Mathematical Methods
for Physics
and Engineering

Mathematical Methods
for Physics
and Engineering

Mattias Blennow

CRC Press
Taylor & Francis Group
Boca Raton London New York

CRC Press is an imprint of the
Taylor & Francis Group, an **informa** business

CRC Press
Taylor & Francis Group
6000 Broken Sound Parkway NW, Suite 300
Boca Raton, FL 33487-2742

Printed on acid-free paper
Version Date: 20171121

International Standard Book Number-13: 978-1-138-05690-9 (Hardback)

Visit the Taylor & Francis Web site at
http://www.taylorandfrancis.com

and the CRC Press Web site at
http://www.crcpress.com

To Ana

Contents

Preface

Mathematics and mathematical methods hold a central role in the scientific pursuit. It is the language we use in order to describe Nature and make quantifiable and testable predictions, the cornerstone of the scientific principle. Compiling all of the mathematics used in physics today would be a daunting task. Instead, this book focuses on common mathematical methods needed to confront modern theoretical physics that are also used in many branches of engineering. The target audience of the book is physics or engineering students that have already taken basic courses in mathematics including linear algebra, multi-variable calculus, and introductory differential equations, typically in the third year of university or later. Unlike many other textbooks on mathematical methods in physics that often use examples from quantum mechanics or other advanced subjects, I have tried to use examples from subjects that should be more familiar to the student as much as possible. With this approach, the student will hopefully be able to see these examples in a new light and prepare his or her intuition for using the new tools in advanced topics at the same time. I have also attempted to introduce some new concepts as needed for the discussion.

It should be noted that this book is not intended to be mathematically rigorous. Instead, the idea is to convey mathematics as a tool for describing physical systems and therefore focus more on the intuition behind the mathematics and its application to physics. You will therefore not find the typical theorem-proof-corollary structure that you might expect from a textbook in mathematics.

The topics covered in the different chapters of this book and their applications are:

1. *Scalars and Vectors*: A basic treatment of vector analysis, which is a fundamental piece of mathematics used in almost every branch of modern physics. The chapter introduces the concepts of fields, their derivatives, and integration. Furthermore, we will work through both non-Cartesian coordinate systems and potential theory, which are often useful tools for solving particular problems. Examples of theories relying heavily on vector analysis include Newtonian mechanics and Maxwell's theory of electromagnetism.

2. *Tensors*: Tensor analysis builds upon the concepts introduced in Chapter 1 and expands them to involve general linear transformations between vectors (and tensors). The approach taken in this chapter is slightly different from what is usually found in textbooks. Instead of starting out with tensors in Cartesian coordinates, we will start by introducing tensors in general coordinates systems already from the beginning as this provides a deeper insight to what tensors are all about. We later treat the special case of Cartesian coordinates and discuss tensor integration and how tensors are applied in solid mechanics, electromagnetism, and classical mechanics. Apart from this, tensor analysis also plays a central role in, for example, fluid mechanics and the special and general theories of relativity.

3. *Partial Differential Equations and Modelling*: Modelling is at the core of physics. Without a mathematical model for a system, no amount of mathematical techniques will help us make predictions about it. The chapter covers the most basic ideas in mathematical modelling using differential equations, starting from the continuity equation,

and discusses techniques important for dealing with them. We will also discuss dimensional analysis, which is central to modelling any physical system and lets us draw conclusions about how they scale, and how delta functions may be used to provide idealised models.

4. *Symmetries and Group Theory*: Symmetry arguments are a powerful tool for simplifying and solving many physical systems. In some particular applications, fundamental symmetries even lie at the heart of the theory itself. The natural language for describing symmetries is that of group theory and this chapter gives a basic introduction to these concepts. The resulting mathematical theory is useful for drawing conclusions about the behaviour of any physical system and is applicable across a wide range of topics.

5. *Function Spaces*: This chapter introduces and treats function spaces as a more abstract form of vector space. We discuss operators acting on these function spaces and their eigenvalues. In particular, we look at Sturm–Liouville theory and its applications. By using separation of variables, we arrive at a large number of different Sturm–Liouville problems whose solutions will be important when approaching the models constructed in Chapter 3. These functions will turn out to appear also in many other physics applications as they are the solutions to very particular differential equations. The framework developed here will be particularly useful in the study of continuous mechanical systems as well as in basic quantum mechanics.

6. *Eigenfunction Expansions*: The methods of Chapter 5 are applied in order to solve many of the models introduced in Chapter 3. In particular, we apply eigenfunction expansions in order to reduce partial differential equations to ordinary differential equations for the expansion coefficients. This is used to treat both homogeneous and inhomogeneous problems in general with additional treatment of critical systems and resonances in driven systems. We also discuss the effects of terminating series solutions using only a finite number of terms to approximate the solution. Furthermore, we discuss the case of infinite domains, where series expansions are replaced by transforms. Methods of this sort are used in the treatment of continuous mechanical systems.

7. *Green's Functions*: The use of Green's functions to solve differential equations is based on the principle of superposition. In particular, it is useful as a method for expressing the solution to a general inhomogeneous linear differential equation in terms of the solution to the fundamental case with a delta function inhomogeneity. We discuss different ways of finding Green's functions for specified problems and applying them to models. Finally, we apply them to the case of non-linear systems by introducing and using perturbation theory. These methods are applicable in a wide range of different fields, from automated control to quantum field theory.

8. *Variational Calculus*: Functionals are maps from sets of functions to real or complex numbers. Variational calculus deals with finding stationary values of such functionals and is applicable to finding the shortest path between two points in a general space or the stable configurations of different physical systems. Many physical principles can be formulated in terms of variational problems and we will use Fermat's principle for light propagation and Hamilton's principle in classical mechanics as examples. Variational methods are also useful as a tool for modelling systems based upon energy methods rather than finding differential equations by analysing infinitesimal elements of a system.

9. *Calculus on Manifolds*: Many physical systems cannot be described in terms of Euclidean spaces. We here introduce the concept of manifolds and describe how calculus

works on them, discussing both derivatives and integrals. The generalisation of the vector and tensor concepts to manifolds holds a central role in the discussion and we treat general curved spaces. The calculus on manifolds is applicable to generalisations of several theories, including classical mechanics, and essential in the study of general relativity.

10. *Classical Mechanics and Field Theory*: The main aim of this chapter is not to introduce new mathematical tools. Instead, it is intended to serve as an example of the application of several of the concepts treated throughout the book. The Newtonian, Lagrangian, and Hamiltonian approaches are introduced and discussed to some extent. In particular, the realisation of the connection between symmetries and conserved quantities through Noether's theorem is fundamental to modern theoretical physics. We also introduce the Lagrangian approach to field theory, starting by taking a continuous classical mechanical system as an example.

To the student

This book will introduce you to many of the fundamental mathematical methods used in modern physics. The range of topics covered is rather broad and the difficulty level will gradually increase. Throughout the book, you will find many examples that are intended to illustrate applications of the more abstract notions that are discussed in the main text. In particular, I strongly recommend reading through the examples if you feel that you are in need of more concrete realisations in order to shape your understanding.

Note that reading this material will only get you so far. In reality, there is no substitute for actually working through the material yourself. I therefore recommend reading the material with pen and paper readily available in order to fill in any steps you do not find immediately evident. Obtaining an intuition for the different topics also requires you to apply the theory to different problems. Each chapter in this book therefore comes with a large number of problems, listed at the end of the chapter, intended to illustrate the application of the theory. In order to master the material, I therefore also suggest that you work through the problems rather than continuing straight on to the next topic.

You should also note that there is a lot of material in this book. You should not be surprised if your course only covers part of the material or your instructor excludes some of the more advanced topics in any given chapter depending on the length of your course. This is intentional and the more advanced topics are also intended to serve as additional input for students who find the material interesting and want to go deeper. If your instructor excludes parts of the material, pay close attention to what problems he or she recommends, as they will likely be representative of what is included in your particular course.

I hope you will enjoy reading this book as much as I have enjoyed writing it.

To the instructor

The material covered in this book is extensive and likely sufficient to last for several courses. Depending on your intentions with your course, you may want to select different parts of the material to present. In many cases, understanding of the earlier chapters is not necessary, but often helpful. Be mindful when you select the material and also make sure to recommend problems that are appropriate and representative of the material that you wish to cover. You can always recommend parts of the book that you do not cover to interested students who wish to obtain a deeper knowledge. For reference, the approximate prerequisites for each chapter are listed below. Obviously, the knowledge of the content for any given chapter may also be acquired elsewhere.

1. *Scalars and Vectors*: Being the first chapter in the book, no knowledge apart from the prerequisites mentioned earlier should be necessary.

2. *Tensors*: Understanding the first chapter is necessary. In particular, the notation introduced in the discussion of non-Cartesian coordinate systems will be used extensively. The potential theory part of the first chapter is not crucial.

3. *Partial Differential Equations and Modelling*: The first chapter is required knowledge as many of the tools of vector analysis will be applied. The second chapter is helpful as some models will be introduced in their general form using tensors, but it is not crucial for most of the basic understanding.

4. *Symmetries and Group Theory*: Again, the first chapter is crucial for a basic understanding. The second chapter is mainly used for some of the discussion on tensor product representations.

5. *Function Spaces*: The first chapter is recommended, in particular the discussion on different coordinate systems. However, much of the material stands on its own and mainly requires basic linear algebra. The discussion of group representations on function spaces requires that the representation theory part of Chapter 4 has been covered.

6. *Eigenfunction Expansions*: Chapters 1, 3, and 5 should be considered as prerequisites as well as basic knowledge on the solution of ordinary differential equations.

7. *Green's Functions*: Chapter 3 is a prerequisite, mainly for the introduction of the models that we will solve. The discussion on distributions in Chapter 5 is helpful, but not crucial. The series method for finding the Green's function of a problem also requires the theory developed in Chapter 5, but may generally be skipped without losing the main message of the chapter.

8. *Variational Calculus*: Requires the knowledge from Chapter 1 and the discussion about the metric tensor in Chapter 2 is helpful. The modelling part refers back to discussions in Chapter 3.

9. *Calculus on Manifolds*: Chapters 1 and 2 are crucial and cannot be skipped. Chapter 8 is necessary for the discussion on geodesics and the Levi-Civita connection. Chapter 3 is necessary for the discussion of the continuity equation in a general manifold.

10. *Classical Mechanics and Field Theory*: Chapter 1 is crucial for all of this chapter. For the discussions on tensors such as the moment of inertia, Chapter 2 is also necessary. The Lagrangian and Hamiltonian mechanics parts, as well as the field theory part, require the knowledge from Chapter 8 and the discussion on the use of manifolds in classical mechanics requires the methods developed in Chapter 9.

Acknowledgments

This book started as a set of lecture notes and supplementary material in a course essentially covering Chapters 3, 5, 6, 7 and 8 as I could not find any textbook that suited the specific needs for that course, which covered mathematical methods with focus on their application to physical systems. In this course, the focus was not on the mathematics, but on the application of the mathematics to physics. In particular, many of the available textbooks used examples from quantum mechanics, which had not yet been covered by the target audience, and did not include any part on mathematical modelling. If not for this course, this book would never have been written.

I would also like to take the opportunity to thank my teaching assistants Stefan Clementz, Per Moosavi, and Mattias Olla for their help in eliminating many typos as well as all students that have pointed out errors and unclear passages.

I have also drawn some inspiration from discussions with the internet community Physics Forums (`www.physicsforums.com`) regarding my mode of presentation and for fine-tuning my formulation of the subject matter to the target audience. Surprisingly often, people have been posting questions regarding the very same material I have been writing about, in some cases leading me to rethink the approach taken.

Since it was written on my free time during evenings and weekends, this book would also never have been completed without the support of my wife. Admittedly, she has rolled her eyes at my devotion to the text more than once, but without her understanding and acceptance, the time necessary to complete the book would not have been available. The book is dedicated to her.

Mattias Blennow
Stockholm, Sweden, 2017

CHAPTER 1

Scalars and Vectors

Essentially everyone who has come into contact with physics at a higher education level is familiar with the concepts of scalars and vectors. The concept of a scalar is easy to grasp as a quantity that has a particular value, such as the kinetic energy of an object, the air pressure at sea level, or the temperature in your oven. The concept of a vector is also relatively straightforward, generally being presented as a directional quantity that has magnitude as well as direction. The typical examples of vector quantities include kinematic quantities such as velocities and accelerations. For example, when flying from New York to Paris, not only the speed (being the magnitude of the velocity) but also the direction of the aircraft velocity is of importance. Flying in the wrong direction we might end up in Anchorage instead, which is not necessarily a bad thing, but probably not what we intended when boarding the flight.

While a scalar quantity can be represented by a single number and a suitable unit of dimension (such as meters, feet, or light years for a distance), a vector needs to be described by several numbers. The most convenient way of doing this is to define a number of linearly independent directions by choosing a set of basis vectors $\vec{e}_i$. The number of such vectors should be equal to the dimension of the space we want to describe, normally three when discussing classical physics, but sometimes less if some directions can be ignored or more in the case where not only the spatial position of a single object is of interest. Some of the concepts in this text, such as the vector cross product, are particularly constructed for three-dimensional space, while others, such as the scalar product or divergence of a vector field, have straightforward generalisations to more or fewer dimensions. The requirement of linear independence of the basis vectors states that it is impossible to write a basis vector as a linear combination of other basis vectors. As a consequence, we can write any vector $\vec{v}$ in an N-dimensional space as

$$\vec{v} = \sum_{i=1}^{N} v^i \vec{e}_i. \tag{1.1}$$

In particular, in three dimensions, $\vec{v} = v^1 \vec{e}_1 + v^2 \vec{e}_2 + v^3 \vec{e}_3$ (note that the superscripts here are indices, not powers!). The N numbers v^i uniquely define the vector $\vec{v}$ and it is common to select basis vectors in such a way that they have magnitude one and are orthogonal. Such a set of basis vectors is referred to as an *orthonormal basis*.

1.1 VECTORS AND ARITHMETICS

The concepts of scalars and vectors come with some basic arithmetic rules. Scalar quantities may be multiplied together, resulting in a new scalar quantity of a different physical dimension, or added (if they have the same physical dimension), resulting in a new scalar quantity of the same dimension. Naturally, such multiplications or additions do not necessarily have a physical meaning and theories of physics are essentially based on how to apply mathematical operations in order to describe the world we live in and make useful predictions.

When it comes to vectors, there is a natural definition of multiplication with a scalar quantity, we multiply the magnitude of the vector by the same amount. In fact, we have already used this in Eq. (1.1) when decomposing the vector $\vec{v}$ into a sum of scalar multiples of the unit vectors $\vec{e}_i$. The same goes for a sum of two vectors, where we can add the coefficients multiplying each of the basis vectors in order to obtain the coefficient of the corresponding basis vector for the vector sum. As in the case of a scalar addition, vector addition also requires the vectors to have the same physical dimension in order to produce meaningful results. In terms of the basis vectors, we therefore have

$$a\vec{v} = a\sum_{i=1}^{N} v^i \vec{e}_i = \sum_{i=1}^{N} (av^i)\vec{e}_i, \tag{1.2a}$$

$$\vec{v} + \vec{w} = \sum_{i=1}^{N} (v^i + w^i)\vec{e}_i. \tag{1.2b}$$

There is no way of defining an addition of a scalar and a vector, but there are different possibilities of creating vector products and these are useful tools when constructing physical theories.

The *scalar product* (also called *dot product* or *inner product*) $\vec{v} \cdot \vec{w}$ of two vectors $\vec{v}$ and $\vec{w}$ is a scalar and is linear in both vectors. It has an intuitive geometrical interpretation (see Fig. 1.1) as the product of the magnitudes of the vectors multiplied by the cosine of the angle between their directions. The vectors are orthogonal if $\vec{v} \cdot \vec{w} = 0$. Thus, if the basis vectors $\vec{e}_i$ are chosen to be orthonormal, i.e., $\vec{e}_i \cdot \vec{e}_j$ equals one if $i = j$ and zero otherwise, then

$$\vec{v} \cdot \vec{w} = \sum_{i=1}^{N}\sum_{j=1}^{N} v^i \vec{e}_i \cdot w^j \vec{e}_j = \sum_{i=1}^{N}\sum_{j=1}^{N} v^i w^j (\vec{e}_i \cdot \vec{e}_j) = \sum_{i=1}^{N} v^i w^i. \tag{1.3}$$

We can also use the scalar product in order to define the *magnitude* (or *norm*) of a vector as

$$|\vec{v}| = \sqrt{\vec{v} \cdot \vec{v}} = \sqrt{\sum_{i=1}^{N} (v^i)^2}. \tag{1.4}$$

It is very common to denote $\vec{v} \cdot \vec{v} = \vec{v}^2$. Note that this is compatible with the definition of a scalar vector multiplication yielding a new vector with a magnitude which is the original vector magnitude multiplied by the scalar. The geometrical interpretation of the scalar product can now be written in the form

$$\vec{v} \cdot \vec{w} = |\vec{v}|\,|\vec{w}|\cos(\alpha), \tag{1.5}$$

where α is the angle between the vector directions. It is also common to denote the magnitude of a vector using a scalar with the same letter, but omitting the vector arrow, i.e., $|\vec{v}| = v$.

In three dimensions, we can also define the anti-symmetric *cross product* (or *vector product*) $\vec{v} \times \vec{w} = -\vec{w} \times \vec{v}$ of two vectors as a new vector. Just as the scalar product, the cross product is linear in both of the vectors. This means that we can completely define it

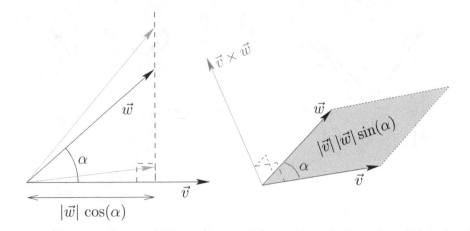

Figure 1.1 Visual of the geometrical interpretation of the scalar (left) and cross (right) products. In the scalar product, the value of the scalar product is the length of $\vec{v}$ multiplied by $|\vec{w}|\cos(\alpha)$. Note that this value does not depend on whether we project $\vec{w}$ on $\vec{v}$ or the other way around. The shaded vectors have the same scalar product with $\vec{v}$ as $\vec{w}$ has. For the cross product, the shaded area is equal to the modulus of the cross product $\vec{v} \times \vec{w}$, which is also orthogonal to both $\vec{v}$ and $\vec{w}$.

in terms of how it acts on an orthonormal set of basis vectors in a *right-handed basis* as

$$\vec{e}_1 \times \vec{e}_2 = \vec{e}_3, \tag{1.6a}$$
$$\vec{e}_2 \times \vec{e}_3 = \vec{e}_1, \tag{1.6b}$$
$$\vec{e}_3 \times \vec{e}_1 = \vec{e}_2, \tag{1.6c}$$
$$\vec{e}_i \times \vec{e}_i = 0, \quad \text{for all } i. \tag{1.6d}$$

This defines the ordering of the basis vectors. For a *left-handed basis*, the definitions come with a minus sign on one side. In terms of the vector components, we obtain

$$\vec{v} \times \vec{w} = (v^2 w^3 - v^3 w^2)\vec{e}_1 + (v^3 w^1 - v^1 w^3)\vec{e}_2 + (v^1 w^2 - v^2 w^1)\vec{e}_3. \tag{1.7}$$

The squared magnitude of $\vec{v} \times \vec{w}$ can now be found using the scalar product and after simplification

$$|\vec{v} \times \vec{w}| = |\vec{v}|\,|\vec{w}|\sin(\alpha), \tag{1.8}$$

where α again is the angle between the vectors $\vec{v}$ and $\vec{w}$. Furthermore, the cross product $\vec{v} \times \vec{w}$ is orthogonal to both $\vec{v}$ and $\vec{w}$ (see Problem 1.5). Combining these properties, the cross product also has a geometrical interpretation as a vector orthogonal to both of the arguments and a magnitude equal to the area spanned by them, see Fig. 1.1.

In addition to the scalar and cross products, there is an additional product called the outer product that is defined regardless of the number of dimensions. This product is a *second rank tensor* and we will return to it in Chapter 2.

1.2 ROTATIONS AND BASIS CHANGES

The choice of basis vectors $\vec{e}_i$ is not unique. In fact, we can select any three linearly independent vectors as the basis and work from there. However, as we have done, it is often convenient to work in terms of an orthonormal basis, where the basis vectors are orthogonal

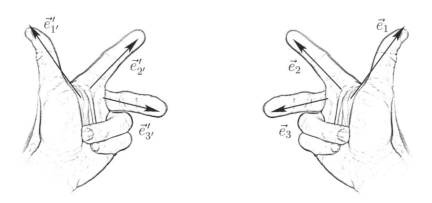

Figure 1.2 A left- and a right-handed set of vectors. When held as in the image, the vectors $\vec{e}_1$, $\vec{e}_2$, $\vec{e}_3$, pointing in the directions of the thumb, index finger, and middle finger of the right hand, respectively, and ordered in that specific order, constitute a right-handed system. On the left part of the image, the vectors $\vec{e}_{1'}'$, $\vec{e}_{2'}'$, $\vec{e}_{3'}'$ constitute a left-handed system.

and have magnitude one. Still, there is a freedom of choice of different orthonormal bases. In general, an orthonormal basis $\vec{e}_i$ can be related to a different orthonormal basis $\vec{e}_{i'}'$ by means of a rotation, assuming that the two bases have the same handedness. Note here that we have chosen to use primed indices for the primed basis. This is a notational convention that is useful to keep track of what indices belong to which basis. As mentioned previously, the handedness of a basis is based on the ordering of the basis vectors as illustrated in Fig. 1.2. In order to transform a right-handed basis into a left-handed one, or vice versa, a reflection is needed in addition to the rotation.

The properties of a vector do not depend on the basis chosen. Regardless of the chosen basis, a vector pointing to the Moon will always point at the Moon, independent of whether one of our basis vectors point to the Sun or not. As mentioned in the beginning of this chapter, the vector $\vec{v}$ may be written in the basis $\vec{e}_i$ according to Eq. (1.1). By the fact that we could just as well use the basis $\vec{e}_{i'}'$, we must therefore have

$$\vec{v} = \sum_{i=1}^{N} v^i \vec{e}_i = \sum_{i'=1'}^{N'} v^{i'} \vec{e}_{i'}'. \tag{1.9}$$

Here we have chosen to use a prime only in the index of $v^{i'}$ to denote that this component belongs to the primed basis. Alternatives to this include using a prime only for the symbol itself v'^i (and at the same time using $\vec{e}_i'$) or using double primes $v'^{i'}$. However, using one prime only is sufficient and our choice is coherent with using primed indices for the primed basis vectors.

With orthonormal bases, the coefficients v^i and $v^{i'}$ can be found through the scalar product with the basis vectors themselves

$$v^i = \vec{e}_i \cdot \vec{v}, \quad v^{i'} = \vec{e}_{i'}' \cdot \vec{v}. \tag{1.10}$$

As we can express $\vec{v}$ using either basis, this leads to

$$v^{i'} = \vec{e}_{i'}' \cdot \sum_{i=1}^{N} v^i \vec{e}_i = \sum_{i=1}^{N} v^i \vec{e}_{i'}' \cdot \vec{e}_i \equiv \sum_{i=1}^{N} v^i a_i^{i'}, \tag{1.11}$$

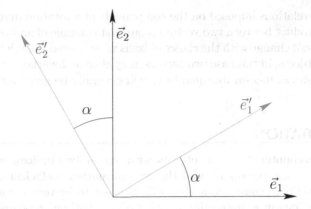

Figure 1.3 Two possible bases in two dimensions related by a rotation with an angle α.

where we have defined the *transformation coefficients* $a_i^{i'} = \vec{e}\,'_{i'} \cdot \vec{e}_i$. From Eq. (1.10), it follows that $a_i^{i'}$ is the component of $\vec{e}_i$ in the $\vec{e}\,'_{i'}$ direction. In a similar fashion, we have

$$v^i = \sum_{i'=1'}^{N'} v^{i'} \vec{e}_i \cdot \vec{e}\,'_{i'} \equiv \sum_{i'=1'}^{N'} v^{i'} a_{i'}^i, \tag{1.12}$$

where $a_{i'}^i = \vec{e}_i \cdot \vec{e}\,'_{i'}$. Due to the symmetry of the scalar product, the transformation coefficients are related as

$$a_{i'}^i = a_i^{i'}. \tag{1.13}$$

They may be thought of as the elements of an $N \times N$ rotation matrix.

Example 1.1 With $N = 2$, we may choose bases according to Fig. 1.3. The transformation coefficients are given by

$$\begin{pmatrix} a_{1'}^1 & a_{2'}^1 \\ a_{1'}^2 & a_{2'}^2 \end{pmatrix} = \begin{pmatrix} \cos(\alpha) & -\sin(\alpha) \\ \sin(\alpha) & \cos(\alpha) \end{pmatrix} = \begin{pmatrix} a_1^{1'} & a_2^{1'} \\ a_1^{2'} & a_2^{2'} \end{pmatrix}^T. \tag{1.14}$$

This is the general expression for a rotation in two dimensions. In three dimensions, a general rotation is given by three angles (e.g., the Euler angles) and in N dimensions, the number of rotation angles is $N(N-1)/2$.

Scalars are just numbers and their values, unlike the vector components, are the same regardless of the choice of basis. From the requirement that the scalar product $\vec{v} \cdot \vec{w}$ is a scalar quantity, i.e., independent of the choice of basis, we obtain the following relations for the transformation coefficients

$$\sum_{i'=1}^N a_{i'}^i a_j^{i'} = \begin{cases} 1, & (i = j) \\ 0, & (i \neq j) \end{cases}, \quad \sum_{i=1}^N a_i^{i'} a_{j'}^i = \begin{cases} 1, & (i' = j') \\ 0, & (i' \neq j') \end{cases}. \tag{1.15}$$

In words, this is the requirement that the magnitude of a vector as well as the angle between two vectors cannot depend on the choice of basis. These relations for the transformation coefficients are the relations imposed on the components of a rotation matrix. Apart from a scalar, the scalar product between two vectors is our first example of an *invariant*, which is a quantity that does not change with the choice of basis vectors even though the representation of the underlying objects, in this case two vectors, may change. Invariants are of fundamental importance to physics as measurable quantities will generally be invariant under changes of basis.

1.3 INDEX NOTATION

We have already encountered several objects with one or two indices, namely the vector components v^i, the basis vectors $\vec{e}_i$, and the transformation coefficients $a_i^{i'}$. It is time to introduce some notational conventions that will turn out to be very powerful and allow us to write our expressions in a more minimalistic fashion, without repeated sums like those in Eq. (1.3). The first convention is the *Einstein summation convention*, which states that whenever an index is repeated twice, it should be summed over from 1 to N. Thus, with this convention, Eq. (1.10) can be expressed as

$$\vec{v} = v^i \vec{e}_i \equiv \sum_{i=1}^{N} v^i \vec{e}_i \tag{1.16}$$

and Eq. (1.3) as

$$\vec{v} \cdot \vec{w} = v^i w^j \vec{e}_i \cdot \vec{e}_j = v^i w^i. \tag{1.17}$$

Indices that appear twice, and thus are summed over, are called *summation indices* or *dummy indices*. The latter name comes from the fact that it generally does not matter what letter is used to represent such an index as it is only a summation variable (as long as the same letter is used for both occurrences of course). For example, the expressions $v^i \vec{e}_i$ and $v^j \vec{e}_j$ are equivalent. However, it is important to make sure that each summation index is used for one sum only. Thus, we should be suspicious if an index appears more than twice in an expression. If a particular set of basis vectors is chosen or inferred, then it is sometimes common to refer to the components v^i and the vector $\vec{v}$ interchangeably as they represent the same information. This will be particularly true when dealing with tensors later on.

Unlike summation indices, *free indices* are indices that appear only once. If an expression in an equation has a free index, then the expression on the other side of the equal sign must have the same free index. Such an expression is valid for all possible values of the free index. For example, the vector components v^i can be expressed as

$$v^i = \vec{e}_i \cdot \vec{v} \tag{1.18}$$

regardless of whether i is equal to one, two, or three. In this equation, we are not allowed to change the i on one side of the equation into a j. On the other hand, we are allowed to change the i to a j on both sides simultaneously, provided that there is not already a different index that has been denoted by j. In the remainder of the book, we will use the Einstein summation convention unless explicitly stated otherwise, e.g., by stating "no sum" in connection to an equation where an index is repeated twice, or by explicitly writing the sum in case of a sum where the index does not appear two times.

1.3.1 The Kronecker delta and the permutation symbol

We have already seen the transformation coefficients as examples of objects with two indices and, in Chapter 2, we will encounter tensors that have several indices. However, before we get there, it is convenient to introduce two objects with two and three indices, respectively. These are the *Kronecker delta* and the *permutation symbol*. The Kronecker delta δ_{ij} is defined by

$$\delta_{ij} = \begin{cases} 1, & (i = j) \\ 0, & (i \neq j) \end{cases}. \tag{1.19}$$

Whenever one of the indices of the Kronecker delta appears as a summation index, the sum can be easily performed as it is just selecting the term of the sum where the summation index is equal to the other index of the Kronecker delta. Thus, this is equivalent to replacing the other occurrence of the summation index with the other index from the Kronecker delta and removing the δ_{ij}, e.g.,

$$\delta_{ij} v^j = v^i. \tag{1.20}$$

Note that the i in this equation is a free index and therefore appears on both sides of the equation.

The Kronecker delta is particularly useful in describing some relations in orthonormal bases. In particular, the requirement for a basis to be orthonormal can be expressed as

$$\vec{e}_i \cdot \vec{e}_j = \delta_{ij}, \tag{1.21}$$

leading to a more shorthand version of Eq. (1.3)

$$\vec{v} \cdot \vec{w} = v^i w^j \vec{e}_i \cdot \vec{e}_j = v^i w^j \delta_{ij} = v^i w^i, \tag{1.22}$$

while Eq. (1.15) takes the form

$$a^i_{i'} a^{i'}_j = \delta_{ij}, \quad a^{i'}_i a^i_{j'} = \delta_{i'j'}. \tag{1.23}$$

The permutation symbol in N dimensions, also called the *Levi-Civita symbol*, has N indices and is denoted $\varepsilon_{ij\ldots}$. It is defined in such a way that $\varepsilon_{12\ldots N} = 1$ and it is anti-symmetric in all indices. This means that if any two indices change positions, the permutation symbol changes sign. In particular, this means that if any two indices are the same, the permutation symbol takes the value zero. In three dimensions, the permutation symbol ε_{ijk} has three indices and six non-zero components

$$\varepsilon_{123} = \varepsilon_{231} = \varepsilon_{312} = -\varepsilon_{132} = -\varepsilon_{321} = -\varepsilon_{213} = 1, \tag{1.24}$$

corresponding to the six possible permutations of the indices 1, 2, and 3.

Much like how the Kronecker delta can be used to express the scalar product $\vec{v} \cdot \vec{w} = v^i w^j \delta_{ij}$, the permutation symbol can be used to express the cross product. In particular, we note that Eqs. (1.6) take the form

$$\vec{e}_j \times \vec{e}_k = \varepsilon_{ijk} \vec{e}_i \tag{1.25}$$

and linearity then gives

$$\vec{v} \times \vec{w} = v^j w^k (\vec{e}_j \times \vec{e}_k) = \vec{e}_i \varepsilon_{ijk} v^j w^k. \tag{1.26}$$

1.3.2 Vector algebra using index notation

The fact that δ_{ij} and ε_{ijk} can be used to express the scalar and cross products can be used in order to write different vector expressions in terms of vector components starting from the basic definitions. For example, the triple product $\vec{u} \cdot (\vec{v} \times \vec{w})$ can be rewritten as

$$\vec{u} \cdot (\vec{v} \times \vec{w}) = u^i \varepsilon_{ijk} v^j w^k. \tag{1.27}$$

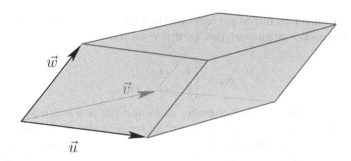

Figure 1.4 The volume of the parallelepiped spanned by the three vectors $\vec{u}$, $\vec{v}$, and $\vec{w}$ can be computed using the triple product $\vec{u} \cdot (\vec{v} \times \vec{w})$.

Although this may not seem as much, we can use the cyclic property of $\varepsilon_{ijk} = \varepsilon_{jki}$, which follows from the anti-symmetry, to change the expression to

$$\vec{u} \cdot (\vec{v} \times \vec{w}) = v^j \varepsilon_{jki} w^k u^i = \vec{v} \cdot (\vec{w} \times \vec{u}). \tag{1.28}$$

From this follows directly that the triple product is invariant under cyclic permutations of the three vectors $\vec{u} \to \vec{v} \to \vec{w} \to \vec{u}$.

Example 1.2 We have earlier seen that the cross product $\vec{v} \times \vec{w}$ has a magnitude which is equal to the area spanned by the vectors $\vec{v}$ and $\vec{w}$ and in addition is orthogonal to both of them. Noting that the volume of a parallelepiped spanned by $\vec{u}$, $\vec{v}$, and $\vec{w}$ is equal to the area spanned by $\vec{v}$ and $\vec{w}$ multiplied by the projection of $\vec{u}$ on the normal direction of that area (see Fig. 1.4), the volume of the parallelepiped is given by

$$V = \vec{u} \cdot (\vec{v} \times \vec{w}). \tag{1.29}$$

There is a subtlety here, which is that this volume might turn out to be negative. This happens if the vector ordering is chosen such that it describes a left-handed system rather than a right-handed one. If we are only interested in the actual volume of the parallelepiped and not the handedness of our selected ordering, we may take the absolute value of the resulting scalar.

Understanding the above example in terms of the properties of the permutation symbol turns out to be helpful, not only in three but in an arbitrary number of dimensions. Assume that our space has N dimensions and that we have N vectors $\vec{v}_n$, with $n = 1, 2, \ldots, N$. Let us study the object

$$V_N = \varepsilon_{i_1 i_2 \ldots i_N} v_1^{i_1} v_2^{i_2} \ldots v_N^{i_N}, \tag{1.30}$$

where we notice that we can always make the replacement

$$\vec{v}_n \to \vec{v}_n + \sum_{m \neq n} \alpha_m \vec{v}_m \tag{1.31}$$

without changing the value of V_N due to the anti-symmetry of the permutation symbol. For example, in two dimensions we have, after changing $\vec{v}_2 \to \vec{v}_2 - \alpha_1 \vec{v}_1$,

$$\varepsilon_{ij} v_1^i (v_2^j + \alpha_1 v_1^j) = \varepsilon_{ij} v_1^i v_2^j - \alpha_1 \varepsilon_{ij} v_1^i v_1^j. \tag{1.32}$$

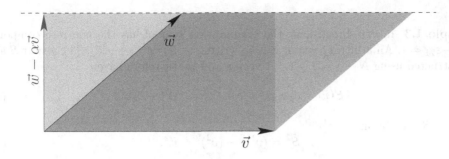

Figure 1.5 A two-dimensional example of changing the vectors spanning a parallelogram without changing the total volume. The original area spanned by the vectors $\vec{v}$ and $\vec{w}$ is composed of the two darker regions. Upon replacing $\vec{w}$ by $\vec{w} - \alpha\vec{v}$, the lighter of the two is replaced by the light shaded area between $\vec{w}$ and $\vec{w} - \alpha\vec{v}$, leaving the total area unchanged.

The anti-symmetric property of $\varepsilon_{ij} = -\varepsilon_{ji}$ now gives

$$\varepsilon_{ij}v_1^i v_1^j = -\varepsilon_{ji}v_1^i v_1^j = -\varepsilon_{ij}v_1^i v_1^j, \tag{1.33}$$

where in the last step we have renamed the dummy indices $i \leftrightarrow j$ and used the fact that v_1^i and v_1^j are numbers that commute. Therefore $\varepsilon_{ij}v_1^i v_1^j = 0$ and

$$\varepsilon_{ij}v_1^i(v_2^j - \alpha_1 v_1^j) = \varepsilon_{ij}v_1^i v_2^j = V_2. \tag{1.34}$$

It follows that if the vectors $\vec{v}_n$ are not linearly independent, then $V_N = 0$. We can take the argument even further and replace all vectors $\vec{v}_n$ with $n < N$ by $\vec{v}_n \to \vec{v}_n - \vec{v}_N(\vec{v}_N \cdot \vec{v}_n)/\vec{v}_N^2$. These new vectors fulfil

$$\vec{v}_N \cdot \left(\vec{v}_n - \vec{v}_N \frac{\vec{v}_N \cdot \vec{v}_n}{\vec{v}_N^2} \right) = \vec{v}_N \cdot \vec{v}_n - \vec{v}_N^2 \frac{\vec{v}_N \cdot \vec{v}_n}{\vec{v}_N^2} = 0 \tag{1.35}$$

and are thus orthogonal to $\vec{v}_N$ just changing the $\vec{v}_n$ by a vector parallel to $\vec{v}_N$. It follows that the new vectors still span the same total volume, see Fig. 1.5 for a two-dimensional example. We can now perform the same procedure using the new $\vec{v}_{N-1}$, then $\vec{v}_{N-2}$, and so forth, until we are left with a set of N orthogonal vectors spanning the same total volume as the original set. Selecting a right-handed system such that $\vec{v}_n = \vec{e}_n |\vec{v}_n|$ (no sum), it follows that V_N is the volume spanned by the original vectors.

In addition, given $N - 1$ vectors $\vec{v}_n$, we can form the new vector

$$\vec{S} = \vec{e}_j \varepsilon_{ji_1 i_2 \ldots i_{N-1}} v_1^{i_1} v_2^{i_2} \ldots v_{N-1}^{i_{N-1}}, \tag{1.36}$$

as a generalisation of the cross product as it maps $N-1$ vectors linearly to a new vector and is completely anti-symmetric. By the same line of argumentation we used for the volume above, it is easy to show (see Problem 1.22) that $\vec{S}$ is orthogonal to all of the $\vec{v}_n$ with a magnitude which is equal to the $N - 1$ dimensional volume spanned by the $\vec{v}_n$ (i.e., in the case of $N = 3$, the two-dimensional area). The vector $\vec{S}$ can therefore be interpreted as an $N - 1$ dimensional volume multiplied by its normal vector in N dimensions.

Example 1.3 In two dimensions, the permutation symbol has the non-zero components $\varepsilon_{12} = -\varepsilon_{21} = 1$. An arbitrary vector can be written as $\vec{v} = v^1\vec{e}_1 + v^2\vec{e}_2$. The vector $\vec{S}$ above is constructed using $N - 1 = 2 - 1 = 1$ vector and we therefore have

$$\vec{S}(\vec{v}) = \vec{e}_1\varepsilon_{12}v^2 + \vec{e}_2\varepsilon_{21}v^1 = \vec{e}_1v^2 - \vec{e}_2v^1. \tag{1.37}$$

Squaring $\vec{S}$, we obtain

$$\vec{S}^2 = (v^1)^2 + (v^2)^2 = \vec{v}^2, \tag{1.38}$$

which is to be expected since the $N - 1$ dimensional volume is a line element spanned by $\vec{v}$. Furthermore, we have

$$\vec{v} \cdot \vec{S} = (v^1\vec{e}_1 + v^2\vec{e}_2) \cdot (\vec{e}_1v^2 - \vec{e}_2v^1) = v^1v^2 - v^2v^1 = 0, \tag{1.39}$$

satisfying the claim that $\vec{S}$ is orthogonal to $\vec{v}$.

Another important relation in vector algebra is the *bac-cab rule*

$$\vec{a} \times (\vec{b} \times \vec{c}) = \vec{b}(\vec{a} \cdot \vec{c}) - \vec{c}(\vec{a} \cdot \vec{b}). \tag{1.40}$$

In order to arrive at this result, we use the *ε-δ-relation*

$$\varepsilon_{ijk}\varepsilon_{k\ell m} = \delta_{i\ell}\delta_{jm} - \delta_{im}\delta_{j\ell}. \tag{1.41}$$

Proving this relation and using it to deduce the bac-cab rule is left as an exercise for the reader (see Problem 1.9). Similar arithmetic can also be applied to rewrite expressions containing more than two or three vectors.

It is also worthwhile to check that the components of the cross product $\vec{v} \times \vec{w}$ actually transform as we would expect a vector to under changes of basis. If we have chosen the new basis vectors $\vec{e}'_{i'}$ such that they form an orthonormal set of unit vectors, then the volume they span should be given by

$$\vec{e}'_{i'} \cdot (\vec{e}'_{j'} \times \vec{e}'_{k'}) = \varepsilon_{i'j'k'} = a^i_{i'}a^j_{j'}a^k_{k'}\varepsilon_{ijk}, \tag{1.42}$$

where we have used that the $\vec{e}_i$ component of $\vec{e}'_{i'}$ is $a^i_{i'}$. Under the assumption that the components of $\vec{v} \times \vec{w}$ should transform as the components of a vector, we then have

$$\vec{e}'_{i'} \cdot (\vec{v} \times \vec{w}) = a^{i'}_i\varepsilon_{ijk}v^jv^k = a^i_{i'}\varepsilon_{ijk}a^j_{j'}v^{j'}a^k_{k'}v^{k'} = \varepsilon_{i'j'k'}v^{j'}w^{k'}, \tag{1.43}$$

which is what we would expect if writing down the cross product in the primed basis from the beginning.

1.4 FIELDS

A central concept to many applications in physics is the notion of a *field*. We will often deal with quantities that are defined throughout space, but may vary from point to point. The usage of fields is found in all branches of physics, from electromagnetism and gravitation to fluid dynamics and material science. The quantised version of classical field theory, quantum field theory (QFT) is the basis for the Standard Model of particle physics, one of the more successful physical theories to date.

In order to define a field, we must first consider the *base space*, i.e., the space that

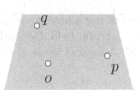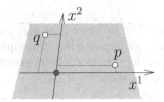

Figure 1.6 In the left figure, a part of a two dimensional affine space with the points p, q, and o marked. In the right figure, we have imposed a coordinate system on the space by selecting the earlier point o as the origin as well as a set of basis vectors. The coordinates of p and q can be deduced by projecting the points on the coordinate axes.

supplies the points where the field is defined. For our current purposes, we will take this to be a part of an *affine space*, which is a space that has the same geometrical structure as $\mathbb{R}^N$ without a fixed origin. As an example, we can consider three-dimensional space, where we, if so willing, can define an origin and base a selection of coordinates on the displacement vector from the origin and a fixed set of basis vectors.

Example 1.4 A two-dimensional affine space is an infinitely extended plane without a fixed origin, see Fig. 1.6. We can define a set of coordinates on this space by selecting an origin and a set of basis vectors $\vec{e}_1$ and $\vec{e}_2$ and using the vector components x^1 and x^2 of the displacement $\vec{x}$ from the origin as coordinates.

Once the base space has been defined, a field assigns a quantity to each point in the base space. The nature of this quantity is always the same for a given field but may be different for different fields. A *scalar field* assigns a scalar value to each point, while a *vector field* assigns a vector value. In the same fashion, a *tensor field* assigns a tensor value to each point, but for the time being we will restrict ourselves to scalar and vector fields. Thus, a field is technically a map from the base space B to a space V (which, for example, may be a scalar or vector space) and we use the notations $\phi(p)$, $\vec{v}(p)$, where p is an point in the base space, for scalar and vector fields, respectively. Naturally, we can call our fields whatever we like and if an origin and/or coordinate system is chosen, we may replace p by the *position vector* $\vec{x} = x^i \vec{e}_i$ or the coordinates x^i. Until Section 1.6, we will work in a Cartesian coordinate system with coordinates x^i. It is also very common to instead use x, y, and z to denote the Cartesian coordinates in three dimensions, but we will keep the x^i notation to reduce the number of conventions used.

While the above may sound slightly intimidating, you are probably already familiar with a large number of different fields. In all of the below examples, the base space is our ordinary three dimensional space

Example 1.5 The *temperature field* $T(\vec{x})$ is a scalar field that assigns a temperature to each point in space. For each point it takes the value of the temperature at that point. The temperature may vary throughout a room or a city, assigning different values to different locations, such as the south pole and the Sahara desert. A typical value of the temperature field may be $T(\text{position of my chair}) = 300$ K.

Example 1.6 The *pressure field* $p(\vec{x})$ takes the value of the ambient pressure of the given spatial point. Since pressure is a scalar quantity, $p(\vec{x})$ is a scalar field. Generally, the pressure will increase in the direction of the gravitational field. A typical value of the pressure field is $p(\text{point in outer space}) = 0$ Pa.

Example 1.7 The *gravitational field* is a vector field and associates a gravitational acceleration $\vec{g}(\vec{x})$ to each point in space. Acceleration has a magnitude and a direction, which is the reason why $\vec{g}(\vec{x})$ is a vector field. The gravitational field outside a spherically symmetric mass distribution has the value

$$\vec{g}(\vec{x}) = -\vec{e}_r \frac{GM}{r^2}, \tag{1.44}$$

where G is Newton's *gravitational constant* (i.e, $G \simeq 6.67 \cdot 10^{-11}$ Nm^2/kg^2), M the total mass, r the distance from the center of the body, and $\vec{e}_r$ is a unit vector directed away from the center of the body.

Upon choosing a coordinate system with coordinates x^i, we can write a scalar field as a single function of these coordinates

$$\phi(x^1, x^2, \ldots) = \phi(p(x^1, x^2, \ldots)) \tag{1.45a}$$

and a vector field as a collection of N functions

$$v^i(x^1, x^2, \ldots) = v^i(p(x^1, x^2, \ldots)) = \vec{e}_i \cdot \vec{v}(p(x^1, x^2, \ldots)). \tag{1.45b}$$

We have here assumed that the basis vectors $\vec{e}_i$ do not depend on the spatial point p. However, it is worth already noting that this will not be generally true once we reach our discussion of curvilinear coordinate systems in Section 1.6. In these settings, the basis vectors will instead be replaced by a set of vector fields that are linearly independent at each point p, but more on that later.

1.4.1 Locality

Just as arithmetic operations can be performed on scalars and vectors, they may also be performed on scalar and vector fields. This is done in such a fashion that the operation is performed for the field values in each point separately. For example, we may multiply a scalar field $\phi(p)$ with a vector field $\vec{v}(p)$. The result of this multiplication will be a vector field $\vec{w}(p)$ that fulfils

$$\vec{w}(p) = \phi(p)\vec{v}(p) \tag{1.46}$$

for all points p. In particular, the value of the product field $\vec{w}$ at any given point p does not depend on the values of ϕ or $\vec{v}$ at points different from p. This property is called *locality* and is a cornerstone in many physical theories.

Example 1.8 If we have a point mass m and know the gravitational field $\vec{g}(p)$, the force on the mass as a function of the spatial point p is given by $\vec{F}(p) = m\vec{g}(p)$. As long as we know the gravitational field, we can compute the force without knowing anything else about the distribution of gravitational sources.

There is a conceptual pitfall here, namely that the gravitational field $\vec{g}(p)$ itself seemingly depends on quantities that are not local. For example, the gravitational field outside a spherically symmetric mass distribution of mass M was given in Eq. (1.44). Clearly, the mass M is not located at the point we are considering the field at. Does this indicate non-locality? The answer is that it does not and that the value of the field follows from physical relations in the form of differential equations that can be written on local form. However, the solution to these equations may and will depend on the distribution of gravitational sources and is obtained by solving the differential equations, a task that we will soon set out to do.

1.4.2   Field integrals

There are several possibilities for integrating fields over the entire or parts of the base space. The integrals may be line integrals, which are integrations along a curve in the base space, surface integrals, which are integrations over a surface in the base space, or volume integrals that integrate over a volume of the base space. In general N dimensional spaces, the integrals can be a subset of the full base space where the subset has any fixed dimension $n \leq N$. We will here mostly restrict our discussion to a three-dimensional base-space.

1.4.2.1   Volume integrals

Volume integrals are relatively straightforward. The volume element dV at $\vec{x}_0$ is the volume enclosed in the parallelepiped described by the coordinates $x_0^i < x^i < x_0^i + dx^i$. This parallelepiped is spanned by the infinitesimal vectors

$$d\vec{x}^i \equiv \frac{\partial \vec{x}}{\partial x^i} dx^i. \quad \text{(no sum)} \qquad (1.47)$$

Note that $d\vec{x}^1$ is the infinitesimal difference vector $\vec{x}(x^1 + dx^1, x^2, x^3) - \vec{x}(x^1, x^2, x^3)$ and similarly for $d\vec{x}^2$ and $d\vec{x}^3$. As described in Example 1.2, the volume element will be given by

$$dV = \left| d\vec{x}^1 \cdot (d\vec{x}^2 \times d\vec{x}^3) \right| = dx^1 dx^2 dx^3, \qquad (1.48)$$

assuming a Cartesian coordinate system such that $\vec{x} = x^i \vec{e}_i$. The integration domain can be specified by defining the set Ω of coordinates that describe the points in the base space that are included in it and the resulting integral is a triple integral over this set of coordinates. The integral I of a scalar field $\phi(\vec{x})$ is therefore given by

$$I = \int_\Omega \phi(\vec{x}) \, dx^1 dx^2 dx^3. \qquad (1.49)$$

This integral is a single number and is independent of the Cartesian coordinate system chosen to represent Ω. Note that when we will deal with curvilinear coordinate systems, this will change slightly due to a more general form of the volume element. In a similar fashion, the integral $\vec{I}$ of a vector field $\vec{v}(\vec{x})$ is given by

$$\vec{I} = \int_\Omega \vec{v}(\vec{x}) \, dx^1 dx^2 dx^3. \qquad (1.50)$$

As long as we restrict ourselves to a Cartesian coordinate system, $\vec{v}(\vec{x})$ can be written in terms of the constant basis vectors $\vec{e}_i$ according to

$$\vec{v}(\vec{x}) = v^i(\vec{x})\vec{e}_i, \tag{1.51}$$

where the v^i are functions of the coordinates. Since the $\vec{e}_i$ are constant, they can be moved out of the integral to obtain

$$\vec{I} = \vec{e}_i \int_\Omega v^i(\vec{x}) \, dx^1 dx^2 dx^3, \tag{1.52}$$

which means that the components of the vector may be integrated separately to provide the components of the integrated field. This is another aspect that will no longer be true once we deal with curvilinear coordinate systems as the basis will then depend on the coordinates.

Example 1.9 The mass of a sphere of radius R depends on the density $\rho(\vec{x})$, which is a scalar field defined in the sphere. The mass of a small volume dV at a point $\vec{x}$ is

$$dm = \rho(\vec{x})dV \tag{1.53}$$

and the total mass of the sphere can be computed by summing the mass of all such elements, i.e., by performing the integral

$$m = \int_\Omega dm = \int_{\vec{x}^2 \leq R^2} \rho(\vec{x}) \, dx^1 dx^2 dx^3. \tag{1.54}$$

Here, Ω is the sphere $\vec{x}^2 \leq R^2$, which assumes that the origin is chosen to be at the center of the sphere.

Example 1.10 The gravitational force on the sphere in the previous example can be computed by summing up the forces on each small volume dV. The force on each volume element is

$$d\vec{F} = \vec{g}(\vec{x})dm = \rho(\vec{x})\vec{g}(\vec{x})dV \equiv \vec{f}(\vec{x})dV, \tag{1.55}$$

where $f(\vec{x}) = \rho(\vec{x})\vec{g}(\vec{x})$ is the *force density*. The total gravitational force on the sphere is therefore

$$\vec{F} = \int_\Omega d\vec{F} = \int_{\vec{x}^2 \leq R^2} \vec{f}(\vec{x}) \, dx^1 dx^2 dx^3. \tag{1.56}$$

1.4.2.2 Surface integrals

Surface integrals appear whenever we have a surface and we need to sum up a set of infinitesimal contributions to a quantity that is defined on that surface. A two dimensional surface may be parametrised by a set of two parameters s and t such that $\vec{x}(s,t)$ describes a point on the surface and the coordinates $x^i(s,t)$ are functions of s and t. The infinitesimal parallelogram described by the parameters $s_0 < s < s_0 + ds$ and $t_0 < t < t_0 + dt$ is spanned by the vectors

$$d\vec{x}_s = \frac{\partial \vec{x}}{\partial s} ds, \quad d\vec{x}_t = \frac{\partial \vec{x}}{\partial t} dt \tag{1.57}$$

evaluated at $(s,t) = (s_0, t_0)$ and its area dS is thus given by

$$dS = |d\vec{x}_s \times d\vec{x}_t| = \left| \frac{\partial \vec{x}}{\partial s} \times \frac{\partial \vec{x}}{\partial t} \right| ds\, dt. \tag{1.58}$$

In many applications, the surface normal $\vec{n}$ is also of importance. Remembering that the cross product is normal to both of the arguments, we will therefore define the *directed surface element*

$$d\vec{S} = \vec{n}\, dS = \frac{\partial \vec{x}}{\partial s} \times \frac{\partial \vec{x}}{\partial t} ds\, dt. \tag{1.59}$$

It is important to remember that there are two possible directions (related to each other by a minus sign) of the surface normal and in each application we must order our parameters s and t in such a way that the normal points in the intended direction. This can be done by remembering that, with the definition above, $\vec{x}_s$, $\vec{x}_t$, and $\vec{n}$ (in that order) form a right-handed system. However, it may be possible to compute $\vec{n}$ through other means than the cross product, in which case it may be more convenient to use $\vec{n}\, dS$ rather than the expression in terms of the cross product without the absolute value. The final double integral is performed over the set of parameters s and t that describe the surface of integration.

Example 1.11 The pressure on a small part of the surface of an object results in a force

$$d\vec{F} = -p(\vec{x})\vec{n}\, dS, \tag{1.60}$$

where the normal direction $\vec{n}$ is chosen to point out of the object. In order to compute the total force on the object, we evaluate the integral

$$\vec{F} = \int_S d\vec{F} = -\int_S p(\vec{x})\, d\vec{S}, \tag{1.61}$$

where S is the surface of the object. For instance, if we wish to compute the force exerted on the half-space $x^3 < 0$, the surface $x^3 = 0$ defines the integration domain and we can use $s = x^1$ and $t = x^2$ as coordinates. We would find that

$$d\vec{x}_s = \frac{\partial \vec{x}}{\partial x^1} ds = \vec{e}_1 ds, \quad d\vec{x}_t = \frac{\partial \vec{x}}{\partial x^2} dt = \vec{e}_2 dt \tag{1.62}$$

leading to

$$d\vec{S} = (\vec{e}_1 \times \vec{e}_2) ds\, dt = \vec{e}_3\, ds\, dt. \tag{1.63}$$

On the chosen surface, this vector points out of the half-space on which we wish to compute the force (see Fig. 1.7) as intended. The total force is thus given by

$$\vec{F} = -\vec{e}_3 \int_{s,t=-\infty}^{\infty} p(\vec{x}(s,t))\, ds\, dt. \tag{1.64}$$

Had we instead chosen $s = x^2$ and $t = x^1$, we would still have parametrised the surface, but obtained a surface normal pointing in the opposite direction and would have had to account for this.

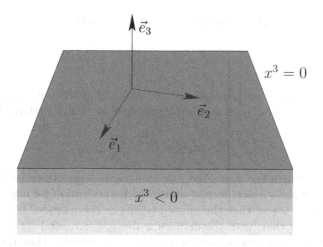

Figure 1.7 The vector $\vec{e}_3$ points out of the half-space $x^3 < 0$ on the surface $x^3 = 0$. This defines the surface normal in the correct direction.

Example 1.12 The convective flow of a substance may be described by the current $\vec{j}(\vec{x}) = \rho(\vec{x})\vec{v}(\vec{x})$ where $\rho(\vec{x})$ is the density of the substance and $\vec{v}(\vec{x})$ its velocity field. The amount of substance $d\Phi$ that flows through an infinitesimal area per unit time is given by the component of the current in the direction of the surface normal multiplied by the infinitesimal area, i.e.,

$$d\Phi = \vec{j}(\vec{x}) \cdot \vec{n}\, dS = \vec{j}(\vec{x}) \cdot d\vec{S} = \rho\vec{v} \cdot d\vec{S}. \tag{1.65}$$

The reason for this is that only the flow in the direction perpendicular to the surface results in a flow through the surface. This $d\Phi$ is thus the amount of substance flowing through the surface in the direction of the surface normal $\vec{n}$. The total flow through a surface is obtained through the integral

$$\Phi = \int_S \vec{j}(\vec{x}) \cdot d\vec{S}, \tag{1.66}$$

which is a scalar quantity as expected.

In general, for any given vector field $\vec{v}(\vec{x})$, the surface integral

$$\Phi = \int_S \vec{v}(\vec{x}) \cdot d\vec{S} \tag{1.67}$$

is called the *flux* of the field $\vec{v}(\vec{x})$ through the surface S, see Fig. 1.8.

1.4.2.3 Line integrals

Line integrals are in many ways reminiscent of surface integrals with the exception that the integration curve is one-dimensional rather than two-dimensional and therefore only needs one parameter. Calling this parameter t, the infinitesimal displacement between $\vec{x}(t)$ and $\vec{x}(t + dt)$ is given by

$$d\vec{x} = \frac{d\vec{x}}{dt}dt = \dot{x}^i \vec{e}_i dt, \tag{1.68}$$

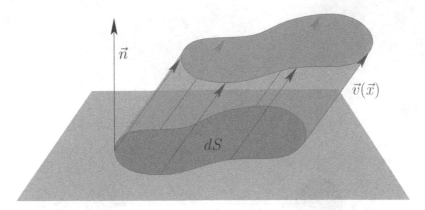

Figure 1.8 The flux of the vector field $\vec{v}(\vec{x})$ through the small surface element $d\vec{S}$ can be geometrically thought of as the volume spanned by $\vec{v}(\vec{x})$ and the surface element. As such it is given by $\vec{v}(\vec{x}) \cdot d\vec{S}$. Integrating over a larger surface we find the flux integral of Eq. (1.67).

where the˙denotes the derivative with respect to the curve parameter t and the last equality holds as long as the basis vectors $\vec{e}_i$ are fixed. As in the case of the surface integral, where we might only need to consider dS rather than $d\vec{S}$, we may be interested only in the magnitude of $d\vec{x}$. However, there are several physical applications where the directionality of $d\vec{x}$, which by construction is in the tangent direction of the curve, is of importance. As in the previous cases, once the curve is parametrised, the integral may be performed as any ordinary integral.

Example 1.13 The most prominent use of line integrals in physics is related to the amount of work done by a force field on an object travelling through it. As discussed previously, the gravitational force on a mass at a position $\vec{x}$ is given by $\vec{F}(\vec{x}) = m\vec{g}(\vec{x})$. The work performed by the force on the object over an infinitesimal displacement $d\vec{x}$ is given by (see Fig. 1.9)

$$dW = \vec{F} \cdot d\vec{x}. \tag{1.69}$$

Summing up the total work done on the object while travelling through the gravitational field results in the integral

$$W = \int_{\Gamma} dW = m \int_{\Gamma} \vec{g}(\vec{x}) \cdot d\vec{x}, \tag{1.70}$$

where Γ is the path taken by the object.

1.4.3 Differential operators and fields

Differential equations are at the very heart of physics. They describe how physical quantities evolve and change with space and time, and formulating and solving them will be a major theme in this book. There are a number of differential operators that will be of importance as we go on to model physical systems and they appear in many different contexts.

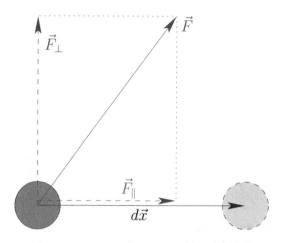

Figure 1.9 The work on an object by a force $\vec{F}$ over a displacement $d\vec{x}$ is given by $dW = \vec{F} \cdot d\vec{x}$. As such, the work done only depends on the force component $\vec{F}_\parallel$, which is parallel to the displacement $d\vec{x}$ and independent of the perpendicular component $\vec{F}_\perp$.

1.4.3.1 The gradient

Any scalar field ϕ is a function of the coordinates x^i. If we want to consider the change of this scalar field as we make small changes to the coordinates, we can consider the difference

$$\delta\phi = \phi(\vec{x} + \varepsilon\vec{a}) - \phi(\vec{x}), \tag{1.71}$$

where ε is a small parameter and $\vec{a}$ is any finite vector. Assuming that ϕ is a differentiable function of the coordinates, we can Taylor expand it to first order in ε and obtain

$$\delta\phi = \phi(\vec{x}) + \varepsilon a^i \partial_i \phi(\vec{x}) - \phi(\vec{x}) + \mathcal{O}(\varepsilon^2) = \varepsilon a^i \partial_i \phi(\vec{x}) + \mathcal{O}(\varepsilon^2), \tag{1.72}$$

where we are using the notation $\partial_i f = \partial f / \partial x^i$. Taking the limit as $\varepsilon \to 0$, we obtain

$$\lim_{\varepsilon \to 0} \frac{\delta\phi}{\varepsilon} = a^i \partial_i \phi = \vec{a} \cdot \nabla\phi. \tag{1.73}$$

The quantity

$$\nabla\phi = \vec{e}_i \partial_i \phi \tag{1.74}$$

is the *gradient* of the scalar field ϕ and completely describes the change of the field for small changes in the coordinates as changing the coordinates by $x^i \to x^i + dx^i$ for small dx^i will result in a change of the field value $d\phi = d\vec{x} \cdot \nabla\phi$.

As evident from its definition, the gradient is a vector field, assigning a vector to each point in space based on the behaviour of the scalar field ϕ in its vicinity. Consequently, the gradient has both a magnitude and a direction. The direction can be interpreted as the direction in which the scalar field grows faster, while the magnitude can be interpreted as how fast it grows in this direction.

The gradient is also important in terms of descriptions of different surfaces. In particular, a scalar field $\phi(\vec{x})$ defines *level surfaces* according to

$$\phi(\vec{x}) = c, \tag{1.75}$$

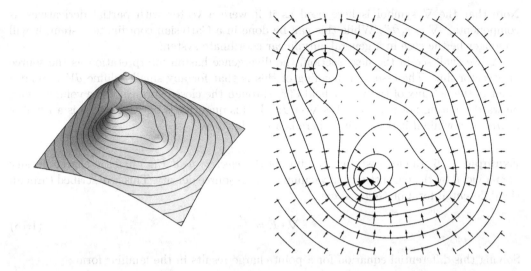

Figure 1.10 The gradient of the height above sea level in a mountain range describes the steepness and direction of the slope. The left figure shows the mountain range and the right shows the gradient field and the level curves. Closer level curves imply a larger gradient.

where c is a constant. This imposes a single constraint on the N coordinates x^i and therefore describes an $N - 1$ dimensional surface. By definition, the value of ϕ does not change if we move within the level surface. This implies that any infinitesimal displacement $d\vec{x}$ such that $\vec{x}$ and $\vec{x} + d\vec{x}$ both lie within the surface must fulfil

$$d\varphi = d\vec{x} \cdot \nabla\phi = 0. \tag{1.76}$$

It follows that $\nabla\phi$ is orthogonal to all infinitesimal displacement within the surface and therefore is orthogonal to the surface itself and thus proportional to the surface normal vector $\vec{n}$, which is defined to be a unit vector orthogonal to the surface.

Example 1.14 In a mountain range, the height above sea level h can be described as a function of the position on the Earth surface (and thus is a scalar field on the Earth surface). The magnitude of the gradient describes the steepness of the slope at each given point and is pointing in the direction of the slope, see Fig. 1.10. A cartographer making a map of the mountain range draws the level curves of the field h, which correspond to the curves with constant height above sea level. The gradient is orthogonal to these curves.

1.4.3.2 The divergence

While the gradient is a differential operator mapping a scalar field to a vector field, the *divergence* does the opposite and maps a vector field to a scalar field. The divergence of the vector field $\vec{v}$ expressed in Cartesian coordinates is defined as

$$\nabla \cdot \vec{v} = \partial_i v^i. \tag{1.77}$$

Note that the ∇ symbol is here used as if it were a vector with partial derivatives as components, "$\nabla = \vec{e}_i \partial_i$". While this can be done in a Cartesian coordinate system, it will no longer be the case in a general curvilinear coordinate system.

As we shall see in the next section, the divergence has an interpretation as the source of a vector flux. The essential meaning of this is that for any small volume dV around a point p, the net flux of the vector field $\vec{v}(\vec{x})$ through the closed surface of the volume (with outward pointing normal vector) is $\nabla \cdot \vec{v} \, dV$. If this quantity is positive, there is a net flux *source* at p while if it is negative, there is a net flux *sink*.

Example 1.15 In electrostatics, the flux of the *electric field* $\vec{E}(\vec{x})$ has the charge density $\rho(\vec{x})$ divided by the permittivity in vacuum ε_0 as its source density. This is described through the differential form of *Gauss's law*

$$\nabla \cdot \vec{E} = \frac{\rho}{\varepsilon_0}. \tag{1.78}$$

Solving this differential equation for a point charge results in the familiar form

$$\vec{E} = \frac{q\vec{e}_r}{4\pi\varepsilon_0 r^2} = k\frac{q\vec{e}_r}{r^2}, \tag{1.79}$$

where q is the charge, $\vec{e}_r$ is a unit vector in the direction away from the charge to the point where the field is evaluated, r is the distance from the charge, and k is *Coulomb's constant*, i.e., $k \simeq 8.99 \cdot 10^9 \text{ Nm}^2/\text{C}^2$.

Example 1.16 We can compute the divergence for the position vector $\vec{x}$ in N dimensions by noting that its ith component is x^i. It follows that

$$\nabla \cdot \vec{x} = \partial_i x^i = \delta_{ii} = N. \tag{1.80}$$

We have here used the fact that $\partial_i x^j = \delta_{ij}$ and the results of Problem 1.7. Note that the Einstein summation convention indicates that the i in δ_{ii} is summed over. Thus, the divergence of the position vector in any vector space is equal to the dimension of the vector space.

1.4.3.3 The curl

Yet another differential operator of importance is the *curl*, which in Cartesian coordinates is defined as

$$\nabla \times \vec{v}(\vec{x}) = \vec{e}_i \varepsilon_{ijk} \partial_j v^k(\vec{x}), \tag{1.81}$$

where again the notation comes from the similarity with a cross product of the vector differential operator ∇ and the vector field on which it acts. The curl completes our set of differential operators acting on fields and maps a vector field to a new vector field.

The interpretation of the curl is somewhat more subtle than that of the gradient and divergence operators. The *circulation* of a vector field with respect to a closed curve is defined as the line integral around the curve

$$C = \oint_\Gamma \vec{v}(\vec{x}) \cdot d\vec{x}. \tag{1.82}$$

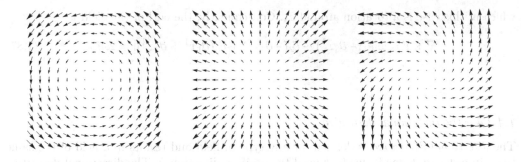

Figure 1.11 The velocity field of a rotating solid object (left). With the origin in the center of rotation, the field is given by $\vec{v} = \omega(-x^2\vec{e}_1 + x^1\vec{e}_2)$, where ω is the angular velocity of the rotation. The middle part of the figure shows a curl free vector field with constant divergence, while the right figure shows a vector field that is the sum of the other two.

In the same fashion as the divergence describes the net flux through the surface of a small volume, the curl describes the circulation of a vector field around a small loop. The circulation integral for the small loop is proportional to the area of the surface spanned by the loop and to the scalar product between the surface normal and the curl, i.e.,

$$dC = (\nabla \times \vec{v}) \cdot d\vec{S}. \tag{1.83}$$

Note that this circulation depends on both the area of the infinitesimal surface as well as its orientation, reaching a maximum when the surface normal aligns with the direction of the curl.

Example 1.17 The force on a point mass m when going around a small loop Γ is given by the force field $\vec{F}(\vec{x})$. The total work done on the mass for one pass through the loop is given by

$$dW = \oint_\Gamma \vec{F} \cdot d\vec{x} \tag{1.84}$$

but is also given by the curl of $\vec{F}$ and the area spanned by the loop

$$dW = (\nabla \times \vec{F}) \cdot d\vec{S}. \tag{1.85}$$

If this quantity is positive, it means that the net force along the loop on the mass is adding energy to the mass and if it is negative, energy is taken away. If $dW \neq 0$, the force field is therefore called *non-conservative*.

Example 1.18 An example of a vector field with a non-zero curl is the local velocity field $\vec{v}(\vec{x})$ of the matter inside a rotating solid object. Such a velocity field is given by (see Fig. 1.11)

$$\vec{v} = \omega(-x^2\vec{e}_1 + x^1\vec{e}_2), \tag{1.86}$$

which in this case is a rotation around $\vec{e}_3$. Consequently, the curl of this field is

$$\nabla \times \vec{v} = \omega[\vec{e}_1(-\partial_3 x^1) + \vec{e}_2(-\partial_3 x^2) + \vec{e}_3(\partial_1 x^1 + \partial_2 x^2)] = 2\omega\vec{e}_3. \tag{1.87}$$

1.4.3.4 The directional derivative

The *directional derivative* is defined for any type of field and results in a field of the same type. In order to define it, we first need to specify a direction $\vec{n}$. The directional derivative of any field F at $\vec{x}$ is then defined as

$$\frac{dF}{d\vec{n}} = \lim_{\varepsilon \to 0} \frac{F(\vec{x} + \varepsilon\vec{n}) - F(\vec{x})}{\varepsilon}. \tag{1.88}$$

Similar to what we did for the gradient, we can Taylor expand $F(\vec{x} + \varepsilon\vec{n})$ and obtain

$$\frac{dF}{d\vec{n}} = n^i \partial_i F \equiv (\vec{n} \cdot \nabla)F. \tag{1.89}$$

In complete analogy, the directional derivative with respect to $\vec{n}$ of a scalar field is just the scalar product of $\vec{n}$ with the gradient of the scalar field. The difference here is that F may be any type of field. When we discuss tensors, this will also hold for a generalised form of the gradient that works for any type of tensor field. It is sometimes preferred to define the the directional derivative only for unit vectors, i.e., for $\vec{n}$ such that $|\vec{n}| = 1$.

1.4.3.5 Second order operators

Given the properties of the gradient, divergence, and curl, there is a finite number of ways we could think of combining them into second order differential operators. With ϕ an arbitrary scalar field and $\vec{v}$ an arbitrary vector field, we could construct the following second order differential operations

$$\nabla \cdot \nabla\phi, \quad \nabla \times \nabla\phi, \quad (\nabla \cdot \nabla)\vec{v}, \quad \nabla(\nabla \cdot \vec{v}), \quad \nabla \cdot (\nabla \times \vec{v}), \quad \nabla \times (\nabla \times \vec{v}). \tag{1.90}$$

For the second and fifth of these, we can argue that

$$\nabla \times \nabla\phi = \vec{e}_i \varepsilon_{ijk} \partial_j \partial_k \phi = 0, \tag{1.91a}$$

$$\nabla \cdot (\nabla \times \vec{v}) = \varepsilon_{ijk} \partial_i \partial_j v^k = 0, \tag{1.91b}$$

just by the assumption that the partial derivatives commute and using the anti-symmetric property of the permutation symbol. Although this may make these second order operators seem trivial and rather uninteresting, these relations will play a crucial role when we discuss potential theory in Section 1.7. The first and third of the listed operators are very similar and only differ in that the first acts on a scalar field and the third acts on a vector field. In both cases, they return a field of the same type and in a Cartesian coordinate system the third operator may be considered to be the first acting on each of the components of $\vec{v}$ separately. This operator is known as the *Laplace operator* and is usually written in one of the following ways

$$\nabla \cdot \nabla = \partial_i \partial_i = \nabla^2 = \Delta. \tag{1.92}$$

The Laplace operator appears in many applications in physics and is often the generalisation of a second derivative in one dimension to similar problems in several dimensions.

Example 1.19 The electric field of a point charge q at the origin may be written as

$$\vec{E} = \frac{q\vec{e}_r}{4\pi\varepsilon_0 r^2},\tag{1.93}$$

where r is the distance from the origin and ε_0 is the permittivity in vacuum. It may also be expressed in terms of the electrostatic potential ϕ by use of the gradient

$$\vec{E} = -\nabla\phi, \quad \phi = \frac{q}{4\pi\varepsilon_0 r}.\tag{1.94}$$

Since the electric field satisfies the relation $\nabla \cdot \vec{E} = 0$ away from the point charge, we find that the potential satisfies

$$\nabla^2\phi = 0\tag{1.95}$$

for $r > 0$. For a general charge distribution, the relation instead becomes

$$\nabla^2\phi = -\nabla \cdot \vec{E} = -\frac{\rho}{\varepsilon_0}.\tag{1.96}$$

The fourth and sixth operators of Eq. (1.90) are related by

$$\nabla \times (\nabla \times \vec{v}) = \nabla(\nabla \cdot \vec{v}) - \nabla^2\vec{v}.\tag{1.97}$$

Thus, they can both be written in terms of each other and the Laplace operator and in total we have two independent second order differential operators.

1.4.3.6 Coordinate independence

So far, we have only discussed the form of the differential operators in one set of Cartesian coordinates. Let us therefore take some time to discuss their expression in a different set of Cartesian coordinates $x'^{i'}$. In a general affine space, we can select new coordinates by choosing a new origin as well as a new set of basis vectors. If the difference vector between the origin of the x' and x coordinates is denoted by $\vec{d}$, any point may be described by either

$$\vec{x} = \vec{e}_i x^i \quad \text{or} \quad \vec{x}' = \vec{e}'_{i'} x'^{i'},\tag{1.98}$$

where $\vec{x}' = \vec{x} + \vec{d}$. We can now use the chain rule to find out how the partial derivatives transform and the result is

$$\partial_i = \frac{\partial}{\partial x^i} = \frac{\partial x'^{i'}}{\partial x^i}\frac{\partial}{\partial x'^{i'}} = \frac{\partial x'^{i'}}{\partial x^i}\partial_{i'}.\tag{1.99}$$

From the relation between the coordinates we have

$$\frac{\partial x'^{i'}}{\partial x^i} = \frac{\partial}{\partial x^i}[\vec{e}'_{i'} \cdot (\vec{x} + \vec{d})] = \frac{\partial}{\partial x^i}(x^j \vec{e}'_{i'} \cdot \vec{e}_j) = a^{i'}_j \delta_{ij} = a^{i'}_i\tag{1.100}$$

and therefore

$$\partial_i = a^{i'}_i \partial_{i'}.\tag{1.101}$$

It now follows that

$$\vec{e}_i\partial_i = a^{i'}_i \vec{e}'_{i'} a^{j'}_i \partial_{j'} = \vec{e}'_{i'}\delta_{i'j'}\partial_{j'} = \vec{e}'_{i'}\partial_{i'}\tag{1.102}$$

which means that ∇ is expressed in the same way regardless of which Cartesian coordinate system is imposed on the base space. This holds for any expression that ∇ appears in, i.e., for the gradient, divergence, and curl, respectively. The crucial step is the realisation that the partial derivatives ∂_i follow the exact same transformation rules as any vector component.

1.5 INTEGRAL THEOREMS

A central part of many physical considerations deals with rewriting the integral of a derivative of a scalar or vector over a region in terms of an integral of the scalar or vector itself over the region boundary or vice versa. The corresponding relation in one dimension is

$$\int_a^b \frac{df}{dx} dx = f(b) - f(a), \qquad (1.103)$$

where the integral of the derivative of f is related to the values of f at the boundary points $x = a$ and $x = b$. In this section we will introduce theorems that are the higher-dimensional equivalents of this relation.

1.5.1 Line integral of a gradient

Assume that we have a scalar field $\phi(\vec{x})$ and wish to compute the line integral

$$L = \int_\Gamma \nabla\phi \cdot d\vec{x}, \qquad (1.104)$$

where Γ is a curve starting in $\vec{x}_0$ and ending in $\vec{x}_1$. This curve may be parametrised by a parameter t such that the position vector is a continuous function $\vec{x}(t)$ with $\vec{x}(0) = \vec{x}_0$ and $\vec{x}(1) = \vec{x}_1$. Rewriting the line integral in terms of this parameter, we obtain

$$L = \int_0^1 \frac{d\vec{x}}{dt} \cdot \nabla\phi \, dt. \qquad (1.105)$$

From the chain rule, we have the relation $(d\vec{x}/dt) \cdot \nabla = (dx^i/dt)\partial_i = d/dt$, leading to

$$L = \int_0^1 \frac{d\phi(\vec{x}(t))}{dt} dt = \phi(\vec{x}(1)) - \phi(\vec{x}(0)) = \phi(\vec{x}_1) - \phi(\vec{x}_0) \qquad (1.106)$$

by virtue of Eq. (1.103). We note here that the integral is only dependent on the values of ϕ at the endpoints and does not depend on the path Γ. In fact, we could have chosen any path with the same endpoints and ended up with the same result. This result is the defining feature of the line integral of a *conservative vector field* and by the relation we have just proven, any vector field which is a gradient of a scalar field is conservative. As we shall see in Section 1.7, for any conservative vector field $\vec{v}$, it is always possible to find a scalar field ϕ such that $\vec{v} = \nabla\phi$. The scalar field ϕ is then called the *potential* of $\vec{v}$.

Example 1.20 The gravitational field outside of a spherical mass distribution with total mass M is given by

$$\vec{g} = -\frac{GM\vec{e}_r}{r^2} = \nabla\left(\frac{GM}{r}\right) \equiv \nabla\phi, \qquad (1.107)$$

where ϕ is the *gravitational potential*. The corresponding force on a test mass m is given by $\vec{F} = m\vec{g}$ and the work done on the test mass when moving it from $\vec{x}_A$ to $\vec{x}_B$ is therefore given by

$$W = \int dW = \int_{\vec{x}_A}^{\vec{x}_B} \nabla m\phi \cdot d\vec{x} = m\phi(\vec{x}_B) - m\phi(\vec{x}_A). \tag{1.108}$$

In other terms, the work done in order to move the test mass is given by the difference in the gravitational potential energy $V = m\phi$ between $\vec{x}_B$ and $\vec{x}_A$.

1.5.2 The divergence theorem

The *divergence theorem* (also known as *Gauss's theorem* or *Ostogradsky's theorem*) relates the flux of a vector field through the closed boundary surface of a volume to the volume integral of the divergence of the vector field over the same volume. Given a volume V with a boundary surface S, the theorem states that

$$\oint_S \vec{v} \cdot d\vec{S} = \int_V \nabla \cdot \vec{v}\, dV. \tag{1.109}$$

The divergence theorem is fundamental in many areas of physics and holds in any number of dimensions. However, we will derive the theorem in three dimensions for concreteness and we start doing so by considering the volume integral

$$\int_V \nabla \cdot \vec{v}\, dV = \int_V (\partial_1 v^1 + \partial_2 v^2 + \partial_3 v^3) dV, \tag{1.110}$$

where we have explicitly written out all of the terms contributing to the divergence. We start by studying the first term in this integral, which can be written as

$$\int_V \partial_1 v^1 dV = \int_{P_{23}V} \left(\sum_n \int_{x_{n,-}^1}^{x_{n,+}^1} \partial_1 v^1 dx^1 \right) dx^2 dx^3. \tag{1.111}$$

Here we start by performing the integral in the x^1 direction and the sum is over the disjoint intervals in x^1 coordinate which are part of the volume V for fixed x^2 and x^3 (see Fig. 1.12). We assume that the lower x^1 boundary in interval n is $x_{n,-}^1$ and the upper $x_{n,+}^1$. The region $P_{23}V$ is the projection of the volume V onto the x^2-x^3-plane. The integral in the x^1 direction is trivial as the integrand is $\partial_1 v^1$ and we thus obtain

$$\int_V \partial_1 v^1 dV = \int_{P_{23}V} \sum_n [v^1(x_{n,+}^1) - v^1(x_{n,-}^1)] dx^2 dx^3. \tag{1.112}$$

Let us consider the surface element of the volume. Around the point with coordinates $(x_{n,+}^1, x^2, x^3)$, the surface may be parametrised with the x^2 and x^3 coordinates and we can write $x_{n,+}^1 = \sigma(x^2, x^3)$. In addition, we know that in order for $x_{n,+}^1$ to be the upper bound of interval n in the x^1 direction, the surface element must have a normal with a positive component in that direction. Using Eq. (1.59), we obtain

$$\begin{aligned} d\vec{S} &= \left(\frac{\partial \vec{x}}{\partial x^2} \times \frac{\partial \vec{x}}{\partial x^3} \right) dx^2 dx^3 \\ &= (\vec{e}_1 \partial_2 \sigma + \vec{e}_2) \times (\vec{e}_1 \partial_3 \sigma + \vec{e}_3) dx^2 dx^3 \\ &= (\vec{e}_1 - \vec{e}_2 \partial_2 \sigma - \vec{e}_3 \partial_3 \sigma) dx^2 dx^3. \end{aligned} \tag{1.113}$$

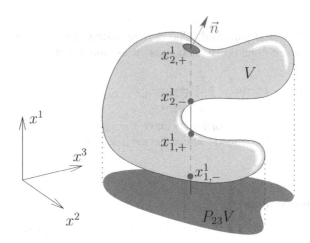

Figure 1.12 In the derivation of the divergence theorem, the integral over the volume V is first performed in the x^1 direction (vertical line in figure for fixed x^2 and x^3) for the term containing $\partial_1 v^1$, resulting in disjoint intervals, in this case $(x^1_{1,-}, x^1_{1,+})$ and $(x^1_{2,-}, x^1_{2,+})$. When integrating x^2 and x^3 over $P_{23}V$, the resulting contributions build up the surface integral for the terms $v^1 dS_1$. The corresponding argument for the other terms of the divergence give the remaining contributions to the flux integral.

From this follows that

$$v^1 dS_1 = v^1 n_1 dS = v^1 dx^2 dx^3 \tag{1.114}$$

at $x^1_{n,+}$. Similar arguments can be made for the surface at $x^1_{n,-}$, but with the difference that the surface normal must have a negative component in the x^1 direction in order to point out of the volume. Thus, we have

$$\oint_S v^1 n_1 dS = \int_{P_{23}V} \sum_i [v^1(x^1_{n,+}) - v^1(x^1_{n,-})] dx^2 dx^3 = \int_V \partial_1 v^1 dV. \tag{1.115}$$

Making the exact same argument for the other terms, i.e., first integrating in the variable that the term has a derivative with respect to, we obtain

$$\oint_S \vec{v} \cdot d\vec{S} = \oint_S v^j n_j dS = \int_V \partial_j v^j dV = \int_V \nabla \cdot \vec{v} \, dV, \tag{1.116}$$

which is the divergence theorem.

There is a minor gap in the proof above, namely that the surface only can be parametrised using x^2 and x^3 only in the regions where $\vec{e}_1$ is not tangent to the surface. While the very existence of $x^1_{n,+}$ implies that this is not the case for the volume integral, we do need to add those surfaces to the surface integral in order to complete the closed surface. However, for such surfaces $v^1 dS_1 = 0$ due to $\vec{n}$ being orthogonal to $\vec{e}_1$ and we can safely add the contribution from such surfaces to the surface integral without changing its value.

As an alternative to the above, we can start by proving the divergence theorem for a

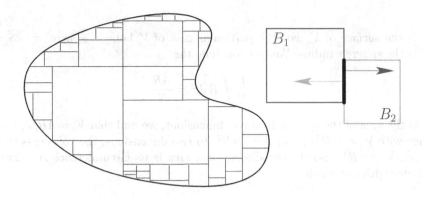

Figure 1.13 The volume V could be built up out of a collection of boxes (left). For the shared surfaces, the flux out of one box is cancelled by the flux into another, since the boxes have surface normals in opposite directions (right). The flux into the box B_1 through the shared surface is exactly the flux out of B_2 through the same surface. This leaves only the outer box surfaces, over which the flux must equal the integral of $\nabla \cdot \vec{v}$ over the volume of all boxes.

very small box B with $x_0^i < x^i < x^i + dx^i$. For the volume integral, we obtain

$$\int_B \partial_1 v^1 dV = \int_{x_0^2}^{x_0^2+dx^2} dx^2 \int_{x_0^3}^{x_0^3+dx^3} dx^3 [v^1(x^1 + dx^1) - v^1(x^1)]$$

$$= \int_{P_{23}B} [v^1 n_1|_{x^1=x_0^1+dx^1} + v^1 n_1|_{x^1=x_0^1}]dS, \qquad (1.117)$$

where we have used that $\vec{n} = \vec{e}_1$ on the end surface $x^1 = x_0^1 + dx^1$ and $\vec{n} = -\vec{e}_1$ on the end surface $x^1 = x_0^1$. The remaining terms in the volume integral contribute with the surface integrals on the other sides, resulting in the divergence theorem for a box. Starting from this, we can build any volume V out of such boxes of different sizes and end up with the divergence theorem for any volume, see Fig. 1.13. For any internal surface between two boxes, the flux out of one box equals the flux into the other since the surface normals have opposite directions (and equal surface). We may therefore cancel these fluxes, leaving only the flux through the unpaired outer surfaces of the boxes, which is equal to the volume integral of the divergence taken over all of the boxes.

This line of argumentation is essentially equivalent to the one we used already, with the advantage that we do not have to worry about disjoint intervals when performing the first integration for a box.

Example 1.21 The divergence theorem can help us compute the relation between the area and the volume of a sphere. In Example 1.16, we saw that $\nabla \cdot \vec{x} = N$, which gives us

$$\int_V \nabla \cdot \vec{x}\, dV = \int_V N\, dV = NV, \qquad (1.118)$$

where V is any volume. By the divergence theorem, it follows that

$$V = \frac{1}{N} \oint_S \vec{x} \cdot d\vec{S}, \qquad (1.119)$$

where S is the surface of V. For the particular case of V being a sphere, $\vec{x} \cdot d\vec{S} = R\,dS$, where R is the sphere's radius. We can conclude that

$$V = \frac{1}{N} \oint R\,dS = \frac{AR}{N}, \tag{1.120}$$

where A is the area of the sphere. In three dimensions, we find that $V = AR/3$, which is in accordance with $V = 4\pi R^3/3$ and $A = 4\pi R^2$. In two dimensions, the volume is the area of a circle, i.e., $V = \pi R^2$, and the one-dimensional area is its circumference $A = 2\pi R$, again in accordance with our result.

Example 1.22 The central piece of electrostatics is *Gauss's law*, which states that the total charge Q_V within a volume V equals the permittivity multiplied by the flux of the electric field through the volume's boundary surface S

$$Q_V = \int_V \rho\,dV = \varepsilon_0 \oint_S \vec{E} \cdot d\vec{S}, \tag{1.121}$$

where ρ is the charge density. By applying the divergence theorem to the flux integral, we obtain

$$\int_V \rho\,dV = \varepsilon_0 \int_V \nabla \cdot \vec{E}\,dV. \tag{1.122}$$

In order for this to hold for *any* volume V, this leads to the conclusion that

$$\nabla \cdot \vec{E} = \frac{\rho}{\varepsilon_0}, \tag{1.123}$$

which is *Gauss's law on differential form*. This is perhaps the most well-known of *Maxwell's equations*.

Example 1.23 A common application of the divergence theorem is the derivation of *Green's identities*, which are identities for integrals involving several fields. Let us start with a scalar field φ and a vector field $\vec{v}$ and the application of the divergence theorem to the volume integral

$$\int_V \nabla \cdot (\varphi \vec{v})dV = \oint_S \varphi \vec{v} \cdot d\vec{S}. \tag{1.124}$$

Applying the divergence to the field $\varphi \vec{v}$ results in

$$\int_V (\varphi \nabla \cdot \vec{v} + \vec{v} \cdot \nabla \varphi)\,dV = \oint_S \varphi \vec{v} \cdot d\vec{S}. \tag{1.125}$$

Green's first identity is derived from this equation by letting $\vec{v} = \nabla \psi$, where ψ is a scalar field

$$\int_V \left[\varphi \nabla^2 \psi + (\nabla \varphi) \cdot (\nabla \psi)\right] dV = \oint_S \varphi \nabla \psi \cdot d\vec{S}. \tag{1.126}$$

Green's second identity is obtained from applying the divergence theorem to the integral

$$\int_V \nabla \cdot (\varphi \nabla \psi - \psi \nabla \varphi) dV = \int_V (\varphi \nabla^2 \psi - \psi \nabla^2 \varphi) dV = \oint_S (\varphi \nabla \psi - \psi \nabla \varphi) \cdot d\vec{S}. \quad (1.127)$$

1.5.3   Green's formula

An important special case of the divergence theorem is *Green's formula*, which is the result of applying the divergence theorem to the two-dimensional case. In this setting, the surface is a curve and the surface element is given by

$$d\vec{S} = \vec{e}_i \varepsilon_{ij} \frac{\partial x^j}{\partial t} dt = \vec{e}_i \varepsilon_{ij} dx^j = \vec{e}_1 dx^2 - \vec{e}_2 dx^1, \quad (1.128)$$

where t is a parameter which parametrises the curve. The volume element is the area element $dx^1 dx^2$ and the divergence theorem now states

$$\int_S \partial_i v^i dx^1 dx^2 = \int_S \left(\frac{\partial v^1}{\partial x^1} + \frac{\partial v^2}{\partial x^2} \right) dx^1 dx^2 = \oint_\Gamma \vec{v} \cdot d\vec{S} = \oint_\Gamma \left(v^1 dx^2 - v^2 dx^1 \right), \quad (1.129)$$

where S is some area and Γ its bounding curve. In order to state this on the form that is usually quoted for Green's formula, we consider the vector field $\vec{v} = Q(\vec{x})\vec{e}_1 - P(\vec{x})\vec{e}_2$, which leads to

$$\int_S \left(\frac{\partial Q}{\partial x^1} - \frac{\partial P}{\partial x^2} \right) dx^1 dx^2 = \oint_\Gamma (P \, dx^1 + Q \, dx^2). \quad (1.130)$$

In order for the normal to point out of the area S, the direction of the curve Γ should be counter clockwise, assuming the usual ordering of the basis vectors $\vec{e}_1$ and $\vec{e}_2$, see Fig. 1.14.

1.5.4   The curl theorem

Just as the divergence theorem connects the volume integral of a divergence with a flux integral over the bounding surface, the *curl theorem* (or *Stokes' theorem*) connects the flux integral of a curl over a surface with a line integral around the bounding curve. If we let S be a surface and Γ its boundary, then

$$\int_S (\nabla \times \vec{v}) \cdot d\vec{S} = \oint_\Gamma \vec{v} \cdot d\vec{x}. \quad (1.131)$$

It is important to note that the direction of the surface normal dictates the direction of the curve Γ, as illustrated in Fig. 1.15. To show this, we parametrise the two-dimensional surface using the parameters s and t such that the surface element is given by

$$d\vec{S} = \frac{\partial \vec{x}}{\partial s} \times \frac{\partial \vec{x}}{\partial t} ds \, dt \quad (1.132)$$

and the surface S corresponds to some area S_p in the s-t parameter space. The sought flux integral is now given by

$$\int_S (\nabla \times \vec{v}) \cdot d\vec{S} = \int_{S_p} (\nabla \times \vec{v}) \cdot \left(\frac{\partial \vec{x}}{\partial s} \times \frac{\partial \vec{x}}{\partial t} \right) ds \, dt. \quad (1.133)$$

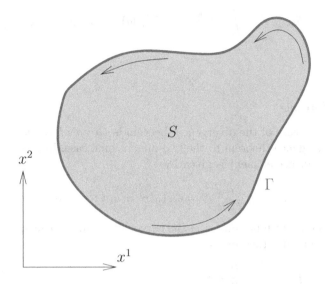

Figure 1.14 In Green's formula, the integral over the boundary curve Γ should be taken counter clockwise.

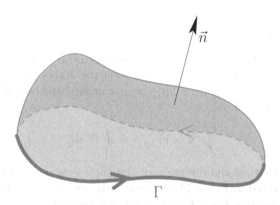

Figure 1.15 The relation between the normal vector of the surface and the direction of the curve for the corresponding line integral based on the curl theorem. Looking through the surface in the normal direction, the bounding curve should be in the clockwise direction.

Using the cyclicity of the vector triple product, the integrand may be rewritten as

$$(\nabla \times \vec{v}) \cdot \left(\frac{\partial \vec{x}}{\partial s} \times \frac{\partial \vec{x}}{\partial t} \right) = \left[\left(\frac{\partial \vec{x}}{\partial s} \times \frac{\partial \vec{x}}{\partial t} \right) \times \nabla \right] \cdot \vec{v} = \left[\frac{\partial \vec{x}}{\partial t} \left(\frac{\partial \vec{x}}{\partial s} \cdot \nabla \right) - \frac{\partial \vec{x}}{\partial s} \left(\frac{\partial \vec{x}}{\partial t} \cdot \nabla \right) \right] \cdot \vec{v}$$

$$= \frac{\partial \vec{x}}{\partial t} \cdot \frac{\partial \vec{v}}{\partial s} - \frac{\partial \vec{x}}{\partial s} \cdot \frac{\partial \vec{v}}{\partial t} = \frac{\partial}{\partial s} \left(\vec{v} \cdot \frac{\partial \vec{x}}{\partial t} \right) - \frac{\partial}{\partial t} \left(\vec{v} \cdot \frac{\partial \vec{x}}{\partial s} \right), \qquad (1.134)$$

where we have used the chain rule $(\partial \vec{x}/\partial \tau) \cdot \nabla = (\partial x^i/\partial \tau)\partial_i = \partial/\partial \tau$. From Green's formula (taking $x^1 = s$ and $x^2 = t$ in Eq. (1.130)), it now follows that

$$\int_S (\nabla \times \vec{v}) \cdot d\vec{S} = \int_{S_p} (\nabla \times \vec{v}) \cdot \left(\frac{\partial \vec{x}}{\partial s} \times \frac{\partial \vec{x}}{\partial t} \right) ds \, dt$$

$$= \int_{S_p} \left[\frac{\partial}{\partial s} \left(\vec{v} \cdot \frac{\partial \vec{x}}{\partial t} \right) - \frac{\partial}{\partial t} \left(\vec{v} \cdot \frac{\partial \vec{x}}{\partial s} \right) \right] ds \, dt = \oint_{\Gamma_p} \left(\vec{v} \cdot \frac{\partial \vec{x}}{\partial t} dt + \vec{v} \cdot \frac{\partial \vec{x}}{\partial s} ds \right)$$

$$= \oint_\Gamma \vec{v} \cdot d\vec{x}, \qquad (1.135)$$

which is the curl theorem.

As an important consequence of the curl theorem, it follows that for any vector field $\vec{w}$ that can be written as the curl of another vector field $\vec{v}$, i.e., $\vec{w} = \nabla \times \vec{v}$, the flux

$$\int_S \vec{w} \cdot d\vec{S} = \oint_\Gamma \vec{v} \cdot d\vec{x}. \qquad (1.136)$$

Thus, the flux of $\vec{w}$ through S is independent of the shape of the actual surface S as long as it has Γ as a boundary curve. It also follows that the flux of $\vec{w} = \nabla \times \vec{v}$ through a closed surface must be zero, as it does not have a boundary curve.

Example 1.24 In magnetostatics, we will be dealing with the magnetic field $\vec{B}$, which may be written as the curl of the vector potential $\vec{A}$

$$\vec{B} = \nabla \times \vec{A}. \qquad (1.137)$$

The flux Φ of the magnetic field through a surface S can then be expressed as

$$\Phi = \int_S \vec{B} \cdot d\vec{S} = \int_S (\nabla \times \vec{A}) \cdot d\vec{S} = \oint_\Gamma \vec{A} \cdot d\vec{S}, \qquad (1.138)$$

where Γ is the boundary curve of S, where we have used the curl theorem. Because of this relation, the magnetic flux through a surface only depends on the surface boundary and not on the surface itself. Any surface with the same boundary as S will result in the same flux.

1.5.5 General integral theorems

Both the divergence and curl theorems may be generalised in such a way that they are not only valid for the special cases when the result of the integral is a scalar. The generalisations are valid for any function $f(\vec{x})$, be it a scalar or a component of a vector or tensor field, and given by

$$\int_V \partial_i f \, dV = \oint_S f \, dS_i, \tag{1.139a}$$

$$\int_S \varepsilon_{ijk} \partial_k f \, dS_j = \oint_\Gamma f \, dx^i, \tag{1.139b}$$

where, as in the case of the divergence and curl theorems, S is the boundary of V for the first equation, while Γ is the boundary of S in the second. The proofs of these relations are completely analogous to those of the earlier theorems. In the case of the divergence theorem, the only modification is that we only need to consider one term, and we integrate in the x^i direction first. In fact, we could prove the divergence theorem by first proving the more general form and afterwards letting f be the jth component of a vector $\vec{v}$. We would obtain

$$I_i^j = \int_V \partial_i v^j \, dV = \oint_S v^j \, dS_i \tag{1.140}$$

and upon letting $i = j$ and summing

$$I_i^i = \int_V \partial_i v^i \, dV = \int_V \nabla \cdot \vec{v} \, dV = \oint_S v^i \, dS_i = \oint_S \vec{v} \cdot d\vec{S}. \tag{1.141}$$

Proving the generalised curl theorem starting from the regular curl theorem is left as Problem 1.44.

1.6 NON-CARTESIAN COORDINATE SYSTEMS

While Cartesian coordinate systems provide a solid basis for dealing with scalars, vectors, and later on tensors, in affine spaces, we will encounter problems that are easier to describe and solve using a different set of coordinates that are not necessarily linear or even orthogonal. In several cases, we will also be able to apply symmetry arguments in a particular set of coordinates that are not as easily applicable in other coordinate systems.

1.6.1 General theory

A coordinate system of an N dimensional affine space is a labelling of all the points in the space by N numbers, which we call *coordinates*. Each combination of possible coordinates should uniquely define a point in the space. Each y^a is a continuous function called a *coordinate function* from the N dimensional space to the real numbers. From this point on, we will adopt the convention of using early letters in the alphabet (a, b, c, ...) as indices for general coordinate systems and reserve letters from the middle of the alphabet (i, j, k, ...) for indices in Cartesian coordinates.

Example 1.25 Given an affine N dimensional space, we can always define a Cartesian coordinate system as we have done previously. We do so by selecting one point in the affine space as the origin O and selecting a set of orthonormal basis vectors $\vec{e}_i$. The coordinates are then defined by mapping the coordinates x^i onto the point P such that

$$\overrightarrow{OP} = \vec{x} = \vec{e}_i x^i, \tag{1.142}$$

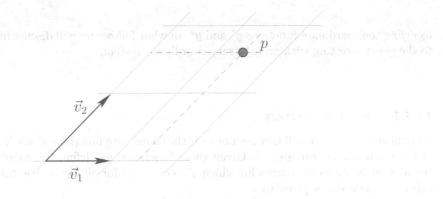

Figure 1.16 A coordinate system for a two-dimensional affine space may be based on the linearly independent, but not orthogonal, set of vectors $\vec{v}_1$ and $\vec{v}_2$. In this case, the coordinates of the point p would be $y^1 = 1.4$, $y^2 = 1.6$.

where $\overrightarrow{OP}$ is the difference vector between P and O. The coordinate functions are given by

$$x^i(P) = \vec{e}_i \cdot \overrightarrow{OP}. \tag{1.143}$$

Since we can always start by writing down a Cartesian coordinate system for an affine space, the coordinates of a more general coordinate system may be written in terms of these Cartesian coordinates for simplicity.

Example 1.26 We can create a coordinate system that is not based on orthogonal basis vectors in a similar way to how we created the Cartesian coordinate system. Instead of a set of orthonormal basis vectors, it suffices to pick a set of linearly independent vectors $\vec{v}_a$. The mapping of the coordinates y^a to the point P is given by

$$\overrightarrow{OP} = \vec{x} = \vec{v}_a y^a. \tag{1.144}$$

However, note that the coordinate functions are not given by $y^a = \vec{v}_a \cdot \overrightarrow{OP}$ in this case as the vectors $\vec{v}_a$ are not necessarily orthogonal or even normalised. For concreteness, let us take a two-dimensional space parametrised using the basis $\vec{v}_1 = \vec{e}_1$ and $\vec{v}_2 = \vec{e}_1 + \vec{e}_2$, where $\vec{e}_1$ and $\vec{e}_2$ form an orthonormal basis, see Fig. 1.16. We obtain

$$\vec{x} = y^1\vec{e}_1 + y^2(\vec{e}_1 + \vec{e}_2) = \vec{e}_1(y^1 + y^2) + \vec{e}_2 y^2. \tag{1.145}$$

From this we can deduce that the coordinate functions are given by

$$y^1 = (\vec{e}_1 - \vec{e}_2) \cdot \vec{x} = x^1 - x^2 \neq \vec{v}_1 \cdot \vec{x}, \tag{1.146a}$$

$$y^2 = \vec{e}_2 \cdot \vec{x} = x^2 \neq \vec{v}_2 \cdot \vec{x}. \tag{1.146b}$$

Equivalently, we could have used

$$y^1 = x^1 - x^2, \tag{1.147a}$$

$$y^2 = x^2, \tag{1.147b}$$

to *define* the coordinate functions y^1 and y^2. In what follows we will discuss how this applies to the general setting with an arbitrary coordinate system.

1.6.1.1 Tangent vector basis

As indicated above, we will now assume that the coordinate functions y^a are known functions of the Cartesian coordinates x^i. Given these functions, we define the *coordinate lines* for the y^a coordinate as the curves for which y^b is constant for all $b \neq a$. The *tangent vector* of this coordinate line is given by

$$\vec{E}_a = \frac{\partial \vec{x}}{\partial y^a} = \frac{\partial x^i}{\partial y^a} \vec{e}_i, \tag{1.148}$$

where $\vec{e}_i$ are the Cartesian basis vectors, which are constant. The partial derivative in this expression is the partial derivative of the Cartesian coordinate functions expressed as functions of the y coordinates. In order for the coordinate system to uniquely define a point, the tangent vectors $\vec{E}_a$ must be linearly independent. We can use this fact in order to use the $\vec{E}_a$ vectors as a set of basis vectors and express any vector as a linear combination of these basis vectors. This presents us with a conundrum as the vectors $\vec{E}_a$ may generally depend on the coordinates. The way forward here is to realise that most vectors are defined only at a given point in space. In particular, for vector fields, it is natural to use the basis defined at a particular spatial point as the basis for the fields at that point. In terms of components, we would write

$$\vec{v}(\vec{x}) = v^a(\vec{x})\vec{E}_a(\vec{x}). \tag{1.149}$$

In what follows, we will often suppress the explicit mention of the spatial dependence of both vector components and basis vectors, i.e., we will assume that $v^a \equiv v^a(\vec{x})$ and $\vec{E}_a \equiv \vec{E}_a(\vec{x})$.

The set of vectors $\vec{E}_a$ is called a *tangent vector basis* and the vector components v^a are called *contravariant vector components*. The reason for this nomenclature is that a vector is not dependent upon the basis used to represent it and in order to ensure that this is the case, the vector components must transform in the opposite way as compared to the basis vectors under a coordinate change. In particular for a vector $\vec{v}$

$$\vec{v} = v^a \vec{E}_a = v^a \frac{\partial x^i}{\partial y^a} \vec{e}_i = v^i \vec{e}_i. \tag{1.150}$$

Scalar multiplication with $\vec{e}_j$ now leads to

$$v^j = \frac{\partial x^j}{\partial y^a} v^a \quad \Longrightarrow \quad v^b = \frac{\partial y^b}{\partial x^j} \frac{\partial x^j}{\partial y^a} v^a = \frac{\partial y^b}{\partial x^j} v^j. \tag{1.151}$$

Here we have used the chain rule and the fact that

$$\frac{\partial y^b}{\partial x^j} \frac{\partial x^j}{\partial y^a} = \delta_a^b. \tag{1.152}$$

Note that we have here introduced a slightly new notation for the δ symbol, namely that it has one index up and one index down. As we shall see, the position of the indices (up or down), that we have so far been quite flexible with, plays a crucial role for non-Cartesian coordinate systems. In Eq. (1.151), we have found that the vector components v^a transform with the partial derivative of the y coordinates with respect to the x coordinates instead of vice versa, as the tangent vector basis does.

We also note that the form of Eq. (1.151) is not restricted to when one of the coordinate systems is Cartesian. If we have two coordinate systems with coordinates y^a and $y'^{a'}$, respectively, we obtain

$$v^{a'} = \frac{\partial y'^{a'}}{\partial x^j} v^j = \frac{\partial y'^{a'}}{\partial x^j} \frac{\partial x^j}{\partial y^a} v^a = \frac{\partial y'^{a'}}{\partial y^a} v^a, \tag{1.153}$$

where we have again applied the chain rule for the derivatives.

Example 1.27 In Example 1.26, we used the vectors $\vec{v}_1$ and $\vec{v}_2$ to define a new coordinate system on a two-dimensional affine space. Towards the end of the example, we noted that the coordinate system could also be defined by specifying the coordinate functions as in Eqs. (1.147). Inverting these equations leads to

$$x^1 = y^1 + y^2, \quad x^2 = y^2. \tag{1.154}$$

It follows directly that the tangent vector basis is given by

$$\vec{E}_1 = \frac{\partial \vec{x}}{\partial y^1} = \vec{e}_1 = \vec{v}_1, \quad \vec{E}_2 = \frac{\partial \vec{x}}{\partial y^2} = \vec{e}_1 + \vec{e}_2 = \vec{v}_2. \tag{1.155}$$

Thus, the tangent vector basis is the basis we used to define the coordinate system in Example 1.26. Note that this is a particular result for the situation when the coordinate functions are linear, we call this type of coordinates *affine coordinates*.

1.6.1.2 Dual basis

Apart from the tangent vector basis, there is a different set of basis vectors which we can define using a given coordinate system. As we move on to more advanced topics, the need shall arise to separate the following concept further from the concept of tangent vectors, but for now let us define a set of vector fields by taking the gradient of the coordinate functions

$$\vec{E}^a = \nabla y^a. \tag{1.156}$$

These vector fields also form a basis at each point in the base space and their relation to the Cartesian basis is

$$\vec{e}_i \cdot \vec{E}^a = \frac{\partial y^a}{\partial x^i} \quad \Longrightarrow \quad \vec{E}^a = \frac{\partial y^a}{\partial x^i} \vec{e}_i. \tag{1.157}$$

Note that the vectors $\vec{E}^a$ follow the very same transformation rule as the contravariant vector components and thus transforms in the opposite fashion as compared to the tangent vector basis. This new set of basis vectors is called the *dual basis* and has a set of interesting properties. The first one that we note is that the vectors in the dual basis are not necessarily parallel to the corresponding tangent vectors.

Example 1.28 Returning again to Example 1.26, Eqs. (1.147) directly gives us

$$\vec{E}^1 = \vec{e}_1 - \vec{e}_2, \quad \vec{E}^2 = \vec{e}_2, \tag{1.158}$$

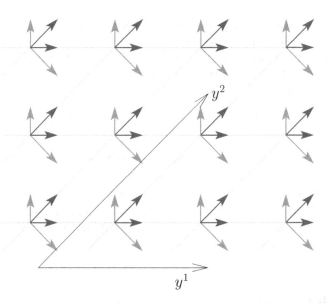

Figure 1.17 The two-dimensional coordinate system $y^1 = x^1 - x^2$, $y^2 = x^2$ along with the associated tangent vector basis (dark) and dual basis (light). The coordinate lines of y^1 are at the same time the level surfaces of y^2. By definition, the tangent vector field $\vec{E}_1$ is tangent to the y^1 coordinate lines while the dual field $\vec{E}^2$ is normal to the y^2 level surfaces. Thus, we must have $\vec{E}_1 \cdot \vec{E}^2 = 0$. The corresponding argument can be made for $\vec{E}_2$ and $\vec{E}^1$.

which are clearly not parallel to $\vec{E}_1$ and $\vec{E}_2$, respectively. The two different bases are illustrated in Fig. 1.17. However, as shown in the figure, the tangent vector field $\vec{E}_1$ is orthogonal to the dual vector field $\vec{E}^2$.

The fact that $\vec{E}_1 \cdot \vec{E}^2 = 0$ in the above example is by no means a coincidence. From the properties of the tangent vector and dual fields, we generally obtain

$$\vec{E}_a \cdot \vec{E}^b = \vec{e}_i \cdot \vec{e}_j \frac{\partial x^i}{\partial y^a} \frac{\partial y^b}{\partial x^j} = \frac{\partial x^i}{\partial y^a} \frac{\partial y^b}{\partial x^i} = \frac{\partial y^b}{\partial y^a} = \delta_a^b. \tag{1.159}$$

Thus, although $\vec{E}^a$ is not generally parallel with $\vec{E}_a$, it is the only dual basis vector that has a non-zero scalar product with $\vec{E}_a$ and this scalar product is equal to one.

Example 1.29 While the tangent vector basis and dual basis are not necessarily parallel to each other, there is an important class of coordinate systems in which they are. An example of this is the coordinate system

$$s = \sinh(x^1), \quad t = \sinh(x^2), \tag{1.160}$$

where we have called the new coordinates s and t rather than y^1 and y^2 in order to avoid confusion in later examples. This coordinate system has the same coordinate lines and coordinate level surfaces as the original Cartesian system as it only rescales the coordinates.

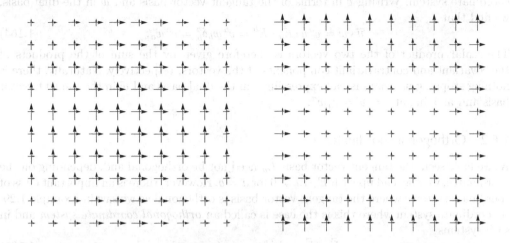

Figure 1.18 The tangent vector basis (left) and dual basis (right) for the coordinate system $s = \sinh(x^1)$, $t = \sinh(x^2)$ with $s = t = 0$ in the middle of the figures (not to scale between the figures). When the magnitude of the tangent vector basis is large, then the magnitude of the dual basis is small and vice versa.

We find that

$$\vec{E}_s = \vec{e}_1 \frac{1}{\sqrt{1+s^2}}, \quad \vec{E}_t = \vec{e}_2 \frac{1}{\sqrt{1+t^2}} \tag{1.161a}$$

$$\vec{E}^s = \vec{e}_1 \sqrt{1+s^2}, \quad \vec{E}^t = \vec{e}_2 \sqrt{1+t^2}, \tag{1.161b}$$

where we have taken the habit of using the coordinate names rather than numbers when speaking about specific components. In Fig. 1.18, we show the tangent vector and dual basis fields for this coordinate system. Another difference from our previous example is that the basis vectors depend on the coordinates, something which is often the case for non-Cartesian coordinate systems.

When it comes to decomposing an arbitrary vector $\vec{v}$ into a linear combination of $\vec{E}^a$ we find that, with $\vec{v} = v_a \vec{E}^a$,

$$\vec{E}_a \cdot \vec{v} = \vec{E}_a \cdot v_b \vec{E}^b = \delta_a^b v_b = v_a. \tag{1.162}$$

Using the expression for $\vec{E}_a$ in Cartesian coordinates, we find that

$$v_a = \frac{\partial x^i}{\partial y^a} \vec{e}_i \cdot \vec{v} = \frac{\partial x^i}{\partial y^a} v^i. \tag{1.163}$$

The components v_a thus transform in the same way as the tangent vector basis under coordinate changes and are therefore called *covariant vector components*.

Before specialising to coordinate systems that fulfil an additional orthogonality criterion, let us examine how to express the scalar product of two general vectors $\vec{v}$ and $\vec{w}$ in a general coordinate system. Writing $\vec{v}$ in terms of the tangent vector basis and $\vec{w}$ in the dual basis, we find that

$$\vec{v} \cdot \vec{w} = v^a w_b \vec{E}_a \cdot \vec{E}^b = v^a w_b \delta^a_b = v^a w_a. \tag{1.164}$$

The scalar product of the two vectors is therefore given by the sum of the products of the covariant and contravariant components of the vectors, respectively. Naturally, there is nothing stopping us from instead expressing $\vec{v}$ in the dual basis and $\vec{w}$ in the tangent vector basis and arriving at $\vec{v} = \vec{w} = v_a w^a$.

1.6.2 Orthogonal coordinates

As we have seen, the tangent vector basis $\vec{E}_a$ need not be orthogonal and, depending on the coordinates, we may end up with $\vec{E}_a \cdot \vec{E}_b \neq 0$ for $a \neq b$. However, there is an important class of coordinate systems where the tangent vector basis is orthogonal as we saw in Example 1.29. A coordinate system where this is the case is called an *orthogonal coordinate system* and in such systems

$$\vec{E}_a \cdot \vec{E}_b = h_a^2 \delta_{ab}, \qquad \text{(no sum)} \tag{1.165}$$

where we have introduced the *scale factors* $h_a = |\vec{E}_a|$, which are the moduli of the tangent vector basis. Since the vectors $\vec{E}_a$ are orthogonal, we can define an *orthonormal basis* $\vec{e}_a$ according to

$$\vec{e}_a = \frac{1}{h_a}\vec{E}_a = \frac{1}{h_a}\frac{\partial \vec{x}}{\partial y^a}. \qquad \text{(no sum)} \tag{1.166}$$

The normalisation condition gives

$$h_a = \sqrt{\sum_i \left(\frac{\partial x^i}{\partial y^a}\right)^2}. \tag{1.167}$$

For our purposes, it is not relevant to work with covariant and contravariant indices once we have established the orthonormal basis $\vec{e}_a$. For the remainder of this chapter, we will therefore use a notation similar to that used for Cartesian coordinates, with the only difference being in the indices used. Naturally, any vector can be written as a linear combination of the $\vec{e}_a$ basis

$$\vec{v} = \tilde{v}_a \vec{e}_a = \sum_a \tilde{v}_a \frac{1}{h_a}\vec{E}_a, \tag{1.168}$$

which leads to the relation $\tilde{v}_a = h_a v^a$ (no sum), where $\tilde{v}_a$ are the *physical components* of the vector $\vec{v}$ in the orthonormal basis, since they are directly related to the projection of the vector onto the basis unit vectors.

Example 1.30 In Example 1.29 we introduced an orthogonal coordinate system with

$$\vec{E}_s = \frac{1}{\sqrt{1+s^2}}\vec{e}_1, \quad \vec{E}_t = \frac{1}{\sqrt{1+t^2}}\vec{e}_2. \tag{1.169}$$

Taking the modulus of these two tangent vectors gives us

$$h_s = \frac{1}{\sqrt{1+s^2}}, \quad h_t = \frac{1}{\sqrt{1+t^2}} \tag{1.170}$$

and thus

$$\vec{e}_s = \vec{e}_1, \quad \vec{e}_t = \vec{e}_2. \tag{1.171}$$

In the above example, the orthonormal basis turned out to be the same as the Cartesian one, but the $\vec{e}_a$ generally depend on the point in the base space, just as $\vec{E}_a$ does, due to the possibility of $\vec{E}_a$ changing direction from point to point.

Let us now find the relation between the dual basis $\vec{E}^a$ and the orthonormal basis $\vec{e}_a$. In order to do this, we expand $\vec{E}^a$ in the orthonormal basis as $\vec{E}^a = \tilde{E}_b^a \vec{e}_b$ and obtain

$$\tilde{E}_b^a = \vec{e}_b \cdot \vec{E}^a = \frac{1}{h_b} \vec{E}_b \cdot \vec{E}^a = \frac{1}{h_b} \delta_b^a. \qquad \text{(no sum)} \qquad (1.172)$$

It follows directly that

$$\vec{e}_a = \frac{1}{h_a} \vec{E}_a = h_a \vec{E}^a. \qquad \text{(no sum)} \qquad (1.173)$$

Thus, not only is the tangent vector basis orthogonal, it is also parallel to the dual basis, with the only difference between the two being a factor of h_a^2. This also means that we could just as well construct the orthonormal basis starting from the dual basis.

1.6.2.1 Integration in orthogonal coordinates

It is worth spending some time and effort on discussing line, surface and volume integrals in orthogonal coordinate systems. In all cases, the relevant quantity will be

$$\frac{\partial \vec{x}}{\partial y^a} = \vec{E}_a = h_a \vec{e}_a. \qquad \text{(no sum)} \qquad (1.174)$$

The form of a line integral does not change significantly, we only need to note that

$$\int_\Gamma \vec{v} \cdot d\vec{x} = \int_\Gamma \vec{v} \cdot \vec{E}_a \dot{y}^a d\tau, \qquad (1.175)$$

where $\dot{y}^a = dy^a/d\tau$ and τ is a curve parameter. This lets us express the integral as

$$\int_\Gamma \vec{v} \cdot d\vec{x} = \int_\Gamma \sum_a \tilde{v}_a h_a \dot{y}^a d\tau. \qquad (1.176)$$

In particular, if integrating along the y^b coordinate line, we would use y^b as a parameter and thus have

$$\int_\Gamma \vec{v} \cdot d\vec{x} = \int_\Gamma \sum_a \tilde{v}_a h_a \frac{\partial y^a}{\partial y^b} dy^b = \int_\Gamma \tilde{v}_b h_b dy^b. \qquad \text{(no sum)} \qquad (1.177)$$

Similar to the above, if we have a surface integral over a coordinate level surface S_c with $y^c = \text{constant}$ and the other coordinates being y^a and y^b, which we can use for parametrising the surface, then the surface element will be given by

$$d\vec{S} = \frac{\partial \vec{x}}{\partial y^a} \times \frac{\partial \vec{x}}{\partial y^b} dy^a dy^b = \vec{E}_a \times \vec{E}_b \, dy^a dy^b$$
$$= h_a h_b (\vec{e}_a \times \vec{e}_b) dy^a dy^b = h_a h_b \vec{e}_c dy^a dy^b, \qquad \text{(no sum)} \qquad (1.178)$$

assuming that $\vec{e}_a$, $\vec{e}_b$, and $\vec{e}_c$ form a right-handed triplet. Since $\vec{e}_c$ is a unit vector, it is the unit normal and the surface area is $dS = h_a h_b dy^a dy^b$ (no sum). A flux integral over this surface is therefore given by

$$\int_{S_c} \vec{v} \cdot d\vec{S} = \int_{S_c} \tilde{v}_c h_a h_b dy^a dy^b. \qquad \text{(no sum)} \qquad (1.179)$$

Finally, parametrising a volume with the coordinates, the volume element dV may be expressed as

$$dV = \frac{\partial \vec{x}}{\partial y^1} \cdot \left(\frac{\partial \vec{x}}{\partial y^2} \times \frac{\partial \vec{x}}{\partial y^3} \right) dy^1 dy^2 dy^3 = \vec{E}_1 \cdot (\vec{E}_2 \times \vec{E}_3) dy^1 dy^2 dy^3$$

$$= h_1 h_2 h_3 dy^1 dy^2 dy^3, \tag{1.180}$$

with a similar generalisation of $dV = \prod_a h_a dy^a$ in an arbitrary number of dimensions. For ease of notation, we introduce the *Jacobian determinant*

$$\mathcal{J} = \prod_a h_a = h_1 h_2 h_3, \tag{1.181}$$

where the last equality holds in three dimensions, which is the product of all of the scale factors. However, it should be noted that the Jacobian determinant also appears in non-orthogonal coordinates, although it does not take the simple form shown above. In general coordinates, we would have

$$\mathcal{J} = \vec{E}_1 \cdot (\vec{E}_2 \times \vec{E}_3), \tag{1.182}$$

again with a straightforward generalisation to an arbitrary number of dimensions. As a result, a volume integral expressed in an arbitrary coordinate system takes the form

$$\int_\Omega \ldots dV = \int_{\Omega_p} \ldots \mathcal{J} \, dy^1 dy^2 dy^3, \tag{1.183}$$

where Ω_p is the set of coordinates corresponding to the volume Ω.

Example 1.31 There is an intuitive interpretation of the integral expressions derived above. The tangent vector $\vec{E}_a = h_a \vec{e}_a$ (no sum) is the change in the position vector associated to an infinitesimal change in the coordinate y^a. A line element therefore becomes directly proportional to $\vec{E}_a dy^a$. Due to the orthogonality of these changes, a surface element spanned by two of the $\vec{E}_a$ will be equal to the product of the scale factors in magnitude and directed in the third direction and the volume spanned by all three of them will be the product of all of the side lengths of the spanned rectangular box.

As an important final comment on integration in orthogonal coordinate systems, it must be noted that if the integrand is a vector quantity, then the result cannot be obtained by integrating the components of the orthonormal basis $\vec{e}_a$ one by one. The reason for this is that the basis vectors themselves are not constant and may therefore not be taken outside of the integral as in the case of Cartesian coordinates.

1.6.2.2 *Differentiation in orthogonal coordinates*

Just as integration takes a slightly different form in general coordinate systems as compared to Cartesian coordinates, the differential operators acting on scalar and vector fields take on a slightly modified form, with the appearance of scale factors slightly simplifying the results. The operator that is easiest to transform to orthogonal coordinates is the gradient of a scalar field φ, which will be given by

$$\nabla \varphi = \vec{e}_i \partial_i \varphi = \vec{e}_i \frac{\partial y^a}{\partial x^i} \partial_a \varphi = \vec{E}^a \partial_a \varphi, \tag{1.184}$$

which holds in any coordinate system. It follows that the partial derivatives $\partial_a \varphi$ are naturally covariant vector components as they will follow the covariant transformation rules. Writing this in terms of the orthonormal basis, the *gradient in orthogonal coordinates* is

$$\nabla \varphi = \sum_a \frac{1}{h_a} \vec{e}_a \partial_a \varphi. \tag{1.185}$$

Due to the appearance of scale factors, expressing the divergence and curl in terms of the orthonormal basis $\vec{e}_a$ and its vector components poses a bit more of a challenge and we will be greatly helped by the following argumentation: Let us consider the dual basis $\vec{E}^a = \nabla y^a = \vec{e}_a/h_a$ (no sum). Taking the cross product between two of these vectors, we obtain

$$\vec{E}^a \times \vec{E}^b = \frac{1}{h_a h_b} \vec{e}_a \times \vec{e}_b = \frac{1}{h_a h_b} \sum_c \varepsilon_{abc} \vec{e}_c = \sum_c \frac{h_c}{\mathcal{J}} \varepsilon_{abc} \vec{e}_c, \tag{1.186}$$

where ε_{abc} has the same properties as the permutation symbol in Cartesian coordinates, i.e., it is completely anti-symmetric and equal to one when $\vec{e}_a$, $\vec{e}_b$, and $\vec{e}_c$ form a right-handed system. Taking the divergence of the original expression leads to

$$\nabla \cdot (\vec{E}^a \times \vec{E}^b) = \nabla \cdot (\nabla y^a \times \nabla y^b) = \nabla y^b \cdot (\nabla \times \nabla y^a) - \nabla y^a \cdot (\nabla \times \nabla y^b) = 0, \tag{1.187}$$

due to $\nabla \times \nabla \varphi = 0$ for all functions φ, and thus also for the coordinate functions. Inserting the expression for $\vec{E}^a \times \vec{E}^b$ in terms of the orthonormal basis into this gives the result

$$\sum_c \nabla \cdot \frac{h_c}{\mathcal{J}} \varepsilon_{abc} \vec{e}_c = 0. \tag{1.188}$$

Multiplying this with ε_{dab}, summing over a and b, and making use of the ε-δ-relation we finally arrive at

$$\sum_{abc} \varepsilon_{dab} \varepsilon_{abc} \nabla \cdot \frac{h_c}{\mathcal{J}} \vec{e}_c = 2 \nabla \cdot \frac{h_d}{\mathcal{J}} \vec{e}_d = 0. \tag{1.189}$$

Equipped with this relation, we move on to expressing the *divergence in orthogonal coordinates*. By expanding the vector $\vec{v}$ as $\vec{v} = \tilde{v}_a \vec{e}_a$, we obtain

$$\nabla \cdot \vec{v} = \nabla \cdot \tilde{v}_a \vec{e}_a = \sum_a \nabla \cdot \left(\frac{\tilde{v}_a \mathcal{J}}{h_a} \frac{h_a}{\mathcal{J}} \vec{e}_a \right) = \sum_a \left[\frac{h_a}{\mathcal{J}} \vec{e}_a \cdot \nabla \left(\frac{\tilde{v}_a \mathcal{J}}{h_a} \right) + \frac{\tilde{v}_a \mathcal{J}}{h_a} \nabla \cdot \left(\frac{h_a}{\mathcal{J}} \vec{e}_a \right) \right]$$

$$= \sum_a \frac{h_a}{\mathcal{J}} \vec{e}_a \cdot \nabla \left(\frac{\tilde{v}_a \mathcal{J}}{h_a} \right) = \frac{1}{\mathcal{J}} \sum_a \partial_a \left(\frac{\tilde{v}_a \mathcal{J}}{h_a} \right), \tag{1.190}$$

where in the last step we have used Eq. (1.185) to express the gradient in terms of the orthonormal basis.

Before moving on with the curl, we note that the *Laplace operator in orthogonal coordinates* can be expressed by combining Eqs. (1.185) and (1.190) into

$$\nabla^2 \varphi = \nabla \cdot \nabla \varphi = \frac{1}{\mathcal{J}} \sum_a \partial_a \left(\frac{\mathcal{J}}{h_a^2} \partial_a \varphi \right). \tag{1.191}$$

Finally, the *curl in orthogonal coordinates* can be expressed using the scale factors and vector components $\tilde{v}_a$ by noting that

$$\nabla \times \vec{v} = \nabla \times v_a \vec{E}^a = -\vec{E}^a \times \nabla v_a + v_a \nabla \times \vec{E}^a = -\vec{E}^a \times \nabla v_a. \tag{1.192}$$

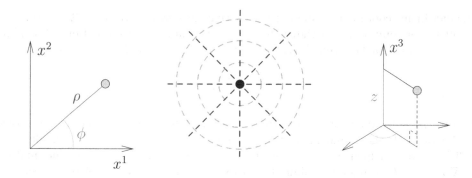

Figure 1.19 The left figure shows the relation between polar coordinates ρ and ϕ and a Cartesian coordinate system in two dimensions. The middle figure shows coordinate lines of ρ (dark) and ϕ (light) based on a polar coordinate system with the black point as origin. A cylinder coordinate system can be created by adding a third direction perpendicular to the plane as shown in the right figure.

Expressing $\vec{E}^a$ in terms of $\vec{e}_a$ and using Eq. (1.185), we find

$$\nabla \times \vec{v} = \sum_{abc} \vec{e}_a \varepsilon_{abc} \frac{1}{h_b h_c} \partial_b \left(h_c \tilde{v}_c \right) = \frac{1}{\mathcal{J}} \sum_{abc} \vec{e}_a \varepsilon_{abc} h_a \partial_b (h_c \tilde{v}_c). \tag{1.193}$$

In matrix determinant notation, this can schematically be written as

$$\nabla \times \vec{v} = \frac{1}{\mathcal{J}} \begin{vmatrix} h_1 \vec{e}_1 & \partial_1 & h_1 \tilde{v}_1 \\ h_2 \vec{e}_2 & \partial_2 & h_2 \tilde{v}_2 \\ h_3 \vec{e}_3 & \partial_3 & h_3 \tilde{v}_3 \end{vmatrix}, \tag{1.194}$$

where the partial derivatives are taken to act on the quantities $h_a \tilde{v}_a$ (no sum) in the third column.

1.6.3 Polar and cylinder coordinates

With the general theory in place, we will now turn our attention to some specific coordinate systems that are highly relevant in several physical applications. The first such system is the set of *polar coordinates* in two dimensions, which are defined by

$$x^1 = \rho \cos(\phi), \tag{1.195a}$$
$$x^2 = \rho \sin(\phi), \tag{1.195b}$$

see Fig. 1.19. This set of coordinates is uniquely defines any point in a Cartesian coordinate system based on the coordinates ρ and ϕ with $0 \leq \rho < \infty$ and ϕ being 2π periodic. We may also extend this to a three dimensional space by mapping the third Cartesian coordinate x^3 to a third coordinate z according to

$$x^3 = z, \tag{1.195c}$$

which along with the earlier definitions of ρ and ϕ defines *cylinder coordinates*. Below we will assume that we are working in three dimensions with cylinder coordinates, remembering that the polar coordinates may be recovered by omitting the z direction, with the obvious exception of the curl, where three dimensions are necessary in order for it to be defined. Using the definition of the tangent vector basis in Eq. (1.148), we find that

$$\vec{E}_\rho = \vec{e}_1 \cos(\phi) + \vec{e}_2 \sin(\phi), \tag{1.196a}$$

$$\vec{E}_\phi = -\vec{e}_1 \rho \sin(\phi) + \vec{e}_2 \rho \cos(\phi), \tag{1.196b}$$

$$\vec{E}_z = \vec{e}_3. \tag{1.196c}$$

It is straightforward to verify that these vectors are orthogonal and therefore we can apply the full framework we have developed for orthogonal coordinate systems. In particular, taking the modulus of these vectors, we find the scale factors for cylinder coordinates

$$h_\rho = 1, \quad h_\phi = \rho, \quad h_z = 1 \tag{1.197}$$

and the orthonormal basis is given by

$$\vec{e}_\rho = \vec{E}_\rho = \vec{e}_1 \cos(\phi) + \vec{e}_2 \sin(\phi), \tag{1.198a}$$

$$\vec{e}_\phi = \frac{1}{\rho}\vec{E}_\phi = -\vec{e}_1 \sin(\phi) + \vec{e}_2 \cos(\phi), \tag{1.198b}$$

$$\vec{e}_z = \vec{E}_z = \vec{e}_3. \tag{1.198c}$$

It should be noted that the position vector $\vec{x}$ can be expressed in cylinder coordinates as

$$\vec{x} = \rho\vec{e}_\rho + z\vec{e}_z. \tag{1.199}$$

Applying the theory developed in Section 1.6.2, we can immediately express the gradient, divergence, curl, and Laplace operator in cylinder coordinates

$$\nabla f = \vec{e}_\rho \partial_\rho f + \vec{e}_\phi \frac{1}{\rho} \partial_\phi f + \vec{e}_z \partial_z f, \tag{1.200a}$$

$$\nabla \cdot \vec{v} = \frac{1}{\rho}(\partial_\rho \rho \tilde{v}_\rho + \partial_\phi \tilde{v}_\phi) + \partial_z \tilde{v}_z, \tag{1.200b}$$

$$\nabla \times \vec{v} = \vec{e}_\rho \left(\frac{1}{\rho} \partial_\phi \tilde{v}_z - \partial_z \tilde{v}_\phi \right) + \vec{e}_\phi (\partial_z \tilde{v}_\rho - \partial_\rho \tilde{v}_z) + \frac{1}{\rho} \vec{e}_z (\partial_\rho \rho \tilde{v}_\phi - \partial_\phi \tilde{v}_\rho), \tag{1.200c}$$

$$\nabla^2 f = \frac{1}{\rho} \partial_\rho(\rho \partial_\rho f) + \frac{1}{\rho^2} \partial_\phi^2 f + \partial_z^2 f. \tag{1.200d}$$

Example 1.32 An important vector field which is most easily described in cylinder coordinates is

$$\vec{v} = \frac{1}{\rho}\vec{e}_\phi. \tag{1.201}$$

It can be used to describe the magnetic field strength around an infinite conductor carrying a current in magnetostatics, but for our current purposes it is sufficient to notice that its divergence and curl both vanish as

$$\nabla \cdot \vec{v} = \frac{1}{\rho} \partial_\phi \frac{1}{\rho} = 0, \tag{1.202}$$

$$\nabla \times \vec{v} = -\vec{e}_\rho \partial_z \frac{1}{\rho} + \vec{e}_z \frac{1}{\rho} \partial_\rho \left(\frac{\rho}{\rho} \right) = 0. \tag{1.203}$$

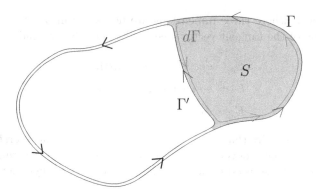

Figure 1.20 The change of contour from Γ to Γ' does not change the value of the circulation integral of $\vec{v}$ as long as $\nabla \times \vec{v} = 0$ on the area S.

These relations are true for all $\rho > 0$, but may not be generally true at the line $\rho = 0$, where the cylinder coordinate system is singular and this vector field will require special treatment. Taking the circulation integral of $\vec{v}$ around a closed curve Γ, we may rewrite it as a circulation integral around Γ' according to Fig. 1.20 as long as $\nabla \times \vec{v} = 0$ on an area with the difference contour as a border, in this case as long as the z-axis does not pass through the loop $d\Gamma$. We may therefore change the shape of the contour as long as we do not change the number of times it winds around the z-axis. For a curve that winds around the z-axis once, it may be deformed into a circle of radius R in the plane $z = 0$. We parametrise this circle by $\rho = R$, $\phi = t$, and $z = 0$, with $0 < t < 2\pi$ and the circulation integral becomes

$$\oint_\Gamma \vec{v} \cdot d\vec{x} = \int_0^{2\pi} \frac{R}{R} dt = 2\pi. \tag{1.204}$$

In the above example, we might have been tempted to apply the curl theorem and ended up with

$$\oint_\Gamma \vec{v} \cdot d\vec{x} = \int_S (\nabla \times \vec{v}) \cdot d\vec{S} = 0, \tag{1.205}$$

where S is the disc with the circle Γ as border. Obviously, $0 \neq 2\pi$ and it would seem that something is amiss. However, this is precisely an effect of the vector field $\vec{v}$ being singular at $\rho = 0$, implying that $\nabla \times \vec{v}$ is in some way related to a δ function along $\rho = 0$, since everywhere else the relation $\nabla \times \vec{v} = 0$ holds true.

1.6.4 Spherical coordinates

The second coordinate system we will have a closer look at is that of *spherical coordinates*. It is most useful whenever a problem exhibits complete rotational symmetry in three dimensions, as is the case for a point charge or a spherically symmetric mass distribution, as we will see in the examples of this section. While the cylinder coordinates in three dimensions introduced polar coordinates in a plane and left the third coordinate untouched, spherical coordinates introduces two angles θ and φ, which uniquely define the direction to a point from the origin, and a radius r, which uniquely defines the distance. The spherical

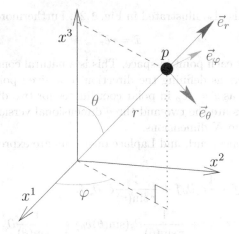

Figure 1.21 The definition of the spherical coordinates in relation to a Cartesian coordinate system. Two angles and a distance uniquely identifies the point p. The unit vectors $\vec{e}_r$, $\vec{e}_\theta$, and $\vec{e}_\varphi$ represent the directions in which p would move when the corresponding coordinates change.

coordinates are defined by

$$x^1 = r\sin(\theta)\cos(\varphi), \tag{1.206a}$$
$$x^2 = r\sin(\theta)\sin(\varphi), \tag{1.206b}$$
$$x^3 = r\cos(\theta), \tag{1.206c}$$

and they are graphically represented in Fig. 1.21. The angle θ is the angle the position vector $\vec{x}$ makes with the x^3-axis, while φ measures the angle between the projection of $\vec{x}$ onto the x^1-x^2-plane and the x^1-axis, much like the polar coordinate ϕ.

From the definition of spherical coordinates, we can find the tangent vector basis in terms of the original Cartesian basis vectors

$$\vec{E}_r = \vec{e}_1\sin(\theta)\cos(\varphi) + \vec{e}_2\sin(\theta)\sin(\varphi) + \vec{e}_3\cos(\theta), \tag{1.207a}$$
$$\vec{E}_\theta = r[\vec{e}_1\cos(\theta)\cos(\varphi) + \vec{e}_2\cos(\theta)\sin(\varphi) - \vec{e}_3\sin(\theta)], \tag{1.207b}$$
$$\vec{E}_\varphi = r\sin(\theta)[-\vec{e}_1\sin(\varphi) + \vec{e}_2\cos(\varphi)]. \tag{1.207c}$$

As in the case of cylinder coordinates, it is straightforward to show that this set of basis vectors is orthogonal and the scale factors

$$h_r = 1, \quad h_\theta = r, \quad \text{and} \quad h_\varphi = r\sin(\theta) \tag{1.208}$$

are found by computing their moduli. Dividing the tangent vector basis by the scale factors provides us with the orthonormal basis

$$\vec{e}_r = \vec{E}_r = \vec{e}_1\sin(\theta)\cos(\varphi) + \vec{e}_2\sin(\theta)\sin(\varphi) + \vec{e}_3\cos(\theta), \tag{1.209a}$$
$$\vec{e}_\theta = \frac{1}{r}\vec{E}_\theta = \vec{e}_1\cos(\theta)\cos(\varphi) + \vec{e}_2\cos(\theta)\sin(\varphi) - \vec{e}_3\sin(\theta), \tag{1.209b}$$
$$\vec{e}_\varphi = \frac{1}{r\sin(\theta)}\vec{E}_\varphi = -\vec{e}_1\sin(\varphi) + \vec{e}_2\cos(\varphi). \tag{1.209c}$$

This set of basis vectors is also illustrated in Fig. 1.21. Furthermore, we obtain the relation

$$\vec{x} = r\vec{e}_r \tag{1.210}$$

for the position vector at each point in space. This is a natural consequence of the selection of the angular parameters as defining the direction to a given point and the coordinate r giving the distance, just as $\vec{x} = \rho\vec{e}_\rho$ in polar coordinates for two dimensions. In fact, polar and spherical coordinates are the two and three dimensional versions of a more general set of spherical coordinates in N dimensions.

The gradient, divergence, curl, and Laplace operators are expressed in spherical coordinates according to

$$\nabla f = \vec{e}_r \partial_r f + \frac{1}{r}\vec{e}_\theta \partial_\theta f + \frac{1}{r\sin(\theta)}\vec{e}_\varphi \partial_\varphi f, \tag{1.211a}$$

$$\nabla \cdot \vec{v} = \frac{1}{r^2}\partial_r(r^2\tilde{v}_r) + \frac{1}{r\sin(\theta)}\partial_\theta(\sin(\theta)\tilde{v}_\theta) + \frac{1}{r\sin(\theta)}\partial_\varphi\tilde{v}_\varphi, \tag{1.211b}$$

$$\nabla \times \vec{v} = \frac{1}{r\sin(\theta)}\vec{e}_r(\partial_\theta \sin(\theta)\tilde{v}_\varphi - \partial_\varphi\tilde{v}_\theta) + \frac{1}{r}\vec{e}_\theta\left(\frac{1}{\sin(\theta)}\partial_\varphi\tilde{v}_r - \partial_r r\tilde{v}_\varphi\right)$$
$$+ \frac{1}{r}\vec{e}_\varphi(\partial_r r\tilde{v}_\theta - \partial_\theta\tilde{v}_r), \tag{1.211c}$$

$$\nabla^2 f = \frac{1}{r^2}\left[\partial_r\left(r^2\partial_r f\right) + \frac{1}{\sin(\theta)}\partial_\theta\left(\sin(\theta)\partial_\theta f\right) + \frac{1}{\sin^2(\theta)}\partial_\varphi^2 f\right]. \tag{1.211d}$$

Example 1.33 An important vector field in physics is the field of a *point source* in three dimensions, which is most easily expressed in spherical coordinates as

$$\vec{v} = \frac{1}{r^2}\vec{e}_r. \tag{1.212}$$

Inserting $\tilde{v}_r = 1/r^2$ into the expressions for the divergence and curl, respectively, we obtain

$$\nabla \cdot \vec{v} = \frac{1}{r^2}\partial_r\frac{r^2}{r^2} = 0, \tag{1.213}$$

$$\nabla \times \vec{v} = \frac{1}{r}\vec{e}_\theta\left(\frac{1}{\sin(\theta)}\partial_\varphi\frac{1}{r^2}\right) - \frac{1}{r}\vec{e}_\varphi\partial_\theta\frac{1}{r^2} = 0. \tag{1.214}$$

As for the field in Example 1.32, this field is both divergence and curl free everywhere where the coordinate system is not singular, in this case at $r = 0$, where the field has a singularity.

If we perform the flux integral over a closed surface S with outward pointing normal vector, the divergence theorem now tells us that the flux Φ through S is given by

$$\Phi = \oint_S \vec{v} \cdot d\vec{S} = \int_V \nabla \cdot \vec{v}\, dV = 0, \tag{1.215}$$

where V is the volume enclosed by S as long as the point $r = 0$ is not part of this volume. If $r = 0$ is enclosed by S, then we can add and subtract the flux integral over a small sphere S_ε of radius ε to the original integral without changing its value, since we are adding and subtracting the same thing, see Fig. 1.22. As a result, the flux through S is given by

$$\Phi = \left(\oint_S \vec{v} \cdot d\vec{S} - \oint_{S_\varepsilon} \vec{v} \cdot d\vec{S}\right) + \oint_{S_\varepsilon} \vec{v} \cdot d\vec{S} = \int_{V_\varepsilon} \nabla \cdot \vec{v}\, dV + \oint_{S_\varepsilon} \vec{v} \cdot d\vec{S}, \tag{1.216}$$

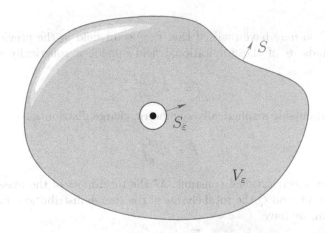

Figure 1.22 By adding and subtracting the flux through the small sphere S_ε, we can rewrite the flux integral over S as the flux integral over S_ε and a volume integral over the volume V_ε in between the S and S_ε.

where V_ε is the volume in between S and S_ε. We have here used the fact that the negative of the integral over S_ε is equal to the integral over S_ε with reversed normal vector, i.e., a normal vector pointing out of the volume V_ε. Since $r = 0$ is not in V_ε, the field $\vec{v}$ is divergence free in this volume and we therefore obtain

$$\Phi = \oint_{S_\varepsilon} \vec{v} \cdot d\vec{S}. \tag{1.217}$$

The surface element $d\vec{S}$ is given by Eq. (1.178) as

$$d\vec{S} = \vec{e}_r h_\theta h_\varphi d\theta \, d\varphi = \vec{e}_r \varepsilon^2 \sin(\theta) \, d\theta \, d\varphi, \tag{1.218}$$

where we have evaluated the scale factors at $r = \varepsilon$ in the last step. We find that

$$\Phi = \oint_{S_\varepsilon} \vec{v} \cdot d\vec{S} = \int_0^\pi \sin(\theta) d\theta \int_0^{2\pi} d\varphi = 4\pi. \tag{1.219}$$

Thus, the flux of the field $\vec{v}$ through a closed surface S is zero if it does not contain the point $r = 0$ and 4π if it does. In order for the divergence theorem to hold, this implies that

$$\nabla \cdot \vec{v} = 4\pi \delta^{(3)}(\vec{x}), \tag{1.220}$$

where $\delta^{(3)}(\vec{x})$ is the three dimensional *delta function* defined by

$$\int_V \delta^{(3)}(\vec{x}) \, dV = \begin{cases} 1, & (\vec{x} = 0 \text{ is in } V) \\ 0, & (\text{otherwise}) \end{cases}. \tag{1.221}$$

For the time being, we will just accept that this is the case, but we shall return to this relation when we discuss Green's functions in Chapter 7.

Example 1.34 You may have realised that the vector field in the previous example has a form that reminds us of the gravitational field outside a spherically symmetric mass distribution

$$\vec{g} = -\vec{e}_r \frac{GM}{r^2} \tag{1.222}$$

or the electric field outside a spherically symmetric charge distribution

$$\vec{E} = \vec{e}_r \frac{kQ}{r^2}, \tag{1.223}$$

where G is Newton's gravitational constant, M the total mass of the mass distribution, k is Coulomb's constant, and Q the total charge of the charge distribution. From Gauss's law on differential form, we have

$$\nabla \cdot \vec{E} = \frac{\rho}{\varepsilon_0}. \tag{1.224}$$

Taking the electric field to be given by Eq. (1.223) in all of space, the resulting charge density must be

$$\rho = \varepsilon_0 \nabla \cdot \vec{E} = 4\pi \varepsilon_0 k Q \delta^{(3)}(\vec{x}) = Q \delta^{(3)}(\vec{x}), \tag{1.225}$$

where the last step applies the fact that $k = 1/4\pi\varepsilon_0$. Thus, the charge density is zero everywhere except for in the origin and the total charge, obtained by integrating the charge density over all of space, is Q. The charge distribution therefore corresponds to a single point charge Q in the origin. Similar arguments can be made for the gravitational field.

Even if the charge distribution is not confined to the origin, but still spherically symmetric, the fields above are still solutions to the field equations outside the charge distribution. This can be derived by using the rotational symmetry of the problem. If the charge distribution is rotated around any axis through the center, then the new charge distribution is equivalent to the original one. However, if we take the rotation axis to be the x^3-axis, then the rotation would have rotated the components of the field orthogonal to $\vec{e}_3$, leading to a different solution for the same charge distribution unless those components are zero, see Fig. 1.23. From this we can draw the conclusion that the field will be of the form $\vec{E} = E(r)\vec{e}_r$, since the solution must be unique and the rotational symmetry also implies that the magnitude of the field cannot depend on the angles θ and φ. If we now take a sphere S_r of radius r such that the entire charge distribution is contained within it, then the integral form of Gauss's law states

$$\oint_{S_r} \vec{E} \cdot d\vec{S} = E(r) \oint_{S_r} dS = E(r) 4\pi r^2 = \frac{Q}{\varepsilon_0}. \tag{1.226}$$

Solving for $E(r)$ results in

$$E(r) = \frac{1}{4\pi\varepsilon_0} \frac{Q}{r^2} = \frac{kQ}{r^2} \tag{1.227}$$

as expected. Solving vector equations by using symmetry arguments in this fashion is not always straightforward and will generally involve special cases or complicated integrals. This will be somewhat simplified by the use of potential theory, which is introduced in the next section.

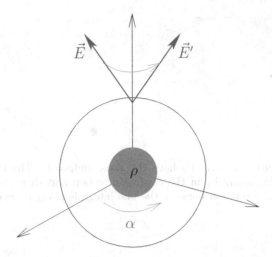

Figure 1.23 Under a rotation by an angle α, if there are components of the electric field $\vec{E}$ orthogonal to the axis of rotation, they will be changed by the rotation to $\vec{E}'$, even if the charge distribution remains the same. This would lead to a different solution unless the orthogonal components are identically zero.

1.7 POTENTIALS

It is sometimes very useful to consider vector fields that can be constructed as derivatives of other fields, both in order to solve differential equations involving fields and in order to make statements of physical importance without having to perform longer calculations. In this section, we will discuss such vector fields, the underlying theory, and some of the physical applications. There are essentially two ways of writing a vector field $\vec{v}$ as a derivative of a different field, either

$$\vec{v} = -\nabla\phi, \tag{1.228a}$$

where ϕ is a scalar field, or

$$\vec{v} = \nabla \times \vec{A}, \tag{1.228b}$$

where $\vec{A}$ is a vector field. The fields ϕ and $\vec{A}$ are known as the *scalar potential* and *vector potential*, respectively. The minus sign in the equation for the scalar potential is purely conventional, but it will appear in this fashion in many physics applications, which is why we adhere to this convention.

1.7.1 Scalar potentials

Vector fields with scalar potentials are typically encountered at very early stages of physics education although it will generally not be stated in the language of vector analysis.

Example 1.35 The gravitational potential Φ is the scalar potential of the gravitational field $\vec{g}$

$$\vec{g} = -\nabla\Phi. \tag{1.229}$$

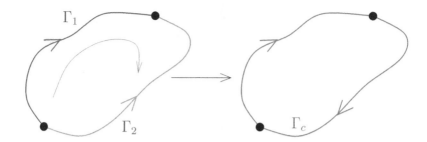

Figure 1.24 The two curves Γ_1 and Γ_2 have the same endpoints. The closed curve Γ_c can be constructed by first going along Γ_1 in the original direction and then along Γ_2 in the opposite direction. This results in a relation between the line integrals along Γ_1 and Γ_2 to the circulation integral along Γ_c.

From Eq. (1.222), we find that, outside a spherical mass distribution,

$$\vec{e}_r \cdot \vec{g} = -\frac{GM}{r^2} = -\partial_r \Phi, \tag{1.230}$$

leading to

$$\Phi = -\frac{GM}{r} \tag{1.231}$$

up to an integration constant that is chosen to be zero by convention, such that the gravitational potential approaches zero as $r \to \infty$.

Coming back to the results of Section 1.4.3, we immediately find that any vector field $\vec{v}$ with a scalar potential ϕ is curl free, as follows from

$$\nabla \times \vec{v} = -\nabla \times \nabla \phi = 0, \tag{1.232}$$

where we have used that $\nabla \times \nabla$ is zero for all scalar fields. It turns out that being curl free is also a sufficient condition for having a scalar potential. Before showing this by explicitly constructing a scalar potential for a curl free vector field, let us examine the properties and physical applications of these fields.

A *conservative vector field* is a vector field $\vec{v}$ for which the line integral

$$L = \int_\Gamma \vec{v} \cdot d\vec{x} \tag{1.233}$$

only depends on the endpoints of the curve Γ. In particular, if we take two different curves Γ_1 and Γ_2 with the same endpoints (see Fig. 1.24), then we can construct a closed curve Γ_c by first going along Γ_1 in the original direction and then along Γ_2 in the opposite direction. This gives us the relation

$$\oint_{\Gamma_c} \vec{v} \cdot d\vec{x} = \int_{\Gamma_1} \vec{v} \cdot d\vec{x} - \int_{\Gamma_2} \vec{v} \cdot d\vec{x} \equiv L_1 - L_2, \tag{1.234}$$

for any vector field $\vec{v}$. Using the curl theorem, we obtain the relation

$$L_1 - L_2 = \int_S (\nabla \times \vec{v}) \cdot d\vec{S}, \tag{1.235}$$

where S is a surface with Γ_c as boundary. If $\nabla \times \vec{v} = 0$, it therefore holds that $L_1 = L_2$, meaning that $\vec{v}$ is conservative if it is curl free. As we have seen that all vector fields with a scalar potential are curl free, they are therefore also conservative. This is also reflected in Eq. (1.106), where we saw that the line integral of a gradient was only dependent on the endpoints of the curve.

Example 1.36 A homogeneous electric field $\vec{E}$ is independent of the position. Naturally this implies that

$$\nabla \times \vec{E} = 0 \tag{1.236}$$

and the field is therefore conservative. The line integral between the two points $\vec{x}_A$ and $\vec{x}_B$ is therefore independent of the path taken and we can take a straight path parametrised by $\vec{x}(t) = \vec{x}_A + t(\vec{x}_B - \vec{x}_A)$, where $t = 0$ represents the start of the curve and $t = 1$ represents the end. Along this curve, $d\vec{x}/dt = \vec{x}_B - \vec{x}_A$, leading to

$$\int_{\vec{x}_A}^{\vec{x}_B} \vec{E} \cdot d\vec{x} = \int_0^1 \vec{E} \cdot (\vec{x}_B - \vec{x}_A)\, dt = \vec{E} \cdot (\vec{x}_B - \vec{x}_A). \tag{1.237}$$

Since $\vec{E}$ is conservative, this will be the result for the line integral of any curve starting in $\vec{x}_A$ and ending in $\vec{x}_B$. We also note that if we define $V = -\vec{E} \cdot (\vec{x} - \vec{x}_A)$, then

$$-\nabla V = \nabla E^i(x^i - x_A^i) = \vec{e}_j E^i \partial_j (x^i - x_A^i) = \vec{e}_j E^i \delta_{ij} = \vec{E}. \tag{1.238}$$

Therefore, the constant field $\vec{E}$ has the potential V.

The above example provides a hint to how we can construct a potential for any conservative field. Assuming a conservative vector field $\vec{v}$, we can construct the scalar field

$$\varphi(\vec{x}) = -\int_{\vec{x}_0}^{\vec{x}} \vec{v}(\vec{\xi}) \cdot d\vec{\xi}, \tag{1.239}$$

which is a line integral along any curve connecting $\vec{x}_0$ and $\vec{x}$. This is well defined due to $\vec{v}$ being conservative, meaning that the integral does not depend on the particular curve chosen. In order to compute the gradient of this scalar field, we need to express the partial derivatives $\partial_i \varphi$ in terms of $\vec{v}$. By definition, the partial derivatives are given by

$$\partial_i \varphi = \lim_{\varepsilon \to 0} \left(\frac{\varphi(\vec{x} + \varepsilon \vec{e}_i) - \varphi(\vec{x})}{\varepsilon} \right). \tag{1.240}$$

We now note that

$$\varphi(\vec{x} + \varepsilon \vec{e}_i) - \varphi(\vec{x}) = -\int_{\vec{x}_0}^{\vec{x} + \varepsilon \vec{e}_i} \vec{v} \cdot d\vec{x} + \int_{\vec{x}_0}^{\vec{x}} \vec{v} \cdot d\vec{x} = -\int_{\vec{x}}^{\vec{x} + \varepsilon \vec{e}_i} \vec{v} \cdot d\vec{x}, \tag{1.241}$$

where we have used that we can select a curve first passing through $\vec{x}$ and then continuing to $\vec{x} + \varepsilon \vec{e}_i$ when computing $\varphi(\vec{x} + \varepsilon \vec{e}_i)$. Again, due to $\vec{v}$ being conservative, we can select the curve $\vec{x}(t) = \vec{x} + t\varepsilon \vec{e}_i$, with t going from 0 to 1, when evaluating the remaining integral. We find that

$$-\int_{\vec{x}}^{\vec{x} + \varepsilon \vec{e}_i} \vec{v} \cdot d\vec{x} = -\varepsilon \int_0^1 \vec{v} \cdot \vec{e}_i dt = -\varepsilon \int_0^1 v^i dt = -\varepsilon v^i(\vec{x} + \varepsilon t_* \vec{e}_i), \tag{1.242}$$

where $0 \leq t_* \leq 1$, and we have used the mean value theorem of integration in the last step. Taking the limit, we obtain

$$\partial_i \varphi = - \lim_{\varepsilon \to 0} \left(v^i(\vec{x} + \varepsilon t_* \vec{e}_i) \right) = -v^i(\vec{x}). \tag{1.243}$$

It follows that

$$-\nabla \varphi = -\vec{e}_i \partial_i \varphi = \vec{e}_i v^i = \vec{v} \tag{1.244}$$

and thus φ is a scalar potential of $\vec{v}$, meaning that a scalar potential of $\vec{v}$ exists as long as $\vec{v}$ is conservative.

Collecting our results, we find that the following three statements are equivalent:

1. The vector field $\vec{v}$ is curl free, $\nabla \times \vec{v} = 0$.

2. The vector field $\vec{v}$ is conservative.

3. The vector field $\vec{v}$ has a scalar potential φ such that $\vec{v} = -\nabla \varphi$.

Example 1.37 A *conservative force field* is a force field $\vec{F}$ for which the work W done by moving an object from $\vec{x}_A$ to $\vec{x}_B$, given by the line integral

$$W = \int_{\vec{x}_A}^{\vec{x}_B} \vec{F} \cdot d\vec{x}, \tag{1.245}$$

does not depend on the path taken. By definition, this force field is a conservative vector field, meaning that it has a scalar potential V and is curl free. The scalar potential is the *potential energy* of the force field and we have

$$\vec{F} = -\nabla V. \tag{1.246}$$

Conservative force fields play a central role in classical mechanics.

In the construction of the scalar potential of a conservative vector field, we introduced the arbitrary point $\vec{x}_0$. As any choice of $\vec{x}_0$ results in a valid scalar potential, the resulting scalar potential is not unique. For the conservative vector field $\vec{v}$, we can define two scalar potentials

$$\varphi_1 = - \int_{\vec{x}_1}^{\vec{x}} \vec{v} \cdot d\vec{x} \quad \text{and} \quad \varphi_2 = - \int_{\vec{x}_2}^{\vec{x}} \vec{v} \cdot d\vec{x}. \tag{1.247}$$

Both of these scalar fields have the property $-\nabla \varphi_i = \vec{v}$ and are therefore scalar potentials of $\vec{v}$. However, they differ by

$$\varphi_2(\vec{x}) - \varphi_1(\vec{x}) = \int_{\vec{x}_1}^{\vec{x}_2} \vec{v} \cdot d\vec{x} \tag{1.248}$$

which is a constant independent of $\vec{x}$. In general, if $\vec{v}$ has the scalar potential φ, then $\phi = \varphi + k$, where k is a constant, will fulfil

$$-\nabla \phi = -\nabla(\varphi + k) = -\nabla \varphi - \nabla k = \vec{v} \tag{1.249}$$

and also be a scalar potential of $\vec{v}$. Any scalar potential is therefore only defined up to a constant and scalar potentials differing by a constant will give the same vector field.

It is also worth noting that the construction of the scalar potential as an integral of the vector field may not always be the easiest way of computing the potential. In general, finding the potential amounts to solving a set of differential equations

$$\partial_i \varphi = -v^i, \tag{1.250}$$

which can often be done in a straightforward fashion, as in Example 1.35.

Example 1.38 In Example 1.36 we found the potential of a constant electric field $\vec{E}$ by integration along a selected curve. We could also have integrated the equation $\partial_i V = -E^i$ directly. Restricting ourselves to two dimensions and starting with $i = 1$, we find that

$$\partial_1 V = -E^1 \quad \Longrightarrow \quad V = -E^1 x^1 + f(x^2), \tag{1.251}$$

where f is only a function of x^2 and thus $\partial_1 f(x^2) = 0$. Inserting this into $\partial_2 V = -E^2$ gives us

$$\partial_2(-E^1 x^1 + f(x^2)) = f'(x^2) = -E^2 \quad \Longrightarrow \quad f(x^2) = -E^2 x^2 + V_0, \tag{1.252}$$

where V_0 is a constant. It follows that

$$V = -E^1 x^1 - E^2 x^2 + V_0 = -\vec{E} \cdot \vec{x} + V_0, \tag{1.253}$$

where V_0 is an arbitrary integration constant. This example has a straightforward generalisation to any number of dimensions.

1.7.2 Vector potentials

Unlike scalar potentials, vector potentials are generally not encountered in physics education until the study of magnetostatics.

Example 1.39 A static magnetic field $\vec{B}$ has a vector potential $\vec{A}$ such that

$$\vec{B} = \nabla \times \vec{A}. \tag{1.254}$$

As we shall see, this follows from *Gauss's law of magnetism*, which on differential form is given by

$$\nabla \cdot \vec{B} = 0. \tag{1.255}$$

In analogy to curl free vector fields having scalar potentials, we shall see that divergence free vector fields have vector potentials.

For any vector field $\vec{v}$ with a vector potential $\vec{A}$, we find that

$$\nabla \cdot \vec{v} = \nabla \cdot (\nabla \times A) = 0. \tag{1.256}$$

All fields with vector potentials are therefore divergence free. This is analogous to what we found in the scalar potential case, where all fields with scalar potentials were found to be curl free. Furthermore, just as vector fields with scalar potentials having line integrals depending only on the curve endpoints, the flux of a field with a vector potential through a surface S only depends on the boundary curve, which follows directly from the curl theorem (see Eq. (1.136)).

In order to show that $\nabla \cdot \vec{v} = 0$ implies the existence of a vector potential, we will proceed with an explicit construction of the potential $\vec{A}$. Unfortunately, the construction cannot proceed in a way similar to the construction of a scalar potential based on a curl free field, mainly due to the loss of a path independent line integral. Instead, we start by considering the scalar field

$$\phi(\vec{x}) = \frac{1}{r}, \tag{1.257}$$

where r is the spherical coordinate. By taking the gradient of this field, we obtain

$$-\nabla\phi = -\vec{e}_r \partial_r \frac{1}{r} = \frac{1}{r^2}\vec{e}_r, \tag{1.258}$$

which is exactly the point source vector field considered in Example 1.33. Using the results from this example, we find that

$$-\nabla^2\phi = 4\pi\delta^{(3)}(\vec{x}). \tag{1.259}$$

By making the translation $\vec{x} \to \vec{x} - \vec{x}_0$, we obtain the relation

$$-\nabla^2 \frac{1}{|\vec{x} - \vec{x}_0|} = 4\pi\delta^{(3)}(\vec{x} - \vec{x}_0). \tag{1.260}$$

Furthermore, for any vector field $\vec{v}$, we can use the defining property of $\delta^{(3)}$ in order to obtain

$$\vec{v}(\vec{x}) = \int \delta^{(3)}(\vec{x} - \vec{x}_0)\vec{v}(\vec{x}_0)dV_0 = -\frac{1}{4\pi}\int \vec{v}(\vec{x}_0)\left(\nabla^2 \frac{1}{|\vec{x} - \vec{x}_0|}\right)dV_0, \tag{1.261}$$

where the integral is taken over all of space and the fields will be assumed to go to zero sufficiently fast as $r \to \infty$. By moving the Laplace operator out of the integral, we obtain

$$\vec{v}(\vec{x}) = -\frac{1}{4\pi}\nabla^2 \int \frac{\vec{v}(\vec{x}_0)}{|\vec{x} - \vec{x}_0|}dV_0 = -\frac{1}{4\pi}\nabla^2\vec{w}, \tag{1.262}$$

where

$$\vec{w} = \int \frac{\vec{v}(\vec{x}_0)}{|\vec{x} - \vec{x}_0|}dV_0. \tag{1.263}$$

From Eq. (1.97), it follows that

$$\vec{v}(\vec{x}) = \frac{1}{4\pi}\nabla \times (\nabla \times \vec{w}) - \frac{1}{4\pi}\nabla(\nabla \cdot \vec{w}). \tag{1.264}$$

We can now rewrite $\nabla \cdot \vec{w}$ according to

$$\nabla \cdot \vec{w} = \int \nabla \cdot \frac{\vec{v}(\vec{x}_0)}{|\vec{x} - \vec{x}_0|}dV_0 = \int \vec{v}(\vec{x}_0) \cdot \nabla \frac{1}{|\vec{x} - \vec{x}_0|}dV_0. \tag{1.265}$$

We now note that $\nabla f(\vec{x} - \vec{x}_0) = -\nabla_0 f(\vec{x} - \vec{x}_0)$, where ∇_0 is taken to be the gradient with respect to $\vec{x}_0$, which leads to

$$\nabla \cdot \vec{w} = -\int \vec{v}(\vec{x}_0) \cdot \nabla_0 \frac{1}{|\vec{x} - \vec{x}_0|}dV_0 = -\int \nabla_0 \cdot \left(\frac{\vec{v}(\vec{x}_0)}{|\vec{x} - \vec{x}_0|}\right)dV_0 = 0, \tag{1.266}$$

where we have used the fact that $\vec{v}$ was taken to be divergence free and again assumed that the $\vec{v}$ goes to zero sufficiently fast as $r \to \infty$ for the surface integral in the divergence theorem to vanish as we integrate over the entire space. It follows that

$$\vec{v} = \nabla \times \vec{A}, \tag{1.267}$$

where

$$\vec{A} = \frac{1}{4\pi} \nabla \times \vec{w} = \frac{1}{4\pi} \nabla \times \int \frac{\vec{v}(\vec{x}_0)}{|\vec{x} - \vec{x}_0|} dV_0, \tag{1.268}$$

and thus $\vec{v}$ has $\vec{A}$ as a vector potential.

While the above tells us of the existence of a vector potential, it is not necessarily the most convenient way of computing it. Again in analogy to the scalar potential case, it is often more advantageous to solve the set of differential equations obtained when demanding that $\vec{v} = \nabla \times \vec{A}$.

Example 1.40 Following Example 1.18, we know that the constant vector field $\vec{\omega} = 2\omega\vec{e}_3$ has the vector potential $\vec{A} = \omega(-x^2\vec{e}_1 + x^1\vec{e}_2)$ (note that $\vec{A}$ was called $\vec{v}$ in Example 1.18) and we now set out to reproduce this result. Starting from $\nabla \cdot \vec{\omega} = 0$, which follows directly from $\vec{\omega}$ being constant, we know that the vector potential exists. We therefore write

$$\nabla \times \vec{A} = \left(\partial_2 A^3 - \partial_3 A^2\right)\vec{e}_1 + \left(\partial_3 A^1 - \partial_1 A^3\right)\vec{e}_2 + \left(\partial_1 A^2 - \partial_2 A^1\right)\vec{e}_3$$
$$= 2\omega\vec{e}_3. \tag{1.269}$$

The vector field components A^i are generally too many to solve for in a unique way and we therefore try to find a solution such that $A^3 = 0$. It follows that

$$\partial_3 A^1 = 0 \implies A^1 = f^1(x^1, x^2), \tag{1.270a}$$

$$\partial_3 A^2 = 0 \implies A^2 = f^2(x^1, x^2), \tag{1.270b}$$

$$\partial_1 f^2 - \partial_2 f^1 = 2\omega. \tag{1.270c}$$

The last of these equations also has several possible solutions, one of which is

$$f^1 = -2\omega x^2, \quad f^2 = 0, \tag{1.271a}$$

which is different from the vector potential we already saw. However, another solution is clearly

$$f^1 = -\omega x^2, \quad f^2 = \omega x^1, \tag{1.271b}$$

which leads to precisely the potential already obtained in Example 1.18. Both of these solutions are perfectly valid vector potentials for $\vec{\omega}$ and it follows that the vector potential is not unique. This is a general result that will hold for any divergence free vector field.

The procedure in the above example can be generally applied to any divergence free vector field $\vec{v}$. We can start by making the ansatz that the vector potential $\vec{A}$ is of the form

$$\vec{A} = A^1\vec{e}_1 + A^2\vec{e}_2, \tag{1.272}$$

which is assuming that the $\vec{e}_3$ component is identically zero. The first component of the relation $\vec{v} = \nabla \times \vec{A}$ is now given by

$$-\partial_3 A^2 = v^1 \implies A^2 = -\int_{x_0^3}^{x^3} v^1(x^1, x^2, z)\, dz + f^2(x^1, x^2), \tag{1.273}$$

for each set of x^1 and x^2. Similarly, we obtain

$$A^1 = \int_{x_0^3}^{x^3} v^2(x^1, x^2, z)\, dz + f^1(x^1, x^2). \tag{1.274}$$

The third component will now be of the form

$$v^3 = \partial_1 A^2 - \partial_2 A^1 = -\int_{x_0^3}^{x^3} [\partial_1 v^1(\vec{x}_z) + \partial_2 v^2(\vec{x}_z)]\, dz + \partial_1 f^2 - \partial_2 f^1, \tag{1.275}$$

where $\vec{x}_z = x^1 \vec{e}_1 + x^2 \vec{e}_2 + z\vec{e}_3$ and f^1 and f^2 are functions of x^1 and x^2. However, since $\vec{v}$ is divergence free, we can replace $\partial_1 v^1 + \partial_2 v^2$ with $-\partial_3 v^3$. We now obtain

$$v^3(\vec{x}) = \int_{x_0^3}^{x^3} \partial_3 v^3(\vec{x}_z)\, dz + \partial_1 f^2 - \partial_2 f^1 = v^3(\vec{x}) - v^3(x^1, x^2, x_0^3) + \partial_1 f^2 - \partial_2 f^1 \tag{1.276}$$

or, cancelling $v^3(\vec{x})$ from both sides,

$$v^3(x^1, x^2, x_0^3) = \partial_1 f^2 - \partial_2 f^1. \tag{1.277}$$

As in the example, this differential equation can be solved by

$$f^1 = -\int_{x_0^2}^{x^2} v^3(x^1, y, x_0^3)\, dy, \quad f^2 = 0 \tag{1.278}$$

and we obtain a vector potential

$$\vec{A} = \vec{e}_1 \left[\int_{x_0^3}^{x^3} v^2(x^1, x^2, z)\, dz - \int_{x_0^2}^{x^2} v^3(x^1, y, x_0^3)\, dy \right] + \vec{e}_2 \int_{x_0^3}^{x^3} v^1(x^1, x^2, z)\, dz. \tag{1.279}$$

Apart from giving us an alternative way of showing that the vector potential exists, this explicit construction gives us a straightforward recipe for computing a vector potential for any divergence free vector field without applying our original construction.

Note that this explicit construction of a vector potential only depends on the local properties of the vector field $\vec{v}$ and does not require it to go to zero sufficiently fast as $r \to \infty$.

Example 1.41 Another example of a divergence free vector field can be found in fluid dynamics. For any fluid, the local velocity field $\vec{v}$ tells us how fast the fluid is moving at each point in space. For an *incompressible fluid*, i.e., a fluid for which the density is constant regardless of external pressure or forces, we must have $\nabla \cdot \vec{v} = 0$, implying the existence of a vector potential for the fluid flow. While this may not be exactly true for any fluid, it is often a good approximation and we will return to the rationale behind this condition for an incompressible fluid in Section 3.9.3.

As we have seen, the vector potential, like the scalar potential for curl free fields, is not unique. In the scalar potential case, we found that any scalar fields that differed by a constant had the same gradient. In the case of a field with a vector potential, we can clearly also add a constant vector field to the potential without affecting the field itself. However, this is not the most general way of obtaining equivalent vector potentials. Let us assume that the vector field $\vec{v}$ is divergence free and that it has two different vector potentials $\vec{A}_1$ and $\vec{A}_2$. By taking the curl of the difference of these vector potentials, we can deduce that

$$\nabla \times (\vec{A}_1 - \vec{A}_2) = \nabla \times \vec{A}_1 - \nabla \times \vec{A}_2 = \vec{v} - \vec{v} = 0. \tag{1.280}$$

Thus, the difference vector field $\delta\vec{A} = \vec{A}_1 - \vec{A}_2$ is curl free. From the previous section, it directly follows that $\delta\vec{A}$ has a scalar potential φ such that $\delta\vec{A} = -\nabla\varphi$. Therefore, for any two vector potentials of the same field, we must have

$$\vec{A}_2 = \vec{A}_1 + \nabla\varphi, \tag{1.281}$$

for some φ. In addition, if we know one vector potential $\vec{A}$, then also

$$\nabla \times (\vec{A} + \nabla\varphi) = \nabla \times \vec{A} + \nabla \times \nabla\varphi = \vec{v} \tag{1.282}$$

for any scalar field φ. It follows that any difference between two vector potentials of the same divergence free field is a gradient and that adding a gradient to the vector potential of a vector field will give a new vector potential of the same field.

1.7.3 Scalar and vector potentials

So far we have seen that curl free vector fields have scalar potentials and that divergence free vector fields have vector potentials. The natural question arises whether or not anything can be said for fields that are neither curl nor divergence free. We shall find that any vector field, regardless of its curl and divergence, may be written using a scalar and a vector potential. In fact, we will be able to decompose any vector field into two parts, where one is curl free and the other divergence free according to

$$\vec{v} = \vec{v}_s + \vec{v}_v, \quad \text{where} \quad \vec{v}_s = -\nabla\varphi, \quad \vec{v}_v = \nabla \times \vec{A}. \tag{1.283}$$

The existence of such a decomposition is known as *Helmholtz's theorem* (or the *fundamental theorem of vector analysis*) and the decomposition itself is called a *Helmholtz decomposition*.

Helmholtz's theorem can be stated in the following way: If V is a volume in three dimensional space, a vector field $\vec{v}$ may be decomposed in a curl free and a divergence free component. We will show that this is true by explicitly constructing such a decomposition in a similar fashion to when we showed that any divergence free field has a vector potential. Starting from $\phi = 1/|\vec{x} - \vec{x}_0|$, we follow the same steps as in Eqs. (1.258) to (1.264), but with the integration domain being V rather than all of space. The scalar and vector potentials are given by $\varphi = \nabla \cdot \vec{w}/4\pi$ and $\vec{A} = \nabla \times \vec{w}/4\pi$, respectively. We find that

$$\varphi = \frac{1}{4\pi} \int_V \vec{v}(\vec{x}_0) \cdot \nabla \frac{1}{|\vec{x} - \vec{x}_0|} dV_0 = -\frac{1}{4\pi} \int_V \vec{v}(\vec{x}_0) \nabla_0 \frac{1}{|\vec{x} - \vec{x}_0|} dV_0$$

$$= -\frac{1}{4\pi} \int_V \nabla_0 \left(\vec{v}(\vec{x}_0) \frac{1}{|\vec{x} - \vec{x}_0|} \right) dV_0 + \frac{1}{4\pi} \int_V \frac{1}{|\vec{x} - \vec{x}_0|} \nabla_0 \cdot \vec{v}(\vec{x}_0) dV_0$$

$$= \frac{1}{4\pi} \int_V \frac{1}{|\vec{x} - \vec{x}_0|} \nabla_0 \cdot \vec{v}(\vec{x}_0) dV_0 - \frac{1}{4\pi} \oint_S \frac{\vec{v}(\vec{x}_0)}{|\vec{x} - \vec{x}_0|} \cdot d\vec{S}_0, \tag{1.284}$$

$$\vec{A} = -\frac{1}{4\pi} \int_V \vec{v}(\vec{x}_0) \times \nabla \frac{1}{|\vec{x} - \vec{x}_0|} dV_0 = \frac{1}{4\pi} \int_V \vec{v}(\vec{x}_0) \times \nabla_0 \frac{1}{|\vec{x} - \vec{x}_0|} dV_0$$

$$= -\frac{1}{4\pi} \int_V \nabla_0 \times \left(\frac{\vec{v}(\vec{x}_0)}{|\vec{x} - \vec{x}_0|} \right) dV_0 + \frac{1}{4\pi} \int_V \frac{1}{|\vec{x} - \vec{x}_0|} \nabla_0 \times \vec{v}(\vec{x}_0) dV_0$$

$$= \frac{1}{4\pi} \int_V \frac{1}{|\vec{x} - \vec{x}_0|} \nabla_0 \times \vec{v}(\vec{x}_0) dV_0 + \frac{1}{4\pi} \oint_S \frac{\vec{v}(\vec{x}_0)}{|\vec{x} - \vec{x}_0|} \times d\vec{S}_0. \tag{1.285}$$

Note that in order to compute these potentials, we only need to know the divergence and curl of $\vec{v}$ in V as well as the value of $\vec{v}$ on the surface S, which is the boundary surface of V.

Just as for scalar and vector potentials, the Helmholtz decomposition is generally not unique. Given a decomposition

$$\vec{v} = -\nabla\varphi_1 + \nabla \times \vec{A}_1, \tag{1.286}$$

we can try to find another pair of potentials φ_2 and $\vec{A}_2$ satisfying the same relation. Such a pair of potentials must fulfil

$$\vec{v} - \vec{v} = -\nabla(\varphi_1 - \varphi_2) + \nabla \times (\vec{A}_1 - \vec{A}_2) = 0. \tag{1.287}$$

Taking the divergence of this relation gives

$$-\nabla^2(\varphi_1 - \varphi_2) = 0, \tag{1.288}$$

implying that the difference of the scalar potentials $f = \varphi_1 - \varphi_2$ must fulfil *Laplace's equation*

$$\nabla^2 f = 0. \tag{1.289}$$

For the difference of the vector potentials, we find that

$$\nabla \times (\vec{A}_1 - \vec{A}_2) = \nabla f. \tag{1.290}$$

Since ∇f is divergence free, it has a vector potential and this equation therefore has a solution. We can therefore add any function satisfying Laplace's equation to the scalar potential as long as we modify the vector potential accordingly.

Example 1.42 In the Helmholtz decomposition above, we will end up with a scalar potential

$$\varphi = -\frac{1}{4\pi} \oint_S \frac{\vec{v}(\vec{x}_0)}{|\vec{x} - \vec{x}_0|} \cdot d\vec{S}_0 \tag{1.291}$$

even in the case when $\nabla \cdot \vec{v} = 0$. However, by the arguments of the previous section, we should be able to write the divergence free vector field $\vec{v}$ as $\vec{v} = \nabla \times \vec{A}$ for some vector potential $\vec{A}$. We note that

$$\nabla^2\varphi = -\frac{1}{4\pi} \oint_S \left(\nabla^2 \frac{1}{|\vec{x} - \vec{x}_0|} \right) \vec{v}(\vec{x}_0) \cdot d\vec{S}_0 = \oint_S \delta^{(3)}(\vec{x} - \vec{x}_0)\vec{v}(\vec{x}_0) \cdot d\vec{S}_0 = 0, \tag{1.292}$$

where we have assumed $\vec{x}$ is an interior point of V so that $\delta^{(3)}(\vec{x} - \vec{x}_0)$ is zero on the boundary S. Since φ fulfils $\nabla^2\varphi = 0$, we can choose a new scalar potential $\varphi_2 = 0$ which fulfils $\nabla^2(\varphi_2 - \varphi) = 0$ and find a corresponding vector potential $\vec{A}_2$ such that

$$\nabla \times (\vec{A}_2 - \vec{A}) = -\nabla\varphi, \quad \nabla \times \vec{A}_2 = -\nabla\varphi + \nabla \times \vec{A} = \vec{v}. \tag{1.293}$$

Thus, we can still construct a vector potential of $\vec{v}$ as per the previous section.

Example 1.43 We have seen that a constant electric field $\vec{E} = E^i \vec{e}_i$ has a scalar potential $V = -\vec{E} \cdot \vec{x}$, the existence of which was implied by $\nabla \cdot \vec{E} = 0$. However, since $\vec{E}$ is constant, we also have

$$\nabla \cdot \vec{E} = \partial_i E^i = 0, \tag{1.294}$$

implying that $\vec{E} = \nabla \times \vec{A}$ for some vector potential $\vec{A}$. Such a vector potential can be constructed by letting

$$\vec{A} = \frac{1}{2} \vec{E} \times \vec{x}, \tag{1.295}$$

which gives us

$$\nabla \times \vec{A} = \frac{1}{2} \nabla \times (\vec{E} \times \vec{x}) = \frac{1}{2} [\vec{E}(\nabla \cdot \vec{x}) - (\vec{E} \cdot \nabla)\vec{x}] = \frac{1}{2}(3\vec{E} - \vec{E}) = \vec{E}. \tag{1.296}$$

This is also consistent with $\nabla^2 \varphi = 0$ implying that the scalar potential can be removed by a redefinition of the vector potential.

Often, the potentials are not physically observable quantities, but the vector fields they describe are. In these situations the potentials can be changed freely as long as they give rise to the same fields. This freedom also means that, when solving physical problems, we may impose additional constraints on the potentials in such a way that the problem is as easy to solve as possible. This can be done as long as our imposed conditions are not so stringent that they imply that the problem has no solution.

1.8   PROBLEMS

Problem 1.1. Write down the following vector expressions in component form using the Einstein summation convention:

  a) $\vec{v} \cdot (k\vec{w} + \vec{u})$

  b) $[(k\vec{v}) \times \vec{w}](\ell + m)$

  c) $\vec{v} \times (\vec{w} \times \vec{u})$

  d) $(\vec{v} \times \vec{w}) \times \vec{u}$

Here, $\vec{u}$, $\vec{v}$, and $\vec{w}$ are vectors while k, ℓ, and m are scalars. Your answers may contain at most one permutation symbol ε_{ijk}.

Problem 1.2. Verify that the cross product satisfies Eq. (1.7) when written on component form.

Problem 1.3. Check that the components of a vector $\vec{v}$ can be found through the use of Eq. (1.10).

Problem 1.4. Consider the three vectors

$$\vec{v}_1 = 3\vec{e}_1 - \vec{e}_2, \quad \vec{v}_2 = 2\vec{e}_2, \quad \vec{e}_3 = -\vec{e}_1 + \vec{e}_2 + 5\vec{e}_3. \tag{1.297}$$

Compute the following:

  a) The magnitude of each of the vectors.

b) The inner product between any pair of the vectors.

c) The cross product between any pair of the vectors.

d) The angles between the vectors.

e) The volume of the parallelepiped spanned by the vectors.

Problem 1.5. Show, using only Eq. (1.6), that the cross product $\vec{v} \times \vec{w}$ is orthogonal to both $\vec{v}$ and $\vec{w}$, i.e., show that

$$\vec{v} \cdot (\vec{v} \times \vec{w}) = \vec{w} \cdot (\vec{v} \times \vec{w}) = 0. \tag{1.298}$$

Problem 1.6. Consider the three vectors

$$\vec{e}_{1'}' = \frac{\vec{e}_1}{\sqrt{2}} - \frac{\vec{e}_3}{\sqrt{2}}, \quad \vec{e}_{2'}' = \frac{\vec{e}_1}{\sqrt{3}} + \frac{\vec{e}_2}{\sqrt{3}} + \frac{\vec{e}_3}{\sqrt{3}}, \quad \vec{e}_{3'}' = \frac{\vec{e}_1}{\sqrt{6}} - \sqrt{\frac{2}{3}}\vec{e}_2 + \frac{\vec{e}_3}{\sqrt{6}}. \tag{1.299}$$

Show that this set of vectors is orthogonal and normalised. Determine whether they form a right- or left-handed set and compute the corresponding transformation coefficients.

Problem 1.7. Show that the Kronecker delta fulfils the relation

$$\delta_{ii} = N, \tag{1.300}$$

where N is the dimensionality of the space.

Problem 1.8. Show that Eq. (1.26) turns into Eq. (1.6) when writing out the sums explicitly by making use of Eq. (1.24).

Problem 1.9. Prove the ε-δ-relation Eq. (1.41) and use it to derive the bac-cab rule Eq. (1.40). *Hint:* In order to prove the ε-δ-relation, you do not need to perform the sum explicitly for all possible indices. It is sufficient to do it for some combinations of free indices and then use the symmetries and anti-symmetries of the expression.

Problem 1.10. Use the ε-δ-relation to express the quantity $\varepsilon_{ijk}\varepsilon_{jk\ell}$ in terms of the Kronecker delta.

Problem 1.11. Consider the vector fields $\vec{v} = x^2\vec{e}_1 - x^1\vec{e}_2$ and $\vec{w} = x^3\vec{e}_1 - x^1\vec{e}_3$ in three dimensions. Compute the following quantities:

a) $\vec{v} \cdot \vec{w}$

b) $\vec{v} \times \vec{w}$

c) $\vec{v} \cdot (\vec{w} \times \vec{x})$

where $\vec{x}$ is the position vector $\vec{x} = x^i\vec{e}_i$.

Problem 1.12. Consider the parallelogram spanned by the vectors $\vec{v}$ and $\vec{w}$, see Fig. 1.25. Show that the diagonals $\vec{v} + \vec{w}$ and $\vec{v} - \vec{w}$ of this parallelogram are orthogonal if and only if $\vec{v}$ and $\vec{w}$ have the same magnitude.

Problem 1.13. A cube of side length ℓ is spanned by the three vectors $\vec{v}_1 = \ell\vec{e}_1$, $\vec{v}_2 = \ell\vec{e}_2$, and $\vec{v}_3 = \ell\vec{e}_3$. Compute the angle between two of the cube diagonals by using the inner product of the displacement vectors between opposite corners.

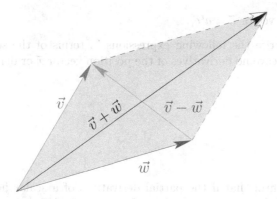

Figure 1.25 A parallelogram spanned by the vectors $\vec{v}$ and $\vec{w}$. The aim of Problem 1.12 is to show that the diagonal vectors $\vec{v}+\vec{w}$ and $\vec{v}-\vec{w}$ are orthogonal only if $\vec{v}$ and $\vec{w}$ have the same magnitude.

Problem 1.14. Find an expression for the squared magnitude of the vector $\vec{v} \times \vec{w}$ in terms of the components of $\vec{v}$ and $\vec{w}$. Also express the magnitudes of $\vec{v}$ and $\vec{w}$ and use your results to find an expression for the sine of the angle between $\vec{v}$ and $\vec{w}$.

Problem 1.15. Consider the two lines given by

$$\vec{x}_1(t) = t\vec{e}_1 + (2-t)\vec{e}_2 \quad \text{and} \quad \vec{x}_2(s) = 2s\vec{e}_1 - \vec{e}_2 - s\vec{e}_3, \qquad (1.301)$$

respectively. Define the difference vector between two points on the lines $\vec{d}(t,s) = \vec{x}_1(t) - \vec{x}_2(s)$ and use it to find the shortest distance between the lines. Show that for the points defining the shortest distance, $\vec{d}$ is orthogonal to the tangent vectors of both lines.

Problem 1.16. An object moves in such a fashion that its position as a function of time t may be described by

$$\vec{x}(t) = r_0 \cos(\omega t)\vec{e}_1 + r_0 \sin(\omega t)\vec{e}_2 + v_0 t\vec{e}_3. \qquad (1.302)$$

a) What is the distance from the origin at an arbitrary time t?

b) What are the velocity and acceleration vectors associated to this motion?

Problem 1.17. Many physical situations involving precession, such as the precession of a spinning top or magnetic spin precession, can be described by the vectorial differential equation

$$\frac{d\vec{L}}{dt} = \vec{v} \times \vec{L}, \qquad (1.303)$$

where $\vec{L}$ is an angular momentum and $\vec{v}$ is a constant vector. Show that the magnitude of the angular momentum $\vec{L}$ and its inner product with $\vec{v}$ are constants in time.

Problem 1.18. Compute the divergence and curl of the following vector fields:

a) $\vec{x} = x^i \vec{e}_i$

b) $\vec{v}_1 = \vec{a} \times \vec{x}$

c) $\vec{v}_2 = x^2 \vec{e}_1 - x^1 \vec{e}_2$

where $\vec{a}$ is a constant vector $\vec{a} = a^i \vec{e}_i$.

Problem 1.19. Rewrite the following expressions in terms of the scalar field ϕ and its derivatives only, i.e., leave no derivatives of the position vector $\vec{x}$ or derivatives of products:

a) $\nabla \cdot (\phi \vec{x})$

b) $\nabla \cdot (\vec{x} \times \nabla \phi)$

c) $\nabla \cdot (\phi \nabla \phi)$

d) $\nabla \times (\vec{x} \times \nabla \phi)$

Problem 1.20. Confirm that if the partial derivatives of a scalar field ϕ commute, i.e., $\partial_i \partial_j \phi = \partial_j \partial_i \phi$, and if the partial derivatives of a vector field $\vec{v}$ commute, then Eqs. (1.91) are satisfied.

Problem 1.21. Express $\nabla \times (\nabla \times \vec{v})$ in terms of the permutation symbol and use the ε-δ-relation to prove Eq. (1.97).

Problem 1.22. Starting from the definition in Eq. (1.36), show that the vector $\vec{S}$ defined in this equation is orthogonal to all $\vec{v}_i$.

Problem 1.23. Consider the following surfaces:

a) $\phi_1(\vec{x}) = x^1 + x^2 + x^3 = 5$

b) $\phi_2(\vec{x}) = (x^1)^2 + (x^2)^2 - x^3 = 0$

c) $\phi_3(\vec{x}) = x^3 - r_0 \cos(kx^1) = -4$

For each surface, parametrise it using the parameters $t = x^1$ and $s = x^2$ and find an expression for the directed area element $d\vec{S}$. Use the fact that the area element is orthogonal to the surface to find a normal vector $\vec{n}$ with magnitude one. Verify that the surface normal in each case is parallel to $\nabla \phi_k$.

Problem 1.24. Show that for a general scalar field $\phi(\vec{x})$, the relation

$$[(\vec{x} \times \nabla) \times (\vec{x} \times \nabla)]\phi = -\vec{x} \times \nabla \phi \tag{1.304}$$

is satisfied.

Problem 1.25. Let $\phi(\vec{x})$ be a scalar field and find the conditions that it has to fulfil in order to satisfy the relation

$$\nabla \times (\nabla \phi \times \vec{a}) = \nabla(\nabla \phi \cdot \vec{a}), \tag{1.305}$$

where $\vec{a}$ is a constant non-zero vector.

Problem 1.26. For a vector field $\vec{v}(\vec{x})$ that fulfils the relation

$$\vec{v}(k\vec{x}) = k^n \vec{v}(\vec{x}), \tag{1.306}$$

where k and n are constants, show that

$$(\vec{x} \cdot \nabla)\vec{v}(\vec{x}) = n\vec{v}(\vec{x}) \tag{1.307a}$$

and compute

$$\nabla \cdot \{\vec{x}[\vec{x} \cdot \vec{v}(\vec{x})]\}. \tag{1.307b}$$

Problem 1.27. For the motion of a rigid body rotating around the point $\vec{x}_0$, the velocity field of the body may be written as

$$\vec{v}(\vec{x}, t) = \vec{\omega}(t) \times (\vec{x} - \vec{x}_0), \tag{1.308}$$

where $\vec{\omega}(t)$ describes the angular velocity of the rotation. Determine an expression for the acceleration $\vec{a}(\vec{x}, t) = d\vec{v}/dt$ and compute the quantities $\nabla \cdot \vec{v}$, $\nabla \times \vec{v}$, $\nabla \cdot \vec{a}$, and $\nabla \times \vec{a}$. *Hint:* When determining the acceleration field, consider the motion $\vec{x}(t)$ of a point in the body that satisfies $d\vec{x}(t)/dt = \vec{v}$.

Problem 1.28. A particle is moving in a force field $\vec{F} = k(x^1\vec{e}_2 - x^2\vec{e}_1)$ from the position $\vec{x} = r_0\vec{e}_1$ to the position $\vec{x} = r_0\vec{e}_2$. Compute the work done by the force when the particle takes the path

a) along a circle of constant radius r_0 and

b) along the straight line between the endpoints.

Determine whether or not the force field is conservative.

Problem 1.29. The *Lorentz force* on a particle with electric charge q in a magnetic field $\vec{B}(\vec{x})$ is given by

$$\vec{F} = q\vec{v} \times \vec{B}, \tag{1.309}$$

where $\vec{v}$ is the particle's velocity $\vec{v} = d\vec{x}/dt$. Show that the total work

$$W = \int \vec{F} \cdot d\vec{x} \tag{1.310}$$

done by the magnetic field on the particle is equal to zero.

Problem 1.30. The half-sphere of radius R defined by $x^3 > 0$ and $\vec{x}^2 = R^2$ can be parametrised by the cylinder coordinates ρ and ϕ as

$$x^1 = \rho\cos(\phi), \quad x^2 = \rho\sin(\phi), \quad x^3 = \sqrt{R^2 - \rho^2}, \tag{1.311}$$

for $\rho < R$. Use this parametrisation to compute the total area of the half-sphere according to

$$A = \int dS. \tag{1.312}$$

Problem 1.31. Consider the vector field

$$\vec{v} = (x^1)^k\vec{e}_1 + (x^2)^k\vec{e}_2 + (x^3)^k\vec{e}_3, \tag{1.313}$$

where k is a positive integer. Compute the flux integral

$$\Phi = \oint_S \vec{v} \cdot d\vec{S}, \tag{1.314}$$

where S is the sphere of radius R with the surface normal pointing away from the origin.

Problem 1.32. In a region where the matter density is given by

$$\rho(\vec{x}) = \frac{\rho_0}{L^2}\vec{x}^2, \tag{1.315}$$

compute the total mass contained in the volumes given by

a) the cube $0 < x^i < L$ $(i = 1, 2, 3)$ and

b) the sphere $r < L$ in spherical coordinates.

Problem 1.33. Consider a fluid with constant density ρ_0 and the velocity field

$$\vec{v} = \frac{v_0}{L}(x^1\vec{e}_2 - x^2\vec{e}_1 + L\vec{e}_3). \tag{1.316}$$

a) Verify that the velocity field is divergence free $\nabla \cdot \vec{v} = 0$.

b) Compute the total momentum inside the cube $0 < x^i < L$ $(i = 1, 2, 3)$.

Problem 1.34. For the fluid in Problem 1.33, compute the mass flux of the fluid

$$\Phi = \int_S \rho_0 \vec{v} \cdot d\vec{S} \tag{1.317}$$

when the surface S in cylinder coordinates is given by

a) the disc $z = z_0$, $\rho < r_0$, $0 < \phi < 2\pi$, with surface normal $\vec{n} = \vec{e}_3$,

b) the cylinder $0 < z < z_0$, $\rho = r_0$, $0 < \phi < 2\pi$, with surface normal $\vec{n} = \vec{e}_\rho$,

c) the ϕ coordinate surface $0 < z < z_0$, $0 < \rho < r_0$, $\phi = \phi_0$, with surface normal $\vec{n} = \vec{e}_\phi$.

Note: We have here used ρ_0 to denote a density, be careful not to confuse it with the cylinder coordinate ρ.

Problem 1.35. A small element $d\vec{x}$ of a conductor carrying a current I in a magnetic field $\vec{B}$ will be subjected to a force

$$d\vec{F} = -I\vec{B} \times d\vec{x}. \tag{1.318}$$

In a constant magnetic field $\vec{B} = \vec{B}_0$

a) compute the total force $\vec{F}$ acting on the closed loop given by

$$x^1 = r_0\cos(t), \quad x^2 = r_0\sin(t), \quad x^3 = 0, \qquad (0 < t < 2\pi) \tag{1.319}$$

b) compute the total torque $\vec{M}$ relative to the origin for the same loop. The torque due to a force $d\vec{F}$ is given by $d\vec{M} = \vec{x} \times d\vec{F}$.

Problem 1.36. For a static electromagnetic field, the electric field $\vec{E}$ has a scalar potential ϕ and the magnetic field $\vec{B}$ is divergence free. Assuming that S is a closed equipotential surface, i.e., that ϕ takes a constant value ϕ_0 on S, compute the volume integral

$$I = \int_V \vec{E} \cdot \vec{B}\, dV, \tag{1.320}$$

where V is the volume enclosed by S, by using the divergence theorem.

Problem 1.37. Compute the closed surface integral

$$\vec{I} = \oint_S \vec{x} \times d\vec{S} \tag{1.321}$$

by rewriting it as an integral over the volume V enclosed by S and simplifying the resulting expression.

Problem 1.38. Use the generalised version of the curl theorem to rewrite the circulation integral

$$\vec{I} = \oint_\Gamma \vec{x} \times d\vec{x} \tag{1.322}$$

around the closed loop Γ as an integral over a surface S bounded by Γ. Simplify your result as far as possible without knowing the loop Γ and then apply your result to the loop given by

$$x^1(t) = r_0 \cos(t), \quad x^2(t) = r_0 \sin(t), \quad x^3(t) = 0. \tag{1.323}$$

Verify that you obtain the same result from the circulation integral as you get from the surface integral.

Problem 1.39. Given that the scalar field $\phi(\vec{x})$ is constant on the boundary curve of a surface S, show that

$$\int_S [(\nabla\phi) \times (\nabla\psi)] \cdot d\vec{S} = 0 \tag{1.324}$$

for any scalar field $\psi(\vec{x})$.

Problem 1.40. Consider again the fluid treated in Problems 1.33 and 1.34. Rewrite the mass flux out of the closed surface S composed of the cylinder area $-z_0 < z < z_0$, $\rho = r_0$, $0 < \phi < 2\pi$ and the two discs $z = \pm z_0$, $\rho < r_0$, $0 < \phi < 2\pi$ as an integral over the enclosed volume and compute the volume integral. Verify that the flux is equivalent to the flux resulting from adding up the fluxes through S (use your results from Problem 1.34).

Problem 1.41. Compute the integral expression

$$I = \int_V \vec{v} \cdot \vec{w} \, dV, \tag{1.325}$$

where $\vec{v}$ is curl free in the volume V and $\vec{w}$ is orthogonal to the surface normal at the boundary of V and divergence free inside V.

Problem 1.42. The surface integral

$$\vec{\Phi} = \oint_S \frac{\vec{x}}{r^5} \vec{x} \cdot d\vec{S} \tag{1.326}$$

can be rewritten as a volume integral

$$\vec{\Phi} = \int_V \vec{x}\phi(\vec{x}) \, dV. \tag{1.327}$$

Determine an expression for the scalar field $\phi(\vec{x})$ assuming that $\vec{x} = 0$ is not within the volume V.

Problem 1.43. Let $\vec{A}$ be a vector field satisfying the relation $\nabla \times (\nabla \times \vec{A}) = 0$ in the volume V and $\vec{n} \cdot [\vec{A} \times (\nabla \times \vec{A})] = 0$ on the boundary surface S of V, where $\vec{n}$ is the surface normal. Show that

$$\int_V (\nabla \times \vec{A})^2 dV = 0. \tag{1.328}$$

Problem 1.44. Use the regular curl theorem to prove the generalised curl theorem, given by Eq. (1.139b).

Problem 1.45. The vector potential $\vec{A}$ of a magnetic dipole with the constant dipole moment $\vec{m}$ is given by

$$\vec{A} = \frac{\mu_0}{4\pi} \frac{\vec{m} \times \vec{x}}{r^3}. \tag{1.329}$$

Compute the corresponding magnetic field $\vec{B} = \nabla \times \vec{A}$ for $r > 0$. Also compute the curl of this magnetic field.

Problem 1.46. With the dual basis at hand, show that the contravariant vector components v^a of a vector $\vec{v}$ may be found as

$$v^a = \vec{E}^a \cdot \vec{v}, \tag{1.330}$$

similar to how the covariant vector components were computed using the tangent vector basis in Eq. (1.162).

Problem 1.47. Verify that the tangent vector bases in cylinder coordinates (see Eqs. (1.196)) and in spherical coordinates (see Eqs. (1.207)) are orthogonal bases.

Problem 1.48. Compute the divergence and curl of the vector field $\vec{v} = \frac{1}{\rho}\vec{e}_\phi$, given in cylinder coordinates, for $\rho > 0$. Use your result to compute the circulation integral

$$I = \oint_\Gamma \vec{v} \cdot d\vec{x}, \tag{1.331}$$

where Γ is the closed curve given by

$$\rho(t) = \rho_0[2 + \sin(t)], \quad \phi(t) = t, \quad z(t) = t(t - 4\pi), \tag{1.332}$$

where $0 < t < 4\pi$.

Problem 1.49. *Hyperbolic coordinates* u and v in two dimensions may be introduced to parametrise the region $x^1, x^2 > 0$ according to

$$x^1 = ve^u, \quad x^2 = ve^{-u}. \tag{1.333}$$

   a) Sketch the coordinate lines corresponding to u and v.

   b) Find the inverse transformation that expresses u and v as functions of x^1 and x^2.

   c) Compute the tangent vector and dual bases.

   d) Determine whether or not hyperbolic coordinates form an orthogonal coordinate system. If they do, compute the scale factors. If they do not, compute the inner products $\vec{E}_u \cdot \vec{E}_v$ and $\vec{E}^u \cdot \vec{E}^v$.

Problem 1.50. The *parabolic coordinates* t and s in two dimensions are defined through the relations

$$x^1 = ts, \quad x^2 = \frac{1}{2}(t^2 - s^2), \tag{1.334a}$$

where $t, s > 0$.

   a) Sketch the coordinate lines corresponding to t and s.

   b) Find the inverse transformation that expresses t and s as functions of x^1 and x^2.

c) Compute the tangent vector and dual bases.

d) Determine whether or not hyperbolic coordinates form an orthogonal coordinate system. If they do, compute the scale factors. If they do not, compute the inner products $\vec{E}_t \cdot \vec{E}_s$ and $\vec{E}^t \cdot \vec{E}^s$.

e) Find the expressions for the gradient, divergence, curl, and Laplace operators in three dimensions when also parametrising the third direction with an additional coordinate z as

$$x^3 = z. \tag{1.334b}$$

f) Express the position vector $\vec{x}$ in parabolic coordinates.

Problem 1.51. A screened point charge at the origin may be described by the Yukawa potential

$$\phi(\vec{x}) = \frac{q}{4\pi r} e^{-kr}, \tag{1.335}$$

where k is a constant and r is the radial spherical coordinate. Compute the flux of the corresponding vector field $\vec{v} = -\nabla\phi$ out of the sphere of radius R both by directly performing the surface integral and by applying the divergence theorem. What can be said about $\nabla^2\phi(\vec{x})$ at the origin?

Problem 1.52. Compute $\nabla \cdot \vec{x}$ and $\nabla \times \vec{x}$ using the expressions for the divergence and curl in cylinder and spherical coordinates. Verify that the results are the same as the result you would find doing the same computation in Cartesian coordinates.

Problem 1.53. Verify that the vector field

$$\vec{v} = \frac{1}{r^2} \vec{e}_r \tag{1.336}$$

is divergence free for $r > 0$. According to our discussion on potentials, this implies that $\vec{v}$ has a vector potential $\vec{A}$ such that

$$\vec{v} = \nabla \times \vec{A}. \tag{1.337}$$

Find a vector potential $\vec{A}$ of $\vec{v}$ that has the form $\vec{A} = A_\varphi \vec{e}_\varphi$ and is divergence free.

Problem 1.54. Assume that the vector fields $\vec{v}(\vec{x})$ and $\vec{w}(\vec{x})$ satisfy

$$\nabla \cdot \vec{v} = \nabla \cdot \vec{w} \quad \text{and} \quad \nabla \times \vec{v} = \nabla \times \vec{w} \tag{1.338a}$$

within a volume V and that

$$\vec{n} \cdot \vec{v} = \vec{n} \cdot \vec{w} \tag{1.338b}$$

on the boundary surface of V. Show that $\vec{v}(\vec{x}) = \vec{w}(\vec{x})$ within the volume V.

Problem 1.55. The force on a charge q at position $\vec{x} \neq 0$ from an electric dipole of dipole moment $\vec{p} = p\vec{e}_3$ at the origin is given by

$$\vec{F} = pq \left(\frac{2\cos(\theta)}{r^3} \vec{e}_r + \frac{\sin(\theta)}{r^3} \vec{e}_\theta \right) \tag{1.339}$$

in spherical coordinates. Compute $\nabla \cdot \vec{F}$ and $\nabla \times \vec{F}$. Does this force field have a corresponding scalar and vector potential?

CHAPTER 2

Tensors

Just as Nature does not care about the set of units we use in order to make measurements, it does not care about what particular set of coordinates we use to describe it. We have seen this already when we discussed vector equations, which must hold regardless of the particular coordinate system. For example, if we have two vectors $\vec{v}$ and $\vec{w}$ whose components fulfil the relation $v^a = w^a$ in one coordinate system, then the components (although different) must be equal in all coordinate systems and we can write down the vector relation $\vec{v} = \vec{w}$ independent of any chosen coordinates. In the same way, we find that scalars, which do not change under coordinate transformations, also fulfil similar statements. For example, if we have a scalar field ϕ and $\phi\vec{v} = \vec{w}$, this is a relation that is also independent of the coordinate system. While we can get pretty far by only using scalars and vectors, there is a generalisation that will turn out to be extremely useful in many areas of physics. This is the realm of tensor analysis, which will allow us to generally write down many different linear relationships in a coordinate independent fashion, just as we have done earlier for scalars and vectors.

In this chapter, we will introduce the general ideas behind tensors, starting by generalising the scalar and vector concepts in order to describe some basic situations in physics. We will make heavy use of the formalism we developed when discussing general coordinate systems in the previous chapter and start from this vantage point to later examine tensor properties in Cartesian coordinates as an important special case. While many texts start by discussing tensors in Cartesian coordinates, the formalism for tensors in arbitrary coordinate systems is not much more difficult and it illuminates the structure in a clearer fashion.

Example 2.1 Let us start with an example taken from rigid body mechanics. Consider a disc of uniformly distributed mass M and radius R that is rotating with angular velocity $\vec{\omega}$, see Fig. 2.1. If the rotation is parallel to the symmetry axis of the disc, then the angular momentum $\vec{L}$ of the disc will also be, and the relationship between the two will be

$$M\frac{R^2}{2}\vec{\omega} = \vec{L}. \tag{2.1a}$$

On the other hand, if the rotation is in a direction perpendicular to the symmetry axis, then the relationship will be

$$M\frac{R^2}{4}\vec{\omega} = \vec{L}. \tag{2.1b}$$

From these considerations, it is clear that, for both directions, the relationship between the

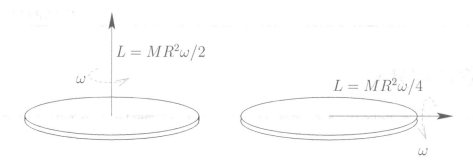

Figure 2.1 A disc of mass M and radius R rotating either around its symmetry axis or around an axis within its plane with angular velocity $\vec{\omega}$. The angular momentum $\vec{L}$ of the disc will generally depend on the direction of the rotation and may not be parallel to $\vec{\omega}$.

angular velocity and the angular momentum is linear. However, since the proportionality constant is different in the different directions, it cannot be a multiplication by a scalar. Instead, we can relate the components of $\vec{\omega}$ and $\vec{L}$ using a Cartesian basis $\vec{e}_i$ according to

$$L^i = I^i_j \omega^j, \tag{2.2}$$

where the I^i_j are a set of nine numbers (one for each possible combination of i and j). This relationship is still linear, but we now have the possibility of ensuring that Eqs. (2.1) hold for both directions, and for any direction of the angular velocity $\vec{\omega}$. The numbers (with dimension of mass×distance^2) I^i_j are the components of the *moment of inertia tensor*.

In general, tensors may have more than two indices, but we will frequently use *matrix notation* for tensors with two indices, this will be indicated by parentheses around the components, i.e.,

$$(I^i_j) = \begin{pmatrix} I^1_1 & I^1_2 & I^1_3 \\ I^2_1 & I^2_2 & I^2_3 \\ I^3_1 & I^3_2 & I^3_3 \end{pmatrix} \tag{2.3}$$

with a straightforward generalisation to cases where the dimension of the base space is different from three.

The above example has so far only showed us that we can write down a linear relationship between two vectors using a set of numbers with two indices. This does not fully illustrate what it means for this object to be a tensor, which will be clearer once we consider what happens if we pick a different set of coordinates with a corresponding basis $\vec{e}'_{i'}$. In general, let us assume that we have a relationship between the vectors $\vec{v}$ and $\vec{w}$ such that $v^i = A^i_j w^j$ in the basis $\vec{e}_i$. However, for this to hold in *any* system, the A^i_j need to follow certain transformation rules. In particular, since both $\vec{v}$ and $\vec{w}$ are vectors, we can express v^i and w^j in terms of their components in the basis $\vec{e}'_{i'}$ and obtain

$$v^i = a^i_{i'} v^{i'} = A^i_j w^j = A^i_j a^j_{j'} w^{j'}. \tag{2.4}$$

Multiplying this equation by $a_i^{k'}$ now gives us

$$a_i^{k'} a_{i'}^i v^{i'} = \delta_{i'}^{k'} v^{i'} = v^{k'} = \underbrace{a_i^{k'} A_j^i a_{j'}^j}_{\equiv A_{j'}^{k'}} w^{j'}, \tag{2.5}$$

where $A_{j'}^{k'}$ are the tensor components relating the components of $\vec{v}$ and $\vec{w}$ in the new basis. We note that the components of A_j^i transform in a similar fashion to how scalars and vectors do. While scalars have zero indices and do not transform under changes of basis, vectors have one index, which is contracted with an index of a transformation coefficient. In the same spirit, the tensor components A_j^i have two indices, which each is contracted with an index of a transformation coefficient. In this case, A_j^i are the components of a *rank two tensor*, which is indicated by the components having two indices. Before long, we will see examples of tensors also of higher rank. For now, we just mention that scalars are tensors of rank zero while vectors are tensors of rank one.

2.1 OUTER PRODUCTS AND TENSOR BASES

So far we have only seen tensors in terms of their components. In many introductory texts, tensors are defined in this fashion, i.e., a set of numbers that follow certain transformation rules under changes of basis and indeed this is often all that is needed from a physics point of view. Once tensors have been introduced in this fashion, we can essentially apply all of the machinery we have learned in vector analysis, i.e., repeated indices are summed over, free indices have to be the same on both sides of an equation, relations take the same form regardless of the basis chosen, equations without free indices represent invariant quantities that may be computed in any basis and so on. Eventually, we will go down the path of referring to tensors in terms of their components, as was indicated already in Chapter 1, where we noted that we can interchangeably refer to the vector components v^a and the vector $\vec{v}$ as long as the vector basis is implicit. However, before doing so, we will explore a few aspects of the construction of tensors.

In Chapter 1 we defined the scalar and cross products of two vectors, the result of which was a scalar and a vector, respectively. We can also define the *outer product* (or *tensor product*) $\vec{v} \otimes \vec{w}$ of the vectors $\vec{v}$ and $\vec{w}$ in the vector space V as bilinear product that takes values in the tensor product space $V \otimes V$. It is important to not mix up the outer product with the cross product, although they have a similar notation; the latter gives a vector and is anti-symmetric, while the former gives an element in the product space and does not have any particular symmetry. For our purposes, it suffices to consider the tensor product space as a linear space to which the outer products of vectors belong. In particular, this means that sums of different outer products, such as

$$T = \vec{v}_1 \otimes \vec{w}_1 + \vec{v}_2 \otimes \vec{w}_2, \tag{2.6}$$

belong to this space. In general, there is no guarantee that an element of this linear space will be an outer product of two vectors. In fact, if $\vec{v}_1$ is linearly independent of $\vec{v}_2$ and $\vec{w}_1$ is linearly independent of $\vec{w}_2$, then it will be impossible to find an outer product of two vectors that equals T.

Taking a vector basis $\vec{e}_i$, the outer product of two vectors may be written

$$\vec{v} \otimes \vec{w} = (v^i \vec{e}_i) \otimes (w^j \vec{e}_j) = v^i w^j (\vec{e}_i \otimes \vec{e}_j) \equiv v^i w^j e_{ij}, \tag{2.7}$$

where we have defined $e_{ij} = \vec{e}_i \otimes \vec{e}_j$. It follows that any outer product may be written as a linear combination of the e_{ij}, which form a basis of the tensor product space. Any element of this space may therefore be written as

$$T = T^{ij}\vec{e}_i \otimes \vec{e}_j = T^{ij}e_{ij} \tag{2.8}$$

and objects of this form are rank two tensors. In a similar fashion, we can define rank three tensors as the linear combinations of tensor products of three vectors

$$T = T^{ijk}\vec{e}_i \otimes \vec{e}_j \otimes \vec{e}_k \equiv T^{ijk}e_{ijk} \tag{2.9a}$$

or rank n tensors as tensor products of n vectors

$$T = T^{i_1 \dots i_n} \bigotimes_{k=1}^{n} \vec{e}_{i_k} = T^{i_1 \dots i_n}e_{i_1 \dots i_n}. \tag{2.9b}$$

Example 2.2 We have seen that the moment of inertia of a disc depends linearly on the direction of the angular velocity $\vec{\omega}$. For angular velocities parallel to the symmetry axis of the disc, the moment of inertia is $I_3 = MR^2/2$, while for directions perpendicular to it, it is given by $I_0 = MR^2/4$. Assuming that the symmetry axis is $\vec{e}_3$ and that $I_j^i \omega^j = I_3 \omega^i$ when $\vec{\omega} = \omega\vec{e}_3$, the moment of inertia tensor needs to have $I_3^3 = I_3$ and $I_3^i = 0$ for $i \neq 3$. This is equivalent to having a contribution $I_3\vec{e}_3 \otimes \vec{e}_3$. The same argument for the orthogonal directions leads us to the conclusion that

$$I = I_0(\vec{e}_1 \otimes \vec{e}_1 + \vec{e}_2 \otimes \vec{e}_2) + I_3\vec{e}_3 \otimes \vec{e}_3, \tag{2.10}$$

i.e., the moment of inertia tensor is the linear combination of three different outer products.

2.1.1 General coordinate bases

In Section 1.6.1, we discussed the tangent vector basis $\vec{E}_a$ together with the dual basis $\vec{E}^a$. As was mentioned, these bases will be fundamentally different once we start discussing more general spaces and not only curvilinear coordinates on a Euclidean space. We will therefore already set up the necessary framework for handling this.

Let us go back to the case where we have a linear relation between two vectors $\vec{v}$ and $\vec{w}$. If we express both of these in tangent vector basis, they may be written as $\vec{v} = v^a\vec{E}_a$ and $\vec{w} = w^a\vec{E}_a$, respectively. The requirement of having a linear relationship between the contravariant vector components gives us

$$v^a = A_b^a w^b \tag{2.11}$$

and in the same way as before, we can obtain the transformation properties of the coefficients A_b^a as

$$A_{b'}^{a'} = \frac{\partial y'^{a'}}{\partial y^a} \frac{\partial y^b}{\partial y'^{b'}} A_b^a \tag{2.12}$$

when changing coordinates from y^a to $y'^{a'}$. In order to create an object with these transformation properties, consider the rank two tensors of the form

$$A = A_b^a(\vec{E}_a \otimes \vec{E}^b) \equiv A_b^a e_a^b. \tag{2.13}$$

Under a change of basis, the basis vectors $\vec{E}_a$ and $\vec{E}^b$ can be expressed in the new bases $\vec{E}'_{a'}$ and $\vec{E}'^{b'}$ according to

$$\vec{E}_a = \frac{\partial y'^{a'}}{\partial y^a}\vec{E}'_{a'} \quad \text{and} \quad \vec{E}^b = \frac{\partial y^b}{\partial y'^{b'}}\vec{E}'^{b'}, \tag{2.14}$$

respectively. Inserting this into the previous equation results in

$$A = A^a_b\left(\frac{\partial y'^{a'}}{\partial y^a}\vec{E}'_{a'}\right) \otimes \left(\frac{\partial y^b}{\partial y'^{b'}}\vec{E}'^{b'}\right) = \frac{\partial y'^{a'}}{\partial y^a}\frac{\partial y^b}{\partial y'^{b'}}A^a_b e'^{b'}_{a'} \equiv A^{a'}_{b'}e'^{b'}_{a'}, \tag{2.15}$$

which is just stating that the tensor components A^a_b transform according to Eq. (2.12). As the distinction between tangent vectors and dual vectors will become important, tensors of this form are *type* $(1,1)$ tensors, whose basis e^b_a is constructed by the outer product of the tangent vector basis with the dual basis.

We may also construct tensors only using the tangent vector basis

$$T = T^{ab}\vec{E}_a \otimes \vec{E}_b = T^{ab}e_{ab} \tag{2.16a}$$

or only using the dual basis

$$T = T_{ab}\vec{E}^a \otimes \vec{E}^b = T_{ab}e^{ab}. \tag{2.16b}$$

The first of these equations defines *contravariant tensors* of rank 2 (also known as type $(2,0)$ tensors), while the second defines *covariant tensors* of rank 2 (type $(0,2)$ tensors). More generally, we can define higher rank tensors using both bases according to

$$T = T^{a_1\ldots a_n}_{b_1\ldots b_m}\bigotimes_{k=1}^{n}\vec{E}_{a_k}\bigotimes_{\ell=1}^{m}\vec{E}^{b_\ell} = T^{a_1\ldots a_n}_{b_1\ldots b_m}e^{b_1\ldots b_m}_{a_1\ldots a_n}, \tag{2.17}$$

i.e., using the outer product of n tangent vectors and m vectors from the dual basis. Such tensors are said to be of *type* (n,m). If $n = 0$ the tensors are *covariant* and if $m = 0$ they are *contravariant*, in both cases referring to the way the tensor components transform under coordinate changes. A tensor that is neither covariant nor contravariant is a *mixed tensor* and its components transform according to the position of the different indices, i.e., upper indices transform contravariantly, while lower indices transform covariantly

$$T^{a'_1\ldots a'_n}_{b'_1\ldots b'_m} = T^{a_1\ldots a_n}_{b_1\ldots b_m}\left(\prod_{k=1}^{n}\frac{\partial y'^{a'_k}}{\partial y^{a_k}}\right)\left(\prod_{\ell=1}^{m}\frac{\partial y^{b_\ell}}{\partial y'^{b'_\ell}}\right). \tag{2.18}$$

Deriving this relation is left as an exercise (see Problem 2.1).

Just as we noted in the case of vectors in general coordinates, there is a one-to-one correspondence between the tensor components in a given coordinate system and the tensor itself. Because of this reason, we will often refer to $T^{a_1\ldots a_n}_{b_1\ldots b_m}$ as a tensor of the appropriate type rather than also writing out the basis.

Example 2.3 We can construct a very important tensor by summing all of the outer products of the tangent vector basis with the corresponding dual basis

$$\delta = \vec{E}^a \otimes \vec{E}_a. \tag{2.19}$$

It is important to note that a appears twice in this definition and therefore should be summed over. This definition can also be written

$$\delta = \delta_b^a e_a^b \qquad (2.20)$$

and it is apparent that this tensor has the components δ_b^a in the chosen coordinate system. Examining the transformation properties of these components, we obtain

$$\delta_{b'}^{a'} = \frac{\partial y'^{a'}}{\partial y^a} \frac{\partial y^b}{\partial y'^{b'}} \delta_b^a = \frac{\partial y'^{a'}}{\partial y^a} \frac{\partial y^a}{\partial y'^{b'}} = \frac{\partial y'^{a'}}{\partial y'^{b'}}, \qquad (2.21)$$

which is one if $a' = b'$ and zero otherwise. Therefore, the components of this tensor are the same regardless of the chosen coordinate system and are non-zero (and equal to one) if the indices match. This shows that the *Kronecker delta* is actually a type (1,1) tensor.

2.2 TENSOR ALGEBRA

The way we have defined tensors as linear combinations of outer products of a number of vectors, it becomes apparent that there are a number of operations that are possible to perform for tensors that result in new tensors. It is worth noting that if tensors are merely introduced as a set of numbers transforming in a particular way under coordinate transformations, which is often the case, then it needs to be checked explicitly that the results of these operations are new tensors. These operations are the operations associated to a linear space, i.e., multiplication by a scalar and addition of tensors of the same type, which follow the distributive properties

$$k_1(T_1 + T_2) = k_1 T_1 + k_1 T_2, \quad (k_1 + k_2)T_1 = k_1 T_1 + k_2 T_1 \qquad (2.22)$$

where the k_i are scalars and the T_i are tensors of the same type, along with the other axioms for a linear space. The linearity of the outer product itself guarantees that the zero tensor is $0 = 0 \otimes \vec{w} = \vec{v} \otimes 0 = 0(\vec{v} \otimes \vec{w})$ for any vectors $\vec{v}$ and $\vec{w}$.

Apart from the operations that naturally follow from tensors belonging to a linear space, there are a number of other operations that may be performed with tensors:

Outer product: Just as the outer product of two vectors, the outer product of two arbitrary tensors, which do not necessarily have to be of the same type, can be defined as

$$T \otimes S = T_{b_1 b_2 \ldots}^{a_1 a_2 \ldots} S_{d_1 d_2 \ldots}^{c_1 c_2 \ldots} (e_{a_1 a_2 \ldots}^{b_1 b_2 \ldots} \otimes e_{c_1 c_2 \ldots}^{d_1 d_2 \ldots}). \qquad (2.23)$$

If the tensor T is of type (n_1, m_1) and the tensor S of type (n_2, m_2), the resulting tensor is of type $(n_1 + n_2, m_1 + m_2)$ and each component is the product of a component of T with a component of S.

Contraction: While the outer product between two tensors always results in a tensor that has a higher rank than either of the arguments (the exception being when one of the tensors is a scalar, in which case the result has the same rank as the other argument), we can define an operation that lowers the rank of a mixed tensor by two. This is the *contraction* (or *trace*), which takes a type (n, m) tensor and maps it to a type $(n-1, m-1)$ tensor and we define it as a linear map acting on the basis according to

$$\mathcal{C}_\mu^\lambda(e_{b_1 \ldots b_n}^{a_1 \ldots a_m}) = \underbrace{(\vec{E}^{a_\lambda} \cdot \vec{E}_c)(\vec{E}_{b_\mu} \cdot \vec{E}^c)}_{=\vec{E}^{a_\lambda} \cdot \vec{E}_{b_\mu}} \bigotimes_{\ell \neq \mu} \vec{E}_{b_\ell} \bigotimes_{k \neq \lambda} \vec{E}^{a_k}$$

$$= \delta_c^{a_\lambda} \delta_{b_\mu}^c e_{b_1 \ldots \cancel{b}_\mu \ldots b_n}^{a_1 \ldots \cancel{a}_\lambda \ldots a_m} = \delta_{b_\mu}^{a_\lambda} e_{b_1 \ldots \cancel{b}_\mu \ldots b_n}^{a_1 \ldots \cancel{a}_\lambda \ldots a_m}, \tag{2.24}$$

where crossed out indices are omitted. This is the contraction between the λth dual basis vector and μth tangent vector in the tensor product. While this may look intimidating, the actual application of this once we suppress the tensor basis will be very straightforward. For example, for a type $(2,3)$ tensor T, the contraction between the first tangent vector and second dual basis vector is given by

$$\mathcal{C}_1^2(T) = T_{cde}^{ab} \mathcal{C}_1^2(e_{ab}^{cde}) = T_{cde}^{ab} \delta_a^d e_{\cancel{a}b}^{c\cancel{d}e} = T_{cae}^{ab} e_b^{ce}, \tag{2.25}$$

which is a type $(1,2)$ tensor with components T_{cae}^{ab}. Similar arguments can be made for any possible contraction of indices and, in component form, making a contraction amounts to setting a covariant index equal to a contravariant one and summing over it. It is left as an exercise (see Problem 2.7) to show that Eq. (2.24) is independent of the chosen basis.

Contracted product: When we were dealing with vectors only, we defined the inner product between the vectors $\vec{v}$ and $\vec{w}$ as

$$\vec{v} \cdot \vec{w} = v_a w^a. \tag{2.26}$$

The generalisation of this to tensors can be constructed by using the two operations defined above. We can construct a product between the tensors T and S by taking their outer product followed by the contraction of one index from T with one index from S (note that there may be several different possibilities here). The resulting tensor has a rank that is two less than the rank of the outer product. It may also be possible to contract more than one pair of indices from the different tensors.

Example 2.4 We saw earlier that the angular momentum for a particular angular velocity could be written as $L^i = I_j^i \omega^j$. It is therefore a contraction of the product between the angular velocity and the moment of inertia tensor. As required from the contracted product of a rank two tensor with a rank one tensor, the resulting tensor has rank one and is therefore a vector.

2.2.1 Tensors and symmetries

For the general tensor, the ordering of the indices is important. Written in component form the type (2,0) tensor T does not necessarily satisfy the relation

$$T^{ab} = T^{ba}. \tag{2.27}$$

Tensors for which this relation does hold are called *symmetric tensors* with the same nomenclature and principle for type (0,2) tensors.

Example 2.5 In Example 2.2, we wrote down the moment of inertia tensor of a disc as a linear combination of the outer product basis. Taking the components from this example, we find that

$$I_1^1 = I_2^2 = I_0, \quad I_3^3 = I_3 \tag{2.28a}$$

for the diagonal elements in the basis where the disc symmetry axis is one of the coordinate axes. The off-diagonal elements are given by

$$I_j^i = 0 = I_i^j \tag{2.28b}$$

for all $i \neq j$. It follows that the moment of inertia tensor for a disc is symmetric under exchange of i and j in this coordinate system. *Note:* In this example we have mixed co-variant and contravariant indices. This is possible since we are using a Cartesian basis, see Section 2.4.

From the definition of a symmetric tensor, it is relatively straightforward to show that if a tensor is symmetric in one coordinate system, then it must be symmetric in all coordinate systems (see Problem 2.9). While both contravariant and covariant rank two tensors can be symmetric, it is not a possibility for a mixed tensor, while it might be possible to find a coordinate system such that T_b^a is equal to T_a^b, this equality being coordinate independent would violate the transformation properties of the tensor components.

Just as some tensors are symmetric, there will also be tensors that are anti-symmetric. This property is similar to the symmetric property, with the only difference being a minus sign

$$A^{ab} = -A^{ba}. \tag{2.29}$$

Any tensor A that satisfies this relation is an *anti-symmetric tensor*. Just as with symmetric tensors, anti-symmetry is a property of the tensor itself and holds in any coordinate system.

Example 2.6 In magnetostatics, the magnetic field $\vec{B}$ fulfils Gauss's law of magnetism $\nabla \cdot \vec{B} = 0$ and thus has a vector potential $\vec{A}$ such that $\vec{B} = \nabla \times \vec{A}$. We therefore find that

$$B^i = \varepsilon_{ijk}\partial_j A_k. \tag{2.30}$$

We can associate a rank two anti-symmetric tensor F to the magnetic field by defining

$$F_{ij} = \varepsilon_{ijk}B^k = \varepsilon_{ijk}\varepsilon_{k\ell m}\partial_\ell A_m = (\delta_{i\ell}\delta_{jm} - \delta_{im}\delta_{j\ell})\partial_\ell A_m = \partial_i A_j - \partial_j A_i. \tag{2.31a}$$

This relation can be inverted according to

$$B^i = \frac{1}{2}\varepsilon_{ijk}F_{jk} = \frac{1}{2}\varepsilon_{ijk}(\partial_j A_k - \partial_k A_j) = \frac{1}{2}(\varepsilon_{ijk}\partial_j A_k + \varepsilon_{ikj}\partial_k A_j) = \varepsilon_{ijk}\partial_j A_k, \tag{2.31b}$$

where in the last step we have renamed the summation indices $j \leftrightarrow k$ in the second term. In fact, the definition of the magnetic field as an anti-symmetric tensor is to some extent more natural as the generalisation to an arbitrary coordinate system becomes $F_{ab} = \partial_a A_b - \partial_b A_a$, while the permutation symbol does not generalise directly, which we will discuss in Section 2.3.3.

As the above example demonstrates, an anti-symmetric rank two tensor in three dimensions has three independent components. In fact, this is the reason why we are able to express the vector $\vec{B}$ in terms of F as well as express F in terms of $\vec{B}$. While a completely general rank two tensor has N^2 independent components, the symmetry and anti-symmetry relations $T^{ab} = \pm T^{ba}$ provide $N(N-1)/2$ constraints for $a \neq b$. While the diagonal relations for $a = b$ are trivially fulfilled for symmetric tensors, they provide an additional N constraints for anti-symmetric tensors. In particular, we find that

$$T^{aa} = -T^{aa} \implies T^{aa} = 0. \quad \text{(no sum)} \quad (2.32)$$

This leaves $N^2 - N(N-1)/2 = N(N+1)/2$ independent components for the symmetric tensor and $N(N+1)/2 - N = N(N-1)/2$ independent components for the anti-symmetric tensor.

Going back to expressing a tensor T using the tensor basis e_{ab}, we can define a linear operator $\mathcal{S}$ by its action on the basis

$$\mathcal{S}e_{ab} = \frac{1}{2}(e_{ab} + e_{ba}) \quad (2.33)$$

mapping any type $(2,0)$ tensor T to its *symmetric part*

$$\mathcal{S}T = T^{ab}\mathcal{S}e_{ab} = \frac{1}{2}T^{ab}(e_{ab} + e_{ba}) = \frac{1}{2}(T^{ab} + T^{ba})e_{ab}, \quad (2.34)$$

i.e., $\mathcal{S}T$ is a tensor with components $(T^{ab}+T^{ba})/2$. In the same spirit, we define the operator $\mathcal{A}$ according to

$$\mathcal{A}e_{ab} = \frac{1}{2}(e_{ab} - e_{ba}), \quad (2.35)$$

which maps T to its *anti-symmetric part* with components $(T^{ab} - T^{ba})/2$. By insertion of the symmetry relations, the symmetric part of an anti-symmetric tensor vanishes and vice versa. It also holds that the symmetric part of a symmetric tensor is equal to the tensor itself, with the equivalent statement also being true for anti-symmetric tensors. The verification of this is the task in Problem 2.10.

Examining the counting of independent components above, we notice that, regardless of the number of dimensions, the number of independent components of a symmetric tensor and that of an anti-symmetric tensor add up to N^2, the number of independent components of a general tensor. This is by no means a coincidence as

$$\mathcal{S}e_{ab} + \mathcal{A}e_{ab} = \frac{1}{2}(e_{ab} + e_{ba} + e_{ab} - e_{ba}) = e_{ab} \quad (2.36)$$

for all elements of the basis e_{ab}. Therefore, any type $(2,0)$ or $(0,2)$ tensor may be decomposed into a symmetric and an anti-symmetric part. While this is true for *any* object with two indices, the real thing to notice here is that the symmetric and anti-symmetric parts are tensors of their own, since the symmetric and anti-symmetric properties are coordinate independent.

The above discussion has focused on rank two tensors only. However, also higher rank tensors may be symmetric or anti-symmetric under the exchange of one or more pairs of indices. For example, the type $(1,3)$ tensor R^a_{bcd} might have the property

$$R^a_{bcd} = R^a_{cbd}, \quad (2.37)$$

which we can refer to by saying that it is symmetric in the first and second covariant indices. Note that if the tensor was also symmetric in the first and third indices, it would also automatically be symmetric in the second and third

$$R^a_{bcd} = \{\text{symmetric in } 12\} = R^a_{cbd} = \{\text{symmetric in } 13\} = R^a_{dbc}$$
$$= \{\text{symmetric in } 12\} = R^a_{bdc}. \tag{2.38}$$

Again, similar considerations can be performed for tensors with anti-symmetries. It may also occur that a tensor is symmetric in one pair of indices and anti-symmetric in another.

Just as we defined the symmetric part of a rank two tensor, we can define the symmetric part of a higher rank tensor under the exchange of the two first indices by

$$\mathcal{S}_{12}e^a_{bcd} = \frac{1}{2}(e^a_{bcd} + e^a_{cbd}) \tag{2.39}$$

with similar definitions for the symmetric part under any other two indices. Even more generally, we can define the symmetrisation with respect to a set of n indices

$$\mathcal{S}_{12\ldots n}e_{a_1 a_2 \ldots a_n} = \frac{1}{n!}\sum_{s \in S_n} e_{s(a_1 a_2 \ldots a_n)}, \tag{2.40}$$

where S_n is the set of all possible permutations of n indices (this set will be discussed in more detail in Section 4.3.3). The tensor components symmetrised with respect to a set of indices is denoted by curly parentheses around the set of indices which are symmetrised, e.g.,

$$T^{\{abc\}} = \frac{1}{6}(T^{abc} + T^{bca} + T^{cab} + T^{bac} + T^{acb} + T^{cba}), \tag{2.41a}$$

$$R_{\{ab\}cd} = \frac{1}{2}(R_{abcd} + R_{bacd}). \tag{2.41b}$$

Similar to the definition of symmetrisation in Eq. (2.40), the anti-symmetrisation can be defined according to

$$\mathcal{A}_{12\ldots n}e_{a_1 a_2 \ldots a_n} = \frac{1}{n!}\sum_{s \in S_n} \text{sgn}(s)e_{s(a_1 a_2 \ldots a_n)}, \tag{2.42}$$

where $\text{sgn}(s)$ is equal to one if the permutation s is even and minus one if it is odd. The components of an anti-symmetrised tensor are denoted by square brackets around the anti-symmetrised indices, e.g.,

$$T^{[ab]c} = \frac{1}{2}(T^{abc} - T^{bac}). \tag{2.43}$$

The notation of symmetrised and anti-symmetrised indices using curly and square brackets is also applicable to any object with indices but, in particular, being symmetric or anti-symmetric in some given indices is a coordinate independent statement for tensors.

2.2.2 The quotient law

There is a relation known as the *quotient law*. It states that not only is the outer product of two tensors a new tensor, but if the components of two tensors T and S are related according to

$$T^{a_1 a_2 \ldots b_1 b_2 \ldots}_{c_1 c_2 \ldots d_1 d_2 \ldots} = Q^{e a_1 a_2 \ldots}_{c_1 c_2 \ldots} S^{b_1 b_2 \ldots}_{e d_1 d_2 \ldots}, \tag{2.44}$$

then $Q_{c_1c_2...}^{ea_1a_2...}$ are also the components of a tensor. In fact, we already used this in the beginning of this chapter to find the transformation rules for the object A_j^i in the relation $w^i = A_j^i v^j$, see Eq. (2.5). Using the transformation properties of T and S, we find that

$$T_{c_1'c_2'...d_1'd_2'...}^{a_1'a_2'...b_1'b_2'...} = \left(\prod_k \frac{\partial y'^{a_k'}}{\partial y^{a_k}}\right)\left(\prod_k \frac{\partial y^{c_k}}{\partial y'^{c_k'}}\right)\left(\frac{\partial y'^{e'}}{\partial y^e}\right) Q_{c_1c_2...}^{ea_1a_2...} S_{e'd_1'd_2'...}^{b_1'b_2'...}$$
$$= Q_{c_1'c_2'...}^{e'a_1'a_2'...} S_{e'd_1'd_2'...}^{b_1'b_2'...}, \tag{2.45}$$

where

$$Q_{c_1'c_2'...}^{e'a_1'a_2'...} = \left(\prod_k \frac{\partial y'^{a_k'}}{\partial y^{a_k}}\right)\left(\prod_k \frac{\partial y^{c_k}}{\partial y'^{c_k'}}\right)\left(\frac{\partial y'^{e'}}{\partial y^e}\right) Q_{c_1c_2...}^{ea_1a_2...} \tag{2.46}$$

needs to hold for Eq. (2.44) to hold for any S.

Example 2.7 In mechanics, there is a linear relationship between the force density per cross sectional area $\vec{f}$ across a small surface dS and the surface element $d\vec{S}$. Denoting the linear coefficients by σ^{ab}, we can write this relation as

$$f^a dS = \sigma^{ab} dS_b = \sigma^{ab} n_b dS. \tag{2.47}$$

Both the normal n_b and the force density f^a are vectors. The quotient law therefore implies that σ^{ab} are the components of a rank two tensor. This tensor is called the *stress tensor* and describes the different forces acting within a continuum.

2.3 TENSOR FIELDS AND DERIVATIVES

The idea behind a *tensor field* is analogous to that of a scalar or vector field, i.e., it associates a tensor of a given type to each point in the base space. Naturally, since scalars and vectors are different types of tensors, scalar and vector fields are the most natural examples of tensor fields. Just as we have used the tangent and dual bases to describe different vectors, a tensor field can be described by expressing its components in the coordinate basis as functions on the base space. For example, a type (2,0) tensor field T would be written as

$$T(p) = T^{ab}(p)e_{ab}(p), \tag{2.48}$$

where we have explicitly written out the dependence on the point p in the base space. Note that the tensor basis e_{ab} is generally also dependent on the point p, which will be important when we consider derivatives of tensor fields. In the following, we will implicitly assume this dependence.

Example 2.8 We have already come into contact with the magnetic field tensor F_{ab} in Example 2.6. Naturally, since the magnetic field $\vec{B}$ is a vector field, the associated tensor $F_{ab} = \partial_a A_b - \partial_b A_a$ is a type (0,2) tensor field.

2.3.1 The metric tensor

Perhaps the most important tensor field for most physics applications is the *metric tensor*, which, as we shall see, generally describes the distance between nearby points. This property will also be used in order to define an inner product between tangent vectors and a natural relationship between contravariant and covariant tensors. We shall start by examining how this works in an affine space with an arbitrary coordinate system. This will suffice in order to get a feeling for what kind of beast the metric tensor is and for interpreting it from a geometrical point of view. In Chapter 9, we generalise this to curved spaces, where the intuition gained from the present discussion will prove useful.

As was discussed in Section 1.6, the tangent vector basis $\vec{E}_a$ need not be orthogonal, meaning that generally $g_{ab} = \vec{E}_a \cdot \vec{E}_b \neq 0$ even if $a \neq b$. For the sake of the argument, let us examine how this quantity transforms under coordinate transformations. By the transformation rules of the individual basis vectors, we find that

$$g_{ab} = \vec{E}_a \cdot \vec{E}_b = \frac{\partial y'^{a'}}{\partial y^a} \frac{\partial y'^{b'}}{\partial y^b} \vec{E}'_{a'} \cdot \vec{E}'_{b'} = \frac{\partial y'^{a'}}{\partial y^a} \frac{\partial y'^{b'}}{\partial y^b} g_{a'b'}. \tag{2.49}$$

Thus, g_{ab} are the components of a type $(0,2)$ tensor

$$g = (\vec{E}_a \cdot \vec{E}_b) e^{ab}, \tag{2.50}$$

which we refer to as the metric tensor. While we have here used the inner product to define the metric, it should be noted that this is backwards when considering more general spaces, where instead the metric will be defining the inner product between vectors. In particular, for any two contravariant vectors v^a and w^a, we have

$$\vec{v} \cdot \vec{w} = v^a w^b \vec{E}_a \cdot \vec{E}_b = g_{ab} v^a w^b, \tag{2.51}$$

where the last step assumes that we have defined the inner product between the tangent vector basis vectors.

Example 2.9 In the case of an orthogonal coordinate system, the tangent vector basis is orthogonal and $\vec{E}_a \cdot \vec{E}_a = h_a^2$ (no sum). It follows that

$$g_{ab} = \begin{cases} h_a^2, & (a = b) \\ 0, & (\text{otherwise}) \end{cases} \tag{2.52}$$

in any orthogonal coordinate system. In particular, for a Cartesian coordinate system, we find that $h_i = 1$ and therefore

$$g_{ij} = \delta_{ij}. \tag{2.53}$$

From the fact that we can remove a summation with the δ, we find

$$\vec{v} \cdot \vec{w} = \delta_{ij} v^i w^j = v^i w^i \tag{2.54}$$

as expected.

Assuming that the coordinates are not singular, i.e., they are a good local description of the space, the metric has the following important properties:

1. *Symmetric*: Due to the inner product being symmetric, it follows that $g_{ab} = \vec{E}_a \cdot \vec{E}_b = \vec{E}_b \cdot \vec{E}_a = g_{ba}$. Therefore, the metric tensor is a symmetric tensor at each point.

2. *Positive definite*: For any vector v^a, it must hold that $g_{ab}v^av^b \geq 0$ with equality implying that $v^a = 0$. This property follows from the corresponding property of the inner product $\vec{v} \cdot \vec{v}$.

Similar to the definition of the metric using the tangent vector basis, we may define the *inverse metric tensor* using the dual basis

$$g^{ab} = \vec{E}^a \cdot \vec{E}^b. \tag{2.55}$$

The nomenclature of referring to g^{ab} as the inverse metric originates from the property

$$g^{ab}g_{bc} = (\vec{E}^a \cdot \vec{E}^b)(\vec{E}_b \cdot \vec{E}_c) = [(\vec{E}^a \cdot \vec{E}^b)\vec{E}_b] \cdot \vec{E}_c = \vec{E}^a \cdot \vec{E}_c = \delta^a_c, \tag{2.56}$$

where we have used the fact that $\vec{v} = (\vec{v} \cdot \vec{E}^b)\vec{E}_b$ for any vector $\vec{v}$. In matrix form, this relationship takes the form

$$(g^{ab})(g_{bc}) = I \iff (g^{ab}) = (g_{bc})^{-1}, \tag{2.57}$$

where I is the identity matrix.

Example 2.10 In Example 1.29, we defined an orthogonal coordinate system with the coordinates $s = \sinh(x^1)$ and $t = \sinh(x^2)$. In later examples, we found that $h_s = 1/\sqrt{1+s^2}$ and $h_t = 1/\sqrt{1+t^2}$. Since the coordinate system is orthogonal, it follows that the metric tensor is diagonal and given by

$$(g_{ab}) = \begin{pmatrix} h_s^2 & 0 \\ 0 & h_t^2 \end{pmatrix} = \begin{pmatrix} \frac{1}{1+s^2} & 0 \\ 0 & \frac{1}{1+t^2} \end{pmatrix}. \tag{2.58}$$

The inverse of this matrix is trivially given by taking the reciprocal of each of the diagonal elements and we obtain

$$(g^{ab}) = (g_{ab})^{-1} = \begin{pmatrix} 1+s^2 & 0 \\ 0 & 1+t^2 \end{pmatrix}. \tag{2.59}$$

2.3.1.1 Distances and the metric tensor

We can express the distance ds between two infinitesimally separated points according to

$$ds^2 = d\vec{x} \cdot d\vec{x}, \tag{2.60}$$

where $d\vec{x}$ is the displacement vector between the points. In Cartesian coordinates, this reduces to $ds^2 = dx^i dx^i$, which is essentially Pythagoras' theorem. However, by applying the chain rule to this relation in a general coordinate system, we find that

$$ds^2 = \left(\frac{\partial \vec{x}}{\partial y^a} \cdot \frac{\partial \vec{x}}{\partial y^b} \right) dy^a dy^b = (\vec{E}_a \cdot \vec{E}_b) dy^a dy^b = g_{ab} dy^a dy^b. \tag{2.61}$$

This *line element* is invariant under coordinate transformations due to the transformation properties of the metric tensor and dy^a and therefore the distance does not depend on the particular set of coordinates chosen. The length of a curve Γ can be computed by summing all of the infinitesimal distances along the curve

$$s_\Gamma = \int_\Gamma ds = \int_0^1 \sqrt{g_{ab}\dot{y}^a \dot{y}^b}\, dt, \tag{2.62}$$

where $\dot{y}^a = dy^a/dt$ and the parameter $t \in [0, 1]$ parametrises the curve by specifying the coordinates $y^a(t)$ as functions of t.

Example 2.11 In the two dimensional plane with polar coordinates, the non-zero components of the metric tensor are given by $g_{\rho\rho} = h_\rho^2 = 1$ and $g_{\phi\phi} = h_\phi^2 = \rho^2$. The length of a curve can therefore be expressed as

$$s_\Gamma = \int_0^1 \sqrt{\dot{\rho}^2 + \rho^2 \dot{\phi}^2}\, dt. \tag{2.63}$$

In particular, a circle of radius R may be parametrised according to $\rho(t) = R$ and $\phi(t) = 2\pi t$, leading to $\dot{\rho} = 0$ and $\dot{\phi} = 2\pi$. As expected, this leads to

$$s_\Gamma = \int_0^1 \sqrt{(2\pi R)^2}\, dt = 2\pi R. \tag{2.64}$$

Naturally, this computation could also be performed in Cartesian coordinates, but it would require a slightly more involved parametrisation of the curve in terms of sines and cosines.

2.3.1.2   Lowering and raising indices

Being a type (0,2) tensor, the metric can be used to relate a tensor of type (n, m) to a tensor of type $(n-1, m+1)$ through a contracted product. In particular, it relates a contravariant vector v^a to a covariant vector v_a according to

$$v_a = g_{ab}v^b. \tag{2.65}$$

For vectors in an affine space, we can recognise this as the relation between the covariant and contravariant components of the vector $\vec{v}$, i.e.,

$$v_a = \vec{E}_a \cdot \vec{v} = \vec{E}_a \cdot (v^b \vec{E}_b) = (\vec{E}_a \cdot \vec{E}_b)v^b = g_{ab}v^b. \tag{2.66}$$

However, the relation will remain true also in more general spaces. Due to the existence of the inverse metric, we can check the consistency according to

$$v^a = g^{ab}v_b = g^{ab}g_{bc}v^c = \delta_c^a v^c = v^a. \tag{2.67}$$

Thus, we may apply the inverse metric tensor to recover the original contravariant vector components.

The above raising and lowering of indices is not restricted to vectors, but may be applied to tensors of any rank apart from scalars, which do not have indices to lower or raise. However, it now becomes important to keep track of the ordering of *any* indices, not only separately for each index type. For example, the type (2,0) tensor T^{ab} may be related to a type (1,1) tensor according to either

$$T^a{}_b = g_{bc}T^{ac} \quad \text{or} \quad T_b{}^a = g_{bc}T^{ca}, \tag{2.68}$$

where it is important to note that the metric tensor is contracted with different indices for the different choices. Unless T^{ab} is symmetric, these two choices refer to different tensors, i.e., $T^a{}_b \neq T_b{}^a$. There is also an associated type (0,2) tensor

$$T_{ab} = g_{ac}g_{bd}T^{cd}, \tag{2.69}$$

where the order of the indices is naturally inferred from the ordering of the indices of the type (2,0) tensor. For the general type (n, m) tensor, we may use the metric and the inverse metric to lower or raise indices as we please, always resulting in a tensor of rank $n + m$, as long as we are careful in keeping track of the index ordering. As mentioned, the exception when we do not need to keep track of the index order is when a tensor is symmetric, resulting in

$$T^a{}_b = g_{bc}T^{ac} = g_{bc}T^{ca} = T_b{}^a. \tag{2.70}$$

As long as we only consider affine spaces, the raising and lowering of indices may be regarded as interchangeably expressing the corresponding vector in the tensor basis as a tangent vector or dual basis vector.

2.3.2 Derivatives of tensor fields

Just as for scalar and vector fields, tensor fields often fulfil different differential equations that can be written on local form. However, we noted that some differential operators take on a more complicated form in a general coordinate system than when written down in Cartesian coordinates. For the case of orthogonal coordinate systems, we solved this by expressing the gradient, divergence, Laplace operator, and curl using the scale factors. An alternative to this, which is applicable to any coordinate system, is to start by considering how a vector field $\vec{v}$ changes with the position by examining the partial derivatives

$$\partial_a\vec{v} = \partial_a v^b \vec{E}_b = \vec{E}_b \partial_a v^b + v^b \partial_a \vec{E}_b, \tag{2.71}$$

where the last step is an application of the product rule. There is a caveat in this line of argumentation: We have assumed that the partial derivative of a vector field makes sense, which a priori requires an unambiguous way of comparing vectors at different points. In an affine space, to which we are currently restricting ourselves, this is straightforward and can be easily done by introducing a Cartesian coordinate system and a fixed set of basis vectors. In more general spaces, this is not as clear and we postpone this discussion for Chapter 9. For now, we just observe that the first term contains a partial derivative of the vector component $\partial_a v^b$ and the second one a partial derivative of the vector basis itself $\partial_a \vec{E}_b$. The latter of these partial derivatives is independent of the vector components v^a and only depends on the coordinate system we have imposed. As it is the derivative of a vector, the result must also be a vector, which we can expand in the tangent vector basis according to

$$\partial_a\vec{E}_b \equiv \Gamma^c_{ab}\vec{E}_c, \tag{2.72}$$

where the *Christoffel symbols* $\Gamma^c_{ab} = \vec{E}^c \cdot \partial_a \vec{E}_b$ are the components of the partial derivative. Using this definition, we can re-express the derivative of the vector field $\vec{v}$ as

$$\partial_a\vec{v} = \vec{E}_b \partial_a v^b + v^b \Gamma^c_{ab}\vec{E}_c = \vec{E}_b(\partial_a v^b + \Gamma^b_{ac}v^c). \tag{2.73}$$

It follows that the components of the partial derivatives of a vector field are not just the partial derivatives of the components themselves, but also has an additional contribution from the change in the basis vectors. In order to avoid confusion when only using components to refer to the fields, we introduce the *covariant derivative* in the y^a direction ∇_a, which acts on contravariant vector components according to

$$\nabla_a v^b = \partial_a v^b + \Gamma^b_{ac} v^c. \tag{2.74}$$

It should be noted (see Problem 2.25) that while $\partial_a v^b$ does not transform as the components of a type (1,1) tensor, the covariant derivative $\nabla_a v^b$ does.

Example 2.12 We introduced a non-orthogonal coordinate system based on the vectors $\vec{v}_1 = \vec{e}_1$ and $\vec{v}_2 = \vec{e}_2 + \vec{e}_1$ in Example 1.26 and found that the tangent vector basis was given by $\vec{E}_a = \vec{v}_a$. Since the vectors $\vec{v}_a$ were taken to be constant, we find that

$$\partial_a \vec{E}_b = \partial_a \vec{v}_b = 0, \tag{2.75}$$

implying that all Christoffel symbols are zero in this coordinate system.

Example 2.13 An orthogonal coordinate system with varying basis vectors was introduced in Example 1.29

$$s = \sinh(x^1), \quad t = \sinh(x^2). \tag{2.76}$$

The tangent vector basis of this coordinate system was found to be

$$\vec{E}_s = \vec{e}_1 \frac{1}{\sqrt{1+s^2}}, \quad \vec{E}_t = \vec{e}_2 \frac{1}{\sqrt{1+t^2}}. \tag{2.77}$$

Differentiating these vectors with respect to the coordinates, we find that

$$\partial_s \vec{E}_s = -\frac{s}{\sqrt{1+s^2}^3} \vec{e}_1 = -\frac{s}{1+s^2} \vec{E}_s, \tag{2.78a}$$

$$\partial_t \vec{E}_t = -\frac{t}{\sqrt{1+t^2}^3} \vec{e}_2 = -\frac{t}{1+t^2} \vec{E}_t, \tag{2.78b}$$

$$\partial_s \vec{E}_t = \partial_t \vec{E}_s = 0. \tag{2.78c}$$

From these relations we can identify that the non-zero Christoffel symbols of this coordinate system are

$$\Gamma^s_{ss} = -\frac{s}{1+s^2}, \quad \Gamma^t_{tt} = -\frac{t}{1+t^2}. \tag{2.79}$$

We shall return to the covariant derivative of a general tensor field shortly, but let us first examine the Christoffel symbols more closely. First of all, we note that it is not only the change in the tangent vector basis that may be expressed in terms of the Christoffel symbols. Using the relation $\vec{E}^a \cdot \vec{E}_b = \delta^a_b$, we find that

$$0 = \partial_c \delta^a_b = \partial_c(\vec{E}^a \cdot \vec{E}_b) = \vec{E}^a \cdot (\partial_c \vec{E}_b) + (\partial_c \vec{E}^a) \cdot \vec{E}_b = \Gamma^a_{cb} + (\partial_c \vec{E}^a) \cdot \vec{E}_b. \tag{2.80a}$$

It directly follows that

$$\partial_c \vec{E}^a = -\Gamma^a_{cb} \vec{E}^b \qquad (2.80b)$$

and so the derivative of the vector field $\vec{v}$ could also be written according to

$$\partial_a \vec{v} = \partial_a(v_b \vec{E}^b) = \vec{E}^b(\partial_a v_b - \Gamma^c_{ab} v_c) \equiv \vec{E}^b \nabla_a v_b, \qquad (2.81)$$

where the action of the covariant derivative on the covariant components v_b is defined as

$$\nabla_a v_b = \partial_a v_b - \Gamma^c_{ab} v_c. \qquad (2.82)$$

As with the action on the covariant vector, $\partial_a v_b$ are not the components of a covariant tensor, but the full covariant derivative $\nabla_a v_b$ is.

Second, we note that the Christoffel symbols are symmetric in the lower indices, i.e., $\Gamma^c_{ab} = \Gamma^c_{ba}$. This follows from the definition of the tangent vector basis and the commutation of the partial derivatives

$$\Gamma^c_{ab} = \vec{E}^c \cdot \partial_a \vec{E}_b = \vec{E}^c \cdot \partial_a \partial_b \vec{x} = \vec{E}^c \cdot \partial_b \partial_a \vec{x} = \Gamma^c_{ba}. \qquad (2.83)$$

We can now express the Christoffel symbols in terms of the metric tensor g_{ab}. Starting from $g_{ab} = \vec{E}_a \cdot \vec{E}_b$ we find that

$$\partial_c g_{ab} = \vec{E}_a \cdot (\partial_c \vec{E}_b) + (\partial_c \vec{E}_a) \cdot \vec{E}_b = \Gamma^d_{cb} \vec{E}_a \cdot \vec{E}_d + \Gamma^d_{ca} \vec{E}_d \cdot \vec{E}_b = \Gamma^d_{cb} g_{ad} + \Gamma^d_{ca} g_{bd}. \qquad (2.84)$$

Contracting this with the inverse metric tensor g^{ea}, it follows that

$$g^{ea} \partial_c g_{ab} = \Gamma^e_{cb} + \Gamma^d_{ca} g^{ea} g_{bd}. \qquad (2.85)$$

Using the symmetry of the Christoffel symbols, these equations can now be rewritten

$$\Gamma^e_{cb} = \frac{1}{2} g^{ea} [\partial_c g_{ab} + \partial_b g_{ac} - (\Gamma^d_{ac} g_{bd} + \Gamma^d_{ab} g_{cd})] = \frac{1}{2} g^{ea} (\partial_c g_{ab} + \partial_b g_{ac} - \partial_a g_{cb}), \qquad (2.86)$$

which can be used to compute the Christoffel symbols directly from the metric tensor. While it is often easier to compute the Christoffel symbols from the partial derivatives of the basis vectors, the expression in terms of the metric will become quite useful later on when we study more general spaces.

2.3.2.1 The covariant derivative

In the case of Cartesian coordinates, we defined the directional derivative $\partial/\partial\vec{n}$ in terms of the gradient $\partial/\partial\vec{n} = \vec{n} \cdot \nabla = n^i \partial_i$. As we noted, it maps a scalar field to another scalar field and a vector field to another vector field. In arbitrary coordinates, we find that

$$\frac{\partial \vec{v}}{\partial \vec{n}} = n^a \partial_a \vec{v} = \vec{E}_b n^a \nabla_a v^b \qquad (2.87)$$

based on our discussion above. It is straightforward to check that the right-hand side is coordinate independent and thus the directional derivative itself is a vector. We will use the notation $\nabla_{\vec{n}} = n^a \nabla_a$ and call this object the covariant derivative in the direction of the vector $\vec{n}$.

Let us examine how the covariant derivative acts on an arbitrary type (n, m) tensor field $T = T^{a_1 \ldots a_n}_{b_1 \ldots b_m} e^{b_1 \ldots b_m}_{a_1 \ldots a_n}$. By the same argumentation as for the vector field, we find

$$\nabla_{\vec{n}} T = n^c \partial_c (T^{a_1 \ldots a_n}_{b_1 \ldots b_m} e^{b_1 \ldots b_m}_{a_1 \ldots a_n}) = n^c (e^{b_1 \ldots b_m}_{a_1 \ldots a_n} \partial_c T^{a_1 \ldots a_n}_{b_1 \ldots b_m} + T^{a_1 \ldots a_n}_{b_1 \ldots b_m} \partial_c e^{b_1 \ldots b_m}_{a_1 \ldots a_n}). \qquad (2.88)$$

As with the derivative acting on the vector field, we have to figure out how the derivative acts on the basis. Let us start by looking at the basis $e_a^b = \vec{E}_a \otimes \vec{E}^b$ and then generalise the result to an arbitrary tensor basis. We find that

$$\partial_c e_a^b = \partial_c(\vec{E}_a \otimes \vec{E}^b) = (\partial_c \vec{E}_a) \otimes \vec{E}^b + \vec{E}_a \otimes (\partial_c \vec{E}^b), \tag{2.89a}$$

where we have used the product rule. This expression only contains derivatives of the tangent vector and dual bases, which can be expressed in terms of the Christoffel symbols

$$\partial_c e_a^b = \Gamma_{ca}^d e_d^b - \Gamma_{cd}^b e_a^d. \tag{2.89b}$$

For the covariant derivative of the tensor $S_b^a e_a^b$, we therefore find

$$\nabla_{\vec{n}} S = n^c [e_a^b \partial_c S_b^a + S_b^a (\Gamma_{ca}^d e_d^b - \Gamma_{cd}^b e_a^d)] = n^c e_a^b (\partial_c S_b^a + \Gamma_{cd}^a S_b^d - \Gamma_{cb}^d S_d^a). \tag{2.90}$$

In the same fashion, the covariant derivative of the arbitrary tensor in Eq. (2.88) is found to be

$$\nabla_{\vec{n}} T = n^c e_{a_1 \ldots a_n}^{b_1 \ldots b_m} \left(\partial_c T_{b_1 \ldots b_m}^{a_1 \ldots a_n} + \sum_{k=1}^{n} \Gamma_{cd}^{a_k} T_{b_1 \ldots b_m}^{a_1 \ldots a_{k-1} d a_{k+1} \ldots a_n} \right.$$
$$\left. - \sum_{\ell=1}^{m} \Gamma_{cb_\ell}^d T_{b_1 \ldots b_{\ell-1} d b_{\ell+1} \ldots b_m}^{a_1 \ldots a_n} \right). \tag{2.91}$$

The above result is quite intuitive, as it includes the derivative of the tensor components together with one term for each factor in the tensor product construction of the basis. If the factor is a tangent vector, the term comes with a positive sign in front of the Christoffel symbol, and if it is a dual basis vector, it comes with a negative sign.

Example 2.14 We should check that the covariant derivative also reduces to the normal directional derivative when acting on scalar fields. Since scalar fields do not come with a basis, we find that for any scalar field ϕ

$$\nabla_{\vec{n}} \phi = n^a \partial_a \phi = \vec{n} \cdot \nabla \phi. \tag{2.92}$$

This agrees with the directional derivative we are already familiar with.

It is also instructive to consider the object $\nabla_c T_{b_1 \ldots b_m}^{a_1 \ldots a_n}$. When contracted with n^c, this object resulted in a tensor of type (n, m), the same type as T. By the quotient rule, it must therefore be the components of a type $(n, m+1)$ tensor, namely

$$\nabla T = \vec{E}^c \otimes \partial_c(T) = \vec{E}^c \otimes e_{a_1 \ldots a_n}^{b_1 \ldots b_m} \nabla_c T_{b_1 \ldots b_m}^{a_1 \ldots a_n} = e_{a_1 \ldots a_n}^{cb_1 \ldots b_m} (\nabla_c T_{b_1 \ldots b_m}^{a_1 \ldots a_n}). \tag{2.93}$$

The resulting tensor field has a rank that is one higher than the original field, a property we recognise from the gradient, which maps a scalar field to a vector field.

Example 2.15 The above construction for a scalar field ϕ results in the vector

$$\nabla \phi = \vec{E}^c \partial_c \phi. \tag{2.94}$$

Not surprisingly, the construction not only has properties similar to the gradient, when applied to a scalar field, it *is* the gradient. It is therefore also no surprise that the contraction with the vector $\vec{n}$ resulted in the directional derivative in the previous example.

With the above in mind, we make the choice of referring to the object as the *gradient of a tensor field*. It will appear in applications such as continuum mechanics.

Example 2.16 Let us apply the gradient to one of the few tensor fields we have come into contact with so far, namely the metric tensor. By insertion, we find that

$$\nabla_c g_{ab} = \partial_c g_{ab} - \Gamma^d_{ca} g_{db} - \Gamma^d_{cb} g_{ad}. \tag{2.95}$$

This result looks strangely familiar. In fact, we can obtain this expression by subtracting the right-hand side of Eq. (2.84) from the left-hand side. Since the sides are equal, the result of this operation is zero and we obtain

$$\nabla_c g_{ab} = 0. \tag{2.96}$$

In particular, this important result allows us to freely raise and lower indices on the tensors that the gradient acts upon. For example, it holds that

$$g_{ab} \nabla_c v^b = \nabla_c (g_{ab} v^b) = \nabla_c v_a. \tag{2.97}$$

2.3.2.2 Divergence

Having constructed generalisations of the directional derivative and the gradient, it stands to reason that there may be straightforward generalisations also of the other differential operators that were discussed in Section 1.4.3. Let us first figure out how to generalise the divergence, which in a heartbeat will also allow us to generalise the Laplace operator, as it is the divergence of the gradient.

For the case of a vector field $\vec{v}$, the divergence was given by

$$\nabla \cdot \vec{v} = \partial_i v^i \tag{2.98}$$

in Cartesian coordinates. As such, it is just the contraction between the two indices of the gradient, since $\nabla_i = \partial_i$ for a Cartesian coordinate system, where the Christoffel symbols vanish. The resulting quantity is a scalar, which is a tensor of rank one less than the original vector field. Generalising to an arbitrary coordinate system, we find that

$$\partial_i v^i = (\partial_i y^a)\partial_a(v^b \partial_b x^i) = (\nabla y^a) \cdot \partial_a(v^b \vec{E}_b) = \vec{E}^a \cdot \vec{E}_b \nabla_a v^b = \nabla_a v^a \tag{2.99a}$$

as expected or, in terms of the Christoffel symbols,

$$\nabla \cdot \vec{v} = \partial_a v^a + \Gamma^a_{ab} v^b. \tag{2.99b}$$

The remaining question is how to generalise this to act on any tensor field. In the case of a contravariant vector, there was only one possible contraction in the gradient, as it was a type (1,1) tensor. If a tensor has a higher rank, we need to select which index the divergence should be taken with respect to. For example, for the tensor T^{ab}, we have the two distinct choices

$$\nabla_a T^{ab} = \partial_a T^{ab} + \Gamma^a_{ac} T^{cb} + \Gamma^b_{ac} T^{ac}, \tag{2.100a}$$

$$\nabla_a T^{ba} = \partial_a T^{ba} + \Gamma^b_{ac} T^{ca} + \Gamma^a_{ac} T^{bc}. \tag{2.100b}$$

These choices will generally be different unless T is symmetric. We may even specify the divergence of a tensor with respect to a covariant index by using the inverse metric tensor to raise the index and then applying the divergence as before. When doing this, it should be noted that

$$\nabla_a(g^{ab}v_b) = g^{ab}\nabla_a v_b = g^{ab}(\partial_a v_b - \Gamma^c_{ab}v_c), \qquad (2.101)$$

due to the covariant derivative of the metric being identically zero.

Having generalised both the gradient and the divergence, we may also generalise the *Laplace operator* to act on any tensor field as

$$\nabla^2 T^{a_1...a_n}_{b_1...b_m} = g^{cd}\nabla_c\nabla_d T^{a_1...a_n}_{b_1...b_m}, \qquad (2.102)$$

i.e., as the divergence of the gradient, where the divergence is taken with respect to the index introduced by the gradient.

2.3.2.3 Generalised curl

In Chapter 1, the curl was defined in Cartesian coordinates as

$$\nabla \times \vec{v} = \vec{e}_i \varepsilon_{ijk}\partial_j v^k. \qquad (2.103)$$

Let us examine some properties of this object and see whether or not we can generalise the concept to apply also to more general tensors and, as a by-product, to spaces that are not necessarily three-dimensional:

1. The curl is a vector with three components. It is composed of ε_{ijk} and $\partial_j v^k$. Remember that we are in Cartesian coordinates, so $\partial_j v^k$ is a tensor as the Christoffel symbols vanish. Since the permutation symbol carries no information about the vector field itself, all of the information on the curl of $\vec{v}$ is contained in the tensor $\partial_j v^k$.

2. Since ε_{ijk} is totally anti-symmetric, only the anti-symmetric part of $\partial_j v^k$ will be relevant to the curl as follows from the argumentation

$$\varepsilon_{ijk}\partial_j v^k = \frac{1}{2}(\varepsilon_{ijk}\partial_j v^k + \varepsilon_{ikj}\partial_k v^j) = \frac{1}{2}\varepsilon_{ijk}(\partial_j v^k - \partial_k v^j), \qquad (2.104)$$

   where we have split the expression into two and renamed $j \leftrightarrow k$ in the second term in the first step and used the anti-symmetry of ε_{ijk} in the second step. Note that the reason that the indices are allowed to fly up and down arbitrarily in these expressions is that they are written using Cartesian coordinates.

3. As demonstrated in Example 2.6, the curl has a one-to-one correspondence to an anti-symmetric rank two tensor F_{jk}. This anti-symmetric tensor is precisely the anti-symmetric part of $\partial_j v^k$. This rank two tensor generalises in a straightforward way to both spaces of different dimensions as well as to other tensors.

Based on the above discussion, we define the curl of the covariant vector field v_a as the anti-symmetric rank two covariant tensor

$$\nabla_a v_b - \nabla_b v_a = \partial_a v_b - \partial_b v_a + (\Gamma^c_{ab} - \Gamma^c_{ba})v_c = \partial_a v_b - \partial_b v_a. \qquad (2.105)$$

Note that we started out by using the covariant derivative ∇_a rather than the partial derivative ∂_a in order to ensure that the result would be a tensor. However, this was not necessary due to the symmetry $\Gamma^c_{ab} = \Gamma^c_{ba}$ of the Christoffel symbols and we might as well have defined it using the partial derivatives.

Just as the divergence of a tensor field of rank higher than one requires us to select the index to contract with the index from the covariant derivative, generalising the curl to such a tensor field requires us to select which index to anti-symmetrise with the index from the covariant derivative. Naturally, there are several different options here. Apart from just choosing one index for anti-symmetrisation with that from the covariant derivative, we may anti-symmetrise an arbitrary number of indices. For example, we might look at the object

$$\nabla_{[a} T_{bcd]} = \partial_{[a} T_{bcd]}. \tag{2.106}$$

Again, the equivalence between the covariant derivative and the partial derivative comes from the anti-symmetrisation. The drawback of anti-symmetrising over all indices is that only the anti-symmetric part of the tensor T_{bcd} will be relevant for the results. However, for a totally anti-symmetric tensor, all options for anti-symmetrisation are equivalent up to a constant. This is a result that plays a profound role in the discussion of differential forms, see Section 9.5.1.

2.3.3   Tensor densities

It is not obvious that scalar functions should be the same regardless of the chosen coordinates. In particular, for quantities such as densities of different sorts, it would also be natural to consider a coordinate density, which would be the mass per coordinate volume rather than the mass per physical volume. This is best illustrated with a one-dimensional example.

Example 2.17   Consider the distribution of kinetic energies of the particles in a monoatomic gas. The number of particles between the energies E and $E + dE$ is then given by $f(E)\,dE$, where $f(E)$ is the distribution function. Changing variables to the momentum p of the particles, related to the energy as $E = p^2/2m$, we find that the number of particles with momentum between p and $p + dp$ is given by

$$f(E)dE = f(E)\frac{dE}{dp}dp = f(p^2/2m)\frac{p}{m}dp \equiv \tilde{f}(p)dp, \tag{2.107}$$

where $\tilde{f}(p)$ is the distribution function in momentum space. Although E and p are directly related and the distribution can be described using either, we find that $\tilde{f}(p) \neq f(E) = f(p^2/2m)$. It is related to the energy distribution function $f(E)$ by the Jacobian dE/dp.

In the same way as in the above example, we could also consider the coordinate mass density $\tilde{\rho}$ with respect to a coordinate volume, i.e., such that the mass dM within a small coordinate box with $y_0^a < y^a < y_0^a + dy^a$ is given by

$$dM = \tilde{\rho}(y_0) \prod_a dy^a. \tag{2.108a}$$

Changing coordinates to $y'^{a'}$, we find that

$$dM = \tilde{\rho}(y_0(y_0')) \mathcal{J} \prod_{a'} dy'^{a'}, \tag{2.108b}$$

where $\mathcal{J}$ is the determinant of the matrix with the partial derivatives $\partial y^a / \partial y'^{a'}$ as its elements. It follows that the coordinate density $\tilde{\rho}'(y_0')$ is given by

$$\tilde{\rho}'(y_0') = \mathcal{J}\tilde{\rho}(y_0(y_0')). \tag{2.109}$$

Since $\tilde{\rho}$ takes different values in different coordinate systems it is not a rank zero tensor. However, the transformation properties of $\tilde{\rho}$ are related to the scalar transformation property by the appearance of the determinant $\mathcal{J}$. We refer to the coordinate density $\tilde{\rho}$ as a rank zero *tensor density* (or *relative tensor*) with weight one. More generally, a tensor density of type (n, m) and weight w is a quantity T whose components transform from a coordinate system y^a to another coordinate system $y'^{a'}$ according to

$$T'^{a_1' \dots a_n'}_{\quad b_1' \dots b_m'} = \mathcal{J}^w \left(\prod_{k=1}^{n} \frac{\partial y'^{a_k'}}{\partial y^{a_k}} \right) \left(\prod_{\ell=1}^{m} \frac{\partial y^{b_\ell}}{\partial y'^{b_\ell'}} \right) T^{a_1 \dots a_n}_{\quad b_1 \dots b_m}. \tag{2.110}$$

Again, this differs from the transformation of tensors (see Eq. (2.12)) only by the appearance of the Jacobian determinant $\mathcal{J}$ raised to the wth power. It follows that a tensor density with weight zero is a normal tensor. However, note that we have here also elected to denote the components of the tensor density in the primed system by a prime, mainly to underline that it is a tensor density and not a tensor. This is most important for scalar densities, which do not have any indices that can be primed, but still take on different values in different coordinates. It is also worth noting that some texts adopt a definition of the tensor density weight w that differs from our convention by a minus sign. Naturally, this may be confusing when consulting several different texts with different conventions and should therefore be kept in mind when doing so.

It is possible to express the Jacobian determinant in terms of the permutation symbol $\varepsilon_{a_1 \dots a_N}$ according to

$$\mathcal{J} = \varepsilon_{a_1' \dots a_N'} \prod_{k=1}^{N} \frac{\partial y^k}{\partial y'^{a_k'}} = \varepsilon_{a_1 \dots a_N} \prod_{k=1}^{N} \frac{\partial y^{a_k}}{\partial y'^{k'}} = \frac{1}{N!} \varepsilon_{a_1' \dots a_N'} \varepsilon_{a_1 \dots a_N} \prod_{k=1}^{N} \frac{\partial y^{a_k}}{\partial y'^{a_k'}}, \tag{2.111}$$

where, as usual, $\varepsilon_{a_1 \dots a_N}$ is totally anti-symmetric and equal to one when all $a_k = k$.

Example 2.18 The permutation symbol $\varepsilon_{a_1 \dots a_N}$ in a general coordinate system turns out to be a tensor density rather than a tensor (which is why it is called the permutation *symbol* and not the permutation *tensor*). Let us consider the result of trying to transform it as a tensor. Starting from a completely anti-symmetric type $(0, N)$ tensor κ, which in the coordinates y^a have the components $\kappa_{a_1 \dots a_N} = \varepsilon_{a_1 \dots a_N}$, we find that

$$\kappa_{a_1' \dots a_N'} = \varepsilon_{a_1 \dots a_N} \prod_{k=1}^{N} \frac{\partial y^{a_k}}{\partial y'^{a_k'}}. \tag{2.112}$$

By construction, it should be clear that both the left- and right-hand sides of this expression are completely anti-symmetric in the free indices. Furthermore, assigning $a_k' = k'$,

$$\kappa_{1'2' \dots N'} = \varepsilon_{a_1 \dots a_N} \prod_{k=1}^{N} \frac{\partial y^{a_k}}{\partial y'^{k'}} = \mathcal{J}. \tag{2.113}$$

Due to the complete anti-symmetry, we conclude that $\kappa_{a_1' \dots a_N'} = \mathcal{J}\varepsilon_{a_1' \dots a_N'}$ and dividing both sides by $\mathcal{J}$ results in

$$\varepsilon_{a'_1...a'_N} = \mathcal{J}^{-1}\varepsilon_{a_1...a_N} \prod_{k=1}^{N} \frac{\partial y^{a_k}}{\partial y^{a'_k}}. \tag{2.114}$$

It follows that the permutation symbol may be interpreted as a type $(0, N)$ tensor density with weight $w = -1$. A priori, sticking to our convention of denoting tensor densities with a prime in the primed coordinates, we might want to put a prime on the transformed ε. However, the permutation symbol will generally be well understood also without the prime.

Just as new tensors can be created from known ones by using different tensor operations, new tensor densities can be created from known ones. In most cases, we may do whatever algebraic manipulations we did to tensors with tensor densities as long as certain requirements are met:

1. *Addition*: Due to the linearity in the construction of the tensor product, there was a natural implementation of addition of tensors. This also generalises to tensor densities, where the linearity of the transformation to the new coordinate system guarantees that the result is a new tensor density. The caveat is that, in order to factor out $\mathcal{J}^w$ after the transformation, the added tensor densities must not only be of the same type, but also the same weight w. Naturally, the resulting tensor density is a tensor density of the same type and weight as the original ones. As an example, the addition of the contravariant vector densities A^a and B^a would have the components

$$(A + B)^a = A^a + B^a, \tag{2.115}$$

   which is a tensor density as long as that A^a and B^a have the same weight.

2. *Multiplication*: In the same fashion as for the construction of the outer product between to tensors, any two tensor densities may be multiplied to yield a new tensor density. If the multiplied tensor densities are of type (n_1, m_1) with weight w_1 and (n_2, m_2) with weight w_2, respectively, then the new tensor density will be of type $(n_1 + n_2, m_1 + m_2)$ with weight $w_1 + w_2$. An example of this operation would be the multiplication of the covariant vector component C_a, i.e., a covariant vector density of weight zero, with the tensor density T^{ab} of weight two. The resulting type $(2, 1)$ tensor density would have the components $C_a T^{bc}$ and have weight two.

3. *Contraction*: The contraction of two free indices in a tensor density works in precisely the same way as for a tensor. The transformation properties of the contracted indices will cancel, leaving the remaining indices to transform as expected. The determinant $\mathcal{J}$ is unaffected by this and the resulting tensor density therefore has the same weight as the original one.

Example 2.19 The *metric determinant* $g = \det(g_{ab})$ can be written in terms of the inverse metric tensor g^{ab} and the permutation symbol $\varepsilon_{a_1...a_N}$ according to

$$g^{-1} = \det(g^{ab}) = \frac{1}{N!}\varepsilon_{a_1...a_N}\varepsilon_{b_1...b_N} \prod_{k=1}^{N} g^{a_k b_k}. \tag{2.116}$$

As the permutations symbols are tensor densities with weight -1 and the inverse metric tensor is a tensor, it follows that the metric determinant g is a scalar density with weight two. Taking the square root of g, we find that $\sqrt{g'} = \sqrt{\mathcal{J}^2 g} = \mathcal{J}\sqrt{g}$, implying that $\sqrt{g}$ is also a scalar density, but with weight one, where we have assumed that $\mathcal{J}$ is positive. This is equivalent to the coordinate bases $\vec{E}_a$ and $\vec{E}'_{a'}$ both being right-handed (or both left-handed).

In Example 2.18, we saw that the permutation symbol may be regarded as a totally anti-symmetric covariant tensor density with weight minus one. We arrived to this conclusion by examining the transformation properties of a covariant tensor $\kappa_{a_1 \ldots a_N}$, which took the values $\varepsilon_{a_1 \ldots a_N}$ in one particular coordinate system. However, the choice of letting the tensor be covariant was arbitrary and we might as well study the contravariant tensor $\tilde{\kappa}^{a_1 \ldots a_N}$. Assume that $\tilde{\kappa}^{a_1 \ldots a_N} = \varepsilon_{a_1 \ldots a_N}$ in a given coordinate system, where we note that the indices on the permutation symbol really are not covariant nor contravariant until we have deduced the transformation properties. We now find that

$$\tilde{\kappa}^{a'_1 \ldots a'_N} = \varepsilon_{a_1 \ldots a_N} \prod_{k=1}^{N} \frac{\partial y'^{a'_k}}{\partial y^{a_k}}, \tag{2.117a}$$

leading to

$$\tilde{\kappa}^{1' \ldots N'} = \varepsilon_{a_1 \ldots a_N} \prod_{k=1}^{N} \frac{\partial y'^{k'}}{\partial y^{a_k}} = \mathcal{J}^{-1}. \tag{2.117b}$$

In the same fashion as in the example, it follows that the permutation symbol may also be considered as a totally anti-symmetric *contravariant* tensor density of weight *one*. It is customary to write this tensor density with the indices raised $\varepsilon^{a_1 \ldots a_N}$ and keep the indices down $\varepsilon_{a_1 \ldots a_N}$ when intending the interpretation as a covariant tensor density. Although this convention is most often convenient, an unfortunate confusion arises when raising and lowering the indices of these tensors using the metric. Since the metric is a tensor, raising the indices of $\varepsilon_{a_1 \ldots a_N}$ using the metric will not result in $\varepsilon^{a_1 \ldots a_N}$, since the result must have weight minus one rather than plus one. In fact, we instead obtain

$$\varepsilon^{a_1 \ldots a_N} \prod_{k=1}^{N} g_{a_k b_k} = g\varepsilon_{b_1 \ldots b_N}, \tag{2.118}$$

where g is the metric determinant discussed above. Since the metric determinant has weight two, the weights of both sides of the equation match. In fact, we could have used this as argumentation for deducing the weight of the metric determinant.

The above relation between the covariant and contravariant interpretations of the permutation symbol incidentally provides us with a way of relating tensor densities of one weight to tensor densities with a different weight. As mentioned earlier, since the metric determinant g is a scalar density with weight two, its square root $\sqrt{g}$ is a scalar density with weight one. For any tensor density $T^{a_1 \ldots a_n}_{b_1 \ldots b_m}$ of weight w, we can construct a tensor

$$\mathcal{T}^{a_1 \ldots a_n}_{b_1 \ldots b_m} = \sqrt{g}^{-w} T^{a_1 \ldots a_n}_{b_1 \ldots b_m}, \tag{2.119}$$

where the factor $\sqrt{g}^{-w}$ compensates the weight of the tensor density to make the result a tensor.

Example 2.20 We can construct a completely anti-symmetric tensor $\eta_{a_1...a_N}$ with components $\varepsilon_{i_1...i_N}$ in any Cartesian coordinate system. Generally, since $\varepsilon_{a_1...a_N}$ is a tensor density of weight minus one, we can do this by assuming that

$$\eta_{a_1...a_N} = k\sqrt{g}\varepsilon_{a_1...a_N}, \tag{2.120}$$

where k is some constant that needs to be determined. Now, in a Cartesian coordinate system, $g_{ij} = \delta_{ij}$, resulting in $g = 1$. It follows that we need $k = 1$ in order to fulfil $\eta_{i_1...i_N} = \varepsilon_{i_1...i_N}$. We also note that, due to the form of the metric in a Cartesian coordinate system, $\eta^{i_1...i_N}$ is also equal to the permutation symbol. We could therefore also have defined it using the contravariant interpretation of the permutation symbol as $\eta^{a_1...a_N} = \sqrt{g}^{-1}\varepsilon^{a_1...a_N}$.

As an application of this example, we can express the cross product of two vectors in a general coordinate system. Since the cross product is a vector and we can express it in terms of the permutation symbol in a Cartesian coordinate system, it must be expressed as

$$(\vec{v} \times \vec{w})^a = \eta^{abc}v_b w_c \tag{2.121}$$

in a general coordinate system.

2.3.4 The generalised Kronecker delta

Since the covariant and contravariant interpretations of the permutation symbol are tensor densities with opposite weight, their product is a tensor of type (N, N) that is completely anti-symmetric in both its covariant and contravariant indices. This tensor is given by

$$\delta^{a_1...a_N}_{b_1...b_N} = \varepsilon^{a_1...a_N}\varepsilon_{b_1...b_N} \tag{2.122}$$

and the use of δ to denote it is not a coincidence as this tensor, and its contractions, turn out to be generalisations of the Kronecker delta tensor. For any $n \leq N$, we define the generalised Kronecker delta of order $2n$ as the type (n, n) tensor

$$\delta^{a_1...a_n}_{b_1...b_n} = \frac{1}{(N-n)!}\delta^{a_1...a_n c_{n+1}...c_N}_{b_1...b_n c_{n+1}...c_N}, \tag{2.123}$$

where the normalisation is such that the elements are either plus or minus one and follows from the $(N-n)!$ possible permutations of the $N-n$ summation indices. Like the Kronecker delta, this generalisation is an *isotropic* tensor, i.e., it has the same components in all coordinate systems

$$\delta^{a_1...a_n}_{b_1...b_n} = \begin{cases} +1, & (a_1...a_n \text{ is an even permutation of } b_1...b_n) \\ -1, & (a_1...a_n \text{ is an odd permutation of } b_1...b_n) \\ 0, & (otherwise) \end{cases} \tag{2.124}$$

Example 2.21 Consider the generalised Kronecker delta of order two. By the definition above, this should be a type $(1, 1)$ tensor that is equal to one if its only covariant index is the same as its only contravariant index and zero otherwise (there is no way to make an odd permutation of one element, there being nothing to permute it with). As such, this should be the Kronecker delta tensor that we are already accustomed to, but let us check this proposition explicitly. We start from the definition and write

$$\delta_b^a = \frac{1}{(N-1)!} \delta_{bc_2\ldots c_N}^{ac_2\ldots c_N}.$$ (2.125)

If $a \neq b$, then this entire expression must equal to zero, since by definition it is necessary that for all c_k, $c_k \neq a$ and $c_k \neq b$. Thus, in order for the expression to be non-zero, we have $N-1$ different c_k, in addition to a and b, which must all take values in the range one to N. However, there is no way of assigning unique values in this range to these $N+1$ different indices and thus $\delta_b^a = 0$.

On the other hand, if $a = b$, we only need to assign unique values to N indices, which is possible. Fixing a, and therefore also b, there are $(N-1)!$ different ways of making this assignment to the $N-1$ indices c_k. Each of those assignments will appear once in the summation and when $a = b$, $ac_2 \ldots c_N$ is the same as $bc_2 \ldots c_N$ and thus an even permutation. This means that each term in the sum contributes with a value of $1/(N-1)!$ due to the normalisation. The sum is therefore equal to one.

Example 2.22 Let us look at the generalised Kronecker delta of order four $\delta_{b_1 b_2}^{a_1 a_2}$. With two covariant and two contravariant indices there are only two possible permutations and we can easily list them. If $a_1 = b_1$ and $a_2 = b_2$, then $a_1 a_2$ is an even permutation of $b_1 b_2$, while if $a_1 = b_2$ and $a_2 = b_2$, it is an odd permutation. We thus obtain

$$\delta_{b_1 b_2}^{a_1 a_2} = \begin{cases} +, & (a_1 = b_1 \neq a_2 = b_2) \\ -1, & (a_1 = b_2 \neq a_2 = b_1) \\ 0, & (otherwise) \end{cases}.$$ (2.126)

Since we just showed that δ_b^a is one if $a = b$, it follows that

$$\delta_{b_1 b_2}^{a_1 a_2} = \delta_{b_1}^{a_1} \delta_{b_1}^{a_1} - \delta_{b_1}^{a_2} \delta_{b_2}^{a_1},$$ (2.127)

which looks stunningly familiar as it is the δ part of the ε-δ-relation. We should not be surprised at this result as, in three dimensions, we also have

$$\delta_{b_1 b_2}^{a_1 a_2} = \frac{1}{(3-2)!} \varepsilon^{a_1 a_2 c} \varepsilon_{b_1 b_2 c} = \varepsilon^{a_1 a_2 c} \varepsilon_{b_1 b_2 c}.$$ (2.128)

This last example is a special case of the more general form

$$\delta_{b_1 \ldots b_n}^{a_1 \ldots a_n} = n! \delta_{b_1}^{[a_1} \ldots \delta_{b_n}^{a_n]},$$ (2.129)

which can be used to express the generalised Kronecker delta in terms of the usual Kronecker delta δ_b^a. Note that the factor $n!$ arises from the definition of the anti-symmetrisation present in this equation, which has an $n!$ in the denominator.

2.3.5 Orthogonal coordinates

Let us briefly examine some of the above ideas applied to an orthogonal coordinate system. The first observation we can make is that, due to the orthogonality, the metric tensor is diagonal by definition

$$(g_{ab}) = \text{diag}(h_1^2, h_2^2, \ldots, h_N^2), \tag{2.130}$$

where the h_a are the scale factors. This follows directly from the relation $g_{ab} = \vec{E}_a \cdot \vec{E}_b$ while keeping in mind that the coordinate system is orthogonal. Inserting this into the expression for the Christoffel symbols in Eq. (2.86), we find that

$$\Gamma_{ab}^c = \delta_{cb}\partial_a \ln(h_b) + \delta_{ca}\partial_b \ln(h_a) - \frac{\delta_{ab}}{2h_c^2}\partial_c h_a^2, \qquad \text{(no sum)} \tag{2.131}$$

where we have used that $(\partial_a h_b^2)/h_b^2 = 2\partial_a \ln(h_b)$. As can be seen from this expression, only Christoffel symbols that contain at least two of the same index can be non-zero. Furthermore, the scale factor belonging to that index must depend on the remaining index. In particular, we obtain

$$\Gamma_{ab}^c = \begin{cases} \partial_b \ln(h_a), & (c = a) \\ -\frac{1}{2h_c^2}\partial_c h_a^2, & (a = b \neq c) \end{cases}. \tag{2.132}$$

Due to the metric being diagonal, the metric determinant is given by the product of the diagonal elements

$$g = \prod_{a=1}^N h_a^2 \implies \sqrt{g} = \prod_{a=1}^N h_a. \tag{2.133}$$

We note that the expression for $\sqrt{g}$ is exactly the same as that for the Jacobian determinant J in Eq. (1.181) for transformations to an orthogonal coordinate system from a Cartesian one. Naturally, this is no coincidence since $\sqrt{g} = 1$ in any Cartesian system and $\sqrt{g}$ is a scalar density of weight one.

Example 2.23 The polar coordinates ρ and ϕ in $\mathbb{R}^2$ constitute an orthogonal coordinate system as discussed in Section 1.6.3, where it was also found that

$$h_\rho = 1, \quad h_\phi = \rho. \tag{2.134}$$

From this follows that

$$g_{\rho\rho} = 1, \quad g_{\phi\phi} = \rho^2, \quad g_{\rho\phi} = g_{\phi\rho} = 0. \tag{2.135}$$

As we only have two coordinates, the argumentation above does not reduce the possible number of non-zero Christoffel symbols. However, since none of the scale factors depend on ϕ, the Christoffel symbols containing derivatives with respect to ϕ are all identically equal to zero. This includes

$$\Gamma_{\phi\phi}^\phi = \partial_\phi \ln(h_\phi) = 0, \tag{2.136a}$$

$$\Gamma_{\rho\rho}^\phi = -\frac{1}{2h_\phi}\partial_\phi h_\rho^2 = 0, \tag{2.136b}$$

$$\Gamma_{\rho\phi}^\rho = \Gamma_{\phi\rho}^\rho = \partial_\phi \ln(h_\rho) = 0. \tag{2.136c}$$

At the same time, the scale factor h_ρ is constant, which gives us

$$\Gamma_{\rho\rho}^\rho = \partial_\rho \ln(h_\rho) = 0. \tag{2.136d}$$

Finally, the non-zero Christoffel symbols are

$$\Gamma^\rho_{\phi\phi} = -\frac{1}{2h_\rho}\partial_\rho h_\phi^2 = -\rho, \tag{2.136e}$$

$$\Gamma^\phi_{\phi\rho} = \Gamma^\phi_{\rho\phi} = \partial_\rho \ln(h_\phi) = \frac{1}{\rho}. \tag{2.136f}$$

These results are in accordance with the derivatives of the tangent vector basis

$$\partial_\rho \vec{E}_\rho = 0, \quad \partial_\phi \vec{E}_\rho = \frac{1}{\rho}\vec{E}_\phi, \quad \partial_\rho \vec{E}_\phi = \frac{1}{\rho}\vec{E}_\phi, \quad \partial_\phi \vec{E}_\phi = -\rho\vec{E}_\rho, \tag{2.137}$$

which can be derived starting from Eq. (1.196). Note that it will often be easier to compute the Christoffel symbols directly from their definition in terms of the partial derivatives of the tangent vector basis.

2.4 TENSORS IN CARTESIAN COORDINATES

As mentioned in the beginning of this chapter, many texts will start the discussion on tensors by considering the special case of tensors in Cartesian coordinates, much in the same way as we first discussed scalars and vectors in Cartesian coordinates in Chapter 1. Instead, we choose to start by discussing the full framework of tensors in arbitrary coordinate systems, using the definition in terms of the outer product in order to build a deeper foundation for our upcoming discussion of more general spaces in Chapter 9, where the discussion on arbitrary coordinate systems will provide us with some additional intuition. However, before jumping into the more general framework, let us take a step back and look at the special case of Cartesian coordinate systems.

We start this endeavour by taking a set of constant orthonormal unit vectors $\vec{e}_i$ and fixing an origin. Any point in space is then described by the coordinates x^i, where the position vector $\vec{x} = x^i\vec{e}_i$ describes the displacement from the origin. The fact that the basis vectors are constant and orthonormal quickly leads us to the conclusion that

$$\vec{e}_i = \vec{E}^i = \vec{E}_i, \tag{2.138}$$

meaning that there is no difference between the tangent vector basis and the dual basis. Thus, in Cartesian coordinates, there is no need to distinguish covariant and contravariant indices and we can write all indices up or down at our leisure. Naturally, there is an argument for keeping the indices where they would belong in a more general setting.

Even when keeping to Cartesian coordinates, we may change to a different Cartesian coordinate system. Selecting the new Cartesian coordinate system $x'^{i'}$, which has the basis vectors $\vec{e}'_{i'}$ and whose origin is displaced with respect to the origin of the x^i coordinates by a vector $\vec{A}$, we find that

$$\vec{x} = \vec{x}' + \vec{A} \implies x'^{i'} = (x^i - A^i)\vec{e}'_{i'} \cdot \vec{e}_i = a^{i'}_i(x^i - A^i), \tag{2.139}$$

where $a^{i'}_i = \vec{e}'_{i'} \cdot \vec{e}_i$ are the transformation coefficients defined in connection to Eq. (1.11). As expected, we find that the more general tensor transformations using partial derivatives reduce to these transformation coefficients as

$$\frac{\partial x'^{i'}}{\partial x^i} = a^{i'}_i = a^i_{i'} = \frac{\partial x^i}{\partial x'^{i'}}. \tag{2.140}$$

Again, it becomes apparent that there is no difference between covariant and contravariant indices in Cartesian coordinates as they transform in the same fashion due to the $i' \leftrightarrow i$ symmetry of the transformation coefficients. With the basis being independent of the position, we may also have tensors that are defined without reference to a particular point in space. In fact, the very first example, Example 2.1 of this chapter, involved such a tensor, the moment of inertia, which is a property of an extended object rather than a single point, although it is expressed with respect to a point (usually the center of mass).

Example 2.24 The Kronecker delta δ_b^a as discussed earlier is generally a tensor. When restricted to Cartesian coordinate systems, we may exchange it for δ_{ij}, which we discussed already in Chapter 1. We have here been able to move the contravariant index down as there is no difference between covariant and contravariant indices in Cartesian coordinates. We also note that, due to the fact that $\vec{e}_i \cdot \vec{e}_j = \delta_{ij}$ and $\vec{E}_i = \vec{e}_i$ in Cartesian coordinates, the metric tensor in Cartesian coordinates becomes

$$g_{ij} = \vec{e}_i \cdot \vec{e}_j = \delta_{ij}. \tag{2.141}$$

In fact, an equivalent definition of transformations between Cartesian coordinate systems is affine transformations that preserve the metric tensor as δ_{ij}.

Naturally, since all basis vectors and components of the metric components are constant, the Christoffel symbols are all identically equal to zero in Cartesian coordinates.

Since the metric tensor for a Cartesian coordinate system is just the Kronecker delta δ_{ij}, it follows directly that the metric determinant is given by $g = 1$ in all such systems. At the same time, we are already aware that the metric determinant is a scalar density with weight two and therefore should transform from frame to frame along with a factor $\mathcal{J}^2$. There is only one way these two can be compatible, which is that any transition between Cartesian coordinate systems must have a Jacobian $\mathcal{J} = \pm 1$, where the positive sign holds for transitions between coordinate systems with the same handedness and the negative sign for transitions between systems with opposite handedness. This also follows from the fact that Cartesian coordinate transformations preserve the δ_{ij} form of the metric tensor, which imposes constraints on the transformation coefficients $a_i^{i'}$. From this insight follows that there is no distinction between tensor densities of different weights as long as we restrict ourselves to right-handed Cartesian coordinates. In particular, all tensor densities can be considered as tensors in right-handed Cartesian coordinates due to the tensor density transformation of Eq. (2.110) reducing to the normal tensor transformation when $\mathcal{J} = 1$, regardless of the weight w.

Example 2.25 Our prime example of a tensor density was the permutation symbol $\varepsilon_{a_1 \ldots a_N}$. When viewed in right-handed Cartesian coordinates, it can be taken as $\varepsilon_{i_1 \ldots i_N} = \varepsilon^{i_1 \ldots i_N} = \eta_{i_1 \ldots i_N}$. Again we note that the positioning of the indices does not matter in this case. The permutation symbol tensor densities, regardless of whether they are viewed as covariant or contravariant, are equivalent to each other and so is the totally anti-symmetric tensor $\eta_{i_1 \ldots i_N}$.

2.5 TENSOR INTEGRALS

Tensor integrals in general coordinates will not necessarily be well defined. This will be particularly true when we discuss curved spaces, where there is no unique way of relating vectors at different points in the base space to each other. In this section, we will therefore start by restricting ourselves to the integration of tensors in Cartesian coordinates and later examine which concepts we can carry with us to integrals in more general spaces.

2.5.1 Integration of tensors in Cartesian coordinates

The integration of vectors discussed in Chapter 1 can be readily extended to the case of tensors in Cartesian coordinates. As was the case for the integration of vectors, the volume, surface, and line elements can be parametrised and computed and the volume element will be an infinitesimal number, while the surface and line elements will be infinitesimal vector quantities. Since the Cartesian bases $e_{i_1 \ldots i_n}$ for any type of tensors are constant, they can be taken out of the integral and each component can be integrated as a separate function.

2.5.1.1 Volume integration

For volume integrals, we first note that the volume element takes on the same expression regardless of the coordinates as long as we use the Cartesian coordinates as integration parameters. In particular, we find that

$$dV = \vec{e}_1 \cdot (\vec{e}_2 \times \vec{e}_3) \, dx^1 dx^2 dx^3 = \varepsilon_{123} dx^1 dx^2 dx^3 = dx^1 dx^2 dx^3. \qquad (2.142)$$

This also follows from the fact that $\mathcal{J} = 1$ for all Cartesian coordinate transformations and we will see the generalisation to arbitrary coordinates when we discuss more general integrations.

Since the volume element is an infinitesimal number, the volume integral of any tensor $T_{i_1 \ldots i_n}$ is a tensor $I_{i_1 \ldots i_n}$ of the same rank and is given by

$$I_{i_1 \ldots i_n} = \int_V T_{i_1 \ldots i_n} \, dV = \int_V T_{i_1 \ldots i_n} dx^1 dx^2 dx^3, \qquad (2.143)$$

where V is the volume over which the integral is taken.

Example 2.26 The moment of inertia tensor of a solid object, which we have discussed earlier, may be expressed using a volume integral. The velocity $\vec{v}$ of the object at a point $\vec{x}$ is given by

$$\vec{v} = \vec{\omega} \times (\vec{x} - \vec{x}_0), \qquad (2.144)$$

where $\vec{x}_0$ is the fixed point of the rotation. The mass within a small volume dV around this point is $dm = \rho \, dV$, where ρ is the mass density of the object. The angular momentum of this mass with respect to $\vec{x}_0$ is therefore given by $d\vec{L} = (\vec{x} - \vec{x}_0) \times dm \, \vec{v}$ and the total angular momentum of the solid can therefore also be written

$$L_i = \vec{e}_i \cdot \int_V \rho \vec{x} \times (\vec{\omega} \times \vec{x}) dV = \omega_j \int_V \rho (x_k x_k \delta_{ij} - x_i x_j) dV, \qquad (2.145)$$

where we have set $\vec{x}_0 = 0$ for simplicity and V is the full extension of the solid. Identification of this expression with Eq. (2.2) shows that the moment of inertia tensor is given by the volume integral of the tensor $\rho(x_k x_k \delta_{ij} - x_i x_j)$

$$I_{ij} = \int_V \rho(x_k x_k \delta_{ij} - x_i x_j) dV. \tag{2.146}$$

In particular, we note that if the volume mass is purely along the rotational axis, i.e., $\vec{\omega} \propto \vec{x}$, then the angular momentum is zero as it should be.

2.5.1.2 Surface integrals

Since the surface element is an infinitesimal vector quantity, we can use it to either increase the rank of the tensor being integrated by one by performing an outer product or decrease it by one by contraction. There is also the possibility of using the scalar surface element, in which case the integral results in a tensor of the same type as the integrand, just as we saw for the volume integral. Naturally, which of these options should be used depends on the physical situation that we wish to describe. Just as for the integrals of vector quantities, the surface element will be given by Eq. (1.59).

Example 2.27 The stress tensor σ_{ij} relates a surface element $d\vec{S}$ to a force acting across the surface as described in Example 2.7. Integrating over the surface S of an object, we can find the total force acting on the object as

$$F_i = \oint_S df_i = \oint_S \sigma_{ij} dS_j. \tag{2.147}$$

It should be noted that this is the external force on the object resulting from contact forces at the surface. There is still the possibility of a force, such as gravity, acting directly on the constituents within the volume. For example, the gravitational force on an object of mass density ρ that takes up the volume V would be given by the volume integral

$$\vec{F}_g = \int_V \rho \vec{g} \, dV, \tag{2.148}$$

where $\vec{g}$ is the external gravitational field.

Example 2.28 Coming back to the moment of inertia tensor, which we saw can be written as a volume integral, some mass distributions will allow us to disregard one dimension of the distribution. In the case we started with, a disc of mass M, we neglected the thickness of the disc. This can be done as long as the thickness is much smaller than the radius R of the disc and the volume integral with the volume density ρ can be replaced by a surface integral with the surface density ρ_s

$$I_{ij} = \int_V \rho(\delta_{ij} x_k x_k - x_i x_j) dV \rightarrow \int_S \rho_s(\delta_{ij} x_k x_k - x_i x_j) dS. \tag{2.149}$$

In our case with the disc, we can compute the moment of inertia tensor about its center by the following consideration. We start by introducing a coordinate system such that the disc is in the x_1-x_2-plane. First of all, the term with $x_i x_j$ will be equal to zero whenever i or j is equal to three as $x_3 = 0$ everywhere on the disc. Furthermore, when $(i,j) = (1,2)$ the integrand for the term $x_1 x_2$ is anti-symmetric with respect to $x_1 \to -x_1$ (and $x_2 \to -x_2$), while the integration domain is symmetric. Thus, the second term will only give a contribution when $i = j = 1$ or 2. Due to the symmetry of the disc, the value will be the same for both these cases and we find

$$\int_S \rho x_1^2 dS = \int_0^R r^3 \, dr \int_0^{2\pi} \rho_s \cos^2(\varphi) d\varphi = \pi \rho_s \frac{R^4}{4}. \tag{2.150}$$

For the first term including the δ_{ij}, we find that

$$\int_S \delta_{ij} \rho_s x_k x_k dS = \delta_{ij} \rho_s \int_0^R r^3 dr \int_0^{2\pi} d\varphi = \pi \rho_s \delta_{ij} \frac{R^4}{2}. \tag{2.151}$$

We now take into account that the mass of the disc is

$$M = \int_S \rho_s dS = \pi R^2 \rho_s, \tag{2.152}$$

resulting in

$$I_{ij} = \frac{MR^2}{4}(\delta_{ij} + \delta_{i3}\delta_{j3}). \tag{2.153}$$

Thus, the non-zero components of the moment of inertia tensor in this coordinate system are $I_{11} = I_{22} = MR^2/4$ and $I_{33} = MR^2/2$. These are precisely the values quoted in the beginning of this chapter. Equivalently, including the tensor basis, this may be expressed as

$$I = \frac{MR^2}{4}(\delta + e_{33}), \tag{2.154}$$

where δ is the rank two Kronecker delta and $e_{33} = \vec{e}_3 \otimes \vec{e}_3$.

2.5.1.3 Line integrals

As was the case with the surface integral, the line element will generally come with a direction. Thus, also for a line integral, the resulting tensor may be of one order lower or higher than the tensor being integrated, depending on whether or not there is a contraction between the integrated tensor and the line element. We may also find situations when the direction of the line element is unimportant, resulting in a tensor of the same order as the integrand.

Example 2.29 The magnetic force on an infinitesimal part $d\vec{x}$ of a conductor carrying a current I is given by $d\vec{F} = I d\vec{x} \times \vec{B}$, where $\vec{B}$ is the external magnetic field. The total force on the conductor may therefore be written as a line integral

$$\vec{F} = I \int_\Gamma d\vec{x} \times \vec{B}, \tag{2.155}$$

where Γ is the curve in space occupied by the conductor. In terms of the magnetic field tensor $F_{ij} = \varepsilon_{ijk}B^k$ introduced earlier, the force can be rewritten as

$$F_i = I \int_\Gamma F_{ij}dx^j, \qquad (2.156)$$

i.e., in terms of the line integral of F_{ij} contracted with the line element. Note that we have here used F_i and F_{ij} to denote different tensors, the first is the total force vector and the second the magnetic field tensor.

2.5.1.4 Integral theorems

As already mentioned, the integral theorems discussed in Section 1.5 may be generalised to the forms

$$\int_V \partial_i T_{j_1 \ldots j_n}\, dV = \oint_S T_{j_1 \ldots j_n}\, dS_i, \qquad (2.157a)$$

$$\int_S \varepsilon_{ijk}\partial_k T_{\ell_1 \ldots \ell_n}\, dS_j = \oint_\Gamma T_{\ell_1 \ldots \ell_n}\, dx^i, \qquad (2.157b)$$

where we have replaced the f of Eqs. (1.139) with the general tensor component $T_{j_1 \ldots j_n}$ as already alluded to when we first discussed these theorems. Contractions of any of the indices of $T_{j_1 \ldots j_n}$ with the free index i are also possible and the integral theorems will still hold since it does not matter whether the contraction is done before or after the integration. This may also be seen by writing the contraction in terms of the constant Kronecker delta, which may be moved out of the integral due to its components being constant, for example

$$\int_V \partial_i T_{ijk}dV = \delta_{i\ell} \int_V \partial_i T_{\ell jk}dV = \delta_{i\ell} \oint_S T_{\ell jk}dS_i = \oint_S T_{ijk}dS_i. \qquad (2.158)$$

Just as the divergence and curl theorems, these generalised integral theorems are powerful tools that may be used in order to simplify several calculations that might otherwise be hard to perform.

Example 2.30 The surface force acting on an object can be found by taking the surface integral

$$F_i = \oint_S \sigma_{ij}dS_j \qquad (2.159)$$

over its entire surface, where σ_{ij} is the stress tensor. By the generalisation of the divergence theorem, we find that

$$F_i = \int_V \partial_j \sigma_{ij}dV. \qquad (2.160)$$

A special application of this formula is *Archimedes' principle*. For an object submerged in a fluid, the contact force from the fluid on the object will be the same as that which the fluid would exert on an equivalent volume of fluid that replaced the object, see Fig. 2.2. Within a stationary fluid of density ρ_0 in a homogeneous gravitational field $\vec{g}$, the stress tensor is given by $\sigma_{ij} = -p\delta_{ij}$, where $p = \rho_0\vec{g}\cdot\vec{x}$ is the pressure in the fluid at $\vec{x}$. It follows that

$$F_i = -\int_V \delta_{ij}\partial_j p\, dV = -\int_V \partial_i \rho_0 g_j x^j dV = -g_i \rho_0 V = -g_i M_V, \qquad (2.161)$$

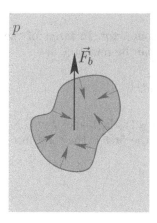

Figure 2.2 According to Archimedes' principle, the buoyant force $\vec{F}_b$ on an object immersed in a fluid with the pressure field p is equal to the force on the corresponding volume if the object was replaced by fluid. The buoyant force can be expressed as an integral over the object's surface using the stress tensor.

where $M_V = \rho_0 V$ is the mass of the displaced fluid. The force is therefore in the opposite direction of the gravitational field and equal in magnitude to the gravitational force on a mass M_V. Including the gravitational force, the total force on the object will therefore be given by

$$\vec{F}_{\text{total}} = \vec{g}(\rho - \rho_0)V, \tag{2.162}$$

where ρ is the density of the object itself. If the object is denser than the fluid, the net force is in the same direction as the gravitational field and this situation is referred to as the object having negative buoyancy. If the fluid is denser than the object, the total force is directed in the opposite direction from the gravitational field and the object has positive buoyancy. In the situation where $\rho = \rho_0$, the net force is zero and the buoyancy is neutral.

2.5.2 The volume element and general coordinates

When discussing volume integrals in general, we have so far only expressed the volume element in Cartesian or orthogonal coordinates, where it takes the form

$$dV = \prod_{a=1}^{N} h_a dx^a. \quad \text{(no sum)} \tag{2.163}$$

This expression was derived from the volume spanned by the tangent vector basis $\vec{E}_a$ and the approach remains valid for a general coordinate system. Wanting to express the volume described by $y_0^a < y^a < y_0^a + dy^a$, the volume is spanned by the vectors $\vec{E}_a dy^a$ (no sum), see Fig. 2.3. In three dimensions, this results in

$$dV = \vec{E}_1 \cdot (\vec{E}_2 \times \vec{E}_3) dy^1 dy^2 dy^3. \tag{2.164}$$

Expressing this triple product in Cartesian coordinates, we find that

$$dV = \varepsilon_{ijk} \frac{\partial x^i}{\partial y^1} \frac{\partial x^j}{\partial y^2} \frac{\partial x^k}{\partial y^3} dy^1 dy^2 dy^3. \tag{2.165}$$

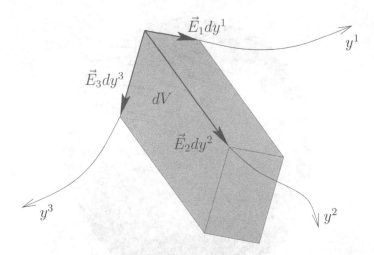

Figure 2.3 In a general coordinate system, the volume dV described by the coordinates $y_0^a < y^a < y_0^a + dy^a$ is spanned by the vectors $\vec{E}_a dy^a$ (no sum). The volume is therefore given by the triple product $dV = \vec{E}_1 \cdot (\vec{E}_2 \times \vec{E}_3) dy^1 dy^2 dy^3$. This is a good approximation as long as the dy^a are small and exact when they are infinitesimal.

The quantity multiplying the differentials dy^a now looks stunningly familiar as the 123 component of a completely anti-symmetric tensor with Cartesian components ε_{ijk}. Fortunately, we already know that this tensor is $\eta_{abc} = \sqrt{g}\,\varepsilon_{abc}$ and it directly follows that

$$dV = \sqrt{g}\,\varepsilon_{123} dy^1 dy^2 dy^3 = \sqrt{g}\,dy^1 dy^2 dy^3. \qquad (2.166)$$

The generalisation of this argumentation to an N dimensional space is straightforward and we find that

$$dV = \sqrt{g} \prod_{k=1}^{N} dy^{a_k} \equiv \sqrt{g}\,d^N y. \qquad (2.167)$$

This form of the volume element should not come as a surprise as we have already seen that the volume element expressed in a general orthogonal coordinate system is $dV = \mathcal{J} d^N y$, where $\mathcal{J}$ is the product of the scale factors, which, as noted earlier, is the square root of the metric determinant in such a coordinate system.

Despite our success in expressing the volume element, integration in a general coordinate system faces another obstacle that is not present in Cartesian coordinates. Unlike the Cartesian basis vectors, the tangent vector basis $\vec{E}_a$ and the dual basis $\vec{E}^a$ are not constant. This results in the complication that we cannot perform integrals of tensor quantities in general coordinates by just integrating the components one by one. For integrals of tensor quantities, we face the problem of which basis to use to express the result. Being an integral over a volume, surface, or curve, the resulting tensor quantity does not belong to a particular point in space, so there can be no preference for any particular basis. In an affine space, we may brush this aside and express the result in a Cartesian coordinate system, but in more general curved spaces, there will not exist a natural way of comparing, much less defining the sum of, tensors at different points in space. However, scalar quantities do not require a basis and integrals of quantities with all indices contracted will therefore continue to make sense even in general spaces.

Figure 2.4 While the flow of a substance on a sphere may be defined as a vector at each point of the sphere, there is no meaningful definition of the vector integral of this flow. However, it is possible to define integrals on the sphere that describe, for example, the total flux into or out of a given region.

Example 2.31 Consider the concentration of a substance on the surface of a sphere. At each point on the sphere, there is a tangent vector J^a describing the flow of the substance, see Fig. 2.4, but the tangent vector spaces at different points are different. There is therefore no relevant way of integrating over the sphere to obtain the average flow of the substance as a vector in any of the tangent spaces of the sphere. Of course, there may be some notion of such an average flow if we are looking at the sphere as embedded in our three dimensional affine space, but with a description that is restricted to the sphere only, this is an impossibility. However, the flux integral

$$\Phi = \int_\Gamma J^a dS_a \tag{2.168}$$

of the vector field J^a over some curve Γ, where dS_a is orthogonal to the curve, is a well defined quantity that describes the amount of substance flowing across Γ per time. These issues will be discussed in more detail in Chapter 9.

2.6 TENSOR EXAMPLES

The last example of the preceding section gives us an appetiser of things to come when we will consider more general spaces. We postpone this discussion for later as the discussion will benefit from material that will be covered in the following chapters, in particular Chapter 8 on variational calculus will be of importance. Instead, we conclude this chapter by giving some more structured examples of the application of tensor analysis within different fields of physics. It should be stressed that these examples are only written for the purpose of

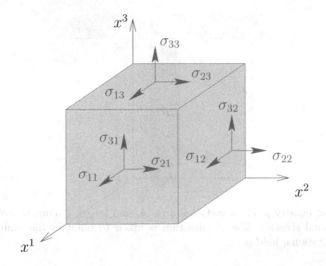

Figure 2.5 The nine components of the stress tensor represented as forces acting on the different faces of a cube. The stress σ_{ij} represents a force in the x^i direction on a surface with a normal $\vec{e}_j$.

discussing the application of tensor analysis in different fields and not as guides or learning material for the subject matters. Each subject is much broader than what can be presented here and will require dedicated texts to understand in full. However, it will provide us with some basics and a context in which we can apply the knowledge acquired throughout this chapter.

2.6.1 Solid mechanics

Solid mechanics is the study of solids and the forces that act within them. In the examples given in this section, we will restrict ourselves to static situations, where any object is in force (and torque) equilibrium. While rank two tensors are relatively common in many areas of physics, as they represent the most general linear relationship between two vectors, solid mechanics presents us with two of the most regularly encountered rank four tensors, the stiffness and compliance tensors.

2.6.1.1 The stress tensor

We are already familiar with one of the main actors of solid mechanics from our earlier examples. The *stress tensor* is a rank two tensor σ_{ij} that relates the directed surface element dS_i to the contact force dF_i acting on that surface element

$$dF_i = \sigma_{ij} dS_j. \tag{2.169}$$

For example, the component σ_{12} describes the force per unit area in the x^1 direction when the area element has a surface normal in the x^2 direction, see Fig. 2.5. For a solid in static equilibrium, we can find a set of differential equations that describe how the stress tensor depends on any volume forces, most commonly a gravitational force. We do this by considering an arbitrary volume V with a boundary surface S. Assuming that the body is subject to a volume force $f_i dV$, where f_i is a vector field, the total force on the volume is

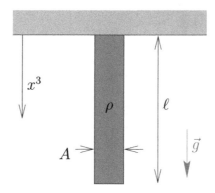

Figure 2.6 A rod of density ρ, cross sectional area A, and length ℓ hanging from the ceiling will be subject to internal stresses. The x^3 direction is taken to point in the same direction as the homogeneous gravitational field $\vec{g}$.

given by

$$F_i = \oint_S \sigma_{ij} dS_j + \int_V f_i dV, \qquad (2.170)$$

where the first term represents the surface forces from the rest of the solid and the second term represents the volume force. Applying the generalised divergence theorem to this relation, we find that

$$F_i = \int_V (\partial_j \sigma_{ij} + f_i) dV = 0, \qquad (2.171)$$

where we have used that in order for the solid to be in static equilibrium, its acceleration, and therefore the total force acting on it, must be equal to zero. Since the volume V was arbitrary, this relation must hold for any volume, which is true only if the integrand is identically equal to zero, i.e.,

$$\partial_j \sigma_{ij} = -f_i. \qquad (2.172)$$

Given appropriate boundary conditions, this differential equation will allow us to solve for the stress tensor for an object subjected to a force density f_i.

Example 2.32 Consider a homogeneous rod of density ρ and square cross section of area A that is hanging from the ceiling as shown in Fig. 2.6. The volume force density acting on this rod due to gravity is given by $f_i = \rho g \delta_{i3}$, where the x^3 direction is defined to be in the same direction as the gravitational field. The resulting differential equation for the stress tensor is

$$\partial_j \sigma_{ij} = -\rho g \delta_{i3}. \qquad (2.173)$$

With no forces acting perpendicular to the x^3 direction, the resulting differential equations for $i = 1, 2$ are solved by $\sigma_{1j} = \sigma_{2j} = 0$. Assuming no forces acting on the free surfaces of the rod, we also find $\sigma_{31} = \sigma_{32} = 0$ as well as the only non-zero component

$$\sigma_{33} = \rho g(\ell - x^3). \qquad (2.174)$$

Note that this gives a force

$$\vec{F} = -\vec{e}_3 \rho g A \ell = -\vec{e}_3 M g, \qquad (2.175)$$

where M is the mass of the rod, from the ceiling acting on the top of the rod. Not surprisingly, this is the exact force needed to cancel the gravitational force of the rod.

Force equilibrium is not sufficient for an object to be in static equilibrium, we also need to impose torque equilibrium in order not to induce rotations. Again looking at a volume V, the torque $\vec{\tau}$ relative to the origin is given by

$$\tau_i = \varepsilon_{ijk}\left(\oint_S x^j \sigma_{k\ell} dS_\ell + \int_V x^j f_k dV\right). \tag{2.176}$$

Note that, due to the force equilibrium that we have already imposed, torque equilibrium around the origin is equivalent to torque equilibrium about any other point. Just as for the force equilibrium, we can apply the generalised divergence theorem to the surface integral, resulting in

$$\tau_i = \int_V \varepsilon_{ijk}(\partial_\ell x^j \sigma_{k\ell} + x^j f_k) dV = \int_V \varepsilon_{ijk}[\sigma_{kj} + x^j(\partial_\ell \sigma_{k\ell} + f_k)] dV$$

$$= \int_V \varepsilon_{ijk}\sigma_{kj} dV, \tag{2.177}$$

where we have applied the requirement of force equilibrium in the last step. Demanding torque equilibrium requires that $\vec{\tau} = 0$ and since this has to hold for any volume V, the integrand must be equal to zero. Multiplying by $\varepsilon_{i\ell m}$, we find that

$$\varepsilon_{i\ell m}\varepsilon_{ijk}\sigma_{jk} = 2\sigma_{[\ell m]} = 0, \tag{2.178}$$

meaning that the anti-symmetric part of the stress tensor σ_{ij} must vanish, i.e., the stress tensor is symmetric.

2.6.1.2 The strain tensor

We have so far only described a situation where we have been concerned with the internal forces inside a solid. If the solid is very stiff, the forces will not cause noticeable deformations of the solid. However, for some applications, it is necessary to also know how the material deforms based on the internal stresses. In order to handle this, we first need a good description of the deformations of the solid. We begin this endeavour by defining the *displacement field* $\vec{u}(\vec{x})$, which describes the displacement of a point in the solid that would be at position $\vec{x}$ if the solid was not strained. It follows that this point is located at

$$\vec{x}' = \vec{x} + \vec{u}(\vec{x}). \tag{2.179}$$

Although this field describes how each point of the solid has been displaced, a non-zero displacement field does not necessarily mean that the solid is strained. Possibilities for such situations include translations and rotations

$$x'^i = R_{ij}x^j + a^i \tag{2.180}$$

of the solid, where R_{ij} are the components of a rotation matrix and $\vec{a}$ is a constant vector. We thus need to relate the displacement field to the strain that is present in the solid.

In the following, we will assume that the solid is relatively stiff, meaning that the partial derivatives of the displacement field are much smaller than one $|\partial_i u_j| \ll 1$. In other words, the difference in the displacement between nearby points in the solid is much smaller than the distance between the points. We define the *linear strain* between two nearby points as

$$\epsilon = \frac{ds' - ds}{ds}, \tag{2.181}$$

where ds is the distance between the points in the unstrained solid and ds' the distance between the points after being displaced. Note that we have here used the symbol ϵ rather than ε for the strain in order to avoid confusion with the permutation symbol when introducing the strain tensor. For the two points $\vec{x}$ and $\vec{x} + \vec{n}\, ds$, where $\vec{n}$ is a unit vector describing the direction of the original spatial separation between the points, we find that the new displacement is given by

$$\vec{x}'(\vec{x} + \vec{n}\, ds) - \vec{x}'(\vec{x}) \simeq [\vec{n} + (\vec{n} \cdot \nabla)\vec{u}]\, ds. \tag{2.182}$$

Squaring this relation we find that, to linear order in u,

$$ds'^2 \simeq ds^2[1 + n_i n_j(\partial_j u_i + \partial_i u_j)] \tag{2.183a}$$

and taking the square root of this expression and expanding in the small derivatives gives us

$$ds' \simeq ds \left[1 + n_i n_j \frac{1}{2}(\partial_j u_i + \partial_i u_j)\right]. \tag{2.183b}$$

It therefore follows that the linear strain is given by

$$\epsilon = n_i n_j \frac{1}{2}(\partial_j u_i + \partial_i u_j) = n_i n_j \epsilon_{ij}, \tag{2.184}$$

where we have defined the symmetric *strain tensor*

$$\epsilon_{ij} = \frac{1}{2}(\partial_j u_i + \partial_i u_j). \tag{2.185}$$

The strain tensor can be used to describe the deformation of the solid in many different manners, but we will leave most of this for dedicated texts on solid mechanics and only provide the following example.

Example 2.33 The *volumetric strain* of a small element of a solid with unstrained volume dV can be quantified as

$$\delta = \frac{dV' - dV}{dV}, \tag{2.186}$$

where dV' is the volume of the element when the solid is strained. Taking an original volume spanned by the coordinate basis vectors $d\vec{x}_i = \vec{e}_i ds_i$ (no sum), we find that

$$d\vec{x}_1' = (\vec{e}_1 + \partial_1 \vec{u})ds_1, \tag{2.187}$$

with the equivalent expressions for $d\vec{x}_2'$ and $d\vec{x}_3'$. The volume dV' is therefore given by

$$dV' = d\vec{x}_1' \cdot (d\vec{x}_2' \times d\vec{x}_3') \simeq (1 + \partial_i u_i)ds_1 ds_2 ds_3 = (1 + \epsilon_{ii})dV. \tag{2.188}$$

Thus, the volumetric strain is given by the trace ϵ_{ii} of the strain tensor.

With the above example in mind, we note that it is always possible to rewrite the strain tensor as

$$\epsilon_{ij} = \frac{1}{3}\theta\delta_{ij} + \underbrace{\left(\epsilon_{ij} - \frac{1}{3}\theta\delta_{ij}\right)}_{\equiv \kappa_{ij}}, \tag{2.189}$$

where θ is an arbitrary scalar field. With the choice of θ equal to the volumetric strain ϵ_{kk}, the tensor κ_{ij} becomes traceless and is known as the *shear strain tensor* (or *deviatoric strain tensor*).

2.6.1.3 The stiffness and compliance tensors

For elastic materials, the internal stresses are linearly related to the strains in the material. For our purposes, we will not consider more complicated situations with large strains and inelastic materials and the most general way of linearly relating two rank two tensors is by a rank four tensor. We define this rank four tensor in the relation between the stress and strain tensors to be the *stiffness tensor* $c_{ijk\ell}$. The corresponding relation, *Hooke's law*, is given by

$$\sigma_{ij} = c_{ijk\ell}\epsilon_{k\ell} \tag{2.190}$$

and the symmetry of the stress tensor implies that the stiffness tensor must be symmetric in the first two indices $c_{ijk\ell} = c_{jik\ell} = c_{\{ij\}k\ell}$. In addition, the fact that the strain tensor is symmetric also implies that any anti-symmetric part with respect to the last two indices will be irrelevant for the relation and we might as well take $c_{ijk\ell} = c_{ij\ell k} = c_{ij\{k\ell\}}$. A general rank four tensor would have $3^4 = 81$ independent components. However, the symmetries of the stiffness tensor brings this number down to $6^2 = 36$.

Example 2.34 If a material is well described only by the deformations in one direction and the deformation only depends on that direction, it can be regarded within a one-dimensional setting with a single coordinate x. As a result, the stress and displacement only have one component and can be written as σ and ε, respectively. Hooke's law now takes the form

$$\sigma = c\varepsilon, \tag{2.191}$$

where c is the single component c_{1111} of the stiffness tensor.

By the assumption of linearity, the stiffness tensor itself is not dependent on the actual strains or stresses in the material, only on the properties of the material itself. Depending on the structure of the material, further simplifications may be performed to reduce the number of independent components. The most constraining case is obtained if we assume that the material is isotropic, meaning that the stiffness tensor must have the same components in all frames. We already know that the rank two Kronecker delta tensor is isotropic and thus any outer product of Kronecker deltas will also be an isotropic rank four tensor. The most general linear combination $C_{ijk\ell}$ of such products is

$$C_{ijk\ell} = c_1\delta_{ij}\delta_{k\ell} + c_2\delta_{ik}\delta_{j\ell} + c_3\delta_{i\ell}\delta_{jk}, \tag{2.192}$$

which has three independent parameters. However, only the first of these terms has the symmetry required from the stiffness tensor and the last two terms display this symmetry only if $c_2 = c_3$. It is possible to show that this is the only isotropic tensor of rank four with the required symmetries and we can therefore write the stiffness tensor of an isotropic material on the form

$$c_{ijk\ell} = \lambda\delta_{ij}\delta_{k\ell} + \mu(\delta_{ik}\delta_{j\ell} + \delta_{i\ell}\delta_{jk}). \tag{2.193}$$

The parameters λ and μ are material constants and we note that the stress is now related to the strain according to

$$\sigma_{ij} = \lambda\delta_{ij}\delta_{k\ell}\varepsilon_{k\ell} + \mu(\varepsilon_{ij} + \varepsilon_{ji}) = \lambda\varepsilon_{kk}\delta_{ij} + 2\mu\varepsilon_{ij}. \tag{2.194}$$

Decomposing the strain tensor into volumetric and shear strains, we find that

$$\sigma_{ij} = \left(\lambda + \frac{2\mu}{3}\right)\varepsilon_{kk}\delta_{ij} + 2\mu\kappa_{ij} \equiv K\varepsilon_{kk}\delta_{ij} + 2G\kappa_{ij}. \tag{2.195}$$

In this last expression we have introduced the *bulk modulus* K and the *shear modulus* G. These are both material constants and can be found tabulated in standard reference texts.

The relation in Eq. (2.190) can also be inverted. We can do this by finding a tensor $s_{ijk\ell}$ such that

$$\epsilon_{ij} = s_{ijk\ell}\sigma_{k\ell}. \tag{2.196}$$

This tensor is known as the *compliance tensor*, which tells us how easy it is to deform the material, with larger components indicating larger deformation for the same stress. By the same argumentation as for the stiffness tensor, the symmetries of the stress and strain tensors imply the very same symmetries for the compliance tensor. The usual form of writing the compliance tensor for an isotropic material is

$$s_{ijk\ell} = \frac{1}{E}\left[\frac{1+\nu}{2}(\delta_{ik}\delta_{j\ell} + \delta_{i\ell}\delta_{jk}) - \nu\delta_{ij}\delta_{k\ell}\right] \tag{2.197a}$$

implying that

$$\epsilon_{ij} = \frac{1}{E}\left[(1+\nu)\sigma_{ij} - \nu\sigma_{kk}\delta_{ij}\right]. \tag{2.197b}$$

The material constants involved here are *Young's modulus* E and *Poisson's ratio* ν. Being the inverse relation of that given by the stiffness tensor, these constants are related to the bulk and shear moduli and are also found in standard references. Relating E and ν to K and G is left as Problem 2.41.

2.6.2 Electromagnetism

Electromagnetism is the theory describing the interactions of electric and magnetic fields with matter based upon charges and currents. For the simplest ideas in the theory, it is sufficient to have a description based on vector analysis using the electric field $\vec{E}$ and magnetic field $\vec{B}$. In free space, these fields satisfy *Maxwell's equations*

$$\nabla \cdot \vec{E} = \frac{\rho}{\varepsilon_0}, \tag{2.198a}$$

$$\nabla \cdot \vec{B} = 0, \tag{2.198b}$$

$$\nabla \times \vec{E} + \frac{\partial\vec{B}}{\partial t} = 0, \tag{2.198c}$$

$$\nabla \times \vec{B} - \frac{1}{c^2}\frac{\partial\vec{E}}{\partial t} = \mu_0\vec{J}, \tag{2.198d}$$

where ρ is the charge density, $\vec{J}$ the current density, c the speed of light in vacuum, ε_0 the permittivity, and μ_0 the permeability. It should be noted that, by definition, $c^2\varepsilon_0\mu_0 = 1$.

Once delving a bit deeper into the theory of electromagnetism, it will become apparent that the use of tensors will help us significantly. In particular, when formulating Maxwell's equations in the framework of special relativity, it will become clear that the electric and magnetic fields are different components of a rank two anti-symmetric tensor in four-dimensional space-time rather than vector fields in space that depend on time. We leave this discussion for dedicated texts and concentrate on other aspects of electromagnetism that may be described using tensor analysis.

2.6.2.1 The magnetic field tensor

We have already encountered the *magnetic field tensor* $F_{ij} = \varepsilon_{ijk}B_k$ in the examples throughout this chapter. The fact that it is an anti-symmetric tensor of rank two is closely related to the relativistic description of electromagnetism we just mentioned. Using the magnetic field tensor instead of the magnetic field vector $\vec{B}$, Maxwell's equations can be recast using the relations

$$\nabla \cdot \vec{B} = \frac{1}{2}\varepsilon_{ijk}\partial_i F_{jk}, \tag{2.199a}$$

$$\vec{e}_i \cdot (\nabla \times \vec{B}) = \frac{1}{2}\varepsilon_{ijk}\partial_j\varepsilon_{k\ell m}F_{\ell m} = \frac{1}{2}(\delta_{i\ell}\delta_{jm} - \delta_{im}\delta_{j\ell})\partial_j F_{\ell m}$$

$$= \frac{1}{2}(\partial_j F_{ij} - \partial_j F_{ji}) = \partial_j F_{ij}. \tag{2.199b}$$

The curl of the magnetic field vector is thus the divergence of the magnetic field tensor and the divergence of the field vector is an anti-symmetric sum of the partial derivatives of the field tensor. Using these expressions, we can rewrite the last two of Maxwell's equations by using the magnetic field tensor as

$$\varepsilon_{ijk}\varepsilon_{k\ell m}\partial_\ell E_m + \varepsilon_{ijk}\frac{\partial B_k}{\partial t} = \partial_i E_j - \partial_j E_i + \frac{\partial F_{ij}}{\partial t} = 0, \tag{2.200a}$$

$$\partial_j F_{ij} - \frac{1}{c^2}\frac{\partial E_i}{\partial t} = J_i. \tag{2.200b}$$

Hardly surprising, the first of these equations states that the time derivative of F_{ij} is an anti-symmetric derivative of the electric field. The third of Maxwell's equations that involves the magnetic field is that of $\nabla \cdot \vec{B}$, which states that the anti-symmetric sum of partial derivatives of the magnetic field tensor in Eq. (2.199a) vanishes.

2.6.2.2 The Maxwell stress tensor

Within electromagnetism, the laws of Newton may seem to be violated unless we also attribute both energy and momentum densities to the electromagnetic field. If the electromagnetic field carries momentum, we may consider the change of momentum within a volume V per time as a force acting on the electromagnetic field within the volume according to Newton's second law

$$\vec{F} = \frac{d\vec{p}}{dt}. \tag{2.201}$$

By Newton's third law, this force needs to be paired with an equal and opposite force. There are two possible types of forces involved here. Adhering to the requirement that physical laws should be local, the electromagnetic field may experience a force from matter

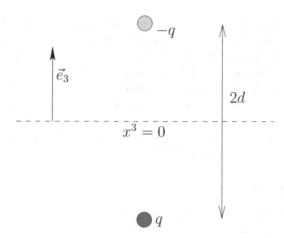

Figure 2.7 Two opposite charges q and $-q$ separated by a distance $2d$. We wish to compute the force of the electromagnetic field in the upper half plane on that in the lower half plane by evaluating the force on the surface $x^3 = 0$ with a surface normal $\vec{e}_3$.

contained in the volume or by a flux of momentum in or out of the volume through the volume's boundary surface. The latter of these options is thus a force acting on the volume's surface, describing the force with which the electromagnetic field outside the volume acts on the field inside the volume. Having read Section 2.6.1, this sounds strangely familiar and it is natural to describe this force with a rank two tensor, the *Maxwell stress tensor*

$$\sigma_{ij} = \varepsilon_0 E_i E_j + \frac{1}{\mu_0} B_i B_j - \frac{1}{2} \left(\varepsilon_0 E^2 + \frac{1}{\mu_0} B^2 \right) \delta_{ij}, \tag{2.202}$$

where E^2 and B^2 are the squares of the electric and magnetic field vectors, respectively. We leave the derivation of the form of the Maxwell stress tensor for a more detailed treatise on electromagnetism and for now only consider the following example.

Example 2.35 Consider a static situation with two charges q and $-q$ separated by a distance $2d$, see Fig. 2.7. In this setting, we know that the force on each of the charges should have the magnitude

$$F = \frac{q^2}{4\pi\varepsilon_0 (2d)^2} \tag{2.203}$$

and be directed in the $\vec{e}_3$ direction for the lower charge q and in the $-\vec{e}_3$ direction for the upper charge $-q$. However, taking the local perspective, these are not forces between the charges themselves, but rather forces with which the fields act on the charges. By Newton's third law, we therefore expect the charges to act on the fields with a force of the same magnitude, but opposite direction. In order for the field to remain static, the force on the field in the $x^3 < 0$ region from that in the $x^3 > 0$ region must therefore precisely balance the force from the charge q.

The electric field from the charges on the surface $x^3 = 0$ are given by

$$\vec{E}_q = \frac{q}{4\pi\varepsilon_0 r^3} (\rho \vec{e}_\rho + d \vec{e}_z), \quad \vec{E}_{-q} = \frac{-q}{4\pi\varepsilon_0 r^3} (\rho \vec{e}_\rho - d \vec{e}_z), \tag{2.204a}$$

where we have introduced polar coordinates and $r^2 = \rho^2 + d^2$. It follows that the total electric field is given by

$$\vec{E} = \vec{E}_q + \vec{E}_{-q} = \frac{qd\vec{e}_z}{2\pi\varepsilon_0 r^3} = \frac{qd\vec{e}_3}{2\pi\varepsilon_0 r^3}. \tag{2.204b}$$

Since there are no currents in the problem, the magnetic field is equal to zero and the surface force on the field in the $x^3 < 0$ region from a small surface element $\vec{e}_z dx^1 dx^2$ is given by

$$dF_i = \sigma_{i3} dx^1 dx^2 = \varepsilon_0 \left(E_i E_3 - \frac{1}{2} E^2 \delta_{i3} \right) dx^1 dx^2. \tag{2.205}$$

The field $\vec{E}_3$ is in the $\vec{e}_3$ direction and we therefore have $E = E_3$ and $E_i = \delta_{i3}E$. It follows that

$$d\vec{F} = \frac{q^2 d^2}{8\pi^2 \varepsilon_0} \frac{1}{(\rho^2 + d^2)^3} \vec{e}_3 dx^1 dx^2. \tag{2.206}$$

Integrating this over the entire surface (the easiest way of doing this is using polar coordinates on the surface), we find that

$$\vec{F} = \frac{q^2}{4\pi\varepsilon_0 (2d)^2} \vec{e}_z, \tag{2.207}$$

exactly the force needed to balance the force from the charge at $x^3 = -d$ on the field. While this example was static and we could have been satisfied with applying Coulomb's law for an interaction between two charges, it demonstrates the ideas of the electromagnetic field being subject to forces as well as the concept of locality.

2.6.2.3 The conductivity and resistivity tensors

You are very likely familiar with *Ohm's law* $V = IR$, where V is the electric potential difference across, I the current through, and R the resistance of a resistor. The resistance R is a physical characteristic of the dimensions of the resistor and the material it is made from. A more fundamental way of writing this relationship is to express the current density $\vec{J}$ as a linear function of the electric field $\vec{E}$. By now, we are familiar with the most general form of a linear relationship between two vectors and know that it is given by a rank two tensor σ_{ij} as

$$J_i = \sigma_{ij} E_j. \tag{2.208}$$

This tensor is the *conductivity tensor* of the material and the notation using σ_{ij} may be unfortunate seeing that we have already used it for the stress tensor of solid mechanics as well as for the Maxwell stress tensor (at least in those cases they were both related to forces). However, these are all very common notations and it should be clear from the context which is intended.

Just as the stiffness tensor we encountered earlier, the conductivity tensor depends only on the material and if the material is isotropic, then the conductivity must also be. The only isotropic rank two tensor is the Kronecker delta and multiples of it and it follows that

$$\sigma_{ij} = \sigma \delta_{ij} \tag{2.209a}$$

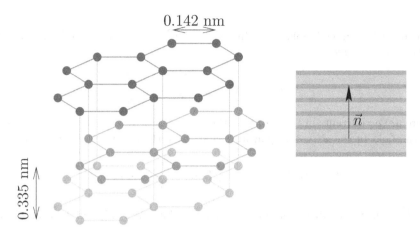

Figure 2.8 Graphite consists of sheets of carbon atoms arranged in a hexagonal grid (left). As a result, the conductivity is high within the sheets, but poor between them, leading to an anisotropic conductivity tensor. The situation is similar to thin copper sheets stacked alternately with an insulator (right). The direction $\vec{n}$ normal to the sheets is singled out as the direction with low conductivity.

for such materials. Since there is only one single number σ that characterises the material, it is known as *the* conductivity of the material. For isotropic materials we therefore have

$$\vec{J} = \sigma \vec{E}, \tag{2.209b}$$

i.e., the current density is directly proportional to the electric field.

Although many conductors are isotropic, at least at the relevant scales, there are materials with an atomic structure that implies high electric conductivity in some directions and low in others. A typical example of such a material is graphite, which consists of several layered sheets of carbon atoms, see Fig. 2.8. The conductivity in the directions of the sheets is relatively high, while transmitting a current perpendicular to the sheets requires a stronger electric field. The resulting relation between the electric field and the resulting current can be written as

$$\vec{J} = \sigma_0 \vec{n}(\vec{E} \cdot \vec{n}) + \sigma_1 \vec{n} \times (\vec{E} \times \vec{n}), \tag{2.210}$$

where $\vec{n}$ is a unit vector normal to the sheets, σ_0 is the conductivity perpendicular to the sheets, and σ_1 the conductivity along the sheets. Writing this in tensor notation, we find that

$$J_i = \sigma_0 n_i n_j E_j + \sigma_1 \varepsilon_{ijk} \varepsilon_{k\ell m} n_j n_\ell E_m = [\sigma_0 n_i n_j + \sigma_1 (\delta_{ij} - n_i n_j)] E_j \tag{2.211a}$$

or in other words

$$\sigma_{ij} = \sigma_0 n_i n_j + \sigma_1 (\delta_{ij} - n_i n_j). \tag{2.211b}$$

A macroscopic analogue of this situation would be stacking sheets of copper alternately with sheets of rubber. On a scale larger than the typical sheet thickness, such a stack would conduct electricity very well along the copper sheets, but poorly in the stacking direction.

In analogy to the inversion of the relation between the stress and strain tensors, the relation between current density and electric field may also be inverted resulting in the relation

$$E_i = \rho_{ij} J_j, \tag{2.212}$$

where the *resistivity tensor* ρ_{ij} satisfies $\rho_{ij}\sigma_{jk} = \delta_{ik}$, i.e., it is the inverse of the conductivity tensor.

2.6.3 Classical mechanics

The use of tensors in more advanced classical mechanics is abundant. In Lagrange's and Hamilton's formulations of mechanics, tensors and curved spaces play a major role for the most general form of the theory. Since we have not yet touched upon curved spaces, our discussion in this section will be limited to some more basic examples, which should be conceptually simpler. Classical mechanics will be treated in a more general setting in Chapter 10.

2.6.3.1 *The moment of inertia tensor*

Our very first example of a tensor was taken from the realm of classical mechanics. The *moment of inertia tensor* was introduced as a description of the linear relationship between the angular velocity $\vec{\omega}$ and the angular momentum $\vec{L}$

$$L_i = I_{ij}\omega_j \tag{2.213}$$

relative to a fixed point of the body. Note that we have here assumed a Cartesian basis. As a result, the index placement is irrelevant and we have moved them all down. In Example 2.26, we showed that the moment of inertia of an object with respect to the origin is given by the integral

$$I_{ij} = \int_V \rho(\delta_{ij} x_k x_k - x_i x_j) dV, \tag{2.214}$$

where V is the volume of the object. In many situations, when considering rotational motion, we are either interested in the momentum of inertia with respect to a point around which the object rotates freely, or around its center of mass. In both cases, we will here only consider the rotational motion. In the case of rotation around the center of mass, any translational motion may be factored out and in the case of a fixed point, the only motion is rotational, meaning that the velocity of a point $\vec{x}$ in the rotating object is given by $\vec{v} = \vec{\omega} \times \vec{x}$. This will be discussed in more detail in Section 10.1.2.

In many senses, the moment of inertia is to rotational motion what mass is to linear motion. As an example of this, the kinetic energy in linear motion is given by the familiar expression $mv^2/2$. Using this to express the kinetic energy of each small part of the object, we can find the object's total kinetic rotational energy

$$E_{\text{rot}} = \frac{1}{2}\int_V \rho(\vec{\omega} \times \vec{x}) \cdot (\vec{\omega} \times \vec{x}) dV = \frac{1}{2}\vec{\omega} \cdot \int_V \rho[\vec{x} \times (\vec{\omega} \times \vec{x})] dV$$

$$= \frac{1}{2}\vec{\omega} \cdot \vec{L} = \frac{1}{2}\omega_i I_{ij}\omega_j. \tag{2.215}$$

In the same way that forces are related to the change in momentum, torques are related to changes in the angular momentum. The torque $\vec{\tau}$ required for an angular acceleration $\vec{\alpha} = \dot{\vec{\omega}}$ is therefore given by

$$\tau_i = \dot{L}_i = \frac{d}{dt}(I_{ij}\omega_j) = \dot{I}_{ij}\omega_j + I_{ij}\alpha_j. \tag{2.216}$$

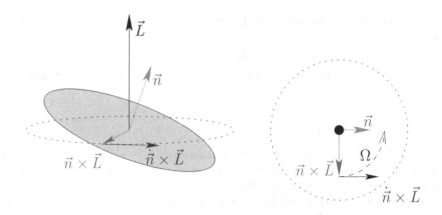

Figure 2.9 For a disc in free rotation, the angular momentum $\vec{L}$ need not be parallel to the principal symmetry direction $\vec{n}$ of the disc. This will lead to the symmetry axis $\vec{n}$ precessing around $\vec{L}$ with an angular velocity Ω, resulting in a wobbling behaviour. We here also show the vectors $\vec{n} \times \vec{L}$, which is orthogonal to both $\vec{n}$ and $\vec{L}$ and its time derivative $\dot{\vec{n}} \times \vec{L}$.

The time derivative $\dot{I}_{ij}$ of the momentum of inertia tensor satisfies the relation

$$\dot{I}_{ij}\omega_j = \varepsilon_{ijk}\omega_j I_{k\ell}\omega_\ell, \tag{2.217}$$

the demonstration of which is left as Problem 2.45. As can be seen from this expression, the angular acceleration $\vec{\alpha}$ may be therefore non-zero even if the torque acting on the object is zero.

Example 2.36 We have previously computed the momentum of inertia tensor for a disc and found that it is given by

$$I_{ij} = I_0(\delta_{ij} + n_i n_j), \tag{2.218}$$

where $I_0 = MR^2/4$ and $\vec{n}$ is a unit vector normal to the disc, see Fig. 2.9. This vector will rotate with the disc and fulfil the relation $\dot{\vec{n}} = \vec{\omega} \times \vec{n}$. Writing $\vec{\omega} = \omega_n \vec{n} + \omega_t \vec{t}$, where $\vec{t}$ is a vector orthogonal to $\vec{n}$, i.e., in the plane of the disc, we find that

$$\vec{L} = I_0(2\omega_n \vec{n} + \omega_t \vec{t}). \tag{2.219}$$

If the disc is freely rotating without an applied torque, then by definition $\dot{\vec{L}} = 0$ and we find that

$$\frac{d(\vec{n} \cdot \vec{L})}{dt} = \dot{\vec{n}} \cdot \vec{L} = I_0(\vec{\omega} \times \vec{n}) \cdot (2\omega_n \vec{n} + \omega_t \vec{t}) = 0. \tag{2.220}$$

Therefore, the projection of $\vec{n}$ onto the angular momentum $\vec{L}$ is constant, which means that it is just rotating around $\vec{L}$, since it is a unit vector. Accordingly, the vector $\vec{n} \times \vec{L}$, which is orthogonal to both $\vec{n}$ and $\vec{L}$ will also be a vector of constant magnitude rotating around $\vec{L}$ with the same angular frequency Ω as $\vec{n}$. We may find this angular frequency by comparing the magnitudes of $\vec{n} \times \vec{L}$ and $d(\vec{n} \times \vec{L})/dt = \dot{\vec{n}} \times \vec{L}$ as shown in the figure. The result for Ω is thus given by

$$\Omega^2 = \frac{[I_0(\vec{\omega} \times \vec{n}) \times (2\omega_n \vec{n} + \omega_t \vec{t})]^2}{[I_0 \vec{n} \times (2\omega_n \vec{n} + \omega_t \vec{t})]^2} = 4\omega_n^2 + \omega_t^2. \tag{2.221}$$

If the disc is rotating mainly around its symmetry axis $\vec{n}$, i.e., if $\omega_n \gg \omega_t$, we obtain $\Omega \simeq 2\omega_n$. This means that the disc will wobble around the axis parallel to the angular momentum with an angular frequency that is twice that with which the disc is spinning. The factor of two arises from the ratio of the disc's moment of inertia along the symmetry axis and the moment of inertia in the other two directions. Objects for which the moment of inertia displays symmetry around one rotational axis, but that have a different ratio between the moment of inertia in the different directions, will lead to another relation between Ω and ω_n.

As a final observation, we note that we can find the power dissipated into rotational energy of the object by differentiating the rotational energy with respect to time

$$\frac{dE_{\text{rot}}}{dt} = \omega_i I_{ij}\alpha_j + \omega_i \dot{I}_{ij}\omega_j = \omega_i \tau_i. \tag{2.222a}$$

Again, we note the resemblance to linear motion, where the power dissipated into an object would be given by $\vec{v} \cdot \vec{F}$. Due to the form of $\dot{I}_{ij}\omega_j$, see Eq. (2.217) the second term in the middle expression vanishes and we also have

$$\frac{dE_{\text{rot}}}{dt} = \omega_i I_{ij}\alpha_j = L_j \alpha_j. \tag{2.222b}$$

The equations of rotational motion being similar to those for linear motion is no coincidence, but a direct result of a more general formalism, where the degrees of freedom of a mechanical system are described using an appropriate number of generalised coordinates.

2.6.3.2 *The generalised inertia tensor*

As we have seen for the moment of inertia tensor, the relations we found were in one-to-one correspondence with similar expressions for linear motion, replacing masses with the moment of inertia and velocities with the angular velocity. It should come as no surprise that this similarity does not appear by accident, but rather results from the application of a more general framework to the special cases of linear motion and rotation.

For a large class of mechanical systems, the *configuration space* is a general space of N-dimensions, i.e., the spatial configuration of the system may be described by N coordinates y^a (see Section 10.2.1). One typical example of this is the *double pendulum* shown in Fig. 2.10, the configuration of which may be fully described by the two angles φ_1 and φ_2. The kinetic energy T in such a general system can be written on the form

$$T = \frac{1}{2}M_{ab}\dot{y}^a \dot{y}^b, \tag{2.223}$$

where $\dot{y}^a = dy^a/dt$ and M_{ab} is the *generalised inertia tensor*. Due to the symmetry of this expression in $\dot{y}^a$ and $\dot{y}^b$, we can take the generalised inertia tensor to be symmetric.

Example 2.37 For a particle of mass m moving in three-dimensional space, the configuration space is a three-dimensional affine space, which we can describe using the Cartesian coordinates x^i or, equivalently, by the position vector $\vec{x}$. Its kinetic energy is given by the expression

$$T = \frac{1}{2}m\vec{v}^2 = \frac{1}{2}m\dot{\vec{x}}^2 = \frac{1}{2}m\delta_{ij}\dot{x}^i \dot{x}^j \tag{2.224}$$

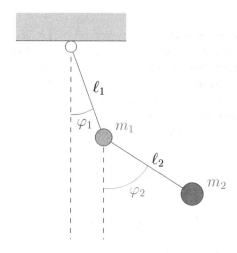

Figure 2.10 A planar double pendulum with lengths ℓ_1 and ℓ_2 and masses m_1 and m_2. The configuration of the double pendulum is completely determined by the two angles φ_1 and φ_2.

and so the inertia tensor is $M_{ij} = m\delta_{ij}$. However, we could also describe the motion of the particle using a curvilinear coordinate system y^a, which would result in the expression

$$T = \frac{1}{2}mg_{ab}\dot{y}^a\dot{y}^b, \tag{2.225}$$

i.e., the inertia tensor would be proportional to the metric g_{ab} with the mass m as the proportionality constant.

Example 2.38 The configuration space of a freely rotating object may be described by the three Euler angles ϕ^a, that parametrise an arbitrary rotation in three dimensions. The components of the angular velocity $\vec{\omega}$ in a Cartesian coordinate system will generally be a linear combination of derivatives of the Euler angles

$$\omega_i = f^i_a(\phi)\dot{\phi}^a. \tag{2.226}$$

As a result, the rotational kinetic energy may be written in the form

$$E_{\text{rot}} = \frac{1}{2}\omega_i I_{ij}\omega_j = \frac{1}{2}f^i_a I_{ij} f^j_b \dot{\phi}^a \dot{\phi}^b. \tag{2.227}$$

We therefore find that the generalised inertia tensor in this coordinate system is $M_{ab} = I_{ij}f^i_a f^j_b$.

Example 2.39 For the *double pendulum* in Fig. 2.10, the positions of the masses may be described in Cartesian coordinates as

$$\vec{x}_1 = \ell_1[-\cos(\varphi_1)\vec{e}_1 + \sin(\varphi_1)\vec{e}_2], \tag{2.228a}$$
$$\vec{x}_2 = \vec{x}_1 + \ell_2[-\cos(\varphi_2)\vec{e}_1 + \sin(\varphi_2)\vec{e}_2]. \tag{2.228b}$$

The kinetic energy in the system is given by

$$T = \frac{1}{2}(m_1\dot{\vec{x}}_1^2 + m_2\dot{\vec{x}}_2^2) = \frac{1}{2}[M\ell_1^2\dot{\varphi}_1^2 + m\ell_2^2\dot{\varphi}_2^2 + 2m\ell_1\ell_2\cos(\varphi_1 - \varphi_2)\dot{\varphi}_1\dot{\varphi}_2], \tag{2.229}$$

where $M = m_1 + m_2$ and $m = m_2$. Identifying with the expression

$$T = \frac{1}{2}M_{ab}\dot{y}^a\dot{y}^b = \frac{1}{2}(M_{11}\dot{\varphi}_1^2 + M_{22}\dot{\varphi}_2^2 + 2M_{12}\dot{\varphi}_1\dot{\varphi}_2), \tag{2.230}$$

where we have explicitly used the symmetry $M_{12} = M_{21}$, we find that

$$M_{11} = M\ell_1^2, \quad M_{22} = m\ell_2^2, \quad M_{12} = m\ell_1\ell_2\cos(\varphi_1 - \varphi_2). \tag{2.231}$$

One thing to note is that the generalised inertia tensor possesses all of the properties normally associated to a metric. In fact, it *defines* a metric on the configuration space, which will often not be a flat space. We will see more of this when we discuss Lagrangian mechanics in Chapter 10. As for the discussion on curved spaces, we postpone this until we have discussed variational calculus in Chapter 8.

2.7 PROBLEMS

Problem 2.1. Derive the transformation properties for the components of a tensor of type (n, m) as quoted in Eq. (2.18).

Problem 2.2. Explicitly write down the tensor basis $e_{ab} = \vec{E}_a \otimes \vec{E}_b$, where $\vec{E}_a$ is the tangent vector basis in polar coordinates, in terms of the Cartesian basis $\vec{e}_i \otimes \vec{e}_j$.

Problem 2.3. Verify that the rank two zero tensor $0 \otimes 0$ satisfies the relation

$$T + 0 \otimes 0 = T \tag{2.232}$$

for any rank two tensor T and that all its components are zero irrespective of the coordinate system used.

Problem 2.4. In an isotropic medium, the polarisation $\vec{P}$ of the medium is directly proportional to the electric field $\vec{E}$ and can be written as

$$\vec{P} = \varepsilon_0 \chi \vec{E}, \tag{2.233}$$

where χ is the *electric susceptibility* of the medium. In a non-isotropic medium, this is no longer true, but the polarisation and electric field are still vectors and their relation is linear such that

$$P^i = \varepsilon_0 \chi^i_j E^j. \tag{2.234}$$

Show that the electric susceptibility χ^i_j is a rank two tensor.

Problem 2.5. Consider Ohm's law, see Eq. (2.208), for an anisotropic medium where the conductivity tensor, in Cartesian coordinates, is given by

$$\sigma_{ij} = \sigma_0 \delta_{ij} + \lambda n_i n_j, \tag{2.235}$$

where λ is a scalar and $\vec{n}$ is a unit vector, which may be taken to be $\vec{n} = \vec{e}_3$.

a) Writing the electric field as $\vec{E} = E_0[\cos(\alpha)\vec{n} + \sin(\alpha)\vec{t}]$, where α is the angle between $\vec{E}$ and $\vec{n}$ and $\vec{t}$ is a unit vector perpendicular to $\vec{n}$, find the corresponding current density $\vec{J}$.

b) Find the angle between $\vec{E}$ and $\vec{J}$ as a function of the angle α. For which values of the angle α are $\vec{E}$ and $\vec{J}$ parallel?

Problem 2.6. Assume that we had instead chosen to define tensors by the transformation properties related to different coordinate systems, i.e., we define that

$$T^{a_1'\dots a_n'}_{b_1'\dots b_m'} = \left(\frac{\partial y'^{a_1'}}{\partial y^{a_1}}\right) \cdots \left(\frac{\partial y'^{a_n'}}{\partial y^{a_n}}\right) \left(\frac{\partial y^{b_1}}{\partial y'^{b_1'}}\right) \cdots \left(\frac{\partial y^{b_m}}{\partial y'^{b_m'}}\right) T^{a_1\dots a_n}_{b_1\dots b_m}. \tag{2.236}$$

Verify explicitly that the following expressions transform in the correct way, and therefore are tensors according to this definition

a) $T^{ab} = cS^{ab}$,

b) $T^a_{bc} = S^a_{bc} + V^a_{bc}$,

c) $T^a_{bcd} = S^a_b V_{cd}$.

Here, c is assumed to be a scalar while the S and V are assumed to be tensors (and thus follow the appropriate transformation rules).

Problem 2.7. Verify that the definition of the contraction in Eq. (2.24) is independent of the coordinates chosen, i.e., that

$$C^\lambda_\mu(e'^{a_1'\dots a_m'}_{b_1'\dots b_n'}) = (\vec{E}'^{a_\lambda} \cdot \vec{E}'_c)(\vec{E}'_{b_\mu} \cdot \vec{E}'^c) \bigotimes_{\ell \neq \mu} \vec{E}'_{b_\ell} \bigotimes_{k \neq \lambda} \vec{E}'^{a_k} \tag{2.237}$$

gives the same result as Eq. (2.24) for the contraction of any tensor.

Problem 2.8. Consider an expression $T_{ab}v^a w^b$ that is invariant under general coordinate transformations for any vectors $\vec{v}$ and $\vec{w}$. Show that T_{ab} are the components of a type $(0,2)$ tensor.

Problem 2.9. Assuming that the tensor T^{ab} is symmetric in the coordinates y^a, show that it must also be symmetric in the coordinates $y'^{a'}$.

Problem 2.10. Verify the following statements:

a) The symmetric part of an anti-symmetric tensor and the anti-symmetric part of a symmetric tensor vanish.

b) The (anti-)symmetric part of a (anti-)symmetric tensor is equal to the tensor itself.

Problem 2.11. Consider the arbitrary tensors T_{ab} and S^{ab}. Show that

$$T_{\{ab\}}S^{[ab]} = T_{[ab]}S^{\{ab\}} = 0, \tag{2.238}$$

i.e., that if we symmetrise one of the tensors and anti-symmetrise the other, the contraction of the symmetrised indices with the anti-symmetrised ones gives zero. Use this result to show that the expression $T_{ab}v^a v^b$ for an arbitrary vector $\vec{v}$ only depends on the symmetric part of T_{ab}.

Problem 2.12. Show that the only tensor T_{abc} that fulfils the symmetry relations

$$T_{abc} = -T_{bac} = T_{acb} \tag{2.239}$$

is necessarily the type $(0, 3)$ zero tensor.

Problem 2.13. The general type $(3, 0)$ tensor has N^3 independent components, where N is the number of dimensions of the space. A general anti-symmetric tensor T^{abc} satisfies the relations $T^{\{ab\}c} = T^{a\{bc\}} = 0$ that restrict the number of independent components. Count the number of independent such relations and conclude what the number of independent components of T^{abc} is. In the same fashion, find the number of independent components of a symmetric tensor S^{abc}.

Problem 2.14. The inverse metric tensor is defined in Eq. (2.55). Starting from this definition, verify that the components g^{ab} transform as the components of a type $(2,0)$ tensor.

Problem 2.15. The magnetic field $\vec{B}$ around a straight wire carrying a current I in the positive z-direction at $\rho = 0$ has the contravariant vector components

$$B^\phi = \frac{\mu_0 I}{2\pi\rho^2}, \quad B^\rho = B^z = 0 \tag{2.240}$$

in cylinder coordinates (note that $\vec{E}_\phi$ is *not* a unit vector). By using the expression for the metric tensor and its inverse in this coordinate system, find the covariant vector components of $\vec{B}$.

Problem 2.16. Verify that for a scalar field ϕ the partial derivatives $\partial_a\phi$ with respect to the coordinates naturally form the covariant components of a vector, i.e., their transformation properties coincide with what you would expect from covariant vector components.

Problem 2.17. Derive the transformation rules for the Christoffel symbols Γ^c_{ab} under changes of coordinate system.

Problem 2.18. Show that the contracted Christoffel symbols Γ^b_{ab} can be written in terms of a partial derivative of the logarithm of the square root of the metric tensor

$$\Gamma^b_{ab} = \partial_a \ln(\sqrt{g}). \tag{2.241}$$

Problem 2.19. Using the results from Problem 2.18, show that the divergence of a general anti-symmetric type $(2,0)$ tensor T^{ab} may be written as

$$\nabla_a T^{ba} = \frac{1}{\sqrt{g}} \frac{\partial}{\partial y^a} \left(T^{ba} \sqrt{g} \right), \tag{2.242}$$

where g is the metric determinant.

Problem 2.20. Starting from the definition of spherical coordinates and using the expressions for the tangent vector basis $\vec{E}_r$, $\vec{E}_\theta$, and $\vec{E}_\varphi$, compute the components of the metric tensor g_{ab} and the Christoffel symbols Γ^c_{ab} in spherical coordinates.

Problem 2.21. Use the expressions for the tangent vector basis in cylinder coordinates in order to find the Christoffel symbols Γ^c_{ab} in these coordinates.

Problem 2.22. A curve in a three-dimensional space may be parametrised by giving the three spherical coordinates as functions $r(t)$, $\theta(t)$, and $\varphi(t)$ of the curve parameter t. Using the expressions for the metric tensor components found in Problem 2.20, write down an integral in terms of these functions that describes the length of a general curve when the curve parameter is given by $0 < t < 1$. Use this expression to find the length of the curve

$$r(t) = R_0, \quad \theta(t) = \theta_0, \quad \varphi(t) = 2\pi t. \tag{2.243}$$

Problem 2.23. In Problems 1.49 and 1.50, we introduced the hyperbolic coordinates u and v

$$x^1 = v e^u, \quad x^2 = v e^{-u} \tag{2.244a}$$

and parabolic coordinates t and s

$$x^1 = ts, \quad x^2 = \frac{1}{2}(t^2 - s^2), \tag{2.244b}$$

respectively, and computed the corresponding tangent vector and dual bases. Use your results from these problems to write down the metric tensor g_{ab} and Christoffel symbols Γ^c_{ab} in these coordinate systems.

Problem 2.24. In Example 1.26 we introduced a non-orthogonal coordinate system in a two-dimensional space based on the coordinates $y^1 = x^1 - x^2$ and $y^2 = x^2$, where x^1 and x^2 are Cartesian coordinates. Compute the components of the metric tensor g_{ab} and its inverse g^{ab} in this coordinate system. Use your result to write down the general expression for the length of a curve given by the functions $y^1(t)$ and $y^2(t)$.

Problem 2.25. Derive the transformation rules for the expression $\partial_a v^b$, where v^b are the contravariant components of a vector field, and verify that it does not transform as the components of a type $(1,1)$ tensor and then verify that $\nabla_a v^b$ does transform as the components of a type $(1,1)$ tensor.

Problem 2.26. Explicitly verify the following identities for the covariant derivative:

a) $\nabla_a v^b T^{cd} = v^b \nabla_a T^{cd} + T^{cd} \nabla_a v^b$

b) $\partial_a v^b w_b = v^b \nabla_a w_b + w_b \nabla_a v^b$

c) $\nabla_a v^a = g^{ab} \nabla_a v_b$

Problem 2.27. Use the expressions for the metric tensor in spherical coordinates found in Problem 2.20 to explicitly write down the divergence $\nabla \cdot \vec{v}$ in terms of the contravariant vector components v^a and functions of the coordinates. Verify that your result is the same as that presented in Eq. (1.211b). *Hint:* You will need to express the physical components of $\tilde{v}_a$ in terms of the contravariant components v^a to make the comparison.

Problem 2.28. Use the generalised expression for the Laplace operator ∇^2 to write down the action of the Laplace operator on a scalar field Φ

$$\nabla^2 \Phi = g^{ab} \nabla_a \nabla_b \Phi \tag{2.245}$$

in a general coordinate system in terms of the Christoffel symbols Γ^c_{ab} and the inverse metric tensor g^{ab}. Use the expressions for the Christoffel symbols and the metric tensor in cylinder coordinates to verify that your expression coincides with Eq. (1.200d).

Problem 2.29. Compute the divergence of the tangent vector bases in cylinder and spherical coordinates.

Problem 2.30. Apply your results from Problem 2.18 and the fact that $\nabla_a g_{bc} = 0$ to verify that the action of the Laplace operator on a scalar field can be written as

$$\nabla^2 \phi = \frac{1}{\sqrt{g}} \partial_a \left(\sqrt{g}\, g^{ab} \partial_b \phi \right), \tag{2.246}$$

where g is the metric determinant.

Problem 2.31. Verify that

  a) the addition of two tensor densities of the same type and weight is a new tensor density of the same type and weight as the original ones,

  b) the multiplication of two tensor densities with weights w_1 and w_2 is a tensor density of weight $w_1 + w_2$, and

  c) the contraction of two free indices (one covariant and one contravariant) in a tensor density results in a new tensor density with the same weight as the original one

by explicitly checking the transformation rules.

Problem 2.32. The Jacobian determinant for a coordinate transformation $y^a \to y'^{a'}$ is given in Eq. (2.111). Assume that we perform a second coordinate transformation $y'^{a'} \to y''^{a''}$ with Jacobian determinant $\mathcal{J}'$. There will also be a corresponding direct coordinate transformation from y^a to $y''^{a''}$ with Jacobian determinant $\mathcal{J}''$. Verify that the Jacobian determinants satisfy the relation

$$\mathcal{J}\mathcal{J}' = \mathcal{J}''. \tag{2.247}$$

In particular, show that $\mathcal{J}' = 1/\mathcal{J}$ when $y''^{a''} = y^a$. Use this result to check that the transformation of a tensor density, defined in Eq. (2.110), is the same when doing the transformations one after the other as when doing the transformation directly from y^a to $y''^{a''}$.

Problem 2.33. Again consider the coordinate system introduced in Example 1.26 (see also Problem 2.24) with the addition of a third dimension, parametrised by the additional orthogonal coordinate $y^3 = 2x^3$. Compute the metric determinant g in this coordinate system and use your result to write down an explicit expression for each of the components of the cross product $\vec{v} \times \vec{w}$ in the y^a coordinates using the relation $\eta^{abc} = \sqrt{g}^{-1} \varepsilon^{abc}$.

Problem 2.34. Show that the divergence of the totally antisymmetric tensor $\eta^{a_1 \cdots a_N} = \varepsilon^{a_1 \cdots a_N}/\sqrt{g}$ vanishes identically in any number of dimensions N. *Hint:* Apply the results of Problem 2.18.

Problem 2.35. Begin from the expression for the curl of a vector field $\nabla \times \vec{v}$ in a Cartesian basis and use a change of coordinates to write down its general expression in an arbitrary coordinate system with the help of the tensor η^{abc} and the metric g_{ab}. Use your result to write down the expression for the covariant components of the quantity $\vec{v} \times (\nabla \times \vec{w}) + \vec{w} \times (\nabla \times \vec{v})$.

Problem 2.36. Using the result from Problem 2.35, compute the components of the curl of the dual basis in cylinder and spherical coordinates. Verify that your final result coincides with what you would obtain if you used Eq. (1.193).

Problem 2.37. Use the relation given in Eq. (2.129) to show that it also holds that

$$\delta^{a_1 \ldots a_n}_{b_1 \ldots b_n} = n! \delta^{a_1}_{[b_1} \ldots \delta^{a_n}_{b_n]}, \tag{2.248}$$

i.e., the same relation holds for anti-symmetrisation of the covariant indices as for the contravariant ones.

Problem 2.38. Start from the definition $x'^{i'} = R^{i'}_i x^i + A^{i'}$ of an affine transformation from a Cartesian coordinate system x^i to a new coordinate system $x'^{i'}$ and the statement that Cartesian coordinate transformations preserve the form $g_{ij} = \delta_{ij}$ of the metric tensor. Show that this requirement implies that

$$R^{i'}_i R^{i'}_j = \delta_{ij} \tag{2.249a}$$

and that

$$x^i = R^{i'}_i x'^{i'} + B^i \tag{2.249b}$$

for some B^i.

Problem 2.39. Consider a fluid with a varying density $\rho(\vec{x})$ and velocity field $\vec{v}(\vec{x})$. Write down integrals describing the following quantities:

a) The total kinetic energy in a volume V.

b) The total momentum of the fluid inside a volume V.

c) The total angular momentum relative to the point $\vec{x} = \vec{x}_0$ of the fluid inside a volume V.

Problem 2.40. Use your results from Problem 2.23 in order to write down an expression for the volume element dV in hyperbolic and parabolic coordinates.

Problem 2.41. Starting from Eqs. (2.195) and (2.197b), express Young's modulus E and Poisson's ratio ν in terms of of the bulk modulus K and the shear modulus G.

Problem 2.42. Use the anti-symmetry and the definition $F_{ij} = \varepsilon_{ijk} B_k$ of the magnetic field tensor to:

a) Show that $F_{ij} F_{jk}$ is symmetric in the free indices i and k.

b) Compute $F_{ij} F_{jk}$ in terms of the magnetic field B_i.

c) Use the result of (b) to express the Maxwell stress tensor in the magnetic field tensor F_{ij} rather than B_k.

Problem 2.43. In Example 2.35, we computed a surface force on the fields in the static situation where two charges q and $-q$ were separated by a distance $2d$. Repeat this computation for the situation when the charges are equal.

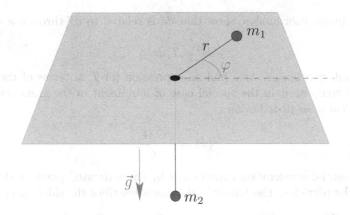

Figure 2.11 The mechanical system in Problem 2.46. The configuration of the system is fully described by the coordinates r and φ. The mass m_1 moves freely in the horizontal plane and the mass m_2 is moving vertically and is affected by the gravitational field $\vec{g}$.

Problem 2.44. Consider a static situation and a volume V in which there are no charges or currents, i.e., Maxwell's equations with $\rho = 0$ and $\vec{J} = 0$ are fulfilled. Show that the total force on the electromagnetic field inside the volume is equal to zero.

Problem 2.45. We just stated that the moment of inertia tensor satisfies the relation

$$\dot{I}_{ij}\omega_j = \varepsilon_{ijk}\omega_j I_{k\ell}\omega_\ell. \tag{2.250}$$

Show that this relation is true by starting from Eq. (2.146) and using the fact that $\vec{v} = \vec{\omega} \times \vec{x}$.

Problem 2.46. A mechanical system consists of a mass m_1 that is free to move in a horizontal plane connected to a mass m_2 by a thread of fixed length that passes through a small hole in the plane, under which the mass m_2 hangs vertically. This system may be given the general coordinates r and φ as shown in Fig. 2.11. The gravitational field $\vec{g}$ is assumed to act vertically. Find the generalised inertia tensor for the system expressed in the r and φ coordinates.

Problem 2.47. In a rotating coordinate system, the fictitious *centrifugal force* on a particle of mass m with a displacement $\vec{x}$ from the rotational center is given by

$$\vec{F}_c = m\vec{\omega} \times (\vec{\omega} \times \vec{x}), \tag{2.251}$$

where ω is the angular velocity. Verify that this may be written in the form

$$F_c^i = T^{ij}x^j, \tag{2.252}$$

where T is a tensor. Express the components of T in terms of the mass m and the components of the angular velocity $\vec{\omega}$. Discuss the requirements for the centrifugal force to vanish.

Problem 2.48. A particle moving in a gravitational potential $\phi(\vec{x})$ is affected by a gravitational force $\vec{F} = m\vec{g}$, where $\vec{g} = -\nabla\phi$ is the gravitational field. Consider the difference in acceleration between a particle at $\vec{x}$ and one at $\vec{x} - d\vec{x}$, where $d\vec{x}$ is a small displacement

$$d\vec{a} = \vec{g}(\vec{x}) - \vec{g}(\vec{x} - d\vec{x}). \tag{2.253}$$

Working in Cartesian coordinates, show that $d\vec{a}$ is related to $d\vec{x}$ through a relation of the form

$$da^i = T^i_j dx^j, \tag{2.254}$$

where T is a rank two tensor. Also find an expression for T in terms of the gravitational potential ϕ and compute it in the special case of movement in the gravitational potential outside a spherical mass distribution

$$\phi(\vec{x}) = -\frac{GM}{r}. \tag{2.255}$$

Note: The differential acceleration experienced by the separated particles defines the *tidal force* between the particles. The tensor T therefore describes the tidal effect.

Problem 2.49. The magnetic force on the small volume dV with a current density $\vec{j}(\vec{x})$ is given by

$$d\vec{F} = \vec{j} \times \vec{B}\, dV, \tag{2.256}$$

where $\vec{B}$ is the magnetic field, which satisfies the conditions

$$\nabla \cdot \vec{B} = 0, \quad \nabla \times \vec{B} = \mu_0 \vec{j}. \tag{2.257}$$

Show that the total magnetic force on a volume V can be written as

$$\vec{F} = \oint_S \vec{e}_i T^{ij} dS_j, \tag{2.258}$$

where S is the boundary of V and T is a rank two tensor. Also find an expression for the components of T.

Problem 2.50. Consider a four-dimensional space where we have introduced an additional orthogonal coordinate $x^0 = ct$ and introduce the completely anti-symmetric tensor $F^{\mu\nu}$ defined by

$$F^{i0} = E^i \quad \text{and} \quad F^{ji} = c\varepsilon_{ijk}B^k, \tag{2.259}$$

where the indices μ and ν run from 0 to 3, the indices i, j, and k from 1 to 3, $\vec{E}$ is the three-dimensional electric field, and $\vec{B}$ the three-dimensional magnetic field. Introducing the four-dimensional vector K^μ with components

$$K^0 = \frac{\rho}{\varepsilon_0} \quad \text{and} \quad K^i = \frac{1}{c\varepsilon_0}J^i, \tag{2.260}$$

where ρ is the charge density and $\vec{J}$ the three-dimensional current density, verify that the differential equation

$$\partial_\mu F^{\mu\nu} = K^\nu \tag{2.261}$$

summarises half of Maxwell's equations. *Note:* This construction will arise very naturally when treating the electromagnetic field in relativity. The four dimensional space we have introduced, which includes the time coordinate t, is the space-time of special relativity. Using the index 0 instead of 4 for the time direction is purely conventional.

CHAPTER 3

Partial Differential Equations and Modelling

Physics is in its essence an experimental science, based on making quantitative predictions for how different systems behave under a set of given conditions and checking experimentally that these predictions are fulfilled. In order to make quantitative predictions, a mathematical model of the system in question must be at hand and a central part of physics is therefore the procedure of accurately describing a given physical system using mathematical tools. In this chapter, we will discuss this procedure and develop tools that can be used for a large variety of different situations. The aim is to familiarise ourselves with physical modelling as well as to introduce some of the differential equations that will be studied in detail in later chapters.

3.1 A QUICK NOTE ON NOTATION

Before delving deep into the business of modelling the world around us using differential equations, it will be beneficial to introduce a handful of different ways of denoting partial derivatives. Each of these notations have their advantages and disadvantages, as some equations will be clearer using one of them and others benefit from a more compact notation.

We have already encountered the notation

$$\frac{\partial}{\partial x^i} = \partial_i \tag{3.1}$$

and we will continue to use it extensively in this and the following chapters. In addition to the spatial coordinates, our partial differential equations will also depend on time and we introduce the similar notation

$$\frac{\partial}{\partial t} = \partial_t. \tag{3.2}$$

Apart from this, it will sometimes be beneficial to write the partial derivatives of a function u as u with an additional subscript, i.e.,

$$\frac{\partial u}{\partial y} = u_y \quad \text{or} \quad \frac{\partial u}{\partial x^i} = u_i, \tag{3.3}$$

where y may be any parameter, i.e., x^i or t. For tensor fields, with indices of their own, we can still use this notation and it will also be found in many textbooks. In order not to mix

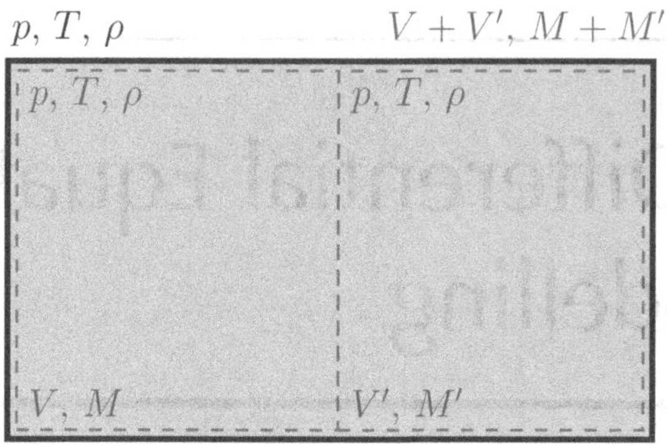

$$p, T, \rho \qquad\qquad V + V', M + M'$$

Figure 3.1 Considering a homogeneous system made up from several subsystems, intensive properties such as pressure, temperature, and density are the same for the system as a whole as well as for the subsystems. Extensive properties such as mass and volume may be different between the subsystems and their value for the whole system is the sum of the values for the subsystems.

the indices of the original tensor with those of the partial derivatives, they are separated by a comma, i.e.,

$$T_{jk,i} = \partial_i T_{jk}. \tag{3.4}$$

In curvilinear coordinates, where there is a difference between the partial derivatives and covariant derivatives, the covariant derivative is instead indicated by switching the comma for a semi-colon $\nabla_a T_{bc} = T_{bc;a}$. We will generally not make extensive use of this notation for tensor fields, but it is good to know that it exists. For second order derivatives, we adopt the similar convention

$$\frac{\partial^2 u}{\partial y^2} = \partial_y^2 u = u_{yy}, \quad \frac{\partial^2 u}{\partial y \partial z} = \partial_y \partial_z u = u_{yz} \tag{3.5}$$

and the generalisation of this to higher order derivatives is straightforward.

3.2 INTENSIVE AND EXTENSIVE PROPERTIES

Considering a physical system, there are generally a number of different physical properties that may be associated with it. These quantities can be subdivided into two different categories depending on how they change when the boundaries of the of the system under study changes. A property that does not change when we redraw the boundaries of a system (assuming the system is homogeneous) is an *intensive property*. As such, an intensive property does not depend on how large part of a system we consider. Typical intensive properties include temperature, pressure, and density. If we consider a given material, these quantities will all be the same regardless of the amount of material we have as long as the conditions are kept the same for the additional material.

Contrary to intensive properties, *extensive properties* do change when we redraw the system boundaries and are proportional to the system size, see Fig. 3.1. Expressed in a different way, an extensive property is additive between system parts. Given a system that can be subdivided into two or more subsystems, the extensive property in the full system is the sum of the same extensive property for each of the subsystems. Some examples of extensive properties are mass, volume, particle number, electrical charge, and momentum. These are all properties that add up between different parts of a system.

Extensive property	System size	Intensive property
Mass [kg]	Volume [m³]	Density [kg/m³]
Charge [C]	Volume [m³]	Charge density [C/m³]
Charge [C]	Area [m³]	Surface charge density [C/m²]
Force [N]	Area [m²]	Pressure [N/m²]
Mass [kg]	Amount [mol]	Molar mass [kg/mol]
Heat capacity [J/K]	Mass [kg]	Specific heat capacity [J/kg K]
Momentum [Ns]	Mass [kg]	Velocity [m/s]

Table 3.1 Some examples of intensive properties related to extensive properties by a given measure of system size. It is worth noting that volume is not the only possible measure of system size, but other extensive properties, such as area or mass, may also be of interest depending on what is being described. The SI units for each property are quoted in square brackets.

Given two different extensive properties, an intensive property may be defined as the quotient between the two extensive ones. For instance, the density is the ratio between a system's total mass and its volume. In fact, when we need to define the size of a system, we can often do so by giving the value of an extensive property such as the volume or mass. With this in mind, there is a natural relation between extensive and intensive properties if a given extensive property is used to denote system size.

If we use the volume of a system as a measure of its size, there will be an intensive property associated to each extensive property. This intensive property is the *concentration*, or *density*, of the extensive property, i.e., the extensive property divided by the system volume. Several examples of intensive properties defined in this fashion can be found in Table 3.1.

As a consequence of extensive properties adding up when combining systems, it is often possible to obtain an extensive property by integrating the corresponding intensive property over the system size. In such integrals, we may also change the measure of system size, e.g., from volume to mass, and under such a change, the resulting Jacobian of the variable change will be an intensive property.

Example 3.1 Taking the air contained in a room as an example, it does not matter whether or not we consider 1 cm³ or 1 dm³ of the air, the intensive properties temperature, pressure, and density of this air will be the same. However, the mass of air contained in 1 cm³ is not the same as the mass contained in 1 dm³. If we wish to compute the heat capacity at constant volume C_V of a given amount of air, we can do so by integrating the specific heat capacity at constant volume c_V over the masses dM of small subsystems

$$C_V = \int c_V \, dM. \tag{3.6a}$$

However, it may be easier to describe the system size in terms of volume, in which case a change of variables leads to

$$C_V = \int c_V \frac{dM}{dV} dV = \int c_V \rho \, dV, \tag{3.6b}$$

which is an integration of the intensive property $c_V \rho$ over the volume.

The world of physics is full of conservation laws for a number of different extensive properties, stating that if a system does not interact with its surroundings in any way, the extensive property will not change. This also means that if we divide this system in two, then any decrease of the extensive property in one part must be accompanied by an equivalent increase of the extensive property in the other part.

Example 3.2 Consider the total momentum of $\vec{p}$ of a system on which there are no external forces. According to Newton's second law $\vec{F} = d\vec{p}/dt$, the momentum in the system is conserved. If we subdivide the system into two parts, one with momentum $\vec{p}_1$ and the other with momentum $\vec{p}_2$, then $\vec{p}_1 + \vec{p}_2 = \vec{p}$ is conserved. This results in

$$\frac{d\vec{p}_1}{dt} + \frac{d\vec{p}_2}{dt} = 0 \quad \Longrightarrow \quad \vec{F}_1 = -\vec{F}_2, \tag{3.7}$$

where $\vec{F}_i = d\vec{p}_i/dt$ is the force acting on each subsystem. Thus, if there is a net force $\vec{F}_1$ acting on the first subsystem, then there is also a net force $-\vec{F}_1$ acting on the second subsystem. This should sound strangely familiar as it is nothing else than Newton's third law of action-reaction forces.

In the following, we will not only consider situations where the total amount is conserved, but also look at processes where there is a net production or loss of a given extensive property, for example if there is an external force acting on our system.

3.3 THE CONTINUITY EQUATION

A central tool in physics modelling is the *continuity equation*, which is a relation among different intensive properties based on an extensive property that may flow from one system to another. For the purpose of the continuity equation, it is most convenient to work with the volume and area of a system as descriptions of the system size. Let us therefore assume that we are looking at a volume V, which may be a part of a larger system and has a boundary surface S. If we are interested in an extensive property Q in this volume, we can relate it to the concentration $q = dQ/dV$ by integration

$$Q = \int dQ = \int_V \frac{dQ}{dV} dV = \int_V q(\vec{x}, t) \, dV. \tag{3.8}$$

After reading the previous chapters, volume integrals of this type should be familiar as it is the volume integral of a scalar field $q(\vec{x}, t)$, which may depend both on the position $\vec{x}$ and the time t. This gives us some hope of being able to apply the machinery already discussed as we work our way forward.

We now wish to describe, in mathematical terms, how Q changes with time. Taking the time derivative, we find that

$$\frac{dQ}{dt} = \int_V \frac{\partial q}{\partial t} dV \tag{3.9}$$

as long as the system boundaries are kept fixed. On its own, this equation does not tell us very much and in order to be useful, we must relate it to other physical quantities. We do this by considering what type of processes could lead to this change. Assuming that Q is continuously transported within a system there are two possibilities, the first of which is a change of Q inside the volume V by production. The *source density* $\kappa(\vec{x}, t)$ is an intensive

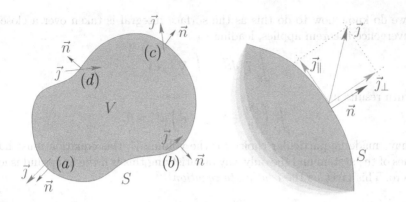

Figure 3.2 The flux out of a volume V through its boundary S depends on the magnitude and direction of the current $\vec{j}$ in relation to the surface normal $\vec{n}$. At (a) the current is directed in the normal direction, providing an outflux proportional to the magnitude of $\vec{j}$, while at (b) the current is parallel to the surface and therefore does not result in a net flux out of the volume. At (c) the current has a component both parallel and perpendicular to the surface, but since the scalar product with $\vec{n}$ is positive, it results in a net outflux. At (d) the scalar product with the normal is negative, resulting in a net influx. In the right part of the figure, we show the decomposition of $\vec{j}$ into components parallel $\vec{j}_\parallel$ and perpendicular $\vec{j}_\perp$ to the surface, i.e., $\vec{j}_\perp$ is parallel to the surface normal. Only the component $\vec{j}_\perp$ is relevant for the flux through the surface.

property related to Q that tells us how much of Q that is produced per volume and time. The corresponding extensive property

$$K = \int_V \kappa(\vec{x}, t)\, dV \tag{3.10}$$

therefore tells us the total production of Q within the volume V per unit time. Naturally, this production will add to Q with time. It should also be noted that destruction of Q may also occur inside the volume. In this case, κ will be negative and correspond to a *sink density*.

The second possibility of changing Q is if there is an in- or outflux through the boundary surface S. In order to model this flux we introduce the current density $\vec{j}(\vec{x}, t)$, which describes the flow of Q with time. In order to change Q inside the volume V, the flow needs to have a component parallel to the surface normal $\vec{n}$, see Fig. 3.2. With the surface normal chosen to point out of the volume, the outflux Φ of Q is given by

$$\Phi = \oint_S \vec{j} \cdot d\vec{S}. \tag{3.11}$$

Just as for the source term, the result of this integral may be negative, which should just be interpreted as a net influx.

Collecting the pieces, the change in Q with time is given by the net production in V minus the outflux through the surface S and we obtain

$$\frac{dQ}{dt} = K - \Phi = \int_V \frac{\partial q}{\partial t}\, dV = \int_V \kappa\, dV - \oint_S \vec{j} \cdot d\vec{S}. \tag{3.12}$$

This relation involves two volume integrals and one surface integral. Would it not be fantastic if we could somehow rewrite the surface integral as a volume integral too? As it so happens, we do know how to do this as the surface integral is taken over a closed surface and the divergence theorem applies, leading to

$$\oint_S \vec{j} \cdot d\vec{S} = \int_V \nabla \cdot \vec{j} \, dV, \tag{3.13}$$

which in turn results in

$$\int_V \left(\frac{\partial q}{\partial t} + \nabla \cdot \vec{j} - \kappa \right) dV = 0. \tag{3.14}$$

Since we have made no particular choice for the volume V, this equation must hold for all sub-volumes of the system and the only way of satisfying this is if the integrand is identically equal to zero. This gives us the *continuity equation*

$$\frac{\partial q}{\partial t} + \nabla \cdot \vec{j} = \kappa \tag{3.15}$$

that relates the change in concentration with time to the source density and the divergence of the current density.

The continuity equation is very general and applies to a large variety of different situations and quantities. However, in order to apply it successfully and actually compute the change in the concentration, we need to provide a model for the current density $\vec{j}$ and the source density κ.

Example 3.3 We can apply the continuity equation to electrical charge inside a conducting material. Since electrical charge is a conserved quantity, there can be no source or sink and we have a situation with $\kappa = 0$. At the same time, Ohm's law for an isotropic and homogeneous conductor tells us that $\vec{j} = \sigma \vec{E}$, where σ is the conductivity and $\vec{E}$ the electric field. Inserting this into the continuity equation, we find that

$$\frac{\partial \rho}{\partial t} = -\nabla \cdot \vec{j} = -\sigma \nabla \cdot \vec{E}. \tag{3.16a}$$

Applying Gauss's law, this relation turns into

$$\frac{\partial \rho}{\partial t} = -\frac{\sigma}{\varepsilon_0} \rho. \tag{3.16b}$$

The only stationary solution to this, assuming non-zero conductivity σ is $\rho = 0$. If the charge density ρ is not zero, it will decrease exponentially with decay constant σ/ε_0.

Example 3.4 Let us assume that we take a handful of sand and throw it into the air such that it disperses without interactions between the grains, which is a good first approximation. Let us also assume that grains that are close have similar velocities so that the velocity of a grain at point $\vec{x}$ can be well approximated by a velocity field $\vec{v}(\vec{x}, t)$. The current density $\vec{j}$ related to the grain concentration will then be equal to $\vec{j} = n\vec{v}$, where n is the number concentration of grains. Since no new grains are produced, the continuity equation becomes

$$\frac{\partial n}{\partial t} + \nabla \cdot (n\vec{v}) = 0. \tag{3.17}$$

This type of equation is relatively common and describes a convective flow without sources.

Figure 3.3 The flow of water out of a faucet is widest at the faucet and becomes narrower as it falls in a gravitational field $\vec{g}$. This is a result of the water speeding up under the influence of gravity.

This last example is a special case of a *convection current*. In convection, the extensive quantity is assumed to be transported with some velocity $\vec{v}$, which depends on position and possibly on time. Convection currents may arise in several different ways as the origin of the velocity field may differ. Typical examples are when particles of a given substance are carried along with the flow of a fluid or when their motion is predetermined based on some initial conditions. The common property of convection currents is that they in all cases take the form

$$\vec{\jmath} = \rho \vec{v}, \tag{3.18}$$

where ρ is the concentration of the extensive quantity. The insertion into the continuity equation yields the *convection equation*

$$\frac{\partial \rho}{\partial t} + \nabla \cdot \rho \vec{v} = \frac{\partial \rho}{\partial t} + \vec{v} \cdot \nabla \rho + \rho \nabla \cdot \vec{v} = \kappa, \tag{3.19}$$

where κ is again the source term. We can also see that if the velocity field can be argued to be divergence free, then the convection equation may be further simplified as the divergence of the current is then equal to the directional derivative of ρ with respect to the velocity $\vec{v}$.

Example 3.5 It is not necessary to consider the volume of the system to be three-dimensional, but we may consider one-dimensional or two-dimensional systems as well, in which case the concentrations become linear or area concentrations. As an example of this, let us consider the steady, non-turbulent flow of water out of a faucet, see Fig. 3.3. In the steady state, the water stream is narrower at the bottom, which most people will be familiar with as the situation can be readily observed in any kitchen with a sink. Taking the linear density in the vertical direction, we find that the flow of water in this direction can be described as

$$j = \rho_\ell v, \tag{3.20}$$

where ρ_ℓ is the linear density and v the velocity of the water. Note that we have dropped the vector arrows here due to the problem being one-dimensional. The velocity, given by the initial velocity v_0 and the acceleration due to gravity g, will be given by

$$v = \sqrt{v_0^2 + 2gx}. \tag{3.21}$$

Assuming a stationary state, i.e., a state where the linear density does not change with time, we find that

$$\frac{dj}{dx} = \frac{d\rho_\ell}{dx}v + \rho_\ell\frac{g}{v} = 0 \quad \Longrightarrow \quad \frac{d\rho_\ell}{dx} = -\rho_\ell\frac{g}{v_0^2 + 2gx} < 0. \tag{3.22}$$

It follows that the linear density will decrease as the water falls and since water is well approximated as having fixed volume density, this means that the cross sectional area of the water stream must decrease. This may be argued in a simpler manner. However, the main point of this example is the realisation that the volume considered need not be three dimensional and how the current can be modelled in terms of the velocity.

3.4 THE DIFFUSION AND HEAT EQUATIONS

Armed with the continuity equation, we are ready to derive one of the more common partial differential equations in classical physics, namely the *diffusion equation*

$$\partial_t u - D\nabla^2 u = \kappa, \tag{3.23}$$

where u is an intensive quantity, often a concentration, D is a constant, and κ a source term. This form of this equation is also known as the *heat equation*, just because it appears also in heat conduction (sometimes referred to as heat diffusion) problems. The diffusion equation and methods for solving it will be a central theme in large parts of the remainder of this book. The methods we will apply will be of interest for solving not only diffusion and heat conduction problems, but will also be applicable in more modern contexts, such as quantum mechanics, due to the similarity to the Schrödinger equation. The mathematics behind diffusion and heat conduction are equivalent, but let us go through each of the phenomena in turn.

3.4.1 Diffusion and Fick's laws

Assume that we wish to describe the diffusion of some extensive property U with a corresponding concentration u. By the argumentation in the previous section, the continuity equation will apply to this setting and what we need to do in order to model the situation is to model the current density $\vec{j}$ in a reasonable fashion. For diffusion, we are not interested in a current which is due to a bulk movement of U and so the current density will not be proportional to the concentration u itself. Instead, we want to describe a situation where U moves from regions of higher concentrations to regions with lower concentration and the direction of the concentration change is naturally given by the gradient ∇u of the concentration. We thus have a situation where we wish to relate the vector field $\vec{j}$ to the vector field ∇u and for our current purposes, we assume that the current is linearly related to ∇u, indicating that the current will be larger if the concentration is changing faster. The most general linear relationship between two vectors is given by a rank two tensor and we therefore write down

$$j_i = -D_{ij}\partial_j u, \tag{3.24}$$

where D_{ij} is the *diffusivity tensor*, which describes how easy it is for the property to move in the volume. The negative sign in this relation is introduced in order for the diffusivity tensor to be positive definite when U is flowing from regions with high concentration to regions with low concentration, meaning that the property will tend to spread out through the material. In the opposite scenario, when U flows from low to high concentration, the resulting differential equation would become unstable and U tend to accumulate in a small region. The form of Eq. (3.24) may be derived from random movements of the property U, which will result in more movement out of high concentration regions than out of low concentration regions. However, this is an argument that we will not go through here and instead we will just state the relation as is.

Inserting the current density into the continuity equation, we obtain

$$\partial_t u + \partial_i j_i = \partial_t u - \partial_i(D_{ij}\partial_j u) = \partial_t u - D_{ij}\partial_i\partial_j u - (\partial_i D_{ij})(\partial_j u) = \kappa, \qquad (3.25)$$

which is the most general form of a partial differential equation describing how the concentration u of a diffusing substance develops in time. The diffusivity tensor is a tensor field that may depend on the spatial position, which is the reason for the appearance of the term involving its divergence $\partial_i D_{ij}$. However, if the material is homogeneous, the tensor field D_{ij} is constant and its divergence is equal to zero. Another simplification occurs if the diffusion is locally isotropic. In such scenarios, the diffusivity tensor is proportional to the Kronecker delta in each point and we may write

$$D_{ij} = D\delta_{ij}, \qquad (3.26)$$

where D is a scalar field. The current density for this case is given by

$$\vec{j} = -D\nabla u, \qquad (3.27a)$$

which is known as *Fick's first law* of diffusion. The corresponding partial differential equation is then given by

$$\partial_t u - D\nabla^2 u - (\nabla D) \cdot (\nabla u) = \kappa. \qquad (3.27b)$$

Finally, if the diffusion is both homogeneous and isotropic, then the scalar field D is constant and we obtain diffusion equation

$$\partial_t u - D\nabla^2 u = \kappa. \qquad (3.28)$$

Even in this case, there remains a large variety of different possibilities for modelling the source term κ, which will lead to different situations with different possible approaches for solutions. The case when there is no source term present is sometimes referred to as *Fick's second law* of diffusion.

Example 3.6 We might want to predict the concentration of a radioactive material within a homogeneous and isotropic substance without convection currents. Assuming that the diffusivity of the radioactive material in the substance is D, the concentration will follow the diffusion equation, but we must ask ourselves what the source term κ will be. The number of radioactive nuclei decaying within a volume V in a short time dt is given by $\lambda N dt$, where N is the total number of radioactive nuclei in the volume and λ the decay constant, which satisfies $\lambda = 1/\tau$, where τ is the half-life of the decaying nuclei. It follows that source density is given by

$$\kappa = -\lambda n, \qquad (3.29)$$

where n is the number density of radioactive nuclei, which therefore satisfies

$$\partial_t n - D\nabla^2 n = -\lambda n. \tag{3.30}$$

Note that this source term is going to be negative and therefore describes a sink. This is hardly surprising as radioactive decay decreases the total amount of radioactive nuclei and we have not assumed any production.

3.4.2 Heat conduction and Fourier's law

In most aspects, the mathematics behind heat conduction is equivalent to that used to derive the diffusion equation. The quantity we are interested in describing is the temperature T in a given volume. In order to accomplish this, we choose to study the *heat* U. The intensive property related to the heat is the heat concentration u, which in turn relates to the temperature through the specific heat capacity c_V and mass density ρ as

$$T = \frac{u}{c_V \rho}. \tag{3.31}$$

It should be noted that both c_V and ρ will generally be material dependent, implying that $c_V \rho$ is a scalar field that may vary in space.

Similar to how a diffusion current is linearly related to the gradient of the concentration, we expect that heat conduction will transfer heat from warmer regions to colder regions, the direction of which is described by the gradient ∇T of the temperature. Assuming the most general linear relationship between this gradient and the heat current, we find that

$$j_i = -\lambda_{ij}\partial_j T, \tag{3.32}$$

where λ_{ij} is the *heat conductivity* tensor. Just as the diffusivity, the heat conductivity is material dependent and may or may not be isotropic or homogeneous, implying different simplifications. Inserted into the continuity equation for heat, the most general form of the differential equation describing heat conduction is given by

$$\frac{\partial u}{\partial t} - \partial_i \lambda_{ij}\partial_j T = \frac{\partial u}{\partial t} - \lambda_{ij}\partial_i\partial_j T - (\partial_i\lambda_{ij})(\partial_j T) = \kappa. \tag{3.33}$$

As for the case of diffusion, the most simplifying assumption is that the material is isotropic and homogeneous, leading to $\lambda_{ij} = \lambda\delta_{ij}$, where λ is a constant. Assuming that the background material properties c_V and ρ do not change with time, we then find that

$$c_V \rho\, \partial_t T - \lambda\nabla^2 T = \kappa, \tag{3.34}$$

which describes how the temperature in the material changes with time. The only difference compared to the diffusion case is the appearance of the factor $c_V \rho$, which may be absorbed into λ if constant.

Example 3.7 We can consider the heat conduction in a material of constant density ρ, heat capacity c_V, and heat conductivity λ. If there is a heat source in the material at the point $\vec{x}_0$, releasing heat at a rate K, the heat production inside a volume V is given by

$$\int_V \kappa(\vec{x})dV = \begin{cases} K, & (\vec{x}_0 \in V) \\ 0, & (\vec{x}_0 \notin V) \end{cases}. \tag{3.35a}$$

or, dividing through by K,

$$\int_V \frac{\kappa(\vec{x})}{K} dV = \begin{cases} 1, & (\vec{x}_0 \in V) \\ 0, & (\vec{x}_0 \notin V) \end{cases}. \tag{3.35b}$$

Comparing to Eq. (1.221), this is the defining property of the three-dimensional delta function $\delta^{(3)}(\vec{x} - \vec{x}_0)$. It follows that the source term κ for a *point source* producing heat at a rate K at the point $\vec{x}_0$ is given by

$$\kappa(\vec{x}) = K\delta^{(3)}(\vec{x} - \vec{x}_0). \tag{3.36}$$

In general, the three-dimensional delta function may be used to describe different point sources or the concentration of any extensive quantity located at a single point in space. Naturally, this is usually a simplification, but it will result in reasonably good results as long as the size of the source is indistinguishable from a point given the precision of measurements.

3.4.3 Additional convection currents

For both diffusion and heat conduction, it may happen that the transportation is not necessarily only due to Fick's or Fourier's laws. Instead, there may also be additional convective currents that are also going to lead to transportation. The perhaps most straightforward example of this is when heat is conducted in a non-stationary medium, resulting in the heat being carried along with the medium in addition to being dispersed through conduction. In a situation such as this, there will be current contributions from both conduction and convection. The total heat current $\vec{j}$ will then be a sum of the different contributions

$$\vec{j} = \vec{j}_{\text{conduction}} + \vec{j}_{\text{convection}} = -\lambda\nabla T + u\vec{v}, \tag{3.37}$$

where we have assumed the medium to be homogeneous and isotropic. The same argumentation can be applied to diffusion with additional convection.

In order to have a solvable model, the velocity field $\vec{v}$ must be modelled just as for pure convection and just as the source term needs to be modelled for diffusion.

Example 3.8 Consider the spread of small particles of density ρ in a stationary fluid of density ρ_0 and assume that the diffusivity of the particles in the fluid is D, see Fig. 3.4. Since the particles and fluid generally have different densities, there will be a net force due to gravity and buoyancy. Particles will therefore be accelerated if they are not also subject to a resistance to motion from the fluid. We assume that force equilibrium between these forces occurs fast and at the terminal velocity $\vec{v}_0$. The corresponding convective current, i.e., the current without diffusion, is therefore

$$\vec{j}_{\text{convection}} = u\vec{v}_0, \tag{3.38}$$

where u is the concentration of particles. Inserting this and the diffusion current into the continuity equation, we end up with

$$\partial_t u - D\nabla^2 u + \vec{v}_0 \cdot \nabla u = 0, \tag{3.39}$$

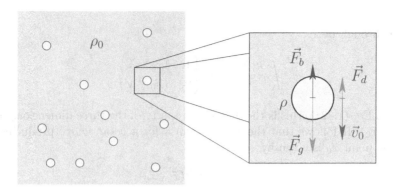

Figure 3.4 Apart from a diffusion current, a convection current may arise for particles of density ρ immersed in a fluid with a different density ρ_0. The gravitational force $\vec{F}_g$ is not balanced by the buoyant force $\vec{F}_b$ alone, but if a particle is moving at terminal velocity $\vec{v}_0$, there is also a drag force $\vec{F}_d$ resulting in force balance. This results in a convection current with velocity $\vec{v}_0$.

since $\vec{v}_0$ is constant and therefore divergence free. In particular, we can here note that in the limit of negligible diffusion, i.e., very small diffusivity, the movement of the particles is dominated by the terminal velocity. On the other hand, if the terminal velocity is very small, which may occur, e.g., if the densities ρ and ρ_0 are very similar, then we instead recover the diffusion equation.

3.5 THE WAVE EQUATION

Another partial differential equation of significance in many physics applications is the *wave equation*

$$\partial_t^2 u - c^2 \nabla^2 u = f, \tag{3.40}$$

where c is the wave velocity and f is a source term. The big difference compared to the diffusion equation is the appearance of a second derivative with respect to time which, as we shall see later on, changes the behaviour of the solutions.

Example 3.9 In one spatial dimension, the source free ($f = 0$) wave equation takes the form

$$\partial_t^2 u - c^2 \partial_x^2 u = (\partial_t + c\partial_x)(\partial_t - c\partial_x)u = 0. \tag{3.41}$$

This will be fulfilled if $\partial_t u = \pm c\partial_x u$, which is solved by

$$u(x,t) = g(x \pm ct). \tag{3.42}$$

The shapes of these solutions remain unchanged with time but travel with a velocity c in the negative x-direction for the positive sign and the positive x-direction for the negative sign. This can be deduced from the fact that at time t_1, the function $g(x_1 \pm ct_1)$ will be the same as $g(x_0)$ when $x_1 \pm ct_1 = x_0$. Thus, in time t_1, the solution has been translated by an amount $\mp ct_1$ as $u(x_1, t_1) = u(x_1 \pm ct_1, 0)$, see Fig. 3.5.

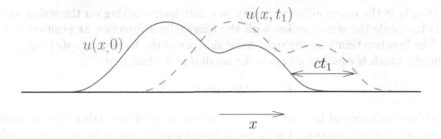

Figure 3.5 A travelling wave of the form $u(x, t) = g(x - ct)$ at times $t = 0$ and $t = t_1$. The wave shape is the same at any time, but is translated in the direction of propagation.

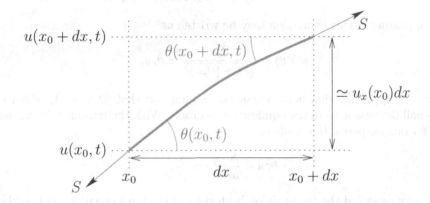

Figure 3.6 The geometry of a small part of a string with tension S along with the forces acting on it as a result of the tension. We are interested in the transversal movement of the string and therefore consider the projection of the forces on the transversal direction only. The tension is assumed to be approximately constant throughout the string.

The wave equation appears in several physical situations, sometimes as an approximative description for small deviations from an equilibrium and sometimes as an exact description given the underlying assumptions. Let us discuss a small subset of these situations in order to use them in examples later on and to see how the wave equation may arise as a result of physical modelling.

3.5.1 Transversal waves on a string

When considering the wave equation, the typical example is often transversal waves on a string with relatively high tension. The modelling that will lead to the wave equation for this string can be based on considering the transversal forces acting on an infinitesimal part of the string with length dx such that the x coordinate is between x_0 and $x_0 + dx$, see Fig. 3.6. The free-body diagram of this part of the string contains a number of forces. We will here only consider the transversal movement and assume that the strain in the string due to the transversal movement is not large enough to significantly change the string tension, which is assumed to be S. The transversal component of the force acting on the string is given by

$$F = S(x_0 + dx) \sin(\theta(x_0 + dx, t)) - S(x_0) \sin(\theta(x_0, t)) + f(x_0, t) dx, \qquad (3.43)$$

where $f(x_0, t)$ is the external force density per unit length acting on the string at x_0 and $\theta(x, t)$ is the angle the string makes with the longitudinal direction at position x, both at time t. The first two terms on the right-hand side are of the form $g(x_0 + dx) - g(x_0)$, where $g = S \sin(\theta)$, which is equal to $g'(x_0) dx$ for small dx. We find that

$$F = [\partial_x(S\sin(\theta)) + f]dx, \tag{3.44}$$

which by Newton's second law must be equal to the mass of the string element multiplied by the transversal acceleration. The mass of the element is given by $m = \rho_\ell dx$, where ρ_ℓ is the linear mass density, and the acceleration is just the second time derivative of the transversal deviation u. We therefore end up with

$$F = ma = \rho_\ell(\partial_t^2 u)dx = [\partial_x(S\sin(\theta)) + f]dx. \tag{3.45}$$

The sine appearing in this expression may be written as

$$\sin(\theta) = \frac{\partial_x u}{\sqrt{1 + (\partial_x u)^2}} \simeq \partial_x u, \tag{3.46}$$

where the last approximation holds under the assumption that $|\partial_x u| \ll 1$, which will be true for small deviations from the equilibrium position. With the tension S being assumed constant for our purposes, this leads to

$$\partial_t^2 u - \frac{S}{\rho_\ell}\partial_x^2 u = \frac{f}{\rho_\ell}, \tag{3.47}$$

where we have cancelled the factor dx on both sides of the force equation. This is the wave equation with a wave speed of $c^2 = S/\rho_\ell$. Thus, the wave speed for propagation of transversal waves on a string increases with increasing string tension and decreases with string density. As we shall see when we solve the wave equation later on, the characteristic frequencies of a guitar or piano string increases with increasing wave speed. That the frequency increases with increasing tension and decreases with string density should be a familiar fact to anyone who is familiar with a string instrument.

Example 3.10 A special case of the sourced wave equation occurs when the string is also subjected to a resistance to movement, such as a damping force due to transferring part of the oscillation energy to a surrounding medium, e.g., in the form of sound waves. Assuming that the resistance from the surrounding medium gives rise to a force density $f = -k\partial_t u$, where k is a non-negative constant, see Fig. 3.7, the wave equation now becomes

$$\partial_t^2 u + k\partial_t u - c^2 \partial_x^2 u = 0. \tag{3.48}$$

This is the *damped wave equation*, which is able to describe oscillations with inherent energy losses. Of course, other external forces may also act on the string, making the right-hand side non-zero.

3.5.1.1 *Wave equation as an application of continuity*

The above derivation may also be constructed as a consequence of the continuity equation for the momentum in the transverse direction. The momentum current $j(x_0, t)$ is the momentum

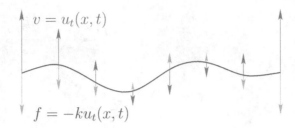

Figure 3.7 For a damped string, we assume that there is a force density f proportional to the local velocity $v = u_t$ of the string. Since the force and the velocity have opposite directions, this will result in a net loss of energy in the string.

transfer per time unit, i.e., force, from the string at $x < x_0$ on the string at $x > x_0$. Using the same logic as before, this can be approximated by $j = -S\partial_x u$. The source term is given by the force density $\kappa = f$, since this is the momentum density added to the string per length and time unit. Finally, the momentum density itself is given by $\rho_\ell v$, where v is the transversal velocity, which may be written as the time derivative of the transversal deviation u. Inserted in the continuity equation, we then obtain the relation

$$\partial_t(\rho_\ell \partial_t u) - \partial_x(S\partial_x u) = \rho_\ell \partial_t^2 u - S\partial_x^2 u = f. \tag{3.49}$$

This is the same as Eq. (3.47) after division by ρ_ℓ. Note that, since we are only dealing with a dependence on one spatial dimension, the x-direction, the gradient and divergence have both reduced to the partial derivative ∂_x.

3.5.2 Transversal waves on a membrane

The very same argumentation as that used for the string may be used for a two-dimensional membrane. While our parameters in the case of the string were the string tension S and the linear density ρ_ℓ, the parameters for the membrane are the membrane surface tension σ and the surface density ρ_A. The membrane tension σ is generally a tensor, but for simplicity we consider the case where the tensor is isotropic and homogeneous. The force across a small section $d\vec{\ell}$ along the surface will also be along the surface but orthogonal to the section, see Fig. 3.8. In order to construct the direction of this force vector, we take the cross product between $d\vec{\ell}$ with the surface normal $\vec{n}$, which will always be a vector in the surface plane, since it is orthogonal to $\vec{n}$, and it will have length $|d\vec{\ell}|$ and be orthogonal also to $d\vec{\ell}$, since $d\vec{\ell}$ is also orthogonal to $\vec{n}$. It follows that the force across the section is given by

$$d\vec{F} = d\vec{\ell} \times \vec{n}\,\sigma \tag{3.50}$$

and in order to find the force we need to find a description for $\vec{n}$.

Let us assume a coordinate system such that the surface extends in the x^1- and x^2-directions when stationary and that the membrane movement occurs in the x^3-direction. As for the case with the string, we will assume that the change of the strain in the membrane is negligible so that the tension σ is constant. If the displacement of the membrane in the x^3-direction is $u(x^1, x^2, t) \equiv u(\vec{x}_2, t)$, where we have introduced $\vec{x}_2$ as the position vector on the two-dimensional surface, then the three dimensional surface describing the membrane shape at time t is given by

$$x^3 = u(\vec{x}_2, t) \quad \Longleftrightarrow \quad f(\vec{x}_3, t) \equiv x^3 - u = 0, \tag{3.51}$$

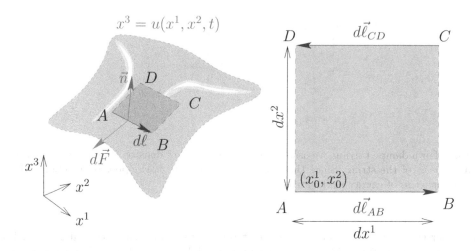

Figure 3.8 We consider the force equation in the transversal direction for a small section of a two-dimensional membrane between the points A, B, C, and D. The force $d\vec{F}$ on the section $d\vec{\ell}$ will be orthogonal to both $d\vec{\ell}$ as well as the surface normal $\vec{n}$ and the surface tension.

where $\vec{x}_3$ is the three-dimensional position vector $\vec{x}_3 = \vec{x}_2 + x^3\vec{e}_3$. In other terms, the membrane at time t is a level surface of $f(\vec{x}_3, t)$. As discussed in Section 1.4.3, the gradient of a function will always be normal to its level surfaces and we must have

$$\vec{n} = \alpha\nabla f = \alpha(-\vec{e}_1\partial_1 u - \vec{e}_2\partial_2 u + \vec{e}_3), \tag{3.52}$$

where α is a normalisation constant. Requiring that $\vec{n}^2 = 1$, we find

$$\alpha = \frac{1}{\sqrt{1 + (\partial_1 u)^2 + (\partial_2 u)^2}} \simeq 1, \tag{3.53}$$

where the approximation holds for oscillations such that $|\partial_i u| \ll 1$. Making this approximation is the membrane equivalent of assuming that the sine of the angle θ which the string makes with the longitudinal direction was given by $\partial_x u$ in Eq. (3.46).

With these considerations in mind, we are ready to discuss the force on a small part of the membrane, given by $x_0^1 < x^1 < x_0^1 + dx^1$ and $x_0^2 < x^2 < x_0^2 + dx^2$. Let us start with the forces acting on the sections with fixed x^2. The section $d\vec{\ell}$ at $x^2 = x_0^2$ is given by

$$d\vec{\ell} = dx^1\vec{e}_1 + \vec{e}_3[u(\vec{x}_{2,0} + \vec{e}_1 dx^1, t) - u(\vec{x}_{2,0}, t)] \simeq dx^1[\vec{e}_1 + \vec{e}_3 u_1(\vec{x}_{2,0}, t)], \tag{3.54}$$

since this is the difference vector between the two corners A and B, see Fig. 3.8. It is here important to keep track of the fact that this section is evaluated at $x^2 = x_0^2$, since the opposite section will be evaluated at $x^2 = x_0^2 + dx^2$. Ignoring this fact, the leading contribution would cancel and we would be left with zero net force. Performing the cross product, we find that the force on the section between A and B is

$$d\vec{F}_{AB} = \sigma\, d\vec{\ell} \times \vec{n} \simeq -\sigma dx^1[\vec{e}_2 + \vec{e}_3 u_2(\vec{x}_{2,0}, t)], \tag{3.55}$$

where terms of second order or higher in $\partial_i u$ have been neglected. In the same way, we can write the force on the section between C and D as

$$d\vec{F}_{CD} = \sigma dx^1[\vec{e}_2 + \vec{e}_3 u_2(\vec{x}_{2,0} + dx^2\vec{e}_2, t)] \tag{3.56}$$

and their sum becomes

$$d\vec{F}_{AB} + d\vec{F}_{CD} = \sigma dx^1 \vec{e}_3 [u_2(\vec{x}_{2,0} + dx^2 \vec{e}_2, t) - u_2(\vec{x}_{2,0}, t)] \simeq \sigma dx^1 dx^2 \vec{e}_3 u_{22}. \qquad (3.57)$$

The equivalent argument for the BC and DA sections results in a similar term, but with u_{22} replaced by u_{11}. With an additional force area density $f(\vec{x}_2, t)$, Newton's second law in the x^3-direction becomes

$$\rho_A(\partial_t^2 u)dx^1 dx^2 = \sigma(u_{11} + u_{22})dx^1 dx^2 + f\, dx^1 dx^2. \qquad (3.58)$$

Cancelling the differentials and dividing through by ρ_A, we again end up with the wave equation

$$\partial_t^2 u - \frac{\sigma}{\rho_A}\nabla^2 u = \frac{f}{\rho_A}. \qquad (3.59)$$

It should be noted that the Laplace operator ∇^2 here only acts on the x^1 and x^2 directions. However, taking it as a three-dimensional Laplace operator is also not the end of the world as u does not depend on x^3 and the three-dimensional Laplace operator therefore reduces to the two-dimensional one.

3.5.3 Electromagnetic waves

While the wave equation for transversal waves on the string and on the membrane were based on approximations, including the small deviation approximation, there are situations where the wave equation is exact as far as the underlying modelling goes. An important example of this is the *wave equation for electromagnetic fields*, which is based on Maxwell's equations (see Eqs. (2.198)). Taking the time derivative of Eq. (2.198c) results in

$$\nabla \times \partial_t \vec{E} + \partial_t^2 \vec{B} = 0, \qquad (3.60)$$

which involves a time derivative of the electric field $\vec{E}$. Luckily, we can express this time derivative in the magnetic field $\vec{B}$ and current density $\vec{J}$ by applying Eq. (2.198d) and doing so leads to

$$\partial_t^2 \vec{B} + c^2 \nabla \times (\nabla \times \vec{B} - \mu_0 \vec{J}) = 0. \qquad (3.61)$$

The $\nabla \times (\nabla \times \vec{B})$ expression may be simplified as

$$\nabla \times (\nabla \times \vec{B}) = \nabla(\nabla \cdot \vec{B}) - \nabla^2 \vec{B} = -\nabla^2 \vec{B}, \qquad (3.62)$$

since Eq. (2.198b) tells us that $\nabla \cdot \vec{B} = 0$. The insertion of this simplification into our differential equation finally results in

$$\partial_t^2 \vec{B} - c^2 \nabla^2 \vec{B} = \frac{1}{\varepsilon_0}\nabla \times \vec{J}, \qquad (3.63)$$

which is a sourced wave equation for the magnetic field $\vec{B}$ with wave velocity c. In fact, Maxwell's equations are usually written in terms of ε_0 and μ_0 only rather than using c^2. It is only once electromagnetic waves are identified with light that the wave equation gives us $c = 1/\sqrt{\varepsilon_0\mu_0}$ as a prediction for the speed of light.

Similar arguments to the one above will also show that the electric field $\vec{E}$ satisfies a sourced wave equation. There are many other aspects to electromagnetic waves, but we leave them for a dedicated discussion on electromagnetism.

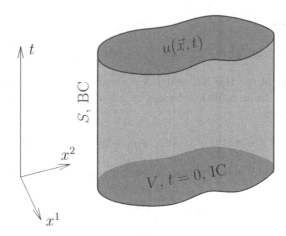

Figure 3.9 The region where we wish to find a solution to partial differential equations is generally bounded in both space and time. The boundary conditions (BC) on the spatial surface S may depend on time as well as on the point on S, while the initial conditions (IC) at $t = 0$ will be functions on the full spatial volume V.

3.6 BOUNDARY AND INITIAL CONDITIONS

People familiar with with ordinary differential equations will know that, in order to find a unique solution, it is necessary to specify a sufficient number of boundary conditions. This does not change when encountering partial differential equations. In the diffusion and heat equations, the parameter space of interest involves not only a spatial volume, but also a time interval, usually from the initial time (often denoted t_0 or defined to be $t = 0$) and forward. Since the diffusion and heat equations are differential equations in both time and space, the boundary of the interesting parameter region is both the spatial boundary of the region we study as well as the entire volume for the initial time t_0, see Fig. 3.9. The exception to this will occur when we search for *stationary states*, i.e., states that do not depend on time.

Physically, the notion of initial and boundary condition relate to the initial state of the system and how the system interacts with the surroundings at the boundaries. For a given sufficient set of initial and boundary conditions, the solution will be unique, a cornerstone in the scientific assumption of a system behaving in the same way under the same conditions, thus ensuring that the underlying theory is predictive.

With the above in mind, we generally assume that we have a problem described by a partial differential equation in the volume V for time $t > t_0$. A *boundary condition* is then of the form

$$f(u(\vec{x}, t), \vec{x}, t) = 0, \tag{3.64}$$

where $\vec{x}$ is a point on the surface S, which is the boundary of V. The function f may depend on both the point on the surface $\vec{x}$ as well as on time t. It may also contain different derivatives of u. An *initial condition* is of the form

$$g(u(\vec{x}, t_0), \vec{x}) = 0, \tag{3.65}$$

where $\vec{x}$ may now be any point in the volume V and, as in the case of the boundary conditions, g may contain derivatives of u.

3.6.1 Boundary conditions

For ordinary differential equations, there are generally as many independent solutions as the order of the differential equation. In the case of partial differential equations that are second order in the spatial derivatives, along with some additional conditions on the form of the differential operator, it is sufficient to give either the function value, normal derivative, or a linear combination thereof, in order for the solution to be unique. In particular, the mentioned conditions are satisfied by the Laplace operator, which we will study extensively. We will get back to this in a short while after discussing some types of boundary conditions that we will encounter.

3.6.1.1 Dirichlet conditions

In many physical applications, it is reasonable to assume that the value of the sought function is known on the boundary. The corresponding boundary condition

$$u(\vec{x}, t) = f(\vec{x}, t), \tag{3.66}$$

valid for all $\vec{x}$ on the boundary of the volume and where f is a known function, is classified as a *Dirichlet boundary condition*.

Example 3.11 A drum may be described as a circular membrane of radius R, which is held fixed in the drum frame at the edges. If the drum frame is rigid, at least in comparison to the membrane, this means that the membrane edge is not subject to transversal motion and thus

$$u(\rho = R, \phi, t) = 0, \tag{3.67}$$

where u is the transversal displacement of the membrane and we have introduced polar coordinates with the center of the membrane as the origin. Since the displacement satisfies the wave equation, which includes the Laplace operator, these boundary conditions will be sufficient to provide a unique solution once sufficient initial conditions have also been specified.

Example 3.12 A sphere of some material is immersed in an oil bath of temperature T_0. If the heat transfer from the oil bath to the sphere is very efficient, then the surface of the sphere will quickly adapt to the oil bath temperature and we will have the boundary condition

$$T(r = R, \theta, \varphi, t) = T_0, \tag{3.68}$$

where we have introduced spherical coordinates centred at the middle of the sphere and assumed that the sphere has radius R. The temperature in the sphere is furthermore assumed to satisfy the heat equation and the boundary condition is therefore sufficient in order to solve the problem if we in addition specify sufficient initial conditions.

In general, Dirichlet boundary conditions apply whenever we keep something fixed at the boundary of the relevant volume.

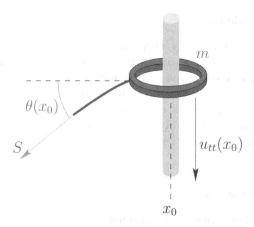

Figure 3.10 A string with tension S ending in a ring of mass m at x_0 that moves freely without friction in the transversal direction. Newton's second law for the ring in the transversal direction gives the boundary condition, which becomes a Neumann boundary condition in the limit $m \to 0$.

3.6.1.2 Neumann conditions

Another commonly encountered type of boundary conditions are *Neumann boundary conditions*, which are conditions on the normal derivative of the function we are solving for, i.e.,

$$\vec{n} \cdot \nabla u = f(\vec{x}, t) \tag{3.69}$$

everywhere on the volume boundary. Here, $\vec{n}$ is the surface normal of the boundary and f is a known function. As we have seen, the gradient often represents a current and physically Neumann conditions therefore tend to appear whenever there is no net flow out of the volume.

Example 3.13 Consider the diffusion of a substance with concentration u within a volume V, which is completely sealed off such that none of the substance may pass the boundary surface S anywhere. In this situation, the flow out of a small part $d\vec{S}$ of the surface will be given by

$$d\Phi = \vec{j} \cdot d\vec{S} = 0 \implies \vec{n} \cdot \vec{j} = 0. \tag{3.70}$$

For diffusion, the current $\vec{j}$ obeys Fick's first law and we obtain

$$\vec{n} \cdot \nabla u = 0, \tag{3.71}$$

where we have divided by the diffusivity D. Thus, the resulting boundary condition is a condition on the normal derivative of u.

Example 3.14 Consider an oscillating string with an end at $x = x_0$, which is attached to a ring of mass m that is moving frictionlessly along a pole, see Fig. 3.10. Since the ring is free to move along the pole, the only force in the transversal direction is that from the string itself. By the same argumentation used in the derivation of the wave equation on the string, this force is given by

$$F = -\frac{Su_x(x_0,t)}{\sqrt{1+u_x(x_0,t)^2}} \simeq -S\partial_x u, \tag{3.72}$$

where the approximation holds for small oscillations. Since the ring's transversal displacement is tied to the string's we find that

$$ma = mu_{tt}(x_0,t) = -Su_x(x_0,t) \tag{3.73}$$

at the string endpoint. If the mass of the ring is negligible, this gives us the Neumann boundary condition

$$u_x(x_0,t) = 0. \tag{3.74}$$

This example may also be derived by considering the transversal momentum transfer from the string to the ring. Since no transversal force may act on the pole, all of the momentum flow from the string at the endpoint will end up in the ring. The change in the ring's momentum is therefore given by the momentum current $j = -S\partial_x u$. When considering that the ring momentum is $mu_t(x_0,t)$, equating the time derivative of this with the momentum current brings us back to the situation just described.

3.6.1.3 Robin boundary conditions

Although it is often a good approximation, it is seldom the case that the actual physical boundary conditions are either Dirichlet or Neumann conditions. Instead, a more accurate description is often given by the *Robin boundary condition*

$$\alpha(\vec{x},t)u(\vec{x},t) + \beta(\vec{x},t)\vec{n} \cdot \nabla u = k, \tag{3.75}$$

where, at every point on the boundary, at least one of the functions α and β is non-zero, and k is a, possibly vanishing, constant. Naturally, the requirement that one of the functions α and β must be non-zero is related to the fact that the boundary condition should be a condition on each point of the surface.

Example 3.15 One important occurrence of Robin boundary conditions is *Newton's law of cooling*, which states that the heat current in the interface between two materials is proportional to the difference in temperature between the materials, with the heat being transferred to the colder one. It may be expressed mathematically as a requirement on the temperature and current at the boundary

$$\vec{n} \cdot \vec{j} = \alpha(T - T_0), \tag{3.76}$$

where $\vec{j}$ is the heat current, α is a *heat transfer coefficient*, T is the temperature within the material for which we wish to solve the heat equation, and T_0 is the temperature of the surrounding medium. Some simplifying assumptions here are to assume that the heat transfer coefficient α is independent of the point on the surface and of the temperature and that T_0 is a constant, i.e., the surrounding material has the uniform temperature T_0. Applying Fourier's law to express the current $\vec{j}$, we find that

$$\alpha T + \lambda \vec{n} \cdot \nabla T = \alpha T_0, \tag{3.77}$$

which is a Robin boundary condition.

Example 3.16 Another situation where Robin conditions apply is in the case where there are both convection and diffusion currents in a material and the transfer out of a given volume must be kept to zero. A possible situation where this occurs is Example 3.8. Since the total current is given by

$$\vec{j} = -D\nabla u + \vec{v}_0 u, \tag{3.78}$$

the condition that the current in the normal direction should be zero turns into the Robin boundary condition

$$\vec{n} \cdot \vec{j} = -D\vec{n} \cdot \nabla u + (\vec{n} \cdot \vec{v}_0)u = 0, \tag{3.79}$$

where we can identify $\alpha = \vec{n} \cdot \vec{v}_0$, $\beta = -D$, and $k = 0$ in Eq. (3.75).

An important class of Robin boundary conditions requires that α and β have the same sign everywhere. This is related to the solution of problems using Sturm–Liouville theory, which we will discuss in Section 5.3. It should also be noted that both Dirichlet and Neumann boundary conditions are special cases of Robin conditions with $\beta = 0$ and $\alpha = 0$, respectively.

3.6.2 Initial conditions

While the boundary conditions specify how our quantity of interest behaves at the surface of the relevant volume, the *initial conditions*, as the name suggests, specify the initial state of the system. The initial conditions are only specified on one boundary, the initial time, and not on every boundary as was the case with the boundary conditions we just discussed. In general, in order to find a unique solution, the number of initial conditions at every point in the volume must be equal to the degree of the highest order time derivative in the differential equation. For ordinary differential equations, this is familiar to anyone who has looked at Newton's second law in the form

$$m\ddot{\vec{x}} = \vec{F}(\vec{x}), \tag{3.80}$$

where m is the mass of an object, $\vec{x}(t)$ its position at time t, and $\vec{F}$ a position dependent force field. In order to fully solve for the motion $\vec{x}(t)$, we need to define the initial position $\vec{x}(0)$ as well as the initial velocity $\dot{\vec{x}}(0)$, which is in line with the equation being a second order differential equation.

Being of first order in the time derivative ∂_t, the diffusion equation requires only one initial condition, namely the initial concentration of the diffusing substance. By contrast, the wave equation is of second order and requires the specification of both the value and rate of change for the sought function at $t = t_0$.

Example 3.17 In the case of a substance diffusing in a medium, a possible initial condition is that the concentration of the substance is constant in the entire relevant volume. In this case, the initial condition

$$u(\vec{x}, t_0) = u_0, \tag{3.81}$$

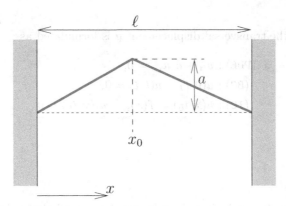

Figure 3.11 A string pulled at x_0 only is released from the resulting stationary state. The initial conditions include the string having the shape shown in this figure and $u_t(x, t_0) = 0$.

where u_0 is the constant concentration, is sufficient to determine the concentration at any later time $t > t_0$, assuming that appropriate boundary conditions are also given.

Example 3.18 A string of length ℓ, which is fixed at its ends, is pulled transversally at $x = x_0$ and takes the shape shown in Fig. 3.11, which may be described by

$$f(x) = \begin{cases} a\frac{x}{x_0}, & (x \le x_0) \\ a\frac{\ell-x}{\ell-x_0}, & (x > x_0) \end{cases}. \tag{3.82}$$

It is then released from rest at time $t = t_0$, which results in the initial conditions

$$u(x, t_0) = f(x), \quad u_t(x, t_0) = 0. \tag{3.83}$$

The latter of these conditions just states that the string is not moving transversally at time t_0 and is necessary in order for a unique solution to exist, since the wave equation is of second order in the time derivative ∂_t.

When writing down a problem in equations, we will from now on explicitly write out the partial differential equation, boundary, and initial conditions, as long as they are available. We will do so including a note on what each equation describes for the model in question, (PDE) for partial differential equation, (BC) for boundary condition, and (IC) for initial condition. Quoting the available information like this is often beneficial, as it will provide structure to the problem to be solved.

Example 3.19 The transversal oscillations on the string in Example 3.18 may be assumed to follow the source free wave equation with wave velocity c. Assuming that the ends of the string are fixed and that the initial conditions are given by the previous example, the full problem for finding the transversal displacement u is formulated as

$$\text{(PDE)} : u_{tt} - c^2 u_{xx} = 0, \tag{3.84a}$$

$$\text{(BC)} : u(0,t) = u(\ell,t) = 0, \tag{3.84b}$$

$$\text{(IC)} : u(x,t_0) = f(x), \quad u_t(x,t_0) = 0. \tag{3.84c}$$

3.6.3 Uniqueness

Let us take a step back and look at how it may be argued that solutions to the partial differential equations described with boundary and initial conditions are unique. We are going to do so by utilising *energy methods*. If we consider a wave equation problem in the volume V with Dirichlet boundary conditions

$$\text{(PDE)} : u_{tt} - c^2 \nabla^2 u = f(\vec{x},t), \qquad (\vec{x} \in V) \tag{3.85a}$$

$$\text{(BC)} : u(\vec{x},t) = g(\vec{x},t), \qquad (\vec{x} \in S) \tag{3.85b}$$

$$\text{(IC)} : u(\vec{x},t_0) = h_1(\vec{x}), \quad u_t(\vec{x},t_0) = h_2(\vec{x}), \qquad (\vec{x} \in V) \tag{3.85c}$$

where S is the boundary of V and f, g, h_1, and h_2 are known functions, we can prove the uniqueness of u by assuming that it is *not* unique and showing that this leads to a contradiction. Let us therefore assume that there exists two different solutions u and v, which both solve the given problem. We can then form the difference function $w = u - v$, which must be a non-zero function if our assumption that u and v are distinct should hold. Using that u and v both fulfil the differential equation as well as boundary and initial conditions, we find that

$$\text{(PDE)} : w_{tt} - c^2 \nabla^2 w = 0, \qquad (\vec{x} \in V) \tag{3.86a}$$

$$\text{(BC)} : w(\vec{x},t) = 0, \qquad (\vec{x} \in S) \tag{3.86b}$$

$$\text{(IC)} : w(\vec{x},t_0) = 0, \quad w_t(\vec{x},t_0) = 0. \qquad (\vec{x} \in V) \tag{3.86c}$$

Introducing the *energy functional*

$$E[u] = \frac{1}{2} \int_V [u_t^2 + c^2 (\nabla u)^2] dV, \tag{3.87}$$

which depends on the function u taken as argument, we find that

$$\frac{dE[w]}{dt} = \int_V [w_t w_{tt} + c^2 (\nabla w_t) \cdot (\nabla w)] dV = \int_V c^2 [w_t \nabla^2 w + (\nabla w_t) \cdot (\nabla w)] dV. \tag{3.88}$$

This expression is exactly Green's first identity, see Eq. (1.126), with $\varphi = w_t$ and $\psi = w$ and we obtain

$$\frac{dE[w]}{dt} = \oint_S w_t \nabla w \cdot d\vec{S} = 0, \tag{3.89}$$

since $w_t = 0$ on the entire surface due to w being identically zero on S, due to the boundary conditions. Since $w(\vec{x},t_0) = 0$ and $w_t(\vec{x},t_0) = 0$, we find that $E[w](t_0) = 0$ and the zero derivative implies that $E[w] = 0$ for all t. However, we must have

$$E[w] \geq 0, \tag{3.90}$$

since all terms in the integral are positive. This inequality becomes an equality only if w is identically zero everywhere and since we have found that $E[w] = 0$ for all t, the only possible conclusion is that $w = 0$, contrary to our starting assumption. It therefore follows that the solution to the partial differential equation with the given boundary and initial conditions is unique.

The reason that this method is called an energy method is that $E[u]$ may be interpreted, up to a multiplicative constant, as the energy in the volume V. Similar considerations may be applied to the diffusion equation as well as to other types of boundary conditions.

3.7 PDES IN SPACE ONLY

One differential equation that we have so far not treated in this chapter is *Poisson's equation*

$$\nabla^2 u(\vec{x}) = -\rho(\vec{x}), \tag{3.91}$$

where u is an unknown function in some volume V and ρ is known. Unlike the diffusion and wave equations, Poisson's equation does not have any time dependence and therefore will not need any initial conditions. We have already seen some examples of situations where Poisson's equation applies in Chapter 1.

Example 3.20 In a static situation, the electric field $\vec{E}$ is curl free, as indicated by Maxwell's equations with $\partial_t \vec{B} = 0$. Thus, in this situation there exists a scalar potential V such that

$$\vec{E} = -\nabla V. \tag{3.92}$$

Inserting this relation into Gauss's law, we find that

$$-\nabla \cdot \vec{E} = \nabla^2 V = -\frac{\rho}{\varepsilon_0}, \tag{3.93}$$

where ρ is the charge density and ε_0 the permittivity. The electrostatic potential V therefore satisfies Poisson's equation.

In many situations Poisson's equation appears as a description of system configurations that minimise the total energy or another property of a system. In particular, an important class of solutions to the diffusion and wave equations are *stationary solutions*, which are solutions that do not depend on time. For these solutions, all time derivatives vanish and the diffusion and wave equations take the forms

$$\nabla^2 u = -\frac{\kappa}{D} \quad \text{and} \quad \nabla^2 u = -\frac{f}{c^2}, \tag{3.94}$$

respectively, i.e., the stationary solutions satisfy Poisson's equation. Naturally, to find stationary states, it is not necessary to specify any initial conditions and only the boundary conditions will be of relevance for solving these equations.

Example 3.21 If an oscillating string is located in a gravitational field of strength g, there will be an external force density $f = -\rho_\ell g$ acting on every part of the string. Inserting this force density into the wave equation, we find that

$$\partial_t^2 u - c^2 \partial_x^2 u = -g. \tag{3.95}$$

For a stationary solution, it holds that $\partial_t^2 u = 0$ and we end up with the differential equation

$$\partial_x^2 u = \frac{g}{c^2} = \frac{\rho_\ell g}{S}, \tag{3.96}$$

which is Poisson's equation in one dimension with a constant source term $-\rho_\ell g/S$. This equation has the general solution $u = \rho_\ell g x^2/2S + Ax + B$, where A and B are arbitrary constants which will be determined by the boundary conditions. Note that this is the solution for a string where the gravitational force is small enough not to violate $|\partial_x u| \ll 1$. This is true as long as $\rho_\ell g L/S \ll 1$, where L is the total length of the string, i.e., for strings with a large tension in comparison to the product of the gravitational field strength, string density, and string length. A word of warning is appropriate here. As argued, this solution is only true for strings under large tension. We will discuss the case when a string is only suspended at its endpoints in Chapter 8.

It is important to note that, while stationary states do not evolve with time, this is not equivalent to stating that there is nothing happening in the system. For example, in a system satisfying Poisson's equation, we may still have $\nabla u \neq 0$, implying that $\vec{j} \neq 0$ for a stationary solution to the diffusion equation. Thus, there can still be currents within the system, but they are arranged in such a way that the quantity u does not change with time. Another way of stating this is reexamining the continuity equation for a stationary solution

$$\nabla \cdot \vec{j} = \kappa, \tag{3.97}$$

since $\partial_t u = 0$. The requirement for a solution to be stationary is therefore that the divergence of the current density is equal to the source density.

Example 3.22 Consider the case of a single point source in three dimensions constantly producing a substance with a rate P, which then diffuses throughout space with diffusivity D. The source term is then given by $\kappa = P\delta^{(3)}(\vec{x})$, where $\delta^{(3)}(\vec{x})$ is the three-dimensional delta function (see Eq. (1.221)), and the differential equation to solve for the stationary solution is

$$\nabla^2 u = -\frac{P\delta^{(3)}(\vec{x})}{D}. \tag{3.98}$$

Assuming the concentration goes to zero at infinity, we have already solved this problem in Example 1.33, where we discussed the point source and found that

$$\vec{j} = \frac{P}{4\pi r^2}\vec{e}_r \tag{3.99}$$

(note that, for Fick's first law, u is a scalar potential for the current density). Thus, in the stationary solution, there is a current in the direction away from the source and with a magnitude decreasing with the square of the distance. As the area of a sphere centred on the production point grows as r^2, the fluxes through all such spheres are the same for the stationary solution.

One special case of Poisson's equation that holds particular importance is the case where the source term is zero, i.e.,

$$\nabla^2 u = 0. \tag{3.100}$$

In fact, it is so special that it has warranted its own name, *Laplace's equation*. Based on the discussion we have had so far, we can conclude that this equation describes stationary states where there are no sources. In order for such a situation to hold any interest, the boundary conditions must be non-trivial. This may occur, e.g., when we have a soap film inside a non-planar wire frame.

Another important situation appears when we consider *steady state* solutions. Unlike stationary solutions, all time derivatives of a steady state solution do not necessarily vanish. Instead, we impose the slightly less constraining requirement that some properties of the solution, for example the amplitude of oscillation, should not change with time. Taking the wave equation as an example, we may consider solutions of the form

$$u(\vec{x}, t) = \cos(\omega t)v(\vec{x}). \tag{3.101}$$

Inserting this into the wave equation, we obtain

$$-(\omega^2 + c^2 \nabla^2)v = \frac{f}{\cos(\omega t)}. \tag{3.102}$$

This is particularly useful if either $f = 0$ or $f = g(\vec{x})\cos(\omega t)$. In the first of these situations, we obtain the *Helmholtz equation*

$$\nabla^2 v + k^2 v = 0, \tag{3.103a}$$

where $k = \omega/c$, and in the second the inhomogeneous Helmholtz equation

$$\nabla^2 v + k^2 v = -\frac{g(\vec{x})}{c^2}. \tag{3.103b}$$

The Helmholtz equation may be regarded as an eigenvalue equation for the operator $-\nabla^2$, a fact that we shall be using extensively later on. It may also be used to solve situations with time varying boundary conditions or external forces using an assumed steady state.

Example 3.23 Consider the drum from Example 3.11 with the modification that instead of fixing its boundary, the boundary is undergoing simple harmonic motion with angular frequency ω. This imposes the boundary conditions

$$\text{(BC)}: \ u(\rho = R, \phi, t) = A\cos(\omega t), \tag{3.104}$$

where A is a constant amplitude. Assuming a steady state $u = v(\vec{x})\cos(\omega t)$, the wave equation results in the Helmholtz equation for v as described above. This is not self consistent unless the boundary conditions for v are time independent, but luckily

$$u(\rho = R, \phi, t) = v(\rho = R, \phi)\cos(\omega t) = A\cos(\omega t), \tag{3.105a}$$

leading to the boundary condition

$$\text{(BC)}: \ v(\rho = R, \phi) = A, \tag{3.105b}$$

which does not depend on time t.

Regarding the uniqueness of the solutions to Poisson's equation, it can be proven much in the same way as for the time-dependent diffusion and wave equations. However, there is one exception, which occurs when all of the boundary conditions are Neumann conditions. Given a solution to Poisson's equation $\nabla^2 u = -\rho$ with Neumann conditions $\vec{n} \cdot \nabla u = g$ on the boundary, we can always shift u by a constant, i.e., $v = u + C$ without affecting either the differential equation or the boundary conditions. Instead, there is a consistency condition between the inhomogeneities ρ and g, which may be obtained by integrating ∇u over the boundary

$$\oint_S g \, dS = \oint_S \nabla u \cdot d\vec{S} = \int_V \nabla^2 u \, dV = -\int_V \rho \, dV, \tag{3.106}$$

where we have used the boundary condition in the first step, the divergence theorem in the second, and Poisson's equation in the third. If this condition is not fulfilled, no solution to the problem exists.

3.8 LINEARISATION

A *linear differential operator* $\hat{L}$ of order m is a differential operator which maps a function to a linear combination of u and its partial derivatives of order m and lower. In particular, it may be written as

$$\hat{L}u = \sum_{|\alpha| \leq m} a_\alpha \partial_\alpha u, \tag{3.107}$$

where $\alpha = (\alpha_1, \ldots, \alpha_N)$ is a *multi-index*, i.e., a combination of N numbers, and $|\alpha| = \alpha_1 + \ldots + \alpha_N$. We have also introduced the notation

$$\partial_\alpha = \partial_1^{\alpha_1} \ldots \partial_N^{\alpha_N}, \tag{3.108}$$

where N is the total dimension of the space. The constants a_α are generally functions that may be different for each value of the multi-index α.

Example 3.24 The wave operator in one spatial dimension and time $\partial_t^2 - c^2 \partial_x^2$ is an example of a linear differential operator of order two. Since the total dimension of the parameter space is two, with the parameters being time t and position x, the multi-indices $\alpha = (\alpha_t, \alpha_x)$ have two entries. The only non-zero coefficients a_α in the linear combination giving the wave operator are

$$a_{(2,0)} = 1 \quad \text{and} \quad a_{(0,2)} = -c^2. \tag{3.109}$$

In a similar fashion, the diffusion operator and Laplace operators are also examples of linear differential operators of order two.

In fact, all of the differential operators that we have encountered so far have been linear differential operators, with the exception of the wave equations on the string and membrane before we made the assumption of small oscillations, which would have turned out to be

$$u_{tt} - c^2 \partial_x \left(\frac{u_x}{\sqrt{1 + u_x^2}} \right) = 0 \tag{3.110}$$

for the the string without an additional force density. However, it should be noted that any of the differential equations we have encountered may turn out to include a non-linear differential operator if the source term is not linear in u. We define a *linear differential equation* to be a differential equation of the form

$$\hat{L}u = f, \tag{3.111}$$

where $\hat{L}$ is a linear differential operator and f is a known function which does not depend on u. Furthermore, if $f = 0$ the differential equation is *homogeneous* and, consequently, if $f \neq 0$ the differential equation is *inhomogeneous*. Likewise, we can classify boundary and initial conditions as linear and homogeneous if on the boundary of the relevant region

$$\hat{B}u = g \quad \text{and} \quad g = 0, \tag{3.112}$$

respectively, where $\hat{B}$ is a linear differential operator and g is a function. A linear problem is a problem where both the partial differential equation as well as the boundary and initial conditions are linear.

Example 3.25 The damped wave equation from Example 3.10

$$(\text{PDE}): \ u_{tt} + k u_t - c^2 u_{xx} = 0 \tag{3.113}$$

is a linear and homogeneous differential equation. If the string is fixed at the endpoints, then the boundary conditions

$$(\text{BC}): \ u(0,t) = u(\ell,t) = 0, \tag{3.114}$$

where ℓ is the string length are also homogeneous. Specifying initial conditions on u and u_t, the problem can be solved but, in order to obtain a non-trivial solution, at least one of the initial conditions must be inhomogeneous.

The last comment in the example above is an important one. If the differential equation as well as the boundary and initial conditions are linear and homogeneous, then $u = 0$ will be a solution. Based on the uniqueness of the solution, this implies that this is the only solution to such a problem, which is rather uninteresting.

On the other hand, an issue which is interesting is that any differential equation may be written as a linear differential equation for small deviations from a known solution. Assume that we have a non-linear differential equation

$$(\text{PDE}): \ \mathcal{F}(\partial_\alpha u) = f(\vec{x}), \tag{3.115}$$

where $\mathcal{F}(\partial_\alpha u)$ is a function of u and its partial derivatives up to some order m, $f(\vec{x})$ is an inhomogeneity, and the boundary conditions are given by

$$(\text{BC}): \ \mathcal{G}(\partial_\alpha u) = g(\vec{x}), \tag{3.116}$$

where $\mathcal{G}$ is also a function of u and its partial derivatives and $g(\vec{x})$ is a function on the boundary. If we have found one solution u_0 to this differential equation with given inhomogeneities in the differential equation $f = f_0$ and boundary condition $g = g_0$, then we can study solutions $u = u_0 + \varepsilon v$, for $f = f_0 + \varepsilon f_1$ and $g = g_0 + \varepsilon g_1$, where ε is assumed to be a small number. Expanding the differential equation to first order in ε results in

$$
\mathcal{F}(\partial_\alpha(u_0 + \varepsilon v)) \simeq \mathcal{F}(\partial_\alpha u_0) + \varepsilon \left. \frac{d\mathcal{F}(\partial_\alpha(u_0 + \varepsilon v))}{d\varepsilon} \right|_{\varepsilon=0}
$$
$$
= f_0(\vec{x}) + \varepsilon \left. \frac{d\mathcal{F}(\partial_\alpha(u_0 + \varepsilon v))}{d\varepsilon} \right|_{\varepsilon=0} = 0. \tag{3.117}
$$

Here, the term $\mathcal{F}(\partial_\alpha u_0) = f_0(\vec{x})$, since u_0 is assumed to be a solution of the differential equation with the inhomogeneity f_0. In order for this equation to hold to linear order in ε, we therefore need the derivative evaluated at $\varepsilon = 0$ to be equal to f_1. By applying the chain rule, this condition may be written as

$$
\left. \frac{d\mathcal{F}(\partial_\alpha(u_0 + \varepsilon v))}{d\varepsilon} \right|_{\varepsilon=0} = \sum_{|\alpha| \leq m} \left. \frac{\partial \mathcal{F}}{\partial(\partial_\alpha u)} \right|_{u=u_0} \left. \frac{d}{d\varepsilon} \partial_\alpha(u_0 + \varepsilon v) \right|_{\varepsilon=0}
$$
$$
= \sum_{|\alpha| \leq m} \left. \frac{\partial \mathcal{F}}{\partial(\partial_\alpha u)} \right|_{u=u_0} \partial_\alpha v = f_1(\vec{x}). \tag{3.118}
$$

Note that the partial derivatives of $\mathcal{F}$ are taken with respect to its arguments, which are different partial derivatives of the function u. Evaluated at $u = u_0$, these partial derivatives are known functions and the resulting differential equation for v is linear and homogeneous. A similar argument is then applied to the boundary condition.

Example 3.26 The perhaps most common application of linearisation occurs for time-dependent differential equations with stationary solutions. For example, imagine that we wish to solve the diffusion equation with a source term that is a function of u, i.e.,

$$
\text{(PDE)}: \quad u_t - D\nabla^2 u = f(u), \tag{3.119}
$$

and that we have homogeneous Dirichlet boundary conditions $u = 0$ on the boundary of the volume of interest V. A stationary solution to this problem $\tilde{u}(\vec{x})$ would then satisfy

$$
-D\nabla^2 \tilde{u} = f(\tilde{u}), \tag{3.120}
$$

and the homogeneous boundary condition. Letting $u = \tilde{u} + v$, where v is assumed to be a small deviation as long as the initial condition only deviates slightly from $\tilde{u}$, we find that

$$
\underbrace{\tilde{u}_t - D\nabla^2 \tilde{u} - f(\tilde{u})}_{=0} + v_t - D\nabla^2 v = f(\tilde{u} + v) - f(\tilde{u}), \tag{3.121}
$$

where we have used the fact that $\tilde{u}$ is a solution to the differential equation with the source term $f(\tilde{u})$. Since v is assumed small, we can use $f(\tilde{u} + v) - f(\tilde{u}) \simeq v f'(\tilde{u})$ and obtain the differential equation

$$
v_t - D\nabla^2 v = v f'(\tilde{u}) \tag{3.122}
$$

for the deviation v. Since $\tilde{u}$ satisfies the homogeneous boundary conditions, we also find that $u = \tilde{u} + v = v = 0$ on the boundary. The inhomogeneity required to have a non-trivial solution for v will arise from any initial deviation from the stationary state, a deviation that must be small in order for our procedure to be valid.

It should be noted that deviations will sometimes grow even if they start out as small deviations. In those situations, the solution to the linearised differential equation will be valid only as long as the deviation may be considered small.

3.9   THE CAUCHY MOMENTUM EQUATIONS

We have seen examples of how the continuity equation may be used to derive several important differential equations. Let us now examine how the continuity equation for different properties may be used in modelling by considering the momentum and mass continuity equations in a streaming fluid. A *fluid* is a medium, which may be, e.g., a gas or a liquid, whose shear strain is not able to generate any forces and therefore flows under any shear stress. Instead, as we will discuss, there may be a force related to the *shear strain rate*. This force is related to the viscosity of the fluid.

In any fluid, we may consider the flow velocity field $\vec{v}(\vec{x}, t)$, which is the velocity of the fluid at a given point in space and time, along with the density field $\rho(\vec{x}, t)$, which describes the density of the fluid. The momentum density $\vec{P}$ is then the intensive property given by

$$\vec{P} = \frac{d\vec{p}}{dV} = \frac{dm}{dV}\frac{d\vec{p}}{dm} = \rho\vec{v}, \tag{3.123}$$

where we have used that $d\vec{p} = \vec{v}\,dm$ is the momentum and dm is the mass enclosed in a small volume dV. Using the continuity equation for momentum, we find that

$$\frac{\partial P_i}{\partial t} + \partial_j j_{ij} = \rho\frac{\partial v_i}{\partial t} + v_i\frac{\partial \rho}{\partial t} + \partial_j j_{ij} = f_i, \tag{3.124}$$

where $\vec{f}$ is a force density acting in the volume dV. It should here be noted that, since the momentum density is a vector, the momentum current density is a rank two tensor with components j_{ij} describing the current density of momentum in the x^i-direction flowing in the x^j-direction. The time derivative $\partial\rho/\partial t$ may be expressed using the continuity equation for mass, which takes the form

$$\frac{\partial \rho}{\partial t} = -\nabla \cdot (\rho\vec{v}), \tag{3.125}$$

where we have used that the mass transport is a convective process given by the flow of the medium to write $\vec{j}_{\mathrm{mass}} = \rho\vec{v}$ and assumed that there is no mass source or sink. The continuity equation for the momentum is now given by

$$\rho\frac{\partial v_i}{\partial t} - v_i\partial_j(\rho v_j) + \partial_j j_{ij} = f_i \tag{3.126}$$

and the remaining task is to model the momentum current density j_{ij}. There are generally two different contributions to this current, the first being the convective transport of momentum by the flow $\vec{v}$. Since the convective current is given by the density multiplied by the flow velocity, we find that

$$j_{ij}^{\mathrm{conv}} = \rho v_i v_j, \tag{3.127}$$

or in other terms, using the outer product notation, $j^{\mathrm{conv}} = \rho\vec{v} \otimes \vec{v}$. This momentum flow is illustrated in Fig. 3.12.

The second contribution to the momentum current density comes from any forces acting on the surface of the volume. If the stress tensor is σ_{ij}, then the contribution from these forces to the momentum current is

$$j_{ij}^{\mathrm{force}} = -\sigma_{ij}. \tag{3.128}$$

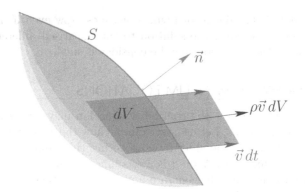

Figure 3.12 In a time dt, the volume $dV = \vec{v}\,dt \cdot d\vec{S}$ flows through a surface $d\vec{S}$. This volume carries with it a momentum $\rho\vec{v}\,dV$, resulting in a momentum current.

Note that the minus sign arises due to σ_{ij} describing the forces acting on the volume and therefore represents a flux of momentum into the volume instead of out of it. Combining these two contributions to the momentum current density, we find that

$$j_{ij} = \rho v_i v_j - \sigma_{ij} \tag{3.129}$$

and inserting this into the continuity equation for momentum yields

$$\rho\frac{\partial v_i}{\partial t} - v_i \partial_j(\rho v_j) + \partial_j(\rho v_i v_j - \sigma_{ij}) = \rho\frac{\partial v_i}{\partial t} + \rho v_j \partial_j v_i - \partial_j \sigma_{ij} = f_i. \tag{3.130}$$

It is common to also write f_i in terms of the force per mass rather than the force per volume, i.e., using that

$$d\vec{F} = \vec{g}\,dm = \vec{g}\frac{dm}{dV}dV = \vec{g}\rho\,dV \equiv \vec{f}\,dV, \tag{3.131}$$

where $\vec{g}$ is the force per mass. This choice is particularly natural in the case when the external force is due to a gravitational field. Doing this replacement and moving the term arising from the stress tensor to the right-hand side of the equation, we finally end up with

$$\frac{\partial v_i}{\partial t} + v_j \partial_j v_i = g_i + \frac{1}{\rho}\partial_j \sigma_{ij}. \tag{3.132}$$

These are the *Cauchy momentum equations*, which apply to the flow of any general fluid. There are several applications in fluid mechanics where additional assumptions may be made and these equations may be rewritten in a more particular form, we shall discuss some of these cases shortly. Let us first remark that the left-hand side of the Cauchy momentum equations may be rewritten as

$$\frac{\partial v_i}{\partial t} + v_j \partial_j v_i = \frac{Dv_i}{Dt}, \tag{3.133}$$

where D/Dt is the *material derivative* defined by

$$\frac{Df}{Dt} = \frac{\partial f}{\partial t} + (\vec{v}\cdot\nabla)f. \tag{3.134}$$

The material derivative represents a change with time of an intensive property f for a point flowing with the medium.

Example 3.27 Consider a fluid that does not support any internal stresses, i.e., there are no internal forces in the fluid such that $\sigma_{ij} = 0$. The Cauchy momentum equations now become

$$\frac{D\vec{v}}{Dt} = \vec{g}, \tag{3.135}$$

with the interpretation that the velocity of a particle flowing with the fluid changes proportionally to the force per mass. For a gravitational field, this is nothing else than the statement that the acceleration for a point flowing with the fluid is equal to the gravitational acceleration.

Example 3.28 In hydrostatics, stationary fluids are considered, implying that the velocity field vanishes identically $\vec{v} = 0$. In this situation, there will be no shear forces in the fluid as they cannot be supported. The stress tensor can then be written as $\sigma_{ij} = -p\delta_{ij}$, where p is the *pressure*. From the Cauchy momentum equations follows that

$$\frac{1}{\rho}\nabla p = \vec{g}. \tag{3.136}$$

This is nothing but a statement on how the pressure varies with respect to a force field $\vec{g}$. In particular, when the force field is a homogeneous gravitational field $\vec{g} = -g\vec{e}_3$, we find that

$$p = -\rho g x^3 + p_0, \tag{3.137}$$

where p_0 is the pressure at $x^3 = 0$. This linear dependence of the pressure on the coordinate x^3 should be familiar to most physics students. We note that, since $\rho\vec{g} = \nabla p$, the vector field $\rho\vec{g}$ must be curl free in order for its scalar potential p to exist. We come to the conclusion that there are no solutions to the Cauchy momentum equations with $\vec{v} = 0$ for which $\rho\vec{g}$ has a non-zero curl.

3.9.1 Inviscid fluids

An *inviscid fluid* is a fluid where there are no shear forces regardless of the movement of the fluid. Therefore, as in Example 3.28, the stress tensor is given by

$$\sigma_{ij} = -p\delta_{ij}. \tag{3.138}$$

This reduces the Cauchy momentum equation to

$$\frac{\partial \vec{v}}{\partial t} + (\vec{v} \cdot \nabla)\vec{v} = -\frac{1}{\rho}\nabla p + \vec{g}, \tag{3.139}$$

which is one of the *Euler equations* describing inviscid flows. The remaining Euler equations are the continuity equations for mass and energy, respectively,

$$\frac{\partial \rho}{\partial t} + \nabla \cdot (\rho\vec{v}) = 0, \tag{3.140a}$$

$$\frac{\partial \varepsilon}{\partial t} + \nabla \cdot [\vec{v}(\varepsilon + p)] = 0, \tag{3.140b}$$

where $\varepsilon = \rho e + \rho \vec{v}^2/2$ is the energy density, the two terms representing the internal energy with specific internal energy per mass e, and the kinetic energy due to the movement of the fluid. The current term $\vec{v}\varepsilon$ represents the convective flow of energy while the $\vec{v}p$ term results from the work done by the pressure forces.

Example 3.29 For a homogeneous *incompressible* fluid, the density in a small volume is constant along the flow, in other words the volume taken up by each part of the fluid remains the same. This can be expressed as the material derivative of the density vanishing, giving us

$$\frac{\partial \rho}{\partial t} + \nabla \cdot \rho \vec{v} = \frac{D\rho}{Dt} + \rho \nabla \cdot \vec{v} = \rho \nabla \cdot \vec{v} = 0 \quad \Longrightarrow \quad \nabla \cdot \vec{v} = 0 \tag{3.141}$$

from the continuity equation for mass. If the fluid is also inviscid and the force per mass $\vec{g}$ has a potential ϕ such that $\vec{g} = -\nabla \phi$, we find that

$$\frac{D\vec{v}}{Dt} = -\frac{1}{\rho}\nabla p - \nabla \phi, \tag{3.142}$$

which we may multiply by the momentum density $\rho\vec{v}$ and simplify to

$$\frac{1}{2}\frac{D(\rho\vec{v}^2)}{Dt} + (\vec{v}\cdot\nabla)(p + \rho\phi) - \phi(\vec{v}\cdot\nabla)\rho = 0. \tag{3.143}$$

If we are considering stationary flows, all quantities involved are time independent and the material derivative is equivalent to the directional derivative in the direction of the flow $D/Dt = \vec{v}\cdot\nabla$ and we find

$$\frac{D}{Dt}\left(\frac{1}{2}\rho\vec{v}^2 + p + \rho\phi\right) - \phi\frac{D\rho}{Dt} = \rho\frac{D}{Dt}\left(\frac{1}{2}\vec{v}^2 + \frac{p}{\rho} + \phi\right) = 0, \tag{3.144}$$

where we have used that $D\rho/Dt = 0$ for an incompressible flow. It directly follows that

$$\frac{1}{2}\vec{v}^2 + \frac{p}{\rho} + \phi = \text{constant} \tag{3.145}$$

along the flow lines, i.e., along the path taken by any small fluid parcel. This relation is known as *Bernoulli's principle*.

Example 3.30 The *isentropic flow* of a gas is both adiabatic and reversible. In particular, this means that an ideal gas obeys the *adiabatic gas law*

$$pV^\gamma = \text{constant} \quad \Longleftrightarrow \quad p = K\rho^\gamma, \tag{3.146}$$

where $\gamma = c_p/c_V$ is the ratio between the heat capacities for fixed pressure and fixed volume and K is a constant. With no force field $\vec{g}$ acting on the gas, which is a good approximation for many cases, there exists a stationary solution $p = p_0$, $\rho = \rho_0$, $\vec{v} = 0$ to the Euler equations. Linearising the general problem around this solution, the momentum and mass continuity equations state that

$$\frac{\partial \vec{v}}{\partial t} = -\frac{1}{\rho_0} \nabla p_1, \tag{3.147a}$$

$$\frac{\partial \rho_1}{\partial t} + \rho_0 \nabla \cdot \vec{v} = 0, \tag{3.147b}$$

where $p = p_0 + p_1$ and $\rho = \rho_0 + \rho_1$ (note that the velocity $\vec{v}$ is expanded around $\vec{v} = 0$ so we do not need to introduce a new variable for the linear term).

Linearising the adiabatic gas law results in

$$p_1 = K\gamma\rho_0^{\gamma-1}\rho_1 \implies \rho_1 = \frac{\rho_0}{\gamma p_0} p_1 \tag{3.148}$$

and inserting this into Eq. (3.147b) leads to

$$\frac{\partial p_1}{\partial t} + \gamma p_0 \nabla \cdot \vec{v} = 0. \tag{3.149}$$

Finally, differentiation of this relation with respect to time t and subsequent insertion of the mass continuity equation yields the wave equation

$$\frac{\partial^2 p_1}{\partial t^2} - c^2 \nabla^2 p_1 = 0, \tag{3.150}$$

where the wave velocity is given by $c^2 = \gamma p_0/\rho_0$. This equation thus describes waves of pressure deviations p_1 in a gas, i.e., *sound waves*.

3.9.2 Navier–Stokes equations

A particular form of the Cauchy momentum equation relies upon a series of quite general assumptions. When we were discussing stress and strain theory in the solid mechanics case (see Section 2.6.1), we made some similar assumptions so some arguments from there will be beneficial also in this case, the big difference being that the stress in a fluid cannot be linearly related to the strain tensor. However, the stress in a fluid may depend on the velocity field $\vec{v}$. As for the solid mechanics case, we make the assumption that the stresses do not depend on fluid movements which are overall translations or rotations, leading us to define the *strain rate tensor*

$$\varepsilon_{ij} = \frac{1}{2}(\partial_i v_j + \partial_j v_i) \tag{3.151}$$

in analogue with the strain tensor defined for the solid case. The next assumption is that of a linear relationship between the strain rate tensor ε_{ij} and the stress tensor σ_{ij}. Again in analogue to the solid case, this generally leads us to define the rank four *viscosity tensor* $C_{ijk\ell}$ which enters in the relation

$$\sigma_{ij} = C_{ijk\ell}\varepsilon_{k\ell}. \tag{3.152}$$

If the stress is symmetric, then it can be argued on the same grounds as for the stiffness tensor that $C_{ijk\ell}$ is symmetric in the first two indices as well as in the last two. This is the case for fluids that cannot carry any significant intrinsic angular momentum, i.e., an angular momentum related to the rotation of the constituents. In addition, fluids are very commonly isotropic, leading to the viscosity tensor necessarily being isotropic as well and based on the symmetries we must have

$$C_{ijk\ell} = \lambda\delta_{ij}\delta_{k\ell} + \mu(\delta_{ik}\delta_{j\ell} + \delta_{i\ell}\delta_{jk}), \tag{3.153}$$

where λ is the *bulk viscosity* and μ the *dynamic viscosity*. Using the Kronecker deltas to perform the contractions, the relation between the viscous stress and the strain rate tensor is given by

$$\sigma_{ij} = \lambda\varepsilon_{kk}\delta_{ij} + 2\mu\varepsilon_{ij} = \left(\lambda + \frac{2}{3}\mu\right)\varepsilon_{kk}\delta_{ij} + 2\mu\left(\varepsilon_{ij} - \frac{1}{3}\varepsilon_{kk}\delta_{ij}\right), \tag{3.154}$$

where the first term is the isotropic stress and the second the traceless deviatoric stress. Sometimes the *second viscosity* $\zeta = \lambda + 2\mu/3$ is introduced to slightly simplify this expression.

In the above, we note that $\varepsilon_{kk} = \partial_k v_k = \nabla \cdot \vec{v}$. This means that for an incompressible fluid there is no isotropic stress due to viscosity. In this situation, the viscous stress is only due to the strain rate that for the isotropic part means a change in volume and there may still be a thermodynamic pressure p providing an isotropic stress, while it is the deviatoric stress that necessarily is zero for a fluid with zero strain rate. Inserted into the Cauchy momentum equation, this leads to

$$\rho\frac{Dv_i}{Dt} = \rho g_i + \partial_i(\zeta\varepsilon_{kk} - p) + 2\mu\partial_j\varepsilon_{ij} - \frac{2}{3}\mu\partial_i\varepsilon_{kk}$$
$$= \rho g_i - \partial_i\tilde{p} + \mu\partial_j\partial_j v_i + \frac{\mu}{3}\partial_i\partial_j v_j \tag{3.155a}$$

or in other terms

$$\rho\frac{D\vec{v}}{Dt} = \rho\vec{g} - \nabla\tilde{p} + \mu\nabla^2\vec{v} + \frac{\mu}{3}\nabla(\nabla \cdot \vec{v}), \tag{3.155b}$$

where we have introduced the mechanical pressure $\tilde{p} = p - \zeta\nabla \cdot \vec{v}$. This is the *Navier-Stokes momentum equation* for a compressible fluid. In general, just as the Euler equation, this equation needs to be supplemented with additional assumptions on the continuity of mass and energy as well as appropriate boundary conditions in order to fully describe the flow. It should also be noted that we have assumed the viscosities λ and μ to be constants. In general, this may not be true as the viscosity may depend on several parameters, including density and pressure.

Example 3.31 Consider an incompressible stationary flow in a fixed direction, which may be taken to be $\vec{e}_3$, see Fig. 3.13. The velocity field may then be written as $\vec{v} = v\vec{e}_3$, where v is a scalar field that, due to the incompressibility condition $\nabla \cdot \vec{v} = \partial_3 v = 0$, cannot depend on x^3 and therefore is a function of x^1 and x^2 only. If the flow is driven by a pressure gradient $\nabla p = \vec{e}_3 p_0/\ell$, then the Navier-Stokes momentum equation takes the form

$$(\text{PDE}): \quad 0 = -\frac{p_0}{\ell} + \mu\nabla^2 v \quad \implies \quad \nabla^2 v = \frac{p_0}{\ell\mu}. \tag{3.156}$$

The velocity v therefore satisfies Poisson's equation with a constant source term. In order to solve this equation, we need to impose boundary conditions. For the fixed direction flow to make sense, the region of interest must be translationally invariant in the x^3 direction and the region in the x^1-x^2-plane is independent of x^3. This is the region in which we wish

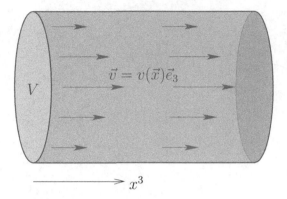

Figure 3.13 A parallel flow through a cylinder along the cylinder symmetry axis, taken to be the $\vec{e}_3$ direction. Assuming that the stationary flow is pressure driven, that the fluid has a shear viscosity μ, and that the flow is incompressible, the flow can be computed.

to solve for v and we refer to it as V. One possible set of boundary conditions, called *no slip conditions*, is given by requiring that the flow velocity is zero at the region boundary S, resulting in the Dirichlet boundary condition

$$(\text{BC}) : \ v(\vec{x} \in S) = 0. \tag{3.157}$$

Taking the flow region to be a cylinder of radius R, the region V is a disc of radius R on which we use polar coordinates ρ and ϕ. Due to the rotational symmetry of the problem, both in the differential equation and the boundary conditions, the solution cannot depend on ϕ and we find that v is only a function of the parameter ρ, $v = v(\rho)$. Inserting this into the differential equation leads to

$$v''(\rho) + \frac{1}{\rho} v'(\rho) = \frac{p_0}{\ell \mu}, \tag{3.158}$$

which may be solved using an integrating factor ρ. The resulting solution is given by

$$v(\rho) = \frac{p_0}{4\ell\mu}(\rho^2 - R^2). \tag{3.159}$$

It should be noted that we have imposed a regularity condition $v(0) < \infty$ in order to fix one of the integration constants to zero. As should be expected, the resulting flow is in the negative $\vec{e}_3$ direction since $\rho \leq R$. Naturally, this corresponds to a flow from higher to lower pressure. Furthermore, the flow is proportional to the driving pressure gradient p_0/ℓ and inversely proportional to the viscosity μ of the fluid.

While this model may be very rudimentary, it does demonstrate some basic dependencies of a pressure driven flow. The actual situation is often not going to result in a flow of this type, even if the container through which a fluid flows would be infinite, there is no guarantee that the solution is going to be stable even if it exists. This must be checked explicitly and is often not the case in fluid dynamics. However, we will continue to use this and flows based on similar assumptions as very rudimentary models mainly for illustrative purposes.

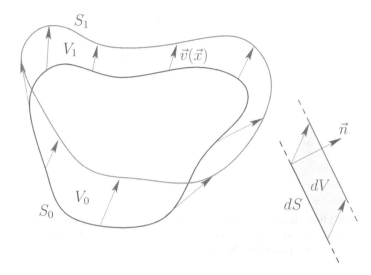

Figure 3.14 During a time dt, the surface S_0 moves to the surface S_1 according to the displacement field $\vec{v}\,dt$. If the flow is incompressible, the enclosed regions V_0 and V_1 will have the same volume, resulting in $\nabla \cdot \vec{v} = 0$. The lower right corner shows the volume dV flowing out through a surface $d\vec{S}$ per time unit.

3.9.3 Incompressible flow

We have already discussed *incompressible flows* and seen the condition $\nabla \cdot \vec{v} = 0$ resulting from requiring the material derivative of the density to be equal to zero. Let us briefly comment on this condition and find its geometrical interpretation. Consider a volume V_0 with boundary S_0 as depicted in Fig. 3.14. The material within this volume will move according to the velocity flow $\vec{v}$, meaning that the material at the point $\vec{x}$ at time t_0 will be at the point $\vec{x} + \vec{v}\,dt$ at time $t_0 + dt$ for small dt. The material at the surface S at t_0 will form a new surface S_1 at $t_0 + dt$, as shown in the figure. By assumption of continuous deformation of the fluid, the surfaces will both enclose the same material and the question we wish to answer is how the volumes of the enclosed volumes relate to each other.

Looking at a small surface $d\vec{S}$, the volume swept out by this surface while moving with velocity $\vec{v}$ during a time dt is given by

$$dV = \vec{v} \cdot d\vec{S}\,dt. \tag{3.160}$$

This is also depicted in the right part of Fig. 3.14. Summing over the entire surface S_0, this swept volume will represent the difference between the volumes enclosed by S_0 and S_1. By applying the divergence theorem, we find that

$$\frac{dV_0}{dt} = \oint_{S_0} \vec{v} \cdot d\vec{S} = \int_{V_0} \nabla \cdot \vec{v}\,dV. \tag{3.161}$$

It follows that if $\nabla \cdot \vec{v} = 0$, then the change in the volume occupied by the material does not change with time.

An alternative way of arriving at the same result is to compute the volume V_1, enclosed by the surface S_1, using the integral

$$V_1 = \int_{V_1} dV. \tag{3.162}$$

Making a change of variables to $\vec{x}' = \vec{x} - \vec{v}\,dt$, we find that

$$V_1 = \int_{V_0} \mathcal{J}\,dV_0, \tag{3.163}$$

where $\mathcal{J}$ is the Jacobian determinant of the transformation, given by

$$\mathcal{J} = \frac{1}{N!}\varepsilon_{i_1\ldots i_N}\varepsilon_{j_1\ldots j_N}\prod_{k=1}^{N}\frac{\partial x_{i_k}}{\partial x'_{j_k}} = \frac{1}{N!}\delta^{i_1\ldots i_N}_{j_1\ldots j_N}\prod_{k=1}^{N}(\delta_{i_k j_k} + dt\,\partial_{j_k}v_{i_k})$$

$$\simeq \frac{1}{N!}\delta^{i_1\ldots i_N}_{i_1\ldots i_N} + \frac{dt}{(N-1)!}\delta^{i_1 i_2\ldots i_N}_{j_1 i_2\ldots i_N}\partial_{j_1}v_{i_1} = 1 + dt\,\delta_{i_1 j_1}\partial_{j_1}v_{i_1} = 1 + dt\,\nabla\cdot\vec{v}, \tag{3.164}$$

where we have used the properties of the generalised Kronecker delta from Section 2.3.4 and kept terms only to linear order in dt. It follows that

$$V_1 = \int_{V_0}(1 + dt\,\nabla\cdot\vec{v})dV_0 = V_0 + dt\int_{V_0}\nabla\cdot\vec{v}\,dV_0, \tag{3.165}$$

which gives the same result for $V_1 - V_0$ as that obtained in Eq. (3.161).

3.10 SUPERPOSITION AND INHOMOGENEITIES

A fundamental tool in the treatment of linear differential equations, partial as well as ordinary, is the *superposition principle*. As we have discussed, a general linear differential equation will be of the form

$$(\text{PDE}): \hat{L}u = f, \tag{3.166a}$$

where $\hat{L}$ is a linear differential operator and f is some function that does not depend on u. If we also have linear boundary and initial conditions of the form

$$(\text{BC/IC}): \hat{B}u = g \tag{3.166b}$$

on the boundary (spatial as well as temporal) where $\hat{B}$ is also a linear operator, then the problem is linear and allows superposition of the solutions. This means that if we have two functions u_1 and u_2 which solve the problem for different inhomogeneities, f_1 and g_1 for u_1 and f_2 and g_2 for u_2, then the sum $u_{1+2} = u_1 + u_2$ is a solution to the problem

$$(\text{PDE}): \hat{L}u_{1+2} = f_{1+2} = f_1 + f_2, \tag{3.167a}$$

$$(\text{BC/IC}): \hat{B}u_{1+2} = g_{1+2} = g_1 + g_2, \tag{3.167b}$$

where again the boundary condition is only valid on the boundary. In order to arrive at this conclusion, we use the linear property of the differential operator $\hat{L}$

$$\hat{L}(u_1 + u_2) = \hat{L}u_1 + \hat{L}u_2 = f_1 + f_2 \tag{3.168}$$

and similarly for the operator $\hat{B}$ specifying the boundary conditions.

Example 3.32 The superposition principle is useful for combining solutions to linear problems into new solutions. A typical example is given by the electrostatic potential of two point charges q_1 and q_2 at points $\vec{x}_1$ and $\vec{x}_2$, respectively. The differential equation satisfied by the electrostatic potential Φ is

$$\nabla^2 \Phi = -\frac{\rho}{\varepsilon_0}, \tag{3.169}$$

which follows from Maxwell's equations. The Laplace operator ∇^2 is linear and the potential for a single point charge q is given by

$$\Phi = \frac{q}{4\pi\varepsilon_0 r}, \tag{3.170}$$

where r is the distance to the charge. From the superposition principle therefore follows that the total potential for both charges is given by

$$\Phi = \frac{1}{4\pi\varepsilon_0} \left(\frac{q_1}{|\vec{x} - \vec{x}_1|} + \frac{q_2}{|\vec{x} - \vec{x}_2|} \right). \tag{3.171}$$

The electric field $\vec{E}$ can be deduced from this potential through the relation $\vec{E} = -\nabla\Phi$.

The above example shows the power of superposition, i.e., we can take the solutions for two different inhomogeneities and use them to construct the solution to a new problem. In fact, this method is very general and we may do the same also for a continuous charge distribution $\rho(\vec{x})$. If we consider the contribution from a small volume dV_0 around $\vec{x}_0$, the contribution from the charge within that volume is given by

$$d\Phi(\vec{x}) = \frac{1}{4\pi\varepsilon_0} \frac{dq}{|\vec{x} - \vec{x}_0|} = \frac{\rho(\vec{x}_0)}{4\pi\varepsilon_0} \frac{dV_0}{|\vec{x} - \vec{x}_0|}. \tag{3.172}$$

Summing the contributions by integration over the entire charge distribution, we find that

$$\Phi(\vec{x}) = \frac{1}{4\pi\varepsilon_0} \int \frac{\rho(\vec{x}_0)}{|\vec{x} - \vec{x}_0|} dV_0. \tag{3.173}$$

This is an example of a solution expressed using an integral with a *Green's function* $G(\vec{x}, \vec{x}_0) = -1/4\pi|\vec{x} - \vec{x}_0|$. We shall return to this method for solving partial differential equations in Chapter 7.

In a similar fashion, superposition may be used to split a problem into several problems, each of which is easier to solve than the original one. In the following, we will discuss some possible approaches.

3.10.1 Removing inhomogeneities from boundaries

In many situations, in particular when solving linear partial differential equations using series methods, see Chapter 6, it is beneficial to work with homogeneous boundary conditions. Consider the linear problem

$$\text{(PDE)} : \hat{L}u = f, \qquad\qquad (\vec{x} \in V) \tag{3.174a}$$

$$\text{(BC)} : \hat{B}u = g, \qquad\qquad (\vec{x} \in S_1) \tag{3.174b}$$

where S_1 is part of the boundary of V. We can exchange the boundary condition for a homogeneous boundary condition by selecting any function u_0 that satisfies the boundary condition, without any regard for the differential equation, and requiring that $u = v + u_0$. Inserting this into the linear problem, we obtain a new linear problem

$$\text{(PDE)} : \hat{L}v = f - \hat{L}u_0, \qquad\qquad (\vec{x} \in V) \qquad\qquad (3.175a)$$

$$\text{(BC)} : \hat{B}v = 0, \qquad\qquad (\vec{x} \in S_1) \qquad\qquad (3.175b)$$

for v. Since we can compute $\hat{L}u_0$, we can find the new inhomogeneity in the differential equation and possibly obtain a problem that is easier to solve. It should be noted that there may also be other boundary conditions and it may not always be the best idea to find a function u_0 that makes all boundary conditions homogeneous.

Example 3.33 Let us take a situation where we are studying transversal oscillations of a string for which one end is fixed and the other is subjected to forced oscillations with angular frequency ω and amplitude A

$$\text{(PDE)} : u_{tt} - c^2 u_{xx} = 0, \qquad\qquad (3.176a)$$

$$\text{(BC)} : u(0,t) = 0, \quad u(\ell, t) = A\sin(\omega t), \qquad\qquad (3.176b)$$

$$\text{(IC)} : u(x,0) = u_t(x,0) = 0. \qquad\qquad (3.176c)$$

We have here also assumed that the string is initially at rest with zero transversal displacement at $t = 0$. A function that fulfils both boundary conditions is $u_0(x,t) = A\sin(\omega t)x/\ell$ and letting $u = u_0 + v$, we find that v must satisfy

$$\text{(PDE)} : v_{tt} - c^2 v_{xx} = A\omega^2 \sin(\omega t)\frac{x}{\ell}, \qquad\qquad (3.177a)$$

$$\text{(BC)} : v(0,t) = 0, \quad v(\ell, t) = 0, \qquad\qquad (3.177b)$$

$$\text{(IC)} : v(x,0) = 0, \quad v_t(x,0) = -A\omega\frac{x}{\ell}. \qquad\qquad (3.177c)$$

At first glance, this may seem like a loss, we just traded a single inhomogeneity in a boundary condition for inhomogeneities in both the differential equation as well as the initial conditions. However, as we will discuss in Chapters 5 and 6, having homogeneous boundary conditions will be a prerequisite for expanding solutions into series using Sturm–Liouville theory. Besides, we can always find the solution for the inhomogeneous differential equation and the inhomogeneous initial condition separately and construct the full solution by superposition.

3.10.2 Using known solutions

The method discussed above is even more powerful if we do care about which function u_0 to choose. In particular, if we have a problem to which we can find a solution, disregarding any initial conditions, using it as u_0 will remove the inhomogeneities from both the boundaries as well as from the differential equation. Consider the linear problem

$$\text{(PDE)} : \hat{L}u = f, \qquad\qquad (3.178a)$$

$$\text{(BC)} : \hat{B}u = g, \qquad\qquad (3.178b)$$

$$\text{(IC)} : \hat{I}u = h, \qquad\qquad (3.178c)$$

where the boundary condition is valid on the spatial boundary of the volume V in which we wish to solve for u and the initial condition is valid at $t = 0$. If it exists, the stationary or steady state solution u_0 must fulfil both the differential equation as well as the boundary conditions, but will generally not satisfy the initial condition. Again taking $u = v + u_0$, we now find that

$$(\text{PDE}) : \hat{L}v = f - \hat{L}u_0 = 0, \tag{3.179a}$$

$$(\text{BC}) : \hat{B}v = g - \hat{B}u_0 = 0, \tag{3.179b}$$

$$(\text{IC}) : \hat{I}v = h - \hat{I}u_0, \tag{3.179c}$$

where the initial condition will generally not be homogeneous. However, both the differential equation and the boundary conditions are homogeneous and after finding the most general solution to these, the final solution may be found by adapting the resulting free parameters to the initial conditions.

Example 3.34 Let us study heat conduction along an insulated rod of length ℓ without a heat source and initially with the temperature T_0 everywhere. Furthermore the ends are being held at temperatures T_1 and T_2, respectively, and we have the linear problem

$$(\text{PDE}) : T_t - aT_{xx} = 0, \tag{3.180a}$$

$$(\text{BC}) : T(0,t) = T_1, \quad T(\ell,t) = T_2, \tag{3.180b}$$

$$(\text{IC}) : T(x,0) = T_0. \tag{3.180c}$$

The stationary solution to this problem is $T^{\text{st}} = T_1 + (T_2 - T_1)x/\ell$ and defining $T = u + T^{\text{st}}$, we find that

$$(\text{PDE}) : u_t - au_{xx} = 0, \tag{3.181a}$$

$$(\text{BC}) : u(0,t) = 0, \quad u(\ell,t) = 0, \tag{3.181b}$$

$$(\text{IC}) : u(x,0) = T_0 - T_1 - (T_2 - T_1)\frac{x}{\ell}. \tag{3.181c}$$

The resulting problem for u is therefore homogeneous in both the differential equation as well as the boundary conditions. In order to solve the problem, we need to find the general solution for the homogeneous differential equation and boundary conditions and adapt it to the inhomogeneous initial conditions.

This approach is the partial differential equation equivalent of first finding a particular solution to an ordinary differential equation and then adding a homogeneous solution to it in order to adapt it to the boundary conditions. In this case, we find a solution that matches both the differential equation as well as the boundary conditions and use it as the particular solution. A solution to the homogeneous problem is then added and the freedom in choosing it is used to adapt the full solution to the original initial conditions. It should also be noted that a similar approach may be used for problems that are not time dependent by instead leaving out one of the spatial boundaries when looking for a particular solution.

3.11 MODELLING THIN VOLUMES

In many physical situations where a volume may be considered thin in one or more directions, it is often possible to rewrite the full model in terms of a lower dimensional one by making an assumption of the solution being approximately constant in the thin dimension. In order to illustrate this, we will consider the heat conduction problem in three dimensions within the volume $0 < x^3 < h$, where h may be considered small, the meaning of which must be interpreted in terms of the temperature diffusivity a and the typical time scales involved in the problem. We will assume a source density $\kappa(\vec{x}, t)$ and that the boundary conditions at $x^3 = 0$ and h are given by Newton's law of cooling with heat transfer coefficient α and external temperature T_0. The resulting linear problem is of the form

$$\text{(PDE)}: T_t - a\nabla^2 T = \kappa(\vec{x}, t), \qquad\qquad (0 < x^3 < h) \qquad (3.182a)$$

$$\text{(BC)}: \alpha T + \lambda \partial_3 T = \alpha T_0, \qquad\qquad (x^3 = h) \qquad (3.182b)$$

$$\alpha T - \lambda \partial_3 T = \alpha T_0, \qquad\qquad (x^3 = 0) \qquad (3.182c)$$

$$\text{(IC)}: T(\vec{x}, 0) = T_1(\vec{x}). \qquad\qquad (3.182d)$$

We now introduce the x^3 average $\bar{f}(x^1, x^2, t)$ of any function $f(\vec{x}, t)$ as

$$\bar{f} = \frac{1}{h} \int_0^h f(\vec{x}, t) dx^3, \qquad\qquad (3.183)$$

with the aim of finding a differential equation that will describe the x^3 averaged temperature $\bar{T}$. Note that the given problem does not have any boundary conditions in the x^1 and x^2 directions and that the resulting differential equation for $\bar{T}$ should not involve any boundary conditions, but be a problem in two dimensions.

Dividing the differential equation for T by h and integrating with respect to x^3, we now find that

$$\bar{T}_t - a\nabla^2 \bar{T} - a\frac{1}{h}\int_0^h \partial_3^2 T\, dx^3 = \bar{\kappa}, \qquad\qquad (3.184)$$

where we have used that

$$\nabla^2 \bar{T} = \frac{1}{h}\int_0^h (\partial_1^2 + \partial_2^2) T\, dx^3, \qquad\qquad (3.185)$$

since the partial derivatives may be moved outside of the integral and $\bar{T}$ does not depend on x^3. Luckily, the integral in Eq. (3.184) has an integrand that is a total derivative and we can perform it without trouble

$$\int_0^h \partial_3^2 T\, dx^3 = [\partial_3 T]_{x^3=0}^{x^3=h} = \alpha(2T_0 - T|_{x^3=h} - T|_{x^3=0}) \simeq 2\alpha(T_0 - \bar{T}), \qquad (3.186)$$

where we have assumed that $T|_{x^3=h} + T|_{x^3=0} \simeq 2\bar{T}$, i.e., that the averaged temperature is roughly the mean value of the boundary temperatures. This is a good approximation as long as h may be considered small. For the initial condition, we average both sides with respect to x^3, resulting in the problem

$$\text{(PDE)}: \bar{T}_t - a\nabla^2 \bar{T} = \bar{\kappa} - 2a\frac{\alpha}{h}(\bar{T} - T_0), \qquad\qquad (3.187a)$$

$$\text{(IC)}: \bar{T}(x^1, x^2, 0) = \bar{T}_1(x^1, x^2). \qquad\qquad (3.187b)$$

This problem is now a two-dimensional approximation of the full three-dimensional problem and the boundary in the thin x^3 direction has been exchanged for an additional temperature dependent source term. It should be noted that this source term corresponds to a source whenever $\bar{T} < T_0$, corresponding to a net influx of heat, and to a sink whenever $\bar{T} > T_0$, corresponding to a net outflux of heat.

This approach works well in the case of Robin or Neumann boundary conditions. If the boundary conditions in the thin direction are of Dirichlet type, we cannot use them to express the normal derivative as done in Eq. (3.186).

Physical dimension	Symbol	Typical units
Length	L	m, mm, yards, light years
Time	T	s, days, weeks, years
Mass	M	kg, g, tonnes
Temperature	Θ	K
Electric charge	Q	C

Table 3.2   The five basic physical dimensions, the symbols used to represent them, and some typical units in which they are measured. Generally, the physical dimension of any quantity may be expressed as a product of these dimensions.

3.12   DIMENSIONAL ANALYSIS

Although often taken almost for granted, let us briefly remind ourselves on the use of *physical dimensions* and their importance when it comes to physics in general and differential equation modelling in particular. All physical quantities are not just numbers to be put into an equation, but are associated with a physical dimension. There are five basic physical dimensions, summarised in Table 3.2, and the physical dimension of any quantity will generally be some product of powers of these. As physical quantities are multiplied together, the physical dimension of the resulting quantity is the product of the physical dimensions of the original quantities. Similarly, when the ratio between two physical quantities is taken the result is the ratio of the physical dimensions. Given a physical quantity q, we will denote its physical dimension as $[q]$.

Example 3.35   An object travels a distance ℓ at constant velocity v during the time t. The velocity may be computed as $v = \ell/t$ and the involved physical dimensions are $[\ell] = \mathsf{L}$, $[t] = \mathsf{T}$ and $[v] = [\ell]/[t] = \mathsf{L}/\mathsf{T}$. As such, ℓ is a length, which is measured using length units, t is a time, which is measured in time units, and v is a velocity, which is measured in units of length per time. This example should already be familiar to most, but illustrates the general idea that is prevalent throughout physics.

While quantities of different physical dimension may be multiplied and divided with each other to yield new quantities of yet a third physical dimension, it makes no sense to add or subtract quantities of different physical dimension as the result would be nonsense, there just is no way of adding 1 m and 0.2 s and getting a meaningful answer. The exception to this general rule is when an underlying equivalence between the two dimensions is assumed, for example by using units such that a normally dimensionful natural constant becomes dimensionless. This is common in both quantum mechanics and relativity, where one will often use units such that Planck's constant $\hbar = 1$ and the speed of light in vacuum $c = 1$, respectively. In these situations, it is always possible to reinsert the natural constants into the resulting expressions using dimensional analysis.

Example 3.36   In relativity with units such that $[c] = \mathsf{L}/\mathsf{T} = 1$, we find that $\mathsf{L} = \mathsf{T}$ and so the number of basic physical dimensions is reduced by one. The dimensionless gamma factor is then given by

$$\gamma = \frac{1}{\sqrt{1-v^2}}. \tag{3.188}$$

If we wish to use a different set of units where $L \neq T$, the subtraction in the denominator must be dimensionless in order for the gamma factor to be dimensionless. It follows that the term v^2 must be multiplied by some power of c in order to have the correct dimensions, i.e., we must make the substitution $v^2 \to c^\alpha v^2$. We can find the power α by ensuring that

$$[c^\alpha v^2] = L^{2+\alpha} T^{-2-\alpha} = 1 \quad \Longrightarrow \quad \alpha = -2. \tag{3.189}$$

The expression for the gamma factor with c reinserted is therefore

$$\gamma = \frac{1}{\sqrt{1 - \frac{v^2}{c^2}}}. \tag{3.190}$$

In particular, the idea that quantities with different physical dimensions may not be added applies to functions, which must receive arguments with the appropriate physical dimension in order to be meaningful. For example, a quantity with physical dimension cannot be the argument of the function $f(x) = x + x^2$ as it would involve adding two quantities with different physical dimension, $[x]$ and $[x]^2$, respectively. On the other hand, for the function $f(x) = x\ell + x^2$, where $[\ell] = L$ the argument must be dimensionful as both terms must have the same dimension. This requirement tells us that

$$[x][\ell] = [x]L = [x]^2 \quad \Longrightarrow \quad [x] = L. \tag{3.191}$$

Common mathematical functions such as the exponential function, the logarithm, sines, and cosines may all be written in the form of series expansions

$$f(x) = \sum_{n=0}^{\infty} a_n x^n, \tag{3.192}$$

where the coefficients a_n are dimensionless. By the discussion above, the argument x must therefore be dimensionless and it is useful to remember that all of these functions must have dimensionless arguments, it is meaningless to talk about the logarithm of a length $\ln(\ell)$. However, it is always possible to take the logarithm of a length divided by some reference length ℓ_0, i.e., $\ln(\ell/\ell_0)$, as the argument is then dimensionless.

Dimensional analysis also provides us with a very powerful tool for estimating the dependence of a physical quantity of interest on other given physical quantities. By knowing the dimension of the target quantity and the given quantities, we can perform the following procedure:

1. Identify the dimension of the target quantity.

2. Write down a general product of powers of the given quantities.

3. Identify the dimension of the product constructed in 2 with the dimension of the target quantity. Each basic physical dimension must come with the correct power.

4. Solve the ensuing linear system of equations to find the appropriate dependence.

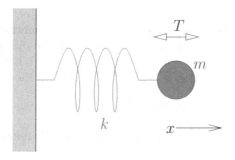

Figure 3.15 A basic spring-mass system. The dependence of the period T of completing one full oscillation in the x-direction on the mass m and spring constant k may be deduced through dimensional analysis.

Example 3.37 We are interested in finding out how the period of oscillation T for a spring-mass system with mass m and spring constant k, see Fig. 3.15, depends on these quantities. The most general product we may write using m and k is $m^\alpha k^\beta$, where α and β are to be determined. We know that the target quantity T has dimensions $[T] = \mathsf{T}$ and that $[m] = \mathsf{M}$. The dimension of k can be deduced from the force relation

$$F = kx \quad \Longrightarrow \quad [F] = \frac{\mathsf{ML}}{\mathsf{T}^2} = [k][x] = [k]\mathsf{L}, \tag{3.193}$$

where x is the displacement from the position where the mass is when the spring is relaxed. It follows that $[k] = \mathsf{M}/\mathsf{T}^2$ and that

$$\mathsf{T} = [m]^\alpha[k]^\beta = \mathsf{M}^{\alpha+\beta}\mathsf{T}^{-2\beta} \quad \Longrightarrow \quad \alpha + \beta = 0, \ -2\beta = 1. \tag{3.194}$$

Solving the system of equations leads to $\alpha = -\beta = 1/2$ and thus $\sqrt{m/k}$ is the only combination of m and k that has the physical dimension of time. In general, this may have to be corrected through multiplication by a dimensionless constant C that must be found by modelling and solving the system, but the final solution must be on the form $T = C\sqrt{m/k}$.

3.12.1 Units

When dealing with variables that have physical dimension, we must always select a set of *units* to work with. A complete set of units is defined by specifying reference values for all of the basic physical dimensions that are of relevance. In the general case, with all the five basic physical dimensions listed in Table 3.2, we would have to specify a reference value u_i for each of them, where the index i denotes the corresponding basic physical dimension, which we will call the *base unit*. The table also lists some common choices of units used for each of the basic physical dimensions.

Once the base units have been selected, any quantity p that has the physical dimension of the base unit X will have a *measured value* λ_p in this set of units given by

$$p = \lambda_p u_\mathsf{X}, \tag{3.195}$$

such that λ_p is dimensionless. As we have seen, a quantity may also have a physical dimension that is the product of powers of all the basic ones. In particular, if we consider a quantity p such that

$$[p] = \prod_i \mathsf{X}_i^{\alpha_i}, \tag{3.196}$$

where X_i is one of the basic physical dimensions, we find that we can use the unit

$$u_p = \prod_i u_{\mathsf{X}_i}^{\alpha_i} \tag{3.197}$$

as a unit for p.

Example 3.38 If we want to consider how a particle moves there are two basic physical dimensions which are of relevance, length L and time T. A common choice for the base units u_L and u_T is given by

$$u_\mathsf{L} = 1 \text{ m} \quad \text{and} \quad u_\mathsf{T} = 1 \text{ s}. \tag{3.198}$$

Any length ℓ can now be written as a number λ_ℓ multiplied by u_L and any time t can be written as a number λ_t multiplied by u_T. When we wish to describe velocities, which have physical dimension L/T, we would obtain $\alpha_\mathsf{L} = 1$ and $\alpha_\mathsf{T} = -1$, indicating that velocities should be measured in terms of the unit

$$u_v = \frac{u_\mathsf{L}}{u_\mathsf{T}} = \frac{1 \text{ m}}{1 \text{ s}} = 1 \, \frac{\text{m}}{\text{s}}. \tag{3.199}$$

As people who have spent a significant amount of time in both countries using *imperial units* (feet, gallons, pounds, etc.) and countries using the *metric system* (meters, kilograms, etc.) will be painfully aware of, the choice of the base units is arbitrary and different default choices have developed in different places. Using different base units will naturally result in different measured values for the same physical quantity, but there is some order to the madness. Consider a physical dimension X for which we could use the base unit u_X or the base unit u'_X. Any quantity p with the physical dimension of X can now be expressed in terms of either of these base units as

$$p = \lambda_p u_\mathsf{X} = \lambda'_p u'_\mathsf{X}, \tag{3.200}$$

where λ_p and λ'_p are the measured values in the respective units. In particular, since u_X has physical dimension X, it can be expressed as $u_\mathsf{X} = x_\mathsf{X} u'_\mathsf{X}$ and doing so we find that

$$\lambda_p x_\mathsf{X} u'_\mathsf{X} = \lambda'_p u'_\mathsf{X}, \tag{3.201}$$

implying that $\lambda'_p = \lambda_p x_\mathsf{X}$. Finding the transformation relation for the composite units, we find that the measured value of a quantity p which has physical dimensions according to Eq. (3.196) changes as

$$\lambda'_p = \lambda_p \prod_i x_{\mathsf{X}_i}^{\alpha_i}. \tag{3.202}$$

Example 3.39 When measuring lengths, we could choose to use with the unit of $u_L = 1$ m or with the unit of $u'_L = 1$ ft. In converting measured values between these two units, we would make use of the relation 1 ft = 0.3048 m, implying that

$$x_L = \frac{1 \text{ m}}{1 \text{ ft}} = \frac{1}{0.3048} \simeq 3.28. \tag{3.203}$$

An area A has physical dimension L^2 and the measured value λ_A in terms of m^2 will be related to the measured value λ'_A in terms of ft^2 as

$$\lambda'_A = x_L^2 \lambda_A \simeq 10.8 \lambda_A. \tag{3.204}$$

As a direct consequence of the above discussion, any dimensionless physical quantity, such as the ratio between two lengths, has all of the $\alpha_i = 0$ and consequently we find that if p is dimensionless, then

$$\lambda_p = \lambda'_p \prod_i x_{X_i}^0 = \lambda'_p, \tag{3.205}$$

i.e., the measured value of a dimensionless quantity does not change when we change the base units.

3.12.2 The Buckingham π theorem

Consider a situation where we have n different physical variables q_i that we know are related and that these have units built upon k different basic physical dimensions. Any physical relation connecting these variables can be written in terms of a dimensionless function

$$f(q_1, \ldots, q_n) = 0, \tag{3.206}$$

but the freedom in how this function can be constructed can be significantly reduced through means of dimensional analysis.

Example 3.40 In the case of constant motion in one dimension starting at the origin, we may want to relate the displacement from the origin x, the time t, and the velocity v. We already know that the resulting relationship can be written as $x = vt$, but let us put it on the form of Eq. (3.206). By dividing with x on both sides and subtracting one, we find that

$$0 = \frac{vt}{x} - 1 \equiv f(x, t, v), \tag{3.207}$$

where $f(x, t, v)$ is the dimensionless function describing the physical relationship between x, t, and v.

With our assumption of there being k different basic physical dimensions, we will have at most k variables that have an independent physical dimension in the sense that all other variables will have the same dimension as some combination of the first k. If this is not the case, we have a degenerate situation and could essentially select a smaller set of basic physical dimensions. For our purposes here, we will assume that there are k variables that have independent physical dimension and that we order those to be the first k of the set of variables q_i. By our assumption, any q_i with $i > k$ now has a physical dimension that is dependent on the dimension of the first k variables. Concretely, we find that

$$[q_i] = \prod_{j=1}^{k} [q_j]^{\beta_{ij}}, \tag{3.208}$$

where the β_{ij} are numbers and $i > k$. Defining the new variable

$$Q_i(q_1, \ldots, q_k) = \prod_{j=1}^{k} q_j^{\beta_{ij}}, \tag{3.209}$$

we therefore find that $[Q_i] = [q_i]$ and we can construct a dimensionless quantity

$$\pi_{i-k} = \frac{q_i}{Q_i} \tag{3.210}$$

that can be used to rewrite the physical relationship between the q_i as

$$0 = f(q_1, \ldots, q_n) = f(q_1, \ldots, q_k, \pi_1 Q_{k+1}, \ldots, \pi_{n-k} Q_n)$$
$$\equiv F(q_1, \ldots, q_k, \pi_1, \ldots, \pi_{n-k}). \tag{3.211}$$

However, by assumption, the q_i with $i \leq k$ have independent physical dimension and it is therefore not possible to construct any dimensionless product of powers of these which depends on them. As a result, the function F must be independent of these variables and we find that

$$F(\pi_1, \ldots, \pi_{n-k}) = f(q_1, \ldots, q_n) = 0, \tag{3.212}$$

where we have removed the dependence on the q_i from F. The physical relationship among all of the q_i can therefore be rewritten as a relationship between the $n - k$ dimensionless variables π_i. This result is known as the *Buckingham π theorem*, where the π refers to the dimensionless variables, and is a central result in dimensional analysis.

Example 3.41 As an application of the Buckingham π theorem, let us reconsider the problem of Example 3.37, where we wished to find the form of the relationship between the spring constant k, the mass m, and the period of oscillation T. These parameters depend on the two basic physical dimensions M and T and so any physical relationship should be possible to express in terms of $3 - 2 = 1$ dimensionless parameter π. Taking $q_1 = T$, $q_2 = m$, and $q_3 = k$, we find that

$$[q_3] = [k] = [T]^{-2}[m]^1 = [q_1]^{-2}[q_2]^1 \implies Q_3 = \frac{m}{T^2}. \tag{3.213}$$

Directly applying the definition of the dimensionless π parameter now results in

$$\pi = \frac{q_3}{Q_3} = \frac{kT^2}{m} \tag{3.214}$$

and the physical relationship between them must take the form

$$F(\pi) = 0 \implies \pi = \frac{kT^2}{m} = F^{-1}(0) \equiv C^2, \tag{3.215}$$

where F^{-1} is the inverse of the function F and C^2 is a dimensionless constant. Solving for the period of oscillation gives us

$$T = C\sqrt{\frac{m}{k}}, \tag{3.216}$$

which is the exact same result as in Example 3.37.

Example 3.42 Let us consider a situation where there will be several dimensionless combinations π_i and we know the actual result, motion with constant acceleration a and initial velocity v starting at the origin and we are looking for the dependence of the displacement x on time t. The basic physical units involved in this situation are length L and time T and the known result is of the form

$$x = vt + \frac{at^2}{2}, \tag{3.217}$$

but we wish to know how far dimensional analysis will get us. The choice of dimensionally independent quantities in this case is arbitrary so let us therefore consider two different choices, starting with x and t. For the remaining two quantities v and a, we find that the corresponding dimensionless variables are

$$\pi_1 = \frac{vt}{x} \quad \text{and} \quad \pi_2 = \frac{at^2}{x}, \tag{3.218}$$

respectively. The physical relationship between the quantities must therefore be of the form

$$F(\pi_1, \pi_2) = 0. \tag{3.219}$$

We note that this is the requirement for a level curve in the two-dimensional π_1-π_2-plane and, as long as this curve is a one-to-one relationship, we can write π_1 as a function of π_2

$$\pi_1 = g_1(\pi_2) \quad \Longrightarrow \quad \frac{vt}{x} = g_1\left(\frac{at^2}{x}\right). \tag{3.220}$$

Comparing to the known result, we find that $g_1(\pi_2) = 1 - \pi_2/2$, but there is no way of figuring this out from dimensional analysis alone. With this choice of dimensionally independent quantities, it is therefore not possible to solve for x, since it appears both on the left-hand side and in the argument of g_1 on the right-hand side.

A different choice of dimensionally independent quantities could have been v and a. In this scenario, we would instead obtain

$$\pi_1' = \frac{xa}{v^2} \quad \text{and} \quad \pi_2' = \frac{at}{v}, \tag{3.221}$$

where we have introduced the primes to distinguish these dimensionless variables from the previous ones. In the same fashion, we conclude that

$$\pi_1' = g_2(\pi_2') \quad \Longrightarrow \quad x = \frac{v^2}{a} g_2\left(\frac{at}{v}\right), \tag{3.222}$$

where it should be noted that g_2 is not the same function as the g_1 we obtained when using different dimensionally independent quantities. This is as far as we will get based on dimensional analysis alone, but comparison with the known relationship would give us $g_2(\pi_2') = \pi_2' + \pi_2'^2/2$.

3.12.3 Dimensional analysis and modelling

Whenever writing down a physical model, in terms of differential equations or otherwise, it is good practice to make sure that it is dimensionally consistent. We can always do this by ensuring that all of the terms in a given expression have the same physical dimension and that any function in the expression is taking arguments with the appropriate dimension. Throughout this chapter, we have modelled different physical situations using differential equations and integrals. It is therefore worth having a look at how these relate to dimensional analysis.

For any derivative, ordinary as well as partial, we are looking at limits of the form

$$\frac{df}{dx} = \lim_{h \to 0} \frac{f(x+h) - f(x)}{h}. \tag{3.223}$$

For the expression inside the limit to be dimensionally consistent, the dimension of h must be equal to $[x]$ for the argument of $f(x+h)$ to be meaningful. As the dimension of the expression does not change with h, it will also be the dimensions of the derivative, i.e.,

$$\left[\frac{df}{dx}\right] = \frac{[f]}{[h]} = \frac{[f]}{[x]}. \tag{3.224}$$

In a more physical approach, the derivative is essentially a small change in f, which must have units of $[f]$, divided by a small change in x, which must have units of $[x]$, thereby giving a dimension $[f]/[x]$ for the derivative df/dx.

Example 3.43 Let us deduce the physical dimension of the diffusivity D, which appears in the diffusion equation

$$u_t - D\nabla^2 u = \kappa. \tag{3.225}$$

For the left-hand side to be meaningful, both terms must have the same physical dimension and it follows that

$$[D] = \frac{[u_t]}{[\nabla^2 u]}. \tag{3.226}$$

The nominator here is the dimension of the derivative of u with respect to time t and we therefore find that $[u_t] = [\partial u / \partial t] = [u]/\mathsf{T}$ and the denominator contains second derivatives with respect to the lengths x^i, e.g., $[\partial^2 u / \partial (x^1)^2] = [u]/[x^1]^2 = [u]/\mathsf{L}^2$. Inserting this into the above equation for $[D]$, we obtain

$$[D] = \frac{[u]/\mathsf{T}}{[u]/\mathsf{L}^2} = \frac{\mathsf{L}^2}{\mathsf{T}}. \tag{3.227}$$

The diffusivity of any quantity u therefore has dimensions of length squared divided by time, regardless of the dimension of u.

A similar approach may be taken to the dimensional analysis of integrals. Considering that an integral is in essence nothing more than the limit of a sum, its dimension will be given by

$$\left[\int f(x)\,dx\right] = [f][dx] = [f][x], \tag{3.228}$$

i.e., it will be the product of the dimensions of the integrand and that of the variable being integrated over. One important thing to remember is that integrals over multiple variables will give a contribution to the dimension from every variable that is being integrated over. In particular, for a three-dimensional volume integral, $[dV] = \mathsf{L}^3$.

3.12.4   Parameters as units

In some situations when a physical problem depends on a number of unknown physical input parameters, it can be useful to use these input parameters to define the base units as much as possible. This is particularly applicable when attempting to solve such a problem numerically as the resulting problem will generally be dimensionless and depend at most on a number of dimensionless parameters. This lets us solve a well defined numerical problem without having to assume anything about the unknown physical variables and at the end the result may be rescaled with the correct physical units in order to produce a meaningful result. The procedure for doing so is best illustrated through an example.

Example 3.44 Consider the one-dimensional heat transport problem in a material

$$(\text{PDE}) : T_t - aT_{xx} = 0, \tag{3.229a}$$

$$(\text{BC}) : T(0,t) = 0, \quad T(\ell,t) = T_0, \tag{3.229b}$$

$$(\text{IC}) : T(x,0) = 0, \tag{3.229c}$$

where $T(x,t)$ is the temperature at position x at time t, ℓ is the length of the one-dimensional region we are interested in, and a is a material constant with physical dimension L^2/T. One end of the region is being held at zero temperature, the other at temperature T_0, and the entire system is initially at zero temperature. This problem contains three basic physical dimensions, length L, time T, and temperature Θ and we have three unknown physical variables T_0, a, and ℓ. Using these to construct base units of the correct physical dimensions, we find that

$$u_\mathsf{L} = \ell, \quad u_\mathsf{T} = \frac{\ell^2}{a}, \quad \text{and} \quad u_\Theta = T_0. \tag{3.230}$$

The remaining physical quantities in the problem may now be written in terms of these units as

$$x = \ell\xi, \quad t = \frac{\ell^2}{a}\tau, \quad \text{and} \quad T = T_0\theta(\xi,\tau), \tag{3.231}$$

where ξ, τ, and θ are all dimensionless quantities. The partial differential equation now takes the form

$$(\text{PDE}) : \theta_\tau - \theta_{\xi\xi} = 0, \tag{3.232a}$$

$$(\text{BC}) : \theta(0,\tau) = 0, \quad \theta(1,\tau) = 1, \tag{3.232b}$$

$$(\text{IC}) : \theta(\xi,0) = 0, \tag{3.232c}$$

which is a differential equation where the numerical values of the input parameters are known and therefore can be put directly into a computer without specifying any of the unknown parameters T_0, a, and ℓ. The resulting numerical solution for $\theta(\xi,\tau)$ is illustrated in Fig. 3.16. Note how we in this figure have selected to use the original quantities as axis labels rather than introducing the dimensionless quantities θ, ξ, and τ, e.g., we have used T/T_0 instead of θ.

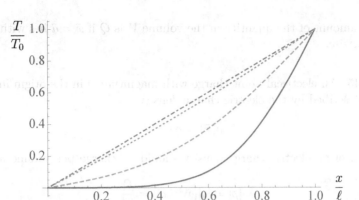

Figure 3.16   The evolution of the temperature in Example 3.44 based on a numerical solution to the resulting dimensionless problem. The solution is shown for $at/\ell^2 = 0.03, 0.1, 0.3$, and 1, respectively (lower to upper curve).

3.13   MODELLING WITH DELTA FUNCTIONS

We have already run into the delta function and the three-dimensional *delta function*. Let us examine them a bit closer and figure out some of their important properties and how they may be used to model densities that may be considered very thin, i.e., points, lines, and surfaces. The defining property of the one-dimensional delta function is

$$\int \delta(x - x_0)f(x)dx = f(x_0) \tag{3.233}$$

as long as x_0 is within the integration boundaries. For any expression of this type, we can examine its physical dimension in order to deduce the physical dimension of the delta function by looking at the dimensions of the integral

$$[\delta(x - x_0)][f][dx] = [f] \quad \Longrightarrow \quad [\delta(x - x_0)] = \frac{1}{[x]}. \tag{3.234}$$

In other words, the one-dimensional delta function has a physical dimension that is the reciprocal of the physical dimension of its argument. This is important to remember when using delta functions in modelling, as the physical dimension of the delta function will be needed to do the proper dimensional consistency checks. In a similar fashion, we can consider the N-dimensional delta function with the defining relation

$$\int \delta^{(N)}(\vec{x} - \vec{a})f(\vec{x})d^N x = f(\vec{a}) \tag{3.235}$$

as long as $\vec{a}$ is within the integration region. As a result, we find that $[\delta^{(N)}] = 1/[x]^N$.

In N dimensions, we now know that $\delta^{(N)}(\vec{x})$ has the physical dimension of the reciprocal of the volume. If we multiply it by an extensive physical quantity Q, such as a charge or a mass, we will have a quantity that has the same dimension as a concentration of that physical quantity. Indeed, the resulting function $Q\delta^{(N)}(\vec{x} - \vec{a})$ will be describing the density if we gather an amount of the extensive quantity at the point $\vec{a}$. This is also in accordance with the fact that

$$\int_V Q\delta^{(N)}(\vec{x} - \vec{a})\, dV = \begin{cases} Q, & \vec{a} \in V \\ 0, & \text{otherwise} \end{cases}, \tag{3.236}$$

i.e., the total amount of the quantity in the volume V is Q if $\vec{x} = \vec{a}$ lies within the volume and zero otherwise.

Example 3.45 An electrical point charge with magnitude q in the origin in three dimensions can be described by the electric charge density

$$\rho(\vec{x}) = q\delta^{(3)}(\vec{x}). \tag{3.237}$$

The dimension of an electric charge density should be charge per volume and indeed we find that

$$[\rho] = [q][\delta^{(3)}(\vec{x})] = \frac{Q}{L^3}. \tag{3.238}$$

In what follows, we will specialise to the case where the extensive quantity is a charge, but it should be kept in mind that the general approach is valid regardless of the type of extensive quantity we are dealing with. In other words, we could just as well be talking about a mass or another extensive quantity gathered at a single point, line, or surface.

3.13.1 Coordinate transformations

Using Cartesian coordinates x^i, we can express the N-dimensional delta function $\delta^{(N)}$ in terms of one-dimensional delta functions of the coordinates as

$$\delta^{(N)}(\vec{x} - \vec{a}) = \prod_{i=1}^{N} \delta(x^i - a^i). \tag{3.239}$$

We can verify that this gives the proper integration property by integrating this along with a function $f(\vec{x}) = f(x^1, \ldots, x^N)$ to obtain

$$\int f(x^1, \ldots, x^N) \prod_{i=1}^{N} \delta(x^i - a^i) dx^i = f(a^1, \ldots, a^N) = f(\vec{a}), \tag{3.240}$$

where we have used the defining property of the one-dimensional delta functions and assumed that $\vec{x} = \vec{a}$ is inside the integration region. This is nothing else than the defining property of $\delta^{(N)}(\vec{x} - \vec{a})$.

In some situations, using Cartesian coordinates may not be the best option and we may then ask ourselves the question of how the N-dimensional delta function can be expressed in terms of one-dimensional delta functions of general coordinates. Since it should be zero everywhere but at the point $\vec{a}$, we can make the ansatz

$$\delta^{(N)}(\vec{x} - \vec{a}) = D(\vec{a}) \prod_{b=1}^{N} \delta(y^b - a^b), \tag{3.241}$$

where a^b are the general coordinates of the point for which $\vec{x} = \vec{a}$. The function $D(\vec{a})$ that precedes the product of one-dimensional delta functions is necessary as the coordinates may not have the appropriate physical dimensions and so this must somehow be ensured.

Example 3.46 In the case of polar coordinates in a two-dimensional space, we find that

$$[\delta^{(2)}] = \frac{1}{L^2} = [D][\delta(r - a^r)][\delta(\phi - a^\phi)] = \frac{[D]}{L}, \tag{3.242}$$

since the angle ϕ is dimensionless $[\phi] = 1$. It follows that the function D must satisfy $[D] = 1/L$.

Writing down the defining relation of the N-dimensional delta function and expressing it in the curvilinear coordinates, we find that

$$f(\vec{a}) = \int f(y^1, \ldots, y^N) D(\vec{a}) \left(\prod_{b=1}^{N} \delta(y^b - a^b) \right) \mathcal{J} \, dy^1 \ldots dy^N = f(\vec{a}) D(\vec{a}) \mathcal{J}(\vec{a}), \tag{3.243}$$

where $\mathcal{J}$ is the Jacobian determinant. It follows that $D(\vec{a}) = 1/\mathcal{J}(\vec{a})$ and therefore

$$\delta^{(N)}(\vec{x} - \vec{a}) = \frac{1}{\mathcal{J}(\vec{a})} \prod_{b=1}^{N} \delta(y^b - a^b). \tag{3.244a}$$

In particular, in orthogonal coordinates the Jacobian determinant may be expressed as the product of the scale factors and we find that

$$\delta^{(N)}(\vec{x} - \vec{a}) = \prod_{b=1}^{N} \frac{\delta(y^b - a^b)}{h_b}. \tag{3.244b}$$

Example 3.47 In spherical coordinates in three dimensions, the scale factors are given by $h_r = 1$, $h_\theta = r$, and $h_\varphi = r \sin(\theta)$. As a consequence, the three-dimensional delta function at the point $\vec{x}_0$ with coordinates $r = r_0$, $\theta = \theta_0$, and $\varphi = \varphi_0$ will be given by

$$\delta^{(3)}(\vec{x} - \vec{x}_0) = \frac{1}{r_0^2 \sin(\theta_0)} \delta(r - r_0) \delta(\theta - \theta_0) \delta(\varphi - \varphi_0). \tag{3.245}$$

Looking at the dimensions of this expression, they are also consistent with $[\delta^{(3)}] = 1/L^3$ as the pre-factor has dimensions of $1/L^2$, the radial delta function $1/L$, and the angular delta functions are dimensionless. Note that it does not matter whether we use r_0 and θ_0 or r and θ in the factor in front of the delta functions as the delta functions are non-zero only when $r = r_0$ and $\theta = \theta_0$.

3.13.2   Lines and surfaces

As we have just seen, the N-dimensional delta function can be used to describe densities of point charges. However, point charges are not the only type of charges where the density becomes singular due to having a finite amount of charge within a region with vanishing volume. In three-dimensional spaces, we can also have an extensive quantity distributed along a line or on a surface, which both have zero volume. Let us therefore demonstrate how these can be described using delta functions.

In the case of a point charge in three dimensions, we found that the density was described by three delta functions, each corresponding to one of the three conditions necessary to specify a single point. In the case of a line or a surface, they will instead be specified using two and one conditions, respectively. The first order of business is therefore to write the density of the line charge and surface charge as

$$\rho(\vec{x}) = \rho_\ell(\vec{x})f_1(\vec{x})\delta(g_1(\vec{x}))\delta(g_2(\vec{x})) \quad \text{and} \quad \rho(\vec{x}) = \sigma(\vec{x})f_2(\vec{x})\delta(h(\vec{x})), \tag{3.246}$$

respectively, where $\rho_\ell(\vec{x})$ is the linear density of the line charge, $\sigma(\vec{x})$ the surface density of the surface charge, $f_i(\vec{x})$ are normalising functions, $g_i(\vec{x}) = 0$ are the conditions for the line, and $h(\vec{x}) = 0$ is the condition for the surface. We can now normalise these densities properly by making sure that whenever we integrate over a volume V containing a given part of the line Γ or part of the area S, the resulting total charge is given by

$$Q = \int_\Gamma \rho_\ell(\vec{x})d\ell \quad \text{and} \quad Q = \int_S \sigma(\vec{x})dS, \tag{3.247}$$

respectively. By definition of the density, this charge should also be given by

$$Q = \int_V \rho(\vec{x})dV. \tag{3.248}$$

Using the delta functions to remove the superfluous integrals, we can find the normalising functions $f_i(\vec{x})$.

Example 3.48 Consider a surface charge density on the sphere of radius R given by $\sigma(\theta, \varphi)$. The defining condition for the sphere is $h(\vec{x}) = r - R = 0$ in spherical coordinates and we therefore make the ansatz

$$\rho(\vec{x}) = \sigma(\theta, \varphi)f_2(\vec{x})\delta(r - R). \tag{3.249}$$

The resulting charge within a volume V will then be given by

$$Q = \int_V \sigma(\theta, \varphi)f_2(\vec{x})\delta(r - R)r^2 \sin(\theta)dr\, d\theta\, d\varphi. \tag{3.250}$$

Using the delta function to perform the integral over r, we find that

$$Q = \int_S \sigma(\theta, \varphi)f_2(\vec{x})R^2 \sin(\theta)d\theta\, d\varphi, \tag{3.251}$$

where S is the surface on the sphere $r = R$ that is contained in the volume V. The surface element on the sphere is given by $dS = R^2 \sin(\theta)d\theta\, d\varphi$ and we end up with

$$Q = \int_S \sigma(\theta, \varphi)f_2(\vec{x})dS = \int_S \sigma(\theta, \varphi)dS. \tag{3.252}$$

Since this should hold for any volume V, we conclude that $f_2(\vec{x}) = 1$ in this case. The density of the surface charge is therefore described by

$$\rho(\vec{x}) = \sigma(\theta, \varphi)\delta(r - R). \tag{3.253}$$

Again, the physical dimensions of this expression are consistent. A surface charge density should have physical dimension Q/L^2 and a charge density Q/L^3. Since the one-dimensional delta function $\delta(r - R)$ has the physical dimension of $1/[r] = 1/L$, the physical dimensions of both sides of the expression match.

Example 3.49 A line charge along the line $\theta = \theta_0$ and $\varphi = \varphi_0$ in spherical coordinates has the linear charge density $\rho_\ell(r)$. The constraining functions are given by $g_1(\theta) = \theta - \theta_0$ and $g_2(\varphi) = \varphi - \varphi_0$, respectively, and we make the ansatz

$$\rho(\vec{x}) = \rho_\ell(r) f_1(\vec{x}) \delta(\theta - \theta_0) \delta(\varphi - \varphi_0). \qquad (3.254)$$

Integrating over a spherical shell with $R < r < R + \varepsilon$, where ε is a small positive number, we find that the total charge within this shell is given by

$$Q = \int_{r=R}^{R+\varepsilon} \int_{\theta=0}^{\pi} \int_{\varphi=0}^{2\pi} \rho_\ell(r) f_1(r, \theta, \varphi) \delta(\theta - \theta_0) \delta(\varphi - \varphi_0) r^2 \sin(\theta) dr\, d\theta\, d\varphi$$

$$= \int_{R}^{R+\varepsilon} \rho_\ell(r) f_1(r, \theta_0, \varphi_0) r^2 \sin(\theta_0) dr \simeq \rho_\ell(R) f_1(R, \theta_0, \varphi_0) R^2 \sin(\theta_0) \varepsilon. \qquad (3.255a)$$

On the other hand, the charge should also be given by

$$Q = \int_{R}^{R+\varepsilon} \rho_\ell(r) dr = \rho_\ell(R) \varepsilon. \qquad (3.255b)$$

As a result, we can identify the normalisation

$$f_1(r, \theta, \varphi) = \frac{1}{r^2 \sin(\theta)}. \qquad (3.256a)$$

and therefore the line charge is described by the charge density

$$\rho(\vec{x}) = \frac{\rho_\ell(r)}{r^2 \sin(\theta)} \delta(\theta - \theta_0) \delta(\varphi - \varphi_0). \qquad (3.256b)$$

Again, we can easily verify that this charge density has dimensions of Q/L^3.

The above examples are special cases of the situation when the described line or surface is a coordinate line or surface in orthogonal coordinates. For an orthogonal coordinate system, let us look at a surface described by $y^1 = a^1$, but remember that the argumentation will be the same regardless of which coordinate we place the constraint on. A surface charge on this surface will be described by the charge density

$$\rho = \sigma f_2 \delta(y^1 - a^1). \qquad (3.257)$$

Integrating over a volume V, expressing it in the orthogonal coordinate system, and using the delta function to perform the y^1 integral results in

$$Q = \int_V \rho\, dV = \int_V \sigma f_2 \delta(y^1 - a^1) h_1 h_2 h_3 dy^1 dy^2 dy^3 = \int_{y^1 = a^1} \sigma f_2 h_1 h_2 h_3 dy^2 dy^3, \qquad (3.258)$$

where the surface integral over $y^1 = a^1$ is taken over the part of the surface contained in V. This charge is also given by

$$Q = \int_{y^1 = a^1} \sigma\, dS = \int_{y^1 = a^1} \sigma \left| \frac{\partial \vec{x}}{\partial y^2} \times \frac{\partial \vec{x}}{\partial y^3} \right| dy^2 dy^3, \qquad (3.259)$$

where we also know that

$$\left| \frac{\partial \vec{x}}{\partial y^2} \times \frac{\partial \vec{x}}{\partial y^3} \right| = h_2 h_3 \left| \vec{e}_2 \times \vec{e}_3 \right| = h_2 h_3. \qquad (3.260)$$

We therefore conclude that

$$\int_{y^1 = a^1} \sigma f_2 h_1 h_2 h_3 dy^2 dy^3 = \int_{y^1 = a^1} \sigma h_2 h_3 dy^2 dy^3 \qquad (3.261)$$

and therefore

$$f_2 = \frac{1}{h_1} \implies \rho(\vec{x}) = \sigma \frac{\delta(y^1 - a^1)}{h_1}. \qquad (3.262)$$

This is also in agreement with the dimensionality of the charge density. In particular, we find that the dimension of h_1 is given by

$$[h_1] = \left[\frac{\partial \vec{x}}{\partial y^1} \right] = \frac{\mathsf{L}}{[y^1]}. \qquad (3.263)$$

This implies that

$$[\rho] = \frac{[\sigma][\delta(y^1 - a^1)]}{[h_1]} = \frac{\mathsf{Q}}{\mathsf{L}^2} \frac{[y_1]}{[y_1]\mathsf{L}} = \frac{\mathsf{Q}}{\mathsf{L}^3}, \qquad (3.264)$$

as expected. Similar argumentation may be applied to a line charge located at a coordinate line, see Problem 3.44.

3.14   PROBLEMS

Problem 3.1. Identify which of the following quantities are extensive and which are intensive properties:

a) Entropy

b) Tension

c) Chemical potential

d) Melting point

e) Angular momentum

f) Kinetic energy

Problem 3.2. Consider a substance that is produced with a source density

$$\kappa(\vec{x}, t) = \kappa_0 \exp\left(-\frac{\rho^2}{r_0^2} \right) \exp\left(-\frac{t}{\tau} \right) \qquad (3.265)$$

within a two-dimensional space, where ρ is the radial polar coordinate and r_0, τ, and κ_0 are constants with the appropriate physical dimensions. Find an expression for the production per unit time within the sub-region $\rho \leq R$ and the total amount produced in the entire space from $t = 0$ to $t \to \infty$.

Problem 3.3. The concentration u of a substance at a given time t_0 is described by

$$u(\vec{x}, t_0) = u_0 \cos\left(\frac{\pi x^1}{L}\right) \sin^2\left(\frac{\pi x^2}{L}\right) \cos^2\left(\frac{\pi x^3}{L}\right), \tag{3.266}$$

where u_0 is a constant. Assume that the current is given by Fick's law with a known diffusivity D and compute the total flux of the substance out of the cube $0 < x^i < L$ at $t = t_0$.

Problem 3.4. Just as for the string in Section 3.5.1, the wave equation for transversal motion of the membrane derived in Section 3.5.2 may also be derived from the continuity equation for the momentum in the x^3 direction. The corresponding momentum current is

$$\vec{j} = -\sigma \nabla u. \tag{3.267}$$

Verify that the wave equation follows from the continuity equation with this current.

Problem 3.5. In the diffusion of a substance that has a different density than the medium in which it is diffusing, the total current will be the sum of the usual diffusive current in Fick's law and a current due to gravitation

$$\vec{j}_g = k(\rho - \rho_0)u\vec{g}, \tag{3.268}$$

where k is a constant, ρ and ρ_0 are the densities of the substance and the medium, respectively, u is the concentration and $\vec{g}$ the gravitational field. Use the continuity equation for the substance to derive a partial differential equation for how the concentration u depends on time and space. Also find the boundary conditions that must be fulfilled in order for no substance to leave or enter the volume V.

Problem 3.6. Consider the homogeneous Robin boundary condition

$$\alpha \vec{n} \cdot \nabla u + \beta u = 0 \tag{3.269}$$

on the boundary surface of some region. Verify that the Dirichlet and Neumann type boundary conditions are special cases and discuss the limits in which Newton's law of cooling, see Example 3.15, turns into a Dirichlet and Neumann boundary condition, respectively. What is the physical interpretation?

Problem 3.7. In Section 3.5.3, it was shown that the magnetic field $\vec{B}$ fulfils the sourced wave equation with the wave speed $c = 1/\sqrt{\varepsilon_0 \mu_0}$. Starting from Maxwell's equations (see Eqs. (2.198)), show that this is also true for the electric field $\vec{E}$.

Problem 3.8. When studying electric fields in matter, it is often convenient to work in terms of the *electric displacement field* $\vec{D} = \varepsilon_0 \vec{E} + \vec{P}$, where $\vec{P}$ is the polarisation density. For weak external fields, the polarisation density can be assumed to depend linearly on the electric field and therefore

$$P^i = \varepsilon_0 \chi^i_j E^j, \tag{3.270}$$

where χ is the rank two electric susceptibility tensor, see also Problem 2.4. From Maxwell's equations, find the divergence of $\vec{D}$. In particular, simplify your result for the case $\chi^i_j = \delta_{ij}$.
Note: The divergence $\nabla \cdot \vec{D}$ is often written on the form $\nabla \cdot \vec{D} = \rho_f$, where ρ_f is the so-called *free charge*.

Problem 3.9. A substance is dissolved in a fluid and well mixed in such a way that its concentration u is constant throughout the fluid. Verify that concentration will not change due to convective currents if the fluid flow is incompressible.

Problem 3.10. Write down the diffusion equation in terms of a Cartesian coordinate system for an anisotropic medium where the diffusivity tensor is given by

$$D_{ij} = D_0\delta_{ij} + D_1 n_i n_j, \tag{3.271}$$

$\vec{n}$ is a constant unit vector, and D_0 and D_1 are constants. *Hint:* You can choose the coordinate system in such a way that $\vec{n}$ is one of the basis vectors.

Problem 3.11. A one-dimensional string with tension S is constructed from two parts with different densities ρ_- and ρ_+, respectively. The point where the two parts meet may be assigned the position $x = 0$ and at this position a point mass m has been attached. Find a partial differential equation for the transversal displacement of the string and specify the conditions the solution must fulfil at $x = 0$. The string may be considered infinitely extended in both directions.

Problem 3.12. Show that the one-dimensional wave equation

$$(\text{PDE}): \ u_{tt}(x,t) - c^2 u_{xx}(x,t) = 0 \tag{3.272a}$$

is solved by the ansatz $u(x,t) = f(x - ct) + g(x + ct)$, where f and g are functions of one variable. Use this ansatz to solve the wave equation on the real line $-\infty < x < \infty$ with the initial conditions

$$(\text{IC}): \ u(x,0) = u_0(x), \quad u_t(x,0) = 0. \tag{3.272b}$$

Problem 3.13. A radioactive material with a mean lifetime of τ is diffusing in a medium. There is a sink density due to the decay given by $\kappa(\vec{x},t) = -u(\vec{x},t)/\tau$, where $u(\vec{x},t)$ is the concentration of the material. This results in the partial differential equation

$$(\text{PDE}): \ u_t - D\nabla^2 u = -\frac{u}{\tau}. \tag{3.273}$$

Show that this can be rewritten as a source free diffusion equation by the substitution $u(\vec{x},t) = v(\vec{x},t)e^{-t/\tau}$.

Problem 3.14. The temperature $T(x,t)$ in an isolated rod of length ℓ that conducts heat with heat conductivity λ can be described by the one-dimensional heat equation

$$(\text{PDE}): T_t - aT_{xx} = \kappa(x), \tag{3.274a}$$

$$(\text{BC}): T(0,t) = T(\ell,t) = T_0, \tag{3.274b}$$

where a is a material constant related to heat capacity, heat conductivity, and density, $\kappa(x)$ is a constant source term, and we have assumed that the rod's ends are being held at a constant temperature T_0. Express the stationary solution to the problem as an integral and compute it for the case of $\kappa(x) = \kappa_0\delta(x - x_0)$.

Problem 3.15. Consider a volatile substance dissolved in a medium where it diffuses with diffusivity D. The medium is being kept in a cylindrical glass that is impenetrable to the volatile substance, but has an exposed top surface, where the evaporation rate is proportional to the concentration of the substance. Write down a model, including boundary and initial conditions, that describes the concentration of the substance in the medium as a function of time. The substance may be assumed to be evenly distributed in the medium at the initial time.

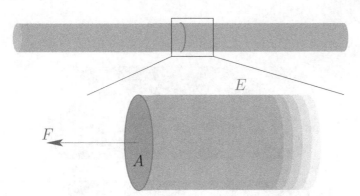

Figure 3.17 The force across a section of a long metal rod depends on the cross sectional area A, the elasticity module E of the metal, and the strain of the rod at the position of the section.

Problem 3.16. A substance is produced within a volume V with a source density that is proportional to the square of its concentration. Furthermore, the substance is assumed to diffuse with a constant diffusivity throughout V and to evaporate quickly on the surface ∂V. Construct a partial differential equation with appropriate boundary conditions that describe how the concentration $u(\vec{x}, t)$ of the substance depends on time and space. Furthermore, assume that there exists a stationary solution $u_0(\vec{x})$ to the problem and linearise the time-dependent partial differential equation around this solution.

Problem 3.17. Consider the situation where heat is conducted away from a sphere of radius R, i.e., look at the heat conduction in the region $r > R$. Assuming that the sphere uniformly produces a total power of P and that the medium conducting the heat is isotropic and homogeneous, find a differential equation describing the spherically symmetric stationary state of the temperature in the surrounding medium. How does the stationary solution depend on the distance r from the center of the sphere?

Problem 3.18. When the wave speed c and the damping factor k are large in the damped wave equation, show that it can be regarded as a diffusion equation. Also discuss what it means for c and k to be large and express the corresponding diffusivity.

Problem 3.19. In a long metal rod with Young's modulus E, the force across a section perpendicular to the rod is given by

$$F = \sigma A = \varepsilon E A \tag{3.275}$$

in the direction of the surface element, where σ is the stress in the longitudinal direction, ε is the strain, and A the cross-sectional area, see Fig. 3.17. Find a model describing the longitudinal displacement of the rod as a function of the position on the rod and time.

Problem 3.20. Consider the derivation of the wave equation for a pressure wave in a cylinder. Using an external pressure of p_0, find the boundary condition for the pressure at one of the cylinder ends if the end is open. When the end is closed by a rigid surface, use the fact that the velocity of the medium at the end must be zero in the normal direction and derive the corresponding boundary condition on the pressure field, see Fig. 3.18.

Problem 3.21. A string's end at coordinate $x = x_0$ is attached to two springs, each with spring constant k, see Fig. 3.19, such that they extend whenever the string deviates

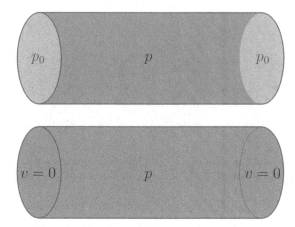

Figure 3.18 The boundary conditions on the pressure p at the ends of a cylinder depend on whether the cylinder is open or closed. If it is open the pressure will adapt to the external pressure p_0 and if it is closed the velocity v of the contained fluid at the boundary must be equal to zero.

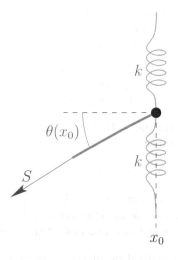

Figure 3.19 The end of a string at $x = x_0$ is attached to two springs with spring constant k that act against transversal deviations from the equilibrium position. The tension in the string is assumed to be S.

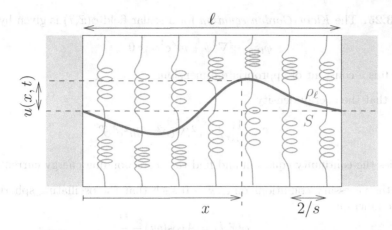

Figure 3.20 A string with length density ρ_ℓ and tension S subjected to a restoring force due to a large number of small springs, each with spring constant k and with s springs per unit length.

transversally from its equilibrium position. Find the boundary condition that the transversal displacement of the string must satisfy at this end if the tension in the string is S.

Problem 3.22. In a quadratic area described by the coordinates $0 < x, y < \ell$, heat is being introduced evenly through the boundary at $y = 0$ at a rate Q while the boundary at $y = \ell$ is heat isolated. At the boundaries in the x direction heat is being removed with a constant current density component in the normal direction. The stationary state for this situation may be described by the partial differential equation

$$\text{(PDE)} : \nabla^2 T(x,y) = -\frac{\kappa(x,y)}{\lambda}, \tag{3.276a}$$

$$\text{(BC)} : T_y(x,0) = -\frac{Q}{\ell\lambda}, \quad T_y(x,\ell) = 0, \quad T_x(0,y) = -T_x(\ell,y) = \frac{j_0}{\lambda}, \tag{3.276b}$$

where $\kappa(x,y)$ is a term describing the time-independent production of heat and λ is the heat conductivity. Find a consistency condition for the constant j_0 in the following cases:

a) There is no heat produced inside the area, $\kappa(x,y) = 0$.

b) The heat production is given by $\kappa(x,y) = \kappa_0 \sin(\pi x/\ell) \sin(\pi y/\ell)$.

Problem 3.23. A two-dimensional membrane with surface tension σ is being suspended in a rigid frame that is described by the curve

$$\rho = R, \quad z = z_0 \cos(3\phi) \tag{3.277}$$

in cylinder coordinates with $z_0 \ll R$. Find a partial differential equation along with appropriate boundary conditions that describes the stationary shape of the membrane.

Problem 3.24. A string with linear density ρ_ℓ is suspended between two walls a distance ℓ apart with tension S. Along the string, a number of small springs with spring constant k are attached to the string in such a way that they are stretched when the string has a transversal displacement, see Fig. 3.20. Assuming that there are s springs per unit length and that s is large enough for the resulting force density to be considered as a continuum, determine the partial differential equation which describes small transversal oscillations of the string and specify the boundary conditions. You do not need to specify the initial conditions.

Problem 3.25. The *Klein–Gordon equation* for a scalar field $\phi(\vec{x}, t)$ is given by

$$\phi_{tt} - c^2 \nabla^2 \phi + m^2 c^4 \phi = 0, \tag{3.278}$$

where $m > 0$ is a constant of appropriate dimensions.

a) Show that the energy density

$$\mathcal{E} = \frac{1}{2}[\phi_t^2 + c^2(\nabla\phi)^2 + m^2 c^4 \phi^2] \tag{3.279}$$

satisfies the continuity equation and find the corresponding energy current $\vec{j}$.

b) Find the necessary conditions on $\omega, k > 0$ such that the oscillating spherically symmetric function

$$\phi(\vec{x}, t) = A\cos(\omega t)\frac{e^{-kr}}{r} \tag{3.280}$$

is a solution to the Klein–Gordon equation with $r > 0$ being the radial spherical coordinate. Can you use this solution to find a stationary state as $\omega \to 0$?

c) Instead of the exponentially decaying solution in (b), consider the spherical wave solution

$$\phi(\vec{x}, t) = A\cos(\omega t)\frac{\cos(kr)}{r} \tag{3.281}$$

and find the corresponding relation between ω and k. In particular, find the frequency ω corresponding to $k \to 0$.

Problem 3.26. Verify that the differential operator

$$\hat{L} = \sum_{|\alpha| \leq m} a_\alpha \partial_\alpha, \tag{3.282}$$

where α is a multi-index is a linear differential operator, i.e., show that $\hat{L}(k_1 f_1 + k_2 f_2) = k_1 \hat{L} f_1 + k_2 \hat{L} f_2$, where the k_i are constants and the f_i are functions.

Problem 3.27. Identify the non-zero coefficients a_α in Eq. (3.282) when $\hat{L}$ is the Laplace operator in Cartesian coordinates. Repeat the same exercise for the diffusion operator in Cartesian coordinates.

Problem 3.28. Imagine a one-dimensional situation where the velocity of a convective flow is given by

$$v = k(u_0 - u), \tag{3.283}$$

k and u_0 are constants, and $u = u(x, t)$ is the concentration, i.e., the velocity of the flow is decreasing as the concentration increases.

a) Find the resulting partial differential equation and show that any constant concentration $u(x, t) = \tilde{u}_0$ gives a particular solution.

b) Linearise the differential equation around the constant concentration solution found in (a) by making the ansatz $u(x, t) = \tilde{u}(x, t) + \tilde{u}_0$ and ignoring terms of order two or higher in $\tilde{u}$.

c) Show that the resulting linearised equation can be solved by the ansatz $\tilde{u}(x, t) = f(x - ct)$, as long as c is chosen appropriately. Find an expression for c and interpret your result physically.

Problem 3.29. The temperature $T(x,t)$ in an isolated rod of length ℓ follows the one-dimensional heat equation

$$(\text{PDE}): \quad T_t - aT_{xx} = \kappa(x,t), \tag{3.284a}$$

where $\kappa(x,t)$ represents a source term. If the end at $x = 0$ is kept at temperature T_0 and a constant flow of heat is introduced into the rod through the boundary at $x = \ell$, the boundary conditions will be of the form

$$(\text{BC}): \quad T(0,t) = T_0, \quad T_x(\ell,t) = q_0. \tag{3.284b}$$

We also assume that the entire rod takes the temperature T_0 at time $t = 0$, leading to the initial condition

$$(\text{IC}): \quad T(x,0) = T_0. \tag{3.284c}$$

Rewrite this problem as two separate problems in the same region where one of the problems only has an inhomogeneity in the differential equation and the other only has an inhomogeneity in one of the boundary conditions.

Problem 3.30. A circular membrane with surface tension σ is attached to a ring of radius R. A sinusoidal motion is imposed on the ring, leading to the transversal displacement $u(\rho, \phi, t)$ of the membrane satisfying the wave equation with inhomogeneous boundary conditions

$$(\text{PDE}): u_{tt} - c^2 \nabla^2 u = 0, \tag{3.285a}$$

$$(\text{BC}): u(R, \phi, t) = A\sin(\omega t), \tag{3.285b}$$

where ρ and ϕ are the polar coordinates on the membrane, A is the amplitude of the sinusoidal motion, and ω the angular frequency. Transfer the inhomogeneity in the boundary condition to the differential equation for a new function $v(\rho, \phi, t)$ by making the ansatz $u(\rho, \phi, t) = v(\rho, \phi, t) + u_0(\rho, \phi, t)$ with an appropriate choice of $u_0(\rho, \phi, t)$. What is the resulting inhomogeneity in the differential equation for v?

Problem 3.31. Consider the functional

$$F[u] = \frac{1}{2} \int_V u^2 dV. \tag{3.286}$$

Show that the diffusion problem with Neumann boundary conditions

$$(\text{PDE}): u_t - D\nabla^2 u = \kappa(\vec{x},t), \qquad (\vec{x} \in V) \tag{3.287a}$$

$$(\text{BC}): \vec{n} \cdot \nabla u = \phi(\vec{x},t), \qquad (\vec{x} \in S) \tag{3.287b}$$

$$(\text{IC}): u(\vec{x},t_0) = u_0(\vec{x}), \qquad (\vec{x} \in V) \tag{3.287c}$$

where S is the boundary of the volume of interest V, has a unique solution by assuming that there exist two solutions u and v and constraining $dF[u-v]/dt$.

Problem 3.32. In Example 3.31 we considered a flow in the direction of the symmetry axis of a cylinder of radius R.

a) Compute the total flow of the fluid through the cylinder by integrating the flux over the cylinder cross section.

b) Compute the strain rate tensor ε_{ij} and use it to deduce the viscous force acting on a small volume. Show that it exactly balances the force due to the pressure gradient.

Problem 3.33. For an inviscid flow that satisfies the Euler equations, show that the total momentum within a volume V

$$\vec{P} = \int_V \rho \vec{v} \, dV \tag{3.288}$$

is conserved if $\vec{v} \cdot \vec{n} = 0$ and $p = p_0$ is constant on the boundary of V and there is no external force per unit mass acting on the fluid, i.e., $\vec{g} = 0$ in Eq. (3.139).

Problem 3.34. Consider a situation where the viscous forces dominate the Navier–Stokes momentum equations, i.e., when the terms proportional to ρ may be neglected. Write down the Navier–Stokes momentum equation in this situation and simplify it as far as possible if the fluid is also incompressible.

Problem 3.35. Assuming that the flow of water out of a garden hose may be considered inviscid. Apply Bernoulli's principle to find an expression for the velocity at which the water exits the nozzle. You can assume that the hose has an inner diameter d, that the flow rate of water (volume per second) is Γ, the water has a density ρ, and that the pressure difference between the hose and the atmosphere is δp. Estimate the numerical value for the velocity given reasonable assumptions on the given parameters. You may assume that there is no change in the gravitational potential along the flow.

Problem 3.36. Consider the three-dimensional heat equation in a thin region $0 < x^3 < h$ with inhomogeneous Neumann boundary conditions

$$(\text{BC}): \quad \partial_3 T(x^1, x^2, 0, t) = -f(x^1, x^2), \quad \partial_3 T(x^1, x^2, h, t) = g(x^1, x^2). \tag{3.289}$$

Use the fact that the region can be considered thin to find a partial differential equation in two spatial dimensions describing the averaged temperature

$$\tilde{T}(x^1, x^2, t) = \frac{1}{h} \int_0^h T(x^1, x^2, x^3, t) dx^3. \tag{3.290}$$

Problem 3.37. For a thin homogeneous spherical shell of radius R and thickness $r_0 \ll R$, construct a two-dimensional partial differential equation describing how the temperature in the shell depends on position and time. Assume that Newton's law of cooling is satisfied at the shell's surfaces and that temperature outside the shell is T_0 and the temperature inside it is T_1.

Problem 3.38. A small ball of radius R is made from a material with a large heat diffusion coefficient a. The ball is placed in a medium with temperature T_0 and Newton's law of cooling

$$\vec{n} \cdot \nabla T + \alpha(T - T_0) = 0, \tag{3.291}$$

where $\vec{n}$ is the surface normal, is assumed to hold at the boundary. Show that if a is large enough for the temperature to be effectively constant within the ball, then the average temperature $\tilde{T}(t)$ of the ball will follow the ordinary differential equation

$$\frac{d\tilde{T}}{dt} = \beta(T_0 - \tilde{T}), \tag{3.292}$$

where β is a constant. Express β in terms of the constant α in Newton's law of cooling.

Problem 3.39. The rear window of a car often comes equipped with a defrosting system consisting of a number of horizontal heating wires in the glass. Assuming that the window has a rectangular shape and that the outside temperature is T_0, construct a two-dimensional model in terms of a partial differential equation for how the temperature in the window depends on the position and time. You may assume that Newton's law of cooling is satisfied at the window's surfaces and that the window is heat-isolated at the edges. The heat production of the wires is q per length and time and the wires otherwise do not affect the heat diffusion inside the window.

Problem 3.40. In a long cylinder of radius R, a physical quantity is assumed to satisfy the wave equation with homogeneous Dirichlet boundary conditions

$$\text{(PDE)} : u_{tt} - c^2 \nabla^2 u = 0, \qquad\qquad (\rho < R) \qquad\qquad (3.293a)$$
$$\text{(BC)} : u(\vec{x}, t) = 0, \qquad\qquad (\rho = R) \qquad\qquad (3.293b)$$

where ρ is the radial cylinder coordinate. For particular solutions on the form $u(\vec{x}, t) = f(\vec{x}) \cos(\omega t)$, find a partial differential equation, including boundary conditions, that $f(\vec{x})$ must satisfy.

Problem 3.41. Consider an isotropic material such that the stiffness tensor is given by

$$c_{ijk\ell} = \lambda \delta_{ij}\delta_{k\ell} + \mu(\delta_{ik}\delta_{j\ell} + \delta_{i\ell}\delta_{jk}) \qquad\qquad (3.294)$$

as discussed in relation to Eq. (2.193).

a) Using Hooke's law $\sigma_{ij} = c_{ijk\ell}\epsilon_{k\ell}$, where $\epsilon_{k\ell} = (\partial_k u_\ell + \partial_\ell u_k)/2$ is the linear strain and $\vec{u}(\vec{x}, t)$ is the displacement field, find an expression for the force $d\vec{F}$ on a small volume dV in terms of λ, μ, and $\vec{u}$ due to external forces on its surface.

b) Newton's second law for the small volume takes the form

$$d\vec{F} = \rho \partial_t^2 \vec{u} \, dV. \qquad\qquad (3.295)$$

Apply your result from (a) to show that the volumetric strain $\delta = \epsilon_{kk}$ satisfies a wave equation of the form

$$\partial_t^2 \delta - c^2 \nabla^2 \delta = 0 \qquad\qquad (3.296)$$

and find an expression for the wave velocity c.

Problem 3.42. Consider the flow of a fluid between two parallel plates separated by a distance ℓ. The fluid may be considered to have dynamic viscosity μ and no slip conditions are applied at the interfaces with the plates. The fluid has a pressure gradient parallel to the plates. Find the stationary flow of the fluid.

Problem 3.43. For a test particle of negligible mass, the orbital period T in a circular orbit in the usual gravitational potential outside of a spherical mass distribution depends on the mass of the central body M, the radius of the orbit R, and Newton's gravitational constant G. Use dimensional analysis to derive *Kepler's third law*

$$T^2 \propto R^3 \qquad\qquad (3.297)$$

for this type of orbit, where the proportionality constant is fixed by M and G. *Hint:* Newton's gravitational law was not yet discovered when Kepler formulated his laws. Thus, Kepler did not have the luxury of being able to use dimensional analysis, but you do!

Problem 3.44. For a line charge $\rho_\ell(y^1)$ localised to a coordinate line with fixed $y^2 = a^2$ and $y^3 = a^3$ in orthogonal coordinates, verify that the charge density is given by the expression

$$\rho(\vec{x}) = \rho_\ell(y^1)\frac{\delta(y^2 - a^2)\delta(y^3 - a^3)}{h_2 h_3}. \tag{3.298}$$

Problem 3.45. A surface charge $\sigma(x^1, x^2) = \sigma_0 e^{-k^2[(x^1)^2+(x^2)^2]}$ is located in the plane $x^3 = 0$. In Cartesian coordinates, the charge density is given by

$$\rho(\vec{x}) = \sigma(x^1, x^2)\delta(x^3). \tag{3.299}$$

Express this charge density in spherical and cylinder coordinates, respectively.

Problem 3.46. An object is thrown horizontally from a height h with velocity v in a constant gravitational field g. Without performing any actual kinematic computations, determine how the following quantities scale with h and g:

a) The time t_0 at which the object hits the ground. You may assume that this does not depend on v.

b) The horizontal distance d at which the object lands. You may assume that this only depends on v and t_0.

Now assume that the object is thrown at an angle θ from the horizontal. Use the Buckingham π theorem to determine the general structure of the relationships between:

c) h, v, g, t_0, and θ.

d) h, v, g, d, and θ.

Solve the kinematic problem exactly and identify the form of the functions of dimensionless parameters you have introduced in (c) and (d).

Problem 3.47. A homogeneous metal sphere of radius R containing a radioactive isotope is constantly heated due to the radioactive decays inside of it, which may be considered as a constant source term κ_0 in the heat equation as long as we are only interested in time scales much shorter than the lifetime of the isotope. Construct a partial differential equation, including boundary conditions, for the temperature inside the sphere. You may assume that the sphere's surface is held at the constant temperature T_0. Find how the stationary temperature in the middle of the sphere scales with the radius R, the source term κ_0, and the heat diffusion coefficient a without performing any explicit computations.

Problem 3.48. Consider the damped one-dimensional wave equation with a periodic source term, homogeneous Dirichlet boundary conditions, and homogeneous initial conditions

$$(\text{PDE}) : u_{tt} + k u_t - c^2 u_{xx} = f_0 \sin(\omega t), \tag{3.300a}$$
$$(\text{BC}) : u(0, t) = u(\ell, t) = 0, \tag{3.300b}$$
$$(\text{IC}) : u(x, 0) = 0, \quad u_t(x, 0) = 0. \tag{3.300c}$$

Introduce dimensionless parameters and rewrite the differential equation as a differential equation for a dimensionless function with dimensionless parameters. Does the solution of the dimensionless problem depend on any dimensionless parameters apart from the dimensionless time and the dimensionless distance you have introduced?

CHAPTER 4

Symmetries and Group Theory

Symmetries play a central role in modern physics and the mathematical language for describing them is found in group theory. Invoking symmetry arguments can aid us in analysing and making simplifying statements regarding many physical systems. As we move on, symmetries are also going to be central in deriving conservation laws in classical mechanics and even further on they are at the very foundation of quantum physics and the Standard Model of particle physics, topics that will not be covered in this book.

In this chapter, we will briefly discuss the very basic foundations of symmetry arguments, how transformations of time, space, and other properties may be described within the group theory language, and how it may help us confront different physical situations. The examples considered will be basic and classical in order to keep the prerequisite physics knowledge at a more fundamental level. The aim is to introduce a rudimentary symmetry thinking and provide the main ideas, while a deeper treatment is left out of our discussion for brevity.

4.1 WHAT IS A SYMMETRY?

A *symmetry* of a physical system is a transformation that leaves the system, or a particular property of the system, *invariant*. In effect, this means that the property is the same before and after the transformation is applied. In this respect, transformations are ways of mathematically rewriting or reparametrising a system with a new set of parameters describing the same physical situation. Although this may sound quite abstract, there are many transformations that should be familiar.

Example 4.1 Let us consider a homogeneous sphere of mass density ρ_0, see Fig. 4.1. Placing the origin of a Cartesian coordinate system in the center of the sphere, the density function $\rho(\vec{x})$ is given by

$$\rho(\vec{x}) = \begin{cases} \rho_0, & (r \leq R) \\ 0, & (r > R) \end{cases}, \tag{4.1a}$$

where $r = \sqrt{\vec{x}^2}$ and R is the radius of the sphere. If we introduce new Cartesian coordinates $x'^{i'} = a_i^{i'} x^i$, which are related to the old coordinates by a *rotation* about the origin, we find that

$$\rho(\vec{x}') = \begin{cases} \rho_0, & (r' \leq R) \\ 0, & (r' > R) \end{cases}, \tag{4.1b}$$

where now $r' = \sqrt{\vec{x}'^2}$. The form of the function ρ is therefore the same in both of the systems

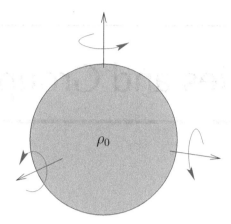

Figure 4.1 A sphere of homogeneous mass density ρ_0 is symmetric under any rotation around an axis passing through its center. Regardless of how many of these rotations are performed, the mass density will be the same after the rotation as it was before.

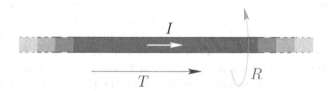

Figure 4.2 An infinite straight conductor carrying a current I displays a translational symmetry along its axis T and a rotational symmetry around its axis R. Any such transformation, or a combination of such transformations, results in the same current density as before the transformation.

and is invariant under rotations. All rotations about the origin are therefore symmetries of the mass distribution.

An important aspect of symmetries is that if two different transformations are both symmetries of a physical system, then also their composition is a symmetry. In the above example, we could consider rotations around a given axis $\vec{n}_1$ as a symmetry and rotations around a different axis $\vec{n}_2$ as another. Performing any rotation around $\vec{n}_1$ followed by a rotation around $\vec{n}_2$, or vice versa, will then also leave the system invariant. As we shall see, this is a fundamental aspect of why group theory is useful in describing transformations and symmetries.

Example 4.2 An infinite straight conductor carries a current I as shown in Fig. 4.2. The resulting current density is symmetric under translations along the conductor direction as well as under rotations around the conductor axis. Any combination of such translations and rotations are also symmetries of the current density.

Symmetries do not need to be continuous as in the case of rotations and they also do not

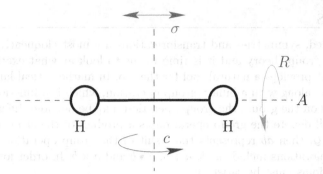

Figure 4.3 The symmetries of the hydrogen molecule H_2. The molecule is symmetric under any rotation around the axis A connecting the two atoms R, rotations by π around any axis perpendicular to it c, and reflections in the plane orthogonal to the connecting axis σ.

need to be spatial transformations of a system. They may also be discrete, for example if a rotation needs to be performed by a finite angle in order to be a symmetry or transformations of other physical variables.

Example 4.3 As an example of discrete symmetries, we may consider the symmetries of a hydrogen molecule as depicted in Fig. 4.3. Apart from exhibiting a rotational symmetry R around the axis A going through both hydrogen atoms, the molecule is symmetric under rotations c by an angle π around any axis through its center point which is orthogonal to A, as well as under spatial reflections σ in the plane orthogonal to A. Any of these transformations will bring us back to a situation with a hydrogen atom in the same position where a hydrogen atom was originally situated. Apart from the rotations R around A, all of these transformations are discrete. For example, rotating by an angle $\pi/3$ around an axis perpendicular to A is not a symmetry of the hydrogen molecule.

Example 4.4 Maxwell's equations in vacuum, i.e., without charge or current densities, are of the form

$$\nabla \cdot \vec{E} = \nabla \cdot \vec{B} = 0, \quad \nabla \times \vec{E} + \partial_t \vec{B} = \nabla \times \vec{B} - \frac{1}{c^2}\partial_t \vec{E} = 0. \tag{4.2}$$

These equations are symmetric under the transformation

$$\vec{E} \to \vec{E}' = \cos(\alpha)\vec{E} + \sin(\alpha)c\vec{B}, \tag{4.3a}$$

$$\vec{B} \to \vec{B}' = -\sin(\alpha)\frac{1}{c}\vec{E} + \cos(\alpha)\vec{B}, \tag{4.3b}$$

where α is a fixed angle. Note that this transformation only involves a transformation of the fields $\vec{E}$ and $\vec{B}$ and does not involve spatial transformations.

4.2 GROUPS

As just mentioned, symmetries and transformations are most eloquently described by the mathematics of group theory and it is time for us to look at what exactly is meant by a *group* and why it provides a natural tool for the job. In mathematical language, a group $\mathcal{G}$ is a set of objects along with a binary group operation, which is a function of two elements in $\mathcal{G}$. Depending on the group, the group operation may be denoted differently, but for our purposes, we will denote the group operation as a product of the elements, i.e., if a and b are elements of $\mathcal{G}$, then ab represents the result of the group operation between a and b. Other common notations include $a \cdot b$, $a + b$, $a \bullet b$ and $a \times b$. In order for $\mathcal{G}$ to be a group, the following axioms must be satisfied:

1. *Closure*: The group must be closed under the group operation. This means that if a and b are elements of $\mathcal{G}$, then so is ab, i.e.,

$$a, b \in \mathcal{G} \implies ab \in \mathcal{G}. \tag{4.4a}$$

2. *Identity*: The group must contain an identity element e that satisfies

$$ea = ae = a \tag{4.4b}$$

for all a in the group.

3. *Inverse*: For every element a in the group, there must exist an inverse element a^{-1} that is also in the group and satisfies

$$aa^{-1} = a^{-1}a = e, \tag{4.4c}$$

where e is the identity element.

4. *Associativity*: For any three elements a, b, c in the group, the group operation must satisfy the relation

$$(ab)c = a(bc) \equiv abc. \tag{4.4d}$$

In words, this means that first applying the group operation between a and b and then applying the group operation between the result and c must give the same result as first applying the group operation between b and c and then applying it to a and the result.

Example 4.5 Considering the group axioms, it is straightforward to check that the set of rotations in three dimensions form a group:

1. As we have already discussed, performing two consecutive rotations results in a new rotation. If a and b are rotations, then we define ab as the transformation obtained by first performing b and then performing a. The order here is chosen by pure convention, but will be the natural choice later on.

2. The identity element is the identity transformation, which may be regarded as a rotation by an angle zero. It is clear that, for any rotation a, doing nothing and then doing a will give the same rotation as first doing a and then doing nothing.

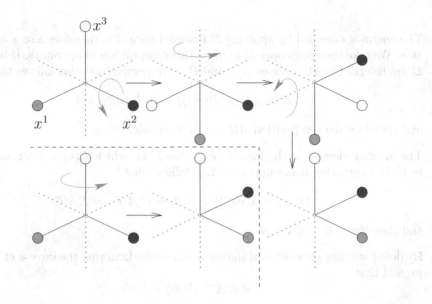

Figure 4.4 An example of consecutive rotations in three dimensions. The axes in the figure should be interpreted as indicated in the upper left, with the x^i denoting the positive x^i-axis. The gray arrows symbolise rotations by $\pi/2$ performed to get to the next image in the sequence, indicated by the straight black arrows. The upper sequence, terminating in the lower right, is the rotation $R_2 R_3 R_1$, where R_i is the rotation by $\pi/2$ around the x^i-axis. We see that the end result is the same as that produced by starting from the same situation and only applying R_3. Since the transformation of a complete set of basis vectors uniquely determines a rotation, we therefore find that $R_2 R_3 R_1 = R_3$.

3. If we have a rotation a that is a rotation by an angle θ about the axis $\vec{n}$, then the inverse of a is the rotation by an angle $-\theta$ about the same axis $\vec{n}$. This corresponds to reversing the transformation and ending up where we started.

4. With any three rotations a, b, and c, the triple product abc will describe first performing c, then b, and finally a.

An example of a group relation from the rotation group is illustrated in Fig. 4.4.

Example 4.6 Given a vector space V, the set of linear operators on V consists of maps A from V to itself such that

$$A(a_1\vec{v}_1 + a_2\vec{v}_2) = a_1 A(\vec{v}_1) + a_2 A(\vec{v}_2) \tag{4.5}$$

for all vectors $\vec{v}_1$ and $\vec{v}_2$ and constants a_1 and a_2. The set of all invertible linear operators on V form a group. The group axioms can be checked explicitly:

1. For any linear operators A and B, we find that

$$A(B(a_1\vec{v}_1 + a_2\vec{v}_2)) = A(a_1 B(\vec{v}_1) + a_2 B(\vec{v}_2)) = a_1 A(B(\vec{v}_1)) + a_2 A(B(\vec{v}_2)). \tag{4.6}$$

The operator obtained by applying B first and then A is therefore also a linear operator. We take the group operation to be this composition of operators. If both A and B are invertible with inverses A^{-1} and B^{-1}, respectively, it also follows that

$$B^{-1}(A^{-1}(A(B(\vec{v})))) = B^{-1}(B(\vec{v})) = \vec{v} \tag{4.7}$$

and therefore the composition AB is also invertible.

2. The identity element is the identity operator E for which $E(\vec{v}) = \vec{v}$ for all vectors $\vec{v}$ in V. For any other linear operator A, it follows that

$$AE(\vec{v}) = A(E(\vec{v})) = A(\vec{v}) = E(A(\vec{v})) = EA(\vec{v}) \tag{4.8}$$

and therefore $AE = EA = A$.

3. By definition, any element A of the set is invertible. Denoting the inverse of A by A^{-1}, we find that

$$\vec{v} = A^{-1}(A(\vec{v})) = E(\vec{v}). \tag{4.9a}$$

Acting on this relation with A and letting $\vec{w} = A(\vec{v})$, it also follows that

$$\vec{w} = A(\vec{v}) = A(A^{-1}(A(\vec{v}))) = A(A^{-1}(\vec{w})). \tag{4.9b}$$

Consequently, we find that $AA^{-1} = A^{-1}A = E$. Furthermore, A^{-1} must also be a linear operator as

$$A^{-1}(a_1 A(\vec{v}_1) + a_2 A(\vec{v}_2)) = A^{-1}A(a_1\vec{v}_1 + a_2\vec{v}_2) = a_1\vec{v}_1 + a_2\vec{v}_2$$
$$= a_1 A^{-1}(A(\vec{v}_1)) + a_2 A^{-1}(A(\vec{v}_2)). \tag{4.10}$$

4. For any two linear operators A and B on V, the composition is defined according to

$$AB(\vec{v}) = A(B(\vec{v})). \tag{4.11}$$

Using this definition for the composition $(AB)C$, we find that

$$(AB)C(\vec{v}) = A(B(C\vec{v})) = A(BC)(\vec{v}) \tag{4.12}$$

and the composition is therefore associative.

The set of invertible linear operators on V therefore form a group with the composition of transformations as the group operation.

An important property that is *not* necessarily satisfied for a group is *commutativity* of the group operation. A commutative operation is one where the order of the arguments is irrelevant, i.e.,

$$ab = ba \tag{4.13}$$

for all a and b. A group for which the group operation is commutative is called a *commutative group* or *Abelian group*. Conversely, if the group is not commutative, it is called non-commutative or non-Abelian.

Even if a group is non-Abelian, there are going to be elements in every group that do satisfy Eq. (4.13). Such elements are said to *commute* with each other and, in particular,

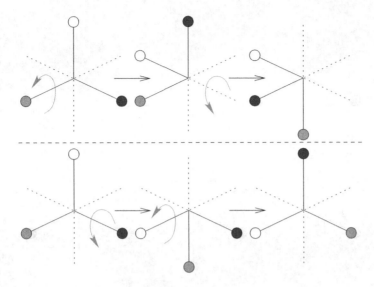

Figure 4.5 Example of non-commutative rotations. In the upper row, we first perform the rotation R_1 by an angle $\pi/2$ arond the x^1-axis followed by the rotation R_2 by $\pi/2$ around the x^2-axis. In the lower row, the order is reversed, which produces a different end result. Therefore, $R_2 R_1 \neq R_1 R_2$.

every element commutes with its inverse and the identity element commutes with all other elements by definition.

Example 4.7 Rotations in three dimensions are an example of a non-Abelian group as rotations about two different different axes generally do not commute. This is illustrated in Fig. 4.5, where we can see that rotating by an angle $\pi/2$ around $\vec{e}_1$ and then around $\vec{e}_2$ is not equivalent to performing the same rotations in the opposite order.

Example 4.8 An example of a group that is Abelian is the set of translations in any dimension. A translation $T_{\vec{v}}$ transforms a position vector $\vec{x}$ to $T_{\vec{v}}\vec{x} = \vec{x} + \vec{v}$, where $\vec{v}$ is a constant vector. It follows that

$$T_{\vec{v}}T_{\vec{w}}\vec{x} = \vec{x} + \vec{v} + \vec{w} = T_{\vec{w}}T_{\vec{v}}\vec{x} \tag{4.14}$$

since vector addition is commutative.

4.2.1 Conjugacy classes

The concept of *conjugacy classes* is based on a classification of the elements of a group into a number of distinct subsets of the group. We start by defining that for two elements a and b of a group, b is conjugate to a if the group contains an element g such that

$$b = gag^{-1}. \tag{4.15}$$

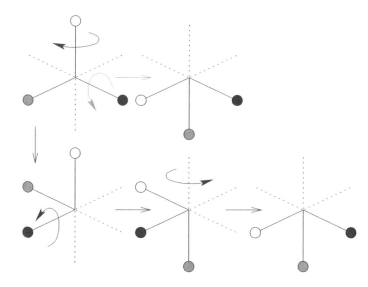

Figure 4.6 An illustration of the conjugacy between a rotation around the x^2-axis to one around the x^1-axis, both by an angle $\pi/2$. Starting from the upper left configuration, we can either perform the rotation around the x^2-axis (light arrows), or first rotate by $-\pi/2$ around the x^3-axis, perform the rotation around x^1, and finally rotate back around the x^3-axis (dark arrows), ending up with the same final configuration.

We write this relation as $b \sim a$ and it is easy to check (see Problem 4.6) that conjugacy is an equivalence relation, meaning that $b \sim a$ implies $a \sim b$ and that $a \sim b$ together with $b \sim c$ implies $a \sim c$. From the perspective of transformations, conjugacy has a very intuitive interpretation. If we can find a transformation g such that a is equivalent to first performing g, then b, and finally applying the inverse transformation of g, then a and b are conjugate to each other. The conjugacy class of a can now be defined as the set of all elements of the group that are conjugate to a.

Example 4.9 For the rotation group that we have looked at in the previous examples, two rotations are conjugate to each other if they rotate by the same angle θ, regardless of the axis of rotation. Considering rotations by an angle θ around the axes $\vec{n}_1$ and $\vec{n}_2$, respectively, the rotation g in the conjugacy relation between them is any rotation that rotates $\vec{n}_1$ into $\vec{n}_2$. As an illustration, the conjugacy between a rotation by $\pi/2$ around $\vec{e}_1$ and one around $\vec{e}_2$ is shown in Fig. 4.6.

Because of the defining properties of the identity element e, there can be no other elements in the conjugacy class of e. Assuming that $a \sim e$ would imply that

$$a = geg^{-1} = gg^{-1} = e. \tag{4.16}$$

It directly follows that there are at least two conjugacy classes for all groups except the *trivial group* containing only the identity element. Similarly, in all Abelian groups, every element forms its own conjugacy class, which follows by using the commutative property on the conjugacy relation between a and b

$$a = gbg^{-1} = bgg^{-1} = be = b. \tag{4.17}$$

Thus, if $a \sim b$ in an Abelian group, then also $a = b$.

Example 4.10 In two dimensions, there is only one possible rotation around a fixed origin. Consequently, all rotations commute, with the composition of (counter clock-wise) rotations by angles θ_1 and θ_2 being a rotation by $\theta = \theta_1 + \theta_2$. It follows that the group of rotations in two dimensions is Abelian and that each element of the group is its own conjugacy class. As for three-dimensional rotations, each conjugacy class consists of all rotations by a given angle around an arbitrary axis. In order to include rotations in both directions in the same conjugacy class in the two-dimensional case, we would have to include a mirror operation σ given by $x^1 \to x^1$, $x^2 \to -x^2$, which would map counter clock-wise rotations to clock-wise rotations. The resulting group will not be Abelian and all conjugacy classes except that of the unit element e and the rotation by and angle π will contain more than one group element.

4.2.2 Subgroups

In physics, it is very common that a system is not completely symmetric under the full set of possible transformations, but only under a subset. For example, we may have a situation where a system is not invariant under all rotations, but is invariant under rotations by an angle $2\pi/m$ (with m being an integer) around some axis $\vec{n}$. This leads us to the concept of a *subgroup*, which is a subset $\mathcal{H}$ of a group $\mathcal{G}$ which is also a group under the group operation in $\mathcal{G}$. For any subset of $\mathcal{G}$, we can check whether it is a subgroup by going through the group axioms:

1. *Closure*: For any a and b in $\mathcal{H}$, the product ab must also be in $\mathcal{H}$. This is usually not true for all subsets of $\mathcal{G}$ and must be checked explicitly.

2. *Identity*: Since the unit element e of $\mathcal{G}$ is the only element which can have the required properties, it must be an element of any subgroup of $\mathcal{G}$.

3. *Inverse*: The subgroup must contain the inverse of all its elements, i.e., if a is an element in the subgroup, then so is a^{-1}.

4. *Associativity*: This one is easy and not necessary to check. It follows directly from the associativity of the original group $\mathcal{G}$.

Example 4.11 For many groups, there are several possible subgroups. Taking three-dimensional translations as an example, some possible subgroups are the set of translations in a given direction and the set of free translations in two given directions together with translations by multiples of ℓ in the third direction. It is worth noting that the set of translations by multiples of ℓ in an arbitrary direction is not a subgroup as it is not closed under the group operation. By performing two consecutive translations by ℓ in different directions, we may end up with a translation by a distance anywhere between 0 and 2ℓ, see Fig. 4.7.

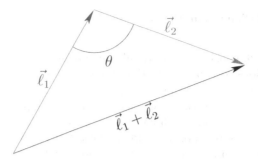

Figure 4.7 The net translation after consecutive translations by $\vec{\ell}_1$ and $\vec{\ell}_2$ is the translation by the sum of the two displacement vectors. Even if $\vec{\ell}_1$ and $\vec{\ell}_2$ both have length ℓ, the length of the final translation is $\ell\sqrt{2 - 2\cos(\theta)}$. The set of translations by a distance which is an integer multiple of ℓ is therefore not a subgroup as it is not closed under the group operation.

A common occurrence when the concept of subgroups will be of importance is when a system is almost symmetric with respect to a set of transformations $\mathcal{G}$, but in reality is only symmetric under a subgroup $\mathcal{H}$ of those. To a first approximation, the system may then be treated as symmetric under the full group and the behaviour of the corrections to this approximation will be well described by the mathematics relating the group and the subgroup. This process is usually referred to as *symmetry breaking* of $\mathcal{G}$ to $\mathcal{H}$.

Example 4.12 The carbon monoxide molecule consists of one carbon atom and one oxygen atom with masses of 12 and 16 atomic mass units, respectively. If the atoms would have the same mass, then the mass distribution of the molecule would have the same symmetries as the hydrogen molecule discussed in Example 4.3. The mass difference between the carbon and oxygen atoms breaks the symmetry to the set of rotations around the axis connecting the molecules only. Rotations by π around an axis perpendicular to this as well as the spatial reflections are no longer part of the symmetry group.

4.2.3 Homomorphisms

The final general concept that we will discuss before having a look at different useful groups is that of group *homomorphisms*. A homomorphism is a mapping h from a group $\mathcal{G}$ to a group $\mathcal{H}$ that preserves the structure of the group operation, i.e.,

$$h(ab) = h(a)h(b), \tag{4.18}$$

where a and b are elements of $\mathcal{G}$. In words, this tells us that taking the product ab in $\mathcal{G}$ and then applying the homomorphism to the result should be the same as first acting with the homomorphism individually on a and b and applying the group operation in $\mathcal{H}$ to the results.

Example 4.13 For all groups, there exists a homomorphism to any other group. If we map all elements of $\mathcal{G}$ to the identity of $\mathcal{H}$, i.e., $h(a) = e$ for all a, then the homomorphism relation is fulfilled as

$$h(ab) = e = e^2 = h(a)h(b). \qquad (4.19)$$

This is the *trivial homomorphism* between $\mathcal{G}$ and $\mathcal{H}$.

From this example, it should be clear that homomorphisms in general are not invertible. However, if a homomorphism is a bijection, then it is called an *isomorphism* and all of the relations valid for the elements of $\mathcal{G}$ are also valid for the corresponding mapped elements in $\mathcal{H}$. If two groups allow an isomorphism, then they are said to be *isomorphic*. It is fairly straightforward to show, see Problem 4.12, that being isomorphic is an equivalence relation on groups and physicists will commonly refer to isomorphic groups as being the same group.

Example 4.14 Naturally, every group $\mathcal{G}$ is isomorphic to itself as the identity mapping $h(a) = a$ is an isomorphism. In general, it is also possible for other mappings from the group to itself to be isomorphisms. Such isomorphisms are referred to as *automorphisms*, while a general homomorphism from a group to itself that is not necessarily an isomorphism is called an *endomorphism*.

4.3 DISCRETE GROUPS

Now that the most important pieces of the framework of group theory are in place, let us go on to discuss some important examples and aspects of different groups. Although many of our examples so far have been in terms of continuous groups, we will start by considering *discrete groups*, which are groups with a countable (but possibly infinite) number of elements. Discrete symmetries will often occur in physical systems and they provide a good place to start getting used to the group framework. In the next section, we shall return to discussing continuous groups.

Any discrete group can be constructed using a set of group *generators* g_i along with a set of relations that they satisfy. In order to have a complete set of generators, it must be possible to write any group element as a product of generators (and their inverses). The relations among the generators will define any additional structure the group may have, thereby allowing the possibility of a single group element to be written in several different ways. In addition, if there exists a finite set of generators with which any element of a group can be written, then the group is *finitely generated*.

Example 4.15 The infinite chain of points of separation ℓ, see Fig. 4.8, has a symmetry group containing translations by a multiple of ℓ and reflections. The group is generated by the translation T_ℓ, for which $T_\ell x = x + \ell$, and the reflection σ, for which $\sigma x = -x$. It also holds that $\sigma^2 = e$ and that $\sigma T_\ell \sigma T_\ell = e$. Any group that is isomorphic to this group may be defined by two generators obeying these relations. We also note that the last relation is equivalent to $T_\ell^{-1} = \sigma T_\ell \sigma$ and it follows that T_ℓ and T_ℓ^{-1} are in the same conjugacy class. Since two generators are enough to write any element in this group, it is finitely generated.

The above example shows us a finitely generated group that has an infinite number of elements. However, some groups contain only a finite number of elements and are therefore

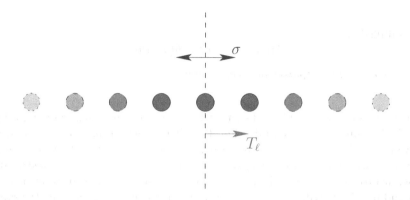

Figure 4.8 An infinite chain of points with fixed separation ℓ has a symmetry group that includes translations by multiples of ℓ along the chain direction and reflections in a plane orthogonal to the chain. These transformations are generated by the translation T_ℓ and the reflection σ at one of the points in the chain.

called *finite groups*. An advantage of finite groups is the possibility of constructing a *group table* (or *Cayley table*) of the form:

e	a_1	a_2	$\cdots$	a_{N-1}
a_1	$a_1 a_1$	$a_1 a_2$	$\cdots$	$a_1 a_{N-1}$
a_2	$a_2 a_1$	$a_2 a_2$	$\cdots$	$a_2 a_{N-1}$
$\vdots$	$\vdots$	$\vdots$	$\ddots$	$\vdots$
a_{N-1}	$a_{N-1} a_1$	$a_{N-1} a_2$	$\cdots$	$a_{N-1} a_{N-1}$

The table is here constructed for a group with N elements, called the *order* of the group. This nomenclature may be slightly confusing, as a group element a is also said to have *order* k if $a^k = e$. It is therefore important to specify if the order of the group or the order of an element is intended. Since the products in the table are group elements, specifying each entry $a_i a_j$ in the table in terms of which element it is, the group is completely defined. In order for the group axioms to be fulfilled, this table must be a magic square, where each row and each column contains every group element exactly once. For any finite Abelian group the relation $a_i a_j = a_j a_i$ implies that the group table is symmetric under the exchange of rows and columns.

Example 4.16 The most basic example of a finite group is the trivial group containing only one element, the identity. Going slightly beyond this, there is only one group (up to isomorphisms) that contains two elements. This group contains the identity element e as well as a second element σ, which has to be its own inverse, i.e., $\sigma^2 = e$. The group table for this group is

e	σ
σ	e

From the table, it is apparent that the group is Abelian, due to the symmetry when exchanging rows and columns. The group consisting of the real numbers 1 and -1 is isomorphic to this group with $e = 1$ and $\sigma = -1$ and with the group operation being regular multiplication.

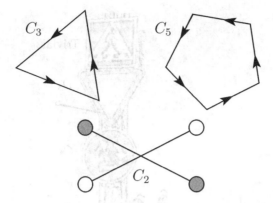

Figure 4.9 Three figures that display the cyclic symmetries C_3, C_5, and C_2, respectively (in two dimensions). The upper two are regular polygons with directed edges, while the lower figure is a bit more intricate, but still displays a cyclic symmetry, in this case a rotation by π.

While the group table in the example above is quite manageable, it starts getting cumbersome for groups with a large number of elements. It is therefore often preferable to just specify the group in terms of its generators and relations among them.

4.3.1 The cyclic group

A common symmetry in physical systems, which we have already alluded to in some examples, is invariance under rotations by a finite angle $2\pi/m$, where m is an integer. A few examples of systems exhibiting such symmetries are shown in Fig. 4.9. Performing a $2\pi/m$ rotation m times results in a rotation by an angle 2π, a full turn, that brings the system back to its original configuration. Based on this, the *cyclic group* of order m can be defined using a single generator c that satisfies the relation

$$c^m = e. \tag{4.20}$$

The cyclic group of order m is generally represented by the symbol C_m.

The elements of the cyclic group are all of the form c^k, where $k < m$. Since the cyclic group only has one generator and every group element commutes with itself, it generally holds that

$$c^k c^\ell = c^{k+\ell} = c^\ell c^k. \tag{4.21}$$

The cyclic group of any order is therefore Abelian. The cyclic group is also the only group, up to isomorphisms, of order m as long as m is a prime, implying that the lowest order of a group that is not cyclic is four.

Example 4.17 The group of integers modulo m is isomorphic to the cyclic group C_m when the group operation is taken to be addition of integers. The identity element of this group is $e = 0$ and the generator is $c = 1$. Alternatively, the generator can be chosen to be c^{m-1}, or c to the power of any integer smaller than m that is coprime with m.

Figure 4.10 Examples of road signs with different symmetries. From top to bottom, the road signs have the symmetry groups: the trivial group, D_3, and C_3, respectively. The bottom two signs are both invariant under rotations by $2\pi/3$ and, unlike the bottom sign, the middle sign is in also invariant under reflection in a vertical line through its center.

4.3.2 The dihedral group

Many physical systems are invariant not only under rotations by a given angle, but also under some spatial reflection. An example of a road sign with such a symmetry is shown in Fig. 4.10 along with a road sign without symmetry and one with cyclic symmetry. It is a special case of the more general symmetry of a regular polygon with m sides that, apart from the rotation by $2\pi/m$, may be reflected in any line passing through its center and of one of its vertices or midpoints of its edges (for odd m, these lines will pass through one vertex and the midpoint of one edge, while for even m they will pass through two vertices or the midpoints of two edges). Although there are m different possible reflections, they may all be written in terms of a single reflection σ and the rotations, which again are generated by a cyclic element c. This group is called the *dihedral group D_m* and is of order $2m$. It is generated by the elements c and σ, which satisfy the relations

$$c^m = e, \quad \sigma^2 = e, \quad \text{and} \quad \sigma c \sigma c = e. \tag{4.22}$$

Example 4.18 The simplest dihedral group that is not isomorphic to the trivial group or C_2 is D_2, which is a group of order four generated by two generators, c and σ, each of which is of order two. The group table is given by

e	c	σ	σc
c	e	σc	σ
σ	σc	e	c
σc	σ	c	e

From this table, it becomes clear that the group is Abelian. This is a consequence of both

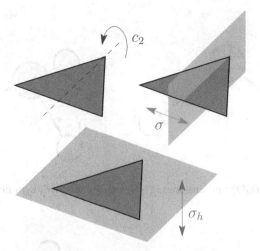

Figure 4.11 The different ways of realising the two-dimensional dihedral reflection in three dimensions, here for the equilateral triangle with the symmetry group D_3. The options are either a three-dimensional rotation c_2 by an angle π or a reflection σ. If both of these are symmetries, the resulting group is the dihedral group D_{3h} and their composition is a reflection σ_h in the plane of the triangle.

generators being their own inverses and the relation $\sigma c \sigma c = e$. Multiplying the latter relation by $c\sigma$ from the right, we find that

$$c\sigma = \sigma c \sigma c c \sigma = \sigma c \sigma^2 = \sigma c, \tag{4.23}$$

stating that the generators, and thus all group elements, commute. It should be noted that this is not true for the general dihedral group D_m.

4.3.2.1 Dihedral groups and three dimensions

The discussion above regarding the dihedral group has focused on the group structure and the example given in terms of the D_3 symmetry of a road sign was implicitly assuming that the symmetry of the sign was considered in two dimensions only. If we wish to realise a symmetry of this type in terms of rotations and reflections in three dimensions, there are two different choices. Starting from the line that defined the two-dimensional reflection, the reflection may be realised either by a rotation c_2 by an angle π around this line, or by a reflection σ in a plane that contains the line and is perpendicular to the two-dimensional plane containing the figure, see Fig. 4.11. These transformations of the three-dimensional space are distinct. Given any vector, they will treat the component of the vector that is perpendicular to the plane differently. Together with the cyclic rotations, including any of these operations will result in a group that is isomorphic to D_n. By convention, when talking about the action of these groups on three-dimensional space, the group including the rotation is usually referred to as D_n, while the one including the reflection is referred to C_{nv}. This distinction is particularly important when considering the symmetry groups of different molecules, which will often be of the type C_{nv} where $n = 2$ or 3 as for water or

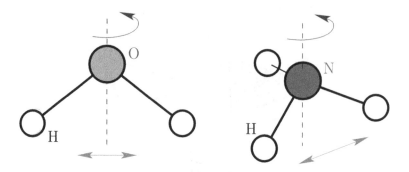

Figure 4.12 The water (H_2O) and ammonia (NH_3) molecules with symmetry groups C_{2v} and C_{3v}, respectively.

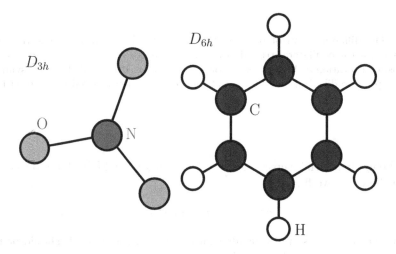

Figure 4.13 The structure of the planar molecules nitrate (NO_3, left) and benzene (C_6H_6, right) which display symmetry under the dihedral groups D_{3h} and D_{6h}, respectively.

ammonia, see Fig. 4.12. The C_{nv} symmetries are often referred to as *pyramidal*, since it is the symmetry group of a pyramid with an n-sided regular polygon as its base.

In addition, some molecules are symmetric under both the rotation by π and the reflection described above. This is particularly true for planar molecules such as benzene or nitrate, see Fig. 4.13. Although they map the same atoms to the same sites, the rotation and reflection are different transformations and the resulting symmetry group, called D_{nh}, contains both of them as separate group elements. The extension of the two-dimensional symmetry D_n to D_{nh} can be made by including both c_2 and σ as group generators. However, there is an option of using the product $\sigma_h = c_2\sigma$ to replace one of these generators, see Fig. 4.11. This is a reflection in the two-dimensional plane containing the original D_n symmetry and commutes with all group elements.

4.3.3 The symmetric group and permutations

From our discussion on the symmetries of tensors in Section 2.2.1, we are already somewhat familiar with the set S_n of all possible permutations of n elements. In fact, this set is a group with the group operation ab given by first performing the permutation b and then performing the permutation a, where a and b are members of S_n. This group is called the *symmetric group* and its subgroups are *permutation groups*.

Since n distinct elements may be ordered in $n!$ different ways, the group S_n is of order $n!$. An element σ of S_n is a rearrangement of the n objects, which we may number from 1 to n. Denoting the new position of the element originally at position k by $\sigma(k)$, we may represent a permutation as

$$\sigma = \begin{pmatrix} 1 & 2 & \cdots & n \\ \sigma(1) & \sigma(2) & \cdots & \sigma(n) \end{pmatrix}, \tag{4.24}$$

where the lower element in a row is the new position of the upper element in the same row after the permutation. However, a more common and short hand notation is given by splitting a permutation into *cycles*. Since $\sigma(k)$ is a bijection from the numbers 1 to n to themselves, starting with 1, we know that σ maps it to $\sigma(1)$. Furthermore, the element that used to be in position $\sigma(1)$ is now mapped to $\sigma(\sigma(1))$ and so on. We can write this as $1 \to \sigma(1) \to \sigma(\sigma(1)) \to \ldots$, but this cycle must eventually return to 1 before it has more than n links. For example, in S_5, we may have the cycle $1 \to 2 \to 5 \to 1$, telling us that under this permutation, 1 maps to 2, which maps to 5, which maps to 1. This cycle can be represented as (125), but does not a priori tell us anything about how the elements 3 and 4 are treated by the permutation. We can specify this by also writing down the cycle or cycles containing the remaining elements. In this case, starting with 3, there are two options: either $\sigma(3) = 3$ or $\sigma(3) = 4$. The first case would be represented as two separate cycles of length one $(3)(4)$, while the latter would be a single cycle of length two (34). Cycles of length one are often omitted when writing down a permutation and so the first case would be written only as (125), while the latter would be written $(125)(34)$.

The cycle notation also makes it easy to perform consecutive permutations. In order to write down the result of consecutively applied cycles, we follow the following algorithm:

1. Select the position of one element that is not yet part of any cycle in the result. If this is the first time this step is taken, position 1 will do nicely. Commit this position to memory and write it down.

2. Considering the cycles in the product from right to left, if the position in memory is part of a cycle, then memorise the next position in the cycle.

3. When arriving at the far left, write down the memorised position if this is not the position written down in step 1, then repeat step 2. Otherwise, write parentheses around the discovered cycle. If there are still positions whose mapping is not known, go back to step 1 and if not, the permutation is now known. You may cancel out any cycles of length one at this stage.

Since they deal with different positions, cycles that do not affect at least one common position always commute.

Example 4.19 Consider the permutations (123) and (143), which are elements of S_4. By application of the algorithm given above, we find that $(123)(143) = (14)(23)$ (see Table 4.1). Since (14) and (23) are cycles containing no common positions, they commute and we find $(14)(23) = (23)(14)$.

Step	In	Step action	Out	Result
1	-	Select from 1234	1	1
2	1	(123)(**143**)	4	
2	4	(123)(143)	4	
3	4	Write down 4	4	14
2	4	(123)(**143**)	3	
2	3	(**123**)(143)	1	
3	1	End of cycle	-	(14)
1	-	Select from **23**	2	(14)2
2	2	(123)(143)	2	
2	2	(**123**)(143)	3	
3	3	Write down 3	3	(14)23
2	3	(123)(**143**)	1	
2	1	(**123**)(143)	2	
3	2	End of cycle	-	(14)(23)

Table 4.1 An example of how to apply the group action in the symmetric group S_4 using the cycle representation. We take the action between the cycles (123) and (143) and perform the action step by step as described in the text. The column labelled "Step" refers to the step of the algorithm in the text. The "In" and "Out" columns refer to the position committed to memory coming into the given step and going out of the given step, respectively. The "Step action" describes the measure taken in the step. For step 2 in the algorithm, the cycle under consideration has been underlined and the incoming and outgoing positions are shown in bold face. Finally, the "Result" column is updated whenever a step dictates that we write something down.

The order of any element of S_n is equal to the least common multiple of the lengths of its cycles. We can also use $n-1$ cycles of length two to generate the full group. Example sets of such generating cycles are the cycles of the form $(1k)$ or the cycles of the form $((k-1)k)$, both for $k \neq 1$. An important homomorphism from S_n to C_2 maps all such generators to -1 (taking the elements of C_2 to be 1 and -1 and the group operation to be multiplication). Elements that contain an even number of generators consequently map to 1, while elements containing an odd number of generators map to -1. These elements are called *even* and *odd* permutations, respectively. Denoting this homomorphism for an element σ by the sign function $\mathrm{sgn}(\sigma)$, we find that the permutation symbol may be written as

$$\varepsilon_{\sigma(1)...\sigma(N)} = \mathrm{sgn}(\sigma). \tag{4.25}$$

Permutations groups are often of relevance when discussing the symmetries and anti-symmetries of different tensors.

4.4 LIE GROUPS

In the case of discrete groups, we found that they could be written in terms of a set of generators and integer powers of these generators. For many physical transformations, such as general translations and rotations, this is not the case. Instead, for any finite translation, there is always the possibility of performing a translation which translates half the distance. Yet, we can try to implement a similar concept for such continuous groups by looking at small transformations, i.e., transformations that are not too far away from the identity transformation. Naturally, the definition of "small" may be in question here, but for now we will stick to a more conceptual approach rather than discussing this in more detail. Groups of this type are known as *Lie groups*.

Suppose we have a group element g that is close to the identity element $e = 1$ (calling the identity 1 is very common and will decrease the possibility for confusion in this section). As for the case of discrete groups, we may obtain different group elements by taking powers of g, i.e., g^n where n is an integer. However, unlike the discrete scenario, there will always be a group element h such that

$$h^2 = g, \tag{4.26a}$$

or, formally,

$$h = g^{1/2}. \tag{4.26b}$$

This is not restricted to $1/2$, or even to rational numbers as the transformation is continuous. Instead, there will generally be elements of the form g^θ, where θ is any real number, that belong to the group.

Example 4.20 The most straightforward Lie group may be the set of translations in one dimension. For any translation T_ℓ given by $x \to T_\ell x = x + \ell$, we may write $T_\ell^\theta x = x + \theta\ell$, giving a translation by a new distance $\theta\ell$, which in general may be any distance, depending on the value of θ.

We shall now introduce a bit of formalism that will be of great use and is central to many group theory considerations in physics. If it seems too abstract or ad hoc, this will hopefully be resolved to some extent when we discuss matrix groups later in this section. We start by formally writing the element g as

$$g(\theta) = e^{\theta J}, \tag{4.27}$$

where θ parametrises the size of the transformation and J is a *generator* of the Lie group. In some cases, in particular in quantum mechanics, it is preferable to introduce the generator on the form $g(\theta) = e^{-i\theta J}$ instead. Of course, this is purely conventional, but for our purposes in this book the definition above will be preferable. As long as we are dealing with group transformations generated by J only, this definition satisfies the rule

$$g(\theta)g(\phi) = e^{\theta J}e^{\phi J} = e^{(\theta+\phi)J} = g(\theta + \phi). \tag{4.28}$$

If a group has several generators J_i and is Abelian, we also have

$$g_i(\theta)g_j(\phi) = e^{\theta J_i}e^{\phi J_j} = g_j(\phi)g_i(\theta) = e^{\phi J_j}e^{\theta J_i}, \tag{4.29}$$

indicating that we may as well write this as $e^{\theta J_i + \phi J_j}$, or more generally $e^{\theta_i J_i}$ applying the summation convention to the exponent. Even in the case of a non-Abelian group, we can formally expand the exponents for small values of θ_i, leading to

$$e^{\theta_1 J_1}e^{\theta_2 J_2} \simeq (1 + \theta_1 J_1)(1 + \theta_2 J_2) \simeq 1 + \theta_1 J_1 + \theta_2 J_2 \simeq e^{\theta_1 J_1 + \theta_2 J_2}, \tag{4.30}$$

in the case of two generators and with a straightforward generalisation to the case of an arbitrary number of generators. Here, the meaning of the addition is not yet clear, but for now we keep it as a formal tool and assume that the normal rules of addition apply. Close to the identity element, the group therefore behaves as a linear vector space with the generators J_i as a basis. This vector space is called the *Lie algebra* of the group.

The occurrence of the word algebra in the name seemingly indicates that there should be some algebraic structure related to the Lie algebra. Indeed, this becomes clearer when we consider the second order terms in the exponential expansion and find that

$$e^{\theta_1 J_1} e^{\theta_2 J_2} \simeq \left(1 + \theta_1 J_1 + \frac{\theta_1^2}{2} J_1^2\right) \left(1 + \theta_2 J_2 + \frac{\theta_2^2}{2} J_2^2\right)$$

$$\simeq 1 + \theta_1 J_1 + \theta_2 J_2 + \frac{1}{2}(\theta_1^2 J_1^2 + \theta_2^2 J_2^2) + \theta_1 \theta_2 J_1 J_2. \tag{4.31}$$

It is necessary to here note that, since the group may be non-Abelian, the different generators J_i do not necessarily commute. By exchanging $\theta_i \to -\theta_i$ in the expression, we also find that

$$e^{-\theta_1 J_1} e^{-\theta_2 J_2} \simeq 1 - \theta_1 J_1 - \theta_2 J_2 + \frac{1}{2}(\theta_1^2 J_1^2 + \theta_2^2 J_2^2) + \theta_1 \theta_2 J_1 J_2, \tag{4.32}$$

which differs only in the signs of the linear terms. Multiplying the two together, again keeping terms only to second order in θ_i, we find that

$$e^{\theta_1 J_1} e^{\theta_2 J_2} e^{-\theta_1 J_1} e^{-\theta_2 J_2} \simeq 1 + \theta_1 \theta_2 \underbrace{(J_1 J_2 - J_2 J_1)}_{\equiv [J_1, J_2]}, \tag{4.33}$$

where $[J_1, J_2]$ is the *Lie bracket* between the generators J_1 and J_2. As all of the elements on the left-hand side are part of the group, the right-hand side must also be and it should still be a small deviation from the identity. To leading order, we obtain

$$e^{\theta_1 J_1} e^{\theta_2 J_2} e^{-\theta_1 J_1} e^{-\theta_2 J_2} = e^{\theta_1 \theta_2 [J_1, J_2]} \tag{4.34}$$

and thus $\theta_1 \theta_2 [J_1, J_2]$ must belong to the Lie algebra. This expression is of the form $g_1 g_2 g_1^{-1} g_2^{-1} = \delta$ and therefore the group element δ describes the mismatch when consecutively performing two transformations after applying their inverses in the same order. We note that, if g_1 and g_2 commute, then

$$\delta = g_1 g_2 g_1^{-1} g_2^{-1} = g_1 g_1^{-1} g_2 g_2^{-1} = 1^2 = 1 \tag{4.35}$$

and it follows that $[J_1, J_2] = 0$. The Lie bracket is the algebraic structure in the Lie algebra and specifying how it acts on the different generators uniquely defines the local behaviour of the group. Furthermore, since the Lie bracket of two elements results in a new element close to the identity, the meaning of the summation should be clearer as we are adding elements of the Lie algebra, which is a vector space, together. For group elements that are further away from the identity, we would need to include higher orders of the expansion, but they will all include Lie brackets and result in adding elements of the Lie algebra together.

In addition to the local group structure, given by the Lie algebra, there may be additional relations that will be satisfied for the group elements, much like the relations we encountered for the discrete groups.

Example 4.21 Locally, the rotations by an angle θ in two dimensions and translations by a distance $\theta\ell$ in one dimension are exactly the same. Their Lie algebras are one-dimensional vector spaces with only one generator J. As the Lie bracket is anti-symmetric, the only possibility is that $[J, J] = 0$ for both groups. However, the groups are different due to the relation

$$e^{2\pi J} = 1 \tag{4.36}$$

that applies to the case of the rotations, while all group elements with different θ are distinct for the translations.

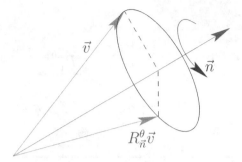

Figure 4.14 The rotation around any vector $\vec{n}$ preserves the projection of the rotated vector $\vec{v}$ on $\vec{n}$ and rotates the perpendicular component of $\vec{v}$ in a circle around it.

Before looking into a the very concrete application found in matrix groups, let us see how this formalism applies to rotations in three dimensions and translations in a general number of dimensions.

4.4.1 Rotations

In order to have a look at the abstract *group of rotations* in three dimensions, we start by considering how a general rotation about an axis defined by the unit vector $\vec{n}$ transforms a vector. We shall later see that the resulting group is isomorphic to a particular group of 3×3 matrices, but for now we keep the treatment more formal.

When rotating a vector $\vec{v}$ by an angle θ about an axis defined by $\vec{n}$, the resulting vector $R_{\vec{n}}^{\theta}\vec{v}$ should have the same projection onto $\vec{n}$ as the original one, see Fig. 4.14. At the same time, the component orthogonal to $\vec{n}$ is rotated by an angle θ. Using the bac-cab rule, the vector $\vec{v}$ may be rewritten as

$$\vec{v} = \vec{n}(\vec{n} \cdot \vec{v}) - \vec{n} \times (\vec{n} \times \vec{v}), \tag{4.37}$$

where the first term on the right-hand side is parallel to and the second is orthogonal to $\vec{n}$. Therefore, the rotation does not change the first term, but the second term is rotated into the vector $\vec{n} \times \vec{v}$, which is orthogonal to both terms. We find that

$$R_{\vec{n}}^{\theta}\vec{v} = \vec{n}(\vec{n} \cdot \vec{v}) - \cos(\theta)\,\vec{n} \times (\vec{n} \times \vec{v}) + \sin(\theta)\,\vec{n} \times \vec{v}$$
$$= \vec{v}\left(1 - \frac{\theta^2}{2}\right) + \frac{\theta^2}{2}\,\vec{n}(\vec{n} \cdot \vec{v}) + \theta\,\vec{n} \times \vec{v} + \mathcal{O}(\theta^3), \tag{4.38}$$

where we have expanded the expression to order θ^2 for small θ in the second step. Using this expression, we can compute $R_1^{\theta_1} R_2^{\theta_2} R_1^{-\theta_1} R_2^{-\theta_2}\vec{v}$, where $R_i^{\theta} = R_{\vec{e}_i}^{\theta}$, by repeatedly inserting the expression obtained from one rotation into the next one and keeping terms only to second order in the angles. After some algebra, we find that

$$R_1^{\theta_1} R_2^{\theta_2} R_1^{-\theta_1} R_2^{-\theta_2}\vec{v} \simeq \vec{v} + \theta_1\theta_2\,\vec{e}_3 \times \vec{v} \simeq R_3^{\theta_1\theta_2}\vec{v}. \tag{4.39}$$

The resulting rotation is therefore a rotation by an angle $\theta_1\theta_2$ around the x^3-axis, which is illustrated in Fig. 4.15. Written in terms of the generators of rotation, we therefore have

$$e^{\theta_1\theta_2[J_1,J_2]} = e^{\theta_1\theta_2 J_3} \tag{4.40a}$$

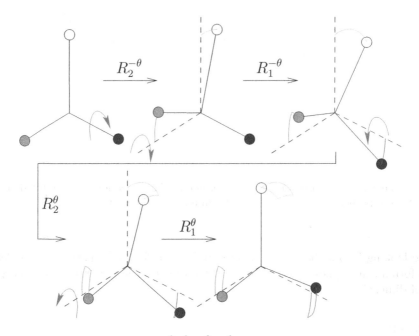

Figure 4.15 The effect of the rotation $R_1^\theta R_2^\theta R_1^{-\theta} R_2^{-\theta}$ on the three-dimensional basis vectors. The shaded curves trace out the path taken by the tips of the vectors during the rotations. In the end, we have obtained an overall rotation by an angle θ^2 around the x^3-axis.

and may identify the Lie bracket relation

$$[J_1, J_2] = J_3. \tag{4.40b}$$

This is one of the defining relations for the Lie algebra of rotations in three dimensions. Cyclic permutation of the basis vectors in the argument above more generally results in the Lie bracket

$$[J_i, J_j] = \varepsilon_{ijk} J_k. \tag{4.41}$$

This is one of the most common Lie bracket relations in physics and it is central in the classical and quantum mechanical treatments of angular momentum. We shall soon see how this relation is realised in a matrix group that is isomorphic to the group of rotations. In addition to the Lie bracket relation, the rotations satisfy the relation

$$R_{\vec{n}}^{2\pi} = 1, \tag{4.42}$$

i.e., rotations by an angle 2π bring back the identity element.

4.4.2 Translations

The abstract *group of translations* in N dimensions is an Abelian Lie group with N generators. The general translation $T_{\vec{\ell}}$ by a vector $\vec{\ell}$ transforms the position vector $\vec{x}$ according to

$$T_{\vec{\ell}} \vec{x} = \vec{x} + \vec{\ell} \tag{4.43}$$

and the generators P_i may be chosen according to $e^{\ell P_i} = T_{\ell \vec{e}_i}$. For two consecutive translations, we find that

$$T_{\vec{\ell}_1} T_{\vec{\ell}_2} \vec{x} = \vec{x} + \vec{\ell}_1 + \vec{\ell}_2 = T_{\vec{\ell}_2} T_{\vec{\ell}_1} \vec{x} = T_{\vec{\ell}_1 + \vec{\ell}_2} \vec{x}, \tag{4.44}$$

from which the Abelian property immediately follows. From the relation

$$e^{\ell_1 \ell_2 [P_i, P_j]} = T_{\ell_1 \vec{e}_i} T_{\ell_2 \vec{e}_j} T_{-\ell_1 \vec{e}_i} T_{-\ell_2 \vec{e}_j} = 1 = e^0 \tag{4.45}$$

for small translations, we also deduce that $[P_i, P_j] = 0$ as expected for an Abelian group.

It is relatively straightforward to also consider the more general group of rotations and translations. Deducing the resulting Lie bracket relations among the generators of translation P_i and the generators of rotations J_j is left as an exercise, see Problem 4.28.

4.4.3 Matrix groups

Some examples of groups that are more hands on than the abstract translations and rotations we just discussed are provided by *matrix groups*. Matrices provide us with some very good group candidates due to the fact that matrix multiplication is associative, giving us one condition less to check. The remaining group axioms provide some rather general constraints that must be fulfilled by a matrix group:

1. *Closure*: To start with, matrix multiplication requires that the group operation exists for any two elements of the group. This constrains the matrices in a matrix group to all be square matrices of a given dimension $n \times n$. Just any set of $n \times n$ matrices will not do; it must be checked explicitly to determine whether or not it is closed under matrix multiplication. However, knowing that we must restrict ourselves to square matrices is a good start.

2. *Identity*: The unit matrix I with ones on the diagonal and zeroes everywhere else is the only matrix that can satisfy the requirement of being an identity element. A possible exception to this would be a projection operator, which projects any matrix on a lower dimensional subspace. While such constructions are possible, they would just correspond to a matrix group of smaller square matrices.

3. *Inverse*: For a the inverse of a matrix A to exist, it is necessary that the determinant $\det(A) \neq 0$. This excludes any set containing matrices with determinant zero from being a group. In addition, we must also check that the inverse A^{-1} belongs to the group, which must be done on a case by case basis.

Example 4.22 The most general group of $n \times n$ matrices is that which includes all of the matrices not explicitly excluded by the considerations above. Working backwards, the existence of an inverse excludes matrices with zero determinants and we are left with matrices A with non-zero determinants. Since the unit matrix has a determinant of one, it is part of this set and for any two matrices A and B with non-zero determinants, we find that

$$\det(AB) = \det(A)\det(B) \neq 0, \tag{4.46}$$

implying that the product AB is a matrix with a non-zero determinant as well, thus verifying the closure of the set under multiplication.

This group of $n \times n$ matrices with non-zero determinants is called the *general linear group* of degree n and is denoted as $GL(n)$. A general $n \times n$ matrix has n^2 independent entries and, considering small deviations from the unit matrix I, we can vary any of these entries independently. The Lie algebra of $GL(n)$ is therefore the n^2-dimensional vector space containing all $n \times n$ matrices.

When we are considering a matrix Lie group, such as $GL(n)$, its Lie algebra will be a different set of $n \times n$ matrices and the exponentiation $e^{\theta J}$ is given meaning in terms of a matrix exponent, given by the standard series definition of the exponential function

$$e^{\theta J} = \sum_{k=0}^{\infty} \frac{\theta^k J^k}{k!}. \tag{4.47}$$

Considering only terms up to linear order in θ, we can find the properties of the elements of the Lie algebra, while if we include terms up to second order in θ, we can express the Lie bracket as

$$[J_i, J_j] = J_i J_j - J_j J_i, \tag{4.48}$$

where the terms on the right-hand side are matrix multiplications and the difference is a regular matrix difference. The Lie bracket for any matrix group is therefore given by this matrix commutator.

Example 4.23 We argued that the Lie algebra of $GL(n)$ is the space of all matrices in the previous example. We may also see this by considering a general small deviation from the unit matrix

$$e^{\theta J} \simeq 1 + \theta J. \tag{4.49}$$

Taking the determinant of this matrix, we find that

$$\det(e^{\theta J}) = e^{\theta \operatorname{tr}(J)} \neq 0, \tag{4.50}$$

where $\operatorname{tr}(J)$ is the trace of J, regardless of J. It follows that J is an arbitrary matrix and the Lie algebra therefore has n^2 independent generators.

4.4.3.1 The orthogonal group

In the examples above using the general linear group $GL(n)$ we essentially considered the largest possible group of $n \times n$ matrices. As many other groups, $GL(n)$ contains a large number of possible subgroups and we will now consider a few of these that are of particular importance in physics, starting with the *orthogonal group* of degree n, denoted by $O(n)$. It is defined as the set of $n \times n$ matrices A that satisfy the orthogonality relation

$$A^T A = A A^T = I, \tag{4.51}$$

where A^T is the transpose of A. In other words, the orthogonal group is the set of matrices whose transpose is their inverse. Considering the group axioms, we find that:

1. *Closure*: If A and B are elements of $O(n)$, then

$$AB(AB)^T = A \underbrace{BB^T}_{=I} A^T = AA^T = I \tag{4.52a}$$

   and it follows that AB is also an element of $O(n)$.

2. *Identity*: The unit matrix I is an element of $O(n)$ as

$$II^T = I^2 = I. \tag{4.52b}$$

3. *Inverse*: If A is an element of $O(n)$, its inverse is A^T. We find that

$$A^T(A^T)^T = A^T A = I \tag{4.52c}$$

as $(A^T)^T = A$ by definition of the transpose. Therefore, A^T is also an element of $O(n)$.

An important property of all elements of $O(n)$ is found by taking the determinant of the defining relation $AA^T = I$. Doing so results in

$$\det(AA^T) = \det(A)^2 = \det(I) = 1 \quad \Longrightarrow \quad \det(A) = \pm 1 \tag{4.53}$$

and tells us that all elements in $O(n)$ have a determinant that is either one or minus one. In fact, based on the relation

$$\det(AB) = \det(A)\det(B), \tag{4.54}$$

we find that the determinant $\det(A)$ is a homomorphism from $O(n)$ to C_2, where C_2 is here represented by the set $\{1, -1\}$ with multiplication as the group operation. This homomorphism also splits $O(n)$ into two subsets, the orthogonal matrices with determinant one and the orthogonal matrices with determinant minus one. The first of these subsets is also a subgroup of $O(n)$, known as the *special orthogonal group* $SO(n)$ of matrices A satisfying

$$AA^T = A^T A = I \quad \text{and} \quad \det(A) = 1. \tag{4.55}$$

Example 4.24 The orthogonal group $O(1)$ is the set of numbers a satisfying $a^2 = 1$, i.e., the set containing 1 and -1. The group operation is regular multiplication and the group is isomorphic to C_2. The special orthogonal group $SO(1)$ additionally requires that $a = 1$, leaving it as the trivial group containing only the identity element.

Example 4.25 The orthogonal group $O(2)$ is the set of 2×2 matrices satisfying $AA^T = I$. We may write a general 2×2 matrix A as

$$A = \begin{pmatrix} A_{11} & A_{12} \\ A_{21} & A_{22} \end{pmatrix}. \tag{4.56}$$

Inserting this into the orthogonality relation, we find that

$$AA^T = \begin{pmatrix} A_{11}^2 + A_{12}^2 & A_{11}A_{21} + A_{12}A_{22} \\ A_{21}A_{11} + A_{22}A_{12} & A_{22}^2 + A_{21}^2 \end{pmatrix} = \begin{pmatrix} 1 & 0 \\ 0 & 1 \end{pmatrix}. \tag{4.57}$$

These are three independent equations, one for each diagonal element and one for the off-diagonal relations that are the same. The first element on the diagonal

$$A_{11}^2 + A_{12}^2 = 1 \tag{4.58}$$

is the equation for a circle of radius one and may be parametrised as

$$A_{11} = \cos(\theta), \quad A_{12} = \sin(\theta). \tag{4.59}$$

In the same fashion, we also conclude that

$$A_{21} = \sin(\varphi), \quad A_{22} = \cos(\varphi) \tag{4.60}$$

from the second diagonal element. Insertion into the off-diagonal relation results in

$$\sin(\varphi)\cos(\theta) + \cos(\varphi)\sin(\theta) = \sin(\varphi + \theta) = 0. \tag{4.61}$$

The solutions to this equation are $\theta = -\varphi + \pi k$, where k is an integer. However, all solutions differing by an integer multiple of 2π are equivalent in terms of the matrix elements A_{ij}. It follows that the two possible solutions are $\theta = -\varphi$ and $\theta = -\varphi + \pi$, corresponding to matrices of the form

$$A_{\pm}(\varphi) = \begin{pmatrix} \pm\cos(\varphi) & \mp\sin(\varphi) \\ \sin(\varphi) & \cos(\varphi) \end{pmatrix}. \tag{4.62}$$

Thus, any matrix in $O(n)$ can be written in one of these forms with a single continuous parameter φ and a choice of sign $\pm$. Taking the determinant, we find that

$$\det(A_{\pm}(\varphi)) = \pm[\cos^2(\varphi) + \sin^2(\varphi)] = \pm 1, \tag{4.63}$$

telling us that the group $SO(2)$ are the matrices on the form A_+.

The product of two matrices in $SO(2)$ is given by

$$A_+(\varphi_1)A_+(\varphi_2) = A_+(\varphi_1 + \varphi_2) \tag{4.64}$$

and evidently $A(2\pi) = A(0) = I$. This looks strangely similar to the relations $R^{\varphi_1}R^{\varphi_2} = R^{\varphi_1+\varphi_2}$ and $R^{2\pi} = e$ satisfied by the group of two dimensional rotations and consequently the mapping $f(R^{\varphi}) = A_+(\varphi)$ is an isomorphism between such rotations and $SO(2)$. Due to this isomorphism, $SO(2)$ is often also referred to as the group of (proper) rotations in two dimensions.

The fact that $SO(2)$ is isomorphic to rotations in two dimensions is neither a coincidence nor a special case. If we consider an n-dimensional vector space with two orthonormal sets of basis vectors $\vec{e}_i$ and $\vec{e}_{i'}'$, they are related by a rotation if both sets are right-handed. The corresponding transformation coefficients

$$a_i^{i'} = \vec{e}_{i'}' \cdot \vec{e}_i = a_{i'}^i, \tag{4.65}$$

which uniquely define the rotation, were found to obey the relation

$$a_{i'}^i a_j^{i'} = \delta_{ij} \tag{4.66a}$$

in Section 1.2 (see also Eq. (1.23)). With the matrix notation $A = (a_{i'}^i)$, this equation takes the form

$$AA^T = I. \tag{4.66b}$$

Thus, the matrix containing all of the transformation coefficients is indeed an orthogonal matrix. Furthermore, if both sets of basis vectors are right-handed, then $\det(A) = 1$ and the matrix belongs to $SO(n)$. If we also include the possibility of having left-handed sets of basis vectors, then the full set of transformations between different sets of basis vectors is described by $O(n)$, with transformations between right- and left-handed bases having determinant minus one. In addition to rotations, such transformations include reflections.

The full group $O(n)$ has the same infinitesimal generators as $SO(n)$, but also includes a finite generator that changes direction of one of the basis vectors as other reflection operators may be constructed from this one by rotations.

The Lie algebra of $SO(n)$ may be found by considering small rotations

$$e^{\theta J} \simeq 1 + \theta J. \tag{4.67}$$

From the orthogonality relation, it follows that

$$e^{\theta J} e^{\theta J^T} \simeq (1 + \theta J)(1 + \theta J^T) \simeq 1 + \theta(J + J^T) = 1 \tag{4.68}$$

to first order in θ. This immediately implies that

$$J^T = -J. \tag{4.69}$$

Therefore, the Lie algebra of $SO(n)$ consists of all real anti-symmetric $n \times n$ matrices.

Example 4.26 Let us check the cases $n = 1$, 2, and 3. For the case of $n = 1$, we have seen that the group $SO(1)$ is isomorphic to the trivial group and thus has no generators. This is perfectly in order as there are no anti-symmetric 1×1 matrices. In the case of $n = 2$, there is only one linearly independent anti-symmetric matrix, namely

$$J = \begin{pmatrix} 0 & -1 \\ 1 & 0 \end{pmatrix}. \tag{4.70}$$

This matrix satisfies the relation $J^2 = -I$ and it follows that the series expansion of the exponent in Eq. (4.47) may be split into terms with even and odd k according to

$$e^{\theta J} = I \sum_{k=0} \frac{(-1)^k \theta^{2k}}{(2k)!} + J \sum_{k=0} \frac{(-1)^k \theta^{2k+1}}{(2k+1)!}. \tag{4.71a}$$

The first sum in this expression is the series expansion of $\cos(\theta)$, while the second sum is the series expansion of $\sin(\theta)$. It immediately follows that

$$e^{\theta J} = I \cos(\theta) + J \sin(\theta) = \begin{pmatrix} \cos(\theta) & -\sin(\theta) \\ \sin(\theta) & \cos(\theta) \end{pmatrix} \tag{4.71b}$$

as expected. Finally, for $SO(3)$, there are three linearly independent real anti-symmetric 3×3 matrices

$$J_1 = \begin{pmatrix} 0 & 0 & 0 \\ 0 & 0 & -1 \\ 0 & 1 & 0 \end{pmatrix}, \quad J_2 = \begin{pmatrix} 0 & 0 & 1 \\ 0 & 0 & 0 \\ -1 & 0 & 0 \end{pmatrix}, \quad \text{and} \quad J_3 = \begin{pmatrix} 0 & -1 & 0 \\ 1 & 0 & 0 \\ 0 & 0 & 0 \end{pmatrix}, \tag{4.72}$$

where J_i is the generator of rotations about the x^i-axis.

4.4.3.2 The unitary group

While the orthogonal group describes general basis changes in a real vector space, the physics of quantum mechanics will require the use of complex vector spaces. While the majority of this book is written using examples from classical physics only, it would be a sin not to include a short discussion on the *unitary group*, which describes basis changes in complex vector spaces.

Much like the orthogonal group is defined by relating the inverse of a matrix to its transpose, an $n \times n$ matrix V belongs to the unitary group $U(n)$ if its Hermitian conjugate $V^\dagger$, i.e., the matrix obtained by transposing the matrix and taking the complex conjugate of each element, is its inverse

$$VV^\dagger = V^\dagger V = I. \tag{4.73}$$

For real matrices, this unitarity relation is equivalent to the defining relation of orthogonal matrices and therefore $O(n)$ is a subgroup of $U(n)$. Unlike $O(n)$, where we were forced to include a reflection to generate the full group, $U(n)$ is completely generated by its Lie algebra. The fact that the elements of the orthogonal group necessarily have determinant ± 1 originated as a result of taking the determinant of the orthogonality relation. Doing this with an element V in the unitary group results in

$$\det(VV^\dagger) = \det(V)\det(V^\dagger) = \det(V)\det(V)^* = |\det(V)|^2 = 1, \tag{4.74}$$

which only states that $\det(V)$ is a complex number of modulus one.

Example 4.27 For $n = 1$, the unitary group $U(1)$ is not a finite group as $O(1)$ (in fact, it is one of the most important Lie groups in physics). Any complex 1×1 matrix is just a complex number z, which may be parametrised as $z = re^{i\theta}$, where r and θ are both real and $r > 0$. The unitarity condition implies that

$$zz^* = r^2 e^{i(\theta-\theta)} = r^2 = 1 \tag{4.75}$$

resulting in $r = 1$ but leaving θ arbitrary. The group $U(1)$ is therefore the set of complex numbers of modulus one, written on the form $z(\theta) = e^{i\theta}$, and its single generator is $J = i$. Furthermore, it holds that $z(2\pi + \theta) = z(\theta)$, which is a very familiar relation and shows that $U(1)$ is isomorphic to $SO(2)$. This should not be surprising as $SO(2)$ is the symmetry group of a circle while complex numbers of the form $e^{i\theta}$ describe the unit circle in the complex plane.

In the case of $O(n)$, we defined the subgroup $SO(n)$ as the matrices in $O(n)$ that have determinant one. In complete analogy, the *special unitary group* $SU(n)$ is defined as the subgroup of $U(n)$ that consists of unitary matrices of determinant one.

The Lie algebra of $U(n)$ can be found by considering the infinitesimal version of the unitarity relation, i.e.,

$$e^{\theta J} e^{\theta J^\dagger} \simeq I + \theta(J + J^\dagger) = I \quad \Longrightarrow \quad J = -J^\dagger. \tag{4.76}$$

The generators of $U(n)$ are therefore the set of anti-Hermitian matrices, i.e., matrices that are equal to the negative of their own Hermitian conjugates. If we restrict ourselves to $SU(n)$, we also have the additional requirement

$$\det(e^{\theta J}) = e^{\theta \operatorname{tr}(J)} = 1, \tag{4.77}$$

implying that $\operatorname{tr}(J) = 0$. The generators of $SU(n)$ are therefore the set of traceless anti-Hermitian matrices.

Example 4.28 Just as $SO(1)$, the requirements of the group $SU(1)$ are so restrictive that the group is trivial and the lowest n for which $SU(n)$ is non-trivial is $n = 2$. For the unitary group $U(2)$ the generators are given by

$$-i\sigma_0 = -iI, \qquad\qquad -i\sigma_1 = -i\begin{pmatrix} 0 & 1 \\ 1 & 0 \end{pmatrix},$$

$$-i\sigma_2 = -i\begin{pmatrix} 0 & -i \\ i & 0 \end{pmatrix}, \qquad\qquad -i\sigma_3 = -i\begin{pmatrix} 1 & 0 \\ 0 & -1 \end{pmatrix}, \qquad (4.78)$$

where we recognise $-i\sigma_2$ as the generator of the $SO(2)$ subgroup we saw previously. The matrices σ_1, σ_2, and σ_3 are called the *Pauli matrices*. In fact, all of the generators could be taken as the generator of a subgroup isomorphic to $SO(2)$. For the special unitary group $SU(2)$, the generators are the same apart from $-i\sigma_0$, which is not traceless and must be removed. The Lie bracket for $SU(2)$ is given by

$$[-i\sigma_i, -i\sigma_j] = 2\varepsilon_{ijk}(-i\sigma_k). \qquad (4.79)$$

This looks similar to the Lie bracket of $SO(3)$ given in Eq. (4.41) and differs only by a factor of two. However, we may equally well generate $SU(2)$ with $S_i = -i\sigma_i/2$ as generators. This makes the Lie bracket exactly the same as that of $SO(3)$ and there is therefore a homomorphism from $SU(2)$ to $SO(3)$ given by

$$f(e^{\theta_i S_i}) = e^{\theta_i J_i}. \qquad (4.80a)$$

It should be stressed that this homomorphism is *not* an isomorphism. In particular, we find that with $\theta_2 = 2\pi$ and $\theta_1 = \theta_3 = 0$,

$$e^{2\pi S_2} = e^{-i\pi\sigma_2} = -I \quad \text{and} \quad e^{2\pi J_2} = I, \qquad (4.80b)$$

where the identity matrix in these relations is 2×2 in the first and 3×3 in the second. It follows that $f(-I) = I$, implying that the homomorphism is not one-to-one as also $f(I) = I$. Indeed, the homomorphism is actually two-to-one and we say that $SU(2)$ is a double cover of $SO(3)$. The group $SU(2)$ is central in the quantum mechanical treatment of angular momentum, just as $SO(3)$ is central to the classical treatment.

4.5 REPRESENTATION THEORY

A fundamental tool in examining how different systems behave under certain transformations is found in representation theory. A *representation* ρ of a group on a vector space V is a homomorphism from the group to the set of linear operators on V. Note that these operators form a group and it therefore makes sense to talk about such a homomorphism, see Example 4.6. The dimension of V is also referred to as the dimension of the representation. If the vector space V is finite dimensional with N dimensions, then the representation can be written down as $N \times N$ matrices. The *action* of a group element a on a vector x in V is given by applying the homomorphism to the vector and obtaining a new vector

$$x' = \rho(a)x. \qquad (4.81)$$

Example 4.29 We have already seen how the abstract group of rotations in three dimensions can be isomorphically mapped to $SO(3)$. The rotation may be written as

$$x'^{i'} = a_i^{i'} x^i \quad \text{or} \quad x' = (a_i^{i'})x = \rho(a)x, \tag{4.82}$$

where x and x' are interpreted as column vectors containing the components of $\vec{x}$ in the last expression, and this is the action of the rotation on the vector $\vec{x}$. The representation $\rho(a) = (a_i^{i'})$ is the *fundamental representation* of rotations. It represents each rotation by a matrix containing the transformation coefficients.

Example 4.30 The matrix group $SU(2)$ may be represented by a three-dimensional representation using the homomorphism between $SU(2)$ and $SO(3)$.

Any representation of a group also naturally gives rise to representations of all its subgroups by restricting the homomorphism ρ to the subgroup. Any element a of the subgroup maps to a subgroup of the representation, since the structure of the group operation is preserved. More precisely, if a and b are in a subgroup of the represented group, then ab is also in the subgroup. This means that

$$\rho(a)\rho(b) = \rho(ab) \tag{4.83}$$

and the group operation between $\rho(a)$ and $\rho(b)$ therefore gives a new element that is the representation of an element in the subgroup. The representation of the subgroup therefore forms a subgroup of the representation of the full group.

Example 4.31 Starting from the two-dimensional representation

$$\rho(R_\alpha) = \begin{pmatrix} \cos(\alpha) & -\sin(\alpha) \\ \sin(\alpha) & \cos(\alpha) \end{pmatrix} \tag{4.84}$$

of rotations in two dimensions, we can also find a two-dimensional representation of any cyclic group C_m, since we know that C_m is isomorphic to rotations by an angle $2\pi/m$. Calling the representation $\tilde{\rho}$, it is given by $\tilde{\rho}(c^k) = \rho(R_{2\pi k/m})$, where c is the generator of C_m. Taking C_3 with the elements e, c and c^2 as an example, the representation becomes

$$\tilde{\rho}(e) = I, \quad \tilde{\rho}(c) = \frac{1}{2}\begin{pmatrix} -1 & -\sqrt{3} \\ \sqrt{3} & -1 \end{pmatrix}, \quad \text{and} \quad \tilde{\rho}(c^2) = \frac{1}{2}\begin{pmatrix} -1 & \sqrt{3} \\ -\sqrt{3} & -1 \end{pmatrix}. \tag{4.85}$$

4.5.1 Tensor products and direct sums

If we have a representation of a group $\mathcal{G}$ on a vector space V, the action of a group element a on a vector $\vec{v}$ can generally be written as

$$v'^{i'} = \rho(a)_i^{i'} v^i, \tag{4.86}$$

where the components $\rho(a)_i^{i'}$ may be seen as the components of a matrix. In the case of rotations, we found that these components were the transformation coefficients $a_i^{i'}$ and putting them into a matrix gave us the matrix group $SO(3)$. Similarly, for any group, we may define the action on a scalar quantity ϕ through

$$\phi' = \rho(a)\phi = \phi, \tag{4.87}$$

i.e., the representation maps all a to the trivial group. According to their definition as linear combinations of products such as

$$T = \vec{v} \otimes \vec{w}, \tag{4.88}$$

where $\vec{v}$ and $\vec{w}$ are elements of V, the set of rank two tensors forms a vector space. We may define a representation $\tilde{\rho}$ of $\mathcal{G}$ on this vector space by the action

$$\tilde{\rho}(a)T = (\rho(a)\vec{v}) \otimes (\rho(a)\vec{w}) \equiv (\rho(a) \otimes \rho(a))(\vec{v} \otimes \vec{w}). \tag{4.89}$$

This is a representation of $\mathcal{G}$ on the tensor product space $V \otimes V$ and we call it the *tensor product representation* (or just *product representation*).

Example 4.32 In three dimensions, consider any of the rank two tensors mentioned in Chapter 2 and call it T_{ij}. This tensor transforms as

$$T_{i'j'} = a_i^{i'} a_j^{j'} T_{ij} \tag{4.90}$$

under rotations. This can be written on the form

$$T_{I'} = A_I^{I'} T_I, \tag{4.91}$$

where $I = (ij)$ and $I' = (i'j')$ are sets of two indices, which each may take $3^2 = 9$ different values, and $A_I^{I'} = a_i^{i'} a_j^{j'}$. Writing down the matrix $(A_I^{I'})$ by labelling each row with the two indices I' and each column with the two indices I, we end up with a 9×9 matrix representation of $SO(3)$. This is a rank two tensor product representation.

In general, a tensor product need not be taken between a vector space and itself. We may also have a situation where we have a representation ρ_V of the group $\mathcal{G}$ on the vector space V and a different representation ρ_W of the same group on the vector space W. We can still construct the tensor product space $V \otimes W$ as the set of linear combinations of objects of the form $\vec{v} \otimes \vec{w}$, where $\vec{v}$ is an element of V and $\vec{w}$ an element of W. The tensor product representation $\rho_{V \otimes W}$ of $\mathcal{G}$ on $V \otimes W$ is then defined by the action

$$\rho_{V \otimes W}(a)(\vec{v} \otimes \vec{w}) = (\rho_V(a)\vec{v}) \otimes (\rho_W(a)\vec{w}), \tag{4.92}$$

in analogy with Eq. (4.89). If V is n-dimensional and W is m-dimensional, the dimension of the tensor product representation is nm.

A different way of constructing a new representation from known ones is the *direct sum*. Assuming that ρ_V and ρ_W are representations of the group $\mathcal{G}$ on the vector spaces V and W, respectively, we can create a new representation on the vector space $V \oplus W$. Rather than having a basis created by the outer product of the basis vectors in each of the spaces as the tensor product space, this vector space is the space of pairs of vectors

$$\vec{u} = (\vec{v}, \vec{w}) \equiv \vec{v} \oplus \vec{w}, \tag{4.93}$$

where $\vec{v}$ is a vector in V and $\vec{w}$ is a vector in W. The addition of vectors and the multiplication of vectors by scalars are given by

$$\vec{u}_1 + \vec{u}_2 = (\vec{v}_1, \vec{w}_1) + (\vec{v}_2, \vec{w}_2) = (\vec{v}_1 + \vec{v}_2, \vec{w}_1 + \vec{w}_2), \tag{4.94a}$$
$$k\vec{u} = k(\vec{v}, \vec{w}) = (k\vec{v}, k\vec{w}), \tag{4.94b}$$

respectively. If the dimensions of V and W are n_V and n_W, respectively, then this vector space has dimension $n_V + n_W$ with a basis on the form

$$\vec{e}_k^{V \oplus W} = \begin{cases} \vec{e}_k^V \oplus 0, & (k \leq n_V) \\ 0 \oplus \vec{e}_{k-n_V}^W, & (k > n_V) \end{cases}. \tag{4.95}$$

A new representation $\rho_{V \oplus W}$ of $\mathcal{G}$ is then given by letting the representations ρ_V and ρ_W act on their respective sub-spaces, i.e.,

$$\rho_{V \oplus W}(a)(\vec{v} \oplus \vec{w}) = (\rho_V(a)\vec{v}) \oplus (\rho_W(a)\vec{w}) \tag{4.96}$$

for any element a in the group $\mathcal{G}$. It is important to note the difference between this representation and the tensor product representation. Although very similar to Eq. (4.92), the representation here is acting on a very different vector space. Creating new representations in this way may not be terribly exciting, since it is really nothing more than studying two separate representations acting on their respective vector spaces. Instead, the process of more relevance is going the other way by rewriting more complicated representations as a direct sum of simpler representations.

Example 4.33 As an illustration of the difference between the tensor product representation and the direct sum representation, let us consider the representation of two-dimensional rotations by $SO(2)$ matrices

$$\rho_m(R_\theta) = \begin{pmatrix} \cos(m\theta) & -\sin(m\theta) \\ \sin(m\theta) & \cos(m\theta) \end{pmatrix}, \tag{4.97}$$

where m is an integer. Since these representations are two-dimensional, both the tensor product $\rho_m \otimes \rho_n \equiv \rho_{m \otimes n}$ and the direct sum $\rho_m \oplus \rho_n \equiv \rho_{m \oplus n}$ will be four-dimensional representations. The tensor product representation is given by

$$\rho_{m \otimes n}(R_\theta) = \begin{pmatrix} c_{m\theta}\rho_n(R_\theta) & -s_{m\theta}\rho_n(R_\theta) \\ s_{m\theta}\rho_n(R_\theta) & c_{m\theta}\rho_n(R_\theta) \end{pmatrix}$$
$$= \begin{pmatrix} c_{m\theta}c_{n\theta} & -c_{m\theta}s_{n\theta} & -s_{m\theta}c_{n\theta} & s_{m\theta}s_{n\theta} \\ c_{m\theta}s_{n\theta} & c_{m\theta}c_{n\theta} & -s_{m\theta}s_{n\theta} & -s_{m\theta}c_{n\theta} \\ s_{m\theta}c_{n\theta} & -s_{m\theta}s_{n\theta} & c_{m\theta}c_{n\theta} & -c_{m\theta}s_{n\theta} \\ s_{m\theta}s_{n\theta} & s_{m\theta}c_{n\theta} & c_{m\theta}s_{n\theta} & c_{m\theta}c_{n\theta} \end{pmatrix}, \tag{4.98}$$

where in the middle step each entry represents a 2×2 block of the 4×4 matrix, given by the matrix $\rho_n(R_\theta)$, and we have introduced the short-hand notations $s_\alpha = \sin(\alpha)$ and $c_\alpha = \cos(\alpha)$. On the other hand, the direct sum representation is given by

$$\rho_{m \oplus n} = \begin{pmatrix} \rho_m(R_\theta) & 0 \\ 0 & \rho_n(R_\theta) \end{pmatrix} = \begin{pmatrix} c_{m\theta} & -s_{m\theta} & 0 & 0 \\ s_{m\theta} & c_{m\theta} & 0 & 0 \\ 0 & 0 & c_{n\theta} & -s_{n\theta} \\ 0 & 0 & s_{n\theta} & c_{n\theta} \end{pmatrix}, \tag{4.99}$$

i.e., a block-diagonal matrix where each block is independent of the other under the group operation.

4.5.2 Reducible representations

Two representations ρ and ρ' of a group on the same vector space V are said to be *equivalent* if there is an invertible linear operator U on the vector space such that

$$\rho'(a) = U\rho(a)U^{-1} \tag{4.100}$$

for all elements a in the group (it is also easy to show that for any given ρ and U, this defines a representation ρ', see Problem 4.39). A representation ρ is said to be *reducible* if it is equivalent to a representation ρ_D of the form

$$U\rho(a)U^{-1} = \rho_D(a) = \begin{pmatrix} \rho_1(a) & \sigma(a) \\ 0 & \rho_2(a) \end{pmatrix}, \tag{4.101}$$

where $\rho_i(a)$ are $n_i \times n_i$ blocks containing an n_i-dimensional representation and σ is an $n_1 \times n_2$ block satisfying $\sigma(ab) = \rho_1(a)\sigma(b) + \sigma(a)\rho_2(b)$. For finite groups, it is always possible to select a basis of the vector space such that $\sigma(a) = 0$ for all a. In this case, Eq. (4.101) becomes

$$\rho_D(a) = \begin{pmatrix} \rho_1(a) & 0 \\ 0 & \rho_2(a) \end{pmatrix}, \tag{4.102}$$

which tells us that the representation may be written as a direct sum of the representations ρ_1 and ρ_2, each acting on a subspace of V. In other terms, ρ_1 and ρ_2 are representations on V_1 and V_2, respectively, $V = V_1 \oplus V_2$, and $\rho_D = \rho_1 \oplus \rho_2$. Since the original representation ρ is equivalent to ρ_D, we will also say that $\rho = \rho_1 \oplus \rho_2$.

Example 4.34 Consider the group of rotations around the x^3-axis. As a subgroup of $SO(3)$, this group may be represented by

$$\rho(R_{\vec{e}_3}^\theta) = \begin{pmatrix} \cos(\theta) & -\sin(\theta) & 0 \\ \sin(\theta) & \cos(\theta) & 0 \\ 0 & 0 & 1 \end{pmatrix}, \tag{4.103}$$

which is block-diagonal and therefore reducible. It is a direct sum of the trivial representation acting on the subspace spanned by $\vec{e}_3$ and the $SO(2)$ representation acting on the subspace spanned by $\vec{e}_1$ and $\vec{e}_2$.

Given an n-dimensional representation ρ on a vector space V, let us consider a reduction of the tensor representation $\rho \otimes \rho$ that is always possible. Introducing a basis $\vec{e}_i$ of V, any tensor in $V \otimes V$ may be written as

$$T = T_{ij}\vec{e}_i \otimes \vec{e}_j = (T_{\{ij\}} + T_{[ij]})\vec{e}_i \otimes \vec{e}_j, \tag{4.104}$$

where $T_{\{ij\}}$ and $T_{[ij]}$ are the symmetric and anti-symmetric components of the tensor, respectively. For any symmetric tensor S, it holds that

$$(\rho_{V \otimes V} S)_{ij} = \rho_{ik}\rho_{j\ell}S_{k\ell} = \rho_{ik}\rho_{j\ell}S_{\ell k} = (\rho_{V \otimes V} S)_{ji} \tag{4.105}$$

and so the action of the group preserves the subspace of symmetric tensors. Similar arguments applied to the anti-symmetric tensors will tell us that also the subspace of anti-symmetric tensors is preserved under the group action. We can always select a basis of symmetric tensors spanning the symmetric subspace $V \otimes_+ V$ and a basis of anti-symmetric tensors spanning the anti-symmetric subspace $V \otimes_- V$ such that

$$(V \otimes_+ V) \oplus (V \otimes_- V) = V \otimes V. \tag{4.106}$$

In this basis, the representation $\rho_{V \otimes V}$ is now of the form

$$\rho_{V \otimes V} = \begin{pmatrix} \rho_{V \otimes_+ V} & 0 \\ 0 & \rho_{V \otimes_- V} \end{pmatrix}, \tag{4.107}$$

where $\rho_{V \otimes_\pm V}$ are the $n(n \pm 1)/2$-dimensional representations on the $V \otimes_\pm V$ subspaces. This is a reduction of the tensor product representation into the symmetric and anti-symmetric tensor product representations.

Example 4.35 Let us get back to the tensor product representation $\rho_1 \otimes \rho_1$ in Example 4.33. The matrix representation in Eq. (4.98) is written in the basis $e_{ij} = \vec{e}_i \otimes \vec{e}_j$, which is neither symmetric nor anti-symmetric. According to the discussion now at hand, we should be able to block diagonalise it by instead using the basis

$$E_1 = e_{11} + e_{22}, \quad E_2 = e_{11} - e_{22}, \quad E_3 = e_{12} + e_{21}, \quad E_4 = e_{12} - e_{21}, \tag{4.108}$$

where the first three basis tensors are explicitly symmetric and E_4 is explicitly anti-symmetric. Checking the action on each of these basis tensors, we find that

$$\rho_{1 \otimes 1} E_1 = E_1, \tag{4.109a}$$
$$\rho_{1 \otimes 1} E_2 = \cos(2\theta) E_2 - \sin(2\theta) E_3, \tag{4.109b}$$
$$\rho_{1 \otimes 1} E_3 = \sin(2\theta) E_2 + \cos(2\theta) E_3, \tag{4.109c}$$
$$\rho_{1 \otimes 1} E_4 = E_4. \tag{4.109d}$$

Writing this in matrix form, we have

$$\rho_{1 \otimes 1} = \begin{pmatrix} 1 & 0 & 0 \\ 0 & \rho_2 & 0 \\ 0 & 0 & 1 \end{pmatrix} = 1 \oplus \rho_2 \oplus 1, \tag{4.110}$$

where ρ_2 is the two-dimensional representation on the subspace spanned by E_2 and E_3 and the ones on the right-hand side denote the trivial representations on E_1 and E_4.

In the example above, we were able to split the tensor product representation $\rho_{1 \otimes 1}$ into not only symmetric and anti-symmetric parts, but we also managed to split the symmetric part into two representations $\rho_{1 \otimes_+ 1} = 1 \oplus \rho_2$. As with many things, this is no mere co-incidence, but rather a consequence of the representation ρ_1 being in terms of orthogonal matrices. For any such representation in terms of $SO(n)$ matrices, we will always have the relation

$$\rho_{V \otimes V} e_{kk} = e_{kk} \tag{4.111a}$$

or, in terms of components

$$\rho_{ik} \rho_{j\ell} \delta_{k\ell} = \rho_{ik} \rho_{jk} = \delta_{ij}, \tag{4.111b}$$

since $e_{kk} = \delta_{ij} e_{ij}$. The interpretation of this is that the Kronecker delta tensor transforms according to the trivial representation under rotations. The remaining part of the symmetric tensor product space is traceless and tensors in it are of the form $T = T_{ij}e_{ij}$ such that

$$T_{ii} = 0. \qquad (4.112)$$

This does not generalise directly to unitary representations on complex vector spaces, but a similar construction is possible using a tensor product representation of the vector space with its dual space, i.e., the space of linear maps from the vector space to complex numbers.

4.6 PHYSICAL IMPLICATIONS AND EXAMPLES

We have now seen several examples of groups and symmetries, but our treatment has been largely abstract and mathematical. It is therefore about time that we discuss the relevance of symmetries and group theory in physics. When we discussed the rotation group, we considered how it transformed a general vector $\vec{v}$. For a physical vector quantity, such a vector will be an element of an N-dimensional vector space and a rotation will transform it into a different vector in a linear fashion. This provides us with a natural representation of the rotation group on this vector space and writing it in terms of a matrix, this representation will be in terms of $SO(N)$. Correspondingly, tensors will transform under representations built up from this representation by tensor products.

4.6.1 Reduction of possible form of solutions

Symmetries can often help in reducing the complexity of a given problem and help us argue why a solution must have a particular form. Consider a situation where a tensor property (of arbitrary rank) of a material depends only on the material, and the material itself displays a symmetry that is a subgroup of rotations, i.e., if we perform a rotation in this subgroup, then the material again looks the same, meaning that the property must have the same components after the rotation. Let us take a step back and consider what this implies. We know that the property should transform under the appropriate tensor product representation, which may be reduced into a direct sum of different representations. These different representations constitute subspaces of the full representation space, but unless the representation on a subspace is trivial, the components in that subspace are generally going to change. The conclusion from this is that the only components that are allowed to be non-zero all correspond to subspaces that transform according to the trivial representation of the symmetry group.

Example 4.36 In Example 1.34 we considered a spherically symmetric charge distribution and deduced that the corresponding electric field necessarily must be directed in the $\vec{e}_r$ direction. Let us reexamine this statement in the language of group theory. Considering the field $\vec{E}(x^3\vec{e}_3)$ along the x^3-axis, the position as well as the charge distribution are invariant under rotations about the x^3-axis. At the same time, the electric field is a vector and transforms according to the $SO(3)$ representation under any rotations. In particular, when restricting ourselves to rotations about the x^3-axis, we saw in Example 4.34 that this representation decomposes into the direct sum of an $SO(2)$ representation and a trivial representation. The trivial representation corresponds to the subspace spanned by $\vec{e}_3$ and so only vectors of the form

$$\vec{E} = E_3\vec{e}_3 \qquad (4.113)$$

are allowed, since the $SO(2)$ representation is not trivial and thus forbids any non-zero components in the subspace spanned by $\vec{e}_1$ and $\vec{e}_2$. Similar arguments may be applied for any point $\vec{x} = r\vec{e}_r$. Taking rotations around $\vec{e}_r$ into account, the resulting field must be of the form $\vec{E} = E\vec{e}_r$.

The above argument is true for any type of representation and in order for a symmetry to allow a non-zero element, the representation must contain at least one subspace for which the representation is trivial. We shall return to these issues in the next section, where we will deduce a general method of finding out whether a given representation contains a trivial subspace. However, since the road to this result is relatively long, let us first consider some options that are more tedious on the computational side, but less theoretically demanding.

The big question we need to answer is whether or not a given representation contains a subspace that transforms under the trivial representation. Let us therefore assume that we have a representation ρ acting on a vector space V. For any vector $\vec{v}$ belonging to a subspace transforming under the trivial representation, we find that for all a in the group

$$\rho(a)\vec{v} = \vec{v} \tag{4.114a}$$

or, writing this in component form

$$\rho(a)_{ij}v^j = v^i = \delta_{ij}v^j. \tag{4.114b}$$

This is nothing but an eigenvector equation for ρ_{ij} with the eigenvalue one. In order for it to be possible to satisfy this equation, a necessary and sufficient condition is that

$$\det(\rho - I) = 0, \tag{4.115}$$

where I is the identity matrix. If this is satisfied, then Eqs. (4.114) allow non-trivial solutions and solving for the corresponding eigenvector will result in a vector transforming under the trivial representation.

Example 4.37 Looking back to the rank two tensor representation of $SO(2)$ in Example 4.35, we might not have made the choice of basis for the symmetric tensor product representation in such a way that the reduction of the symmetric representation into $1 \oplus \rho_2$ was obvious. Instead, a perhaps more natural choice would have been

$$E_1 = e_{11}, \quad E_2 = e_{22}, \quad \text{and} \quad E_3 = \frac{1}{\sqrt{2}}(e_{12} + e_{21}). \tag{4.116}$$

In this basis, the representation $\rho_{1\otimes+1}$ is given by

$$\rho_{1\otimes+1} = \begin{pmatrix} c^2 & s^2 & -sc\sqrt{2} \\ s^2 & c^2 & sc\sqrt{2} \\ -sc\sqrt{2} & sc\sqrt{2} & 2c^2 - 1 \end{pmatrix} \implies \det(\rho_{1\otimes+1} - 1) = 0, \tag{4.117}$$

indicating that the representation has a subspace transforming under the trivial representation. Solving the corresponding eigenvector equation, we find that this subspace is spanned by

$$E_1' = E_1 + E_2 = e_{11} + e_{22}, \tag{4.118}$$

which was found to span the subspace transforming under the trivial representation in Example 4.35.

We have here used the representation of an arbitrary element of the group for our argumentation. In general, this might be cumbersome and it should be noted that it is always sufficient to consider only the generators of the involved symmetry.

It is also possible to argue for a particular functional dependence of a physical quantity based on symmetry arguments. If a physical situation is symmetric under a subgroup of rotations and translations $\vec{x} \to \vec{x}' = R_{\vec{n}}^{\theta}\vec{x} + \vec{a}$ we may consider representations of this group in terms of linear operators on the vector space of functions in the region of interest Ω. Note that, in order to be a symmetry of the system, a necessary condition is that the transformation maps Ω to Ω. That the set of functions on Ω forms a vector space is easy to check (see Chapter 5) with the addition and multiplication by a constant being defined pointwise, i.e., for two functions f and g, we define

$$(f + g)(\vec{x}) = f(\vec{x}) + g(\vec{x}) \quad \text{and} \quad (kg)(\vec{x}) = kg(\vec{x}), \tag{4.119}$$

where k is a constant. It should be noted that this vector space is generally of infinite dimension.

If a system is symmetric with respect to a rotation $R_{\vec{n}}^{\theta}$ followed by a translation by $\vec{a}$, then any physical quantity $Q(\vec{x})$ resulting from solving the system must satisfy

$$Q(R_{\vec{n}}^{\theta}\vec{x} + \vec{a}) = \rho(R_{\vec{n}}^{\theta}, \vec{a})Q(\vec{x}), \tag{4.120}$$

where ρ is a representation of rotations and translations on the space that Q belongs to, i.e., the value of Q at any points that are mapped into each other by the symmetry must be related by this representation. Note that if Q is a scalar, this representation will be the trivial representation. It is here important to note that the system must be completely determined with a unique solution in order to apply this argument. If we consider a partial differential equation in a region that displays a certain symmetry, it is also necessary that the boundary conditions also display this symmetry in order for this argument to apply. If the boundary conditions do not display the symmetry, or are unknown, the argument does not apply.

Example 4.38 Let us again return to Example 1.34 for which we considered the electric field $\vec{E}(\vec{x})$ for a spherically symmetric charge distribution. In Example 4.36, we argued by rotational symmetry that $\vec{E}(\vec{x}) = E(\vec{x})\vec{e}_r$. In order to do so, we considered rotations around an axis connecting the origin with $\vec{x}$. If we now consider general rotations $\vec{x}' = R_{\vec{n}}^{\theta}\vec{x}$, we find that the electric field $\vec{E}'$ after rotation will be given by

$$\vec{E}'(\vec{x}') = R_{\vec{n}}^{\theta}\vec{E}(\vec{x}) = E(\vec{x})R_{\vec{n}}^{\theta}\vec{e}_r(\vec{x}), \tag{4.121}$$

where we have explicitly written out the fact that $\vec{e}_r$ depends on the position $\vec{x}$. We now apply the relation

$$R_{\vec{n}}^{\theta}\vec{e}_r(\vec{x}) = \frac{1}{r}R_{\vec{n}}^{\theta}\vec{x} = \frac{1}{r}\vec{x}' = \vec{e}_r(\vec{x}'), \tag{4.122}$$

which leads to

$$\vec{E}'(\vec{x}') = E(\vec{x})\vec{e}_r(\vec{x}'). \tag{4.123}$$

If the system is symmetric under rotations, this must be equal to the field at $\vec{x}'$ before the rotation, i.e.,

$$\vec{E}'(\vec{x}') = \vec{E}(\vec{x}') = E(R_{\vec{n}}^{\theta}\vec{x})\vec{e}_r(\vec{x}') \implies E(\vec{x}) = E(R_{\vec{n}}^{\theta}\vec{x}). \tag{4.124}$$

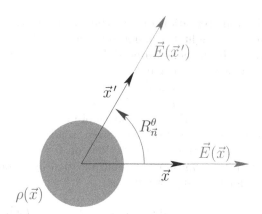

Figure 4.16 The electrical field $\vec{E}$ at positions $\vec{x}$ and $\vec{x}' = R^\theta_{\vec{n}}\vec{x}$ around a spherically symmetric charge distribution can be related by symmetry arguments. The symmetry requires that the field is on the form $\vec{E} = E(r)\vec{e}_r$.

Consequently, the function $E(\vec{x})$ must be equal for all points that can be mapped into each other by rotations, which are all the points with the same radial coordinate r. The function can therefore only be a function of this coordinate, since points with different θ and φ but the same r may be mapped to each other by rotations. We can therefore write $E(\vec{x}) = E(r)$. This situation is illustrated in Fig. 4.16.

4.6.2 Important transformations in physics

Symmetries of physical systems play a central role in modern physics. We have already seen examples of important transformations in the form of rotations and translations. There are some other transformations that are of great importance and we will briefly discuss them here before we go on to dig a bit deeper into representation theory.

4.6.2.1 Time translations and reversal

A common and important question is to consider what happens to a physical system under *time translations and reversal*. Just as spatial translations and rotations are defined by how they act on spatial coordinates, the time translations and reflections $T^s_\pm$ can be defined based on how they act on the time coordinate t

$$T^s_\pm t = \pm t + s, \tag{4.125}$$

where the positive sign corresponds to a pure time translation while the negative sign gives a time translation and reversal. It should be clear that these transformations constitute a group with the identity element T^0_+ for the very same reasons as translations and reversal of any spatial coordinate is. The group has one continuous generator E, which generates translations, and one discrete generator σ, which generates reversal. As we go deeper into the more mathematical formulation of classical mechanics we shall find that invariance under time translations is intimately related to the conservation of energy.

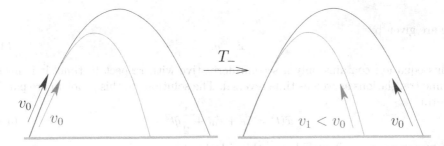

Figure 4.17 The left part of the figure shows a throw parabola (dark) which is the path taken by an object in a homogeneous gravitational field in the absence of air resistance, as well as the path taken by an object subject to air resistance proportional to the velocity (light). The time reversal of the parabola would also be a viable solution to the equations of motion, while the path taken when air resistance is present is not. When including air resistance, the time reversed path would seem to gain energy and speed up.

In order to examine how physical relations transform under time transformations, we should first note that most relations depend only on differences in time and not on the actual value of the time coordinate t. This is in direct analogy to spatial transformations, where it does not matter where we place our origin. As a result, physical relations generally contain derivatives with respect to time and the interesting question becomes how these transform under time transformations. Answering this is a matter of applying the chain rule for derivatives. Taking $t' = T_\pm^s t$, we find that

$$\frac{d}{dt} = \frac{dt'}{dt}\frac{d}{dt'} = \pm\frac{d}{dt'}. \tag{4.126}$$

It directly follows that any differential equation in t is invariant under time translations, while invariance under time reversal occurs when the differential equation contains only even powers of d/dt.

The question is now what invariance under time translations and reversals implies for physics. Starting by looking at time translations, let us assume that we have a system of differential equations in t that displays time translational invariance. If we have a solution to the system, applying a time translation to that solution will give a new solution. Naturally, the solution will generally not fulfil the same initial conditions, but it will be a solution to the differential equations for the time translated initial conditions. This is important in physics as it is related to the repeatability of experiments, i.e., if we repeat an experiment with the same initial conditions, then we expect to obtain the same results. On the other hand, invariance under time reversal means that, given a solution, it would still be a solution to the same differential equations if time was going backwards. If we could make a video recording of a physical process invariant under time reversal, we would not be able to tell whether the tape was being played forward or backward as both options would be physically viable. Systems that are not invariant under time reversal are generally dissipative and playing the tape backwards would show unphysical behaviour such as diffusion systems gathering all substance in one specific location or objects accelerating due to friction.

Example 4.39 Let us consider the motion of a particle under the influence of a homogeneous gravitational field $\vec{g}$. In the absence of air resistance, the equations of motion for this particle are given by

$$\ddot{\vec{x}} = \vec{g}. \tag{4.127}$$

Since this equation contains only a second derivative with respect to time, it is invariant under time translations as well as time reversal. The solutions to this problem are parabolae of the form

$$\vec{x}(t) = \vec{x}_0 + \vec{v}_0 t + \frac{1}{2}\vec{g}t^2 \tag{4.128a}$$

and the transformation $T_\pm^s$ would map this solution to

$$T_\pm^s \vec{x}(t) = \underbrace{\left(\vec{x}_0 + \vec{v}_0 s + \frac{1}{2}\vec{g}s^2\right)}_{=\vec{x}_0'} \pm \underbrace{(\vec{v}_0 + \vec{g}s)}_{=\pm\vec{v}_0'} t + \frac{1}{2}\vec{g}t^2, \tag{4.128b}$$

which is of the same form with $\vec{x}_0 \to \vec{x}_0'$ and $\vec{v}_0 \to \vec{v}_0'$ and therefore a different solution, see Fig. 4.17.

If we introduce air resistance proportional to the speed of the object, the equation of motion will instead read

$$\ddot{\vec{x}} = \vec{g} - k\dot{\vec{x}}, \tag{4.129}$$

where k is a constant. The appearance of the single time derivative in $\dot{\vec{x}}$ breaks the symmetry of the system under time reversal. The most blatant example of this would be a particle falling at the terminal velocity $\dot{\vec{x}} = \vec{g}/k$. Time reversal of this solution would lead to $T_-^0 \dot{\vec{x}} = -\vec{g}/k$, a particle rising with constant velocity against the gravitational field.

4.6.2.2 *Spatial reflections (parity)*

Until now, when looking at different coordinate systems, we have mainly considered coordinate systems where the basis vectors are ordered into a right-handed basis. We have seen that different Cartesian bases are related to each other by rotations isomorphic to the special orthogonal group $SO(N)$. If we take any right-handed Cartesian basis $\vec{e}_i$ and do the transformation $\vec{e}_1 \to -\vec{e}_1$, but keep the rest of the basis vectors fixed, our right-handed basis transforms into a left-handed one by a *spatial reflection* (or *parity* transformation). Naturally, it does not really matter which basis vector is reflected in this fashion, the choice of $\vec{e}_1$ is just one of convenience as it exists for any dimensionality of the space. Using a rotation, this may be transformed in such a way that it corresponds to flipping any of the other basis vectors instead. Additionally, if the number of dimensions is odd, we can perform a rotation resulting in all basis vectors being flipped, i.e., $\vec{e}_i \to -\vec{e}_i$, see Fig. 4.18. In an even number of dimensions, flipping all basis vectors is instead equivalent to a rotation. The group of rotations and spatial reflections is isomorphic to the orthogonal group $O(N)$ with proper rotations forming the $SO(N)$ subgroup.

We have seen and considered how vectors and tensors transform under proper rotations. When also considering spatial reflections, there is an additional classification to be made. Let us start by considering how scalars transform. The set of scalars constitute a one-dimensional representation and generally we have seen that they are in the trivial representation of any group. The additional possibility for a one-dimensional representation in the case of $O(N)$ is the map

$$\rho(a) = \det(a), \tag{4.130}$$

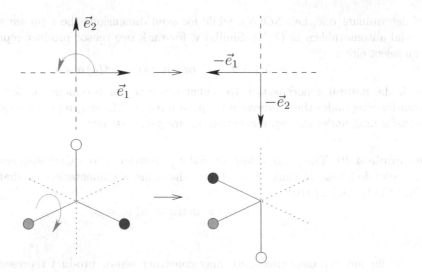

Figure 4.18 In two dimensions (upper figures) it is possible to perform a rotation that maps every basis vector according to $\vec{e}_i \to -\vec{e}_i$. In three dimensions (lower figures) this is no longer possible. Rotating two basis vectors into their negatives, the third will always end up pointing in the same direction as it originally did. In order to do the mapping $\vec{e}_i \to -\vec{e}_i$, an additional spatial reflection is necessary.

which is equal to ± 1 for any element a. We therefore have two possible transformations in a one-dimensional vector space,

$$\rho(a)\phi = \phi \quad \text{and} \quad \rho(a)\phi = \det(a)\phi. \tag{4.131}$$

Quantities transforming according to the first of these transformation properties are called (proper) scalars while quantities transforming according to the second are *pseudo-scalars*. While scalars take the same values regardless of the transformation, i.e., regardless of whether we consider a right- or left-handed coordinate system, pseudo-scalars transform as scalars under rotations but change sign under spatial reflections.

We can handle the possible representations of vectors and general tensors in a very similar fashion. For vectors, the canonical representation of a general rotation and spatial reflection would be

$$v^{i'} = \rho(a)v^i = a_i^{i'} v^i, \tag{4.132}$$

where $a_i^{i'}$ are the transformation coefficients defined in Chapter 1, now forming the components of a general $O(N)$ matrix. However, as for the scalars, this representation can be multiplied by the determinant $\det(a)$ to obtain the representation

$$\rho(a) = \det(a)a, \tag{4.133}$$

where a is an element of $O(N)$. Verifying that this is a homomorphism is left as Problem 4.41. It should be noted that

$$\det(\det(a)a) = \det(a)^N \det(a) = \det(a)^{N+1} = (\pm 1)^{N+1}. \tag{4.134}$$

Thus, in vector spaces of odd dimension, the representation $\det(a)a$ is in terms of matrices of determinant one, i.e., $SO(N)$, while for even dimensions the representation is an nontrivial automorphism of $O(N)$. Similarly, for rank two tensor product representations, we can select either

$$\rho_+(a) = a \otimes a \quad \text{or} \quad \rho_-(a) = \det(a)\, a \otimes a, \tag{4.135}$$

with the natural generalisation to higher order tensor products. As for scalars, objects transforming under the representation ρ_+ are referred to as (proper) tensors, while objects transforming under the representation ρ_- are *pseudo-tensors*.

Example 4.40 The *permutation symbol* ε_{ijk} restricted to a Cartesian coordinate system is a pseudo-tensor of rank three. Using the same argumentation as that leading up to Eq. (2.114), we find that

$$\varepsilon_{i'j'k'} = \det(a)\, a_i^{i'} a_j^{j'} a_k^{k'}\, \varepsilon_{ijk}. \tag{4.136}$$

As for any representations, we may construct tensor product representations also by using the pseudo-tensor representations. In general, due to $(-1)^2 = 1$, we find that the tensor product of two pseudo-tensor representations is a tensor representation, while the tensor product of a pseudo-tensor representation and a tensor representation is a new pseudo-tensor representation.

Let us take a step back and consider the physical implications of a quantity transforming as a tensor or pseudo-tensor. In particular, we will look at vectors and pseudo-vectors and work in $N = 3$ dimensions, where a particular spatial reflection is given by transforming all basis vectors $\vec{e}_i$ to $-\vec{e}_i$ and thus all (proper) vectors $\vec{v}$ are transformed to $-\vec{v}$. From this follows that any product involving two vectors is going to be invariant under this transformation, in particular, the cross product

$$\vec{v} \times \vec{w} \to (-\vec{v}) \times (-\vec{w}) = \vec{v} \times \vec{w} \tag{4.137}$$

and the direct product

$$\vec{v} \otimes \vec{w} \to (-\vec{v}) \otimes (-\vec{w}) = \vec{v} \otimes \vec{w}. \tag{4.138}$$

We learn two things from this consideration. First of all, the cross product $\vec{v} \times \vec{w}$ does not transform as a proper vector as it does not change sign under the reflection. The difference in sign tells us that it is a pseudo-vector. This is also consistent with the cross product being the contraction between two vectors and the permutation symbol, which is a pseudo-tensor. The second thing we learn is that tensors of rank two do not change sign under this transformation and we are starting to see a pattern. Extending this argument to tensors of arbitrary rank, we find that tensors of even rank, such as scalars and rank two tensors, are invariant under $\vec{e}_i \to -\vec{e}_i$, while tensors of odd rank, such as vectors, acquire a change of sign. This is summarised in Table 4.2.

Example 4.41 As we have seen, cross-products between vectors are generally pseudo-vectors. This already gives us a large number of physical examples of pseudo-vectors. One of the more important examples of this is the *angular momentum* of a particle, given by

$$\vec{L} = \vec{x} \times \vec{p}, \tag{4.139}$$

where $\vec{x}$ is the position vector and $\vec{p}$ the (linear) momentum. In the same fashion, the contribution from a small volume dV to the angular momentum of a general mass distribution

Tensor rank	Proper tensor	Pseudo-tensor
Zero (scalar)	$\phi \to \phi$	$\phi \to -\phi$
One (vector)	$\vec{v} \to -\vec{v}$	$\vec{v} \to \vec{v}$
Two	$T \to T$	$T \to -T$
k	$T \to (-1)^k T$	$T \to (-1)^{k+1} T$

Table 4.2   The behaviour of tensors and pseudo-tensors under the spatial reflection $\vec{e}_i \to -\vec{e}_i$ in $N = 3$ dimensions. These properties also hold in any odd number of dimensions.

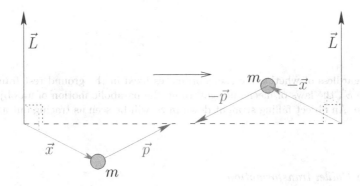

Figure 4.19   The behaviour of the angular momentum $\vec{L} = \vec{x} \times \vec{p}$ of a mass m under the transformation $\vec{e}_i \to -\vec{e}_i$. While the vectors $\vec{x}$ and $\vec{p}$ transform as $\vec{x} \to -\vec{x}$ and $\vec{p} \to -\vec{p}$, the angular momentum pseudo-vector is left invariant under the reflection.

is given by

$$dL = \vec{x} \times d\vec{p} = \rho\, \vec{x} \times \vec{v}\, dV \tag{4.140}$$

and is also a pseudo-vector. Under the spatial reflection $\vec{e}_i \to -\vec{e}_i$, the angular momentum transforms according to

$$\vec{L} \to (-\vec{x}) \times (-\vec{p}) = \vec{x} \times \vec{p} = \vec{L}. \tag{4.141}$$

This is illustrated in Fig. 4.19.

4.6.2.3   Galilei transformations

Already Newtonian mechanics has the notion of *inertial frames* built into it. An inertial frame is a frame of reference in which the laws of physics hold in their simplest form. For example, Newton's laws hold as stated in all inertial frames. All inertial frames are related to each other by a fixed rotation and a linearly time dependent translation. As such, they move with constant velocity with respect to each other. An *event* is specified by giving a time and a point in space, i.e., it specifies when and where, and is described by four coordinates $(t, \vec{x})$ in *space-time* for any inertial frame.

The relation between two inertial frames with coordinates $(t, \vec{x})$ and $(t', \vec{x}')$, respectively,

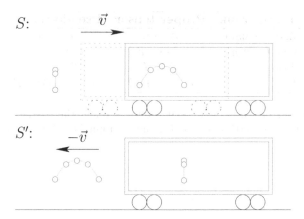

Figure 4.20 Regardless of whether we use coordinates fixed in the ground rest frame S or in the train rest frame S', the laws of mechanics governing the parabolic motion of an object in free fall will be the same. An object falling straight down in S' will be seen as tracing out a parabola in S and vice versa.

is given by the *Galilei transformation*

$$t \to t' = t + s, \tag{4.142a}$$

$$\vec{x} \to \vec{x}' = R_{\hat{n}}^\theta \vec{x} + \vec{v}t + \vec{a}, \tag{4.142b}$$

where $\vec{v}$ is the relative velocity between the frames (in the primed frame). The translations in time and space are specified by s and $\vec{a}$, while the rotation is given by $R_{\hat{n}}^\theta$. It is often common to restrict the Galilei transformations to inertial frames with a shared space-time origin, i.e., $t = 0$ and $\vec{x} = 0$ is the same event as $t' = 0$ and $\vec{x}' = 0$. A Galilei transformation with a non-zero velocity $\vec{v}$ is called a *boost*.

Example 4.42 Consider a passenger performing physical experiments on board a train moving at constant velocity $\vec{v}$ relative to the ground. For this passenger, it is convenient to use a set of coordinates S' fixed relative to the train in order to describe the surroundings. An external observer standing on the ground might describe the same physics using a frame S that is fixed relative to the ground and will be able to describe the same experiments using the very same physical laws as the passenger but using the alternative coordinate system. The two coordinate systems are related by a Galilei transformation, see Fig. 4.20. In this case, the transformation has been chosen such that the velocity $\vec{v} = -v\vec{e}_1$ and such that $y = y'$ and $z = z'$. The resulting transformation

$$t' = t, \tag{4.143a}$$

$$x' = x - vt, \tag{4.143b}$$

$$y' = y, \tag{4.143c}$$

$$z' = z, \tag{4.143d}$$

is the Galilei transformation in *standard configuration*.

The set of Galilei transformations form the *Galilean group* and are perhaps not as iconic as their relativistic counterpart, the *Lorentz transformations*, that describe the relativistic relations between space-time coordinates in different inertial frames sharing the origin. The relativistic counterpart of the more general Galilei transformation, without a common origin, is called the *Poincaré transformation*.

4.7 IRREPS AND CHARACTERS

We shall now turn to a branch of group theory that will require a more theoretical treatment, which will take us some time to work through. However, the rewards in terms of the insights we will be able to draw from the symmetries in a physical system will be well worth the effort in the end. We will mainly work with representations of finite groups in terms of unitary matrices. The framework also readily extends to compact Lie groups, but that discussion requires a notion of integration over the group, which we will not cover here. It should be noted that the restriction to unitary representations is not constraining, since any matrix representation (seen as a representation on a complex vector space) is equivalent to a unitary representation.

4.7.1 Irreducible representations

We have already discussed how some representations are reducible. In particular, we have seen how the tensor product representation $\rho \otimes \rho$ may be decomposed into the direct sum of its symmetric and anti-symmetric parts

$$\rho \otimes \rho = (\rho \otimes_+ \rho) \oplus (\rho \otimes_- \rho). \tag{4.144}$$

As we also saw, it is sometimes possible to further reduce these representations into direct sums of representations of lower dimension. Naturally, this cannot go on forever. At the very least, the process must stop when we reach one-dimensional representations, since one dimensional vector spaces cannot be split into direct sums of lower dimensional ones. We call representations that cannot be reduced *irreducible representations* or, the more common short-hand, *irreps*. For every finite group, we will find that there are only a finite number of (inequivalent) irreps and it will be possible to write any representation as a direct sum of these.

Example 4.43 The finite symmetric group S_3 of permutations of three elements can be represented by the three-dimensional representation given by the mapping of the generators (12) and (13)

$$\rho((12)) = \begin{pmatrix} 0 & 1 & 0 \\ 1 & 0 & 0 \\ 0 & 0 & 1 \end{pmatrix} \quad \text{and} \quad \rho((13)) = \begin{pmatrix} 0 & 0 & 1 \\ 0 & 1 & 0 \\ 1 & 0 & 0 \end{pmatrix}. \tag{4.145}$$

It is straightforward to check that this representation is unitary and that the vector space V_1 spanned by

$$v_1 = \begin{pmatrix} 1 \\ 1 \\ 1 \end{pmatrix} \tag{4.146}$$

is invariant under this representation, since v_1 is an eigenvector of both $\rho((12))$ and $\rho((13))$

with eigenvalue one. We may therefore write this representation as a direct sum of representations on V_1 and V_2, where V_2 is the set of vectors orthogonal to v_1. In V_2, it is impossible to find any vector that is a simultaneous eigenvector of the representations of both generators and the representation on V_2 is therefore also irreducible.

By the definition of an irrep ρ on a vector space V, for any non-zero vector $\vec{v} \in V$, the entire vector space V must be spanned by vectors of the form $\rho(a)\vec{v}$, where a are the elements of the represented group. If this is not the case, then the subspace spanned by those vectors is invariant under the representation, which therefore is reducible. We will use this fact extensively in the upcoming discussion.

4.7.2 Schur's lemmas and the orthogonality theorem

The classification of irreps of any finite group rests upon two mathematical facts, known as *Schur's lemmas*, which restrict the forms of linear operators on the vector spaces on which irreducible representations act. Schur's lemmas state that:

1. Given an irrep ρ on a vector space V, any linear operator A that commutes with $\rho(a)$ for all group elements a is proportional to the identity operator on V. In other words, if for all vectors $\vec{v}$ in V and all group elements a

$$\rho(a)A\vec{v} = A\rho(a)\vec{v}, \tag{4.147}$$

then $A\vec{v} = \lambda\vec{v}$, where λ is a constant.

2. Given two *inequivalent* irreps ρ^1 and ρ^2 acting on the vector spaces V_1 and V_2, respectively, the only linear transformation B from V_1 to V_2 fulfilling

$$B\rho^1(a) = \rho^2(a)B \tag{4.148}$$

for all group elements a is the transformation mapping all elements of V_1 to zero.

The first of these lemmas follows from considering an eigenvector $\vec{v}$ of A, which always exists due to the fundamental theorem of algebra guaranteeing the existence of an eigenvalue. By definition, this vector satisfies $A\vec{v} = \lambda\vec{v}$ for some λ. Acting with A on $\rho(a)\vec{v}$ leads to

$$A\rho(a)\vec{v} = \rho(a)A\vec{v} = \rho(a)\lambda\vec{v} = \lambda\rho(a)\vec{v} \tag{4.149}$$

and it follows that $\rho(a)\vec{v}$ is also an eigenvector of A with the same eigenvalue λ. Since vectors of the form $\rho(a)\vec{v}$ spans the entire vector space V, it must hold that $A\vec{v} = \lambda\vec{v}$ for all vectors $\vec{v}$ in V and thus the operator A is proportional to the identity operator with proportionality constant λ.

The argumentation behind the second lemma is slightly more involved and requires us to split it into three different cases based on the dimensionality of the two irreps. If the dimensions of the irreps are n_1 and n_2, respectively, then we can start by considering the case when $n_2 < n_1$. The subspace K of V_1 such that

$$B\vec{v} = 0 \tag{4.150}$$

if $\vec{v}$ is in K, i.e., the *kernel* of B, then has a dimension of at least $n_1 - n_2 \geq 1$. Taking any vector $\vec{v}$ in K, we find that

$$B\rho^1(a)\vec{v} = \rho^2(a)B\vec{v} = \rho^2(a)0 = 0, \tag{4.151}$$

implying that all of the vectors of the form $\rho^1(a)\vec{v}$ are in K, implying that the kernel K is actually the full space V_1, since ρ^1 was assumed to be irreducible. Therefore, $B\vec{v} = 0$ for all $\vec{v}$ in V_1, which proves the lemma for $n_2 < n_1$.

If we instead consider $n_2 \geq n_1$, then we can study the set of vectors $\vec{w}$ in V_2 for which

$$\vec{w} = B\vec{v}, \tag{4.152}$$

where $\vec{v}$ is an element of V_1, i.e., we are looking at the *image* I of the linear transformation B. By acting on $\vec{w}$ with the representation ρ^2, we find that

$$\rho^2(a)\vec{w} = \rho^2(a)B\vec{v} = B\rho^1(a)\vec{v}. \tag{4.153}$$

Since $\rho^1(a)\vec{v}$ is also an element of V_1, $\rho^2(a)\vec{w}$ is an element of I. It follows that either I is the full vector space V_2 or $\vec{w} = 0$. If the dimension of I is smaller than n_2, the only option is that $\vec{w} = 0$, again implying that $B = 0$. This is always the case if $n_1 < n_2$, since the image of B cannot have a dimension larger than n_1. If $n_1 = n_2$, then it is possible for I to have dimension n_2. However, in this case B is a bijection and therefore invertible, which leads to

$$\rho^1(a) = B^{-1}\rho^2(a)B. \tag{4.154}$$

This would mean that ρ^1 and ρ^2 were equivalent representations, breaking the assumption that they are inequivalent. The only remaining possibility is therefore that the dimension of I is smaller than n_2, completing the proof of the lemma by implying that $B = 0$ also in this case.

Let us use Schur's lemmas to derive an orthogonality relation among different irreps. For any group $\mathcal{G}$, we can number the possible inequivalent irreps ρ^μ by numbers $\mu = 1, 2, \ldots$. We denote the vector spaces the irreps act on by V_μ and their dimensionalities by n_μ. Given any two irreps ρ^μ and ρ^ν as well as a linear operator T from V_ν to V_μ, we can construct the linear operator

$$B = \sum_a \rho^\mu(a)T\rho^\nu(a^{-1}), \tag{4.155}$$

where the sum runs over all elements a in $\mathcal{G}$. For any element b of the group, we now find that

$$\rho^\mu(b)B = \sum_a \rho^\mu(b)\rho^\mu(a)T\rho^\nu(a^{-1}) = \sum_a \rho^\mu(ba)T\rho^\nu(a^{-1}). \tag{4.156a}$$

Introducing $c = ba$, the sum may be rewritten as a sum over c, leading to

$$\rho^\mu(b)B = \sum_c \rho^\mu(c)T\rho^\nu(c^{-1}b) = \sum_c \rho^\mu(c)T\rho^\nu(c^{-1})\rho^\nu(b) = B\rho^\nu(b). \tag{4.156b}$$

This is the very same type of relation that Schur's lemmas deal with, the first lemma for $\mu = \nu$ and the second lemma for $\mu \neq \nu$. In the latter case, the irreps are not equivalent and $B = 0$, while in the former case we find that $B = \lambda_T^\mu \mathbf{1}$, where $\mathbf{1}$ is the identity operator on V_μ and λ_T^μ generally depends on the representation ρ^μ as well as on the operator T. This may be summarised as

$$B = \sum_a \rho^\mu(a)T\rho^\nu(a^{-1}) = \lambda_T^\mu \delta_{\mu\nu}\mathbf{1}, \tag{4.157a}$$

where there is no sum over the repeated index μ on the right-hand side (this will be true for the rest of this section). Taking the ij component of this operator in any basis on V_μ and V_ν, we find that

$$\sum_a \rho_{ik}^\mu(a)T_{k\ell}\rho_{\ell j}^\nu(a^{-1}) = \lambda_T^\nu \delta_{\mu\nu}\delta_{ij}. \tag{4.157b}$$

While the discussion so far has been valid for a general operator T, we now select the operator T to be the operator where the component $T_{pq} = 1$ for fixed p and q is the only non-zero component, leading to $T_{ij} = \delta_{ip}\delta_{jq}$. For this operator, we also define $\lambda_T^\mu = \lambda_{pq}^\mu$. The previous equation now takes the form

$$\sum_a \rho_{ip}^\mu(a)\rho_{qj}^\nu(a^{-1}) = \lambda_{pq}^\mu \delta_{\mu\nu}\delta_{ij}. \tag{4.158}$$

Setting $\mu = \nu$ and $i = j$ in this equation and summing over this index results in

$$\lambda_{pq}^\mu n_\mu = \sum_a \rho_{qi}^\mu(a^{-1})\rho_{ip}^\mu(a) = \sum_a \rho_{qp}^\mu(e) = \delta_{qp}\sum_a 1 = \delta_{qp}N_{\mathcal{G}}, \tag{4.159}$$

where we have used that $\delta_{ii} = n_\mu$ for the representation ρ^μ, that the representation $\rho^\mu(e)$ of the unit element e is the identity matrix, and $N_{\mathcal{G}}$ is the order of the group $\mathcal{G}$. Reinserting this into Eq. (4.158), we finally arrive at the *fundamental orthogonality theorem* of irreducible representations

$$\sum_a \rho_{ip}^\mu(a)\rho_{qj}^\nu(a^{-1}) = \frac{N_{\mathcal{G}}}{n_\mu}\delta_{\mu\nu}\delta_{ij}\delta_{pq}. \tag{4.160}$$

4.7.3 Characters

As we have argued in Chapter 2, the physics of any situation should not depend on any particular choice of basis. Extending this argument to group representations, equivalent representations will correspond to different choices of basis in the vector space on which the representations act. In order to classify the representations, we therefore look for properties of representations that are the same for all equivalent representations. Since the trace has the cyclic property

$$\text{tr}(AB) = \text{tr}(BA) \tag{4.161}$$

for any two matrices A and B, we find that for any two equivalent representations ρ and $\rho' = U\rho U^{-1}$, it holds that

$$\text{tr}(\rho'(a)) = \text{tr}(U\rho(a)U^{-1}) = \text{tr}(U^{-1}U\rho(a)) = \text{tr}(\rho(a)) \tag{4.162}$$

and it directly follows that the *character*

$$\chi^\rho(a) = \text{tr}(\rho(a)) \tag{4.163}$$

of the representation ρ is a function from the group to the complex numbers and is the same for all equivalent representations.

Example 4.44 Consider the representation

$$\rho(R^\theta) = \begin{pmatrix} \cos(\theta) & -\sin(\theta) \\ \sin(\theta) & \cos(\theta) \end{pmatrix} \tag{4.164}$$

of rotations in two dimensions. The character of this representation is given by

$$\chi^\rho(R^\theta) = \text{tr}(\rho(R^\theta)) = 2\cos(\theta). \tag{4.165}$$

In particular, we can note that $\chi^\rho(R^0) = 2$, $\chi^\rho(R^{\pi/2}) = 0$, and $\chi^\rho(R^\pi) = -2$.

Example 4.45 The trivial representation $\rho(a) = 1$ also has a trivial character

$$\chi^\rho(a) = \text{tr}(\rho(a)) = \text{tr}(1) = 1 \qquad (4.166)$$

independent of the group element a. This will be of great value to us when we consider the decomposition of general representations into irreps and in particular when we are looking to reduce the possible forms of different tensors.

Since the identity element e is always represented by the unit matrix, we find that

$$\chi^\rho(e) = n_\rho, \qquad (4.167)$$

where n_ρ is the dimension of the representation ρ. Another important observation is that if a and b belong to the same conjugacy class, then we find that

$$\chi^\rho(a) = \chi^\rho(cbc^{-1}) = \text{tr}(\rho(c)\rho(b)\rho(c^{-1})) = \text{tr}(\rho(b)) = \chi^\rho(b). \qquad (4.168)$$

In other words, the characters of any elements of the same conjugacy class are equal and it suffices to find the character's value for one element of each conjugacy class to fully know the character.

If a representation can be written as a direct sum of other representations

$$\rho = \bigoplus_\mu \rho^\mu, \qquad (4.169)$$

it directly follows that

$$\chi^\rho(a) = \sum_\mu \chi^{\rho^\mu}(a), \qquad (4.170)$$

since the trace over the direct product vector reduces to the sum of the traces over each subspace.

Example 4.46 In Example 4.35, we decomposed the tensor product representation of $SO(2)$ into the direct sum $1 \oplus \rho_2 \oplus 1$. Taking the trace of Eq. (4.110), we find

$$\chi^{\rho_1 \otimes 1}(R^\theta) = 1 + \text{tr}(\rho_2(R^\theta)) + 1 = 2 + \chi^{\rho_2}(R^\theta) = 2[1 + \cos(2\theta)]. \qquad (4.171)$$

The ones in the middle step result from the trivial representation having the character one for all elements.

It is useful to write down the characters for all irreps of a given group in terms of a *character table*, where each row represents an irrep and the columns represent the conjugacy classes of the group. The entries in the table contain the values of the character of the representation for the given conjugacy class. It is customary to write down the conjugacy classes together with the number of elements in them as the top row. We will also make sure to remember which group the character table belongs to by writing it down in the upper left entry of the table.

S_3	e	$2C$	3σ
ρ^1	1	1	1
ρ^2	2	−1	0
ρ^3	1	1	−1

Table 4.3 The character table for the irreps of the group S_3. The upper row shows the group followed by its conjugacy classes along with the number of elements in each conjugacy class (when different from one). Each subsequent row starts with a representation followed by the value of its character for each conjugacy class.

Example 4.47 Let us again look at the group S_3, for which we split a three-dimensional representation into the trivial representation and a two-dimensional representation in Example 4.43. The conjugacy classes of this group are

$$e = \{e\}, \quad C = \{(123), (132)\}, \quad \text{and} \quad \sigma = \{(12), (23), (13)\}, \tag{4.172}$$

see Problem 4.22, where e is the identity element, C is the set of cyclic permutations, and σ the set of odd permutations. The representation given in Example 4.43 gives

$$\chi^\rho(e) = 3, \quad \chi^\rho(C) = \chi^\rho((123)) = 0, \quad \text{and} \quad \chi^\rho(\sigma) = \chi^\rho((12)) = 1. \tag{4.173}$$

Denoting the trivial representation by ρ^1 and the two-dimensional representation in the example by ρ^2, we find that $\chi^1(a) = 1$ for all a and consequently that $\chi^2 = \chi^\rho - \chi^1$ is given by

$$\chi^2(e) = 2, \quad \chi^2(C) = -1, \quad \text{and} \quad \chi^2(\sigma) = 0. \tag{4.174}$$

Both of these representations are irreducible, but there also exists one more one-dimensional irreducible representation, which we will here call ρ^3, given by

$$\rho^3(e) = \rho^3(C) = -\rho^3(\sigma) = 1. \tag{4.175}$$

Since the representation is one-dimensional, this directly gives the characters of the representation. All irrep characters for S_3 are summarised as a character table in Table 4.3.

A word of warning is necessary at this point. Since character tables are of great importance to many different fields, the notation may vary across different sources and irreps may have several different names as well. However, the basic structure of the character table remains the same.

4.7.3.1 Orthogonality of characters

The characters of irreps have a peculiar and very useful property. Using the fundamental orthogonality theorem, we can sum over $i = p$ and over $j = q$ in Eq. (4.160) and obtain

$$N_G \delta_{\mu\nu} = \sum_a \rho_{ii}^\mu(a)\rho_{jj}^\nu(a^{-1}) = \sum_a \chi^\mu(a)\chi^\nu(a^{-1}), \tag{4.176}$$

where $\chi^\mu(a) = \chi^{\rho^\mu}(a)$ is the character of the irrep ρ^μ. We now introduce the *character inner product*

$$\langle \chi, \chi' \rangle = \frac{1}{N_\mathcal{G}} \sum_a \chi(a)\chi'(a^{-1}) = \frac{1}{N_\mathcal{G}} \sum_a \chi(a)\chi'(a)^*, \tag{4.177}$$

where the last equality holds assuming that the representations are unitary, i.e., $\rho(a^{-1}) = \rho(a)^{-1} = \rho(a)^\dagger$, and we obtain

$$\langle \chi^\mu, \chi^\nu \rangle = \delta_{\mu\nu} \tag{4.178}$$

for all irreps. In other words, the irreps are orthonormal under the inner product. Since the character of any elements in the same conjugacy class are equal, the inner product may also be written as

$$\langle \chi, \chi' \rangle = \frac{1}{N_\mathcal{G}} \sum_{i=1}^m k_i \chi(C_i)\chi'(C_i)^*, \tag{4.179}$$

where C_i is the ith conjugacy class, k_i the number of elements in the class, and m is the number of conjugacy classes.

Example 4.48 Consider the representations ρ^2 and ρ^3 from Example 4.47. The inner product of the characters of these representations is given by

$$\langle \chi^2, \chi^3 \rangle = \frac{1}{6}[\chi^2(e)\chi^3(e)^* + 2\chi^2(C)\chi^3(C)^* + 3\chi^2(\sigma)\chi^3(\sigma)^*]$$

$$= \frac{1}{6}[2 \cdot 1 + 2 \cdot (-1) \cdot 1 - 3 \cdot 0 \cdot (-1)] = 0, \tag{4.180}$$

in accordance with the orthogonality of characters of inequivalent irreps.

4.7.3.2 Decomposition into irreps

We have arrived at the final aim of this section, the decomposition of any representation into a direct sum of irreps. As mentioned earlier, any reducible representation may be written as a direct sum of irreps

$$\rho = \bigoplus_k \rho^k, \tag{4.181a}$$

where $\rho^k = \rho^{\mu_k}$ is an irrep. In general, the direct sum may contain several copies of the same irrep and we may instead write this as

$$\rho = \bigoplus_\mu k_\mu \rho^\mu, \tag{4.181b}$$

where the sum is now taken over all irreps and k_μ is the number of times the irrep ρ^μ occurs in ρ. We find that the character of this representation is given by

$$\chi(a) = \sum_\mu k_\mu \chi^\mu(a) \tag{4.182}$$

and taking the inner product with the character χ^ν we find that

$$\langle \chi^\nu, \chi \rangle = \sum_\mu k_\mu \langle \chi^\nu, \chi^\mu \rangle = \sum_\mu \delta_{\mu\nu} k_\mu = k_\nu. \tag{4.183}$$

This is a wonderful result of deep significance that is worth all of the trouble we went through to arrive at it. Given the character χ of any representation, it tells us that we can find the number of times the representation contains an irrep ρ^ν by taking the inner product of χ with χ^ν.

Example 4.49 Again returning to Example 4.43, we found the character of the representation mentioned in that example to be

$$\chi(e) = 3, \quad \chi(C) = 0, \quad \text{and} \quad \chi(\sigma) = 1 \tag{4.184}$$

in Example 4.47. We could have found this result independent of any knowledge of the decomposition of the representation into irreps by taking the traces of the corresponding 3×3 matrices. Using Table 4.3, we can compute the inner products with the irreducible representations

$$\langle \chi^1, \chi \rangle = \frac{1}{6}[3 + 0 + 3] = 1, \tag{4.185a}$$

$$\langle \chi^2, \chi \rangle = \frac{1}{6}[6 + 0 + 0] = 1, \tag{4.185b}$$

$$\langle \chi^3, \chi \rangle = \frac{1}{6}[3 + 0 - 3] = 0, \tag{4.185c}$$

and thereby find that $\rho = \rho^1 \oplus \rho^2$.

Example 4.50 If we consider a tensor product representation $\rho \otimes \rho$, we find that the character is given by

$$\chi^{\rho \otimes \rho}(a) = \text{tr}(\rho(a) \otimes \rho(a)) = \rho_{ii}(a)\rho_{jj}(a) = \chi(a)^2. \tag{4.186}$$

Therefore, the character of the tensor product may be easily computed as long as we know the character of ρ. As an example, let us take the tensor representation $\rho_1 \otimes \rho_1$ of $SO(2)$ from Example 4.35 and use it to represent the finite subgroup of rotations by an angle $2\pi/3$. As mentioned earlier, we could apply similar considerations also to compact Lie groups, but we have here restricted ourselves to finite groups for simplicity. The group elements are e, c, and c^2, where the generator c is the counter-clockwise rotation by $2\pi/3$ and each element forms its own conjugacy class. The character of the two-dimensional representation

$$\rho_1(c) = \begin{pmatrix} \cos(2\pi/3) & -\sin(2\pi/3) \\ \sin(2\pi/3) & \cos(2\pi/3) \end{pmatrix} = \frac{1}{2}\begin{pmatrix} -1 & -\sqrt{3} \\ \sqrt{3} & -1 \end{pmatrix} \tag{4.187}$$

is given by

$$\chi(e) = 2, \quad \chi(c) = -1, \quad \text{and} \quad \chi(c^2) = -1. \tag{4.188}$$

Consequently, the character of the tensor product representation $\rho \otimes \rho$ is given by

$$\chi^{\rho_1 \otimes \rho_1}(e) = 4, \quad \chi^{\rho_1 \otimes \rho_1}(c) = 1, \quad \text{and} \quad \chi^{\rho_1 \otimes \rho_1}(c^2) = 1. \tag{4.189}$$

Taking the inner product with the character of the trivial representation, we find that

$$\langle \chi^1, \chi^{\rho_1 \otimes \rho_1} \rangle = \frac{1}{3}(4 + 1 + 1) = 2. \tag{4.190}$$

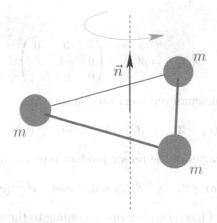

Figure 4.21 The symmetry of three equal masses m placed in the corners of an equilateral triangle may be used to argue for the corresponding moment of inertia tensor being of the form $I_0 \delta + I_1 \vec{n} \otimes \vec{n}$.

The tensor product representation therefore contains two copies of the trivial representation. This is exactly in accordance with the result we found in Example 4.35.

4.7.4 Physical insights

In the Section 4.6, we already discussed the physical implications of symmetries and group theory to some extent. The new tool that we have at hand allows us to take these implications one step further. Let us start by considering the possible degrees of freedom in a physical system with a given symmetry. We argued that if a physical system has a particular symmetry, then any property of the system should be invariant under the action of the symmetry, i.e., the property must transform according to the trivial representation of the symmetry group. As we have just seen, given any representation we can easily find out how many times the representation contains the trivial representation and therefore reduce the possible number of free parameters the property has in a system displaying the given symmetry.

Example 4.51 Let us consider the moment of inertia for the configuration shown in Fig. 4.21, consisting of three masses m at the vertices of an equilateral triangle. This system has a subgroup of rotations in $SO(3)$ as a symmetry and this subgroup is isomorphic to S_3, corresponding to a permutation of the masses. We know that the moment of inertia I_{ij} for any object is a rank two symmetric tensor and therefore transforms according to the representation $\rho \otimes_+ \rho$ under rotations, where ρ is the fundamental representation of $SO(3)$. Taking $\vec{e}_3$ as the axis perpendicular to the triangle, this representation is given by

$$\rho(R_{\vec{e}_3}^{2\pi/3}) = \rho(C) = \begin{pmatrix} -\frac{1}{2} & -\frac{\sqrt{3}}{2} & 0 \\ \frac{\sqrt{3}}{2} & -\frac{1}{2} & 0 \\ 0 & 0 & 1 \end{pmatrix}, \tag{4.191a}$$

and

$$\rho(R_{\tilde{e}_1}^{\pi}) = \rho(\sigma) = \begin{pmatrix} 1 & 0 & 0 \\ 0 & -1 & 0 \\ 0 & 0 & -1 \end{pmatrix}. \tag{4.191b}$$

The character of the fundamental representation is therefore given by

$$\chi^{\rho}(e) = 3, \quad \chi^{\rho}(C) = 0, \quad \text{and} \quad \chi^{\rho}(\sigma) = -1 \tag{4.192}$$

and consequently the character of the tensor product representation $\rho \otimes \rho$ is

$$\chi^{\rho \otimes \rho}(e) = 9, \quad \chi^{\rho \otimes \rho}(C) = 0, \quad \text{and} \quad \chi^{\rho \otimes \rho}(\sigma) = 1. \tag{4.193}$$

However, the momentum of inertia transforms according to the symmetric $\rho \otimes_+ \rho$ representation so we must figure out a way to find its characters. We could do this by introducing a basis on the space of symmetric tensors, but this would involve constructing 6×6 matrices and taking their trace. Instead, we can use the fact that

$$\rho \otimes \rho = (\rho \otimes_+ \rho) \oplus (\rho \otimes_- \rho) \tag{4.194a}$$

and consequently

$$\chi^{\rho \otimes_+ \rho} = \chi^{\rho \otimes \rho} - \chi^{\rho \otimes_- \rho}, \tag{4.194b}$$

meaning that we may find the characters of $\rho \otimes_+ \rho$ by constructing the characters of $\rho \otimes_- \rho$. This task is slightly less daunting, as we will be working with 3×3 matrices rather than 6×6 ones.

For the anti-symmetric product space, we take the basis

$$E_1 = e_{23} - e_{32}, \quad E_2 = e_{31} - e_{13}, \quad \text{and} \quad E_3 = e_{12} - e_{21}. \tag{4.195}$$

In this basis, the resulting 3×3 matrix representation is given by

$$(\rho \otimes_- \rho)(C) = \begin{pmatrix} -\frac{1}{2} & \frac{\sqrt{3}}{2} & 0 \\ -\frac{\sqrt{3}}{2} & -\frac{1}{2} & 0 \\ 0 & 0 & 1 \end{pmatrix}, \quad (\rho \otimes_- \rho)(\sigma) = \begin{pmatrix} 1 & 0 & 0 \\ 0 & -1 & 0 \\ 0 & 0 & -1 \end{pmatrix}, \tag{4.196}$$

leading to the characters

$$\chi^{\rho \otimes_- \rho}(e) = 3, \quad \chi^{\rho \otimes_- \rho}(C) = 0, \quad \text{and} \quad \chi^{\rho \otimes_- \rho}(\sigma) = -1 \tag{4.197}$$

and therefore

$$\chi^{\rho \otimes_+ \rho}(e) = 6, \quad \chi^{\rho \otimes_+ \rho}(C) = 0, \quad \text{and} \quad \chi^{\rho \otimes_+ \rho}(\sigma) = 2. \tag{4.198}$$

Taking the inner product with the trivial representation, we find that

$$\langle \chi^1, \chi^{\rho \otimes_+ \rho} \rangle = \frac{1}{6}(6 + 0 + 3 \cdot 2) = 2, \tag{4.199}$$

indicating that the moment of inertia of an object with the given symmetry has two possible degrees of freedom. What this symmetry argument does not tell us is the form or values of these, but we can reduce the amount of computation necessary in order to find them and be sure that we have found them all once we have done so.

Another physics application of characters and representation theory is *symmetry breaking*. There are many situations in physics that have an approximate symmetry, meaning that a property does not change significantly under a certain group transformation. As a first approximation, such a system and its most important characteristics may be found by treating the system as if it was symmetric. Any corrections to the results will then be proportional to the level at which the symmetry is broken. If the symmetry is broken to a subgroup of the approximate symmetry, the possible corrections due to the symmetry can also be deduced from group theory.

Example 4.52 Let us again consider the moment of inertia for an object with the S_3 symmetry we encountered in the previous example, but this time we attach the three masses on a heavy spherical shell of mass $M \gg m$. Since the masses m are small in comparison, the system is approximately symmetric under the full group of $SO(3)$ rotations. Its moment of inertia, being a rank two symmetric tensor, then transforms according to the symmetric tensor product representation $\rho \otimes_+ \rho$, which we have seen can be reduced to a one dimensional trivial representation ρ^1 and a five-dimensional one ρ^5 consisting of symmetric traceless tensors of rank two, which also happens to be an irrep of $SO(3)$, i.e.,

$$\rho \otimes_+ \rho = \rho^1 \oplus \rho^5. \tag{4.200}$$

Since the representation contains the trivial representation only once, only this invariant subspace may have a non-zero contribution in the first approximation. We find that

$$I_{ij} = k_0 M R^2 \delta_{ij} + m R^2 T_{ij}, \tag{4.201}$$

where k_0 is a constant of order one, R is the radius of the sphere, and the traceless second order correction T_{ij} is still to be determined, but will generally have components of order one at most. The first term, proportional to δ_{ij}, represents the subspace that transforms under the trivial representation, which may be seen by applying an arbitrary rotation.

In the second order correction, the system will still be symmetric under the S_3 subgroup. Since the ρ^1 irrep is invariant under the action of $SO(3)$, it will also be invariant under the action of S_3. However, the representation ρ^5 may now contain subspaces that were not invariant under the general $SO(3)$ action, but are invariant when the action is restricted to that of S_3. As in the previous example, we find that the representation $\rho \otimes_+ \rho$ contains the trivial representation ρ^1 two times. Naturally, the subspace invariant under the full $SO(3)$ rotations is one of these, but there is an additional trivial subspace that was earlier part of the ρ^5 irrep of $SO(3)$. It is relatively straightforward to conclude that this subspace corresponds to

$$(T_{ij}) = k_1 \begin{pmatrix} -1 & 0 & 0 \\ 0 & -1 & 0 \\ 0 & 0 & 2 \end{pmatrix}. \tag{4.202}$$

The full second order correction to the moment of inertia is therefore of the form

$$(T_{ij}) = k_1 \begin{pmatrix} -1 & 0 & 0 \\ 0 & -1 & 0 \\ 0 & 0 & 2 \end{pmatrix} + k_1'(\delta_{ij}). \tag{4.203}$$

As for the example above, it is generally true that the action of any subgroup will leave any irrep of the full group invariant. A direct implication of this is that if we have already split the original representation into irreps of the full group, we only need to consider how the different irreps break into irreps of the subgroup.

Let us finally discuss the vibrational modes of molecules that exhibit certain symmetries. Assume that we have a molecule with a number M different atoms. We can write down the total energy of the molecule as

$$E = \sum_{k=1}^{M} \frac{m_k}{2} \dot{\vec{x}}_i^2 + V, \tag{4.204}$$

where $\vec{x}_k$ is the displacement from the equilibrium position of the kth atom, m_k is the mass of the kth atom, and V is the potential energy of the system, which generally depends on all relative displacements. In order to write this in a somewhat more manageable form, we introduce the $3M$-dimensional vector $\vec{X}$, where

$$\vec{X} = \sum_{k=1}^{M} \frac{1}{\sqrt{m_k}} (x_k^1 \vec{e}_{3k-2} + x_k^2 \vec{e}_{3k-1} + x_k^3 \vec{e}_{3k}), \tag{4.205}$$

and $\vec{e}_i$ are the basis vectors of the $3M$-dimensional space (not of the three-dimensional one!). In other words, the first three components of $\vec{X}$ contain the displacements of the first atom divided by the square root of its mass and so on. In this notation, the total energy may be written as

$$E = \frac{1}{2} \dot{X}_\mu \dot{X}_\mu + \frac{1}{2} X_\mu V_{\mu\nu} X_\nu \tag{4.206}$$

for small oscillations. The form of the second term, where $V_{\mu\nu}$ is symmetric and semi-positive definite, follows from the assumption that $\vec{X}$ is the displacement from the equilibrium, where the potential has a minimum.

Differentiating the energy with respect to time gives the differential equation

$$\ddot{X}_\mu + V_{\mu\nu} X_\nu = 0 \tag{4.207}$$

and since V is symmetric, it has an orthogonal set of eigenvectors $\vec{Y}^q$ such that $V_{\mu\nu} Y_\nu^q = \omega_q^2 Y_\mu^q$ (no sum). The differential equations in the eigenvector directions decouple and we find the general solution

$$\vec{X} = \sum_{q=1}^{3M} A_q \vec{Y}^q f(\omega_q, t, \phi_q), \tag{4.208}$$

where

$$f(\omega_q, t, \phi_q) = \begin{cases} \cos(\omega_q t + \phi_q), & (\omega_q > 0) \\ t + \phi_q, & (\omega_q = 0) \end{cases}, \tag{4.209}$$

A_q is the amplitude of the vibration in the $\vec{Y}^q$ direction, and ϕ_q its phase when $\omega_q > 0$ and a translation in time if $\omega_q = 0$.

Example 4.53 Since the above discussion has been rather abstract, let us discuss an example in one dimension using only two atoms of mass m connected by a potential $V = k(d - d_0)^2/2$, where k is a constant, d is the distance between the atoms, and d_0 the distance at the equilibrium, see Fig. 4.22. If x_k is the displacement of particle k from the equilibrium position, then the energy is given by

$$E = \frac{m}{2}(\dot{x}_1^2 + \dot{x}_2^2) + \frac{k}{2}(x_1 - x_2)^2. \tag{4.210}$$

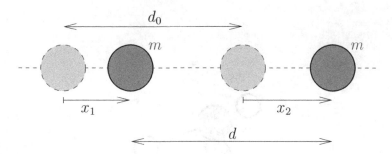

Figure 4.22 We consider two masses m with a potential such that a minimum is obtained whenever the masses are a distance d_0 apart. The mass displacements from one of the configurations where this is satisfied are x_1 and x_2, respectively. The potential for any displacements may be written as a function of the distance $d = d_0 + x_2 - x_1$ between the masses.

The vector $\vec{X}$ is now two-dimensional with components $X_k = x_k/\sqrt{m}$, leading to

$$E = \frac{1}{2}(\dot{X}_1^2 + \dot{X}_2^2) + \frac{k}{2m}(X_1^2 - 2X_1X_2 + X_2^2). \tag{4.211}$$

From this relation, we find that

$$(V_{\mu\nu}) = \frac{k}{m}\begin{pmatrix} 1 & -1 \\ -1 & 1 \end{pmatrix} \tag{4.212}$$

with $\omega_1 = 0$ and $\omega_2 = \sqrt{2m/k}$. The first eigenvalue corresponds to the eigenvector $\vec{Y}^1 = (\vec{e}_1 + \vec{e}_2)/\sqrt{2}$, which gives overall translations of the molecule, while the second case with the eigenvector $\vec{Y}^2 = (\vec{e}_1 - \vec{e}_2)/\sqrt{2}$ corresponds to the vibrational mode of the molecule with angular frequency $\omega_2 = \sqrt{2m/k}$.

So what does all of this have to do with symmetries and group theory? While it cannot help us deduce the actual angular frequencies ω_q, group theory will help us in determining the degeneracies in the spectrum of ω_q. If we consider a molecule with some symmetry group $\mathcal{G}$, then the action of the symmetry group on the displacement vector $\vec{X}$ will define a $3M$-dimensional representation ρ of $\mathcal{G}$. Given a symmetry represented by $\rho(a)$, then $\rho(a)\vec{X}(t)$ is a solution to the equations of motion as long as $\vec{X}(t)$ is and it follows that

$$\rho(a)\ddot{\vec{X}}(t) + V\rho(a)\vec{X}(t) = 0. \tag{4.213}$$

In addition, we can always act on the equation of motion for $\vec{X}(t)$ by $\rho(a)$ in order to obtain

$$\rho(a)\ddot{\vec{X}}(t) + \rho(a)V\vec{X}(t) = 0, \tag{4.214}$$

leading to the relation $V\rho(a) = \rho(a)V$ for all $\vec{X}(t)$. By Schur's first lemma, it follows that V is proportional to the identity on all irreps into which the representation ρ decomposes. In particular, this means that all states in the same irrep correspond to solutions with the same angular frequency. By decomposing the $3M$-dimensional representation into irreps, we can therefore find out how many different vibrational frequencies the molecule allows as

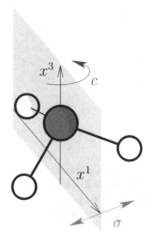

Figure 4.23 The ammonia molecule has a symmetry group generated by the rotation c by $2\pi/3$ around the x^3-axis and the reflection σ in the x^1-x^3-plane. Symmetry arguments may be used to find the form of its vibrational spectrum.

well as the degeneracy of each frequency, i.e., the number of linearly independent vibrations with the same frequency.

It should be noted that the above procedure will also include overall translational and rotational modes. These must be removed from the vibrational spectrum, but this can be performed by finding the corresponding representations, decomposing them into irreps, and removing the irreps from the final direct sum.

Example 4.54 Consider the ammonia molecule consisting of one nitrogen atom and three hydrogen atoms. The ammonia molecule has a symmetry group isomorphic to S_3 generated by the $2\pi/3$ rotation c around the x^3-axis and the reflection σ in the x^1-x^3-plane, see Fig. 4.23. The action of the symmetry on the resulting 12-dimensional representation is a permutation of the displacements for the three hydrogen atoms together with the $O(3)$ matrix corresponding to the transformation acting on each of the three-dimensional displacement vectors. The resulting representation is therefore a tensor product between the two representations describing the permutations of vertices and the rotations of the displacements at each vertex. The representation of S_3 described in Example 4.43, which we will here call ρ_S, describes the permutation of the hydrogen atoms, while the nitrogen atom is always at the same vertex and therefore transforming under the trivial representation ρ^1. Furthermore, the displacements transform according to the fundamental representation ρ_V of $O(3)$ constrained to the S_3 subgroup. The full representation is therefore given by

$$\rho = (\rho_S \oplus \rho^1) \otimes \rho_V \equiv \rho_p \otimes \rho_V. \tag{4.215}$$

In order to produce the characters of this representation, we could find the explicit 12×12 matrix representation, but a much more straightforward way is to use our knowledge of how the characters of direct sums and tensor products behave. We find that the character of the direct sum $\rho_p = \rho_S \oplus \rho^1$ is given by

$$\chi_{\rho_p} = \chi_{\rho_S} + \chi^1, \tag{4.216a}$$

leading to

$$\chi_{\rho_P}(e) = 4, \quad \chi_{\rho_P}(C) = 1, \quad \text{and} \quad \chi_{\rho_P}(\sigma) = 2. \tag{4.216b}$$

Similarly, for the tensor product representation the character is the product of the characters of the factors

$$\chi_\rho(a) = \chi_{\rho_P \otimes \rho_V}(a) = \chi_{\rho_P}(a)\chi_{\rho_V}(a) \tag{4.217a}$$

and we find that

$$\chi_\rho(e) = 12, \quad \chi_\rho(C) = 0, \quad \text{and} \quad \chi_\rho(\sigma) = 2. \tag{4.217b}$$

Taking the inner product with the irreps ρ^1, ρ^2, and ρ^3, whose characters are given in Table 4.3, we find that

$$\langle \chi^1, \chi_\rho \rangle = \frac{1}{6}(12 \cdot 1 + 2 \cdot 0 \cdot 1 + 3 \cdot 2 \cdot 1) = 3, \tag{4.218a}$$

$$\langle \chi^2, \chi_\rho \rangle = \frac{1}{6}(12 \cdot 2 + 2 \cdot 0 \cdot (-1) + 3 \cdot 2 \cdot 0) = 4, \tag{4.218b}$$

$$\langle \chi^3, \chi_\rho \rangle = \frac{1}{6}(12 \cdot 1 + 2 \cdot 0 \cdot 1 + 3 \cdot 2 \cdot (-1)) = 1. \tag{4.218c}$$

The decomposition of ρ into irreps is therefore given by

$$\rho = 3\rho^1 \oplus 4\rho^2 \oplus \rho^3. \tag{4.219}$$

As already mentioned, this representation also contains the translational and rotational modes of the molecule, which must be removed if the representation should contain only the vibrational modes. The translational modes are described by a single displacement vector, which transforms under the fundamental representation ρ_V, while the rotational displacements are described by vectors of the form $\theta\vec{n} \times \vec{x}$, which transform according to the pseudo-vector representation ρ_A (we use the A to denote the pseudo-vector representation as pseudo-vectors are also referred to as *axial vectors*). With our knowledge of characters and decomposition into irreps, it is straightforward (see Problem 4.50) to show that these representations reduce into $\rho_V = \rho^1 \oplus \rho^2$ and $\rho_A = \rho^2 \oplus \rho^3$, respectively. Consequently, we find that the vibrational degrees of freedom transform according to the six-dimensional representation

$$\rho_{\text{vib}} = 2\rho^1 \oplus 2\rho^2. \tag{4.220}$$

In other words, the vibrational spectrum will generally contain four different frequencies, two of which have a twofold degeneracy.

4.8 OUTLOOK

We have barely just begun to scrape on the top of the use of symmetries and group theory in physics, but we have laid a fairly solid foundation for things to come. Let us briefly discuss the applications of group theory that will appear in more advanced physics examples.

The perhaps most iconic statement within classical physics is *Noether's theorem*, which relates the existence of continuous symmetries in physical systems to conserved quantities. All of the conservation laws we are familiar with from elementary classical mechanics are related to a corresponding symmetry. To give some examples, energy conservation relates to a system being invariant under time translations, while conservation of momentum relates

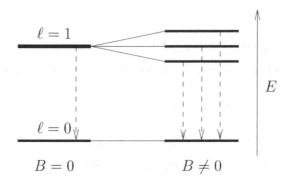

Figure 4.24 With zero external magnetic field, atomic states with the same angular momentum ℓ are degenerate in energy with degeneracy $2\ell+1$. As such, there is only one possible energy difference between states of different angular momentum. Once a magnetic field is applied, the states with the same angular momentum separate in energy, allowing for different energy transitions between angular momentum levels. Here, we illustrate this effect for the transitions between the $\ell = 1$ states to the $\ell = 0$ state.

to spatial translations. We shall discuss this in greater detail in Section 10.2.4, when we have developed the necessary framework to prove the theorem in one of its simpler forms.

In the study of quantum mechanics, we will also encounter our fair share of group theory. Quantum mechanics is a theory built upon linear operators on vector spaces and the properties of these operators and the states they act on will often be clearer when adding a seasoning of group theory. In particular, the concept of symmetry breaking will appear again and allow us to discuss how different degenerate energy eigenstates split into states with different energies when a symmetry is broken. A good example of such a symmetry breaking is the Zeeman effect, where the splitting of degenerate atomic energy levels due to an external magnetic field is considered. Before the introduction of the external magnetic field, the situation for the atom is symmetric under the full $SO(3)$ rotation group, where all states of the same total angular momentum form an irrep and consequently have the same energy. With the introduction of an external magnetic field, the $SO(3)$ symmetry is broken, splitting the representations of equal angular momentum into irreps of the new symmetry group, in this case one-dimensional irreps with a given component of the angular momentum in the magnetic field direction, see Fig. 4.24 for an illustration.

But this is not all, one of the most successful theories in modern theoretical physics, the Standard Model of particle physics, is based upon symmetries of the force carrying fields of the electromagnetic, weak, and strong interactions.

4.9 PROBLEMS

Problem 4.1. Consider the following surfaces:

a) an infinite two-dimensional plane $x^3 = 0$ and

b) an infinite cylinder $(x^1)^2 + (x^2)^2 = R^2$ in three dimensions.

Find the symmetry groups of each surface (as a subgroup of translations and rotations). Which is the largest group that is a subgroup of both of these symmetry groups?

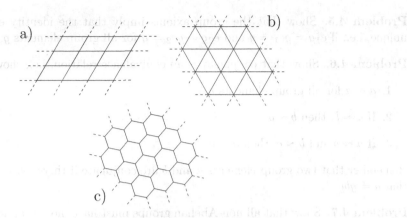

Figure 4.25 Three infinitely repeated grids based on (a) a parallelogram, (b) a triangular grid, and (c) a hexagonal honeycomb grid. In Problem 4.3, you are asked to find the corresponding symmetry groups.

Problem 4.2. Consider the density functions

$$\rho_1(\vec{x}) = \rho_0 e^{-k\vec{x}^2} \quad \text{and} \quad \rho_2(\vec{x}) = \rho_0 e^{-\vec{k}\cdot\vec{x}} \tag{4.221}$$

in three dimensions, where k and $\vec{k}$ are a scalar and a constant vector, respectively. Determine if they are invariant under the following transformations:

a) Rotations about the origin $\vec{x} = 0$.

b) Translations $\vec{x} \to \vec{x} + \vec{\ell}$.

c) Reflections $\vec{x} \to -\vec{x}$.

Problem 4.3. Find the symmetries that leave the infinite grids displayed in Fig. 4.25 invariant. How many generators are required for each of the symmetry groups? What relations among the generators can you find?

Problem 4.4. Two of Maxwell's equations are

$$\nabla \cdot \vec{B} = 0, \quad \nabla \times \vec{E} + \frac{\partial \vec{B}}{\partial t} = 0 \tag{4.222}$$

and imply that the electric and magnetic fields may be written in terms of the scalar potential $\phi(\vec{x}, t)$ and the vector potential $\vec{A}(\vec{x}, t)$ as

$$\vec{E} = -\nabla\phi - \frac{\partial \vec{A}}{\partial t}, \quad \vec{B} = \nabla \times \vec{A}. \tag{4.223}$$

Verify that the *gauge transformations*

$$\phi \to \phi - \frac{\partial \alpha}{\partial t}, \quad \vec{A} \to \vec{A} + \nabla\alpha, \tag{4.224}$$

where $\alpha = \alpha(\vec{x}, t)$ is a scalar field, leave the electric and magnetic fields invariant.

Problem 4.5. Show that the group axioms imply that the identity element e must be unique, i.e., if $e_1 g = g e_1 = g$ and $e_2 g = g e_2 = g$ for all group elements g, then $e_1 = e_2$.

Problem 4.6. Show that conjugacy is an equivalence relation, i.e., show that:

1. $a \sim a$ for all group elements a.

2. If $a \sim b$, then $b \sim a$.

3. If $a \sim b$ and $b \sim c$, then $a \sim c$.

Remember that two group elements a and b are conjugate if there exists an element g such that $a = gbg^{-1}$.

Problem 4.7. Show that all non-Abelian groups must have more than four elements based on the requirement that for any two elements x and y not to commute, it is necessary that xy and yx are different elements and that a non-Abelian group must contain at least two elements that do not commute.

Problem 4.8. Determine which of the following Cayley tables define a group:

1	a	b	c	1	a	b	c
a	b	c	1	a	c	1	b
b	1	a	c	b	1	c	a
c	b	1	a	c	b	a	1

1	a	b	c	d	1	a	b	c	d
a	b	c	d	1	a	1	d	b	c
b	c	d	1	a	b	c	1	d	a
c	d	1	a	b	c	d	a	1	b
d	1	a	b	c	d	b	c	a	1

If any of these Cayley tables does not define a group, state which group axiom is violated. The identity element has been denoted 1 in this problem.

Problem 4.9. We have argued that every element of rotations in two dimensions forms its own conjugacy class due to the group being Abelian. Expanding this group by including reflections, the group is no longer Abelian. Find the conjugacy classes of the resulting group of rotations and reflections.

Problem 4.10. Given two groups $\mathcal{G}_1$ and $\mathcal{G}_2$, the *direct product* $\mathcal{G}_1 \times \mathcal{G}_2$ can be defined as the set of pairs $a = (a_1, a_2)$, where a_1 is an element of $\mathcal{G}_1$ and a_2 an element of $\mathcal{G}_2$, with a group operation defined by

$$ab = (a_1, a_2)(b_1, b_2) = (a_1 b_1, a_2 b_2). \tag{4.225}$$

Verify that $\mathcal{G}_1 \times \mathcal{G}_2$ is a group.

Problem 4.11. Assume that a homomorphism ϕ is invertible, i.e., that there exists an inverse mapping ϕ^{-1} such that $\phi^{-1}(\phi(a)) = a$ for all a. Show that the inverse mapping ϕ^{-1} is also a homomorphism.

Problem 4.12. Show that being isomorphic is an equivalence relation, i.e., show that:

1. Any group $\mathcal{G}$ is isomorphic to itself.

2. If $\mathcal{G}_1$ is isomorphic to $\mathcal{G}_2$, then $\mathcal{G}_2$ is isomorphic to $\mathcal{G}_1$.

3. If $\mathcal{G}_1$ is isomorphic to $\mathcal{G}_2$ and $\mathcal{G}_2$ is isomorphic to $\mathcal{G}_3$, then $\mathcal{G}_1$ is isomorphic to $\mathcal{G}_3$.

Remember that a group $\mathcal{G}$ is isomorphic to another group $\mathcal{G}'$ if there exists a one-to-one homomorphism from $\mathcal{G}$ to $\mathcal{G}'$.

Problem 4.13. For a homomorphism h from $\mathcal{G}_1$ to $\mathcal{G}_2$, verify that the elements a of $\mathcal{G}_1$ for which

$$h(a) = e, \tag{4.226}$$

i.e., the set of elements that map to the identity of $\mathcal{G}_2$, is a subgroup of $\mathcal{G}_1$.

Problem 4.14. Verify that for a fixed group element g, the map $f_g(a) = gag^{-1}$ from the group to itself is an automorphism.

Problem 4.15. Show that the direct product $C_p \times C_q$ (see Problem 4.10) is isomorphic to C_{pq} if and only if p and q are coprime, i.e., if they do not have any common prime factors.

Problem 4.16. Determine which of the dihedral groups D_n are Abelian and find their conjugacy classes.

Problem 4.17. For the symmetry group of a set of points along a line with equidistant spacing ℓ discussed in Example 4.15, we found that translations T_ℓ by a distance ℓ and reflections σ generated the group and that $\sigma T_\ell \sigma = T_\ell^{-1}$. Verify that this implies $\sigma T_\ell^n \sigma = T_\ell^{-n}$.

Problem 4.18. Determine the conjugacy classes of the symmetry group discussed in Problem 4.17.

Problem 4.19. Perform the following group operations in the symmetric group S_4:

a) $(1234)(324)$

b) $(12)(23)(34)(14)$

c) $(123)(34)(123)$

d) $(24)(13)(12)(24)(13)$

Express your answers in terms of commuting cycles.

Problem 4.20. Determine the order of the following elements of the symmetric group S_6:

a) $(145)(236)$

b) $(1356)(24)$

c) $(162)(34)$

d) (12346)

Problem 4.21. In the symmetric group S_n, find an expression for the inverse of the element $(a_1 a_2 \ldots a_k)$. Use your expression to explicitly write down the inverse of your results in Problem 4.19.

Problem 4.22. The symmetric group S_3, which is isomorphic to the dihedral group D_3, contains the six elements e, (12), (23), (13), (123), and (132). Find the conjugacy classes of this group.

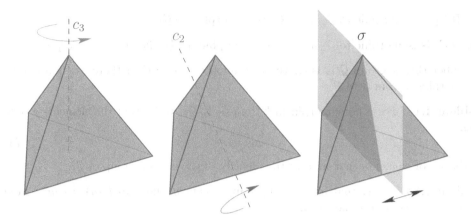

Figure 4.26 The different generators of the tetrahedral group T_d consist of the cyclic generators c_2 and c_3 of order two and three, respectively, and the reflection σ.

Figure 4.27 Regular six-sided dice have their sides numbered 1 through 6. In Problem 4.24, this is to be exploited in order to find the symmetry group of the cube as a subgroup of S_6.

Problem 4.23. The tetrahedral group T_d is the symmetry group of a regular tetrahedron, see Fig. 4.26. It is generated by a rotation c_3 through an angle $2\pi/3$ around an axis passing through one of its vertices and the center of the opposite face, a rotation c_2 by an angle π around an axis passing the center of two opposite edges, and a reflection σ in a plane containing one of the edges and the center of the opposite edge. Show that T_d is isomorphic to the symmetric group S_4 and find the conjugacy classes. Interpret the type of transformation each conjugacy class corresponds to.

Problem 4.24. Construct the symmetry group of a cube by numbering its sides 1 through 6 (as normal six-sided dice, see Fig. 4.27) and writing down the resulting transformations as a subgroup of S_6. What is the order of the group? By labelling the diagonals 1 through 4 and considering how the transformations act on these diagonals, find the complete set of symmetries of the cube.

Problem 4.25. Verify that the set of real numbers forms a group with the addition of numbers as the group operation. Also show that they do not form a group with multiplication as the group operation. What is the largest subset of the real numbers that forms a group with multiplication as the group operation?

Problem 4.26. Show that for the matrix group $SO(2)$ of rotations in two dimensions, where we can denote the rotation by an angle θ as R_θ, the map $h_1(R_\theta) = R_{-\theta}$ is an automorphism. Is the map $h_2(R_\theta) = R_{-2\theta}$ an automorphism?

Problem 4.27. Consider the group of translations and rotations in two dimensions. Determine a complete set of generators for this group and derive the corresponding Lie bracket relations.

Problem 4.28. We have shown the Lie bracket relations

$$[J_i, J_j] = \varepsilon_{ijk} J_k, \tag{4.227a}$$
$$[P_i, P_j] = 0, \tag{4.227b}$$

for rotations and translations, respectively. For the combined group of rotations and translations, deduce the form of the Lie bracket $[P_i, J_j]$.

Problem 4.29. The special linear group of complex 2×2 matrices $SL(2, \mathbb{C})$ is defined as the subset of complex 2×2 matrices that have determinant one. Verify that this set forms a subgroup of $GL(2, \mathbb{C})$ and determine the constraints on the matrices in its Lie algebra.

Problem 4.30. Starting from translations and rotations in three dimensions, show that translations in a direction $\vec{n}$ commute with rotations about $\vec{n}$ and that the set of combinations of those transformations form a subgroup. Also show that any two such subgroups, defined by different vectors $\vec{n}_1$ and $\vec{n}_2$ are isomorphic to each other.

Problem 4.31. Determine whether the matrices

$$A_1 = \frac{1}{\sqrt{2}} \begin{pmatrix} 1 & 1 \\ -1 & -1 \end{pmatrix}, \quad A_2 = \frac{1}{2} \begin{pmatrix} \sqrt{3} & 1 \\ 1 & -\sqrt{3} \end{pmatrix}, \quad A_3 = \frac{1}{5} \begin{pmatrix} 3 & 4 \\ -4 & 3 \end{pmatrix} \tag{4.228}$$

are elements of the group $O(2)$ or not. For those that are, determine the corresponding angle φ and sign in Eq. (4.62).

Problem 4.32. The generators J_i of the matrix group $SO(3)$ are given in Eq. (4.72).

a) Verify that the generators satisfy the relations

$$J_1^2 = - \begin{pmatrix} 0 & 0 & 0 \\ 0 & 1 & 0 \\ 0 & 0 & 1 \end{pmatrix}, \quad J_2^2 = - \begin{pmatrix} 1 & 0 & 0 \\ 0 & 0 & 0 \\ 0 & 0 & 1 \end{pmatrix}, \quad J_3^2 = - \begin{pmatrix} 1 & 0 & 0 \\ 0 & 1 & 0 \\ 0 & 0 & 0 \end{pmatrix}. \tag{4.229}$$

b) The above relations may be written as $J_i^2 = -P_i$, where P_i is a projection matrix with the special property that $P_i^k = P_i$, where k is a positive integer. Use this property along with the series expansion of the exponential function to show that

$$e^{\theta J_1} = \begin{pmatrix} 1 & 0 & 0 \\ 0 & \cos(\theta) & -\sin(\theta) \\ 0 & \sin(\theta) & \cos(\theta) \end{pmatrix}. \tag{4.230}$$

c) Repeat the exercise in (b) to deduce the forms of $e^{\theta J_2}$ and $e^{\theta J_3}$.

Problem 4.33. We have seen that, for any matrix group, taking the determinant $\det(A)$ of a matrix A in the group is a homomorphism to the set of real or complex numbers, depending on the matrix group, with multiplication as the group operation. What are the subgroups of the real or complex numbers corresponding to the determinant homomorphism for the groups $GL(N,\mathbb{C})$, $O(N)$, $SO(N)$, $U(N)$, and $SU(N)$, respectively?

Problem 4.34. Show that the set of continuously differentiable functions $f(x)$ that satisfy $f'(x) > 0$ everywhere and for which $f(x) \to \pm\infty$ as $x \to \pm\infty$ forms a group with the composition of functions

$$(fg)(x) = f(g(x)) \tag{4.231}$$

as the group operation and the identity function $e(x) = x$ as the identity element.

Problem 4.35. Verify that the matrix representations of the rotation generators J_i in Eq. (4.72) satisfy the Lie bracket relation

$$[J_i, J_j] = \varepsilon_{ijk} J_k, \tag{4.232}$$

where the Lie bracket is taken to be the matrix commutator $[A, B] = AB - BA$.

Problem 4.36. Verify that the Pauli matrices in Eq. (4.78) satisfy the commutation relation

$$[\sigma_i, \sigma_j] = 2i\varepsilon_{ijk}\sigma_k. \tag{4.233}$$

Problem 4.37. Check that the tensor product action defined in Eq. (4.92) constitutes a representation, i.e., verify that the mapping from the group to these linear operators is a homomorphism.

Problem 4.38. In Eq. (4.145), we introduced a three-dimensional representation of the symmetric group S_3. Check that the representation of the group generators (12) and (13) satisfy the same product rules as the generators themselves. In particular, check that

$$\rho((12))^2 = I, \quad \rho((13))^2 = I, \tag{4.234a}$$

and that

$$\rho((123))^3 = \rho((13)(12))^3 = [\rho((13))\rho((12))]^3 = I. \tag{4.234b}$$

Problem 4.39. Given a representation ρ of a group $\mathcal{G}$ on some vector space V and an invertible linear operator U acting on V, show that if we define

$$\rho'(a) = U\rho(a)U^{-1} \tag{4.235}$$

for all elements a of $\mathcal{G}$, then ρ' is also a representation of $\mathcal{G}$.

Problem 4.40. We have earlier seen that the moment of inertia for a disc of mass M and radius R has two distinct eigenvalues. Show that any object (in three dimensions) whose symmetry group is the group of rotations around a single axis has at most two distinct eigenvalues by applying symmetry arguments.

Problem 4.41. While the fundamental representation of an $O(3)$ matrix is the identity automorphism mapping any matrix a to a, the pseudo-vector representation is given by

$$\rho_A(a) = \det(a)a. \tag{4.236}$$

Verify that this mapping is a homomorphism.

Problem 4.42. The fundamental representation of the matrix group $SU(2)$ is in terms of its action on a two-dimensional complex vector space, which we can call V. We can construct a new representation by taking the tensor product of two such representations, which is a representation of dimension four. Verify that this new representation is reducible and can be split into a one-dimensional and a three-dimensional representation. Show that the three-dimensional representation resulting from this reduction is irreducible.

Problem 4.43. Construct an isomorphism between the dihedral group D_n and a subgroup of $O(2)$.

Problem 4.44. The Galilei transformations are defined in Eq. (4.142). For Galilei transformations with fixed origin, i.e., $s = 0$ and $\vec{a} = 0$, derive the composition law for the transformation given by first applying the transform $(R_{\vec{n}_1}^{\theta_1}, \vec{v}_1)$ followed by an application of the transform $(R_{\vec{n}_2}^{\theta_2}, \vec{v}_2)$. Verify that the block matrix

$$\rho(R_{\vec{n}}^{\theta}, \vec{v}) = \begin{pmatrix} A & v \\ 0 & 1 \end{pmatrix}, \tag{4.237}$$

where A is the $SO(3)$ representation of $R_{\vec{n}}^{\theta}$ and v is a column matrix containing the components of $\vec{v}$, is a representation of the Galilei group.

Problem 4.45. The heat equation and wave equation are given by

$$u_t(\vec{x}, t) - a\nabla^2 u(\vec{x}, t) = 0 \quad \text{and} \quad u_{tt}(\vec{x}, t) - c^2\nabla^2 u(\vec{x}, t) = 0, \tag{4.238}$$

respectively. Find out whether these equations are invariant under time reversal and parity transformations.

Problem 4.46. The Lie algebra of the restricted Galilei transformations in the Problem 4.44 contains the usual generators of rotation J_i and the generators of boosts C_i such that $e^{v_i C_i}$ is a boost by a velocity $\vec{v}$. Derive the Lie bracket relations for these transformations.

Problem 4.47. For a set of point masses m_i subjected only to internal forces, Newton's laws of motion may be summarised as

$$\ddot{\vec{x}}_i = -\frac{1}{m_i}\nabla_i V, \quad \text{(no sum)} \tag{4.239}$$

where $\vec{x}_i$ is the position of particle i, V is the potential energy of the system, which may be assumed to depend only on the distances between the particles, and ∇_i is the gradient with respect to $\vec{x}_i$. Verify that this set of equations is invariant under Galilei transformations.

Problem 4.48. In Eq. (4.145), a representation of the symmetric group S_3 is given. Verify that the vector v_1 given in the corresponding example is the only vector that is an eigenvector to the action of all elements of S_3.

Problem 4.49. Consider the representations of D_{3h} with characters given by the character table in Table 4.4. Determine the irreps that are contained in these representations.

Problem 4.50. In Example 4.54, we restricted the $O(3)$ group of rotations and reflections to the S_3 symmetry group of the ammonia molecule. Compute the characters of the different group elements in S_3 for the vector and pseudo-vector representations of $O(3)$. Use the resulting characters to split the representations into irreps of S_3.

D_{3h}	e	$2C_3$	$3C_2$	σ_h	$2S_3$	$3\sigma_v$
R_1	3	0	1	3	0	1
R_2	3	0	−1	−1	2	−1
R_3	4	1	0	0	−3	0
R_4	3	3	1	−1	−1	−3

Table 4.4   The character table for some representations of D_{3h}, to be used in Problem 4.49.

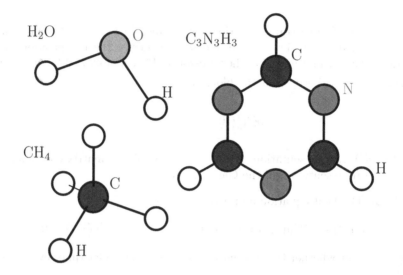

Figure 4.28   The structures of the planar water (H_2O) and 1,3,5-triazine ($C_3N_3H_3$) molecules and the tetrahedral methane (CH_4) molecule.

Problem 4.51. In Example 3.31 we stated without further argumentation that the rotational symmetry of the problem ensured that the solution could not depend on the angle ϕ. Use symmetry arguments to verify this statement.

Problem 4.52. Find the number of distinct vibrational frequencies allowed by the symmetries of the following molecules and state their degeneracies:

a) Water (H_2O)

b) Methane (CH_4)

c) 1,3,5-triazine ($C_3N_3H_3$)

See Fig. 4.28 for the molecular structures.

Problem 4.53. The vibrational spectrum of a molecule is changed if one of the atoms is replaced by a different isotope. Qualitatively describe what would happen to the vibrational spectrum of the methane molecule in Problem 4.52 if one of the hydrogen nuclei was replaced by a deuterium nucleus, i.e., instead of a proton there would now be a proton and a neutron, resulting in a particle with a larger mass. What would happen to the vibrational spectrum of the triazine molecule if one of the carbon nuclei was replaced by a different isotope?

CHAPTER 5

Function Spaces

A *function space* is the general name for essentially any set of functions satisfying some particular criteria, which may differ between different function spaces. In physics, we will often be dealing with very particular function spaces as classes of functions to which we expect that the solutions to our physical problems belong. The main rationale for referring to these sets of functions as spaces rather than just calling them function sets is the fact that they will often turn out to be abstract vector spaces, which are generally of infinite dimension, and we will therefore start this chapter by examining some properties of these.

In Chapter 6, we will use the theory discussed in this chapter to approach a large number of physics problems, in particular the partial differential equations discussed in Chapter 3. The methods in this chapter will also be highly relevant to studies of quantum mechanics, as different quantum states will be represented by functions that are eigenfunctions to a particular linear operator, the Hamiltonian.

Example 5.1 The concentration $u(\vec{x}, t)$ of a substance is a function of both space and time. If we wish to describe this concentration within a finite volume V for a fixed time $t = t_0$, we know that the total amount U in the volume will be given by

$$U(t_0) = \int_V u(\vec{x}, t_0)\, dV. \tag{5.1}$$

A reasonable assumption on the possible functions u is that, in order to describe a physical process, the total amount should be finite. This requires that the integral satisfies $|U(t_0)| < \infty$, i.e., the function $u(\vec{x}, t)$ must be *integrable* on the domain V. Regardless of what differential equation the concentration will need to satisfy, we should therefore look for solutions that fulfil this condition and the function space of integrable functions on V is the space describing the possible concentrations.

For the case of a concentration, it is often unphysical to consider negative concentrations. In such cases, the function space of interest would be the set of non-negative integrable functions on V.

5.1 ABSTRACT VECTOR SPACES

When discussing vectors in Chapter 1, we essentially restricted ourselves to vectors in finite-dimensional vector spaces where each basis vector could be interpreted as defining a physical direction. Holding on to a number of properties of these vector spaces, we can construct a more abstract notion of a *general vector space*. In order for a set V to form a vector space, we require that there are two operations defined involving elements of V, *addition* of two elements and *multiplication* by scalars. The requirement on the addition of elements in V is that V should be an Abelian group with the addition as the group operation. As such, the addition of elements in V must satisfy the relations

$$v_1 + v_2 = v_2 + v_1, \quad \text{(commutativity)} \tag{5.2a}$$

$$v_1 + (v_2 + v_3) = (v_1 + v_2) + v_3, \quad \text{(associativity)} \tag{5.2b}$$

where the v_i can be any elements of V. Furthermore, it is necessary that the additive identity 0, for which $v + 0 = v$, belongs to V and that for each element v in V, its additive inverse $-v$, for which $v + (-v) = 0$ is also an element of V. For brevity of notation, when addition of the additive inverse is intended, it is common to denote this with a minus sign, as for the real numbers, i.e.,

$$v_1 - v_2 \equiv v_1 + (-v_2). \tag{5.3}$$

The second operation we demand from a vector space is the multiplication of an element of the vector space and a scalar, which may be a real or a complex number and which we will denote av, where a is the scalar and v the element of the vector space. Depending on whether this operation is defined only for real numbers or also for general complex numbers, the vector space is referred to as a real or complex vector space, respectively. We require that the multiplication fulfils the following conditions:

1. *Compatibility*: Performing several multiplications with two different scalars a and b should be equivalent to multiplying by the product of the scalars:

$$a(bv) = (ab)v. \tag{5.4a}$$

2. *Identity*: Multiplication with one should give back the original vector:

$$1v = v. \tag{5.4b}$$

3. *Distributivity*: The multiplication must be distributive under addition and vice versa:

$$a(v_1 + v_2) = av_1 + av_2 \quad \text{and} \quad (a_1 + a_2)v = a_1 v + a_2 v. \tag{5.4c}$$

In general, real and complex vector spaces have many things in common and can be treated in essentially the same fashion. The big difference is that when we talk about scalars in the context of a vector space, we will generally mean a complex number when the vector space is complex and a real number when the vector space is real. Since the treatment is similar, future definitions made will generally be written down for complex vector spaces and particularly commented on when special situations occur for real vector spaces.

Example 5.2 We recognise all of the above properties from the usual addition and multiplication of numbers as well as from finite-dimensional vector spaces, where any vector $\vec{v}$ can be written in terms of the basis vectors $\vec{e}_i$ as $\vec{v} = v^i \vec{e}_i$. As the vector components add individually under vector addition in such a vector space, the required properties of the vector addition follow directly from the corresponding properties of the usual addition of real or complex numbers. Likewise, multiplication by a scalar can be done component by component and the required properties again follow in a straightforward fashion from those relating to the multiplication of numbers.

Example 5.3 The subject of this chapter is function spaces and let us therefore consider the space of all real polynomials of order $N - 1$. Any such polynomial $f(x)$ may be written as

$$f(x) = \sum_{k=0}^{N-1} a_k x^k, \tag{5.5}$$

which is a linear combination of the N monomials x^k for $0 \leq k < N$. Identifying $\vec{e}_k = x^{k-1}$ as a basis, the situation for these polynomials is exactly equivalent to that in the previous example and this function space is a vector space of dimension N. Addition of two polynomials is done by adding the coefficients a_k pair-wise and multiplication by a scalar is equivalent to multiplying all of the coefficients by this scalar.

5.1.1 Inner products and completeness

The vector spaces considered in Chapter 1 were all equipped with a Cartesian inner product, which in turn could be used to define the length of a vector by taking the square root of its inner product with itself. As such, we seek to find a similar construction for an abstract vector space and define an *inner product* on a vector space V to be a mapping taking two elements v_1 and v_2 in V to a number $\langle v_1, v_2 \rangle$, which has the following properties:

1. *Conjugate symmetry*: When exchanging the arguments, the inner product is complex conjugated

$$\langle v_1, v_2 \rangle = \langle v_2, v_1 \rangle^*. \tag{5.6a}$$

2. *Linearity*: The inner product must be linear in the second argument

$$\langle w, a_1 v_1 + a_2 v_2 \rangle = a_1 \langle w, v_1 \rangle + a_2 \langle w, v_2 \rangle. \tag{5.6b}$$

3. *Positive definite*: The inner product must be positive definite, i.e.,

$$\langle v, v \rangle \geq 0, \tag{5.6c}$$

with equality implying that $v = 0$.

Note that the inner product is not linear in the first argument. Instead, using the first two properties, we find

$$\langle a_1 v_1 + a_2 v_2, w \rangle = \langle w, a_1 v_1 + a_2 v_2 \rangle^* = a_1^* \langle v_1, w \rangle + a_2^* \langle v_2, w \rangle, \tag{5.7}$$

a property known as *anti-linearity*. A word of warning is appropriate in connection to this: The notation presented here is that most commonly used in physics. On the contrary, in mathematics it is more common to define the inner product to be linear in the *first* argument, resulting in it being anti-linear in the *second* argument. Naturally, this is a pure notational issue and either may be used without problems as long as it is done consistently and the reader is aware of which is intended. In this book, we will keep to the physics convention with the inner product being linear in the second argument but the issue should be kept in mind, in particular when reading mathematics oriented texts. Naturally, this is not an issue for real vector spaces for which there is no difference between linearity and anti-linearity. Furthermore, the formulation of the positive definiteness is relying on the fact that $\langle v, v \rangle$ is a real number due to the conjugate symmetry

$$\langle v, v \rangle = \langle v, v \rangle^*. \tag{5.8}$$

A vector space with an associated inner product is generally called an *inner product space*.

Example 5.4 Let us again look at the finite vector spaces of dimension N spanned by the basis vectors $\vec{e}_i$ with $\vec{v} = v^i \vec{e}_i$. The scalar product

$$\vec{v} \cdot \vec{w} = v^i w^i \tag{5.9}$$

is symmetric, linear, and positive definite as long as we consider only real linear combinations of the $\vec{e}_i$, which we restricted ourselves to in Chapter 1. In order to generalise this definition to complex vector spaces, we must require that

$$\vec{v} \cdot \vec{w} = (v^i)^* w^i, \tag{5.10}$$

which has all of the required properties of an inner product.

Example 5.5 The choice of inner product on a vector space is not unique. Just as we can use the Cartesian scalar product, given a real base space, any rank two tensor g_{ab} will define an inner product

$$\left\langle \vec{E}_a, \vec{E}_b \right\rangle = g_{ab} \tag{5.11}$$

as long as it is symmetric and positive definite. This should be familiar as the very same properties we discussed in connection to the metric tensor. Thus, rather than defining the metric tensor using the scalar product in a Euclidean base space, we can define the inner product by first defining a metric tensor.

5.1.1.1 Geometry in inner product spaces

It is hardly surprising that a general inner product shares a lot of properties with the scalar product that we are already familiar with, as the scalar product on a finite dimensional real vector space is a special case of an inner product. Many of the properties of the scalar product have a geometrical interpretation, which may therefore also be extended to a general inner product. The first of these properties is related to the length of a vector, which we have seen expressed as the square root of the inner product of a vector with itself (cf. Eq. (1.4)). As a generalisation of this concept, we define the *norm* of a vector v in an inner product space as

$$\|v\| = \sqrt{\langle v, v \rangle}. \tag{5.12}$$

The inner product also defines a concept of *orthogonality* in an abstract vector space. In analogy to having two vectors orthogonal whenever their scalar product is zero, two vectors v and w in an abstract vector space are said to be orthogonal if

$$\langle v, w \rangle = 0. \tag{5.13}$$

This concept of orthogonality also extends *Pythagoras' theorem* to abstract vector spaces. If v and w are orthogonal and $z = v + w$, then

$$\|z\|^2 = \langle z, z \rangle = \langle v, v \rangle + \langle w, w \rangle + \underbrace{\langle v, w \rangle + \langle w, v \rangle}_{=0} = \|v\|^2 + \|w\|^2. \tag{5.14}$$

In addition, the *Cauchy–Schwarz* and *triangle inequalities*

$$|\langle v, w \rangle|^2 \le \|v\|^2 \|w\|^2 \quad \text{and} \quad \|v + w\| \le \|v\| + \|w\| \tag{5.15}$$

follow directly from the requirements on an inner product, see Problem 5.3.

It should be stressed that these properties depend not only on the vector space, but also on the chosen inner product. It is perfectly possible that two vectors that are orthogonal under one inner product are not orthogonal under a different inner product on the same vector space.

5.1.1.2 Convergence of series

The notion of a distance in a vector space implied by the vector norm directly provides us with a tool for defining the convergence of any infinite series $\{v_n\} = \{v_1, v_2, \ldots\}$ of vectors in the vector space. We start by reminding ourselves that a *Cauchy sequence* $\{v_n\}$ of elements in any space with a well defined distance function $d(x, y)$ is a sequence such that for every $\varepsilon > 0$, there exists an integer N such that

$$d(v_n, v_m) < \varepsilon \tag{5.16a}$$

for all $n, m > N$. In our case, the distance function is $d(x, y) = \|x - y\|$ and the requirement is therefore given by

$$\|v_n - v_m\| < \varepsilon. \tag{5.16b}$$

In addition, we also say that a sequence *converges* to v if for all $\varepsilon > 0$, there exists an N such that

$$d(v, v_n) = \|v - v_n\| < \varepsilon \tag{5.17}$$

for all $n > N$. Although these definitions are similar, a Cauchy sequence is not necessarily converging to an element within the space itself.

Example 5.6 The set of rational numbers $\mathbb{Q}$ of the form n/m, where n and m are integers, has a well defined distance function $d(x, y) = |x - y|$. However, not all Cauchy sequences in this set converge to a rational number. If we consider the sequence

$$\{x_n\} = \{3, 3.1, 3.14, 3.141, 3.1415, 3.14159, \ldots\}, \tag{5.18}$$

we find that taking any $\varepsilon > 0$, the assignment $N = -\log_{10}(\varepsilon)$ results in $|x_n - x_m| < \varepsilon$ for all $n, m > N$ and the sequence is therefore a Cauchy sequence. However, there is no rational number x to which the sequence converges. The sequence does converge if instead considered as a sequence of real numbers with the limit $x = \pi$, which is an irrational number.

Unlike the rational numbers in the example above, the real numbers $\mathbb{R}$ do have the property that all Cauchy sequences converge to an element in $\mathbb{R}$. Sets with this property are called *complete spaces* and we define a *Hilbert space* to be an inner product space that is complete. Although they may be of finite dimension, Hilbert spaces are generally infinite dimensional vector spaces and are fundamental to many physics applications. The entire foundation of quantum mechanics is ultimately based upon Hilbert spaces and linear operators acting on them.

As just mentioned, the dimension of a Hilbert space need not be finite. On the contrary, it is not even necessary that a countable set of basis vectors $\{\vec{e}_1, \vec{e}_2, \vec{e}_3, \ldots\}$ exists. However, in the cases where such a basis does exist, the Hilbert space is referred to as *separable*.

5.1.2 Function spaces as vector spaces

As advertised in the title, the subject of this chapter is *function spaces* and all of our discussion on abstract vector spaces would therefore be moot unless it was applicable to those. Luckily, function spaces are often vector spaces and we have already seen an example of this in the form of real polynomials of a given degree in Example 5.3. The natural way of turning a space of functions on some domain D into a vector space is through pointwise addition and multiplication of functions, i.e., we define addition of functions f_1 and f_2 as

$$(f_1 + f_2)(x) = f_1(x) + f_2(x) \tag{5.19a}$$

and multiplication of a function f with a scalar a as

$$(af)(x) = af(x) \tag{5.19b}$$

for all x in D. It is easy to check that all of the requirements on the operations in an abstract vector space are satisfied by these definitions as long as the vector space is closed under them, i.e., the sum of two functions and the product of a scalar and a function should also be in the function space.

Example 5.7 The set of all differentiable functions on the real numbers $C^1(\mathbb{R})$ forms a vector space under the addition and multiplication just defined. For any functions f_1 and f_2 in this set, we find that $f_1 + f_2$ as well as af_i are also differentiable functions, in particular

$$(f_1 + f_2)' = f_1' + f_2' \quad \text{and} \quad (af_i)' = af_i'. \tag{5.20}$$

Example 5.8 The domain D does not need to be a continuous set. In fact, we could select a finite set $\{x_n\}$ of N points in $\mathbb{R}$ as the domain and the set of all functions on this set will also be a vector space. In particular, we can write down the the functions

$$e_m(x_n) = \delta_{mn} \tag{5.21}$$

for all $m \leq N$ and consequently describe any function f on this domain by specifying the N numbers f_m in the sum

$$f = \sum_{m=1}^{N} f_m e_m. \tag{5.22}$$

It follows that the functions e_m constitute a finite basis of the vector space and the value of the function f for any point x_n is given by

$$f(x_n) = \sum_{m=1}^{N} f_m e_m(x_n) = \sum_{m=1}^{N} f_m \delta_{nm} = f_n. \tag{5.23}$$

As a result, the vector space is of finite dimension. We can also do the same thing for a countably infinite set of points, in which case the resulting vector space is separable.

Example 5.9 It is also important to note that we cannot just take any set of functions and expect it to be a vector space. An example of this is the set of functions that take the value $f(x_0) = a \neq 0$ at some point x_0 in the domain. For the sum of two functions in this set, we would obtain

$$(f_1 + f_2)(x_0) = f_1(x_0) + f_2(x_0) = 2a \neq a \tag{5.24}$$

and as a result the set is *not* closed under addition, which is one of the requirements for the set to form a vector space. However, if $a = 0$, the set does form a vector space (unless other conditions are violated).

5.1.2.1 Inner products on function spaces

Let us consider what types of inner products we can construct for functions on a continuous domain D, which generally may be all of an N-dimensional space or a subset of it. We will restrict our discussion to the vector space of functions that are square integrable on D, i.e., functions f that satisfy

$$\int_D |f(x)|^2 \, dV < \infty, \tag{5.25}$$

where dV is the volume element in the N-dimensional space. We can easily check that this set of functions forms a vector space since

$$\int_D |f_1 + f_2|^2 \, dV \leq \int_D (|f_1| + |f_2|)^2 \, dV \leq \int_D \left(|f_1|^2 + |f_2|^2 + 2\max(|f_1|^2, |f_2|^2) \right) dV$$

$$\leq 3 \int_D \left(|f_1|^2 + |f_2|^2 \right) dV < \infty, \tag{5.26}$$

where we have used that $|f_1| \, |f_2| \leq \max(|f_1|^2, |f_2|^2)$ and that $\max(|f_1|^2, |f_2|^2) \leq |f_1|^2 + |f_2|^2$ for all values of the arguments. In order to construct an inner product on this vector space, we need to find a mapping that takes functions f_1 and f_2 to scalars and satisfies the inner product requirements. Among these requirements is the requirement of being linear in the second argument. Noting that any integral of the form

$$I_2[f_2] = \int_D W(x) f_2(x) dV, \tag{5.27a}$$

where W is an arbitrary function that does not spoil the convergence of the integral, is linear in f_2, a strong candidate would be a mapping that behaves in this fashion. Similarly, the inner product should be anti-linear in the first argument, which is a property of the integral

$$I_1[f_1] = \int_D f_1^*(x) W(x) dV. \tag{5.27b}$$

Taking $W = f_1$ in the first of these relations and $W = f_2$ in the second results in the same expression and we therefore have a candidate inner product on the form

$$\langle f_1, f_2 \rangle = \int_D f_1^*(x) f_2(x) dV. \tag{5.28}$$

This definition indeed satisfies all of the requirements on an inner product and is commonly encountered in physics. It should be noted that, as for other vector spaces, there are several other possibilities of defining the inner product. A very general form of an inner product would be of the form

$$\langle f_1, f_2 \rangle = \int_D \int_D f_1^*(x) f_2(x') \tilde{w}(x, x') \, dV \, dV', \tag{5.29}$$

where the function $\tilde{w}$ must satisfy a number of constraints to keep the inner product positive definite and well defined on the function space. This form of an inner product does not appear very often, but the special case of $\tilde{w}(x, x') = w(x)\delta(x - x')$ does, leading to

$$\langle f_1, f_2 \rangle = \int_D w(x) f_1^*(x) f_2(x) \, dV. \tag{5.30}$$

The function $w(x)$ must here satisfy $w(x) > 0$ and is called a *weight function*. In the rest of this chapter, this is the only type of inner product on function spaces that we will encounter.

5.2 OPERATORS AND EIGENVALUES

Linear operators on different vector spaces in general, and on function spaces in particular, will appear in a large variety of different physical situations. We have already discussed linear differential operators in connection to the modelling of physical systems in Chapter 3, but let us now discuss the theory of linear operators on abstract vector spaces in a little bit more detail. Our discussion will start with linear operators on finite vector spaces and we shall gradually work our way towards the limit of separable Hilbert spaces, and finally see how the theory generalises to Hilbert spaces in general.

5.2.1 Application of operators in finite spaces

Linear operators appear already in basic mechanics, although usually introduced as matrices. Looking at a mechanical system with a N degrees of freedom $\vec{X}$ around an equilibrium, which may be taken to be $\vec{X} = 0$, its equations of motion can be written on the form

$$\ddot{\vec{X}} = -\hat{V}\vec{X}, \tag{5.31}$$

where $\hat{V}$ is a linear operator taking a vector in the N dimensional configuration space to a different vector in the configuration space. This operator will generally depend on the inertia and potential energy of the system. Taking the usual scalar product on the configuration space, we find that

$$\frac{1}{2}\frac{d\dot{\vec{X}}^2}{dt} = \dot{\vec{X}} \cdot \ddot{\vec{X}} = -\dot{\vec{X}} \cdot (\hat{V}\vec{X}). \tag{5.32}$$

If $\hat{V}$ has the property

$$\vec{Y}_1 \cdot (\hat{V}\vec{Y}_2) = (\hat{V}\vec{Y}_1) \cdot \vec{Y}_2 \tag{5.33}$$

for all $\vec{Y}_i$, it is said to be *symmetric*, which for this finite vector space reduces to the relation $V_{ij} = V_{ji}$ in any basis on the vector space, where $V_{ij} = \vec{e}_i \cdot \hat{V}\vec{e}_j$. We then find that

$$\frac{1}{2}\frac{d\dot{\vec{X}}^2}{dt} = -\frac{1}{2}[\dot{\vec{X}} \cdot (\hat{V}\vec{X}) + \vec{X} \cdot (\hat{V}\dot{\vec{X}})] = -\frac{1}{2}\frac{d(\vec{X} \cdot \hat{V}\vec{X})}{dt} \qquad (5.34)$$

and therefore

$$\frac{dE}{dt} = 0, \quad \text{where} \quad E = \frac{1}{2}\left(\dot{\vec{X}}^2 + \vec{X} \cdot \hat{V}\vec{X}\right) \qquad (5.35)$$

is related to the total energy of the system. Note that the symmetry of $\hat{V}$ will generally be a consequence of it being related to the second derivative of the potential energy $V(\vec{X})$ of the system, which in any coordinate basis satisfies

$$\partial_i\partial_j V = \partial_j\partial_i V. \qquad (5.36)$$

In order to solve Eq. (5.31), we look for eigenvectors of $\hat{V}$, which are vectors $\vec{X}_k$ such that

$$\hat{V}\vec{X}_k = \lambda_k\vec{X}_k, \qquad \text{(no sum)} \qquad (5.37)$$

where the eigenvalue λ_k is a scalar. In general, symmetric operators will have a set of eigenvectors that are orthogonal with real eigenvalues, see Problem 5.14. As the eigenvectors may be normalised, we take such a set of orthonormal eigenvectors $\vec{X}_k$ as a basis and, since any vector may be expanded in terms of this basis, expand the solution in terms of them

$$\vec{X} = \sum_{k=1}^{N} f_k(t)\vec{X}_k, \qquad (5.38)$$

where the $f_k(t)$ are functions of t to be determined. Inserting this into Eq. (5.31) and taking the inner product of the entire equation with $\vec{X}_\ell$, we find that

$$\vec{X}_\ell \cdot \ddot{\vec{X}} = \sum_{k=1}^{N} f_k''(t)\vec{X}_\ell \cdot \vec{X}_k = \sum_{k=1}^{N} f_k''(t)\delta_{\ell k} = f_\ell''(t) \qquad (5.39a)$$

for the left-hand side and

$$-\vec{X}_\ell \cdot \hat{V}\vec{X} = -\sum_{k=1}^{N} f_k(t)\vec{X}_\ell \cdot \hat{V}X_k = -\sum_{k=1}^{N} f_k(t)\vec{X}_\ell \cdot \lambda_k\vec{X}_k$$

$$= -\sum_{k=1}^{N} f_k(t)\lambda_k\delta_{\ell k} = -\lambda_\ell f_\ell(t), \qquad (5.39b)$$

for the right-hand side. In other words, we find that

$$f_\ell''(t) = -\lambda_\ell f_\ell(t), \qquad (5.39c)$$

which is a linear ordinary differential equation for every value of ℓ. We will later apply precisely the same argumentation for linear operators on infinite dimensional function spaces, but we shall first discuss these operators in some detail. For now, let us give an example in a finite dimensional setting in order to substantiate the theory.

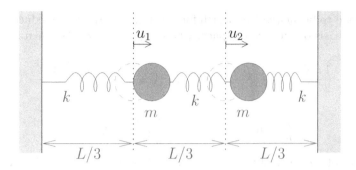

Figure 5.1 Two masses m moving between two fixed points separated by L under the influence of three springs of spring constant k. The displacement of the masses from their equilibrium positions are u_1 and u_2, respectively.

Example 5.10 Consider two masses m moving in one dimension and subjected to the forces of three springs of rest length $L/3$ with spring constant k, see Fig. 5.1. We denote the displacement of the masses by u_k ($k = 1, 2$) and the configuration space of the system is two-dimensional and we take a basis $\vec{e}_1$, which corresponds to $u_1 = 1$, $u_2 = 0$, and $\vec{e}_2$, which corresponds to $u_1 = 0$, $u_2 = 1$. A general displacement is therefore

$$\vec{u} = u_1 \vec{e}_1 + u_2 \vec{e}_2, \tag{5.40a}$$

which may also be represented as a column vector

$$u = \begin{pmatrix} u_1 \\ u_2 \end{pmatrix} = u_1 \underbrace{\begin{pmatrix} 1 \\ 0 \end{pmatrix}}_{=\vec{e}_1} + u_2 \underbrace{\begin{pmatrix} 0 \\ 1 \end{pmatrix}}_{=\vec{e}_2}. \tag{5.40b}$$

Applying Newton's second law to the first of the masses, we find that

$$\ddot{u}_1 = \frac{k}{m}[-u_1 + (u_2 - u_1)], \tag{5.41}$$

where the term containing only u_1 arises due to the spring to the left of the mass and the term containing $u_2 - u_1$ due to the spring to the right. A similar equation, but with $u_1 \leftrightarrow u_2$ can be found for the second mass and the full set of equations of motion for the system may be written

$$\ddot{u} = -\hat{V}u, \tag{5.42}$$

where $\hat{V}$ is the symmetric 2×2 matrix

$$\hat{V} = \frac{k}{m} \begin{pmatrix} 2 & -1 \\ -1 & 2 \end{pmatrix}. \tag{5.43}$$

This matrix has the normalised eigenvectors

$$X_1 = \frac{1}{\sqrt{2}} \begin{pmatrix} 1 \\ 1 \end{pmatrix} \quad \text{and} \quad X_2 = \frac{1}{\sqrt{2}} \begin{pmatrix} 1 \\ -1 \end{pmatrix} \tag{5.44}$$

with eigenvalues $\lambda_1 = k/m$ and $\lambda_2 = 3k/m$, respectively. Writing the general solution to the problem as $u(t) = f_1(t)X_1 + f_2(t)X_2$, we find that

$$f_k(t) = A_k \cos(\omega_k t + \phi_k), \tag{5.45}$$

where $\omega_k = \sqrt{\lambda_k}$ and A_k and ϕ_k are integration constants. The general solution to the problem is therefore

$$u(t) = \left[A_1 \cos(t\sqrt{k/m} + \phi_1)X_1 + A_2 \cos(t\sqrt{3k/m} + \phi_2)X_2 \right]. \tag{5.46}$$

The constants A_k and ϕ_k need to be fixed using initial conditions on the positions and velocities of the two masses.

5.2.2   Operators on inner product spaces

When considering operators acting on general vector spaces, we want to carry several of the properties we encountered in the finite dimensional setting with us. As for the finite setting, we define an operator $\hat{V}$ to be *symmetric* with respect to the inner product if the relation

$$\left\langle u, \hat{V}v \right\rangle = \left\langle \hat{V}u, v \right\rangle \tag{5.47}$$

is satisfied for all vectors u and v in a real vector space. The corresponding relation in a complex vector space defines *Hermitian operators*. Just as symmetric operators may be represented by symmetric matrices in finite-dimensional real vector spaces, Hermitian operators may be represented by Hermitian matrices in finite-dimensional complex vector spaces.

Example 5.11 Let us consider the linear differential operator ∂_x along with the inner product

$$\langle f, g \rangle = \int_{-\infty}^{\infty} f(x)g(x)dx \tag{5.48}$$

on a suitable space of functions in one dimension that decay sufficiently fast as $x \to \pm\infty$. Using partial integration, we find that

$$\langle f, \partial_x g \rangle = \int_{-\infty}^{\infty} f(x)g'(x)dx = -\int_{-\infty}^{\infty} f'(x)g(x)dx = -\langle \partial_x f, g \rangle. \tag{5.49}$$

Because of this, the operator ∂_x is not a symmetric operator (due to the appearance of the minus sign, it is *anti*-symmetric). However, this also means that

$$\langle f, \partial_x^2 g \rangle = -\langle \partial_x f, \partial_x g \rangle = \langle \partial_x^2 f, g \rangle, \tag{5.50}$$

implying that the operator ∂_x^2 is symmetric.

As for operators on finite vector spaces, operators on more general vector spaces may have eigenvectors that satisfy

$$\hat{V}u = \lambda_u u, \tag{5.51}$$

where λ_u is the eigenvalue associated with the eigenvector u of the operator $\hat{V}$. For any eigenvector u, a very useful relation to keep in mind is

$$\left\langle v, \hat{V} u \right\rangle = \langle v, \lambda_u u \rangle = \lambda_u \langle v, u \rangle, \tag{5.52}$$

which follows directly from the fact that λ_u is a number and the linear property of the inner product. In particular, this property may be used to show that any symmetric operator must have real eigenvalues, see Problem 5.14. At this point, it should also be noted that linear operators on function spaces are not necessarily differential operators, although we will deal with these to a larger extent than others. The same ideas with eigenvalues and eigenvectors apply equally well to any type of linear operator.

Example 5.12 Let us again consider the linear operator ∂_x. In order to be an eigenvector of this operator, a function f must satisfy

$$f'(x) = \lambda_f f(x) \tag{5.53}$$

for some constant λ_f. This ordinary differential equation has the solution

$$f(x) = e^{\lambda_f x} \tag{5.54}$$

regardless of the value of λ_f. All of the functions on this form are therefore eigenfunctions of ∂_x. Whether or not these eigenfunctions are actually a part of the function space depends on whether or not they satisfy the requirements of belonging to it.

Example 5.13 As an example of a linear operator that is not a differential operator, let us consider pointwise multiplication by the function

$$\pi_{(a,b)}(x) = \begin{cases} 1, & a < x < b \\ 0, & \text{otherwise} \end{cases}. \tag{5.55}$$

This defines the linear operator $\hat{\pi}_{(a,b)}$ such that

$$(\hat{\pi}_{(a,b)} f)(x) = \pi_{(a,b)}(x) f(x). \tag{5.56}$$

Any function that is non-zero only in the interval $a < x < b$ is an eigenfunction of this operator with eigenvalue one, while any function that is zero in the interval is an eigenfunction with eigenvalue zero.

5.2.2.1 Differential operators and discretisation

It is sometimes useful to compare problems in finite dimensional vector spaces with problems in infinite dimensional ones. While the mathematical structure in terms of operators may be more evident in the finite dimensional spaces, as they are being represented by matrices, problems with many degrees of freedom are often easier to solve if we approximate the number of degrees of freedom as infinite while keeping macroscopic quantities fixed. In

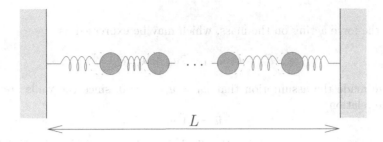

Figure 5.2 The same as Fig. 5.1, but for an increasing number of masses and springs. Allowing the number of masses and springs to grow while keeping some macroscopic quantities such as the total mass, the overall spring constant, and total length fixed, the problem will be well approximated by a continuum.

performing this sort of comparisons, it is useful to keep the finite difference approximations of different derivatives in mind, most notably

$$u_x(x_k) \simeq \frac{u(x_{k+1}) - u(x_k)}{\Delta x} \quad \text{and} \tag{5.57a}$$

$$u_{xx}(x_k) \simeq \frac{u(x_{k+1}) - 2u(x_k) + u(x_{k-1})}{(\Delta x)^2}, \tag{5.57b}$$

for the first and second derivatives, respectively, where Δx is the distance between the equidistantly spaced points x_k.

If we are faced with an infinite dimensional problem, such as the oscillations of a string of a given length, where the function $u(x, t)$ describing the shape of the string at time t belongs to an infinite dimensional function space, the problem may be approximated by considering the function u only at a finite discrete set of points x_k. The function space on such a set is finite dimensional and any derivatives may be approximated by the finite difference approximations. As such, we will be able to approximate the differential operator in terms of a matrix, much like the one that showed up in Example 5.10.

On the other hand, if we are faced with a system with a very large number of degrees of freedom, the solution will generally involve the diagonalisation of a very large matrix, i.e., an $N \times N$ matrix for N degrees of freedom. Instead of performing this diagonalisation explicitly, it may be easier to approximate the system with an infinite dimensional one. This is illustrated in the following example.

Example 5.14 We consider a situation that is very similar to Example 5.10, but instead of having two masses and three springs, we consider the case where we have N masses and $N + 1$ springs, while we keep the total distance of the chain fixed at L, see Fig. 5.2. The distance between consecutive masses is given by $a = L/(N + 1)$ and the total mass of the chain is $M = Nm$, where m is the mass of each small mass. We can also define the macroscopic spring constant $k_0 = k/(N+1)$, where k is the spring constant for each of the small springs. The constant k_0 is the spring constant of $N + 1$ springs with spring constant k connected in series. If we consider how the deviation u_n of the nth mass depends on time, we apply Newton's second law and find

$$\ddot{u}_n = \frac{F_n}{m}, \tag{5.58}$$

where F_n is the force acting on the mass, which may be expressed as

$$F_n = k(u_{n+1} - 2u_n + u_{n-1}). \tag{5.59}$$

We have here made the assumption that $u_0 = u_{N+1} = 0$, since the walls are fixed. This results in the relation

$$\ddot{u} = -\hat{V}u, \tag{5.60}$$

where u is a column vector containing the displacements u_n and $\hat{V}$ is the $N \times N$ matrix

$$\hat{V} = \frac{k}{m} \begin{pmatrix} 2 & -1 & 0 & 0 & \cdots & 0 \\ -1 & 2 & -1 & 0 & \cdots & 0 \\ 0 & -1 & 2 & \ddots & \ddots & \vdots \\ 0 & 0 & \ddots & \ddots & -1 & 0 \\ \vdots & \vdots & \ddots & -1 & 2 & -1 \\ 0 & 0 & \cdots & 0 & -1 & 2 \end{pmatrix}. \tag{5.61}$$

Although this matrix is sparse and can be easily diagonalised by a computer, we would like to find some analytical results as well. In order to do so, we instead consider the limit where $N \to \infty$ while keeping the macroscopic quantities M, k_0, and L fixed. In the continuum limit, we consider a function $u(x,t)$, implying that $u_n = u(x_n, t)$, where $x_n = an$. Rewriting Eq. (5.58) in terms of the macroscopic variables, we find that

$$u_{tt}(x_n) \simeq \frac{k_0}{M} N^2 (u_{n+1} - 2u_n + u_{n-1}) \simeq \frac{k_0}{M} N^2 a^2 u_{xx}(x_n) = \frac{k_0 L^2}{M} u_{xx}(x_n). \tag{5.62}$$

Introducing $c^2 = k_0 L^2 / M$, this is the wave equation with wave velocity c. If N is very large, we should therefore find a reasonable approximation to the solution if we solve the continuous problem instead.

5.3 STURM–LIOUVILLE THEORY

A very important class of differential operators in physics is that of the so-called *Sturm–Liouville operators*. A Sturm–Liouville operator is a differential operator that can be written as

$$\hat{L} = -\frac{1}{w(x)} \left[\frac{\partial}{\partial x} p(x) \frac{\partial}{\partial x} - q(x) \right], \tag{5.63}$$

where $w(x)$, $p(x)$, and $q(x)$ are all positive functions. The reason these are so important is that it is common for operators of the form

$$\hat{L} = -\frac{1}{w(\vec{x})} \left[\nabla \cdot p(\vec{x}) \nabla - q(\vec{x}) \right], \tag{5.64}$$

to appear in different physical situations. Writing this operator down in component form in Cartesian coordinates, we find that

$$\hat{L} = -\frac{1}{w} \left[\partial_i p \partial_i - q \right] \equiv \sum_{i=1}^{N} \hat{L}_i - \frac{q}{w} \tag{5.65}$$

and the operator is therefore a sum of several Sturm–Liouville operators $\hat{L}_i$ in different variables.

Example 5.15 The negative of the Laplace operator $-\nabla^2$ is of the form

$$-\nabla^2 = -\partial_i \partial_i. \tag{5.66}$$

Each term in this sum is a Sturm–Liouville operator with $w = p = 1$ and $q = 0$.

5.3.1 Regular Sturm–Liouville problems

Most of the useful properties of Sturm–Liouville operators are apparent already for one-dimensional problems in which a single Sturm–Liouville operator appears. For this reason, we shall start our discussion regarding these properties for one-dimensional problems and later use separation of variables to generalise the discussion to problems in higher dimensions. For now, we assume that we have a Sturm–Liouville operator that is of the form

$$\hat{L} = -\frac{1}{w(x)} \left[\partial_x p(x) \partial_x - q(x) \right]. \tag{5.67}$$

Since this is a differential operator of order two, we will need two boundary conditions when solving differential equations involving it. Assume that we are studying a problem in the interval $a \leq x \leq b$ and we are looking at boundary conditions of the form

$$\alpha_a f(a) + \beta_a f'(a) = 0, \quad \alpha_b f(b) + \beta_b f'(b) = 0. \tag{5.68}$$

Since the boundary conditions are homogeneous, it is relatively easy to show that functions fulfilling them form a vector space on which the Sturm–Liouville operator can act. The problem of finding the eigenvalues and eigenvectors of this problem is called a *regular Sturm–Liouville problem* based on the fact that the boundary conditions are regular. Explicitly, the problem involves finding functions $f(x)$ such that

$$\hat{L}f(x) = \lambda f(x), \tag{5.69}$$

where $f(x)$ satisfies the boundary conditions and λ is a constant.

Example 5.16 Let us solve the most basic regular Sturm–Liouville problem we may imagine, namely that of $\hat{L} = -\partial_x^2$ on the interval $[0,1]$ with homogeneous Dirichlet boundary conditions

$$(\text{ODE}) : \ -f''(x) = \lambda f(x), \tag{5.70a}$$
$$(\text{BC}) : f(0) = f(1) = 0. \tag{5.70b}$$

We can solve the differential equation by looking at the three different cases $\lambda < 0$, $\lambda = 0$, and $\lambda > 0$ separately:

$\lambda < 0$: In this case, we can introduce $k > 0$ such that $k^2 = -\lambda$. The general solution to the differential equation is then given by

$$f(x) = Ae^{kx} + Be^{-kx} = A' \cosh(kx) + B' \sinh(kx), \tag{5.71a}$$

where A, B, A', and B' are constants to be determined by the boundary conditions. With the homogeneous Dirichlet boundary conditions the only possible solution is $A = B = 0$, i.e., the trivial solution that is the zero vector in the function space.

$\lambda = 0$: The differential equation is now $f''(x) = 0$ with the solution

$$f(x) = Ax + B. \tag{5.71b}$$

Again, the boundary conditions leave only the trivial solution $A = B = 0$.

$\lambda > 0$: Similar to the first case, we may introduce $k > 0$, but this time such that $\lambda = k^2$. The general solution to the differential equation is now

$$f(x) = A\cos(kx) + B\sin(kx). \tag{5.71c}$$

Again, the boundary condition at $x = 0$ implies that $f(0) = A = 0$, but the boundary condition at $x = 1$ is given by

$$f(1) = B\sin(k) = 0, \tag{5.72}$$

which allows for B to be arbitrary whenever $\sin(k) = 0$. This occurs whenever $k = \pi n$, where n is a positive integer (n must be positive since k is).

From the above discussion follows that there is a countable set of eigenfunctions

$$f_n(x) = \sin(k_n x), \tag{5.73}$$

where $k_n = \pi n$, with corresponding eigenvalues $\lambda_n = k_n^2 = \pi^2 n^2$. We will continue to use this example to illuminate the concepts throughout this section.

In general, we could ask ourselves if we should not also consider the possibility of having complex eigenvalues λ in the example above. However, if we introduce the inner product

$$\langle f, g \rangle = \int_0^1 f(x)g(x)dx \tag{5.74}$$

on the function space, we find that $-\partial_x^2$ is a symmetric operator and therefore must have real eigenvalues. As should be verified (see Problem 5.18), any Sturm–Liouville operator in a regular Sturm–Liouville problem is symmetric with respect to the inner product

$$\langle f, g \rangle = \int_D f(x)g(x)w(x)dx, \tag{5.75}$$

where D is the domain of the function space and $w(x)$ is the function appearing in the definition of the Sturm–Liouville operator, which is here taken as the weight function of the inner product.

Another important property of the eigenfunctions belonging to a Sturm–Liouville problem is that they, as the eigenvectors of a symmetric operator, are orthogonal to each other as long as they have distinct eigenvalues. Deriving this property is left as an exercise, see Problem 5.14.

Example 5.17 We can check explicitly that the eigenfunctions we found in Example 5.16 are orthogonal with respect to the inner product of Eq. (5.74). We find that

$$\langle f_n, f_m \rangle = \int_0^1 \sin(\pi n x) \sin(\pi m x) dx, \tag{5.76}$$

where the integrand may be rewritten using the trigonometric identity

$$\sin(\pi n x) \sin(\pi m x) = \frac{1}{2} \left[\cos(\pi(n-m)x) - \cos(\pi(n+m)x) \right]. \tag{5.77}$$

Integration for $n \neq m$ now results in

$$\langle f_n, f_m \rangle = \frac{1}{2\pi} \left[\frac{\sin(\pi(n-m)x)}{n-m} - \frac{\sin(\pi(n+m)x)}{n+m} \right]_0^1 = 0. \tag{5.78a}$$

On the other hand, if $n = m$, we find that

$$\langle f_n, f_n \rangle = \frac{1}{2} \left[x - \frac{\sin(\pi(n+m)x)}{\pi(n+m)} \right]_0^1 = \frac{1}{2} \tag{5.78b}$$

due to $\cos(\pi(n-n)x) = 1$. We therefore also find that the norm of f_n is given by $\|f_n\| = 1/\sqrt{2}$.

5.3.1.1 Sturm–Liouville's theorem

We have just seen that, from a Sturm–Liouville operator being symmetric with respect to a particular inner product, its eigenfunctions must be orthogonal with respect to that inner product (as long as they have distinct eigenvalues). In Example 5.16, we also showed that the eigenfunctions f_n of the Sturm–Liouville operator $-\partial_x^2$ acting on functions that satisfy homogeneous Dirichlet boundary conditions at $x = 0$ and $x = 1$ all had distinct eigenvalues $\lambda_n = \pi^2 n^2$, implying that they are all orthogonal to each other. These properties are not coincidental, but follow directly from *Sturm–Liouville's theorem*, which states that, for any regular Sturm–Liouville problem:

1. There exists an infinite set of eigenfunctions f_n with distinct eigenvalues $\lambda_1 < \lambda_2 < \lambda_3 < \ldots$.

2. This set of eigenfunctions forms a *complete basis* on the function space, i.e., any function $g(x)$ in the function space satisfying the boundary conditions may be written as

$$g(x) = \sum_{n=1}^{\infty} g_n f_n(x), \tag{5.79}$$

where the g_n are the expansion coefficients in the basis $f_n(x)$.

The proof of this theorem is relatively involved and requires more mathematics than what is available to us at this point. However, since the theorem is fundamental for the upcoming discussion, we accept it as is for now.

Example 5.18 In the case of Example 5.16, Sturm–Liouville's theorem tells us that any function $g(x)$ that satisfies $g(0) = g(1) = 0$ may be written as a linear combination of the sine functions $\sin(\pi n x)$

$$g(x) = \sum_{n=1}^{\infty} g_n \sin(\pi n x). \qquad (5.80)$$

This is the *Fourier series* for functions of this type.

For any given function $g(x)$ in the function space, the coefficients of the expansion into eigenfunctions of the Sturm–Liouville operator may be found by using the properties of the inner product. In particular, by taking the inner product of Eq. (5.79) with f_m, we find that

$$\langle f_m, g \rangle = \sum_n g_n \langle f_m, f_n \rangle = \sum_n g_n \delta_{nm} \langle f_m, f_m \rangle = g_m \langle f_m, f_m \rangle, \qquad (5.81)$$

where we have used the fact that the eigenfunctions are orthogonal. Dividing by $\|f_m\|$, the expansion coefficient g_m is given by

$$g_m = \frac{\langle f_m, g \rangle}{\langle f_m, f_m \rangle} = \frac{\langle f_m, g \rangle}{\|f_m\|^2}. \qquad (5.82a)$$

In order to simplify this expression, it is often convenient to work in a normalised basis where $\|f_n\| = 1$, where this equation reduces to

$$g_m = \langle f_m, g \rangle. \qquad (5.82b)$$

In such a basis, the full eigenfunction expansion of the function g is of the form

$$g(x) = \sum_n \langle f_n, g \rangle f_n(x) \qquad (5.83a)$$

while, in the case where the eigenfunctions are not normalised, we instead have

$$g(x) = \sum_n \frac{\langle f_n, g \rangle}{\|f_m\|^2} f_n(x). \qquad (5.83b)$$

It should be noted that this expansion is exactly equivalent to the expression of a general vector in a finite dimensional vector space in terms of an orthonormal vector basis, see Eqs. (1.1) and (1.10).

Example 5.19 In the space of functions satisfying $g(0) = g(1) = 0$, we have seen that finding the series expansion of a function $g(x)$ amounts to finding the coefficients of its Fourier series. This is generally done by the use of *Fourier's trick*, i.e., multiplying by one of the sines $\sin(\pi m x)$ and integrating from 0 to 1. For example, if we wish to find the Fourier coefficients g_n for the function $g(x) = x(1 - x)$, we would compute the integral

$$\int_0^1 x(1 - x) \sin(\pi m x)dx = \frac{2}{\pi^3 m^3}[1 - (-1)^m]. \qquad (5.84a)$$

This must be equal to

$$\sum_n g_n \int_0^1 \sin(\pi m x) \sin(\pi n x)dx = \sum_n g_n \delta_{nm} \frac{1}{2} = \frac{g_m}{2}, \qquad (5.84b)$$

where we have used the results of Example 5.17, implying that

$$g_m = \frac{4}{\pi^3 m^3}[1 - (-1)^m] = \begin{cases} \frac{1}{\pi^3 k^3}, & (m = 2k) \\ 0, & (m = 2k + 1) \end{cases}. \tag{5.84c}$$

The Fourier expansion therefore tells us that

$$x(1 - x) = \sum_{k=1}^{\infty} \frac{\sin(2\pi kx)}{\pi^3 k^3}. \tag{5.85}$$

Although it pre-dates the Sturm–Liouville theory, Fourier's trick is of course nothing else than taking the inner product between the functions as described above.

5.3.2 Periodic and singular Sturm–Liouville problems

Not all Sturm–Liouville problems of physical interest are regular and so we need to consider a few other situations as well. Luckily, the most important aspect of Sturm–Liouville's theorem, the existence of a complete set of eigenfunctions, will remain true also for the problems discussed in this section.

A *periodic Sturm–Liouville problem* is the problem of finding the eigenfunctions of a Sturm–Liouville operator on a function space of periodic functions f such that

$$f(x) = f(x + L), \tag{5.86}$$

where L is the period of the functions. Due to the periodicity, we may restrict the study of this set of functions to the interval $0 \leq x \leq L$, or any other interval of length L, with the imposed boundary conditions

$$f(0) = f(L) \quad \text{and} \quad f'(0) = f'(L). \tag{5.87}$$

Using these boundary conditions, it is again relatively easy to show that the Sturm–Liouville operator will be symmetric with respect to the inner product using the $w(x)$ appearing in its definition as the weight function (naturally, the functions in the definition of the Sturm–Liouville operator need to be periodic for a periodic Sturm–Liouville problem) and integrating over a full period

$$\langle f, g \rangle = \int_0^L f(x)g(x)w(x)dx. \tag{5.88}$$

The consequences of the Sturm–Liouville operator being symmetric, such as the eigenvalues being real, remain true in this case.

Example 5.20 Let us again consider the Sturm–Liouville operator $-\partial_x^2$, this time restricted to 2π periodic functions $f(x) = f(x + 2\pi)$. As in Example 5.16, we find that the eigenfunctions must satisfy the differential equation

$$-f''(x) = \lambda f(x) \tag{5.89}$$

and there are no periodic solutions for $\lambda < 0$. For $\lambda = 0$ we find that

$$f(x) = Ax + B, \tag{5.90}$$

which is periodic only if $A = 0$, resulting in a single non-trivial solution $f(x) = B$ that is periodic. In the case $\lambda = k^2 > 0$, the solutions are of the form

$$f(x) = Ae^{ikx} + Be^{-ikx}. \tag{5.91}$$

The condition of periodicity requires that

$$A + B = Ae^{i2\pi k} + Be^{-i2\pi k} \tag{5.92}$$

which is fulfilled if k is an integer. In general, we therefore find that the eigenfunctions are $f_n = e^{ikx}$, where k is an arbitrary integer that may be positive, negative, or zero (when including the result from the $\lambda = 0$ case, which is $f_0 = 1 = e^{i0x}$). However, apart from the case $k = 0$, we find that the solutions are degenerate, i.e., for each solution e^{ikx} with $k \neq 0$, there exists a linearly independent solution e^{-ikx} with the same eigenvalue $\lambda = k^2$.

Finally, we note that if we want our eigenfunctions to be real, e.g., if we are describing a real function space, then we can introduce the linear combinations

$$c_n(x) = \frac{e^{inx} + e^{-inx}}{2} = \cos(nx) \quad \text{and} \quad s_n(x) = \frac{e^{inx} - e^{-inx}}{2i} = \sin(nx) \tag{5.93}$$

for $n > 0$. These functions are linear combinations of the eigenfunctions $e^{\pm inx}$ and are therefore also eigenfunctions with eigenvalue n^2.

The example above shows that the eigenfunctions of periodic Sturm–Liouville problems are not necessarily non-degenerate. However, the set of all eigenfunctions remains complete and it is always possible to construct a complete orthonormal basis using the eigenfunctions by taking any complete set of eigenfunctions and applying an orthogonalisation procedure, such as the Gram–Schmidt procedure, to it.

In many physical applications, there will also be situations where the function $p(x)$ that appears in the definition of the Sturm–Liouville operator goes to zero at some point x_0. Such cases may be treated as long as this point is on the boundary of the domain of the function space and we replace the boundary conditions at x_0 with the requirement that the functions are finite, i.e.,

$$|f(x_0)| < \infty. \tag{5.94}$$

It may also happen that the interval is not finite, but that we are studying functions on an infinite interval. In this case, the requirement on the function space is that the functions should have finite norms according to the inner product. The problems of searching for eigenfunctions and eigenvalues of Sturm–Liouville operators in these situations are called *singular Sturm–Liouville problems*. In general, it will still be possible to find a complete set of eigenvectors to a singular Sturm–Liouville problem, but the eigenvalues may be degenerate and the *spectrum* of the Sturm–Liouville operator, i.e., the set of eigenvalues, may contain a continuous part. This occurs frequently for several potentials in quantum mechanics, where a potential may result in a spectrum with a discrete part, corresponding to bound states, and a continuous part, corresponding to free states (or scattering states).

Example 5.21 Consider the radial part of the Laplace operator in spherical coordinates in three dimensions

$$\hat{L} = -\frac{1}{r^2}\partial_r r^2 \partial_r. \tag{5.95}$$

If we are studying a region where the origin $r = 0$ is included in our domain, then the problem of finding eigenvalues and eigenfunctions to this operator will be a singular Sturm–Liouville problem, as will the problem of finding eigenvalues and eigenvectors when the domain includes $r \to \infty$. A typical problem where this will be an issue would be finding the electrostatic potential $V(r)$ of a spherically symmetric charge distribution $\rho(r)$ in free space, for which we would need to solve the differential equation

$$-\frac{1}{r^2}\partial_r r^2 \partial_r V = \frac{\rho(r)}{\varepsilon_0} \tag{5.96}$$

between $r = 0$ and $r \to \infty$. This problem is singular both due to the singularity in $p(r) = r^2$ at $r = 0$ and due to the domain not having an upper boundary in r.

5.4 SEPARATION OF VARIABLES

Separation of variables is a technique that may be used to find solutions to some differential equations. The separation we will be mainly concerned with here is the separation of partial differential equations, which essentially amounts to rewriting a differential equation in several variables as an ordinary differential equation in one of the variables and a partial differential equation in the remaining variables. Naturally, if there are only two variables in the problem, the separation will result in two ordinary differential equations. The idea is to reduce the problem to a set of simpler problems that may be solved individually. If we have a partial differential equation in the variables x, y, and z, we can attempt to separate the z coordinate by looking for solutions of the form

$$f(x, y, z) = g(x, y)Z(z). \tag{5.97a}$$

If we are successful, we may continue to try to separate the x and y coordinates by the ansatz

$$g(x, y) = X(x)Y(y) \implies f(x, y, z) = X(x)Y(y)Z(z). \tag{5.97b}$$

If we have a homogeneous linear partial differential equation

$$\hat{L}f(x, y, z) = 0, \tag{5.98}$$

where $\hat{L}$ is a general linear differential operator, the separation of the z variable proceeds in the following manner:

1. Make the ansatz $f(x, y, z) = g(x, y)Z(z)$ (c.f., Eq. (5.97a)).

2. Insert the ansatz into the differential equation and express the derivatives in terms of derivatives of g and Z, respectively.

3. Attempt to rewrite the resulting differential equation as an equation where one of the sides depend only on z and the other side only depend on x and y.

4. Since each side depends on different parameters, both sides must be equal to the same constant. Assume this constant is arbitrary and solve the resulting differential equations separately.

It should be noted that the separation ansatz is not guaranteed to work, nor is it guaranteed to provide solutions that may be adapted to the appropriate boundary conditions when it does. However, as we shall see when we apply separation of variables to multidimensional problems involving Sturm–Liouville operators, we will obtain the most general solutions by virtue of the properties of one-dimensional Sturm–Liouville problems.

Example 5.22 Let us look at the partial differential equation

$$(\partial_x + \partial_y)f(x,y) = 0 \tag{5.99}$$

and search for separated solutions of the form $f(x,y) = X(x)Y(y)$. Inserting the separation ansatz into the differential equation, we find that

$$X'(x)Y(y) + X(x)Y'(y) = 0 \implies \frac{X'(x)}{X(x)} = -\frac{Y'(y)}{Y(y)}. \tag{5.100}$$

Since the left-hand side depends only on x and the right-hand side depends only on y, they must both be equal to the same constant λ and we may write

$$X'(x) = \lambda X(x), \quad Y'(y) = -\lambda Y(y), \tag{5.101}$$

with the solutions

$$X(x) = Ae^{\lambda x}, \quad Y(y) = Be^{-\lambda y} \implies f(x,y) = Ce^{\lambda(x-y)}, \tag{5.102}$$

where $C = AB$. Note that all functions of this form are solutions to the original differential equation, but not all solutions are of this form. In general, we may create superpositions of these solutions that will not be a product of one function of x and one of y. In fact, all functions of the form $f(x,y) = g(x-y)$ solve the differential equation, as may be seen by inserting this into the differential equation

$$(\partial_x + \partial_y)g(x-y) = g'(x-y)(\partial_x + \partial_y)(x-y) = g'(x-y)(1-1) = 0. \tag{5.103}$$

5.4.1 Separation and Sturm–Liouville problems

The separation of variables has a special application when the problem in question involves one or more Sturm–Liouville operators. Imagine that we are given a problem of the form

$$(\text{PDE}): \ [f(y)\hat{L}_x + \hat{L}_y]u(x,y) = g(x,y), \tag{5.104a}$$

where $\hat{L}_x$ is a Sturm–Liouville operator in x, $\hat{L}_y$ is a linear differential operator in y and $f(y)$ and $g(x,y)$ are known functions of their arguments. Furthermore, assume that we have homogeneous boundary conditions on coordinate surfaces of x

$$(\text{BC}): \ \alpha_\pm u(x,y) + \beta_\pm \partial_x u(x,y) = 0. \quad (x = x_\pm) \tag{5.104b}$$

Since the partial differential equation involves a Sturm–Liouville operator, we can look for solutions to the one-dimensional eigenvalue problem

$$(\text{ODE}) : \hat{L}_x X(x) = \lambda X(x), \tag{5.105a}$$

$$(\text{BC}) : \alpha_\pm X(x_\pm) + \beta_\pm X'(x_\pm) = 0. \tag{5.105b}$$

Due to this being a Sturm–Liouville problem, we know that this problem has a set of solutions X_n with corresponding eigenvalues λ_n and that this set forms a complete basis of the function space satisfying the given boundary conditions. In particular, we note that, for any fixed y, the function $u(x, y)$ is a function of x that satisfies the given boundary conditions and therefore it is possible to write it as a linear combination of the functions X_n

$$u(x, y) = \sum_n X_n(x) Y_n(y), \tag{5.106a}$$

where the expansion coefficients $Y_n(y)$ will generally depend on which value we choose to fix y at. The equivalent argumentation can be made for the function $g(x, y)$, which we write

$$g(x, y) = \sum_N X_n(x) g_n(y), \tag{5.106b}$$

where again the expansion coefficients g_n generally depend on y. Inserting this into the original partial differential equation, we find that

$$[f(y)\hat{L}_x + \hat{L}_y]u(x, y) = \sum_n \left[f(y) Y_n(y) \hat{L}_x X_n(x) + X_n(x) \hat{L}_y Y_n(y) \right]$$

$$= \sum_n X_n(x) \left[f(y) Y_n(y) \lambda_n + \hat{L}_y Y_n(y) \right] = \sum_n X_n(x) g_n(y). \tag{5.107}$$

The last two steps of this equation may be rewritten in the form

$$\sum_n X_n(x) \left[f(y) Y_n(y) \lambda_n + \hat{L}_y Y_n(y) - g_n(y) \right] = 0 \tag{5.108}$$

and we can observe that the left-hand side is a linear combination of the functions X_n. The important thing to note is that, since they are the solutions to a Sturm–Liouville problem, the X_n are linearly independent and for a linear combination of X_n to be equal to zero, all of the coefficients must be equal to zero, i.e.,

$$f(y) Y_n(y) \lambda_n + \hat{L}_y Y_n(y) - g_n(y) = 0. \tag{5.109}$$

This is an ordinary differential equation in the variable y that may be solved separately. We have thus used the fact that the differential operator in our problem involved a Sturm–Liouville operator in the x coordinate in order to separate it away. The only remainder of it in the differential equation for $Y_n(y)$ is the eigenvalue λ_n. Naturally, the coordinate y may be replaced by a more general set of coordinates in a higher dimensional space and if $\hat{L}_y$ contains more one-dimensional Sturm–Liouville operators, this may be used to separate the problem even further. We will use this technique extensively in Chapter 6 when applying the concepts of this chapter to physical problems.

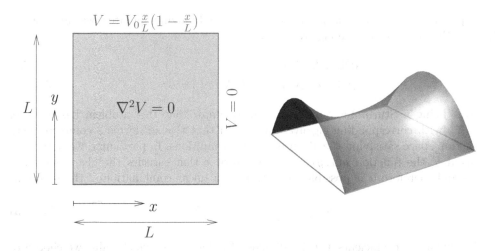

Figure 5.3 The definition of and solution to a problem where we know the electric potential on the boundaries of a square and wish to know the potential inside it.

Example 5.23 Consider the situation shown in Fig. 5.3, where we wish to find the electrostatic potential V in two dimensions on a square $0 < x < L$, $0 < y < L$ with boundary conditions

$$(\text{BC}): \quad V(x,0) = V(x,L) = V_0\frac{x}{L}\left(1 - \frac{x}{L}\right), \quad V(0,y) = V(L,y) = 0 \qquad (5.110a)$$

and where there is no charge in the square itself, i.e., the electrostatic potential satisfies Laplace's equation

$$(\text{PDE}): \quad -\nabla^2 V(x,y) = (-\partial_x^2 - \partial_y^2)V(x,y) = 0. \qquad (5.110b)$$

Since we have homogeneous boundary conditions on the coordinate surfaces $x = 0$ and $x = L$, we can expand the potential in the eigenfunctions $X_n(x) = \sin(\pi n x/L)$ to the Sturm–Liouville problem

$$(\text{ODE}): \quad -X_n''(x) = \lambda_n X_n(x), \qquad (5.111a)$$
$$(\text{BC}): \quad X_n(0) = X_n(L) = 0. \qquad (5.111b)$$

The expansion will be of the form

$$V(x,y) = \sum_{n=1}^{\infty} Y_n(y)\sin(k_n x), \qquad (5.112)$$

where $k_n = \pi n/L$, which inserted into Laplace's equation results in

$$\sum_{n=1}^{\infty} \underbrace{\left[k_n^2 Y_n(y) - Y_n''(y)\right]}_{=0} \sin(k_n x) = 0. \qquad (5.113)$$

The coefficient in front of each sine in the sum must be zero since the sines are linearly independent. We therefore find that the expansion coefficients $Y_n(y)$ must satisfy the ordinary differential equation

$$\text{(ODE)}: \quad k_n^2 Y_n(y) - Y_n''(y) = 0 \tag{5.114}$$

for all n. The boundary conditions for this differential equation can be found by expanding the boundary conditions in the y direction in the sines as well

$$V_0 \frac{x}{L}\left(1 - \frac{x}{L}\right) = \sum_n \frac{4V_0}{\pi^3 n^3}[1 - (-1)^n]\sin(k_n x). \tag{5.115}$$

Note that this is the same expansion as that in Example 5.19 with a rescaling $x \to x/L$. Comparing to the expansion of $V(x,0) = V(x,L)$, we find that

$$\text{(BC)}: \quad Y_n(0) = Y_n(L) = \frac{4V_0}{\pi^3 n^3}[1 - (-1)^n]. \tag{5.116}$$

The resulting problem is easily solved in terms of exponential functions and the solution is shown in the figure as well.

As a special case of the procedure above, we may be interested in eigenvalue problems such as

$$[f(y)\hat{L}_x + \hat{L}_y]u(x,y) = \lambda u(x,y), \tag{5.117}$$

where both $\hat{L}_x$ and $\hat{L}_y$ are Sturm–Liouville operators and λ is an eigenvalue to be determined. We also assume that we have homogeneous boundary conditions on both the x and y coordinate surfaces. Separating out the x variable, we find that

$$[\hat{L}_y + f(y)\lambda_n]Y_n(y) = \lambda Y_n(y), \tag{5.118}$$

where λ_n is the eigenvalue of $X_n(x)$ in the expansion of u. Since $\hat{L}_y$ is a Sturm–Liouville operator, the combination $\hat{L}_y + f(y)\lambda_n$ is also a Sturm–Liouville operator. The resulting problem is therefore a new Sturm–Liouville problem that generally depends on n and has a complete set of orthogonal eigenfunctions $\Upsilon_{nm}(y)$ with corresponding eigenvalues λ_{nm} as solutions. Note that we generally will need two indices, n and m, to denote these solutions, one that numbers the eigenvalue in the x-direction and one that numbers the solution to the differential equation in y for that eigenvalue.

The main thing to note with this situation is that any function $u(x,y)$ satisfying the boundary conditions in the x direction may be written as a linear combination of the eigenfunctions X_n of $\hat{L}_x$

$$u(x,y) = \sum_n Y_n(y)X_n(x) \tag{5.119a}$$

due to the eigenfunctions being a complete basis in the x direction. Furthermore, since $\Upsilon_{nm}(y)$ for a fixed n is a complete basis for functions of y that satisfy the boundary conditions in the y direction, the coefficients $Y_n(y)$ may be expanded in terms of these functions

$$Y_n(y) = \sum_m A_{nm}\Upsilon_{nm}(y), \tag{5.119b}$$

where A_{nm} are the expansion coefficients. As a result, any function $u(x,y)$ that satisfies the boundary conditions of the problem may be written as

$$u(x,y) = \sum_{n,m} A_{nm}X_n(x)\Upsilon_{nm}(y). \tag{5.120}$$

It follows that the functions $f_{nm}(x, y) = X_n(x)\Upsilon_{nm}(y)$ form a complete basis of variable separated eigenfunctions to the two-dimensional problem that we started with in Eq. (5.117). Again, this readily generalises to problems of higher dimensions. For example, in three-dimensions, we may find complete sets of eigenfunctions of the form

$$f_{nm\ell}(x^1, x^2, x^3) = X_n^1(x^1)X_{nm}^2(x^2)X_{nm\ell}^3(x^3). \tag{5.121}$$

The next section will be dedicated to finding functions of this type for several different special cases of physical importance.

5.5   SPECIAL FUNCTIONS

With the theory developed in the previous sections of this chapter at hand, let us turn to some particular cases and find some solutions of relevance for physics applications. We will mainly be working with the Laplace operator ∇^2 in different geometries in which it may be separated into a sum of Sturm–Liouville operators in different coordinate directions.

Example 5.24 Let us consider the Laplace operator in two dimensions in a region $0 < x < L_x$, $0 < y < L_y$, and homogeneous boundary conditions. By separating the x coordinate in eigenvalue problem $-\nabla^2 u = \lambda u$, we find that

$$X_n(x) = A_n \cos(k_n x) + B_n \sin(k_n x), \tag{5.122}$$

where the value of k_n and any constraints on the constants A_n and B_n will be implied by the boundary conditions. For example, with homogeneous Dirichlet conditions, we would find $k_n = \pi n / L_x$ and $A_n = 0$. Expanding the function u in terms of these functions, the differential equation for Y_n becomes

$$Y_n''(y) + (\lambda - k_n^2)Y_n(y) = 0, \tag{5.123}$$

which is a Sturm–Liouville problem for the operator $-\partial_y^2$ with eigenvalue $\lambda - k_n^2$. The solution to this problem will be of the form

$$\Upsilon_m(y) = C_m \cos(\ell_m y) + D_m \sin(\ell_m y), \tag{5.124}$$

with eigenvalue ℓ_m^2, which is independent of n. The possible values of ℓ_m and constraints on the constants will be deduced from the boundary conditions in the y direction. The eigenvalue of the product solution $X_n(x)\Upsilon_m(y)$ is given by the relation

$$\lambda_{nm} - k_n^2 = \ell_m^2 \quad \Longrightarrow \quad \lambda_{nm} = k_n^2 + \ell_m^2, \tag{5.125}$$

i.e., it is the sum of the eigenvalues of the solutions in either direction. With homogeneous Dirichlet boundary conditions on all boundaries, we would obtain $\ell_m = \pi m / L_y$ and the eigenfunctions

$$f_{nm}(x, y) = \sin(k_n x) \sin(\ell_m x). \tag{5.126}$$

These eigenfunctions are shown for $1 \leq n, m \leq 3$ in Fig. 5.4.

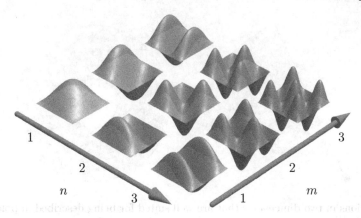

Figure 5.4 The eigenfunctions $\sin(k_n x)\sin(\ell_m y)$ for $1 \leq n, m \leq 3$ in the case of a square where $L_x = L_y$ (and hence $k_n = \ell_n$).

5.5.1 Polar coordinates

For solving problems in regions where the boundaries are coordinate surfaces in polar or cylinder coordinates, i.e., circle segments and radial lines in two dimensions, with the addition of planes parallel to the x^1-x^2-plane in three dimensions, it will be beneficial to work with these coordinates. The Laplace operator in polar coordinates is given by

$$\nabla^2 = \frac{1}{\rho}\partial_\rho \rho \partial_\rho + \frac{1}{\rho^2}\partial_\phi^2. \tag{5.127}$$

We will assume that there are homogeneous boundary conditions and that we are looking for eigenfunctions of the Laplace operator such that $-\nabla^2 f(\rho, \phi) = \lambda f(\rho, \phi)$.

The first thing to notice about the Laplace operator in polar coordinates is the appearance of the operator ∂_ϕ^2 in the angular part. We are already accustomed to this operator after working with the Laplace operator in Cartesian coordinates and know that it has eigenfunctions on the form

$$\Phi_m(\phi) = A_m \cos(k_m \phi) + B_m \sin(k_m \phi). \tag{5.128}$$

There are now two options for a region of the sort we have described, either it has radial lines as boundaries, or it is rotationally symmetric, see Fig. 5.5. In the case where there are boundaries in the ϕ-direction, the problem of finding eigenvalues of $-\partial_\phi^2$ is a regular Sturm–Liouville problem, which we have already discussed. On the other hand, the rotationally symmetric domain is of particular interest and, since ϕ corresponds to the same spatial point as $\phi + 2\pi$, we must impose cyclic boundary conditions, from which we obtain $k_m = m$. In this case, it is also often more convenient to work with the eigenfunctions

$$\Phi_m(\phi) = A_m e^{im\phi} \tag{5.129}$$

and let m be any integer (negative as well as positive and zero, see Example 5.20). Regardless of the boundary conditions, regular or cyclic, the separated eigenfunction $f(\rho, \phi)$ of the Laplace operator will be given by

$$f(\rho, \phi) = R_m(\rho)\Phi_m(\phi), \tag{5.130}$$

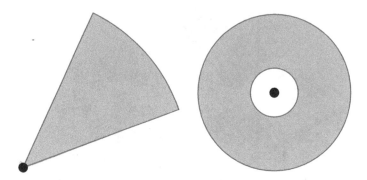

Figure 5.5 Regions in two dimensions that are well suited for being described in polar coordinates with the black dots as the origin. The wedge on the left has boundaries that are coordinate surfaces in both ρ and ϕ, while the rotationally symmetric region on the right has coordinate surfaces in ρ as boundaries, but is cyclic in ϕ.

which leads to the Sturm–Liouville problem

$$-\frac{1}{\rho}\partial_\rho\rho\partial_\rho R_m + \frac{k_m^2}{\rho^2}R_m = -R_m'' - \frac{1}{\rho}R_m' + \frac{k_m^2}{\rho^2}R_m = \lambda R_m. \tag{5.131}$$

This differential equation appears in many situations in physics, not only when separating the Laplace operator in polar coordinates, and is known as *Bessel's differential equation*. Since our boundary conditions were assumed to be homogeneous, we will also have homogeneous boundary conditions for Bessel's differential equation and the problem of finding the eigenfunctions will be a regular Sturm–Liouville problem as long as $\rho = 0$ is not part of the domain and the domain is bounded. On the other hand, if $\rho = 0$ is part of the domain, then the Sturm–Liouville operator is singular and for the boundary at $\rho = 0$ we instead impose that the solutions must be regular $|R_m(0)| < \infty$.

Example 5.25 A physical situation where we may be faced with a problem best expressed in polar coordinates is when we wish to describe the transversal motion of a circular membrane of radius r_0 with a given isotropic tension S that is held fixed at the borders, i.e., essentially a drum skin. The motion can be modelled using the wave equation in two spatial dimensions, which takes the form

$$\frac{1}{c^2}\partial_t^2 u - \nabla^2 u = \left(\frac{1}{c^2}\partial_t^2 - \partial_\rho^2 - \frac{1}{\rho}\partial_\rho - \frac{1}{\rho^2}\partial_\phi^2\right)u = 0 \tag{5.132}$$

in polar coordinates, where $u = u(\rho, \phi, t)$ is the transversal displacement of the membrane and the wave velocity c is given by the tension and the surface density ρ_S as $c^2 = S/\rho_S$, see Section 3.5.2. The boundary conditions on the membrane border are given by

$$u(r_0, \phi, t) = 0. \tag{5.133}$$

For solving this problem, it will be beneficial to find the eigenfunctions of the Laplace operator with the same boundary conditions in order to expand the solution in a series. The eigenfunction problem follows the same steps as outlined above in order to separate the angular part and we are left with looking for radial functions $R_m(\rho)$ solving the Sturm–Liouville problem

$$(\text{PDE}): \ -R_m''(\rho) - \frac{1}{\rho}R_m'(\rho) + \frac{m^2}{\rho^2}R_m(\rho) = \lambda R_m(\rho), \tag{5.134a}$$

$$(\text{BC}): \ R_m(r_0) = 0, \ |R_m(0)| < \infty. \tag{5.134b}$$

The first boundary condition results from requiring that $R_m(r_0)\Phi_m(\phi) = 0$ and the second is the requirement that the solutions are regular due to the Sturm–Liouville problem being singular at $\rho = 0$.

5.5.1.1 Bessel functions

Solving Bessel's differential equation is a rather complicated task and we will therefore discuss a way of obtaining solutions when m is an integer. We start by considering the fact that we wish to find eigenfunctions to Laplace's equation

$$\nabla^2 f + \lambda f = 0. \tag{5.135}$$

One solution to this problem, for the moment disregarding any boundary conditions, is given by

$$f(x^1, x^2) = e^{i\sqrt{\lambda}x^2}, \tag{5.136}$$

which can be seen by inserting this function into the eigenvalue equation. Since we wish to work in polar coordinates, we now express x^2 as $x^2 = \rho\sin(\phi)$ and find that

$$f(\rho, \phi) = e^{i\sqrt{\lambda}\rho\sin(\phi)}. \tag{5.137}$$

Again, it can be checked explicitly by insertion into the eigenfunction equation that this is indeed an eigenfunction with eigenvalue λ. For any fixed ρ, we can expand this function as a function of ϕ in terms of the functions $e^{im\phi}$, as they form a complete basis of 2π periodic functions

$$f(\rho, \phi) = \sum_m f_m(\rho)e^{im\phi}, \tag{5.138}$$

where we have introduced the ρ-dependent expansion coefficients $f_m(\rho)$. By insertion into the eigenfunction equation, these coefficients must satisfy

$$\nabla^2 f + \lambda f = \sum_m \left[f_m''(\rho) + \frac{1}{\rho}f_m'(\rho) - \frac{m^2}{\rho^2}f_m(\rho) + \lambda f_m(\rho) \right] e^{im\phi} = 0. \tag{5.139}$$

Since this is a linear combination of the functions $e^{im\phi}$ that are linearly independent, each term in the sum must be zero and it follows that $f_m(\rho)$ satisfies Bessel's differential equation for $k_m = m$. We can obtain an integral expression for f_m by multiplying Eq. (5.136) by $e^{-im\phi}$ and integrating, i.e., taking the inner product with one of the basis functions of the 2π periodic functions, which results in

$$f_m(\rho) = \frac{1}{2\pi} \int_{-\pi}^{\pi} e^{i(\sqrt{\lambda}\rho\sin(\phi) - m\phi)}d\phi \equiv J_m(\sqrt{\lambda}\rho), \tag{5.140}$$

where we have introduced the *Bessel functions of the first kind* J_m. In themselves, the Bessel functions are independent of the eigenvalue λ as $\sqrt{\lambda}$ is part of its argument. The eigenvalue is instead implied by how the argument relates to the variable in the differential equation. This is in direct analogy with how the definitions of the sine and cosine functions do not specify an eigenvalue in Cartesian coordinates (or rather, normalise the eigenvalue to one), but the eigenvalue is instead given by the factor multiplying the coordinate in the argument, which can be determined from the boundary conditions.

In the more general setting where $k_m = \nu$ is not necessarily an integer, the Bessel function J_ν may be defined through its series expansion

$$J_\nu(x) = \sum_{k=0}^{\infty} \frac{(-1)^k}{k!\Gamma(k+\nu+1)} \left(\frac{x}{2}\right)^{2k+\nu}, \tag{5.141}$$

where Γ is the *gamma function*

$$\Gamma(t) = \int_0^\infty x^{t-1} e^{-x} dx. \tag{5.142}$$

This definition is valid for both positive and negative ν and it should be noted that J_ν and $J_{-\nu}$ solve the same differential equation, which is fortunate as we expect that there should be two independent solutions to Bessel's differential equation, since it is a differential equation of order two. Near $x = 0$, we find that the Bessel functions behave as

$$J_\nu(x) \simeq \frac{1}{\Gamma(\nu+1)} \frac{x^\nu}{2^\nu} + \mathcal{O}(x^{2+\nu}), \tag{5.143}$$

implying that they are singular at $x = 0$ for negative ν unless $\Gamma(\nu+1) \to \infty$, which occurs when ν is a negative integer. This behaviour will allow us to disregard the solutions with negative ν as solutions to the singular Sturm–Liouville problem when $\rho = 0$ is part of the domain and we impose the regularity condition that the eigenfunctions must be regular at the origin.

When $\nu = m$ is an integer, J_m and J_{-m} are no longer independent, as may be derived directly from the integral representation

$$J_{-m}(x) = \frac{1}{2\pi} \int_{-\pi}^{\pi} e^{im\phi + ix\sin(\phi)} d\phi = \{\tau = -\phi + \pi\} = \frac{1}{2\pi} e^{im\pi} \int_0^{2\pi} e^{-im\tau - ix\sin(\tau-\pi)} d\tau$$

$$= (-1)^m \frac{1}{2\pi} \int_{-\pi}^{\pi} e^{-im\tau + ix\sin(\tau)} d\tau = (-1)^m J_m(x), \tag{5.144}$$

where we have done a variable substitution in the first step and used the fact that both $e^{-im\tau}$ and $e^{ix\sin(\tau)}$ are 2π periodic functions of τ to change the interval of integration in the last step. With only the Bessel functions of the first kind, we are therefore missing one independent solution for integer values of ν. This may be remedied by the introduction of the *Bessel functions of the second kind* (also known as *Weber functions* or *Neumann functions*) defined as

$$Y_\nu(x) = \frac{J_\nu(x)\cos(\pi\nu) - J_{-\nu}(x)}{\sin(\pi\nu)} \tag{5.145a}$$

for non-integer values of ν and

$$Y_m(x) = \lim_{\nu \to m} Y_\nu(x) \tag{5.145b}$$

when m is an integer. For non-integer ν this is just a rearrangement of the general solution, similar to rewriting e^{ix} and e^{-ix} in terms of sine and cosine, while for integer m, we obtain a second linearly independent solution to Bessel's differential equation. That it is possible to obtain a linearly independent function in this way is based on the fact that $J_{-m} = (-1)^m J_m$ only implies that $J_{-\nu}$ behaves as

$$J_{-\nu}(x) \simeq (-1)^m J_m(x) + (\nu - m) f_m(x) \tag{5.146}$$

for $\nu \simeq m$, where $f_m(x) = dJ_{-\nu}(x)/d\nu|_{\nu=m}$ is not proportional to J_m. This is not unique to function spaces, but we can have similar situations also in finite dimensional vector spaces.

Example 5.26 As an example of such a situation, let us consider the two dimensional case with

$$\vec{v}_1(\theta) = \cos(2\theta)\vec{e}_1 + \sin(2\theta)\vec{e}_2, \quad \vec{v}_2(\theta) = \cos(\theta)\vec{e}_1 + \sin(\theta)\vec{e}_2. \tag{5.147}$$

The vectors $\vec{v}_1$ and $\vec{v}_2$ are linearly independent for all values of θ that are not integer multiples of π, for which

$$\vec{v}_1(n\pi) = \vec{e}_1, \quad \vec{v}_2(n\pi) = (-1)^n \vec{e}_1. \tag{5.148}$$

For non-integer multiples of π, we can exchange $\vec{v}_2$ for the vector

$$\vec{u}(\theta) = \frac{\vec{v}_2(\theta)\cos(\theta) - \vec{v}_1(\theta)}{\sin(\theta)}, \tag{5.149}$$

which is linearly independent from $\vec{v}_1$. In the limit of $\theta \to \pi n$, we find that

$$\vec{v}_1 \simeq \vec{v}_1(\pi n) + (\theta - \pi n)\vec{v}_1'(\pi n) = \vec{e}_1 + 2(\theta - \pi n)\vec{e}_2, \tag{5.150a}$$

$$\vec{v}_2 \simeq \vec{v}_2(\pi n) + (\theta - \pi n)\vec{v}_2'(\pi n) = (-1)^n[\vec{e}_1 + (\theta - \pi n)\vec{e}_2], \tag{5.150b}$$

to linear order in $\theta - \pi n$. Along with $\sin(\pi n) \simeq (-1)^n(\theta - \pi n)$, this implies that

$$\vec{u} \simeq \frac{[\vec{e}_1 + (\theta - \pi n)\vec{e}_2] - \vec{e}_1 - 2(\theta - \pi n)}{(-1)^n(\theta - \pi n)} = (-1)^{n+1}\vec{e}_2, \tag{5.151}$$

which is linearly independent from $\vec{v}_1(\pi n) = \vec{e}_1$, see Fig. 5.6.

The Bessel functions of the second kind $Y_\nu(x)$ all have the property of being singular in the origin. Graphs of the Bessel functions of the first and second kind for a selection of integer ν are shown in Figs. 5.7 and 5.8, respectively. There is a property of the Bessel functions that is general and apparent from the figures, which is that they are oscillating back and forth around zero. Consequently, the equations $J_\nu(x) = 0$ and $Y_\nu(x) = 0$ have an infinite number of roots, as do the corresponding equations $J_\nu'(x) = 0$ and $Y_\nu'(x) = 0$ for the derivatives. This property is essential in finding non-trivial solutions to the eigenfunction equation with given homogeneous boundary conditions. Generally, the zeros of the Bessel functions are not integer multiples of each other as the zeros of the sine and cosine functions and they have to be found through approximative means or through numerical computations. Luckily, this is an exercise which has been performed and documented in tables for the occasions where numerical values are necessary. We will denote the kth zero of the Bessel function J_m by α_{mk} and the kth zero of its derivative by α_{mk}'. The values of the first couple of zeros are listed in Table 5.1. It is always true that

$$\alpha_{mk} < \alpha_{nk} \tag{5.152}$$

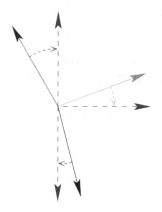

Figure 5.6 The situation where the two black vectors become linearly dependent in a limit. However, the properly scaled difference (gray vector) has non-zero limit which may be taken as a second linearly independent basis vector.

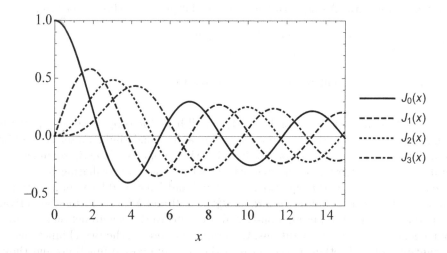

Figure 5.7 The first four Bessel functions of the first kind $J_m(x)$ for integer m. Notable characteristics are the oscillatory behaviour and that only J_0 is non-zero at $x = 0$.

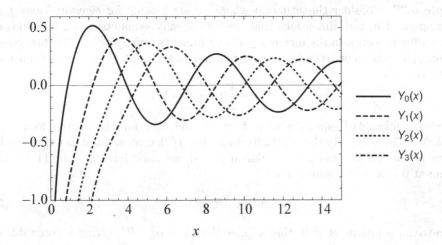

Figure 5.8   The first four Bessel functions of the second kind $Y_m(x)$ for integer m. Notable characteristics are the oscillatory behaviour and that all of the functions diverge in the limit $x \to 0$.

α_{mk}		m			α'_{mk}		m		
k	0	1	2	3	k	0	1	2	3
1	2.40	3.83	5.14	6.38	1	0	1.84	3.05	4.20
2	5.52	7.02	8.42	9.76	2	3.83	5.33	6.71	8.02
3	8.65	10.17	11.62	13.02	3	7.02	8.54	9.97	11.36
4	11.79	13.32	14.80	16.22	4	10.17	11.71	13.17	14.59
5	14.93	16.47	17.96	19.41	5	13.32	14.86	16.35	17.79

Table 5.1   Tables containing the first five zeros of the first four Bessel functions and their derivatives, i.e., α_{mk} and α'_{mk}, respectively. The values have been given to a precision of two decimals but are generally transcendental. Note that zero is sometimes omitted as α'_{01} and all of the α'_{0k} are then moved up one step in the table. We have chosen to use this convention as $J_0(0)$ is non-zero, implying that $J_0(\alpha'_{01}x)$ is a non-trivial solution to Bessel's differential equation with $m = 0$, i.e., the constant solution, for the cases where it is relevant. This will save some amount of bookkeeping in those problems, but the convention should be kept in mind when consulting any table of Bessel function zeros. For the Bessel functions with $m > 1$, the derivative in $x = 0$ is also equal to zero. However, we do not need to include zero as α'_{m1} as the functions $J_m(x)$ are also equal to zero in $x = 0$ and therefore $J_m(0x)$ would correspond to a trivial solution.

for all $m < n$.

Example 5.27 Consider the situation where we are looking for eigenfunctions f of the Laplace operator in two dimensions that are rotationally symmetric, i.e., functions of the radial coordinate ρ only, in the region $\rho \leq R$ and fulfil the homogeneous Dirichlet boundary condition $f(R) = 0$. In order to be an eigenfunction of the Laplace operator, f must satisfy the equation

$$-\nabla^2 f = -\frac{1}{\rho}\partial_\rho \rho \partial_\rho f = \lambda f, \qquad (5.153)$$

i.e., Bessel's differential equation with $k = 0$, and therefore be of the form $f(\rho) = AJ_0(\sqrt{\lambda}\rho) + BY_0(\sqrt{\lambda}\rho)$. By the regularity condition $f(0) < \infty$ and the fact that the Bessel functions of the second kind are singular at $\rho = 0$, we must have $B = 0$. The boundary condition at $\rho = R$ now requires that

$$AJ_0(\sqrt{\lambda}R) = 0. \qquad (5.154)$$

For non-trivial solutions $A \neq 0$, this implies that $\lambda = \alpha_{0k}^2/R^2$, giving a countable set of eigenfunctions

$$f_k(\rho) = J_0(\alpha_{0k}\rho/R) \qquad (5.155)$$

that form a complete basis for the space of rotationally symmetric functions satisfying the homogeneous boundary condition.

A large number of useful integral and derivative relations of the Bessel functions may be derived from Eq. (5.140). Some of these are listed in Appendix A.3.2 and their derivation is left as problems at the end of this chapter (see Problems 5.34 and 5.35). For our purposes, the most important relation is the integral relation

$$\int_0^1 xJ_m(\alpha_{mk}x)J_m(\alpha_{m\ell}x)dx = \frac{\delta_{k\ell}}{2}J_m'(\alpha_{mk})^2 = \frac{\delta_{k\ell}}{2}J_{m+1}(\alpha_{mk})^2, \qquad (5.156)$$

which is related to the orthogonality and normalisation of the different eigenfunctions of Bessel's differential equation.

Returning to the problem of finding eigenfunctions of the two-dimensional Laplace operator, having found the eigenfunctions of Bessel's differential equation in terms of the Bessel functions, the two-dimensional eigenfunctions will be of the form

$$f_{km}(\rho, \phi) = \left[A_{mk}J_m(\sqrt{\lambda_{mk}}\rho) + B_{mk}Y_m(\sqrt{\lambda_{mk}}\rho)\right]\Phi_m(\phi), \qquad (5.157)$$

where $\Phi_m(\phi)$ is given by Eq. (5.128) and the relations between A_{mk} and B_{mk} as well as the possible values of λ_{mk} are governed by the radial boundary conditions. Due to the orthogonality relations for each of the individual one-dimensional Sturm–Liouville problems, these functions will be orthogonal under the inner product

$$\langle f, g \rangle = \int_D f(\rho, \phi)g(\rho, \phi)dA = \int_D f(\rho, \phi)g(\rho, \phi)\rho \, d\rho \, d\phi, \qquad (5.158)$$

where D is the domain of the eigenfunctions.

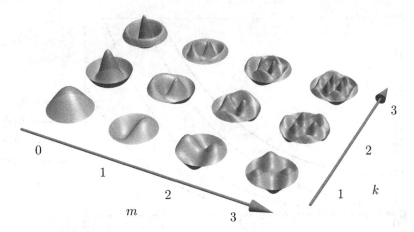

Figure 5.9 The shape of the first few eigenfunctions $f^c_{mk}(\rho, \phi) = J_m(\alpha_{mk}\rho/R)\cos(m\phi)$ for $m = 0, 1, 2$ and 3 of the Laplace operator on a disc $\rho \leq R$ with the Dirichlet boundary condition $f_{mk}(R, \phi) = 0$.

Example 5.28 A commonly encountered problem that will involve expansion in Bessel functions is that of finding the eigenfunctions of the Laplace operator on a disc of radius R with homogeneous Dirichlet boundary conditions. We have already concluded that these functions must be of the form given in Eq. (5.157) and since the domain is the full disc, we find that

$$\Phi_m(\phi) = C_m e^{im\phi} + D_m e^{-im\phi}. \tag{5.159}$$

Since $\rho = 0$ is part of our domain, we must also have all $B_{mk} = 0$ in order for our eigenfunctions to fulfil the criterion of being regular at the origin. This leaves us with the eigenfunctions

$$f_{mk}(\rho, \phi) = J_m(\sqrt{\lambda_{mk}}\rho)(C_{mk}e^{im\phi} + D_{mk}e^{-im\phi}), \tag{5.160}$$

where the constant A_{mk} in front of the Bessel function has been absorbed in C_{mk} and D_{mk}, respectively. We now require that the boundary condition $f(R, \phi) = 0$ is satisfied and that the solution is non-trivial, leading to

$$J_m(\sqrt{\lambda_{mk}}R) = 0 \implies \lambda_{mk} = \frac{\alpha^2_{mk}}{R^2}. \tag{5.161}$$

The resulting functions f_{mk} are degenerate with two solutions for each fixed m and k, except for when $m = 0$, corresponding to different choices of C_{mk} and D_{mk}. If we wish to have explicitly real eigenfunctions, we may choose the functions

$$f^c_{mk}(\rho, \phi) = J_m(\alpha_{mk}\rho/R)\cos(m\phi) \quad \text{and} \quad f^s_{mk}(\rho, \phi) = J_m(\alpha_{mk}\rho/R)\sin(m\phi) \tag{5.162}$$

rather than those using the exponentials $e^{\pm im\phi}$. Some of these functions are illustrated in Fig. 5.9.

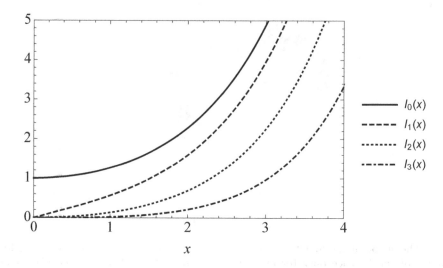

Figure 5.10 The first four modified Bessel functions of the first kind $I_m(x)$ for integer m. Like the Bessel functions of the first kind J_m, the modified Bessel functions of the first kind are finite at $x = 0$. However, they grow exponentially as $x \to \infty$ and do not have any zeros.

5.5.1.2 Modified Bessel functions

At times, we will encounter *Bessel's modified differential equation*

$$-\frac{1}{\rho}\partial_\rho\rho\partial_\rho f + \frac{\nu^2}{\rho^2}f = -f'' - \frac{1}{\rho}f' + \frac{\nu^2}{\rho^2}f = -\lambda f, \tag{5.163}$$

with λ being a fixed positive number. The difference from Bessel's differential equation is the sign of λ and the modified equation will appear mainly when we have inhomogeneous boundary conditions in the radial direction as a result of separating out other coordinates from the problem. Formally, the equation will be satisfied by functions of the form

$$f(\rho) = AJ_\nu(i\sqrt{\lambda}\rho) + BY_\nu(i\sqrt{\lambda}\rho). \tag{5.164}$$

The functions $J_\nu(ix)$ and $Y_\nu(ix)$ will generally be complex valued, but may be turned into real functions by multiplication with $i^{-\nu}$. It is therefore convenient to introduce the *modified Bessel functions* of the first and second kind according to

$$I_\nu(x) = i^{-\nu}J_\nu(ix) \quad \text{and} \quad K_\nu(x) = \frac{\pi}{2}\frac{I_{-\nu}(x) - I_\nu(x)}{\sin(\pi\nu)}, \tag{5.165}$$

respectively. For integer m, K_m is defined by the limit $\nu \to m$ as was the case for the Bessel functions of the second kind. Like the Bessel functions of the second kind, the modified Bessel functions of the second kind diverge when the argument goes to zero. The first few modified Bessel functions of the first and second kind for integer values of ν are shown in Figs. 5.10 and 5.11, respectively.

The modified Bessel functions will play the same role in polar and cylinder coordinates as the hyperbolic sine and cosine (alternatively the exponential functions) play in Cartesian coordinate systems and they do share some characteristics with them. For example, in Cartesian coordinates, we find that

$$-\partial_x^2 \sinh(x) = -\sinh(x), \tag{5.166a}$$

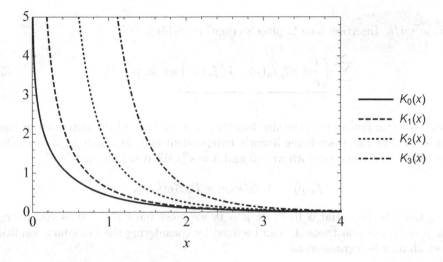

Figure 5.11   The first four modified Bessel functions of the second kind $K_m(x)$ for integer m. Like the Bessel functions of the second kind Y_m, the modified Bessel functions of the second kind diverge at $x = 0$. However, they decrease exponentially as $x \to \infty$ and do not have any zeros.

while in polar coordinates

$$\left(-\frac{1}{\rho}\partial_\rho \rho \partial_\rho - \frac{m^2}{\rho^2}\right) I_m(x) = -I_m(x). \tag{5.166b}$$

In both cases, the differential operator on the left-hand side is the differential operator resulting from separating out other coordinates from the Laplace operator and rather than having a positive eigenvalue, these functions therefore have a negative eigenvalue and grow exponentially as the argument $x \to \infty$.

Example 5.29   The modified Bessel functions will appear in situations when we are solving Laplace's equation in cylinder coordinates. Consider the situation where we wish to compute the electric potential inside a cylinder of radius R and height h where the cylindrical surface is being kept at a potential

$$(\text{BC}): \ V(R, z) = V_0 \frac{z}{h}\left(1 - \frac{z}{h}\right) \tag{5.167a}$$

while the end caps are grounded

$$(\text{BC}): \ V(\rho, 0) = V(\rho, h) = 0. \tag{5.167b}$$

Assuming there are no charges inside the cylinder, the potential inside the cylinder satisfies Laplace's equation

$$(\text{PDE}): \ \nabla^2 V = 0. \tag{5.168}$$

Since the problem is rotationally symmetric, the potential will be a function of the coordinates ρ and z only. Separating out the z coordinate and expanding in the eigenfunctions $\sin(k_n z)$ of $-\partial_z^2$, we find that

$$V(\rho, z) = \sum_{n=1}^{\infty} f_n(\rho) \sin(k_n z), \tag{5.169}$$

where $k_n = \pi n/h$. Insertion into Laplace's equation yields

$$\sum_{n=1}^{\infty} \underbrace{\left(\frac{1}{\rho}\partial_\rho \rho \partial_\rho f_n(\rho) - k_n^2 f_n(\rho)\right)}_{=0} \sin(k_n z) = 0. \tag{5.170}$$

As earlier, the term multiplying the sine functions in each term of the sum must be equal to zero due to the sine functions being linearly independent and thus each f_n satisfies Bessel's modified differential equation with $m = 0$ and $\lambda = k_n^2$. We conclude that

$$f_n(\rho) = A_n I_0(k_n\rho) + B_n K_0(k_n\rho). \tag{5.171}$$

Requiring that the potential is finite at $\rho = 0$, we must have $B_n = 0$ as K_0 diverges in the limit $\rho \to 0$. The constants A_n can be fixed by considering the boundary condition at $\rho = R$, which may be expressed as

$$V(R, z) = \sum_{n=1}^{\infty} A_n I_0(k_n R) \sin(k_n z) = V_0 \frac{z}{h}\left(1 - \frac{z}{h}\right)$$

$$= \sum_{n=1}^{\infty} \frac{4V_0}{\pi^3 n^3}[1 - (-1)^n]\sin(k_n z), \tag{5.172}$$

where we have used the results of Example 5.19 to expand the boundary condition in terms of sine functions. Solving for A_n now yields

$$A_n = \frac{4V_0}{\pi^3 n^3 I_0(k_n R)}[1 - (-1)^n] \tag{5.173a}$$

and consequently

$$V(\rho, z) = \sum_{n=1}^{\infty} \frac{4V_0}{\pi^3 n^3}[1 - (-1)^n]\frac{I_0(k_n\rho)}{I_0(k_n R)}\sin(k_n z). \tag{5.173b}$$

5.5.2 Spherical coordinates

Just as we have considered regions where the boundaries are coordinate surfaces in polar or cylinder coordinates, we may be interested in doing the same for regions in three dimensions whose boundaries are coordinate surfaces in spherical coordinates. In particular, this applies to any rotationally symmetric regions such as the interior of a sphere or the region between two concentric spherical shells. As a by-product, we are going to develop a basis for functions on a sphere, much like the functions $e^{im\phi}$ form a basis for functions on a circle, i.e., 2π periodic functions. Let us start by recalling the form of the Laplace operator in spherical coordinates

$$\nabla^2 = \frac{1}{r^2}\partial_r r^2 \partial_r + \frac{1}{r^2}\left(\frac{1}{\sin(\theta)}\partial_\theta \sin(\theta)\partial_\theta + \frac{1}{\sin^2(\theta)}\partial_\varphi^2\right) = \frac{1}{r^2}\partial_r r^2 \partial_r - \frac{1}{r^2}\hat{\Lambda}, \tag{5.174}$$

where

$$\hat{\Lambda} = -\frac{1}{\sin(\theta)}\partial_\theta \sin(\theta)\partial_\theta - \frac{1}{\sin^2(\theta)}\partial_\varphi^2 \tag{5.175}$$

is a differential operator acting only on the angular coordinates θ and φ. In particular, we note that this expression contains Sturm–Liouville operators in all of the different coordinates and is well suited for variable separation. Starting by separating the coordinate φ, the operator $-\partial_\varphi^2$ is equivalent to the operator $-\partial_\phi^2$ encountered when we considered polar coordinates. In the same fashion as for polar coordinates, we therefore obtain the eigenfunctions

$$\Phi_m(\varphi) = A_m e^{im\varphi} + B_m e^{-im\varphi} = C_m \cos(m\varphi) + D_m \sin(m\varphi). \tag{5.176}$$

Again, we underline the fact that m will be an integer whenever we consider a domain that is rotationally symmetric around the x^3-axis, but in general may take on different values if this is not the case. For the purposes of this section, we will assume that the domain does display this rotational symmetry, but this subtlety should be kept in mind in other situations (see, e.g., Problem 5.50).

5.5.2.1   Legendre polynomials and associated Legendre functions

Solving the periodic Sturm–Liouville problem in the φ coordinate and assuming the eigenvalue of Φ_m with respect to the operator $-\partial_\varphi^2$ to be m^2, the separated eigenfunctions of $-\nabla^2$ take the form

$$f_m(r,\theta,\phi) = R_m(r)\Theta_m(\theta)\Phi_m(\varphi) \tag{5.177}$$

resulting in the differential equation

$$-\frac{1}{r^2}\partial_r r^2 \partial_r R_m \Theta_m + \frac{1}{r^2}\hat{\Lambda}_m R_m \Theta_m = \lambda R_m \Theta_m, \tag{5.178}$$

where

$$\hat{\Lambda}_m = -\frac{1}{\sin(\theta)}\partial_\theta \sin(\theta)\partial_\theta + \frac{m^2}{\sin^2(\theta)} \tag{5.179}$$

is a Sturm–Liouville operator in the θ direction and we therefore look for its eigenfunctions and eigenvalues.

Let us start our search for the eigenfunctions of $\hat{\Lambda}_m$ by making the transformation of variables

$$\xi = \cos(\theta) \quad \Longrightarrow \quad \partial_\theta = \frac{\partial\xi}{\partial\theta}\partial_\xi = -\sin(\theta)\partial_\xi. \tag{5.180}$$

Upon performing this transformation, the eigenfunction equation for $\hat{\Lambda}_m$ turns into

$$\hat{\Lambda}_m = -\partial_\xi(1-\xi^2)\partial_\xi P_m(\xi) + \frac{m^2}{1-\xi^2}P_m(\xi) = \mu_m P_m(\xi), \tag{5.181}$$

where $\Theta_m(\theta) = P_m(\cos(\theta)) = P_m(\xi)$ and μ_m is the sought eigenvalue. This differential equation is known as the *general Legendre equation*, with the special case of $m = 0$ being referred to as *Legendre's differential equation*. We shall begin by dealing with this special case and finding the eigenfunctions of $\hat{\Lambda}_0$ and later generalising the solutions to arbitrary m. For the first part, we shall suppress the use of the index m as it is implicitly set to zero.

If we are studying the Laplace operator in the full range of θ, i.e., 0 to π, the variable ξ ranges from -1 to 1 and the Sturm–Liouville operator appearing in Legendre's differential equation is singular at both ends, indicating that we are dealing with a singular Sturm–Liouville problem and looking for solutions that are finite at $\xi = \pm 1$ as our only boundary conditions. Looking at Legendre's differential equation

$$-\partial_\xi(1 - \xi^2)\partial_\xi P(\xi) = \mu P(\xi), \tag{5.182}$$

we may expand $P(xi)$ as a Maclaurin series

$$P(\xi) = \sum_{n=0}^{\infty} p_n \xi^n. \tag{5.183}$$

Inserting this into the differential equation results in the recursion relation

$$p_{n+2} = \left(\frac{n}{n+2} - \frac{\mu}{(n+2)(n+1)} \right) p_n, \tag{5.184}$$

implying that the entire series is defined by fixing μ, p_0, and p_1. As $n \to \infty$, $p_{2n} \sim 1/n$, and thus the series

$$P(1) = \sum_{n=0}^{\infty} p_n \tag{5.185}$$

diverges unless it terminates at a finite $n = \ell < \infty$. If the series diverges, then $P(\xi)$ is singular in $\xi = 1$, but we are explicitly looking for solutions that are not singular in $\xi = \pm 1$. The conclusion from this is that the series must terminate and that $P(\xi)$ is therefore a polynomial of some degree ℓ. In order for $p_{\ell+2} = 0$ to hold, we find that

$$\ell - \frac{\mu}{\ell+1} = 0 \quad \Longrightarrow \quad \mu = \ell(\ell+1), \tag{5.186}$$

providing us with the corresponding eigenvalues μ. Since the recursion relations for odd and even n are independent and Eq. (5.186) can only hold for $n = \ell$, it follows that for odd ℓ all p_n with even n are equal to zero and vice versa. Consequently, the polynomials are all either odd or even functions of ξ, depending on their degree. The polynomial of this form of degree ℓ is denoted by $P_\ell(\xi)$ and is the *Legendre polynomial* of degree ℓ. As usual, the eigenfunctions are defined only up to a constant and the customary normalisation of the Legendre polynomials is to require that $P_\ell(1) = 1$. The polynomial coefficients may be found through the recursion relation for any given ℓ or may be computed through *Rodrigues' formula*

$$P_\ell(\xi) = \frac{1}{2^\ell \ell!} \frac{d^\ell}{d\xi^\ell} (\xi^2 - 1)^\ell. \tag{5.187}$$

Clearly, this is a polynomial of degree ℓ and it is easily verified by insertion into Legendre's differential equation that it is an eigenfunction with the appropriate eigenvalue. The Legendre polynomials up to a degree of $\ell = 5$ are given in Table 5.2 and those up to a degree $\ell = 4$ are shown in Fig. 5.12.

Being a second order differential equation, there should exist a second linearly independent solution $Q_\ell(\xi)$ to Legendre's differential equation for every fixed eigenvalue and the general solution should be of the form

$$P(\xi) = A P_\ell(\xi) + B Q_\ell(\xi). \tag{5.188}$$

However, the solutions $Q_\ell(\xi)$ are inevitably singular and since we are searching for non-singular solutions, we will largely disregard them.

ℓ	$P_\ell(\xi)$	$\ell(\ell+1)$
0	1	0
1	ξ	2
2	$\frac{1}{2}(3\xi^2 - 1)$	6
3	$\frac{1}{2}(5\xi^3 - 3\xi)$	12
4	$\frac{1}{8}(35\xi^4 - 30\xi^2 + 3)$	20
5	$\frac{1}{8}(63\xi^5 - 70\xi^3 + 15\xi)$	30

Table 5.2   The first six Legendre polynomials $P_\ell(\xi)$ along with the corresponding eigenvalues $\ell(\ell+1)$ of Legendre's differential equation.

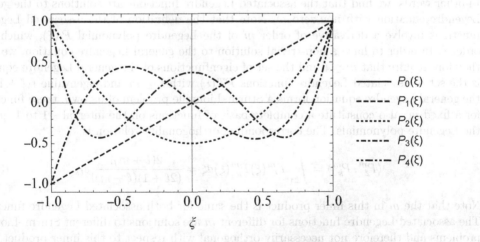

Figure 5.12   The first five Legendre polynomials $P_\ell(\xi)$. Note how all of the Legendre polynomials satisfy the property $P_\ell(-\xi) = (-1)^\ell P_\ell(\xi)$.

As the Legendre polynomials are eigenfunctions with different eigenvalues of a Sturm–Liouville operator whose weight function is one, they must be orthogonal under the inner product

$$\langle f, g \rangle = \int_{-1}^{1} f(\xi) g(\xi) d\xi \tag{5.189a}$$

or, in terms of the original coordinate θ,

$$\langle f, g \rangle = \int_{0}^{\pi} f(\cos(\theta)) g(\cos(\theta)) \sin(\theta) d\theta, \tag{5.189b}$$

which may be found by performing the change of variables in reverse. In general, the orthogonality relation between the Legendre polynomials can be summarised as

$$\langle P_\ell, P_k \rangle = \int_{-1}^{1} P_\ell(\xi) P_k(\xi) d\xi = \frac{2}{2\ell + 1} \delta_{\ell k}. \tag{5.190}$$

Introducing the *associated Legendre functions*

$$P_\ell^m(\xi) = (-1)^m (1 - \xi^2)^{m/2} \frac{d^m}{d\xi^m} P_\ell(\xi), \tag{5.191}$$

applying the Legendre differential operator $\partial_\xi (1 - \xi^2) \partial_\xi$, and using the fact that $P_\ell(\xi)$ satisfies Legendre's differential equation results in

$$\partial_\xi (1 - \xi^2) \partial_\xi P_\ell^m(\xi) = \frac{m^2}{1 - \xi^2} P_\ell^m(\xi) - \ell(\ell + 1) P_\ell^m(\xi). \tag{5.192}$$

In other words, we find that the associated Legendre functions are solutions to the general Legendre equation with non-zero m. Note that the definition of the associated Legendre functions involve a derivative of order m of the Legendre polynomial $P_\ell(\xi)$, which is of order ℓ. In order to have a non-trivial solution to the general Legendre equation, we must therefore require that $m \leq \ell$ and the set of eigenfunctions of the general Legendre equation is the set of associated Legendre functions $P_\ell^m(\xi)$ with $\ell \geq m$ and eigenvalue $\ell(\ell + 1)$. As the general Legendre equation forms a Sturm–Liouville problem of its own, these functions, for a fixed m, also constitute a complete basis of functions on the interval -1 to 1, just as the Legendre polynomials. The corresponding orthogonality relation is

$$\langle P_\ell^m, P_k^m \rangle = \int_{-1}^{1} P_\ell^m(\xi) P_k^m(\xi) d\xi = \frac{2(\ell + m)!}{(2\ell + 1)(\ell - m)!} \delta_{\ell k}. \tag{5.193}$$

Note that the m in this inner product is the same for both associated Legendre functions. The associated Legendre functions for different m are solutions to different Sturm–Liouville problems and therefore not necessarily orthogonal with respect to this inner product.

Example 5.30 Consider the electric potential $V(r, \theta)$ between two concentric shells with radii r_1 and r_2, respectively, where $r_1 < r_2$. If there is no charge density in the region $r_1 < r < r_2$, then the potential satisfies Laplace's equation

$$\text{(PDE)}: \quad \nabla^2 V(r, \theta) = 0 \tag{5.194a}$$

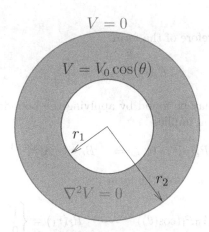

Figure 5.13 A system with two concentric shells with given electrical potentials. The figure shows a section through the center of the shells with the inner shell having a radius r_1 and the outer shell a radius r_2. The potential in the region between the spheres satisfies Poisson's equation.

in this region. The inner shell is held at a potential

$$(\text{BC}): \ V(r_1, \theta) = V_0 \cos(\theta), \tag{5.194b}$$

while the outer is kept grounded

$$(\text{BC}): \ V(r_2, \theta) = 0 \tag{5.194c}$$

(see Fig. 5.13). Since the boundary conditions are invariant under rotations around the x^3-axis, the solution will not depend on the spherical coordinate φ. Expanding the solution for fixed r in the Legendre polynomials $P_\ell(\cos(\theta))$, we obtain

$$V(r, \theta) = \sum_{\ell=0}^{\infty} R_\ell(r) P_\ell(\cos(\theta)) \tag{5.195}$$

and insertion into Laplace's equation yields

$$\nabla^2 V(r, \theta) = \sum_{\ell=0}^{\infty} \underbrace{\left[R_\ell''(r) + \frac{1}{r} R_\ell'(r) - \frac{\ell(\ell+1)}{r^2} R_\ell(r) \right]}_{=0} P_\ell(\cos(\theta)) = 0, \tag{5.196}$$

where the factor in front of the Legendre polynomial in each term must be equal to zero due to the linear independence of the Legendre polynomials. The resulting differential equation

$$R_\ell''(r) + \frac{2}{r} R_\ell'(r) - \frac{\ell(\ell+1)}{r^2} R_\ell(r) = 0 \tag{5.197}$$

is of the Cauchy–Euler type and can be solved by the ansatz $R_\ell(r) = r^k$, giving

$$k(k-1) + 2k - \ell(\ell+1) = 0 \quad \Longrightarrow \quad k = \ell \text{ or } k = -(\ell+1). \tag{5.198}$$

The general solution is therefore of the form

$$R_\ell(r) = A_\ell r^\ell + \frac{B_\ell}{r^{\ell+1}}. \tag{5.199}$$

The constants A_ℓ and B_ℓ may be found by applying the boundary conditions. First of all, the outer boundary condition implies

$$R_\ell(r_2) = 0 \quad \Longrightarrow \quad B_\ell = -A_\ell r_2^{2\ell+1}. \tag{5.200}$$

The inner boundary conditions is of the form

$$V(R_1,\theta) = V_0 P_1(\cos(\theta)) \quad \Longrightarrow \quad R_\ell(r_1) = \begin{cases} V_0, & (\ell = 1) \\ 0, & (\ell \neq 1) \end{cases}, \tag{5.201}$$

where the inference is made by noting that the Legendre polynomials are linearly independent and identifying the terms in the series solution with the boundary condition. This implies $A_\ell = 0$ for all $\ell \neq 1$ and

$$A_1 = \frac{V_0 r_1^2}{r_1^3 - r_2^3} \quad \Longrightarrow \quad R_1(r) = \frac{V_0 r_1^2}{r^2} \frac{r_2^3 - r^3}{r_2^3 - r_1^3}. \tag{5.202}$$

The full solution for the potential between the spherical shells is therefore

$$V(r,\theta) = \frac{V_0 r_1^2}{r^2} \frac{r_2^3 - r^3}{r_2^3 - r_1^3} \cos(\theta). \tag{5.203}$$

Combining Rodrigues' formula with the definition of the associated Legendre functions, we find that

$$P_\ell^m(\xi) = \frac{(-1)^m}{2^\ell \ell!} (1 - \xi^2)^{m/2} \frac{d^{\ell+m}}{d\xi^{\ell+m}} (\xi^2 - 1)^\ell. \tag{5.204}$$

This also allows the extension of associated Legendre functions to negative values of m as long as $|m| \leq \ell$, which is the range for which the associated Legendre functions exist. This definition results in

$$P_\ell^{-m}(\xi) = (-1)^m \frac{(\ell - m)!}{(\ell + m)!} P_\ell^m(\xi). \tag{5.205}$$

Naturally, this means that P_ℓ^m and P_ℓ^{-m} are linearly dependent, so we can choose to include either, but not both or none, in the complete bases and they only differ in the normalisation.

5.5.2.2 Spherical harmonics

We may combine our solutions for the Sturm–Liouville operators in the θ and φ directions to construct a set of eigenfunctions of the operator $\hat{\Lambda}$ that form a complete basis for the functions on the unit sphere, i.e., functions $f(\theta,\varphi)$. For a fixed θ, we expand the function f in terms of the functions $e^{im\varphi}$ and obtain

$$f(\theta,\varphi) = \sum_{m=-\infty}^{\infty} \Theta_m(\theta) e^{im\varphi}. \tag{5.206a}$$

The functions $\Theta_m(\theta)$ may further be expanded in terms of the complete basis of associated Legendre functions $P_\ell^m(\cos(\theta))$ to yield

$$f(\theta, \varphi) = \sum_{m=-\infty}^{\infty} \sum_{\ell \geq |m|} A_{\ell m} P_\ell^m(\cos(\theta)) e^{im\varphi} = \sum_{\ell=0}^{\infty} \sum_{m=-\ell}^{\ell} A_{\ell m} P_\ell^m(\cos(\theta)) e^{im\varphi}, \qquad (5.206b)$$

where all we have done in the second step is to change the order of the sums. The set of functions $P_\ell^m(\cos(\theta)) e^{im\varphi}$ thus forms a complete basis for the functions on the sphere which is known as the *spherical harmonics* Y_ℓ^m. We may choose to normalise this basis as we see fit and this normalisation tends to differ between different disciplines. We will use the normalisation

$$Y_\ell^m(\theta, \varphi) = (-1)^m \sqrt{\frac{(2\ell + 1)}{4\pi} \frac{(\ell - m)!}{(\ell + m)!}} P_\ell^m(\cos(\theta)) e^{im\varphi}, \qquad (5.207)$$

which is commonly used in quantum physics, but the normalisation should be checked whenever it is of importance for an application.

All of the spherical harmonics are orthogonal under the inner product

$$\langle f, g \rangle = \int_{S^2} f(\theta, \varphi)^* g(\theta, \varphi) dA = \int_{\varphi=0}^{2\pi} \int_{\theta=0}^{\pi} f(\theta, \varphi)^* g(\theta, \varphi) \sin(\theta) \, d\theta \, d\varphi, \qquad (5.208)$$

where S^2 denotes the sphere, which is the domain of the functions. The orthogonality of spherical harmonics with different m follows from the orthogonality of the eigenfunctions $e^{im\varphi}$, while the orthogonality of spherical harmonics with the same m, but different ℓ, follows from the orthogonality of the associated Legendre functions. With our chosen normalisation, the spherical harmonics are orthonormal

$$\left\langle Y_\ell^m, Y_{\ell'}^{m'} \right\rangle = \delta_{\ell\ell'} \delta_{mm'}. \qquad (5.209)$$

By design, the spherical harmonics are also the eigenfunctions of the operator $\hat{\Lambda}$ and we find that

$$\hat{\Lambda} Y_\ell^m(\theta, \varphi) = \ell(\ell + 1) Y_\ell^m(\theta, \varphi). \qquad (5.210)$$

Since the eigenvalue does not depend on m and $-\ell \leq m \leq \ell$, there are $2\ell + 1$ different spherical harmonics with eigenvalue $\ell(\ell+1)$, i.e., all of the eigenvalues except zero correspond to more than one eigenfunction. A selection of different spherical harmonics are shown in Fig. 5.14. The functions with different values of ℓ in an expansion into spherical harmonics are referred to as different *multipoles*, e.g., the $\ell = 0$ spherical harmonic leads to a monopole, the $\ell = 1$ solutions dipoles, the $\ell = 2$ solutions quadrupoles, etc. An expansion in terms of the spherical harmonics is often referred to as a multipole expansion.

Example 5.31 As they form a complete basis for functions on a sphere, any function on a sphere may be expressed as a series expansion in terms of the spherical harmonics. An example of such an expansion is given in Fig. 5.15, where we show the function $r(\theta, \varphi)$ given by the sum

$$r(\theta, \varphi) = \sum_{\ell=0}^{\ell_{max}} \sum_{m=-\ell}^{\ell} C_{\ell m} Y_\ell^m(\theta, \varphi) \qquad (5.211)$$

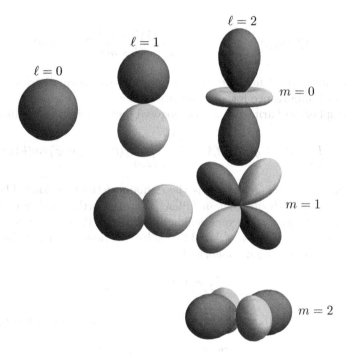

Figure 5.14 The spherical harmonics with $\ell \leq 2$. For each spherical harmonic, we show the absolute value of the real part as the distance from the center. The real part is positive where the figure is dark and negative where it is light. Note how the angles between the maxima of these spherical harmonics are $2\pi/\ell$.

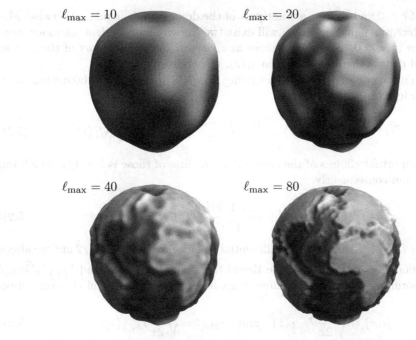

$\ell_{\max} = 10$ $\ell_{\max} = 20$

$\ell_{\max} = 40$ $\ell_{\max} = 80$

Figure 5.15 Earth's topography as given by its series expansion into spherical harmonics. Each successive image includes all terms with $\ell \leq \ell_{\max}$. As more terms are added to the series, the series expansion becomes more and more accurate. Note that the variations have been severely exaggerated. Data source: *C. Hirt, M. Kuhn, W.E. Featherstone, F. Göttl*, Topographic/isostatic evaluation of new generation GOCE gravity field models, *Journal of Geophysical Research – Solid Earth 117 (2012) B05407.*

for different values of $\ell_{\max}$ as a function on the sphere. The constants $C_{\ell m}$ are taken to be the expansion coefficients of the Earth's topography. As seen in the figure, including only the terms corresponding to lower values of ℓ results in a function that gives a very rough idea about the main features. The sharper features in the function are reproduced once terms with higher ℓ are included as well.

5.5.2.3 Spherical Bessel functions

Continuing the search for separated eigenfunctions of the Laplace operator in spherical coordinates, we may now write such a function as

$$f(r,\theta,\varphi) = R(r)Y_\ell^m(\theta,\varphi). \tag{5.212}$$

Inserting this into the eigenfunction equation and using the fact that Y_ℓ^m is an eigenfunction of $\hat{\Lambda}$, we obtain

$$-\frac{1}{r^2}\partial_r r^2 \partial_r R(r) + \frac{\ell(\ell+1)}{r^2}R(r) = \lambda R(r). \tag{5.213}$$

This equation is the spherical coordinate equivalent of Bessel's differential equation and with appropriate homogeneous boundary conditions it will become a Sturm–Liouville problem that is regular if $r = 0$ and $r \to \infty$ are not part of the domain and singular otherwise. Much like Bessel's differential equation, there will exist two linearly independent solutions, one of which is singular at $r = 0$. We refer to these as *spherical Bessel functions* of the first and second kind and denote them by $j_\ell(r)$ and $y_\ell(r)$, respectively.

There are a number of advantages of rewriting Eq. (5.213) using the substitution $R(r) = f_k(r)r^{-k}$, which leads to

$$-f_k''(r) - \frac{2(1-k)}{r}f_k'(r) + \frac{\ell(\ell+1) - k(k-1)}{r^2}f_k(r) = \lambda f_k(r). \tag{5.214}$$

There are two important choices of the constant k, the first of those is $k = 1/2$, which leads to $2(1-k) = 1$ and consequently

$$-f_{\frac{1}{2}}''(r) - \frac{1}{r}f_{\frac{1}{2}}'(r) + \frac{(\ell+1/2)^2}{r^2}f_{\frac{1}{2}}(r) = \lambda f_{\frac{1}{2}}(r), \tag{5.215}$$

i.e., the function $f_{\frac{1}{2}}(r)$ satisfies Bessel's differential equation with $\nu = \ell + 1/2$ and we already know the independent solutions to be the Bessel functions $J_{\ell+\frac{1}{2}}(\sqrt{\lambda}r)$ and $Y_{\ell+\frac{1}{2}}(\sqrt{\lambda}r)$. As per usual, the normalisation of the eigenfunctions needs to be chosen and the conventional choice is

$$j_\ell(x) = \sqrt{\frac{\pi}{2x}}J_{\ell+\frac{1}{2}}(x) \quad \text{and} \quad y_\ell(x) = \sqrt{\frac{\pi}{2x}}Y_{\ell+\frac{1}{2}}(x). \tag{5.216}$$

The second important choice of k is $k = 1$, which leads to the vanishing of the term with the first derivative

$$-f_1''(r) + \frac{\ell(\ell+1)}{r^2}f_1(r) = \lambda f_1(r). \tag{5.217}$$

In particular, this implies that the equation for $\ell = 0$ reduces to $f''(r) + \lambda f(r) = 0$, which we know to have a solution in terms of sines and cosines. Consequently, we find that

$$j_0(x) = \frac{\sin(x)}{x} \quad \text{and} \quad y_0(x) = -\frac{\cos(x)}{x}, \tag{5.218}$$

where the normalisation is chosen to be compatible with Eq. (5.216). For general ℓ, Eq. (5.217) can be solved by

$$j_\ell(x) = (-x)^\ell \left(\frac{1}{x}\frac{d}{dx}\right)^\ell j_0(x) \quad \text{and} \quad y_\ell(x) = (-x)^\ell \left(\frac{1}{x}\frac{d}{dx}\right)^\ell y_0(x). \tag{5.219}$$

Performing these derivatives for the first few spherical Bessel functions of the first kind, we find that

$$j_0(x) = \frac{\sin(x)}{x}, \tag{5.220a}$$

$$j_1(x) = \frac{\sin(x)}{x^2} - \frac{\cos(x)}{x}, \tag{5.220b}$$

$$j_2(x) = \left(\frac{3}{x^2} - 1\right)\frac{\sin(x)}{x} - \frac{3\cos(x)}{x}, \tag{5.220c}$$

$$j_3(x) = \left(\frac{15}{x^3} - \frac{6}{x}\right)\frac{\sin(x)}{x} - \left(\frac{15}{x^2} - 1\right)\frac{\cos(x)}{x}. \tag{5.220d}$$

Similar expressions may be derived for the spherical Bessel functions of the second kind in the same fashion. The spherical Bessel functions of the first and second kind for $\ell \leq 3$ are

$\beta_{\ell k}$		ℓ		
k	0	1	2	3
1	π	4.49	5.76	6.99
2	2π	7.73	9.10	10.42
3	3π	10.90	12.32	13.70
4	4π	14.07	15.51	16.92
5	5π	17.22	18.69	20.12

$\beta'_{\ell k}$		ℓ		
k	0	1	2	3
1	0	2.08	3.34	4.51
2	4.49	5.94	7.29	8.58
3	7.73	9.21	10.61	11.97
4	10.90	12.40	13.85	15.24
5	14.07	15.58	17.04	18.47

Table 5.3 Tables containing the first five zeros of the first four spherical Bessel functions of the first kind j_ℓ and their derivatives, i.e., $\beta_{\ell k}$ and $\beta'_{\ell k}$, respectively. The values have been given to a precision of two decimals but are generally transcendental. As for the Bessel functions, zero is sometimes omitted as β'_{01}. Note that $\beta_{1k} = \beta'_{0(k+1)}$ as $j'_0(x) = -j_1(x)$.

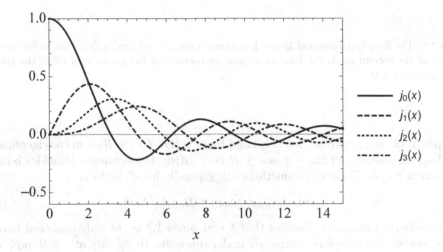

Figure 5.16 The first four spherical Bessel functions of the first kind $j_\ell(x)$. Just as for the Bessel functions of the first kind, we note an oscillatory behaviour and that only j_0 is non-zero at $x = 0$.

shown in Figs. 5.16 and 5.17, respectively. Like the Bessel functions, the spherical Bessel functions and their derivatives have an infinite number of zeros. We will denote the kth zero of j_ℓ by $\beta_{\ell k}$ and the kth zero of its derivative by $\beta'_{\ell k}$. From the form of the spherical Bessel function j_0, it is evident that $\beta_{0k} = \pi k$. The values of $\beta_{\ell k}$ for general ℓ will not be as simple to express, but may be computed numerically or found in tables such as Table 5.3.

Being solutions to a Sturm–Liouville problem, the eigenfunctions that are expressed in terms of j_ℓ and y_ℓ, properly adapted to the boundary conditions, will be orthogonal under the inner product

$$\langle f, g \rangle = \int_a^b f(r)g(r)r^2 dr, \tag{5.221}$$

since the weight function of the Sturm–Liouville operator is $w(r) = r^2$. In particular, the spherical Bessel functions satisfy the relation

$$\int_0^1 j_\ell(\beta_{\ell k}x)j_\ell(\beta_{\ell k'}x)x^2 dx = \frac{\delta_{kk'}}{2}j_{\ell+1}(\beta_{\ell k})^2, \tag{5.222}$$

see Problem 5.45.

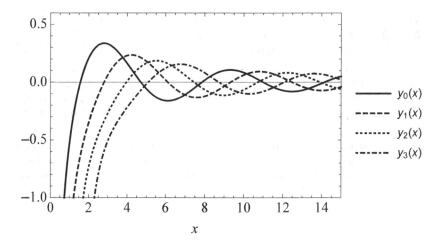

Figure 5.17 The first four spherical Bessel functions of the second kind $y_\ell(x)$. Just as for the Bessel functions of the second kind, the functions show an oscillatory behaviour and all of the functions are singular at $x = 0$.

Example 5.32 As an exercise, let us expand the function $f(r) = R - r$ in the eigenfunctions of the Laplace operator in the region $r \leq R$ that satisfy homogeneous Dirichlet boundary conditions at $r = R$. These eigenfunctions will generally be of the form

$$u_k(r) = A_k j_0(\sqrt{\lambda_k}r) + B_k y_0(\sqrt{\lambda_k}r), \tag{5.223}$$

since the spherical symmetry implies that $\ell = 0$, since Y_0^0 is the only spherical harmonic that is constant and therefore symmetric under rotations. Regularity at $r = 0$ implies that $B_k = 0$ and the homogeneous Dirichlet condition at $r = R$ consequently results in

$$\lambda_k = \left(\frac{\pi k}{R}\right)^2 \tag{5.224}$$

in order to have non-trivial solutions. The expansion in terms of the eigenfunctions $j_0(\pi kr/R)$ is given by

$$f(r) = \sum_{k=1}^{\infty} A_k j_0(\pi kr/R) = \sum_{k=1}^{\infty} C_k \frac{\sin(\pi kr/R)}{r}, \tag{5.225}$$

where $C_k = RA_k/\pi k$. Taking the inner product of f with $\sin(\pi k'r/R)/r$ now results in

$$\frac{1}{2}C_{k'} = \sum_{k=1}^{\infty} \int_0^R \sin(\pi k'r/R)\sin(\pi kr/R)dr$$

$$= \int_0^R (R - r)\sin(\pi k'r/R)r\,dr = \frac{2R^3}{\pi^3 k'^3}[1 - (-1)^{k'}]. \tag{5.226a}$$

Multiplying this equation by two, we find that

$$C_k = \frac{4R^3}{\pi^3 k^3}[1 - (-1)^k] \tag{5.226b}$$

and consequently

$$R - r = \sum_{k=1}^{\infty} \frac{4R^3}{\pi^3 k^3}[1 - (-1)^k]\frac{\sin(\pi k r/R)}{r} = \sum_{k=1}^{\infty} \frac{4R^2}{\pi^2 k^2}[1 - (-1)^k]j_0(\pi k r/R). \quad (5.227)$$

It should be noted that, multiplying the entire expansion by r, this is actually the same series expansion as that found in Example 5.19.

Wrapping up the problem of finding the eigenfunctions of the Laplace operator with homogeneous boundary conditions on a spherical surface, the general eigenfunction will be of the form

$$f_{n\ell m}(r, \theta, \varphi) = [A_{\ell n}j_\ell(\sqrt{\lambda_{\ell n}}r) + B_{\ell n}y_\ell(\sqrt{\lambda_{\ell n}}r)]Y_\ell^m(\theta, \varphi). \quad (5.228)$$

As always, the relation between the constants $A_{\ell n}$ and $B_{\ell n}$ must be determined through the boundary conditions, with $B_{\ell n} = 0$ due to the requirement of regularity if $r = 0$ is part of the domain. In particular, if we are studying the domain $r \leq R$ with homogeneous Dirichlet conditions, the eigenfunctions are given by

$$f_{n\ell m}(r, \theta, \varphi) = j_\ell(\beta_{\ell n}r/R)Y_\ell^m(\theta, \varphi), \quad (5.229)$$

with corresponding eigenvalues $\lambda_{\ell n} = \beta_{\ell n}^2/R^2$. Note that, since the eigenvalue does not depend on m and $-\ell \leq m \leq \ell$, there will be $2\ell + 1$ different eigenfunctions with the same eigenvalue $\lambda_{\ell n}$.

Regardless of the boundary conditions, the eigenfunctions of the form presented above will be orthogonal under the inner product

$$\langle f, g \rangle = \int_{r=a}^b \int_{\theta=0}^\pi \int_{\varphi=0}^{2\pi} f(r, \theta, \varphi)^* g(r, \theta, \varphi)r^2 \sin(\theta)dr\, d\theta\, d\varphi$$

$$= \int_{a \leq r \leq b} f(r, \theta, \varphi)^* g(r, \theta, \varphi)dV, \quad (5.230)$$

where we have assumed the domain of the Sturm–Liouville problem to be $0 \leq a < r < b$ and dV is the regular three-dimensional volume element. The orthogonality of functions with different ℓ or different m follows from the orthogonality of the spherical harmonics under the angular integrals, while the orthogonality of functions with the same ℓ and m, but different n, follows from the orthogonality relations for the spherical Bessel functions.

5.5.3 Hermite functions

The Laplace operator does not always appear on its own as a Sturm–Liouville operator. In many cases that will be encountered in quantum mechanics, we will need to solve Sturm–Liouville problems of the form

$$-\psi''(x) + V(x)\psi(x) = 2E\psi(x), \quad (5.231a)$$

where $V(x)$ is a function describing the potential energy at position x, $\psi(x)$ is the quantum mechanical wave function, and E is an energy eigenvalue. While discussing the methodology of solving partial differential equations by using series and transform solutions, we will not consider problems from quantum mechanics, but as they will be fundamental when pursuing further studies in physics, we will briefly discuss the special case $V(x) = x^2$, corresponding to a quantum mechanical harmonic oscillator, for future reference. It should be noted that performing dimensional analysis on the ensuing differential equation

$$-\psi''(x) + x^2\psi(x) = 2E\psi(x), \tag{5.231b}$$

the coordinate x and the energy eigenvalue E are necessarily dimensionless. However, this may be adapted to a physical situation by an appropriate scaling to a physical distance along with the introduction of the appropriate conversion factors. We will not discuss these here as we are mainly interested in the Sturm–Liouville problem itself.

The weight function of the Sturm–Liouville operator $-\partial_x^2 + x^2$ is equal to one and consequently, the eigenfunctions will be orthogonal with respect to the inner product

$$\langle \psi_1, \psi_2 \rangle = \int_{-\infty}^{\infty} \psi_1(x)^* \psi_2(x) dx \tag{5.232}$$

and we restrict the operator to the function space containing functions of finite norm, i.e., functions that are square integrable on the real line. The problem may be rewritten in a different form by performing the substitution

$$\psi(x) = e^{-x^2/2} H(x), \tag{5.233a}$$

resulting in the Sturm–Liouville problem

$$-e^{x^2}\partial_x e^{-x^2}\partial_x H(x) + x^2 H(x) = (2E-1)H(x), \tag{5.233b}$$

which has the weight function $w(x) = e^{-x^2}$ and whose eigenfunctions are orthogonal under the inner product

$$\langle f, g \rangle_{e^{-x^2}} = \int_{-\infty}^{\infty} f(x)^* g(x) e^{-x^2} dx. \tag{5.234}$$

The function space in question is the space of functions with finite norm under this inner product. Note that we have here denoted the inner product with weight function $w(x)$ by $\langle \cdot, \cdot \rangle_{w(x)}$ to separate it from the inner product with $w(x) = 1$. Applying the differential operator of the Sturm–Liouville problem to $H(x)$ results in the differential equation

$$-H''(x) + 2xH'(x) = (2E-1)H(x) = 2\lambda H(x). \tag{5.235}$$

This equation is known as the *Hermite equation* and the first thing to note is that, as the left-hand side only contains derivatives of the function $H(x)$, any constant will solve the equation with $\lambda = 0$ and therefore $E = 1/2$. We can define the eigenfunction

$$H_0(x) = 1. \tag{5.236}$$

Defining the differential operator $\hat{a}_+$ on the function space by

$$\hat{a}_+ f(x) = 2xf(x) - \partial_x f(x) \tag{5.237}$$

the Hermite equation may be written as

$$\hat{a}_+ \partial_x H(x) = (2E-1)H(x). \tag{5.238}$$

and we find that, for any function $f(x)$,

$$\partial_x \hat{a}_+ f(x) = 2f(x) + (2x - \partial_x)\partial_x f(x) = 2f(x) + \hat{a}_+ \partial_x f(x). \tag{5.239}$$

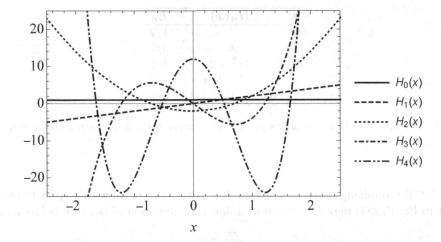

Figure 5.18 The first five Hermite polynomials $H_n(x)$. Note that the Hermite polynomials of even degree are even functions while those of odd degree are odd functions. The Hermite polynomials form a complete basis for functions on the real line that are square integrable with the weight function e^{-x^2}.

In particular, if $H_n(x)$ is an eigenfunction with eigenvalue corresponding to $E = E_n$, using $f(x) = H_n(x)$ and acting on the equation with $\hat{a}_+$ leads to

$$\hat{a}_+\partial_x\hat{a}_+H_n(x) = 2\hat{a}_+H_n(x) + \hat{a}_+\hat{a}_+\partial_xH_n(x) = 2\hat{a}_+H_n(x) + (2E_n - 1)\hat{a}_+H_n(x)$$
$$= [2(E_n + 1) - 1]\hat{a}_+H_n(x). \tag{5.240}$$

Consequently, we find that

$$H_{n+1}(x) \equiv \hat{a}_+H_n(x) = (2x - \partial_x)H_n(x) \tag{5.241}$$

is also an eigenfunction of $\hat{a}_+\partial_x$, but with an eigenvalue corresponding to $E = E_{n+1} = E_n + 1$. Applying the operator $\hat{a}_+$ to a known eigenfunction therefore results in a new eigenfunction with the eigenvalue E increased by one and the operator is therefore called a *raising operator* (also *creation operator*). Starting from $H_0(x)$ as defined above, recursion results in

$$H_n(x) = \hat{a}_+^n H_0(x) = (2x - \partial_x)^n 1. \tag{5.242}$$

Since $E_{n+1} = E_n + 1$ and $E_0 = 1/2$, we find that the eigenvalue corresponding to $H_n(x)$ is $E_n = n + 1/2$. When acting on a polynomial of degree n, the multiplication by $2x$ will raise the polynomial degree to $n + 1$ and the differentiation lowers the polynomial degree to $n - 1$, overall resulting in an addition of two polynomials of degree $n + 1$ and $n - 1$, respectively, which is a polynomial of degree $n + 1$. A direct implication of this is that $H_n(x)$ is a polynomial of degree n as H_0 is a polynomial of degree zero, i.e., a constant. These polynomials are known as the *Hermite polynomials*. It should also be noted that if $H_n(x)$ is an even function of x, then $H_{n+1}(x)$ will be an odd function and vice versa. Since $H_0(x)$ is an even function, all Hermite polynomials of even degree are even functions and all Hermite polynomials of odd degree are odd functions. The five first Hermite polynomials are shown in Fig. 5.18 and listed in Table 5.4. Together, the set of Hermite polynomials form a complete basis of the function space which is the domain of the Sturm–Liouville operator

n	$H_n(x)$	E_n
0	1	1/2
1	$2x$	3/2
2	$4x^2 - 2$	5/2
3	$8x^3 - 12x$	7/2
4	$16x^4 - 48x^2 + 12$	9/2

Table 5.4 The first four Hermite polynomials $H_n(x)$ along with their corresponding eigenvalues $E_n = n + 1/2$.

in Eq. (5.233b), implying that any function $f(x)$ with finite norm with respect to the inner product in Eq. (5.234) may be written as a linear combination of Hermite polynomials

$$f(x) = \sum_{n=0}^{\infty} A_n H_n(x). \tag{5.243}$$

The normalisation of the Hermite polynomials is given by the inner product relation

$$\langle H_n, H_m \rangle_{e^{-x^2}} = \int_{-\infty}^{\infty} H_n(x) H_m(x) e^{-x^2} dx = 2^n n! \sqrt{\pi} \delta_{nm}, \tag{5.244}$$

where the inner product is using the weight function $w(x) = e^{-x^2}$, as defined in Eq. (5.234).

For the original application we had in mind, the solution to the quantum mechanical harmonic oscillator, see Eq. (5.231b), the eigenfunctions are given by

$$\psi_n(x) = \frac{e^{-x^2/2}}{\sqrt{2^n n! \sqrt{\pi}}} H_n(x), \tag{5.245}$$

where the normalisation is chosen in such a way that they are orthonormal with respect to the inner product

$$\langle \psi_n, \psi_m \rangle = \int_{-\infty}^{\infty} \psi_n(x) \psi_m(x) dx = \delta_{nm}. \tag{5.246}$$

The functions ψ_n defined in this manner are called the *Hermite functions* and differ from the Hermite polynomials only by the multiplication with the exponential factor $e^{-x^2/2}$ and a normalisation constant. The Hermite functions form a complete basis for the square integrable functions on the real line. The shapes of the first five Hermite functions are shown in Fig. 5.19.

5.6 FUNCTION SPACES AS REPRESENTATIONS

In Chapter 4, we discussed the representation of groups on vector spaces. With this in mind, we may realise that given any symmetry transformation of the domain of a given function space will provide us with a natural representation of the symmetry group. In particular, if the transformation T is a symmetry transformation that maps $x \to x' = Tx$ on the domain D of a function $f(x)$, where x is in general a set of N coordinates describing the domain, then the mapping

$$f(x) \to (\rho(T)f)(x) = f(Tx) \tag{5.247}$$

is a representation of the corresponding symmetry group as long as $f(Tx)$ belongs to the function space.

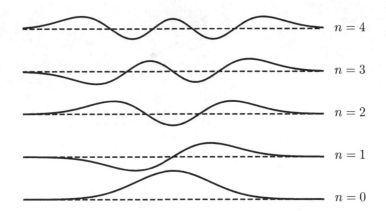

Figure 5.19 The shapes of the first five Hermite functions $\psi_n(x)$ obtained by multiplying the Hermite polynomials by $e^{-x^2/2}$ and normalising. The dashed lines represent the value zero.

Example 5.33 Consider the set of linear functions $f(x) = ax + b$ on the real line. This is a two-dimensional vector space where each vector may be specified by the coefficients a and b, cf. Example 5.3 with $N = 2$. The real line has the reflection $x \to Rx = -x$ as a symmetry and the corresponding representation is the map $f \to \rho(R)f$ such that

$$(\rho(R)f)(x) = f(-x) = -ax + b \equiv a'x + b', \tag{5.248}$$

where a' and b' are the coefficients of the function $\rho(R)f$. Since the vector space is two-dimensional, we may write this down on matrix form by specifying the transformations of the coefficients a and b using a 2×2 matrix

$$\begin{pmatrix} a \\ b \end{pmatrix} \to \begin{pmatrix} a' \\ b' \end{pmatrix} = \begin{pmatrix} -1 & 0 \\ 0 & 1 \end{pmatrix} \begin{pmatrix} a \\ b \end{pmatrix} = \begin{pmatrix} -a \\ b \end{pmatrix}. \tag{5.249}$$

In general, function spaces may be infinite dimensional and therefore not correspond to representations that may be written down in matrix form.

5.6.1 Reducibility

From Schur's first lemma, we know that if an operator commutes with all elements of an irrep, then that operator's action on that irrep must be proportional to that of the identity operator. If we have an operator L and a representation ρ on a vector space V such that

$$L\rho(g)\vec{v} = \rho(g)L\vec{v} \tag{5.250}$$

for all elements g of the represented group and vectors $\vec{v}$ in V, then L is said to commute with the representation ρ, which we may write as the commutator relation

$$L\rho(g) - \rho(g)L = [L, \rho(g)] = 0. \tag{5.251}$$

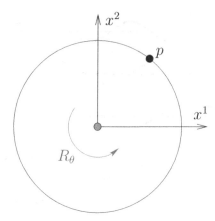

Figure 5.20 Two-dimensional rotations R_θ can transform the point p into any point on the circle that has the distance from p to the origin as its radius. The circle is the simplest subset of the two-dimensional plane which is invariant under rotations.

By Schur's first lemma, this implies that the representation is reducible if L has a spectrum containing more than one eigenvalue and may at least be reduced into representations composed of the subspaces spanned by eigenvectors that have the same eigenvalue.

Example 5.34 Looking at Example 5.33, we have a representation of the group containing the identity transformation e and the reflection of the real line R on the vector space of linear functions. The operator $L = x\partial_x$ acts on this vector space as

$$(Lf)(x) = x\partial_x(ax + b) = ax \tag{5.252}$$

and we therefore find that

$$(L\rho(R)f)(x) = x\partial_x(-ax + b) = -ax = \rho(R)ax = (\rho(R)Lf)(x) \tag{5.253}$$

for all functions $f(x) = ax + b$ in the vector space. That L commutes with $\rho(e)$ is trivial as $\rho(e)$ is the identity transformation and we therefore have $[L, \rho(g)] = 0$ for both $g = e$ and $g = R$. We can draw the conclusion that any eigenfunctions of L with different eigenvalues will be in different irreps. In particular, from Eq. (5.252) it is clear that the eigenfunctions of L are $f(x) = b$ and $f(x) = ax$ with eigenvalues zero and one, respectively. The two-dimensional representation in terms of the linear functions is therefore reducible and, since both eigenvalues are non-degenerate, the functions $f(x) = 1$ and $f(x) = x$ each form one-dimensional irreps.

With this in mind, we turn our focus to the rotation groups. We will start by examining the rotation group in two dimensions and later make the same considerations for three dimensions. In order to study rotations in two dimensions, we ask ourselves what the simplest rotationally symmetric set of points is. Since any given point may be transformed to any other point at the same distance from the origin, see Fig. 5.20, the simplest possible such set is a circle, which may be parametrised by an angle θ. The vector space of functions f on a circle, i.e., functions that are 2π periodic, therefore provides a representation ρ of the group of two-dimensional rotations. The action of this representation is given by

$$(\rho(R_\alpha)f)(\theta) = f(\theta + \alpha), \tag{5.254}$$

where R_α is a rotation by an angle α.

By studying the periodic Sturm–Liouville problem posed by the operator $-\partial_\theta^2$ on 2π periodic functions (see Example 5.20), we found that this vector space has a complete basis consisting of the functions

$$f_n(\theta) = e^{in\theta}, \tag{5.255}$$

where n may take any integer value. These functions are all eigenfunctions of $-\partial_\theta^2$ with eigenvalue n^2. For any function f on the circle and any rotation R_α, we find that

$$-\partial_\theta^2 \rho(R_\alpha)f(\theta) = -\partial_\theta^2 f(\theta + \alpha) = -f''(\theta + \alpha), \tag{5.256a}$$

which follows directly from the chain rule. At the same time, we also have

$$\rho(R_\alpha)(-\partial_\theta^2)f(\theta) = -\rho(R_\alpha)f''(\theta) = -f''(\theta + \alpha) = -\partial_\theta^2 \rho(R_\alpha)f(\theta) \tag{5.256b}$$

leading to

$$[-\partial_x^2, \rho(R_\alpha)] = 0. \tag{5.256c}$$

As a direct consequence, the representation is reducible to subspaces spanned by the eigenfunctions of $-\partial_\theta^2$ that have the same eigenvalues, i.e., to function spaces of the form

$$g(\theta) = Ae^{in\theta} + Be^{-in\theta} = C\sin(n\theta) + D\cos(n\theta) \tag{5.257}$$

for $n \geq 0$ (the function spaces with $n < 0$ are equivalent to the corresponding function space with positive n). These function spaces are two-dimensional representations of the rotation group except for when $n = 0$, for which the representation is one-dimensional. The action of the representation on the function $g(\theta)$ is given by

$$\begin{aligned} \rho(R_\alpha)g(\theta) &= C\sin(n(\theta + \alpha)) + D\cos(n(\theta + \alpha)) \\ &= \underbrace{[C\cos(n\alpha) - D\sin(n\alpha)]}_{\equiv C'}\sin(n\theta) + \underbrace{[C\sin(n\alpha) + D\cos(n\alpha)]}_{\equiv D'}\cos(n\theta). \end{aligned} \tag{5.258}$$

Since the representation is two-dimensional, we may write down the transformation of the components C and D in matrix form

$$\begin{pmatrix} C' \\ D' \end{pmatrix} = \rho(R_\alpha)\begin{pmatrix} C \\ D \end{pmatrix} = \begin{pmatrix} \cos(n\alpha) & -\sin(n\alpha) \\ \sin(n\alpha) & \cos(n\alpha) \end{pmatrix}\begin{pmatrix} C \\ D \end{pmatrix}. \tag{5.259}$$

Comparing with the representations in Example 4.33, we therefore find that these are exactly the representations defined in Eq. (4.97).

It should be pointed out that, although the representation of the rotation group on 2π periodic functions is reducible in this way, the two-dimensional representations we have found are not irreducible when considering them as complex representations. Indeed, we find that due to

$$\rho(R_\alpha)e^{in\theta} = e^{in\alpha}e^{in\theta}, \tag{5.260}$$

both functions $e^{\pm in\theta}$ in Eq. (5.257) span their own complex functional space that is invariant under rotations and the irreps are therefore all one-dimensional. Not only are these representations irreps, but together they form the set of all irreps of $SO(2)$.

With the above in mind, we turn to the representation of the rotation group $SO(3)$.

Just as the elements of the group $SO(2)$ maps a point in two dimensions to points on a circle, the transformations of $SO(3)$ will map a point to points on a sphere. We therefore look at the set of functions on the unit sphere, which we know may be expanded in terms of the spherical harmonics Y_ℓ^m. Furthermore, for any function $f(\theta, \varphi)$, it can be shown that

$$\hat{\Lambda}\rho(R_{\hat{n}}^\alpha)f = \rho(R_{\hat{n}}^\alpha)\hat{\Lambda}f \implies [\hat{\Lambda}, \rho(R_{\hat{n}}^\alpha)] = 0, \tag{5.261}$$

i.e., the representation of the rotation group on the function space of functions on a sphere commutes with the angular part of the Laplace operator. Consequently, the representation is reducible to representations consisting of the subspaces spanned by spherical harmonics that have the same eigenvalue of $\hat{\Lambda}$. As the spherical harmonic Y_ℓ^m corresponds to the eigenvalue $\ell(\ell + 1)$, it is degenerate with all other spherical harmonics with the same ℓ. Furthermore, since $-\ell \leq m \leq \ell$, there are $2\ell + 1$ such spherical harmonics in total. The $SO(3)$ representation of functions on a sphere therefore reduces to representations of all odd dimensions.

Unlike in the $SO(2)$ case, the $SO(3)$ representations in terms of the eigenfunctions with degenerate eigenvalues of $\hat{\Lambda}$ are irreducible. They also form the set of all possible irreps of $SO(3)$. It will turn out that these irreps are also irreps of the unitary group $SU(2)$ based on its double cover mapping to $SO(3)$, but in this case there will also be additional irreps.

5.7 DISTRIBUTION THEORY

Although we generally assume all functions in physics to be smooth, there will be several situations that we may want to model using approximations that are not actually part of the function space we should be considering. In these situations, we may start worrying about whether or not our differential equations are applicable. The resolution of this issue is to consider a generalisation of functions called *distributions*. The mathematical field of distribution theory is very deep, but we will here restrict ourselves to providing the general idea as well as what we need for our peace of mind when we apply differential equations to non-differentiable functions later on.

Example 5.35 One of the more common examples of a distribution that we encounter in physics is the *delta distribution* or, as we have previously known it, the delta function. It is generally introduced as a function that is zero everywhere except for in $x = 0$, but has the property

$$\int \delta(x - a)f(x)dx = f(a). \tag{5.262}$$

Clearly, the delta distribution $\delta(x)$ is not really a function as it does not have a particular value in $x = 0$, but instead hand-wavingly is said to be infinite.

The delta distribution is easily generalised to more than one dimension by making the equivalent assumptions on the behaviour outside of the origin and the behaviour of the integral. It may be used to model several different phenomena in physics, as discussed in Chapter 3, but is perhaps most useful as a description of the density of a quantity that is contained at a single point in space or occurs at a single instant in time, either by approximation or by having it as an underlying assumption of the theory. For example, we may approximate the concentration of a substance by a delta distribution if it is contained in a volume so small that we may not resolve it in experiments or such that it is a good enough approximation for the solution to the problem we are considering. On the other hand, having point charges is usually a basic assumption in classical electromagnetism when describing electrons and protons. These point charges would then be described by delta distributions.

As should be evident from its defining properties, the delta distribution is not really a function, and definitely not a differentiable one as would be required to directly apply some differential equations, such as the diffusion equation, in which the second derivative of the concentration appears.

Rather than referring to mappings from a space of functions to numbers as functions, they are instead called to as *functionals* and the value that the function $\varphi(x)$ maps to under the functional f is usually denoted $f[\varphi]$ with the square brackets to underline the fact that it is a functional. With the delta distribution in mind, we define a distribution to be a linear functional from a set of sufficiently nice functions to the real or complex numbers. In other words, in order to be a distribution, the functional must satisfy the relations

$$f[a\varphi_1] = af[\varphi_1], \tag{5.263a}$$
$$f[\varphi_1 + \varphi_2] = f[\varphi_1] + f[\varphi_2], \tag{5.263b}$$

where a is a number and φ_1 and φ_2 are functions. Naturally, a mathematician would likely encourage us to define what "sufficiently nice" means, but there are many dedicated textbooks on the subject and we will not use many of the properties or go very deep into the theory. Instead, for now we just assume that everything we do with these functions is well defined.

Example 5.36 Since we have already concluded that the delta distribution is not a function, let us define the delta distribution at $x = \ell$ as the functional $\delta_\ell[\varphi]$ for which

$$\delta_\ell[\varphi] = \varphi(\ell). \tag{5.264}$$

This definition is a functional since it maps any function $\varphi(x)$ to its value at $x = \ell$ and we can also check that it is linear by considering

$$\delta_\ell[a_1\varphi_1 + a_2\varphi_2] = a_1\varphi_1(\ell) + a_2\varphi_2(\ell) = a_1\delta_\ell[\varphi_1] + a_2\delta_\ell[\varphi_2]. \tag{5.265}$$

We shall soon see how this relates to the integral property that is usually quoted when referring to the delta distribution as a delta function.

Any locally integrable function $f(x)$, whether differentiable or not, defines a distribution $f[\varphi]$ according to

$$f[\varphi] = \int f(x)\varphi(x)dx, \tag{5.266}$$

where the integration is taken over the full domain of the argument functions φ. This is a functional since it will give a number for any φ and it is linear due to the possibility of moving constants out of the integral and splitting an integral of a sum into the sum of the integrals. It should be noted that a function does not define a unique functional. With the definition of two functionals f_1 and f_2 being equivalent if

$$f_1[\varphi] = f_2[\varphi] \tag{5.267}$$

for all φ, Eq. (5.266) will result in two functions defining the same functional if they only

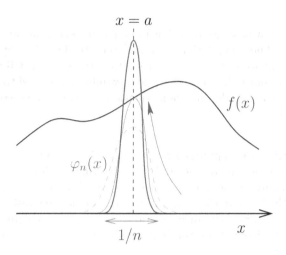

Figure 5.21 We can recover the value at $x = a$ of a function $f(x)$ used to define a distribution by letting it act on a set of functions φ_n with decreasing support of size $1/n$ around $x = a$ and that integrate to one. Note that the limit $\lim_{n\to\infty} \varphi_n$ is not a function, but rather converges towards the delta distribution.

differ in a countable set of points. As long as f is continuous at x, the value of $f(x)$ can be recovered from the functional by taking the limit

$$f(x) = \lim_{n\to\infty} f[\varphi_n], \tag{5.268}$$

where $\varphi_n(x)$ is a series of functions that integrate to one and are non-zero only in an interval of size $1/n$ around x, see Fig. 5.21.

Example 5.37 This definition of a functional given a function largely agrees with the way of thinking about the delta distribution as having the property

$$\delta_a[\varphi] = \int \delta(x - a)\varphi(x)dx = \varphi(a) \tag{5.269}$$

and calling $\delta(x - a)$ a delta function in $x = a$. If we try to use the inversion

$$\delta(x - a) = \lim_{n\to\infty} \delta_a[\varphi_n] = \lim_{n\to\infty} \varphi_n(a) \tag{5.270}$$

we will find that if $a \neq x$, there will exist a number N such that φ_n is zero at a and consequently $\delta(x - a) = 0$ if $n > N$. However, for $x = a$ we cannot use this inversion, as the limit will depend on the set of functions chosen.

5.7.1 Distribution derivatives

As promised in the beginning of this section, we want to introduce a notion of derivatives of distributions in order to justify their use in different physics problems. Looking at the distribution defined by a differentiable function $f(x)$, a natural expectation is that the derivative of the corresponding distribution should be

$$f'[\varphi] = \int f'(x)\varphi(x)dx, \tag{5.271}$$

i.e., the distribution defined by the derivative of $f(x)$. Applying partial integration to this, we find that

$$f'[\varphi] = -\int f(x)\varphi'(x)dx = -f[\varphi'], \tag{5.272}$$

where the vanishing of the boundary term can be attributed to the functions φ being sufficiently nice. For distributions that are not defined in terms of a differentiable function, we instead take the above equation as a definition of the *distribution derivative*. This definition readily generalises to distributions in several dimensions and partial derivatives as

$$\partial_i f[\varphi] = -f[\partial_i \varphi]. \tag{5.273}$$

It should be noted that since $\partial_i f$ is a distribution, it may again be differentiated with respect to any of the coordinates by applying the very same definition of the derivative again.

Example 5.38 Consider the *Heaviside function* $\theta(x)$ on the real numbers, which is defined as

$$\theta(x) = \begin{cases} 1, & (x \geq 0) \\ 0, & (x < 0) \end{cases}. \tag{5.274}$$

The Heaviside distribution $\theta[\varphi]$ is therefore given by

$$\theta[\varphi] = \int_{-\infty}^{\infty} \theta(x)\varphi(x)dx = \int_{0}^{\infty} \varphi(x)dx. \tag{5.275}$$

By definition, the derivative of the Heaviside distribution is given by $\theta'[\varphi] = -\theta[\varphi']$ and we therefore find

$$\theta'[\varphi] = -\int_{0}^{\infty} \varphi'(x)dx = -[\varphi(x)]_{x=0}^{\infty} = \varphi(0), \tag{5.276}$$

where one of the niceness properties of φ is assumed to be that it vanishes as $|x| \to \infty$. Since $\theta'[\varphi] = \varphi(0)$ for all φ, we conclude that

$$\theta'[\varphi] = \delta_0[\varphi]. \tag{5.277}$$

In other words, as distributions are considered equivalent whenever they act the same way on the functions, the derivative of the Heaviside distribution is the delta distribution δ_0. It should be noted that away from $x = 0$, the Heaviside is a constant and therefore has zero as the derivative, which is perfectly compatible with $\delta(x)$ being equal to zero for these values of x.

Throughout this book, whenever there are derivatives acting on functions that are not differentiable, the proper way of making sense of the derivatives is to consider the functions as distributions instead. However, we will not mention this fact explicitly but rather use any resulting properties freely. This will be particularly true in Chapters 6 and 7, where we will expand distributions in terms of eigenfunctions of different Sturm–Liouville operators and solve differential equations where the inhomogeneities may be approximated by delta distributions.

Example 5.39 As a small teaser, let us consider the expansion of the delta distribution $\delta_{L/2}$ in the eigenfunctions of the Sturm–Liouville operator $-\partial_x^2$ on the interval $0 < x < L$ with homogeneous Dirichlet boundary conditions. We know that the corresponding eigenfunctions are $f_n(x) = \sin(k_n x)$, where $k_n = \pi n/L$ and we wish to perform an expansion

$$\delta(x - L/2) = \sum_n d_n \sin(k_n x). \tag{5.278}$$

Solving for d_n, we find that

$$d_n = \frac{2}{L} \int_0^L \sin(k_n x)\delta(x - L/2)dx = \frac{2}{L} \sin\left(\frac{k_n L}{2}\right) = \frac{2}{L} \sin\left(\frac{\pi n}{2}\right). \tag{5.279}$$

Note that $|d_n|$ for odd n is always equal to one and therefore the convergence of the sum in Eq. (5.278) may be put in question and that $\delta(x - L/2)$ is not normalisable since the inner product of $\delta(x - L/2)$ with itself would formally give $\delta(0)$ as the result. Thus, like $\delta(x - L/2)$, its series expansion may not be regarded as an element of the function space, but must be regarded as a distribution.

5.8 PROBLEMS

Problem 5.1. Discuss the requirements that a function must fulfil in order to describe the transversal motion of a string that is held fixed between two points a distance ℓ apart. Does the set of functions satisfying these requirements form a vector space?

Problem 5.2. Verify that the Lie algebra of a matrix group forms a vector space. Find complete bases for the Lie algebras of the groups $SO(2)$, $SU(2)$, and $SU(3)$.

Problem 5.3. Start from the definition of an inner product and derive the Cauchy–Schwarz and triangle inequalities

$$|\langle v, w\rangle|^2 \le \|v\|^2 \|w\|^2 \quad \text{and} \quad \|v + w\| \le \|v\| + \|w\|. \tag{5.280}$$

Hint: Consider the norm of the vector $z = \|w\|^2 v - \langle w, v\rangle w$.

Problem 5.4. Assume that the series $\{v_n\}$ converges to v. Show that the sequence is a Cauchy sequence, i.e., for any $\varepsilon > 0$, there exists an N such that

$$\|v_n - v_m\| < \varepsilon \tag{5.281}$$

if $n, m > N$.

Problem 5.5. Verify that the expression

$$\langle f, g\rangle_w = \int f(x)^* g(x)w(x)dx, \tag{5.282}$$

where $w(x) > 0$, defines an inner product on the space of functions in which $\langle f, f\rangle$ is well defined.

Problem 5.6. In a real vector space, the Cauchy–Schwarz inequality implies that

$$\langle v, w \rangle = \|v\| \, \|w\| \cos(\alpha) \qquad (5.283)$$

for some angle α. This angle may be interpreted geometrically as the angle between the vectors. Compute the angle between the functions $f(x) = 1$ and $g(x) = x$ in the space of functions on the interval $[-1, 1]$ under the inner products

a) $\langle f, g \rangle = \displaystyle\int_{-1}^{1} f(x)g(x)dx$,

b) $\langle f, g \rangle = \displaystyle\int_{-1}^{1} f(x)g(x)(1 - x^3)dx$, and

c) $\langle f, g \rangle = \displaystyle\int_{-1}^{1} f(x)g(x)[1 + \sin^2(\pi x)]dx$.

Problem 5.7. For the Lie algebra of $SU(2)$, verify that if the matrices A and B are in the Lie algebra, then

$$\langle A, B \rangle = -\operatorname{tr}(AB) \qquad (5.284)$$

defines an inner product.

Problem 5.8. We have seen that the domain D of the functions in a function space can be a discrete set in Example 5.8, where D was a discrete set of points $\{x_k\}$. Define an inner product on this function space as

$$\langle f, g \rangle = \sum_{k=1}^{N} w_k f(x_k)^* g(x_k). \qquad (5.285)$$

Find the requirements that the numbers w_k have to satisfy.

Problem 5.9. Determine the condition that $\tilde{w}(x, x')$ in

$$\langle f, g \rangle = \int_{D} \int_{D} \tilde{w}(x, x') f(x)^* g(x') \, dx \, dx' \qquad (5.286a)$$

needs to satisfy in order for the conjugate symmetry

$$\langle f, g \rangle = \langle g, f \rangle^* \qquad (5.286b)$$

to be fulfilled.

Problem 5.10. In the vector space of all polynomials of degree $N - 1$ or less:

a) Verify that $\hat{L} = d/dx$ is a linear operator on the space.

b) Is $\hat{L}$ invertible, i.e., does an operator $\hat{L}^{-1}$ such that $\hat{L}^{-1}\hat{L} = \hat{L}\hat{L}^{-1} = 1$ exist?

c) Is $\hat{L}$ nilpotent, i.e., is $\hat{L}^n = 0$ for some n?

Problem 5.11. Consider the different sets of functions on the interval $0 \leq x \leq 1$ defined by

a) $f(0) = 1$, $f(1) = 0$,

b) $f'(0) = 0$, $f(1) = 0$,

c) $f''(0) = 1$, $f(1) = 1$, and

d) $f(0) = f(1)$,

respectively. Which of these sets form vector spaces with the addition and multiplication by scalar defined pointwise?

Problem 5.12. The functions $f(x)$ and $g(x)$ are both elements of the same function space. Which of the following combinations of $f(x)$ and $g(x)$ must necessarily belong to the function space as well?

a) $\pi f(x) + g(x)$

b) $2f(x) - 5$

c) $f(x)g(x)$

d) $f(2x)$

e) $f(x) - 3g(x)$

Problem 5.13. Let $\hat{V}$ be a symmetric operator on a real Hilbert space with a countable basis $\vec{e}_i$. Let the action of $\hat{V}$ define the numbers V_{ji} through the relation

$$\hat{V}\vec{e}_i = \vec{e}_j V_{ji}. \tag{5.287}$$

Show that $V_{ji} = V_{ij}$.

Problem 5.14. Let $\hat{V}$ be a symmetric operator on some Hilbert space. Prove that:

a) Any eigenvalue of $\hat{V}$ is necessarily real.

b) Eigenvectors with different eigenvalues are necessarily orthogonal.

Problem 5.15. A particle moves randomly between N different sites, which can be labelled by $k = 1, \ldots, N$. For the site k at time t, we can denote the probability of the particle being at the site by $p_k(t)$. We consider a situation where the particle moves from the site k to the sites $k + 1$ and $k - 1$ with the same rate λ, resulting in the differential equation

$$\dot{p}_k = \lambda(p_{k+1} + p_{k-1} - 2p_k). \tag{5.288}$$

For the site $k = 1$, moving to $k - 1$ is replaced by moving to $k = N$, and for $k = N$ moving to $k + 1$ is replaced by moving to $k = 1$. Define the column matrix

$$P(t) = \begin{pmatrix} p_1(t) \\ p_2(t) \\ \vdots \\ p_N(t) \end{pmatrix} \tag{5.289}$$

and find an expression for the matrix A such that $\dot{P} = AP$. For the case of $N = 3$, find the eigenvectors and eigenvalues of A and use them to write down the general solution to the problem. Adapt this solution to the initial condition $p_1(0) = p_3(0) = 0$ and $p_2(0) = 1$.

Problem 5.16. In Problem 5.15 we considered a particle moving randomly between N different sites. For large N, find an approximation to this problem in terms of a differential equation on functions $p(x,t)$ with period ℓ on the form $p_t = \hat{L}p$, where $\hat{L}$ is a linear differential operator in x. Find the corresponding eigenvectors and eigenvalues.

Problem 5.17. The harmonic oscillator can be described by its position $x(t)$ and momentum $p(t)$, which satisfy the differential equations

$$\dot{x}(t) = \frac{p(t)}{m} \quad \text{and} \quad \dot{p}(t) = -kx(t). \tag{5.290}$$

Write this as a matrix equation

$$\begin{pmatrix} \dot{x}(t) \\ \dot{p}(x) \end{pmatrix} = A \begin{pmatrix} x(t) \\ p(t) \end{pmatrix} \tag{5.291}$$

and solve it by finding the eigenfunctions and eigenvalues of A.

Problem 5.18. For any regular Sturm–Liouville problem, show that the Sturm–Liouville operator is symmetric with respect to the corresponding inner product.

Problem 5.19. Verify that any Sturm–Liouville operator with periodic $p(x)$, $q(x)$, and $w(x)$ is symmetric on the space of functions that satisfy periodic boundary conditions with the same period as long as the inner product is taken with the appropriate weight function.

Problem 5.20. For the inner product space of complex square integrable functions on the real line $-\infty < x < \infty$ with the inner product defined by

$$\langle f, g \rangle = \int_{-\infty}^{\infty} f(x)^* g(x) \, dx, \tag{5.292}$$

verify that the following operators are linear and determine whether they are Hermitian or not (assuming that the functions they act on are such that the result is also in the function space):

a) $\hat{L}_1 f(x) = f'(x)$,

b) $\hat{L}_2 f(x) = i f'(x)$,

c) $\hat{L}_3 f(x) = -f''(x)$,

d) $\hat{L}_4 f(x) = x f(x)$.

Problem 5.21. Consider the function space consisting of functions $f(x)$ such that their Taylor expansion

$$f(x) = \sum_{k=0}^{\infty} f_k x^k \tag{5.293}$$

converges for all x. Determine the eigenfunctions and eigenvalues of the derivative operator d/dx on this function space. Is d/dx an invertible operator?

Problem 5.22. For the function space such with elements $f(x)$ that satisfy the Neumann boundary conditions

$$(\text{BC}): \ f'(0) = f'(L) = 0. \tag{5.294}$$

Compute the eigenfunctions of the Sturm–Liouville operator $-\partial_x^2$ on this function space and their corresponding eigenvalues. Verify that these eigenfunctions are orthogonal with respect to the inner product

$$\langle f, g \rangle = \int_0^L f(x)g(x)dx \tag{5.295}$$

and find their normalisation. Finally, express the function

$$f(x) = \sin^2(\pi x/L) \tag{5.296}$$

in terms of the eigenfunction basis.

Problem 5.23. The *Wronskian* of two functions f and g is defined as the function

$$W_{f,g}(x) = f(x)g'(x) - f'(x)g(x). \tag{5.297}$$

Let f and g be eigenfunctions of a regular Sturm–Liouville problem on the interval $a < x < b$ satisfying the homogeneous boundary conditions

$$(\text{BC}): \ f'(a) = \alpha f(a), \quad g'(a) = \alpha g(a). \tag{5.298}$$

Assume that f and g have the same eigenvalue λ with respect to the Sturm–Liouville operator and:

a) Compute the value of the Wronskian at $x = a$.

b) Show that $p(x)W_{f,g}(x)$ is a constant function.

c) Use your result from (a) and (b) to show that $g(x) = Cf(x)$, where C is a constant.

The result of this problem shows that the eigenvalues of a regular Sturm–Liouville problem necessarily are non-degenerate as eigenfunctions with the same eigenvalue differ by the multiplication of a constant.

Problem 5.24. A *projection operator* $\hat{P}$ is an operator satisfying the relation $\hat{P}^2 = \hat{P}$. Show that if a projection operator $\hat{P}$ is invertible, then it is necessarily the identity operator and determine which of the following operators are projection operators:

a) the operator $\hat{\pi}_{(a,b)}$ defined in Eq. (5.55),

b) $\hat{L}f(x) = \dfrac{2}{L} \displaystyle\int_0^L \sin\left(\dfrac{\pi x}{L}\right) \sin\left(\dfrac{\pi x'}{L}\right) f(x')\, dx'$,

c) $\hat{L}f(x) = f'(x)$.

Problem 5.25. Consider two projection operators (see Problem 5.24) $\hat{P}_1$ and $\hat{P}_2$. Show that the sum $\hat{P}_1 + \hat{P}_2$ is a projection operator if $\hat{P}_1\hat{P}_2 = \hat{P}_2\hat{P}_1 = 0$.

Problem 5.26. Write down the two-dimensional Laplace operator ∇^2 in parabolic coordinates (see Problem 1.50). Find the differential equations describing the separated solutions to the Helmholtz equation $\nabla^2 u = -k^2 u$.

Problem 5.27. For the Sturm–Liouville operator $\hat{L} = -d^2/dx^2$, find the normalised eigenfunctions and eigenvalues for the boundary conditions $f'(0) = f(\ell) = 0$.

Problem 5.28. Consider the one-dimensional heat conduction problem

$$(\text{PDE}) : T_t - aT_{xx} = \kappa_0 \delta(x - x_0), \tag{5.299a}$$
$$(\text{BC}) : T_x(0, t) = 0, \quad T(\ell, t) = 0. \tag{5.299b}$$

Determine the stationary solution to this problem by expanding it in the appropriate eigenfunctions of the Sturm–Liouville operator $-\partial_x^2$.

Problem 5.29. Use the expressions for the sine and cosine functions in terms of exponential functions in order to expand $f(x) = \sin(\varphi)\cos^2(\varphi)$ in terms of the exponential functions $e^{in\varphi}$.

Problem 5.30. Find the full solution of the problem in Example 5.23 in terms of the hyperbolic sine and cosine functions. Explicitly verify that your solution satisfies the differential equation as well as the boundary conditions.

Problem 5.31. Consider the functions

a) $u(x, y) = xy$,

b) $u(x, y) = 1 - x^2 - y^2$, and

c) $u(x, y) = e^{x+y}$

in the square $0 < x, y < 1$. For each of the functions, find the functions $Y_n(y)$ in the series expansion

$$u(x, y) = \sum_{n=1}^{\infty} Y_n(y) \sin(\pi n x). \tag{5.300}$$

Note: The functions do not satisfy the same boundary conditions at $x = 0$ and $x = 1$ as the sine functions. However, they may be arbitrarily well approximated by functions that do.

Problem 5.32. Find the possible eigenvalues of the operator $-\nabla^2$ in the square $0 < x, y < L$ when the boundary conditions are given by

a) $u(0, y) = u(L, y) = u(x, 0) = u(x, L) = 0$,

b) $u(0, y) = u(L, y) = u_y(x, 0) = u_y(x, L) = 0$,

c) $u_x(0, y) = u(L, y) = u(x, 0) = u(x, L) = 0$,

d) $u_x(0, y) = u(L, y) = u(x, 0) = u_y(x, L) = 0$.

In each of the cases, identify which eigenvalues are degenerate.

Problem 5.33. The modified Bessel functions of the first and second kind are defined in Eq. (5.165). From the series expansion of the Bessel functions given in Eq. (5.141), verify that the modified Bessel functions are real functions.

Problem 5.34. Starting from the integral form given in Eq. (5.140), show that the Bessel functions satisfy the following relations:

a) $J_{-m}(x) = (-1)^m J_m(x)$,

b) $J_m'(x) = \frac{1}{2}[J_{m-1}(x) - J_{m+1}(x)]$,

c) $J_m(x) = \dfrac{x}{2m}[J_{m-1}(x) + J_{m+1}(x)]$,

d) $xJ'_m(x) = mJ_m(x) - xJ_{m+1}(x)$,

e) $xJ_0(x) = \dfrac{d(xJ_1(x))}{dx}$.

Note that relation (d) implies that $J'_m(\alpha_{mk}) = -J_{m+1}(\alpha_{mk})$ for all zeros α_{mk} of $J_m(x)$.

Problem 5.35. For a fixed m, the Bessel functions $J_m(\alpha_{mk}x)$ are orthogonal with respect to the inner product

$$\langle f, g \rangle = \int_0^1 x f(x) g(x) dx. \tag{5.301}$$

To show this, start by computing the integral

$$(\beta^2 - \alpha^2) \langle J_m(\alpha x), J_m(\beta x) \rangle = (\beta^2 - \alpha^2) \int_0^1 x J_m(\alpha x) J_m(\beta x) dx, \tag{5.302a}$$

where α and β are arbitrary constants, i.e., not necessarily zeros of J_m. Your result should contain the constants α and β as well as the Bessel function J_m and its derivative. As a direct result, show that the integral is equal to zero if α and β are both zeros of J_m or if both are zeros of J'_m.

The normalisation of the inner product for the case when $\alpha = \beta = \alpha_{mk}$, i.e., a zero of J_m may be deduced from the integral you just computed as well. Let $\beta = \alpha + \varepsilon$ and compute the integral

$$\int_0^1 x J_m(\alpha_{mk}x) J_m((\alpha_{mk} + \varepsilon)x) dx \tag{5.302b}$$

in the limit when $\varepsilon \to 0$.

Hint: You may find some of the relations derived in Problem 5.34 to be useful along with the fact that the Bessel functions satisfy Bessel's differential equation.

Problem 5.36. For the two-dimensional domain $\rho < r_0$ in polar coordinates, find the eigenfunctions to the Laplace operator that satisfy homogeneous Neumann boundary conditions at $\rho = r_0$.

Problem 5.37. A wedge shaped two-dimensional region is given by $0 < \rho < r_0$ and $0 < \phi < \phi_0$ in polar coordinates. Find the eigenfunctions of the Laplace operator that satisfy homogeneous Dirichlet conditions on the boundary of this region, i.e., solve the eigenvalue problem

$$(\text{PDE}) : \nabla^2 u(\rho, \phi) + \lambda u(\rho, \phi) = 0, \tag{5.303a}$$

$$(\text{BC}) : u(r_0, \phi) = u(\rho, 0) = u(\rho, \phi_0) = 0. \tag{5.303b}$$

Your result may contain the zeros of any Bessel functions.

Problem 5.38. Consider the two-dimensional region $r_1 < \rho < r_2$ in polar coordinates. If there is no charge density within this region, the electric potential $V(\rho, \phi)$ satisfies the Laplace equation

$$(\text{PDE}) : \frac{1}{\rho}\partial_\rho \rho \partial_\rho V + \frac{1}{\rho^2}\partial_\phi^2 V = 0. \tag{5.304a}$$

Expand the function $V(\rho, \phi)$ in the eigenfunctions of the Sturm–Liouville operator $-\partial_\phi^2$

with appropriate boundary conditions and insert the result into the Laplace equation to obtain an ordinary differential equation for the expansion coefficients, which are generally functions of ρ. Solve this partial differential equation for the boundary conditions

$$(\text{BC}): \quad V(r_1, \phi) = 0, \quad V(r_2, \phi) = V_0 \sin^2(\phi). \tag{5.304b}$$

Problem 5.39. In cylinder coordinates, a function $V(\rho, z)$ satisfies the Laplace equation

$$(\text{PDE}): \quad \nabla^2 V(\rho, z) = \frac{1}{\rho} \partial_\rho \rho \partial_\rho V + \partial_z^2 V = 0 \tag{5.305a}$$

in the region $r_1 < \rho < r_2$, $0 < z < h$ with the boundary conditions

$$(\text{BC}): \quad V(r_1, z) = V_0, \quad V(r_2, z) = V(\rho, 0) = V_z(\rho, h) = 0. \tag{5.305b}$$

Expand $V(\rho, z)$ in terms of the eigenfunctions to the operator $-\partial_z^2$ that satisfy appropriate boundary conditions at $z = 0$ and h. Insert the resulting expansion in the Laplace equation to obtain a set of ordinary differential equations for the expansion coefficients, which are generally functions of ρ. Solve the resulting differential equations.

Problem 5.40. Explicitly verify that the eigenfunctions

$$u_{nm}(\rho, \phi) = J_m(\alpha_{mn} \rho / r_0) e^{im\phi} \tag{5.306}$$

of the Laplace operator in the region $\rho < r_0$ with homogeneous Dirichlet boundary conditions are orthogonal under the inner product

$$\langle f, g \rangle = \int_{\phi=0}^{2\pi} \int_{\rho=0}^{r_0} f(\rho, \phi)^* g(\rho, \phi) \rho \, d\rho \, d\phi. \tag{5.307}$$

Problem 5.41. Find the expansion of the delta distribution $\delta^{(2)}(\vec{x} - \vec{x}_0)$ in the two dimensional disc $\rho < r_0$ in terms of the eigenfunctions of the Laplace operator satisfying homogeneous Dirichlet boundary conditions (see Problem 5.40). *Hint:* Note that the inner product on the disc can be rewritten as

$$\langle f, g \rangle = \int_{|\vec{x}| < r_0} f^*(\vec{x}) g(\vec{x}) \, dx^1 dx^2 \tag{5.308}$$

in Cartesian coordinates.

Problem 5.42. Use the recurrence relation for the coefficients of the Legendre polynomials to write down the form of the Legendre polynomials P_6, P_7, and P_8. Verify your result using Rodrigues' formula.

Problem 5.43. Expand the following polynomials $p_k(x)$ in terms of the Legendre polynomials $P_\ell(x)$:

a) $p_1(x) = 2x^2 + 4x - 2$

b) $p_2(x) = 3x - 2$

c) $p_3(x) = x^3 - 3x + 1$

Problem 5.44. Find explicit expressions for the associated Legendre functions P_1^1, P_2^1, and P_2^2.

Problem 5.45. Using the results of Problem 5.35, show that the spherical Bessel functions satisfy the integral relation given in Eq. (5.222).

Problem 5.46. A grounded metal sphere with radius r_0 is placed in an external electric field. The resulting potential $V(r, \theta, \varphi)$, given in spherical coordinates, satisfies the Laplace equation

$$(\text{PDE}): \quad \nabla^2 V = 0 \tag{5.309a}$$

with the boundary conditions

$$(\text{BC}): \quad V(r_0, \theta, \varphi) = 0, \quad \lim_{r \to \infty} V(r, \theta, \varphi) = V_0 \frac{r}{r_0} \cos(\theta). \tag{5.309b}$$

Expand the potential in spherical harmonics in order to derive a number of differential equations in the radial coordinate r and solve them in order to find the potential V.

Problem 5.47. A potential $V(r, \theta, \varphi)$ satisfies the Laplace equation outside a sphere of radius r_0 with the boundary condition

$$(\text{BC}): \quad V(r_0, \theta, \varphi) = V_0[3 \cos^2(\theta) - 1]. \tag{5.310}$$

Find the potential in the region $r > r_0$ if it is assumed to be finite as $r \to \infty$.

Problem 5.48. Express the following functions $f(\theta, \varphi)$ on the unit sphere in terms of the spherical harmonics $Y_\ell^m(\theta, \varphi)$:

a) $\sin(\theta) \sin(\varphi)$

b) $\sin^2(\theta) \cos(2\varphi)$

c) $\cos^2(\theta)$

Problem 5.49. On the interval $0 < r < r_0$, the functions

$$u_n(r) = j_0\left(\frac{\pi n r}{r_0}\right), \tag{5.311}$$

where n is a positive integer, form a complete orthogonal basis with the inner product

$$\langle f, g \rangle = \int_0^{r_0} f(r)^* g(r) r^2 \, dr. \tag{5.312}$$

Find the coefficients in the series expansion of the function $f(r) = \cos(\pi r / 2 r_0)$ in this basis.

Problem 5.50. Find the eigenfunctions and corresponding eigenvalues of the Laplace operator inside the half-sphere

$$0 < r < r_0, \quad 0 < \theta < \pi, \quad \text{and} \quad 0 < \varphi < \pi \tag{5.313}$$

in spherical coordinates, under the assumption that the boundary conditions are homogeneous Dirichlet conditions.

Problem 5.51. Consider the Sturm–Liouville operator in Eq. (5.231b) and its eigenfunctions $\psi_n(x)$ with corresponding eigenvalues $2E_n$. Show that the operators

$$\hat{\alpha}_\pm = x \mp \partial_x \tag{5.314}$$

relate eigenfunctions with different eigenvalues such that $\hat{a}_{\pm}\psi_n$ is an eigenfunction with eigenvalue $2(E_n \pm 1)$.

We know that the eigenfunction with the smallest eigenvalue is given by

$$\psi_0(x) = \frac{e^{-x^2/2}}{\pi^{1/4}}. \tag{5.315}$$

Verify that $\hat{a}_-\psi_0(x) = 0$, i.e., acting on it with $\hat{a}_-$ does not give a non-trivial eigenfunction with a lower eigenvalue.

Problem 5.52. Show that the operator $\hat{a}_+$ defined in Problem 5.51 satisfies the relation

$$\hat{a}_+ p(x) e^{-x^2/2} = e^{-x^2/2}(2x - \partial_x)p(x). \tag{5.316}$$

Use the same kind of argumentation to express $\hat{a}_- p(x) e^{-x^2/2}$ in a similar manner. What does this tell you about the relation between the Hermite polynomials $H_n(x)$ and $H_{n-1}(x)$?

Problem 5.53. In Example 5.33, we considered the representation of spatial reflections on the function space of linear functions on the real line. Find the three-dimensional matrix representation $\rho(R)$ of the spatial reflection on the space of all polynomials of degree two, i.e., $f(x) = ax^2 + bx + c$ such that

$$\begin{pmatrix} a' \\ b' \\ c' \end{pmatrix} = \rho(R) \begin{pmatrix} a \\ b \\ c \end{pmatrix}. \tag{5.317}$$

In addition, also find the representation of the spatial translations $x \to x' = x - \ell$. Check that the resulting matrix representation satisfies the necessary group relations.

Problem 5.54. Consider the Sturm–Liouville operator

$$\hat{L} = -\frac{d^2}{dx^2} + q(x) \tag{5.318}$$

on the interval $-a < x < a$ for functions $q(x)$ that satisfy $q(-x) = q(x)$. Assume that we have boundary conditions at $x = \pm a$ that are homogeneous and symmetric under the transformation $x \to -x$. Show that the eigenfunctions of the corresponding Sturm–Liouville problem are either even or odd under this transformation, i.e., that $\hat{L}X(x) = \lambda X(x)$ implies that

$$X(-x) = \pm X(x). \tag{5.319}$$

Problem 5.55. The symmetries we encountered in Chapter 4 can be used to argue for the possible degeneracies in the spectrum of a Sturm–Liouville operator.

a) Verify that the Laplace operator ∇^2 commutes with the transformations $\hat{c}u(x,y) = u(y, \ell - x)$ and $\hat{\sigma}u(x,y) = u(y,x)$ and that the homogeneous boundary conditions

$$(\text{BC}): \ u(0,y) = u(\ell, y) = u(x,0) = u(x,\ell) = 0 \tag{5.320}$$

are invariant, i.e., that the transformations map functions satisfying the boundary conditions to other functions satisfying the boundary conditions.

b) Show that the group generated by these transformations is isomorphic to the dihedral group D_4.

c) The vector space of functions on the square $0 < x, y < \ell$ with homogeneous Dirichlet boundary conditions therefore provides a representation of D_4 with the transformations $\hat{c}$ and $\hat{\sigma}$ discussed above. Use the fact that the Laplace operator commutes with the transformations to discuss the different possibilities on how many eigenfunctions can share the same eigenvalue. How does this discussion change if the allowed range of y is increased by a small length δ?

d) Compare your argumentation to the actual set of eigenfunctions and eigenvalues.

Problem 5.56. We have seen that in the function space of functions on the unit sphere, the spherical harmonics with a fixed ℓ form an irrep of the rotation group $SO(3)$. Starting from the spherical harmonic $Y_1^0(\theta, \varphi)$, perform a small rotation by an angle α around the x^2-axis given by $x^1 \to x^1 + \alpha x^3$ and $x^3 \to x^3 - \alpha x^1$ and verify that this rotation results in a linear combination of the spherical harmonics $Y_1^m(\theta, \varphi)$, i.e., the set of spherical harmonics with $\ell = 1$ is closed under rotations about the x^2-axis. Repeat the argument for rotations around the x^1- and x^3-axes and the spherical harmonics $Y_1^{\pm 1}(\theta, \varphi)$. Write down a matrix representation for the action of $SO(3)$ on the spherical harmonics $Y_1^m(\theta, \varphi)$.

Problem 5.57. If f is a distribution and g is a sufficiently nice function, we can define a new distribution fg according to

$$fg[\varphi] = f[g\varphi]. \tag{5.321}$$

Find explicit expressions that only refer to g and φ and their derivatives for the following distributions:

a) $\delta_a g[\varphi]$

b) $\delta'_a g[\varphi]$

c) $(\delta_a g)'[\varphi]$

Also verify that $(fg)'[\varphi] = f'g[\varphi] + fg'[\varphi]$.

CHAPTER 6

Eigenfunction Expansions

With the theory of Chapter 5 in place, we are ready to apply it extensively in order to solve several of the problems we encountered when modelling physics problems using partial differential equations. In particular, we are here going to develop methods of using operator eigenbases in which the original partial differential equations reduce to ordinary differential equations that are generally easier to solve. In fact, we have already seen an example of the approach we will take in Example 5.10, where we diagonalised an operator represented by a 2×2 matrix in order to obtain two non-coupled ordinary differential equations rather than two coupled ones. The approach taken in this chapter is going to be the direct equivalent of this approach with the difference that our operators will be differential operators and the vector spaces under study will be function spaces which are generally infinite dimensional.

In addition to presenting the general technique of series and transform solutions, we will also discuss some of the particular physical properties of the solutions.

6.1 POISSON'S EQUATION AND SERIES

Already in Chapter 5, we saw several examples of solutions to Poisson's equation where we have expanded the solution in different eigenfunctions depending on the geometry of the problem (in particular, see Examples 5.29 and 5.30). We shall now have a closer look at how series expansion using the eigenfunctions of Sturm–Liouville operators may be applied in order to solve more general physics problems, as the earlier examples were mainly restricted to cases where the differential equations were homogeneous.

6.1.1 Inhomogeneous PDE

One of the more common partial differential equations in physics is *Poisson's equation*

$$-\nabla^2 \Phi(\vec{x}) = \rho(\vec{x}), \tag{6.1}$$

where $\Phi(\vec{x})$ is a scalar field that we wish to determine and $\rho(\vec{x})$ is a density of some sort. In general, some additional physical constants may be present, but this does not change the mathematical structure of the problem. Poisson's equation appears in Newton's theory of gravity, where $\Phi(\vec{x})$ is the gravitational potential and $\rho(\vec{x})$ the mass density, as well as in electrostatics, where $\Phi(\vec{x})$ is the electric potential and $\rho(\vec{x})$ the charge density.

In general, we will wish to solve Poisson's equation for cases where the the density $\rho(\vec{x})$ is not zero in the region of interest. With Sturm–Liouville theory in mind, let us first assume that we have homogeneous boundary conditions on our region and consequently that there is a complete set of eigenfunctions $f_n(\vec{x})$ to the Laplace operator that satisfy the boundary conditions and the eigenfunction equation

$$-\nabla^2 f_n(\vec{x}) = \lambda_n f_n(\vec{x}). \tag{6.2}$$

As a consequence of this, we can expand any function $g(\vec{x})$ in terms of the eigenfunctions

$$g(\vec{x}) = \sum_n g_n f_n(\vec{x}). \tag{6.3a}$$

In particular, this is true of both the potential $\Phi(\vec{x})$ as well as the density $\rho(\vec{x})$

$$\Phi(\vec{x}) = \sum_n \Phi_n f_n(\vec{x}) \quad \text{and} \quad \rho(\vec{x}) = \sum_n \rho_n f_n(\vec{x}), \tag{6.3b}$$

respectively. Inserting these expansions into Poisson's equation, we find that

$$\rho(\vec{x}) + \nabla^2 \Phi(\vec{x}) = \sum_n \left[\rho_n f_n(\vec{x}) + \Phi_n \nabla^2 f_n(\vec{x}) \right] = \sum_n \left(\rho_n - \lambda_n \Phi_n \right) f_n(\vec{x}) = 0. \tag{6.4}$$

Since all of the functions $f_n(\vec{x})$ are linearly independent and the final expression is equal to zero, the coefficient of each $f_n(\vec{x})$ must be equal to zero and therefore

$$\rho_n - \Phi_n \lambda_n = 0 \quad \Longrightarrow \quad \Phi_n = \frac{\rho_n}{\lambda_n}. \tag{6.5a}$$

This may also be seen by taking the inner product of $f_m(\vec{x})$ with both sides of Poisson's equation

$$\langle f_m, \rho + \nabla^2 \Phi \rangle = \sum_n (\rho_n - \Phi_n \lambda_n) \langle f_m, f_n \rangle = (\rho_m - \Phi_m \lambda_m) \| f_m \|^2 = 0 \tag{6.5b}$$

due to the orthogonality of the eigenfunctions $f_m(\vec{x})$. If we can find the expansion of $\rho(\vec{x})$, it is now trivial to write down the expansion of the solution $\Phi(\vec{x})$ as well, since the expansion coefficients are related by division with the corresponding eigenvalue only.

The big caveat here is that the density $\rho(\vec{x})$ might not satisfy the boundary conditions. However, in these situations, we can find a density that does and that is physically equivalent to the original $\rho(\vec{x})$ and we treat this situation as if $\rho(\vec{x})$ does satisfy the boundary conditions. The real resolution of this problem is considering $\rho(\vec{x})$ to be a distribution rather than a function, resulting in the expansion being well defined. This will also apply when $\rho(\vec{x})$ is best described by any sort of distribution, e.g., a delta distribution to model a point charge.

Example 6.1 Consider the situation shown in Fig. 6.1, where a two-dimensional membrane is clamped into a rectangular frame. We describe the rectangle by the coordinates $0 < x < L_x$, $0 < y < L_y$, where the boundaries are kept fixed at $u(x, y) = 0$. Gravity acts perpendicular to the frame, affecting the membrane with a force density ρg. We can model the stationary shape of the membrane using Poisson's equation and homogeneous Dirichlet boundary conditions

$$(\text{PDE}) : -\nabla^2 u(x, y) = -\frac{g\rho}{\sigma}, \tag{6.6a}$$

$$(\text{BC}) : u(x, 0) = u(x, L_y) = u(0, y) = u(L_x, y) = 0, \tag{6.6b}$$

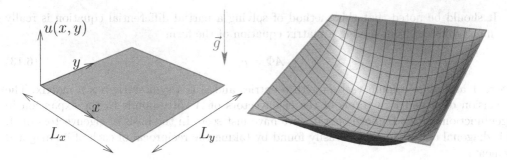

Figure 6.1 The shape of a membrane clamped into a rectangular frame under the influence of gravity may be described using Poisson's equation. The right figure shows the resulting solution, which may be expressed as a series.

where σ is the tension in and ρ the surface density of the membrane. The eigenfunctions of the Laplace operator that satisfy the given boundary conditions are the products of sine functions

$$f_{nm}(x,y) = \sin\left(\frac{\pi n x}{L_x}\right)\sin\left(\frac{\pi m y}{L_y}\right) \tag{6.7}$$

with corresponding eigenvalues

$$\lambda_{nm} = \pi^2\left(\frac{n^2}{L_x^2} + \frac{m^2}{L_y^2}\right). \tag{6.8}$$

Since these eigenfunctions form a complete set of basis functions, we expand the solution $u(x,y)$ in terms of them

$$u(x,y) = \sum_{nm} A_{nm} f_{nm}(x,y) = \sum_{nm} A_{nm}\sin\left(\frac{\pi n x}{L_x}\right)\sin\left(\frac{\pi m y}{L_y}\right). \tag{6.9}$$

Inserting this into the differential equation, we now obtain

$$\sum_{nm} A_{nm}\lambda_{nm} f_{nm}(x,y) = -\frac{g\rho}{\sigma}. \tag{6.10}$$

Multiplying by $f_{nm}(x,y)$ and integrating over the rectangle now gives

$$\begin{aligned}
A_{nm} &= -\frac{4\rho g}{\sigma L_x L_y \lambda_{nm}}\int_{x=0}^{L_x}\int_{y=0}^{L_y}\sin\left(\frac{\pi n x}{L_x}\right)\sin\left(\frac{\pi m y}{L_y}\right)dx\,dy \\
&= -\frac{4g\rho}{\sigma\pi^2 nm\lambda_{nm}}[1-(-1)^n][1-(-1)^m].
\end{aligned} \tag{6.11}$$

Consequently, the shape of the membrane in its stationary state is given by

$$u(x,y) = -\sum_{nm}\frac{4g\rho[1-(-1)^n][1-(-1)^m]}{\sigma\pi^2 nm\lambda_{nm}}\sin\left(\frac{\pi n x}{L_x}\right)\sin\left(\frac{\pi m y}{L_y}\right). \tag{6.12}$$

It should be noted that this method of solving a partial differential equation is really not much different from solving a matrix equation of the form

$$A\Phi = \rho, \tag{6.13}$$

where Φ and ρ are column vectors with n entries and A is a symmetric $n \times n$ matrix. The expansion of Φ and ρ in terms of the eigenvectors of A corresponds to the expansion in eigenfunctions of the Laplace operator we have just seen. In the basis of eigenvectors of A, A is diagonal and its inverse is easily found by taking the reciprocal of each of its diagonal elements.

6.1.2   Inhomogeneous boundary conditions

In many situations, we will be faced with problems where the boundary conditions are not homogeneous. In these cases, the approach described above is not directly applicable. Instead, we can apply an approach similar to that adapted in the previous chapter, where we attempt to find coordinate directions that do have homogeneous boundary conditions, expand the solution in eigenfunctions of Sturm–Liouville operators in these coordinates, and end up with an ordinary differential equation for the remaining coordinate. Although we have already discussed this technique, it is well worth repeating the general structure.

Let us assume that we have a problem of the form

$$(\text{PDE}) : \hat{R}_x f(x,y) + g(x)\hat{L}_y f(x,y) = 0, \tag{6.14a}$$

$$(\text{BC}) : f(x,a_y) = f(x,b_y) = 0, \tag{6.14b}$$

$$f(a_x,y) = h_1(y), \quad f(b_x,y) = h_2(y). \tag{6.14c}$$

Here, $\hat{L}_y$ is a Sturm–Liouville operator in the y-direction while $\hat{R}_x$ is a second order differential operator in the x-direction and may be a Sturm–Liouville operator or not. By using the fact that $\hat{L}_y$ is a Sturm–Liouville operator and the boundary conditions in the y-direction are homogeneous, we can expand $f(x,y)$ in its eigenfunctions $Y_n(y)$ for every value of x. Since $f(x,y)$ depends on x, the expansion coefficients obtained when performing this expansion will generally depend on x

$$f(x,y) = \sum_n X_n(x)Y_n(y). \tag{6.15}$$

Since $Y_n(y)$ is an eigenfunction of $\hat{L}_y$, we know that $\hat{L}_y Y_n = \lambda_n Y_n$, where λ_n is the corresponding eigenvalue. Inserting this expansion into the partial differential equation, we now obtain

$$\hat{R}_x f(x,y) + g(x)\hat{L}_y f(x,y) = \sum_n \left[Y_n(y)\hat{R}_x X_n(x) + g(x)X_n(x)\hat{L}_y Y_n(y) \right]$$

$$= \sum_n \underbrace{\left[\hat{R}_x X_n(x) + \lambda_n g(x)X_n(x) \right]}_{=0} Y_n(y) = 0, \tag{6.16}$$

where the coefficient in front of each $Y_n(y)$ must be zero due to the linear independence of these functions. The resulting ordinary differential equation

$$\hat{R}_x X_n(x) + \lambda_n g(x)X_n(x) = 0 \tag{6.17}$$

may be solved separately for each n in order to obtain the solution to the problem. The

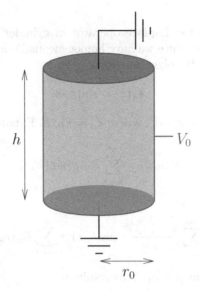

Figure 6.2 A cylinder of radius r_0 and height h where the end-caps are grounded and the cylindrical surface is being held at potential V_0. We wish to find the potential inside the cylinder, which satisfies Laplace's equation with inhomogeneous boundary conditions.

boundary conditions necessary to uniquely solve each differential equation are found by expanding the inhomogeneous boundary conditions $h_i(y)$ in the functions $Y_n(y)$ as well and we find that

$$h_1(x) = \sum_n h_{1n} Y_n(y) = f(a_y, y) = \sum_n X_n(a_y) Y_n(y), \tag{6.18}$$

with the linear independence of the $Y_n(y)$ functions implying that $X_n(a_y) = h_{1n}$ and a similar expression holding at the other boundary.

Sometimes we will not have actual boundary conditions at one of the boundaries and instead be faced with regularity conditions. This will typically occur when we are dealing with operators that are singular at that boundary. The inhomogeneity will then enter at the other boundary and we can apply the regularity condition for each differential equation at the singular boundary.

Example 6.2 Consider a cylinder of radius r_0 and height h where the end-caps are grounded and the cylindrical surface is being held at the constant electric potential $V(r_0, z) = V_0$, see Fig. 6.2. We furthermore assume that there is no charge contained in the cylinder and wish to find the electric potential $V(\rho, z)$ inside it. Note that the potential $V(\rho, z)$ will not depend on the angular coordinate ϕ as the problem is symmetric under rotations around the z-axis. Since the electric potential satisfies Laplace's equation, we therefore need to solve the problem

$$(\text{PDE}): \ -\nabla^2 V(\rho, z) = -\left(\partial_\rho^2 + \frac{1}{\rho}\partial_\rho + \partial_z^2\right) V(\rho, z) = 0, \tag{6.19a}$$

$$(\text{BC}): V(0, z) < \infty, \quad V(r_0, z) = V_0, \quad V(\rho, 0) = V(\rho, h) = 0, \tag{6.19b}$$

where we have written out the Laplace operator in cylinder coordinates and imposed a regularity condition at $\rho = 0$. Since we have homogeneous Dirichlet boundary conditions in the z-direction, we consider the eigenfunctions

$$Z_n(z) = \sin(k_n z) \tag{6.20}$$

of the Sturm–Liouville operator $-\partial_z^2$, where $k_n = \pi n/h$. Expansion of the solution in these eigenfunctions gives

$$V(\rho, z) = \sum_{n=1}^{\infty} R_n(\rho) \sin(k_n z) \tag{6.21a}$$

for the solution and

$$V_0 = 2V_0 \sum_{n=1}^{\infty} \frac{1 - (-1)^n}{\pi n} \sin(k_n z) = \sum_{n=1}^{\infty} R_n(r_0) \sin(k_n z) \tag{6.21b}$$

for the boundary condition at $\rho = r_0$. This results in

$$R_n(r_0) = \frac{2V_0}{\pi n}[1 - (-1)^n] \tag{6.22}$$

due to the linear independence of the sine functions. Furthermore, the regularity condition at $\rho = 0$ implies that $R_n(0) < 0$.

Insertion of the series expansion into the differential equation now leads to

$$\sum_{n=1}^{\infty} \left[-R_n''(\rho) - \frac{1}{\rho} R_n'(\rho) + k_n^2 R_n(\rho) \right] \sin(k_n z) = 0, \tag{6.23}$$

and from the linear independence of the sine functions follows that

$$-R_n''(\rho) - \frac{1}{\rho} R_n'(\rho) + k_n^2 R_n(\rho) = 0 \tag{6.24}$$

for all n. This equation is Bessel's modified differential equation with the modified Bessel functions as solutions

$$R_n(\rho) = A_n I_0(k_n \rho) + B_n K_0(k_n \rho). \tag{6.25}$$

As the modified Bessel functions of the second kind $K_0(x)$ are not regular at $x = 0$, the regularity condition at $\rho = 0$ immediately implies that all $B_n = 0$. We are left with the boundary condition

$$R_n(r_0) = A_n I_0(k_n r_0) = \frac{2V_0}{\pi n}[1 - (-1)^n] \quad \Longrightarrow \quad A_n = \frac{2V_0[1 - (-1)^n]}{\pi n I_0(k_n r_0)}. \tag{6.26}$$

The electric potential inside the cylinder is therefore given by the sum

$$V(\rho, z) = \sum_{n=1}^{\infty} \frac{2V_0[1 - (-1)^n]}{\pi n I_0(k_n r_0)} I_0(k_n \rho) \sin(k_n z). \tag{6.27}$$

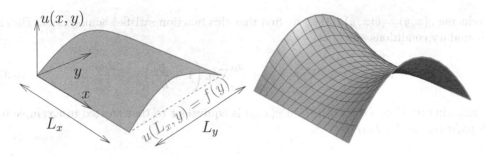

Figure 6.3 The problem of a membrane in a frame that is not flat. Its solution satisfies Laplace's equation with inhomogeneous boundary conditions and the problem may be rewritten as a problem with homogeneous boundary conditions, but with an inhomogeneous differential equation. The solution to the problem for $f(y) = u_0 \frac{y}{L_y}(1 - \frac{y}{L_y})$ is shown in the right figure.

6.1.2.1 Transferring inhomogeneities

There is an alternative to the approach given above and it is to go back to the solution of Poisson's equation with homogeneous boundary conditions, but inhomogeneities in the differential equation. Given a problem of the form

$$\text{(PDE)} : \; -\nabla^2 u(\vec{x}) = 0, \qquad\qquad (\vec{x} \in V) \qquad (6.28\text{a})$$

$$\text{(BC)} : (\alpha + \beta \hat{n} \cdot \nabla)u(\vec{x}) = f(\vec{x}), \qquad (\vec{x} \in S) \qquad (6.28\text{b})$$

where V is the domain in which we wish to solve the differential equation, S its boundary, and $\hat{n}$ the boundary normal, we can transfer the inhomogeneity from the boundary condition to the differential equation itself. We do this by selecting any function $g(\vec{x})$, defined for all points in V, that satisfies $(\alpha + \beta \hat{n} \cdot \nabla)g(\vec{x}) = f(\vec{x})$ on the boundary. Introducing the new function $v(\vec{x}) = u(\vec{x}) - g(\vec{x})$, we find that

$$\text{(PDE)} : \; -\nabla^2 v(\vec{x}) = \nabla^2 g(\vec{x}), \qquad (\vec{x} \in V) \qquad (6.29\text{a})$$

$$\text{(BC)} : (\alpha + \beta \hat{n} \cdot \nabla)v(\vec{x}) = 0. \qquad (\vec{x} \in S) \qquad (6.29\text{b})$$

The action of the Laplace operator on the function $g(\vec{x})$ may be computed and we are then back in the situation where the boundary conditions are homogeneous and the differential equation is inhomogeneous. Whether or not this approach is helpful depends on whether or not the function $g(\vec{x})$ can be chosen in such a way that the resulting problem is easier to solve.

Example 6.3 Consider the same situation as in Example 6.1, but instead of the gravitational pull we introduce a frame that is not flat, see Fig. 6.3. The differential equation describing this situation is

$$\text{(PDE)} : \; -\nabla^2 u(x, y) = 0, \qquad\qquad (6.30\text{a})$$

$$\text{(BC)} : u(x, 0) = 0, \quad u(x, L_y) = 0,$$

$$u(0, y) = u(L_x, y) = f(y) = u_0 \frac{y}{L_y}\left(1 - \frac{y}{L_y}\right). \qquad (6.30\text{b})$$

Introducing $v(x, y) = u(x, y) - f(y)$, we find that this function satisfies homogeneous Dirichlet boundary conditions and that

$$-\nabla^2 v(x, y) = -\nabla^2 u(x, y) + \frac{u_0}{L_y}\nabla^2 y\left(1 - \frac{y}{L_y}\right) = -\frac{2u_0}{L_y^2}. \tag{6.31}$$

The resulting problem for the function $v(x, y)$ is equivalent to that treated in Example 6.1 with $g\rho/\sigma$ replace by $2u_0/L_y^2$.

It should be noted that the problem in the example above has inhomogeneous boundary conditions only in one direction. In general, a problem may have inhomogeneous boundary conditions in several directions and is therefore not directly solvable by expanding in the eigenfunctions in the directions with homogeneous boundary conditions. However, we can then split the problem into several separate problems with each having inhomogeneities in different directions.

6.1.3 General inhomogeneities

Splitting a problem into several simpler ones and adding the results is not an approach unique to the case where we have inhomogeneous boundary conditions in several directions. We can also apply this approach to the situation where we have inhomogeneities not only in the boundary conditions in different directions, but also in the partial differential equation. Let us therefore now consider a problem of the form

$$(\text{PDE}): \ -\nabla^2 u(\vec{x}) = f(\vec{x}), \qquad\qquad (\vec{x} \in V) \tag{6.32a}$$

$$(\text{BC}): (\alpha + \beta\hat{n} \cdot \nabla)u(\vec{x}) = g(\vec{x}), \qquad\qquad (\vec{x} \in S) \tag{6.32b}$$

where S is the boundary of V. In this scenario, we may apply either of the methods discussed above for the case where inhomogeneities appear in the boundary conditions.

6.1.3.1 Superpositions

The perhaps most straightforward way of solving the problem is to consider it as a combination of several different problems. In order to solve the problem in this fashion, we introduce the functions $u_1(\vec{x})$ and $u_2(\vec{x})$ such that

$$(\text{PDE}): \ -\nabla^2 u_1(\vec{x}) = f(\vec{x}), \qquad -\nabla^2 u_2(\vec{x}) = 0, \qquad (\vec{x} \in V) \tag{6.33a}$$

$$(\text{BC}): (\alpha + \beta\hat{n} \cdot \nabla)u_1(\vec{x}) = 0, \qquad (\alpha + \beta\hat{n} \cdot \nabla)u_2(\vec{x}) = g(\vec{x}). \qquad (\vec{x} \in S) \tag{6.33b}$$

Due to the linearity of the problem, we find that $u(\vec{x}) = u_1(\vec{x}) + u_2(\vec{x})$ solves the original problem. If there are several boundaries on which $g(\vec{x})$ is non-zero, the problem for $u_2(\vec{x})$ may be further split into functions that each have inhomogeneous boundary conditions in one coordinate direction only. In this fashion, any inhomogeneous problem may be reduced to solving a number of problems of the forms we have already discussed and adding the solutions, see Fig. 6.4.

Example 6.4 Imagine that we wish to compute the electric potential inside the spherical

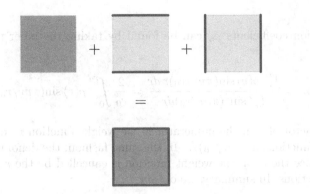

Figure 6.4 The solution to a general linear problem with inhomogeneities in the boundary conditions as well as in the differential equation may be written as a sum of the solutions to problems with inhomogeneities on only one of the coordinate boundaries or only in the differential equation. Here, the light domains and boundaries represent homogeneous differential equations and boundary conditions, respectively, while the dark domain and lines represent inhomogeneities. The solutions to the individual problems with inhomogeneities only in the differential equation or only in one boundary direction can be superposed to construct the solution to the problem with inhomogeneities everywhere.

region $r < r_0$ with a spherically symmetric charge distribution $\rho(r)$ while the surface of the region is held at a potential $V(r_0, \theta) = V_0 \cos(\theta)$. The resulting problem is given by

$$\text{(PDE)} : \ -\nabla^2 V(r, \theta) = \frac{\rho(r)}{\varepsilon_0}, \qquad (r < r_0) \tag{6.34a}$$

$$\text{(BC)} : \ |V(0, \theta)| < \infty, \quad V(r_0, \theta) = V_0 \cos(\theta). \tag{6.34b}$$

We split this problem with inhomogeneities in both the differential equation and boundary conditions into two by the ansatz $V(r, \theta) = V_1(r) + V_2(r, \theta)$, where

$$\text{(PDE)} : \ -\nabla^2 V_1(r) = \frac{\rho(r)}{\varepsilon_0}, \quad -\nabla^2 V_2(r, \theta) = 0, \qquad (r < r_0) \tag{6.35a}$$

$$\text{(BC)} : V_1(r_0) = 0, \quad V_2(r_0, \theta) = V_0 \cos(\theta), \quad |V_i(0, \theta)| < \infty. \tag{6.35b}$$

We have here already concluded that V_1 is a function of the radius r only as both the differential equation and boundary conditions for V_1 are symmetric under rotations.

The problem for V_1 has homogeneous boundary conditions and is independent of the angles θ and φ and it is therefore possible to expand the solution in the functions $R_n = j_0(\beta_{0n} r/r_0) = r_0 \sin(\pi n r/r_0)/\pi n r$. Absorbing the constant $r_0/\pi n$ into the expansion coefficients, we find that

$$V_1(r) = \sum_{n=1}^{\infty} A_n \frac{\sin(\pi n r/r_0)}{r} \quad \text{and} \quad \rho(r) = \sum_{n=1}^{\infty} \rho_n \frac{\sin(\pi n r/r_0)}{r}. \tag{6.36}$$

Insertion into the differential equation now yields

$$A_n = \frac{r_0^2 \rho_n}{\pi^2 n^2} \tag{6.37}$$

and the expansion coefficients ρ_n can be found by taking the inner product of $\rho(r)$ with $\sin(\pi n r/r_0)/r$

$$\rho_n = \frac{\int_0^{r_0} \rho(r)\sin(\pi n r/r_0) r\, dr}{\int_0^{r_0} \sin^2(\pi n r/r_0) dr} = \frac{2}{r_0} \int_0^{r_0} \rho(r) \sin(\pi n r/r_0) r\, dr. \qquad (6.38)$$

Note that the factor of r in the numerator is the weight function r^2 divided by the r that appears in the function $\sin(\pi n r/r_0)/r$. In the same fashion, the denominator does not have a factor of r, since the r^2 in the weight function is cancelled by the r in the denominators of the basis functions. In summary, we obtain

$$A_n = \frac{2r_0}{\pi^2 n^2} \int_0^{r_0} \rho(r) \sin(\pi n r/r_0) r\, dr. \qquad (6.39)$$

In order to have a full solution to the problem, we must also solve the problem for $V_2(r, \theta)$. We do this by expanding the $V_2(r, \theta)$ as well as the inhomogeneity in the boundary condition in the Legendre polynomials $P_\ell(\cos(\theta))$. For the boundary condition, we find that

$$V_0 \cos(\theta) = V_0 P_1(\cos(\theta)) \qquad (6.40)$$

and the only Legendre polynomial of interest is therefore $P_1(\cos(\theta))$, as all others will correspond to completely homogeneous differential equations with trivial solutions. We can therefore directly make the ansatz

$$V_2(r, \theta) = f(r) P_1(\cos(\theta)) = f(r) \cos(\theta). \qquad (6.41)$$

Inserting this into the homogeneous differential equation for $V_2(r, \theta)$ we obtain

$$r^2 f''(r) + 2r f'(r) - 2f(r) = 0. \qquad (6.42)$$

This is a differential equation of Euler type, to which we make the ansatz $f(r) = A r^k$ resulting in $k = 1$ or $k = -2$. The solution for $k = -2$ does not satisfy the condition of being regular at $r = 0$ and the boundary condition at $r = r_0$ results in

$$V_0 = f(r_0) = A r_0 \implies V_2(r, \theta) = \frac{V_0 r}{r_0} \cos(\theta). \qquad (6.43)$$

6.1.3.2 *Transferring inhomogeneities*

As an alternative to splitting the problem into several problems with different inhomogeneities, we can apply the very same approach that was discussed for problems where only the boundary conditions were inhomogeneous and find any function that satisfies the inhomogeneous boundary conditions. As we saw, this generally resulted in an inhomogeneity appearing in the differential equation instead. In the case where the differential is already inhomogeneous, we will find that the new inhomogeneity is just added to the already existing one. We therefore again end up with a problem where there are inhomogeneities only in the differential equation and we may solve it by using the series expansion technique as already discussed.

6.2 STATIONARY AND STEADY STATE SOLUTIONS

When modelling different physical problems in Chapter 3, we discussed *stationary* and *steady state* solutions to time-dependent partial differential equations such as the heat and wave equations. By definition, a stationary state does not depend on time and the time derivative of the quantity of interest u vanishes

$$\partial_t u(\vec{x}, t) = 0. \tag{6.44}$$

For a steady state solution, the requirement is only that the time derivative of some property of the system, such as the amplitude of oscillation or total energy, is equal to zero and the property therefore remains constant.

It is often the case that a time-dependent problem exhibits stationary or steady state solutions. In particular, this may occur when external sources or boundary conditions have a periodic or exponential time dependence. In these situations, the time-dependent problem can usually be reduced to a time-independent one by an appropriate ansatz regarding the time dependence of the solution. In the case of a stationary solution $u(\vec{x}, t) = u(\vec{x})$, both the heat and wave equations reduce to Poisson's equation due to the relations

$$\partial_t u - D\nabla^2 u = -D\nabla^2 u, \quad \partial_t^2 u - c^2\nabla^2 u = -c^2\nabla^2 u. \tag{6.45}$$

This is no longer true for steady state solutions, but it is often the case that the time dependence may be factored out if it is an eigenfunction of the time derivative operator. In particular, if there is a time-dependent inhomogeneity in the heat equation

$$(\text{PDE}): \ \partial_t u - D\nabla^2 u = f(\vec{x})g(t), \tag{6.46}$$

where $g'(t) = kg(t)$, i.e., $g(t) = g_0 e^{kt}$ for some g_0, we can use the ansatz $u(\vec{x}, t) = v(\vec{x})g(t)$ and obtain

$$f(\vec{x})g(t) = v(\vec{x})g'(t) - g(t)D\nabla^2 v(\vec{x}) = g(t)(k - D\nabla^2)v(\vec{x}). \tag{6.47a}$$

We therefore find that

$$-D\nabla^2 v(\vec{x}) + kv(\vec{x}) = f(\vec{x}). \tag{6.47b}$$

The left-hand side contains the negative of the Laplace operator $-\nabla^2$ and if we have homogeneous boundary conditions, we may solve the problem by performing a series expansion in its eigenfunctions, just as we did for Poisson's equation. The same principle also applies for the wave equation when the time dependence of the inhomogeneity can be factored out and satisfies $g''(t) = kg(t)$.

Example 6.5 A homogeneous metallic sphere of radius r_0 and heat diffusion coefficient a is doped with a homogeneously distributed radioactive substance with mean life τ. Due to the radioactive decays, energy will be evenly deposited inside the sphere in proportion to the number of decays. Since this energy will heat up the sphere, the temperature inside the sphere will satisfy the sourced heat equation

$$(\text{PDE}): \ T_t(r, t) - a\nabla^2 T(r, t) = \kappa e^{-t/\tau}, \tag{6.48a}$$

where κ is the value of the source term at $t = 0$. The source decays exponentially with time as the radioactive substance decays and therefore decreases in activity. If the sphere is kept in a surrounding temperature $T = 0$ with a large heat transfer coefficient, the corresponding boundary conditions will be given by

$$\text{(BC)}: \quad T(r_0, t) = 0. \tag{6.48b}$$

We have here assumed that the solution $T(r, t)$ is independent of the angles θ and φ due to the rotational symmetry of the problem. Making the ansatz $T(r, t) = R(r)e^{-t/\tau}$ now leads to the problem

$$\text{(ODE)}: \quad -\frac{1}{\tau} R(r) - a \left[R''(r) + \frac{1}{r} R'(r) \right] = \kappa, \tag{6.49a}$$

$$\text{(BC)}: \quad |R(0)| < \infty, \quad R(r_0) = 0. \tag{6.49b}$$

This problem may be solved by expanding $R(r)$ and κ in the spherical Bessel functions $j_0(\pi k r/r_0)$ according to

$$R(r) = \sum_{k=1}^{\infty} R_k \frac{\sin(\pi k r/r_0)}{r} \quad \text{and} \quad \kappa = \sum_{k=1}^{\infty} \kappa_k \frac{\sin(\pi k r/r_0)}{r}, \tag{6.50}$$

respectively. Insertion into the differential equation and identification of the coefficients in front of each linearly independent spherical Bessel function now yields

$$\left(a \frac{\pi^2 k^2}{r_0^2} - \frac{1}{\tau} \right) R_k = \kappa_k \quad \Longrightarrow \quad R_k = \frac{\kappa_k \tau r_0^2}{a \tau \pi^2 k^2 - r_0^2}. \tag{6.51}$$

The expansion coefficients κ_k may be obtained through the inner product of κ with the spherical Bessel functions.

6.2.1 Removing inhomogeneities

Of course, the stationary or steady state solutions will not always satisfy the initial conditions for a given problem. In fact, since the stationary solution will always be equal to the corresponding initial condition and the initial value for the steady state solution may be found by letting $t = t_0$, where t_0 is the initial time, any other initial condition will lead to a solution that is not stationary or a steady state. However, it is often interesting to expand the general solution around the stationary or steady state solutions, in particular when we are dealing with dissipative systems and any solution approaches these solutions asymptotically.

In the case where we know a particular solution $u_p(\vec{x}, t)$ to the inhomogeneous heat equation

$$\text{(PDE)}: \quad u_t(\vec{x}, t) - a\nabla^2 u(\vec{x}, t) = \kappa(\vec{x}, t), \tag{6.52}$$

we can introduce $v(\vec{x}, t) = u(\vec{x}, t) - u_p(\vec{x}, t)$, resulting in the homogeneous differential equation

$$v_t(\vec{x}, t) - a\nabla^2 v(\vec{x}, t) = 0 \tag{6.53}$$

for the difference $v(\vec{x}, t)$. Setting the initial time to $t = 0$, the initial condition $v(\vec{x}, 0)$ will be given by

$$\text{(IC)}: \quad v(\vec{x}, 0) = u(\vec{x}, 0) - u_p(\vec{x}, 0), \tag{6.54}$$

where the first term will be known due to the initial conditions on the sought function $u(\vec{x}, t)$ and the second term can be computed by letting $t = 0$ in the particular solution $u_p(\vec{x}, t)$. The resulting problem for $v(\vec{x}, t)$ is therefore a homogeneous differential equation with given initial conditions. The inhomogeneity has thus been moved from the differential equation to the initial condition. In particular, stationary and steady state solutions are suitable choices to fill the purpose of $u_p(\vec{x}, t)$ and they can be found through the means we have just discussed.

Example 6.6 In Example 6.5, we computed the temperature inside a sphere with a radioactive substance as the heat source. We found a solution of the form

$$T(r, t) = T_p(r, t) = \sum_{k=1}^{\infty} \frac{\kappa_k \tau r_0^2}{a \tau \pi^2 k^2 - r_0^2} \frac{\sin(\pi k r / r_0)}{r} e^{-t/\tau}, \tag{6.55}$$

where κ_k were the expansion coefficients of the constant function κ. This solution is only the solution to the actual physical problem if the physical problem comes with the initial condition

$$T(r, 0) = T_p(r, 0) = \sum_{k=1}^{\infty} \frac{\kappa_k \tau r_0^2}{a \tau \pi^2 k^2 - r_0^2} \frac{\sin(\pi k r / r_0)}{r}. \tag{6.56}$$

If we instead let the initial condition be given by the sphere being held at a constant temperature T_0, it will be of the form

$$(\text{IC}) : T(r, 0) = T_0. \tag{6.57}$$

We can still use the solution $T_p(r, t)$ to take care of the inhomogeneity in the differential equation by letting $T(r, t) = u(r, t) + T_p(r, t)$, where $u(r, t)$ will be the solution to the problem

$$(\text{PDE}) : u_t(r, t) - a\nabla^2 u(r, t) = 0, \tag{6.58a}$$

$$(\text{BC}) : |u(0, t)| < \infty, \quad u(r_0, t) = 0, \tag{6.58b}$$

$$(\text{IC}) : u(r, 0) = T_0 - T_p(r, 0). \tag{6.58c}$$

Expanding the constant T_0 in terms of the functions $\sin(\pi k r / r_0)/r$, we can write the initial condition for $u(r, t)$ as

$$u(r, 0) = \sum_{k=1}^{\infty} \kappa_k \left[\frac{T_0}{\kappa} - \frac{\tau r_0^2}{a \tau \pi^2 k^2 - r_0^2} \right] \frac{\sin(\pi k r / r_0)}{r}. \tag{6.59}$$

Note that the ratio T_0/κ arises from both κ and T_0 being constants expanded in the same eigenfunctions and the expansion coefficients of T_0 are therefore given by $\kappa_k T_0/\kappa$.

Although we have here used the heat equation as the primary example, it should be noted that the methods presented in this section apply equally well to the wave equation or any other time-dependent partial differential equation that involves a Sturm–Liouville operator in the spatial part.

6.3 DIFFUSION AND HEAT EQUATIONS

We have so far discussed the solution to Poisson's equation and the spatial differential equations resulting from assuming a stationary or steady state solution. It is therefore about time that we turn our attention to the solution of the more general time-dependent problems of physical interest. In particular, we will discuss the solutions and phenomenology of the heat and wave equations we derived in Chapter 3. Starting with the heat equation, we will consider the general problem

$$(\text{PDE}): u_t(\vec{x}, t) - a\nabla^2 u(\vec{x}, t) = \kappa(\vec{x}, t), \qquad (\vec{x} \in V) \qquad (6.60\text{a})$$

$$(\text{BC}): \alpha u(\vec{x}, t) + \beta \vec{n} \cdot \nabla u(\vec{x}, t) = 0, \qquad (\vec{x} \in S) \qquad (6.60\text{b})$$

$$(\text{IC}): u(\vec{x}, 0) = g(\vec{x}), \qquad (\vec{x} \in V) \qquad (6.60\text{c})$$

where V is the domain inside of which we are interested in solving for $u(\vec{x}, t)$ and S is the boundary of V. We have here assumed homogeneous boundary conditions for convenience as it will allow us to expand the solution in eigenfunctions of the operator $-\nabla^2$. For the moment, we will assume that V is a compact volume such that the set of eigenfunctions of $-\nabla^2$ will be countable. Any problem with inhomogeneous boundary conditions may be recast on this form by transferring the inhomogeneity to the differential equation and the initial condition as described earlier.

6.3.1 Initial conditions

Let us first discuss the case where there is no source present, i.e., $\kappa(\vec{x}, t) = 0$, and the only inhomogeneity in the problem is given by the initial condition $g(\vec{x})$. Since our problem contains the operator $-\nabla^2$ and has homogeneous boundary conditions on the surface S, we may express the solution $u(\vec{x}, t)$ and the initial condition $g(\vec{x})$ as linear combinations of a set of eigenfunctions $X_n(\vec{x})$ satisfying

$$-\nabla^2 X_n(\vec{x}) = \lambda_n X_n(\vec{x}), \qquad (6.61)$$

where λ_n is the corresponding eigenvalue. It should be noted that n here is just a dummy variable used for labelling the eigenfunctions. In practice, it may be more convenient to use several dummy variables instead. An example of this is when we have found the eigenfunctions by separation of variables and have one label per direction. The expansions of $u(\vec{x}, t)$ and $g(\vec{x})$ will take the forms

$$u(\vec{x}, t) = \sum_n u_n(t) X_n(\vec{x}), \quad g(\vec{x}) = \sum_n g_n X_n(\vec{x}). \qquad (6.62)$$

The time dependence of the expansion coefficients $u_n(t)$ is necessary as the function $u(\vec{x}, t)$ will generally be different functions of $\vec{x}$ for different t.

Inserting the series expansions of $u(\vec{x}, t)$ and $g(\vec{x})$ into Eqs. (6.60), we find that

$$\sum_n [u'_n(t) + a\lambda_n u_n(t)] X_n(\vec{x}) = 0, \qquad (6.63\text{a})$$

$$\sum_n u_n(0) X_n(\vec{x}) = \sum_n g_n X_n(\vec{x}). \qquad (6.63\text{b})$$

Since the functions $X_n(\vec{x})$ are linearly independent, we may identify the coefficients of each

$X_n(\vec{x})$ on both sides of the equations, leading to an infinite number of decoupled ordinary differential equations

$$(\text{ODE}) : u_n'(t) + a\lambda_n u_n(t) = 0, \tag{6.64a}$$

$$(\text{IC}) : u_n(0) = g_n. \tag{6.64b}$$

Luckily, these equations are all of the same form and are generally solved by

$$u_n(t) = g_n e^{-a\lambda_n t}. \tag{6.65}$$

The solution $u(\vec{x}, t)$ is therefore given by

$$u(\vec{x}, t) = \sum_n g_n X_n(\vec{x}) e^{-a\lambda_n t}. \tag{6.66}$$

This approach to solving the heat equation is exactly equivalent to the way in which we solved Laplace's equation with inhomogeneous boundary conditions earlier. The only difference is that we left out the coordinate direction with the inhomogeneous boundary conditions in the case of Laplace's equation in order to solve the resulting ordinary differential equation in that coordinate. In contrast, for the heat equation we expand in the eigenfunctions of the full Laplace operator, resulting in a set of ordinary differential equations in time rather than in a remaining spatial coordinate.

Example 6.7 In a thin medium placed in a circular Petri dish of radius r_0, a substance diffuses with diffusivity D. An amount Q of the a substance is located at the center of the medium at time $t = 0$ and we are looking to describe the concentration $q(\vec{x}, t)$ of the substance as a function of the position in the medium and of time. Assuming that no substance can pass through the surface of the medium or through the walls of the Petri dish, we may model the time evolution of the concentration by the partial differential equation

$$(\text{PDE}) : q_t(\rho, t) - D\nabla^2 q(\rho, t) = 0, \qquad (\rho < r_0) \tag{6.67a}$$

$$(\text{BC}) : |q(0, t)| < \infty, \quad q_r(r_0, t) = 0, \tag{6.67b}$$

$$(\text{IC}) : q(\vec{x}, 0) = Q\delta^{(2)}(\vec{x}). \tag{6.67c}$$

The homogeneous boundary condition at $\rho = r_0$ is due to the current being zero in the normal direction at this boundary and we have assumed that the concentration only depends on the polar coordinate ρ due to the rotational symmetry of the problem. We have also chosen not to give the initial condition in polar coordinates as the delta function describing the initial condition is at the boundary and coordinate singularity of the problem in polar coordinates.

The eigenfunctions of the operator $-\nabla^2$ satisfying the boundary conditions of Eq. (6.67b) are given by

$$R_m(\rho) = J_0(\alpha_{0m}' \rho/r_0) \tag{6.68}$$

with corresponding eigenvalues $\lambda_m = \alpha_{0m}'^2/r_0^2$. The initial condition may be expanded in these functions and we find that

$$Q\delta^{(2)}(\vec{x}) = Q \sum_{m=1}^{\infty} d_m R_m(\rho), \tag{6.69}$$

where the expansion coefficients d_m may be computed as

$$d_m = \frac{\langle R_m, \delta^{(2)} \rangle}{\|R_m\|^2} \tag{6.70}$$

and the inner product is given by

$$\langle f, g \rangle = \int_{\rho < r_0} f(\vec{x}) g(\vec{x}) dA = 2\pi \int_0^{r_0} \rho f(\rho) g(\rho) d\rho. \tag{6.71}$$

Again it should be noted that the last form of this inner product is not directly applicable to the inner product in the numerator because of the delta function. This inner product is instead easier to compute by applying the defining property

$$\left\langle R_m, \delta^{(2)} \right\rangle = \int_{\rho < r_0} R_m(\rho) \delta^{(2)}(\vec{x}) dA = R_m(0) \tag{6.72}$$

leading to

$$d_m = \frac{R_m(0)}{\|R_m\|^2}. \tag{6.73}$$

Expanding the solution $q(\rho, t)$ in $R_m(\rho)$ and inserting the result into the differential equation, we find that

$$q(\rho, t) = Q \sum_{m=1}^{\infty} \frac{R_m(0)}{\|R_m\|^2} R_m(\rho) e^{-D\alpha_{0m}'^2 t / r_0^2}. \tag{6.74}$$

An important piece of phenomenology is evident from the discussion in this section, the solutions to the homogeneous heat equation with homogeneous boundary conditions are exponentially decaying. The decay rate for each eigenfunction of the Laplace operator is proportional to its eigenvalue and given by $a\lambda_n$, indicating that after a long time has passed, the eigenfunction corresponding to the lowest eigenvalue will completely dominate the solution as the eigenfunctions with higher eigenvalues will decay faster. Since the negative of the Laplace operator is positive definite, none of the solutions will grow exponentially. The only possibility of not having an exponential decay is the case when we have homogeneous Neumann boundary conditions that allow a constant eigenfunction with eigenvalue zero. This is the case in the above example where the eigenfunction for $m = 0$ has zero eigenvalue due to $\alpha_{01}' = 0$.

Any problem involving the heat equation that allows a stationary or steady state solution may be recast on the form we have assumed here by transferring the inhomogeneities to the initial condition as shown in the previous section. The solutions to the heat equation will therefore approach the stationary or steady state solution for large times. When the steady state solution is exponentially decaying as in Example 6.5, the full solution will also decay to zero but the rate at which this occurs will depend on the constant $a\lambda_1\tau$. If $a\lambda_1\tau > 1$, the steady state solution decays slower than all of the contributions from the homogeneous problem, whereas if $a\lambda_1\tau < 1$, the eigenfunction corresponding to the lowest eigenvalue λ_1 will dominate the solution for large times.

6.3.2 Constant source terms

Let us now consider the situation where the source term $\kappa(\vec{x}, t)$ in Eqs. (6.60) is a general function of the spatial variable $\vec{x}$, but does not depend on time, i.e., $\kappa(\vec{x}, t) = \kappa(\vec{x})$. For the sake of the argument, we will work with initial conditions that are homogeneous $g(\vec{x}) = 0$. Since the heat equation is linear, this is not really a restriction as we can always split a problem with several inhomogeneities into one problem with the inhomogeneous source and another one with the inhomogeneous initial condition. The latter of these problems can be solved as just described while for the former we can expand both the inhomogeneous source term and the solution in the eigenfunctions of the Laplace operator as

$$\kappa(\vec{x}) = \sum_n \kappa_n X_n(\vec{x}), \quad u(\vec{x}, t) = \sum_n u_n(t) X_n(\vec{x}). \tag{6.75}$$

Insertion into the differential equation and using the fact that the $X_n(\vec{x})$ are linearly independent to identify the terms in the sums on both sides of the equation now results in

$$(\text{ODE}) : u'_n(t) + a\lambda_n u_n(t) = \kappa_n, \tag{6.76a}$$

$$(\text{IC}) : u_n(0) = 0. \tag{6.76b}$$

This differential equation is solved by

$$u_n(t) = \frac{\kappa_n}{a\lambda_n}(1 - e^{-a\lambda_n t}), \tag{6.77}$$

which may be obtained by first finding the particular solution $u_{n,p}(t) = \kappa_n/a\lambda_n$ and then adapting the homogeneous solution to the initial condition.

In the end, whether we treat any inhomogeneous initial conditions separately or not does not really matter for the solution. If we keep the inhomogeneous initial condition and expand it in the eigenfunctions $X_n(\vec{x})$, we will find the ordinary differential equations

$$(\text{ODE}) : u'_n(t) + a\lambda_n u_n(t) = \kappa_n, \tag{6.78a}$$

$$(\text{IC}) : u_n(0) = g_n. \tag{6.78b}$$

This may be solved in exactly the same fashion as Eqs. (6.76), by finding the particular solution and then adapting the homogeneous solution to the boundary conditions. The resulting solution is now given by

$$u_n(t) = \frac{\kappa_n}{a\lambda_n}(1 - e^{-a\lambda_n t}) + g_n e^{-a\lambda_n t}. \tag{6.79}$$

Note that this is exactly the same result as that which would be obtained by superposing the solution to the problem with an inhomogeneous differential equation and homogeneous boundary conditions with that of the problem with a homogeneous differential equation and inhomogeneous boundary conditions.

Example 6.8 A thin rod of length ℓ with heat diffusion coefficient a is isolated everywhere except for the endpoints, which are kept at a constant temperature T_0, and at the position $x = x_0$, where a constant amount of heat is added per time unit after $t = 0$, see Fig. 6.5. If

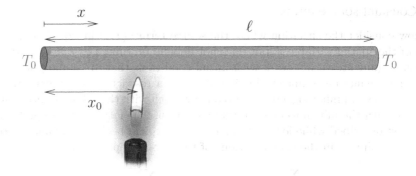

Figure 6.5 A rod of length ℓ that is assumed to be isolated everywhere except the endpoints, which are held at temperature T_0 and at $x = x_0$ where we add a fixed amount of heat per time unit.

we assume that the rod has had time to adapt to the endpoint temperature before $t = 0$, we can model this situation as

$$\text{(PDE)} : T_t(x,t) - aT_{xx}(x,t) = \kappa_0 \delta(x - x_0), \tag{6.80a}$$

$$\text{(BC)} : T(0,t) = T(\ell,t) = T_0, \tag{6.80b}$$

$$\text{(IC)} : T(x,0) = T_0. \tag{6.80c}$$

In order to solve this problem, we start by removing the inhomogeneities in the boundary and initial conditions by introducing $u(x,t) = T(x,t) - T_0$, which will have homogeneous boundary and initial conditions and satisfy the same partial differential equation as $T(x,t)$

$$\text{(PDE)} : u_t(x,t) - au_{xx}(x,t) = \kappa_0 \delta(x - x_0), \tag{6.81a}$$

$$\text{(BC)} : u(0,t) = u(\ell,t) = 0, \tag{6.81b}$$

$$\text{(IC)} : u(x,0) = 0. \tag{6.81c}$$

As we have seen earlier, the eigenfunctions of $-\partial_x^2$ with these boundary conditions are given by the sine functions

$$X_n(x) = \sin\left(\frac{\pi n x}{\ell}\right) \tag{6.82}$$

with corresponding eigenvalues $\lambda_n = \pi^2 n^2/\ell^2 = k_n^2$. Expanding the solution and inhomogeneity in terms of these functions, we find that

$$\kappa_0 \delta(x - x_0) = \frac{2\kappa_0}{L} \sum_{n=1}^{\infty} \sin(k_n x_0) \sin(k_n x) \tag{6.83}$$

and consequently

$$u(x,t) = \frac{2\kappa_0}{aL} \sum_{n=1}^{\infty} \frac{\sin(k_n x_0)}{k_n^2}(1 - e^{-ak_n^2 t})\sin(k_n x). \tag{6.84}$$

An alternative approach to solving a problem with a constant inhomogeneity in the source term is to find the corresponding stationary solution, which will satisfy the boundary conditions and Poisson's equation. This solution will exactly correspond to the constant term in Eq. (6.77) and using this solution we may rewrite the problem as a problem with an inhomogeneity only in the initial condition. This procedure may be favourable if the stationary solution can be found and expressed on a relatively simple form. However, a series expansion may still be necessary in order to find the solution to the remaining problem with the inhomogeneity in the initial condition.

Example 6.9 In the case of Example 6.8, it is relatively easy to find the stationary solution. As $\delta(x - x_0) = 0$ for both $x < x_0$ and for $x > x_0$, the stationary solution in these regions will be a linear function $u_{st}(x)$

$$u_{st}(x) = \begin{cases} A_- x, & (x < x_0) \\ A_+(x - \ell), & (x > x_0) \end{cases}, \tag{6.85}$$

where we have already accounted for the boundary conditions at $x = 0$ and $x = \ell$. In order to satisfy the differential equation at $x = x_0$, the constants $A_\pm$ must be given by

$$A_- = \left(1 - \frac{x_0}{\ell}\right) \kappa_0, \quad A_+ = -\frac{\kappa_0 x_0}{\ell}. \tag{6.86}$$

This can be seen by taking the distribution derivative of the stationary solution, but may also be deduced based on physical reasoning, requiring that the heat transported away must be equal to the heat added in the stationary solution and that temperature is a continuous function.

We may now rewrite the problem as a problem with a homogeneous partial differential equation, homogeneous boundary conditions, and an inhomogeneous initial condition by letting $v(x, t) = u(x, t) - u_{st}(x)$. Due to the homogeneous initial condition for $u(x, t)$, the initial condition for $v(x, t)$ will be given by

$$v(x, 0) = u(x, 0) - u_{st}(x) = -u_{st}(x). \tag{6.87}$$

6.3.3 Critical systems

The cases we have discussed so far have been concerned with situations where the source term has been independent of the actual quantity that we wish to solve for. While this is sufficient to describe many situations, there are also many situations where this is no longer true. Instead, we must often account for situations where the production or destruction of a substance depends on its concentration. Examples of cases where this becomes important include the situation where we are interested in the concentration of a radioactive substance subject to diffusion or the situation where the presence of a substance in a medium acts as a catalyst for its own production, such as for the thermal neutrons in a fission reactor core.

Example 6.10 Consider the situation where we are looking to describe the concentration q of diffusing radioactive atoms as a function of position and time. In this situation, the concentration will not change only due to diffusion, but also drop due to the radioactive decays. The number of decays per unit time inside a volume V is given by the number of atoms of the substance multiplied by the decay rate $1/\tau$, where τ is the mean life of the radioactive decay. We therefore find that the number of decays inside the volume per unit time is given by

$$K = -\frac{1}{\tau}\int_V q(\vec{x},t)dV \tag{6.88}$$

and the source density is therefore

$$\kappa(\vec{x},t) = -\frac{1}{\tau}q(\vec{x},t), \tag{6.89}$$

leading to the partial differential equation

$$q_t(\vec{x},t) - D\nabla^2 q(\vec{x},t) = -\frac{1}{\tau}q(\vec{x},t), \tag{6.90}$$

where D is the diffusivity.

If the presence of a substance instead acts as a catalyst for its own creation, we may have a situation such that the substance is created with a source density proportional to its concentration. This will lead to a differential equation of the form

$$q_t(\vec{x},t) - D\nabla^2 q(\vec{x},t) = \gamma q(\vec{x},t), \tag{6.91}$$

where γ is the rate at which a given amount of substance duplicates.

The only difference between the differential equations derived in the example above is the sign in front of the source term. We can therefore treat both cases in the same fashion by assuming that we have a partial differential equation given by

$$u_t(\vec{x},t) - D\nabla^2 u(\vec{x},t) = \kappa_0 u(\vec{x},t), \tag{6.92}$$

where the constant κ_0 may be taken as either negative or positive depending on the problem in question. By moving the source term to the left-hand side, we find that

$$u_t(\vec{x},t) - (D\nabla^2 + \kappa_0)u(\vec{x},t) = 0. \tag{6.93}$$

Despite the generally non-zero source term, this differential equation is still a homogeneous linear partial differential equation and we will therefore still have the possibility of solving it using superposition. Since κ_0 is taken as a constant, any eigenfunction $X_n(\vec{x})$ of $-\nabla^2$ will satisfy

$$-(D\nabla^2 + \kappa_0)X_n(\vec{x}) = (D\lambda_n - \kappa_0)X_n(\vec{x}), \tag{6.94}$$

where λ_n is the eigenvalue of $X_n(\vec{x})$ with respect to $-\nabla^2$. Because of this, the differential equations obtained when expanding the solution in terms of a linear combination of $X_n(\vec{x})$ will remain decoupled even after the introduction of the source term, but with a shifted eigenvalue. The expansion

$$u(\vec{x},t) = \sum_n u_n(t)X_n(\vec{x}) \tag{6.95}$$

inserted into the differential equation therefore results in the ordinary differential equations

$$u'_n(t) + (D\lambda_n - \kappa_0)u_n(t) = 0 \tag{6.96}$$

and therefore

$$u(\vec{x}, t) = \sum_n u_n(0) X_n(\vec{x}) e^{(\kappa_0 - D\lambda_n)t}. \tag{6.97}$$

The major difference as compared to the solution of the source-free heat equation is the appearance of the constant κ_0 in the exponent. As we discussed in relation to the source free heat equation, its solutions were found to be exponentially decaying with a decay constant proportional to the eigenvalue λ_n. With the introduction of the linear source term, we instead find that the part of the solution proportional to $X_n(\vec{x})$ decays with a decay constant

$$\frac{1}{\tau_n} = D\lambda_n - \kappa_0. \tag{6.98}$$

For the cases where κ_0 is negative, i.e., when the source term describes a sink, the phenomenology does not change significantly. The solutions still decay exponentially, but the process is speeded up by the sink. However, when $0 < \kappa_0 < D\lambda_n$ the decay of the mode $X_n(\vec{x})$ is slowed down and at the critical value

$$\kappa_0 = D\lambda_n \equiv \kappa_{c,n}, \tag{6.99}$$

the coefficient of $X_n(\vec{x})$ is constant in time. When this happens for any λ_n, the solution will no longer approach the stationary or steady state solution for large times. With the ordering $\lambda_1 < \lambda_2 < \dots$ of the eigenfunctions, the critical value $\kappa_{c,n}$ will be lower for smaller values of n and the first mode to be affected in this way will be the $X_1(\vec{x})$ mode.

If we have values of κ_0 which are larger than $\kappa_{c,n}$, the coefficient in front of $X_n(\vec{x})$ will grow exponentially instead of decaying. In this situation, the full solution $u(\vec{x}, t)$ will generally grow exponentially as well. This exponential growth is the underlying reason that a runaway nuclear reaction may occur when a critical mass of fissile material is created. It should be noted that it is important to keep the limitations of the model in mind when we encounter exponentially growing solutions of this kind. In many cases, we may have obtained the model by linearising a non-linear problem in which case the model is only valid for small variations of the solution. With an exponential growth, such approximations will only be valid as long as the solution may be considered a small variation.

The comparison of κ_0 and $D\lambda_n$ has a simple physical interpretation. While κ_0 is the rate at which any amount of the substance duplicates, $D\lambda_n$ is the rate with which any amount of substance in the nth eigenmode diffuses out of the system. Therefore, if κ_0 is larger than $D\lambda_n$, the amount of substance in the nth eigenmode will increase, since it is being produced faster than it diffuses out of the system.

Example 6.11 As a very crude approximation, we describe a nuclear reactor core as a cylinder with radius r_0 and height h. The concentration of thermal neutrons $n(\vec{x}, t)$ in the core follows the partial differential equation

$$\begin{align} \text{(PDE)} : n_t(\vec{x}, t) - D\nabla^2 n(\vec{x}, t) &= \kappa_0 n(\vec{x}, t), & (\vec{x} \in V) \tag{6.100a} \\ \text{(BC)} : n(\vec{x}, t) &= 0, & (\vec{x} \in S) \tag{6.100b} \end{align}$$

where V is the core volume and S its surface. We have here assumed that any neutrons reaching the edge of the core are effectively transported away from it. While this model is extremely crude and not accurate enough to describe an actual reactor core, it serves the purpose of illustrating the mechanism behind a critical system. The constant κ_0 may be thought of as a combination of the source term due to the fission reactions and the sink due to absorption in the control rods.

The eigenfunctions of the Laplace operator satisfying the given boundary conditions are of the form

$$X_{nkm}(\rho, \phi, z) = J_m(\alpha_{mk}\rho/r_0)\sin(\pi n z/h)e^{im\phi} \tag{6.101a}$$

with corresponding eigenvalues

$$\lambda_{nkm} = \frac{\alpha_{mk}^2}{r_0^2} + \frac{\pi^2 n^2}{h^2}. \tag{6.101b}$$

The lowest eigenvalue is therefore obtained for the lowest possible values of α_{mk} and n, which occurs for $m = 0$, $k = 1$, and $n = 1$, resulting in

$$D\lambda_{110} = D\left(\frac{\alpha_{01}^2}{r_0^2} + \frac{\pi^2}{h^2}\right) = \kappa_{c,110}. \tag{6.102}$$

As can be seen from this relation, the physical dimensions of the core directly influence the critical value of κ_0 with a larger core corresponding to a lower critical value. Therefore, a large core will require a much lower value of κ_0 before becoming critical.

In situations where the source term depends on the concentration in a non-linear fashion, the concept of criticality may still be applied to the problem resulting from a linearisation of the problem for small deviations from a stationary solution. Assuming that we have a partial differential equation of the form given in Eqs. (6.60) with a source term

$$\kappa(\vec{x}, t) = f(u(\vec{x}, t)) \tag{6.103}$$

and a stationary solution $u_{\text{st}}(\vec{x})$ satisfying

$$-D\nabla^2 u_{\text{st}}(\vec{x}) = f(u_{\text{st}}(\vec{x})), \tag{6.104}$$

the difference $v(\vec{x}, t) = u(\vec{x}, t) - u_{\text{st}}(\vec{x})$ satisfies

$$v_t(\vec{x}, t) - D\nabla^2 v(\vec{x}, t) = f'(u_{\text{st}}(\vec{x}))v(\vec{x}, t) \tag{6.105}$$

for small $v(\vec{x}, t)$. This is a linear partial differential equation for $v(\vec{x}, t)$ and we can solve it as earlier by finding and expanding in the eigenfunctions to the problem

$$-\left[D\nabla^2 + f'(u_{\text{st}}(\vec{x}))\right]X(\vec{x}) = \lambda X(\vec{x}). \tag{6.106}$$

If all of the eigenvalues λ are positive, the stationary solution is *stable*, indicating that any deviation will tend to decay over time. On the other hand, if any eigenvalue is negative, the solution is unstable and there are deviations from u_{st} that will increase exponentially with time, generally implying that the approximation of small deviations will eventually be violated. If there are eigenvalues that are zero, the time evolution of the corresponding eigenmodes will depend on higher order corrections.

Example 6.12 A bacterial culture is living on the surface of a growth medium in a circular Petri dish of radius r_0, see Fig. 6.6. The bacteria move randomly on this surface resulting in a

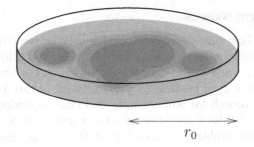

Figure 6.6 The growth of a bacterial culture on the two dimensional surface of a growth medium in a Petri dish of radius r_0 may be described by a non-linear partial differential equation. By linearising the solution around the stationary solutions we can find whether or not they are stable.

diffusive current and their net reproduction is given by $\kappa u(u_0 - u)$, where u is the bacterial concentration, u_0 the bacterial concentration at which there is not enough resources to support more bacteria, and κ is a constant relating to how quickly a bacteria may undergo binary fission. The resulting bacterial concentration is described by the differential equation

$$(\text{PDE}) : u_t(\vec{x}, t) - D\nabla^2 u(\vec{x}, t) = \kappa u(u_0 - u), \tag{6.107a}$$
$$(\text{BC}) : \vec{n} \cdot \nabla u(\vec{x}, t) = 0, \qquad\qquad (\rho = r_0) \tag{6.107b}$$

where D is the diffusivity of the bacteria and $\vec{n}$ is normal to the edge of the Petri dish. There are two stationary solutions to this problem, $u(\vec{x}, t) = 0$ and $u(\vec{x}, t) = u_0$. Both of these solutions result in both sides of the partial differential equation becoming zero.

If we linearise the solution around the stationary solution $u_{\text{st}}(\vec{x}) = 0$ by introducing $u(\vec{x}, t) = u_{\text{st}}(\vec{x}) + v(\vec{x})$ and keeping only terms that are linear in $v(\vec{x})$, we find that

$$v_t(\vec{x}, t) - D\nabla^2 v(\vec{x}, t) = \kappa u_0 v(\vec{x}, t). \tag{6.108}$$

Because of the boundary conditions of the problem, the lowest eigenvalue of the operator $-\nabla^2$ is equal to zero, corresponding to a constant eigenfunction. As a result, the stationary solution $u_{\text{st}}(\vec{x}) = 0$ is unstable as the constant multiplying $v(\vec{x}, t)$ on the right-hand side is positive.

On the other hand, if the system is instead linearised around $u_{\text{st}}(\vec{x}) = u_0$, the linearised differential equation becomes

$$v_t(\vec{x}, t) - D\nabla^2 v(\vec{x}, t) = -\kappa u_0 v(\vec{x}, t). \tag{6.109}$$

In this situation, the constant multiplying $v(\vec{x}, t)$ on the right-hand side is negative, resulting in the linearised source leading to a faster decay of all eigenfunction modes in $v(\vec{x}, t)$. Because of this, the stationary solution $u_{\text{st}}(\vec{x}) = u_0$ is a stable solution. Any small deviation from this solution will result in a solution that approaches $u_{\text{st}}(\vec{x}) = u_0$ for large times.

6.3.4 Time-dependent sources

We have already seen how time-dependent sources may be treated by removing the inhomogeneities from the differential equations and boundary conditions in the cases where it is possible to find a stationary or steady state solution to the problem. However, it may not always be apparent or even possible to find a solution of this form and we may have to take a more general approach for more complicated time-dependent inhomogeneities. One method we may always fall back on is that of series expansions in the eigenfunctions of the Laplace operator. For the problem given by Eqs. (6.60), we may generally consider the case where the source term $\kappa(\vec{x}, t)$ is an arbitrary function of the spatial variables and time. For any fixed t, this function may be expanded as

$$\kappa(\vec{x}, t) = \sum_n \kappa_n(t) X_n(\vec{x}), \tag{6.110}$$

where $X_n(\vec{x})$ are the eigenfunctions of the Laplace operator and the expansion coefficients $\kappa_n(t)$ now generally depend on the time t. With precisely the same argumentation as for the time-independent sources, this eventually leads to the decoupled differential equations

$$u'_n(t) + D\lambda_n u_n(t) = \kappa_n(t). \tag{6.111}$$

Naturally, this differential equation is exactly equivalent with Eq. (6.76a) with the only difference that the right-hand side is now time-dependent. The general solution is given by

$$u_n(t) = u_n(0)e^{-D\lambda_n t} + \int_0^t \kappa_n(\tau)e^{-D\lambda_n(t-\tau)}d\tau. \tag{6.112}$$

While it is straightforward to check that this expression solves the differential equation by differentiating it, its derivation is left for our discussion of Green's functions in Chapter 7.

6.4 WAVE EQUATION

Many of the approaches we have discussed in connection to the heat equation remain valid when attempting to solve the wave equation using series expansions in eigenfunctions of the Laplace operator. The main difference arises from the fact that the wave equation is of second order in the time variable t, whereas the heat equation was of first order in t. Therefore, we will generally require two initial conditions in the case of the wave equation rather than the one that was required for the heat equation. The general wave equation problem that we are interested in will be of the form

$$\text{(PDE)} : u_{tt}(\vec{x}, t) - c^2\nabla^2 u(\vec{x}, t) = f(\vec{x}, t), \qquad (\vec{x} \in V) \tag{6.113a}$$

$$\text{(BC)} : \alpha u(\vec{x}, t) + \beta \vec{n} \cdot \nabla u(\vec{x}, t) = 0, \qquad (\vec{x} \in S) \tag{6.113b}$$

$$\text{(IC)} : u(\vec{x}, 0) = g(\vec{x}), \quad u_t(\vec{x}, 0) = h(\vec{x}). \qquad (\vec{x} \in V) \tag{6.113c}$$

As for the heat equation problem given by Eqs. (6.60), we are here interested in the solution $u(\vec{x}, t)$ within the volume V with boundary S.

6.4.1 Inhomogeneous sources and initial conditions

Let us first discuss the case where we have an inhomogeneous source term $f(\vec{x}, t)$ that does not depend on the function $u(\vec{x}, t)$ as well as inhomogeneous initial conditions $g(\vec{x})$ and

$h(\vec{x})$. Just as we did for the heat equation, the solution $u(\vec{x}, t)$ and the inhomogeneities may be expanded in the eigenfunctions $X_n(\vec{x})$ of the Laplace operator

$$u(\vec{x}, t) = \sum_n u_n(t) X_n(\vec{x}), \qquad\qquad f(\vec{x}, t) = \sum_n f_n(t) X_n(\vec{x}),$$

$$g(\vec{x}) = \sum_n g_n X_n(\vec{x}), \qquad\qquad h(\vec{x}) = \sum_n h_n X_n(\vec{x}), \qquad (6.114)$$

where the functions $f_n(t)$ and the constants g_n and h_n may be determined by taking the inner product of the corresponding functions and the eigenfunctions $X_n(\vec{x})$. Inserting the expansions into the partial differential equation and the initial conditions results in the decoupled ordinary differential equations

$$(\text{ODE}) : u_n''(t) + c^2 \lambda_n u_n(t) = f_n(t), \qquad (6.115a)$$

$$(\text{IC}) : u_n(0) = g_n, \quad u_n'(0) = h_n, \qquad (6.115b)$$

where λ_n is the eigenvalue of $X_n(\vec{x})$ with respect to $-\nabla^2$. This derivation is exactly analogous to what we have seen for the heat equation, with the addition of an extra initial condition due to the wave equation being a differential equation of second order in time. Just as for the heat equation, the resulting ordinary differential equations may be solved separately and the full solution to the problem is given by the superposition of the individual solutions.

In the situation where we do not have any inhomogeneity in the wave equation, i.e., when $f(\vec{x}, t) = 0$, Eq. (6.115a) has the general solution

$$u_n(t) = A_n \sin(\omega_n t) + B_n \cos(\omega_n t), \qquad (6.116)$$

where the *natural angular frequency* ω_n of the nth mode is given by

$$\omega_n = c\sqrt{\lambda_n} \equiv ck_n. \qquad (6.117)$$

As expected, this general solution has two unknown constants A_n and B_n, as the differential equation for $u_n(t)$ is of second order. These two constants are fully determined by the initial conditions, which result in

$$u_n(0) = B_n = g_n, \quad u_n'(0) = \omega_n A_n = h_n. \qquad (6.118)$$

The solution to the homogeneous wave equation is therefore given by

$$u(\vec{x}, t) = \sum_n \left[\frac{h_n}{\omega_n} \sin(\omega_n t) + g_n \cos(\omega_n t) \right] X_n(\vec{x}). \qquad (6.119)$$

Unlike the solutions to the homogeneous heat equation, which were exponentially decreasing with time, the eigenmodes of the wave equation are oscillating with angular frequency ω_n and constant amplitude. Because of this, the solutions to the wave equation will generally not approach a stationary or steady state solution, but instead oscillate around them as long as there are no additional dissipative forces acting on the system.

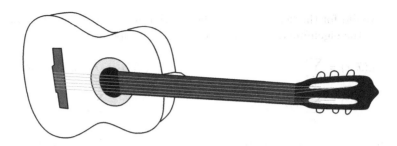

Figure 6.7 The guitar is an example of a plucked string instrument. Sound is generated by plucking the strings and then allowing them to vibrate freely. This differs from instruments such as bowed string instruments that are continuously driven by the rubbing of a bow.

Example 6.13 *Plucked string instruments*, such as a guitar (see Fig. 6.7), are a category of instruments which create sound from the vibrations resulting from a fixed string being plucked, i.e., released from some initial state and then allowed to vibrate under the influence of internal forces only. The wave equation describing the movement of such a string is of the form

$$(\text{PDE}) : u_{tt}(x,t) - c^2 u_{xx}(x,t) = 0, \qquad (0 < x < \ell) \qquad (6.120a)$$
$$(\text{BC}) : u(0,t) = u(\ell,t) = 0, \qquad (6.120b)$$
$$(\text{IC}) : u(x,0) = g(x), \quad u_t(x,0) = h(x), \qquad (6.120c)$$

where $c^2 = S/\rho_\ell$ is the wave speed in the string, ℓ the string length, S the tension in the string, and ρ_ℓ its linear density (see also Example 3.18).

As we have already discussed, the eigenfunctions of the operator $-\partial_x^2$ with homogeneous Dirichlet boundary conditions are given by

$$X_n(x) = \sin(k_n x), \qquad (6.121)$$

where $k_n = \pi n/\ell$. It follows that the string vibrations are going to be described by

$$u(x,t) = \sum_{n=1}^{\infty} \left[\frac{h_n}{\omega_n} \sin(\omega_n t) + g_n \cos(\omega_n t) \right] \sin(k_n x), \qquad (6.122)$$

where $\omega_n = c k_n$ are the natural angular frequencies of the string. The natural frequencies therefore correspond to the sound frequencies that will be produced by the string, with the lowest angular frequency $\omega_1 = c\pi/\ell$, for which $n = 1$, corresponding to the *fundamental frequency* of the string, while the remaining frequencies are *overtones*.

In general, when a string is plucked, all of its eigenmodes will be excited with different amplitudes depending on the exact nature of the initial conditions created. As a result, different ratios of the overtones to the fundamental frequency may arise when plucking the string differently, resulting in different sound sensations.

6.4.2 Damped systems

In the example we have just seen, the plucked string will oscillate forever with constant amplitude. This may be a good model for a short time period after the string is plucked, but it is relatively obvious that it is not for longer times. While a guitar string may continue to produce sound for a some time, the sound eventually fades away and the string comes to a rest due to loss of vibrational energy in the string. Additionally, without the loss of energy in the string, the creation of sound waves would violate the conservation of energy, as energy is needed to create the sound waves. Since infinitely oscillating guitar strings and violation of energy conservation are not particularly pleasing features, we can amend the model by introducing a damping term into the wave equation (see, e.g., Example 3.10)

$$(\text{PDE}): \ u_{tt}(\vec{x}, t) + 2k u_t(\vec{x}, t) - c^2 \nabla^2 u(x, t) = 0. \tag{6.123}$$

This change of the wave equation does not prevent us from expanding the solution and initial conditions in eigenfunctions of the Laplace operator in the same way as earlier. Instead, the main change introduced in the case of the damped wave equation is the appearance of a first order time derivative in each of the resulting ordinary differential equations

$$u_n''(t) + 2k u_n'(t) + \omega_n^2 u_n(t) = 0. \tag{6.124}$$

These differential equations may be solved by making the ansatz

$$u_n(t) = e^{\alpha t}, \tag{6.125}$$

resulting in the characteristic equation

$$\alpha^2 + 2k\alpha + \omega_n^2 = 0 \quad \Longrightarrow \quad \alpha = -k \pm \sqrt{k^2 - \omega_n^2}. \tag{6.126}$$

From these solutions, we can divide the eigenmodes into three different categories:

1. *Underdamped eigenmodes:* In the case where $k < \omega_n$, the roots of Eq. (6.126) are complex and may be written in the form

$$\alpha = -k \pm i\omega_n', \tag{6.127}$$

where $\omega_n' = \sqrt{\omega_n^2 - k^2}$ is a real number. The general solution may therefore be written as

$$u_n(t) = e^{-kt}(A_n e^{i\omega_n' t} + B_n e^{-i\omega_n' t}) = e^{-kt}[C_n \sin(\omega_n' t) + D_n \cos(\omega_n' t)], \tag{6.128}$$

   where we have rewritten the solution in terms of the real sine and cosine functions instead of the complex exponentials in the second step. The solutions are therefore oscillatory with amplitude that decreases exponentially with time. Note that the solutions approach the undamped solutions when $k \to 0$ as the exponential suppression then disappears and $\omega_n' \to \omega_n$.

2. *Overdamped eigenmodes:* For the modes in which the damping is very strong, i.e., when $k > \omega_n$, both roots to Eq. (6.126) are real and given by

$$\alpha_\pm = -k \pm \sqrt{k^2 - \omega_n^2} \equiv -k \pm \delta k. \tag{6.129}$$

   The general solutions for $u_n(t)$ in this regime are of the form

$$u_n(t) = e^{-kt}(A_n e^{\delta k\, t} + B_n e^{-\delta k\, t}). \tag{6.130}$$

   Since $\delta k < k$, these solutions are always exponentially decaying with time and show no oscillatory behaviour. In the limit where $k \gg \omega_n$, we find that $\delta k \to k$ and we are left with a solution that is decaying very slowly and one that decays at the rate $2k$.

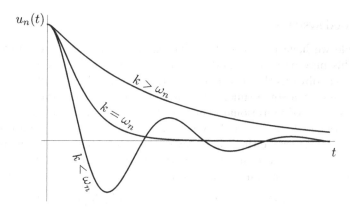

Figure 6.8 Examples of the three possible cases of damping for the same initial conditions.

3. *Critically damped eigenmodes:* Finally, there is a possibility of having $k = \omega_n$ for one eigenmode (or several in the case of degenerate eigenmodes). In this situation, there is only one double root $\alpha = k$ to Eq. (6.126) and the general solution for $u_n(t)$ is given by

$$u_n(t) = e^{-kt}(A_n t + B_n).\qquad(6.131)$$

Much like the overdamped solutions, these solutions do not show any oscillatory behaviour and are exponentially decreasing.

The general behaviour of underdamped, overdamped, and critically damped solutions are shown in Fig. 6.8. It should be noted that a system described by the damped wave equation will contain many eigenmodes and it is possible for a system to simultaneously have some modes that are underdamped, some that are overdamped, and some that are critically damped. In general, it is the eigenmodes with higher frequencies that will display underdamping while the lower frequency modes, if any, will be overdamped.

6.4.3 Driven systems

In many situations, there will be external sources for the wave equation, resulting in a constant or time-dependent inhomogeneity $f(\vec{x}, t)$ that does not depend on the solution $u(\vec{x}, t)$. A case of particular interest occurs in the situation when the external source has a sinusoidal dependence on time

$$f(\vec{x}, t) = F(\vec{x}) \sin(\omega t).\qquad(6.132)$$

In this scenario, we can apply the methods discussed earlier for finding a steady state solution that takes care of this inhomogeneity.

We will consider the general damped situation, which is now of the form

$$(\text{PDE}): \ u_{tt}(\vec{x}, t) + 2k u_t(\vec{x}, t) - c^2 \nabla^2 u(\vec{x}, t) = F(\vec{x}) \sin(\omega t).\qquad(6.133)$$

If we make the ansatz $u(\vec{x}, t) = v(\vec{x}) \sin(\omega t)$, we obtain

$$[-\omega^2 v(\vec{x}) - c^2 \nabla^2 v(\vec{x})] \sin(\omega t) + 2k\omega v(\vec{x}) \cos(\omega t) = F(\vec{x}) \sin(\omega t).\qquad(6.134)$$

In this expression, the time dependent factor $\sin(\omega t)$ only factors out if $k = 0$ and we therefore need to be a little bit more inventive. As series expansions in terms of the eigenfunctions have proven useful before, we express both $F(\vec{x})$ and $u(\vec{x}, t)$ as series expansions in the eigenfunctions $X_n(\vec{x})$ of the Laplace operator with expansion coefficients F_n and u_n, respectively,

$$F(\vec{x}) = \sum_n F_n X_n(\vec{x}), \quad u(\vec{x}, t) = \sum_n u_n(t) X_n(\vec{x}). \tag{6.135}$$

This leads to the ordinary differential equations

$$F_n \sin(\omega t) = u_n''(t) + 2k u_n'(t) + \omega_n^2 u_n(t), \tag{6.136}$$

where $\omega_n^2 = c^2 \lambda_n$ and λ_n is the eigenvalue corresponding to the eigenfunction $X_n(\vec{x})$. These equations have the same problem as the original partial differential equation, i.e., they cannot be solved by the ansatz of a time dependence proportional to $\sin(\omega t)$. Instead, we make the ansatz

$$u_n(t) = A_n \sin(\omega t + \phi_n), \tag{6.137}$$

where A_n is an amplitude and ϕ_n a phase factor. An argument for making this ansatz is that it is again a sinusoidal function of t with the same angular frequency ω as the left-hand side of Eq. (6.136), but has an additional freedom in the phase factor that we can adapt in order for both sides of the equation to have the same time-dependence. Inserting the ansatz into the differential equation results in

$$F_n \sin(\omega t) = [(\omega_n^2 - \omega^2) \sin(\omega t + \phi_n) + 2k\omega \cos(\omega t + \phi_n)] A_n$$
$$= \sqrt{(\omega_n^2 - \omega^2)^2 + (2k\omega)^2} \sin(\omega t + \phi_n + \alpha_n), \tag{6.138}$$

where $\alpha_n = \operatorname{atan}(2k\omega / (\omega_n^2 - \omega^2))$. Identifying the amplitudes and phases of both sides of this equation results in

$$A_n = \frac{F_n}{\sqrt{(\omega_n^2 - \omega^2)^2 + (2k\omega)^2}}, \tag{6.139a}$$

$$\phi_n = -\alpha_n = \operatorname{atan}\left(\frac{2k\omega}{\omega^2 - \omega_n^2}\right). \tag{6.139b}$$

We note that the amplitude A_n is generally maximal for a given source amplitude F_n when the function

$$R_n(\omega) = (\omega_n^2 - \omega^2)^2 + (2k\omega)^2 \tag{6.140}$$

is minimised. If the damping is very small such that $k \to 0$, the corresponding amplitude $A_n \to \infty$ as $\omega \to \omega_n$ while, for an arbitrary k, we find that

$$R_n'(\omega_{rn}) = 0 \quad \Longrightarrow \quad \omega_{rn}^2 = \omega_n^2 - 2k^2 \tag{6.141}$$

defines the *resonant angular frequency* ω_{rn} of the nth eigenmode such that the amplitude A_n is maximal when $\omega = \omega_{rn}$. In the case when $k > \omega_n / \sqrt{2}$, i.e., for sufficiently strong damping in relation to the natural angular frequency ω_n, there is no real solution for the resonant angular frequency and the amplitude always decreases with increasing ω. Note that this occurs already in the underdamped region. The behaviour of the amplitude A_n as a function of the driving angular frequency ω is shown in Fig. 6.9. The phase shift ϕ_n describes when the driven oscillation in the nth mode reaches its maximum in relation to when the driving source does. The correct branch of the atan function compatible with Eq. (6.138) is that for which $-\pi < \phi_n < 0$, such that the shift always represents a phase lag, i.e., the

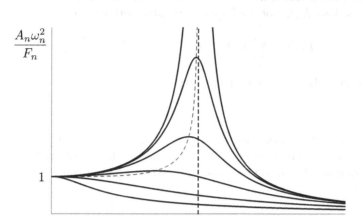

Figure 6.9 The behaviour of the amplitude A_n as a function of the driving angular frequency ω for different values of the damping coefficient k. The maximal amplitude corresponds to the undamped case $k = 0$. The thin dashed line shows how the position and amplitude of the resonant angular frequency ω_{rn} changes as k is varied.

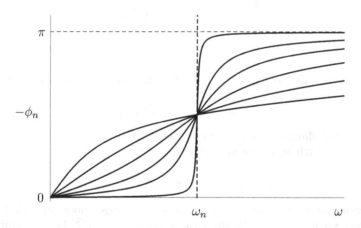

Figure 6.10 The dependence of the phase shift ϕ_n on the driving angular frequency ω for different values of the damping coefficient k. The phase shift is a step function for $k = 0$ and gradually moves away from this shape as k increases.

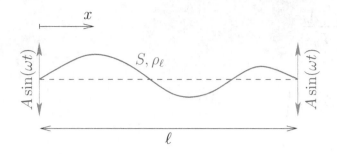

Figure 6.11 Oscillations on a string with tension S and linear density ρ_ℓ can be driven by moving the endpoints to enforce the boundaries moving as $u(0,t) = u(\ell,t) = A\sin(\omega t)$.

maximum of the oscillation amplitude occurs after the maximum of the driving force. The exception is the undamped case with $k = 0$ for which we find that $\phi_n = 0$ whenever $\omega < \omega_n$ and $\phi_n = -\pi$ when $\omega > \omega_n$. The phase shift as a function of the driving angular frequency is shown in Fig. 6.10.

In a system with many different natural angular frequencies ω_n, each one that satisfies $\omega_n > \sqrt{2}k$ has its own resonant angular frequency. This implies that the system may be subjected to resonant driven oscillations at several different values of the driving angular frequency ω. As the damping k is increased, the lower frequency eigenmodes will no longer exhibit a resonant angular frequency, starting with the fundamental mode, which has the lowest natural angular frequency.

Example 6.14 Consider a string with tension S, linear density ρ_ℓ, and length ℓ, see Fig. 6.11. In the absence of an external force, the transversal movement of the string follows the source free wave equation. If we start moving the endpoints in the transversal direction with an angular frequency ω and amplitude A, we can model the movement of the string as

$$\text{(PDE)} : u_{tt}(x,t) - c^2 u_{xx}(x,t) = 0, \tag{6.142a}$$
$$\text{(BC)} : u(0,t) = u(\ell,t) = A\sin(\omega t), \tag{6.142b}$$

where $c^2 = S/\rho_\ell$. We can rewrite this on the form considered above by introducing

$$v(x,t) = u(x,t) - A\sin(\omega t), \tag{6.143}$$

leading to

$$v_{tt}(x,t) - c^2 v_{xx}(x,t) = A\omega^2 \sin(\omega t) \tag{6.144}$$

and homogeneous Dirichlet boundary conditions for $v(0,t)$ and $v(\ell,t)$. The eigenfunctions of the operator $-\partial_x^2$ satisfying the boundary conditions are $X_n(x) = \sin(k_n x)$, where $k_n = \pi n/\ell$ and the resulting steady state solution therefore becomes

$$v(x,t) = 2A\sin(\omega t) \sum_{n=1}^{\infty} \frac{1 - (-1)^n}{\pi n} \frac{\sin(k_n x)}{\omega_n^2/\omega^2 - 1}, \tag{6.145}$$

where $\omega_n = ck_n$. Due to the absence of a damping term, the amplitude of the nth eigenmode diverges as $\omega \to \omega_n$. This may be remedied by introducing a small damping term $k \ll \omega_1$,

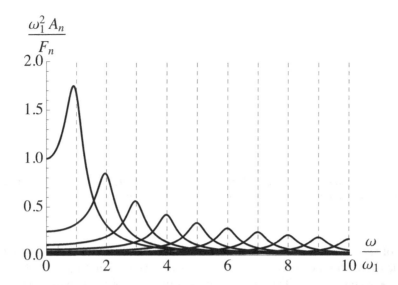

Figure 6.12 The amplitudes of the different modes in the driven one-dimensional wave equation and homogeneous Dirichlet boundary conditions as a function of the driving angular frequency ω. The natural angular frequencies are given by $\omega_n = n\omega_1$ and the damping coefficient is set to $k = 0.3\omega_1$. The vertical dashed lines show the positions of the natural angular frequencies. Note that the resonance appears slightly below the natural angular frequencies for the first few modes and that the relatively large damping results in the resonant amplitude for the first mode being less than a factor two times the amplitude at zero driving angular frequency.

which will result in large, but finite, amplitudes. With the damping term being small, the resonances still occur at $\omega \simeq \omega_n$. We also note that only the eigenmodes with odd n have non-zero amplitude. This is due to the reflection symmetry of the problem around the string's mid-point implying that the even n eigenmodes, whose eigenfunctions are odd with respect to the midpoint, cannot be excited in this fashion. If we instead decide to move only one of the ends, we would excite all of the eigenmodes.

Experimental demonstrations similar to the one described here are often used in high-school classes in order to demonstrate the concept of standing waves. Changing the driving angular frequency ω changes the eigenmode that experiences resonant behaviour of its amplitude, allowing the different eigenmodes to be identified by eye. To be overly picky, the resonant angular frequencies do not occur exactly at the natural angular frequencies ω_n. However, in order for the amplitude to be appreciably larger than the input amplitude at the endpoints, the damping must be so small that the difference between the resonant angular frequency and the natural angular frequency cannot be distinguished, see Figs. 6.12 and 6.13.

6.5   TERMINATING THE SERIES

So far in this chapter, we have mainly encountered solutions in terms of series containing an infinite number of eigenmodes to the Laplace operator or other combinations of Sturm–Liouville operators. The exception to this in the approaches discussed so far would be when

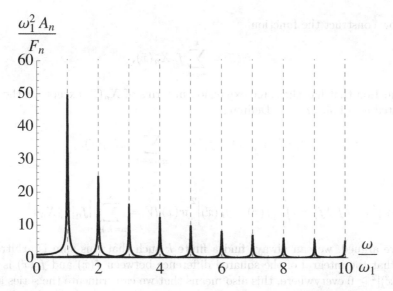

Figure 6.13 The same as Fig. 6.12 but for $k = 0.01\omega_1$. Note how the resonances have become much stronger and that they gain a large factor compared to the amplitude at zero driving angular frequency. By tuning the driving angular frequency to one of the natural angular frequencies, the corresponding mode will completely dominate the solution.

the inhomogeneities are such that only a finite number of the resulting ordinary differential equations are inhomogeneous, which is a rather special case. Since it may be quite cumbersome to deal with infinite series if we wish to compute a numerical value, let us discuss the implications of terminating the series at some finite value of n. We start this discussion by considering a function $f(\vec{x})$ that may be expanded in terms of the eigenfunctions $X_n(\vec{x})$ of some combination of Sturm–Liouville operators

$$f(\vec{x}) = \sum_n f_n X_n(\vec{x}). \tag{6.146}$$

We consider the function space such that $f(\vec{x})$ is square integrable with weight function $w(\vec{x})$ and hence

$$\|f\|^2 = \int_V |f(\vec{x})|^2 \, w(\vec{x}) dV = \langle f, f \rangle, \tag{6.147}$$

where $w(\vec{x})$ is the appropriate weight function for the eigenfunctions $X_n(\vec{x})$ to be orthogonal. Using the expansion of $f(\vec{x})$ in terms of the eigenfunctions, its norm may be rewritten as

$$\langle f, f \rangle = \sum_{nm} f_n^* f_m \langle X_n, X_m \rangle = \sum_{n=1}^{\infty} |f_n|^2 \|X_n\|^2. \tag{6.148}$$

In order for this series to be convergent, for any $\varepsilon > 0$ there must exist a number N such that for all $M > N$

$$\left| \|f\|^2 - \sum_{n=1}^{M} |f_n|^2 \|X_n\|^2 \right| < \varepsilon, \tag{6.149}$$

implying that

$$\sum_{n=M+1}^{\infty} |f_n|^2 \|X_n\|^2 < \varepsilon. \tag{6.150}$$

We can now construct the function

$$\tilde{f}_k(\vec{x}) = \sum_{n=1}^{k} f_n X_n(\vec{x}), \tag{6.151}$$

i.e., the function that has the same expansion in terms of $X_n(\vec{x})$ except for the fact that it is terminated at a finite $n = k$. Defining

$$\delta f(\vec{x}) = f(\vec{x}) - \tilde{f}_k(\vec{x}) = \sum_{n=k+1}^{\infty} f_n X_n(\vec{x}), \tag{6.152}$$

we find that

$$\|\delta f\|^2 = \langle \delta f, \delta f \rangle = \int_V \left| f(\vec{x}) - \tilde{f}_k(\vec{x}) \right|^2 w(\vec{x}) dV = \sum_{n=k+1}^{\infty} |f_n|^2 \|X_n\|^2. \tag{6.153}$$

As we have argued, we can always find a finite k such that this sum is arbitrarily small, implying that the integral of the squared difference between $f(\vec{x})$ and $\tilde{f}_k(\vec{x})$ is as well. As $|f(\vec{x}) - \tilde{f}(\vec{x})|^2 \geq 0$ everywhere, this also means that we can truncate the series at a finite n and still obtain an approximation that should be reasonably close to $f(\vec{x})$. Naturally, the actual n at which we can do this depends on the function $f(\vec{x})$. It should be noted that the above argument does not hold for functions and distributions that are not square integrable as they may formally have infinite norm.

Since the large n eigenfunctions are rapidly oscillating, we are essentially throwing away any small scale structure when approximating a function by truncating a series in this manner. As a result, the smallest features of the approximated function that can be reproduced are given by the highest n eigenfunctions that we choose to keep in the expansion.

Example 6.15 In the simplest case, let us consider the function $f(x) = \theta(1/2 - x)$, where θ is the Heaviside step function, on the interval $0 < x < 1$ and its expansion into the functions $X_n(x) = \sin(\pi n x)$. The infinite series expansion is given by

$$f(x) = \sum_{n=1}^{\infty} \frac{2}{\pi n} [1 - \cos(\pi n/2)] \sin(\pi n x). \tag{6.154}$$

Approximating this with

$$\tilde{f}_k(x) = \sum_{n=1}^{k} \frac{2}{\pi n} [1 - \cos(\pi n/2)] \sin(\pi n x), \tag{6.155}$$

we find that

$$\int_0^1 \left| f(x) - \tilde{f}_k(x) \right|^2 dx = \sum_{n=k+1}^{\infty} \frac{2}{\pi^2 n^2} [1 - \cos(\pi n/2)]^2, \tag{6.156}$$

which can be made arbitrarily small by selecting a large enough k. The function $f(x)$ is shown along with $\tilde{f}_k(x)$ for some choices of k in Fig. 6.14. The typical feature size of the function $\sin(\pi n x)$ is of length $1/n$ and we therefore expect that the size of the region where the sharp feature of $f(x)$ at $x = 1/2$ will not be well reproduced is proportional to $1/k$. The appearance of strong oscillations in this region as the series is terminated is known as *Gibbs' phenomenon*.

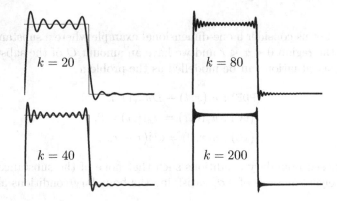

Figure 6.14 The approximations $\tilde{f}_k(x)$ of the step function $\theta(1/2 - x)$ for different values of k. The oscillatory overshooting near the steps is called Gibbs' phenomenon. The appearance of Gibbs' phenomenon in the beginning of the interval is due to the approximation using functions that satisfy homogeneous Dirichlet conditions.

6.5.1 Heat and diffusion equations

In the case of the homogeneous heat equation with homogeneous boundary conditions, we found that the series expansion of the solution was of the form

$$u(\vec{x}, t) = \sum_{n=1}^{\infty} u_n(0) e^{-a\lambda_n t} X_n(\vec{x}). \tag{6.157}$$

Although the entire solution tends to zero, the eigenmodes corresponding to the smaller eigenvalues do so slower than the ones corresponding to larger eigenvalues. If we consider the function

$$v(\vec{x}, t) = e^{a\lambda_1 t} u(\vec{x}, t) = \sum_{n=1}^{\infty} u_n(0) e^{-a(\lambda_n - \lambda_1)t} X_n(\vec{x}), \tag{6.158}$$

this function has one eigenmode that will have a constant expansion coefficient, namely the $n = 1$ eigenmode, while the other eigenmodes show an exponential decay. The function $v(\vec{x}, t)$ will have the same shape as $u(\vec{x}, t)$ for every t and we can consider how important the higher eigenmodes are for this shape by comparing it with

$$\tilde{v}_1(\vec{x}, t) = u_1(0) X_n(\vec{x}), \tag{6.159}$$

which is just the series expansion truncated after the first term. We find that

$$\|v(\vec{x}, t) - \tilde{v}_1(\vec{x}, t)\|^2 = \sum_{n=2}^{\infty} e^{-a(\lambda_n - \lambda_1)t} |u_n(0)|^2 \|X_n\|^2$$

$$\leq e^{-a(\lambda_2 - \lambda_1)t} \sum_{n=2}^{\infty} |u_n(0)|^2 \|X_n\|^2$$

$$= e^{-a(\lambda_2 - \lambda_1)t} \|v(\vec{x}, 0) - \tilde{v}_1(\vec{x}, 0)\|^2 \tag{6.160}$$

and therefore the importance of the eigenmodes with larger eigenvalues for the shape of the solution also decays exponentially. Generally, eigenmodes for which $a(\lambda_n - \lambda_1)t \gg 1$ may be safely ignored.

Example 6.16 Let us consider a one-dimensional example where a substance diffuses with diffusivity D in the region $0 < x < \ell$ and we have an amount Q of the substance at $x = x_0$ at time $t = 0$. This situation can be modelled as the problem

$$\text{(PDE)} : u_t(x,t) - D u_{xx}(x,t) = 0, \tag{6.161a}$$

$$\text{(BC)} : u_x(0,t) = u_x(\ell,t) = 0, \tag{6.161b}$$

$$\text{(IC)} : u(x,0) = Q\delta(x - x_0). \tag{6.161c}$$

We have here chosen boundary conditions such that none of the substance diffuses out of the region. The eigenfunctions of $-\partial_x^2$ satisfying the boundary conditions are given by

$$X_n(x) = \cos(k_n x), \tag{6.162}$$

where $k_n = \pi n/\ell$ and n runs from zero to infinity. The solution to this problem is given in terms of the series

$$u(x,t) = Q \sum_{n=0}^{\infty} \frac{\cos(k_n x_0)\cos(k_n x)}{\|X_n\|^2} e^{-Dk_n^2 t}, \tag{6.163}$$

where

$$\|X_m\|^2 = \begin{cases} \ell, & (n = 0) \\ \frac{\ell}{2}, & (n \neq 0) \end{cases}. \tag{6.164}$$

The approximation

$$\tilde{u}(x,t) = Q\frac{\cos(k_0 x_0)\cos(k_0 x)}{\|X_0\|^2} e^{-Dk_0^2 t} = \frac{Q}{\ell} \tag{6.165}$$

is quite obviously horrible at $t = 0$. However, as time goes by and the eigenmodes with $n \neq 0$ decay away, the entire solution tends to the constant solution where the substance is evenly distributed throughout the region.

6.5.2 Wave equation

Unlike the heat equation, the solutions to the wave equations are oscillatory rather than exponentially decaying. This implies that in order to have a reasonable approximation after truncating the series, we must include all of the terms that were necessary to have a reasonable approximation at $t = 0$. This statement is modified when we instead study the damped wave equation. As we have seen, the introduction of a damping term generally leads to the amplitudes of the eigenmodes being subject to exponential decay just as for the heat equation. All of the eigenmodes with $\omega_n \geq k$ are critically damped or underdamped and therefore exhibit an exponential suppression e^{-kt}. For times $t \gg 1/k$, this suppression is significant and we can truncate the series by excluding all critically damped and underdamped modes to make the approximation

$$u(\vec{x},t) = \sum_n u_n(t) X_n(\vec{x}) \simeq \sum_{\omega_n < k} u_n(t) X_n(\vec{x}). \tag{6.166}$$

Since all of the remaining modes are overdamped, the coefficients $u_n(t)$ may be expressed through Eq. (6.130) and we find

$$u(\vec{x}, t) \simeq \sum_{\omega_n < k} e^{-kt}(A_n e^{\delta k_n t} + B_n e^{-\delta k_n t}) X_n(\vec{x}) \simeq \sum_{\omega_n < k} A_n e^{-(k - \delta k_n)t} X_n(\vec{x}), \quad (6.167)$$

where $\delta k_n = \sqrt{k^2 - \omega_n^2}$ and we have used the fact that the term proportional to B_n decays even faster than the underdamped modes. We now have a situation that is completely analogous to that encountered in the case of the heat equation with $a\lambda_n$ replaced by $k - \delta k_n$. Since δk_n is larger for the modes with lower frequencies, the mode corresponding to the fundamental angular frequency will exhibit the slowest decay and be dominant at large times.

The behaviour of the damped wave equation for large times is very similar to that of the diffusion equation. The eigenmodes with larger eigenvalues all become irrelevant and if we consider the initial conditions of the strongly overdamped modes for which $\omega_n \ll k$, and therefore $\delta k \to k$, we find that

$$A_n = u_n(0) + \frac{u_n'(0)}{2k}, \quad B_n = -\frac{u_n'(0)}{2k}. \quad (6.168)$$

As already mentioned, the term with the coefficient B_n will be strongly suppressed, while the term with the A_n will remain. The evolution of the solution for large times will therefore be equivalent to that of the heat equation with $a\lambda_n \to k - \delta k_n$ and the initial condition

$$u(\vec{x}, 0) = \sum_{\omega_n \ll k} \left[u_n(0) + \frac{u_n'(0)}{2k} \right] X_n(\vec{x}). \quad (6.169)$$

The underlying reason for this is that when $k - \delta k_n$ is small, the time derivative $u_n'(t)$ is suppressed by $k - \delta k_n$ while the second time derivative $u_n''(t)$ is suppressed by $(k - \delta k_n)^2$ in Eq. (6.124). The second time derivative therefore becomes negligible and the damped wave equation reduces to the heat equation for these modes

$$u_n''(t) + 2ku_n'(t) + \omega_n^2 u_n(t) \simeq 2ku_n'(t) + \omega_n^2 u_n(t) = 0. \quad (6.170a)$$

This may be rewritten as

$$u_n'(t) + \frac{\omega_n^2}{2k} u_n(t) = 0, \quad (6.170b)$$

which is the ordinary differential equation describing the evolution of the nth eigenmode of the heat equation with $a\lambda_n = \omega_n^2/2k$. This is consistent with the replacement $a\lambda_n \to k - \delta k_n$ for $\omega_n \ll k$ as

$$k - \delta k_n = k - \sqrt{k^2 - \omega_n^2} \simeq \frac{\omega_n^2}{2k}. \quad (6.171)$$

Naturally, the discussion above requires that there exist modes that are overdamped. If this is not the case, all modes will be exponentially suppressed by the factor e^{-kt} and the solution will tend to zero.

6.6 INFINITE DOMAINS

In this chapter we have so far been concerned with problems defined in a bounded spatial region, leading to the Sturm–Liouville operators of interest having a countable set of eigenfunctions with a corresponding discrete set of eigenvalues. In many situations occurring in physics, it will be motivated to consider unbounded regions such as the real line or the first quadrant of $\mathbb{R}^2$. In such regions, the methods described here will still work to some extent, but the spectrum of the Sturm–Liouville operators will be continuous rather than discrete.

For our purposes in this section, we will consider a problem of the form

$$\text{(PDE)}: \hat{L}u(\vec{x}) = \lambda u(\vec{x}), \qquad\qquad (\vec{x} \in V) \qquad\qquad (6.172a)$$

$$\text{(BC)}: (\alpha + \beta \vec{n} \cdot \nabla)u(\vec{x}) = 0, \qquad\qquad (\vec{x} \in S) \qquad\qquad (6.172b)$$

where $\hat{L}$ is a combination of Sturm–Liouville operators in different coordinates, V is an unbounded region and S its boundary. In addition, we require that the solutions are bounded as $|\vec{x}| \to \infty$

$$\lim_{|\vec{x}| \to \infty} |u(\vec{x})| < \infty. \qquad\qquad (6.172c)$$

The boundary conditions of Eq. (6.172b) will generally not be strong enough to constrain the eigenvalues of $\hat{L}$ to a discrete set as was the case when the region V was bounded. Instead, there will be a continuum of eigenvalues λ that may or may not be degenerate. The corresponding eigenfunctions $X_{\vec{k}}(\vec{x})$ satisfy the eigenfunction relation

$$\hat{L}X_{\vec{k}}(\vec{x}) = \lambda(\vec{k})X_{\vec{k}}(\vec{x}) \qquad\qquad (6.173)$$

as well as the boundary conditions. Here, the vector $\vec{k}$ contains a general set of parameters that uniquely identify the different eigenfunctions. This vector will be restricted to some region $\mathcal{K}$. If the eigenfunctions are not degenerate, we may use the eigenvalue λ itself and $\mathcal{K}$ is then the spectrum of the $\hat{L}$. The functions $X_{\vec{k}}(\vec{x})$ will generally not be square integrable, but will satisfy the inner product relation

$$\langle X_{\vec{k}}, X_{\vec{k}'} \rangle = \int_V X_{\vec{k}}(\vec{x})^* X_{\vec{k}'}(\vec{x})w(\vec{x})dV = 0 \qquad\qquad (6.174)$$

for all $\vec{k} \neq \vec{k}'$, where $w(\vec{x})$ is the weight function associated with the operator $\hat{L}$. In order for this relation to be well defined, it should generally be seen as a distribution on $\mathcal{K}$ and the general form of the inner product will be

$$\langle X_{\vec{k}}, X_{\vec{k}'} \rangle = \int_V X_{\vec{k}}(\vec{x})^* X_{\vec{k}'}(\vec{x})w(\vec{x})dV = N(\vec{k})\delta(\vec{k} - \vec{k}'), \qquad\qquad (6.175)$$

where $N(\vec{k})$ is a function relating to the normalisation of the eigenfunctions. We can always rescale the eigenfunctions as

$$\tilde{X}_{\vec{k}}(\vec{x}) = \frac{X_{\vec{k}}(\vec{x})}{\sqrt{N(\vec{k})}}. \qquad\qquad (6.176)$$

The new eigenfunctions $\tilde{X}_{\vec{k}}(\vec{x})$ are then normalised as

$$\left\langle \tilde{X}_{\vec{k}}, \tilde{X}_{\vec{k}'} \right\rangle = \delta(\vec{k} - \vec{k}'). \qquad\qquad (6.177)$$

This relation is the *orthogonality relation* for the eigenfunctions of the symmetric operator $\hat{L}$.

Just as we could express functions in bounded regions as linear combinations of eigenfunctions to a Sturm–Liouville problem, we can express functions in the unbounded regions as linear combinations of the functions discussed above. The big difference is that the most general linear combination in the discrete case was a sum, where the summation was taken over all of the eigenfunctions, and in order to include all of the eigenfunctions in the unbounded case, we instead need to integrate over all eigenfunctions. In other words, we can write a function $f(\vec{x})$ as

$$f(\vec{x}) = \int_{\mathcal{K}} \tilde{f}(\vec{k}) X_{\vec{k}}(\vec{x}) d\mathcal{K}. \tag{6.178}$$

Comparing with the series expansion

$$f(\vec{x}) = \sum_{n} f_n X_n(\vec{x}) \tag{6.179}$$

in the bounded case, the analogy is made clear by the correspondence of the quantities

$$\vec{k} \longleftrightarrow n, \tag{6.180a}$$

$$\int_{\mathcal{K}} \cdots \longleftrightarrow \sum_{n} \cdots, \tag{6.180b}$$

$$\tilde{f}(\vec{k}) d\mathcal{K} \longleftrightarrow f_n, \tag{6.180c}$$

$$X_{\vec{k}}(\vec{x}) \longleftrightarrow X_n(\vec{x}). \tag{6.180d}$$

Just as for the bounded case, the expansion coefficients $\tilde{f}(\vec{k})$ may be found by taking the inner product between $f(\vec{x})$ and one of the eigenfunctions according to

$$\langle X_{\vec{k}'}, f \rangle = \int_{\mathcal{K}} \tilde{f}(\vec{k}) \langle X_{\vec{k}'}, X_{\vec{k}} \rangle d\mathcal{K} = \int_{\mathcal{K}} \tilde{f}(\vec{k}) N(\vec{k}) \delta(\vec{k} - \vec{k}') d\mathcal{K} = \tilde{f}(\vec{k}') N(\vec{k}'), \tag{6.181a}$$

implying that

$$\tilde{f}(\vec{k}) = \frac{\langle X_{\vec{k}'}, f \rangle}{N(\vec{k})}. \tag{6.181b}$$

The function $\tilde{f}(\vec{k})$ is often referred to as a *transform* of the function $f(\vec{x})$. The two functions convey the same information but one is a function on the spatial domain V, whereas the other is a function on the set $\mathcal{K}$.

Example 6.17 The first example that comes to mind when looking for problems such as the ones described above is letting $\hat{L} = -\partial_x^2$ in one dimension where $-\infty < x < \infty$. This region has no boundaries and therefore no boundary conditions at any finite x. We are left with the singular Sturm–Liouville problem

$$(\text{PDE}) : \ -\partial_x^2 X(x) = \lambda X(x), \tag{6.182a}$$

$$(\text{BC}) : \ |X(\pm\infty)| < \infty. \tag{6.182b}$$

For $\lambda < 0$, we find that

$$X(x) = Ae^{-kx} + Be^{kx}, \tag{6.183}$$

where $k^2 = -\lambda$. None of these solutions are bounded at both infinities and we conclude that $\lambda \geq 0$ for all solutions. For $\lambda = 0$, the general solution is given by

$$X(x) = Ax + B. \tag{6.184}$$

With the requirement of the solution being bounded at the infinities, the constant A must be equal to zero and we are left with a constant function

$$X_0(x) = \frac{1}{2\pi},$$ (6.185)

where the normalisation is purely conventional. Finally, for $\lambda = k^2 > 0$, the general solution is

$$X(x) = Ae^{ikx} + Be^{-ikx}.$$ (6.186)

Both of these solutions are bounded at the infinities and are therefore allowed. We now need to construct $\mathcal{K}$ and we do so by assigning

$$X_k(x) = \frac{1}{2\pi}e^{ikx}$$ (6.187)

for $-\infty < k < \infty$. Again, the normalisation of these functions is arbitrary and the choice of the factor $1/2\pi$ is purely conventional. The case $\lambda = 0$ is here covered by $k = 0$. At the same time, the first term in Eq. (6.186) is covered by $k > 0$, while the second term is covered by $k < 0$. The allowed region $\mathcal{K}$ is therefore the entire real line and we also have

$$\lambda(k) = k^2.$$ (6.188)

The inner product between two of these eigenfunctions is given by

$$\langle X_k, X_{k'} \rangle = \frac{1}{2\pi}\delta(k - k') \quad \Longrightarrow \quad N(k) = \frac{1}{2\pi}.$$ (6.189)

The general expression of a function $f(x)$ in terms of the eigenfunctions is given by

$$f(x) = \frac{1}{2\pi}\int_{-\infty}^{\infty} \tilde{f}(k)e^{ikx}dk$$ (6.190a)

while the function $\tilde{f}(k)$ is found to be

$$\tilde{f}(k) = \frac{2\pi}{2\pi}\int_{-\infty}^{\infty} f(x)e^{-ikx}dx = \int_{-\infty}^{\infty} f(x)e^{-ikx}dx.$$ (6.190b)

Just as the series expansion of a function on a bounded interval in terms of the eigenfunctions of $-\partial_x^2$ was found to be a Fourier series, we may recognise this transform in terms of the eigenfunctions of $-\partial_x^2$ on the entire real line as the *Fourier transform*.

6.6.1 Domains with a boundary

The example we have just seen in the Fourier transform was a special case in that its definition was the entire real line. In many other situations, we will be dealing with a region that has a finite boundary in one direction. In these situations, we are left with a problem of the form

$$(\text{PDE}) : \hat{L}X(x) = \lambda X(x), \tag{6.191a}$$

$$(\text{BC}) : \alpha X(a) + \beta X'(a) = 0, \quad \lim_{x \to \infty} |X(x)| < \infty, \tag{6.191b}$$

where we have put a limit $x > a$ on the domain of the functions $X(x)$. The corresponding problem with an upper boundary and no lower boundary on x is conceptually equivalent so we will concentrate on this formulation. In these situations, the single boundary condition will restrict the number of possible solutions, but the spectrum may still be continuous depending on the operator $\hat{L}$.

6.6.1.1 The Fourier sine and cosine transforms

A special case of the above is obtained when we let $a = 0$ and study the Sturm–Liouville operator $-\partial_x^2$. As was argued in Example 6.17, studying the corresponding problem on the entire real line we ended up with the Fourier transform. For concreteness, let us work with a homogeneous Dirichlet condition at $x = 0$ to obtain

$$(\text{ODE}) : X''(x) + \lambda X(x) = 0, \tag{6.192a}$$

$$(\text{BC}) : X(0) = 0. \tag{6.192b}$$

The requirement of being bounded at $x \to \infty$ along with the boundary condition results in discarding all solutions with $\lambda \leq 0$. We are left with the case $\lambda = k^2$ for which

$$X(x) = A \sin(kx) + B \cos(kx), \tag{6.193}$$

where the boundary condition now implies that $B = 0$. We may therefore take $k > 0$ and consider the non-degenerate solutions

$$X_k(x) = \frac{2}{\pi} \sin(kx), \tag{6.194}$$

where we have chosen the normalisation by convention. The inner product on the functions on the interval $x > 0$ is given by

$$\langle f, g \rangle = \int_0^\infty f(x)^* g(x) dx, \tag{6.195}$$

which implies that

$$\langle X_k, g \rangle = \frac{2}{\pi} \int_0^\infty \sin(kx) g(x) dx = \frac{1}{i\pi} \int_0^\infty (e^{ikx} - e^{-ikx}) g(x) dx$$

$$= \frac{1}{i\pi} \int_{-\infty}^\infty e^{ikx} \bar{g}(x) dx, \tag{6.196}$$

where we have defined $\bar{g}(x) = g(x)$ for $x > 0$ and $\bar{g}(x) = -g(-x)$ for $x < 0$, i.e., $\bar{g}(x)$ is an odd extension of $g(x)$ to the entire real line. In particular, for $g(x) = X_{k'}(x) = 2\sin(k'x)/\pi$ we find

$$\langle X_k, X_{k'} \rangle = -\frac{1}{\pi^2} \int_{-\infty}^\infty e^{ikx} (e^{ik'x} - e^{-ik'x}) dx$$

$$= -\frac{2}{\pi} [\delta(k + k') - \delta(k - k')] = \frac{2}{\pi} \delta(k - k'), \tag{6.197}$$

where the last step follows from both k and k' being positive. Consequently, $N(k) = 2/\pi$ in this case.

Writing down the transform expression, a function on $x > 0$ for which $f(0) = 0$ or, equivalently, an odd function on the entire real line, may be written in terms of the functions $X_k(x)$ as

$$f(x) = \frac{2}{\pi} \int_0^\infty \tilde{f}_s(k) \sin(kx) dx, \tag{6.198a}$$

where we have introduced the *Fourier sine transform* $\tilde{f}_s(k)$ with a subscript s to separate it from the Fourier transform seen in Example 6.17. Taking the inner product with X_k, we can express the Fourier sine transform as

$$\tilde{f}_s(k) = \int_0^\infty \sin(kx) f(x) dx. \tag{6.198b}$$

Expressions similar to these may be obtained for the case of homogeneous Neumann boundary conditions with the exception that the cosine term survives rather than the sine term. In this situation, the proper expansion to the real line will be an even function rather than an odd (see Problem 6.44), leading to the Fourier cosine transform $\tilde{f}_c(k)$ for functions on $x > 0$ satisfying the Neumann conditions or, equivalently, for even functions on the real line. As any function $f(x)$ may be written as a sum of one odd and one even function, the Fourier sine and cosine transforms can be combined such that

$$f(x) = \frac{2}{\pi} \int_0^\infty [\tilde{f}_s(k) \sin(kx) + \tilde{f}_c(k) \cos(kx)] dk. \tag{6.199}$$

This should not be very surprising as the sine and cosine are linearly independent superpositions of the functions $e^{\pm ikx}$.

6.6.1.2 Hankel transforms

In some situations, mainly when dealing with domains that are well described in polar coordinates, we will be faced with problems of the form

$$(\text{ODE}): \; -R''(\rho) - \frac{1}{\rho} R'(\rho) + \frac{\nu^2}{\rho^2} R(\rho) = \lambda R(\rho), \qquad (\rho > 0) \tag{6.200a}$$

$$(\text{BC}): \; |R(0)| < \infty. \tag{6.200b}$$

We recognise the differential equation here as Bessel's differential equation and the boundary condition at $\rho = 0$ implies that the solutions we seek are Bessel functions of the first kind

$$R_k(\rho) = J_\nu(k\rho), \tag{6.201}$$

where $\lambda = k^2$. The Bessel functions satisfy the orthogonality relation

$$\langle R_k, R_{k'} \rangle = \int_0^\infty \rho J_\nu(k\rho) J_\nu(k'\rho) d\rho = \frac{1}{k} \delta(k - k'), \tag{6.202}$$

where the inner product has the weight function $w(\rho) = \rho$ due to the form of Bessel's differential operator. This means that $N(k) = 1/k$ in Eq. (6.175). For a function $f(\rho)$ defined for $\rho \geq 0$ we can now define the *Hankel transform* of order ν as

$$\tilde{f}_\nu(k) = \int_0^\infty \rho f(\rho) J_\nu(k\rho) d\rho \tag{6.203a}$$

with the corresponding inversion formula

$$f(\rho) = \int_0^\infty k\tilde{f}_\nu(k)J_\nu(k\rho)dk. \tag{6.203b}$$

Note that the Hankel transforms for different values of ν are independent and different from each other.

6.7 TRANSFORM SOLUTIONS

Transforms can be used much in the same way as we used series expansions. By expanding a solution in terms of a transform related to the relevant differential operator, we can exchange partial differential operators for a multi-variable function for ordinary differential equations for transform. We can start by studying the homogeneous heat equation

$$\text{(PDE)} : u_t(x,t) - au_{xx}(x,t) = 0, \tag{6.204a}$$

$$\text{(IC)} : u(x,0) = f(x) \tag{6.204b}$$

on the real line. For any fixed value of t, we can express $u(x,t)$ in terms of its Fourier transform

$$u(x,t) = \frac{1}{2\pi} \int_{-\infty}^\infty \tilde{u}(k,t)e^{ikx}dk. \tag{6.205}$$

Inserted into the heat equation, we find that

$$\tilde{u}_t(k,t) + ak^2\tilde{u}(k,t) = 0, \tag{6.206}$$

which is an ordinary differential equation for $u(k,t)$ for every fixed value of k. This is completely analogous to how we obtained ordinary differential equations for the expansion coefficients $u_n(t)$ when we dealt with series solutions, in fact, it is the very same differential equation we obtained when solving the heat equation in that fashion. We already know the solution to this differential is of the form

$$\tilde{u}(k,t) = \tilde{u}(k,0)e^{-ak^2t} = \tilde{f}(k)e^{-ak^2t}, \tag{6.207}$$

where $\tilde{f}(k)$ is the Fourier transform of the initial condition. The solution $u(x,t)$ is therefore given by the integral expression

$$u(x,t) = \frac{1}{2\pi} \int_{-\infty}^\infty \tilde{f}(k)e^{ikx-ak^2t}dx. \tag{6.208}$$

In many cases, the integrals appearing in the transform solutions are analytically solvable either by explicit computation or through knowledge about the Fourier transforms of several template functions. Such knowledge is often listed in tables of Fourier transforms (see Appendix A.4).

Example 6.18 Consider the one-dimensional diffusion of a substance in a medium with diffusivity D, where an amount Q of the substance is originally located at the point $x = x_0$ at time $t = 0$. This situation is described by Eqs. (6.204) with $f(x) = Q\delta(x - x_0)$. We can find the Fourier transform of $f(x)$ as

$$\tilde{f}(k) = \int_{-\infty}^\infty Q\delta(x - x_0)e^{-ikx}dx = Qe^{-ikx_0}. \tag{6.209}$$

By the argumentation above, we find that the Fourier transform of the solution at time t is given by

$$\tilde{u}(k,t) = Q e^{-ak^2 t - ikx_0} \tag{6.210a}$$

and therefore

$$u(x,t) = \frac{Q}{2\pi} \int_{-\infty}^{\infty} e^{-ak^2 t + ik(x-x_0)} dk. \tag{6.210b}$$

This integral can be computed analytically. We start by making the change of variables $s = k\sqrt{at}$, leading to $dk = ds/\sqrt{at}$ and hence

$$u(x,t) = \frac{Q}{2\pi\sqrt{at}} \int_{-\infty}^{\infty} e^{-s^2 + is(x-x_0)/\sqrt{at}} ds. \tag{6.211}$$

Completing the square in the exponent we find that

$$-s^2 + is\frac{x-x_0}{\sqrt{at}} = -\left(s - i\frac{x-x_0}{2\sqrt{at}}\right)^2 - \frac{(x-x_0)^2}{4at} \equiv -z^2 - \frac{(x-x_0)^2}{4at} \tag{6.212}$$

where we have introduced

$$z = s - i\frac{x-x_0}{2\sqrt{at}}. \tag{6.213}$$

Changing variables from s to z, we have the complex integral

$$u(x,t) = \frac{Q}{2\pi\sqrt{at}} e^{-\frac{(x-x_0)^2}{4at}} \int_{\Gamma} e^{-z^2} dz, \tag{6.214}$$

where Γ is the contour in the complex plane where the imaginary part is equal to $-(x - x_0)/2\sqrt{at}$ (see Fig. 6.15). However, the integrand e^{-z^2} is analytic and exponentially suppressed as $\text{Re}(z) \to \pm\infty$ while keeping $\text{Im}(z)$ fixed. The integral along the contour Γ is therefore the same as the integral along the real axis and we are left with the standard Gaussian integral

$$u(x,t) = \frac{Q}{2\pi\sqrt{at}} e^{-\frac{(x-x_0)^2}{4at}} \int_{-\infty}^{\infty} e^{-z^2} dz = \frac{Q}{\sqrt{4\pi at}} e^{-\frac{(x-x_0)^2}{4at}}. \tag{6.215}$$

This is an example of a situation where the transform solution results in an exact analytic expression. The solution of this problem is depicted in Fig. 6.16 for several different times t.

As for the series solutions, there is nothing particular about using this method for solving partial differential equations that are of first order in time, or for only solving partial differential equations that include time as a variable for that matter. We can equally well apply similar methods for solving, e.g., Poisson's equation or the wave equation. Naturally, the use of transform solutions is not restricted to the one-dimensional case but works when an appropriate transform can be written down. A common type of problem where such methods are useful includes problems in all of $\mathbb{R}^N$, where we can make use of an N-dimensional Fourier transform

$$u(\vec{x},t) = \frac{1}{(2\pi)^N} \int_{\mathbb{R}^N} \hat{u}(\vec{k},t) e^{i\vec{k}\cdot\vec{x}} dV. \tag{6.216}$$

Note that the basis functions used to construct this N-dimensional Fourier transform are

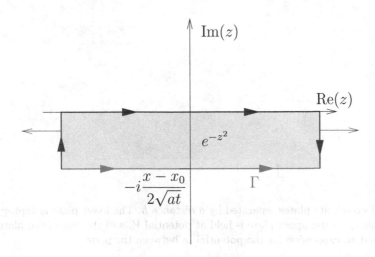

Figure 6.15 The integral in Eq. (6.214) is taken along the curve Γ as the endpoints tend to infinity. As the integrand e^{-z^2} is analytic in the shaded region and exponentially suppressed as $\mathrm{Re}(z) \to \pm\infty$ with fixed imaginary part, the integral may be rewritten as an integral along the real line.

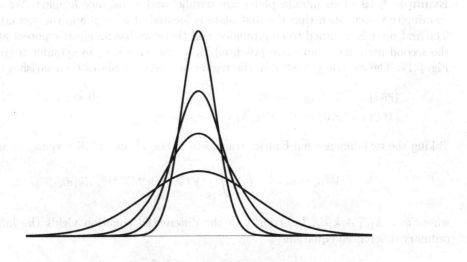

Figure 6.16 The solution to the diffusion equation on the entire real line for a delta function initial condition for different times. As time progresses, the Gaussian shaped solution spreads out.

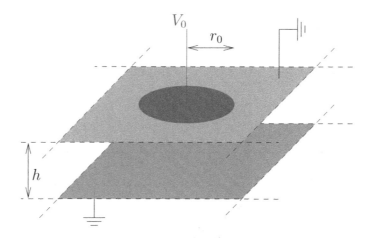

Figure 6.17 Two infinite plates separated by a distance h. The lower plate is kept grounded while a disc of radius r_0 on the upper plate is held at potential V_0 and the rest of the plate is grounded. We wish to find an expression for the potential in between the plates.

of the form

$$X_{\vec{k}}(\vec{x}) = \frac{1}{(2\pi)^N} e^{i\vec{k}\cdot\vec{x}} = \frac{1}{(2\pi)^N} e^{i\sum_j k_j x^j} = \prod_j \frac{e^{ik_j x^j}}{2\pi}. \qquad (6.217)$$

In other words, they are variable separated solutions where each factor corresponds to a basis function from the one-dimensional case.

Example 6.19 Two infinite plates are parallel and a distance h apart. We introduce a coordinate system such that the first plate is located at $x^3 = 0$ and the second at $x^3 = h$. The first plate is assumed to be grounded and therefore has an electric potential zero while the second plate has a non-zero potential V_0 whenever $\rho < r_0$ in cylinder coordinates, see Fig. 6.17. The electric potential in the region between the plates then satisfies

$$\text{(PDE)}: \; -\nabla^2 V(\rho, z) = 0, \qquad\qquad (0 < z < h) \qquad (6.218a)$$
$$\text{(BC)}: \; V(\rho, 0) = 0, \quad V(\rho, h) = V_0 \theta(r_0 - \rho). \qquad\qquad (6.218b)$$

Taking the two-dimensional Fourier transform in the x^1 and x^2 directions, we may write

$$V(\rho, z) = \frac{1}{4\pi^2} \int_{-\infty}^{\infty} \int_{-\infty}^{\infty} \tilde{V}(\vec{k}, z) e^{i(k_1 x^1 + k_2 x^2)} dk_1 dk_2, \qquad (6.219)$$

where $\vec{k} = k_1 \vec{e}_1 + k_2 \vec{e}_2$. Insertion into the differential equation yields the infinite set of ordinary differential equations

$$k^2 \tilde{V}(\vec{k}, z) = \partial_z^2 \tilde{V}(\vec{k}, z), \qquad (6.220)$$

which in turn implies

$$\tilde{V}(\vec{k}, z) = A(\vec{k}) \cosh(kz) + B(\vec{k}) \sinh(kz), \qquad (6.221)$$

where $k = |\vec{k}|$. The boundary condition at $z = 0$ implies $\tilde{V}(\vec{k}, 0) = 0$ and hence $A(\vec{k}) = 0$, leaving us with

$$\tilde{V}(\vec{k}, z) = B(\vec{k}) \sinh(kz). \tag{6.222}$$

In order to find an expression for $B(\vec{k})$, we need to make use of the boundary condition at $z = h$, for which we have the transform

$$\tilde{V}(\vec{k}, h) = \int_{\rho < r_0} V_0 \rho e^{-ik\rho \cos(\phi - \theta_{\vec{k}})} \rho d\rho \, d\phi$$

$$= 2\pi V_0 \int_0^{r_0} \rho J_0(k\rho) d\rho = \frac{2\pi V_0 r_0}{k} J_1(kr_0), \tag{6.223}$$

where $\theta_{\vec{k}}$ is the angle between $\vec{k}$ and the x^1-axis. In the second step we have used the integral expression of Eq. (5.140) for the Bessel function J_0 and in the third we have used the relation $xJ_0(x) = d(xJ_1(x))/dx$ in order to perform the integral. As a consequence, we find that

$$B(\vec{k}) = \frac{2\pi V_0 r_0}{k} \frac{J_1(kr_0)}{\sinh(kh)} \tag{6.224}$$

and therefore

$$V(\rho, z) = \frac{V_0 r_0}{2\pi} \int \frac{J_1(kr_0)}{k} \frac{\sinh(kz)}{\sinh(kh)} e^{i\vec{k} \cdot \vec{r}} dk_1 dk_2$$

$$= V_0 r_0 \int_0^\infty J_1(kr_0) \frac{\sinh(kz)}{\sinh(kh)} J_0(k\rho) dk, \tag{6.225}$$

where we again have used the integral expression of the Bessel function J_0 to perform the angular integral. As expected from the rotational symmetry of the problem, the solution depends only on the coordinates ρ and z. Note that the Fourier transform $\hat{V}(\vec{k}, z)$ also did not depend on the direction of $\vec{k}$, but only on the magnitude k.

This example also demonstrates the close connection between the Fourier transform in two dimensions with the Hankel transform. As our functions did not depend on the angle ϕ, the Fourier transform essentially turned into the Hankel transform of order zero. If there had been an angular dependence, the Fourier transform would have turned into a sum over several different Hankel transforms, essentially replacing one of the Fourier transform integrals by a sum. This is directly related to the different, but equivalent, description of the plane in Cartesian and polar coordinates, respectively.

6.7.1 Mixed series and transforms

In many situations, we will be faced with problems in which the domain of the solution is infinite in some directions and finite in others. In such cases we can combine a transform solution in the infinite directions with a series solution in the finite ones. For concreteness, let us assume that we have a two dimensional eigenvalue problem of the form

$$\text{(PDE)} : \hat{L}u(x, y) = \lambda u(x, y), \tag{6.226a}$$

$$\text{(BC)} : \alpha_a u(a, y) + \beta_a u_x(a, y) = \alpha_b u(b, y) + \beta_b u_x(b, y) = 0, \tag{6.226b}$$

where $\hat{L} = f(y)\hat{L}_x + \hat{L}_y$ is a sum of one-dimensional Sturm–Liouville operators as in Eq. (5.104a). We wish to solve this eigenvalue equation on the domain $a < x < b$ and $-\infty < y < \infty$. Following the same kind of logic as when we separated the eigenfunctions of Sturm–Liouville operators on finite domains, the eigenfunctions $X_n(x)$ of $\hat{L}_x$ have the property

$$\hat{L}_x X_n(x) = \mu_n X_n(x), \tag{6.227}$$

where μ_n is the corresponding eigenvalue. Due to the orthogonality of the functions $X_n(x)$, any function $h(x, y)$ may be expanded in these for a fixed y

$$h(x, y) = \sum_n X_n(x) h_n(y). \tag{6.228}$$

We may now consider the Sturm–Liouville operators $\hat{L}_{y,n} = \hat{L}_y + f(y)\mu_n$ on the infinite domain in the y coordinate with corresponding eigenfunctions $Y_{nk}(y)$, where n is the index belonging to the Sturm–Liouville operator and k a generally continuous parameter denoting the different eigenfunctions of $\hat{L}_{y,n}$. We can now write any function of y, in particular the $h_n(y)$, using the transform

$$h_n(y) = \int_k \tilde{h}_n(k) Y_{nk}(y) dk, \tag{6.229}$$

implying that

$$h(x, y) = \sum_n \int_k \tilde{h}_n(k) X_n(x) Y_{nk}(y) dk. \tag{6.230}$$

Note that we have chosen to expand $h_n(y)$ in terms of the eigenfunctions of the corresponding $\hat{L}_{y,n}$. We thus find an expansion of $h(x, y)$ in terms of the eigenfunction products $X_n(x) Y_{nk}(y)$ that satisfy

$$\hat{L} X_n(x) Y_{nk}(y) = X_n(x) \underbrace{[f(y)\mu_n + \hat{L}_y]}_{=\hat{L}_{y,n}} Y_{nk}(y) = \lambda_{nk} X_n(x) Y_{nk}(y), \tag{6.231}$$

where λ_{nk} is the eigenvalue of $Y_{nk}(y)$ with respect to $\hat{L}_{y,n}$. These product functions are therefore a complete set of eigenfunctions satisfying the eigenvalue equation given in Eq. (6.226a). The expansion functions $\tilde{h}_n(k)$ are the combined transforms and series coefficients.

While the above discussion deals with a two-dimensional situation, the generalisation to more than two dimensions is straightforward. Each of the infinite directions will then correspond to a transform and each of the finite directions to a series.

Example 6.20 In a long canal with stationary water, an amount Q of a substance is released at time $t = 0$ in the middle of the canal. If we assume that the substance diffuses with diffusivity D, that the width of the canal is h, and approximate the canal to be of infinite length, the concentration $u(x, y, t)$ of the substance will satisfy the relations

$$(\text{PDE}) : u_t(x, y, t) - D\nabla^2 u(x, y, t) = 0, \tag{6.232a}$$
$$(\text{BC}) : u_y(x, 0, t) = u_y(x, h, t) = 0, \tag{6.232b}$$
$$(\text{IC}) : u(x, y, 0) = Q\delta(x)\delta(y - h/2). \tag{6.232c}$$

We have here introduced a coordinate system according to Fig. 6.18 and assumed that no

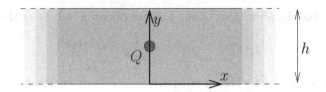

Figure 6.18 An amount Q of a substance is released in the middle of a canal with stationary water and allowed to diffuse. We introduce a coordinate system such that $-\infty < x < \infty$ and $-h/2 < y < h/2$.

substance will flow out of the canal at the boundaries. For each fixed t, we can express the concentration as

$$u(x,y,t) = \frac{1}{2\pi} \sum_n \int_k \tilde{u}_n(k,t) \cos(\kappa_n y) e^{ikx} dk, \tag{6.233}$$

i.e., as an expansion in the eigenfunctions $\cos(\kappa_n y) e^{ikx}/2\pi$ of the Laplace operator, which satisfy

$$-\frac{1}{2\pi} \nabla^2 \cos(\kappa_n y) e^{ikx} = (\kappa_n^2 + k^2) \frac{1}{2\pi} \cos(\kappa_n y) e^{ikx}, \tag{6.234a}$$

with corresponding eigenvalue

$$\lambda_n(k) = \kappa_n^2 + k^2, \tag{6.234b}$$

where $\kappa_n = \pi n/h$. After solving the ordinary differential equations resulting from insertion into the problem, we find that

$$\tilde{u}_n(k,t) = \tilde{u}_n(k,0) e^{-D\lambda_n(k)t}. \tag{6.235}$$

As usual, the initial conditions $\tilde{u}_n(k,0)$ may be found by expressing the original initial condition in terms of the function basis as well. We find that

$$\tilde{u}_n(k,0) = \frac{1}{N_n} \int_{x=-\infty}^{\infty} \int_{y=0}^{h} Q\delta(x)\delta(y - h/2) \cos(\kappa_n y) e^{-ikx} dx dy$$

$$= \frac{Q}{N_n} \cos(\pi n/2), \tag{6.236}$$

where $N_n = (1 + \delta_{0n}) h/2$ is a normalisation factor based on the normalisation of $\cos(0y) = 1$ being different from the normalisation of the other eigenfunctions in the y-direction. All in all, the concentration is given by the function

$$u(x,y,t) = \frac{Q}{2\pi} \sum_n \frac{\cos(\kappa_n y) \cos(\pi n/2)}{N_n} \int_k e^{ikx} e^{-D\lambda_n(k)t} dk. \tag{6.237}$$

The integral part of this solution may be performed in the same way as for Example 6.18.

In some situations, different sets of coordinates on the same domain may lead to different numbers of coordinates giving rise to transforms instead of series. An example of this occurs when we consider an infinite two-dimensional plane. If we introduce Cartesian coordinates

x^i on the plane, the eigenfunctions of the Laplace operator will be of the form

$$f_{\vec{k}}(\vec{x}) = \frac{1}{4\pi^2} e^{i\vec{k}\cdot\vec{x}} = \frac{1}{4\pi^2} e^{i(k_1 x^1 + k_2 x^2)}, \tag{6.238}$$

with eigenvalue $\lambda(\vec{k}) = \vec{k}^2$. The parameters k_i are both continuous, leading to a two-dimensional Fourier transform.

If we instead choose to use polar coordinates ρ and ϕ, we obtain a periodic Sturm–Liouville problem in the angular direction when computing the eigenfunctions, while the radial direction results in different Bessel functions. The eigenfunctions are therefore of the form

$$g_{n,k}(\rho, \phi) = J_n(k\rho)e^{in\phi}, \tag{6.239}$$

where n is any integer and the corresponding eigenvalues are $\lambda_n(k) = k^2$.

An important observation here is that the resulting function bases span the same vector space and are all eigenfunctions of the same operator. In particular, this implies that the subspaces with the same eigenvalues of the Laplace operator must be the same as well. Consequently, it should be possible to express any of the functions $f_{\vec{k}}(x, y)$ solely in terms of the functions $g_{n,k}(\rho, \phi)$ that have the same eigenvalue. In other terms, we should be able to write

$$\frac{1}{4\pi^2} e^{i\vec{k}\cdot\vec{x}} = \sum_n A_n(\vec{k}) J_n(k\rho)e^{in\phi}. \tag{6.240}$$

This looks stunningly familiar. If we look at the special case of $\vec{k} = k\vec{e}_2$ and rewrite the left-hand side in polar coordinates, we find that

$$\frac{1}{4\pi^2} e^{ik\rho\sin(\phi)} = \sum_n A_n(k\vec{e}_2) J_n(k\rho)e^{in\phi}. \tag{6.241}$$

This is nothing but the expansion of the function $e^{i\sqrt{\lambda}x^2}/4\pi^2$ in terms of the polar functions $e^{i\pi n\phi}$ as given in Eq. (5.138), where we found the expansion coefficients to satisfy Bessel's differential equation and giving us the integral representation of the Bessel functions in Eq. (5.140). We therefore find that $A_n(k\vec{e}_2) = 1/4\pi^2$. The value of $A_n(\vec{k})$ for a general vector $\vec{k}_\alpha = k[\cos(\alpha)\vec{e}_1 + \sin(\alpha)\vec{e}_2]$ can be found through the relation

$$\vec{k}_\alpha \cdot \vec{x} = k\rho\cos(\phi - \alpha) = k\rho\sin(\phi - \alpha + \pi/2) \tag{6.242}$$

leading to

$$\frac{1}{4\pi^2} e^{i\vec{k}_\alpha \cdot \vec{x}} = \frac{1}{4\pi^2} \sum_n J_n(k\rho)e^{in(\phi - \alpha + \pi/2)} = \frac{1}{4\pi^2} \sum_n i^n e^{-in\alpha} J_n(k\rho)e^{in\phi} \tag{6.243a}$$

and thus

$$A_n(\vec{k}_\alpha) = \frac{i^n}{4\pi^2} e^{-in\alpha}. \tag{6.243b}$$

The relation may also be inverted in order to express the functions $g_{n,k}(\rho, \phi)$ in terms of the $f_{\vec{k}}(x, y)$, see Problem 6.48.

6.8 DISCRETE AND CONTINUOUS SPECTRA

When we consider operators of the Sturm–Liouville form

$$\hat{L} = -\frac{1}{w(x)} \left[\partial_x p(x) \partial_x - q(x) \right], \tag{6.244}$$

we have seen that the spectrum depends on the functions $p(x)$, $q(x)$, and $w(x)$. In particular, we have seen that the spectrum is discrete for regular Sturm–Liouville problems and that it may be continuous for singular problems, such as the one resulting from studying $\hat{L} = -\partial_x^2$ on the entire real line. However, we have also seen that even if the domain is infinite, this does not necessarily imply that the spectrum is continuous. An example of this is the Hermite functions we encountered in Section 5.5.3.

In addition to the cases where the spectrum is discrete and continuous, respectively, there are some cases in which the spectrum has a continuous as well as a discrete part. These situations occur when the regularity conditions on the eigenfunctions along with the differential equation itself imply that the eigenvalues must take a particular form for some range of eigenvalues.

Example 6.21 The above is best illustrated by an example. We consider the situation where a substance is diffusing in a medium with diffusivity D and there is a source located at $x = 0$. If the source production per time unit is proportional to the concentration of the substance at the source, then the concentration $u(x,t)$ will satisfy the partial differential equation

$$\text{(PDE)}: \quad u_t(x,t) - Du_{xx}(x,t) = \kappa_0 u(x,t)\delta(x), \tag{6.245}$$

where κ_0 is a constant and we also impose the regularity condition that the concentration should be finite at infinity. If we can find the eigenfunctions of the operator

$$\hat{L} = -\partial_x^2 - \frac{\kappa_0}{D}\delta(x), \tag{6.246}$$

then we may solve this problem using the same methods that we have already discussed in this chapter. Taking our usual approach, we start searching for the eigenfunctions by writing down the eigenvalue equation

$$\hat{L}X(x) = -X''(x) - \frac{\kappa_0}{D}\delta(x)X(x) = \lambda X(x). \tag{6.247}$$

For all $x \neq 0$, the second term in the middle step equals zero due to the δ function and we find the solutions

$$X_\pm(x) = A_\pm e^{i\sqrt{\lambda}x} + B_\pm e^{-i\sqrt{\lambda}x}, \tag{6.248}$$

where $X_+(x)$ is the solution in the region $x > 0$ and $X_-(x)$ is the solution in the region $x < 0$. The overall solution may be written as

$$X(x) = X_+(x)\theta(x) + X_-(x)\theta(-x), \tag{6.249}$$

where $\theta(x)$ is the Heaviside step function. Taking the derivatives of the solution, we find that

$$X'(x) = X_+'(x)\theta(x) + X_-'(x)\theta(-x) + \delta(x)[X_+(0) - X_-(0)], \tag{6.250a}$$

$$X''(x) = X_+''(x)\theta(x) + X_-''(x)\theta(-x) + \delta(x)[X_+'(0) - X_-'(0)] \\ + \delta'(x)[X_+(0) - X_-(0)], \tag{6.250b}$$

where we have used the fact that $\delta(x)f(x) = \delta(x)f(0)$. Furthermore, we know that $X_\pm''(x) = -\lambda X_\pm(x)$ and therefore

$$X''(x) = -\lambda X(x) + \delta(x)[X_+'(0) - X_-'(0)] + \delta'(x)[X_+(0) - X_-(0)]. \tag{6.251}$$

Comparing with the original partial differential equation, we must therefore have

$$X_+(0) = X_-(0) \equiv X(0), \tag{6.252a}$$

$$X'_+(0) - X'_-(0) = -\frac{\kappa_0}{D} X(0). \tag{6.252b}$$

For $\lambda = 0$, these equations cannot be satisfied simultaneously as both solutions $X_\pm(x)$ would be constants, while for $\lambda = k^2 > 0$, we find that

$$A_+ + B_+ = A_- + B_-, \tag{6.253a}$$

$$ik\frac{D}{\kappa_0}(A_+ - A_- - B_+ + B_-) = -A_+ - B_+. \tag{6.253b}$$

This system of equations has the non-trivial solutions

$$A_- = A_+ - \frac{i\kappa_0}{2kD}(A_+ + B_+), \tag{6.254a}$$

$$B_- = B_+ + \frac{i\kappa_0}{2kD}(A_+ + B_+), \tag{6.254b}$$

indicating the existence of two independent solutions for each value of k.

So far, the treatment has been analogous to that made for the operator $-\partial_x^2$ in the sense that we have studied the case $\lambda > 0$ and found a continuous spectrum. The real difference appears when we try to find solutions for $\lambda = -k^2 < 0$. In this situation, we found no eigenfunctions of $-\partial_x^2$ that were bounded at the infinities. However, in this case, the solutions are of the form

$$X_\pm(x) = A_\pm e^{kx} + B_\pm e^{-kx}. \tag{6.255}$$

The condition of $X(x)$ being bounded at the infinities now implies that $A_+ = B_- = 0$ and the conditions we found for satisfying the differential equation at $x = 0$ become

$$A_- = B_+, \tag{6.256a}$$

$$-kB_+ - kA_- = -\frac{\kappa_0}{D}B_+. \tag{6.256b}$$

This set of equations has a non-trivial solution only if

$$k = \frac{\kappa_0}{2D} \tag{6.257}$$

and so the only allowed eigenvalue for $\lambda < 0$ is

$$\lambda_0 = -\frac{\kappa_0^2}{4D^2}. \tag{6.258}$$

This represents a single isolated eigenvalue and therefore a discrete part of the spectrum.

The general solution to the problem may now be written as a linear combination of all of the eigenfunctions

$$u(x,t) = A_0(t)e^{-\kappa_0|x|/2D} + \int_0^\infty [A(k,t)X_k(x) + B(k,t)X_{-k}(x)]dk, \tag{6.259}$$

where $X_k(x)$ and $X_{-k}(x)$ represent the two independent eigenfunctions with eigenvalue

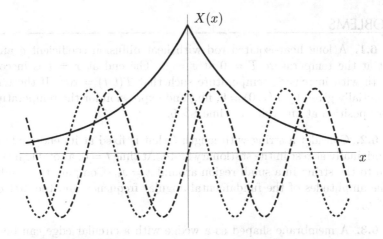

Figure 6.19 The shape of the eigenfunction corresponding to the discrete eigenvalue $\lambda_0 = -\kappa_0^2/4D^2$ is shown as a solid curve. Two eigenfunctions in the continuous spectrum are also shown, one symmetric (dotted) and one anti-symmetric (dashed) both corresponding to the same eigenvalue.

$\lambda = k^2 > 0$. While the amplitudes $A(k,t)$ and $B(k,t)$ correspond to eigenfunctions with positive eigenvalues and will therefore decay exponentially with time

$$A(k,t) = A(k,0)e^{-Dk^2t}, \quad B(k,t) = B(k,0)e^{-Dk^2t}, \tag{6.260}$$

the solution with a negative eigenvalue λ_0 results in an exponentially increasing solution for $A_0(t)$

$$A_0(t) = A_0(0)e^{\kappa_0^2 t/4D}. \tag{6.261}$$

As a consequence, we can draw the conclusion that having a point production of a substance proportional to the concentration of the substance itself will generally lead to the concentration increasing exponentially and that the distribution of the substance will have the shape $e^{-\kappa_0|x|/2D}$. This is illustrated in Fig. 6.19.

In the example above, we found only one eigenfunction in the discrete spectrum, but generally the discrete part of the spectrum may contain more than one eigenfunction (see Problem 6.53). Operators with a combination of continuous and discrete spectra are of high relevance in quantum mechanics, where the discrete spectrum will correspond to *bound states* while the continuous part of the spectrum will correspond to *free states* (or *scattering states*).

In general, for an operator with a discrete spectrum with corresponding eigenfunctions $X_n(\vec{x})$ and a continuous spectrum with corresponding eigenfunctions $X_{\vec{k}}(\vec{x})$, any expansion in terms of the eigenfunctions will include both the discrete and the continuous part of the spectrum. The expansion of a function $f(\vec{x})$ will in this case be of the form

$$f(\vec{x}) = \sum_n f_n X_n(\vec{x}) + \int_{\mathcal{K}} \hat{f}(k) X_{\vec{k}}(\vec{x}) d\mathcal{K}. \tag{6.262}$$

6.9   PROBLEMS

Problem 6.1. A long heat-isolated rod with heat diffusion coefficient a and length ℓ is being kept at the temperature $T = 0$ at $x = 0$. The end at $x = \ell$ is in contact with a thermal bath with increasing temperature such that $T(\ell, t) = \alpha t^2$. If the temperature in the rod is initially given by $T(x, 0) = 0$, find and expression for the temperature $T(x, t)$ for an arbitrary position at an arbitrary time $t > 0$.

Problem 6.2. Consider a string with length ℓ that is fixed at its endpoints at $x = 0$ and $x = \ell$ and originally at rest in the stationary state. At time $t = 0$, a total transversal impulse p_0 is added to the string in a small region around $x = x_0$. Compute the resulting quotient between the amplitudes of the fundamental angular frequency and the first overtone as a function of x_0.

Problem 6.3. A membrane shaped as a wedge with a circular edge can be described by the polar coordinates $0 < \rho < r_0$ and $0 < \phi < \phi_0$. Assume that the membrane is fixed at its borders and placed in an external gravitational field that is orthogonal to the membrane. Determine the transversal displacement of the membrane in the stationary state if its surface density is ρ_A and its surface tension is σ. Your answer may contain a single integral of a Bessel function.

Problem 6.4. Determine the transversal motion $u(x, t)$ of a string with wave velocity c that is fixed at its endpoints at $x = 0$ and $x = \ell$ and at time $t = 0$ satisfies the initial conditions

$$(\text{IC}) : u(x, 0) = \begin{cases} 3\varepsilon \frac{x}{\ell}, & (0 < x < \frac{\ell}{3}) \\ 3\varepsilon \frac{\ell - x}{2\ell}, & (\frac{\ell}{3} \leq x < \ell) \end{cases} \tag{6.263a}$$

$$u_t(x, 0) = 0. \tag{6.263b}$$

Problem 6.5. In a very long cylindrical conductor, the conductivity σ is related to the temperature T of the conductor as

$$\sigma = \sigma_0 - \mu T, \tag{6.264}$$

where σ_0 and μ are positive constants of suitable physical dimension. The heat production κ in the conductor is given by $\kappa = \sigma E^2$, where E is a constant electric field and the cylinder's surface is kept at the constant temperature $T = 0$. Determine the stationary temperature in the conductor, which may be assumed to have a radius r_0.

Problem 6.6. The electric potential $V(r, \theta)$ between two concentric spherical shells of radius r_1 and r_2, respectively, where $0 < r_1 < r_2$, with no charge in between them satisfies the differential equation

$$(\text{PDE}) : \nabla^2 V(r, \theta) = 0, \qquad\qquad (r_1 < r < r_2) \tag{6.265a}$$

$$(\text{BC}) : V(r_1, \theta) = \alpha \frac{3\cos^2(\theta) - 1}{r_1^3}, \quad V(r_2, \theta) = V_0 \cos(\theta), \tag{6.265b}$$

where we have used the rotational symmetry of the problem to conclude that V cannot depend on the azimuthal angle φ. Determine the physical dimension of the constant α and compute the potential as a function of r and θ.

Problem 6.7. A conducting spherical shell of radius r_0 is grounded, resulting in a potential $V = 0$ on the shell. Inside the shell there is a charge distribution given by the charge density

$$\rho = \rho_0 \frac{r(r_0 - r)}{r_0^2} \cos(\theta). \tag{6.266}$$

The potential inside the shell satisfies Poisson's equation

$$(\text{PDE}) : \ -\nabla^2 V = \frac{\rho}{\varepsilon_0}. \tag{6.267}$$

Find an expression for the potential as a function of the position.

Problem 6.8. On a spherical shell of radius r_0 in three dimensions, the electric potential is kept at V_0 when $x^3 > 0$ and at $-V_0$ when $x^3 < 0$. Compute the resulting potential outside of the shell $r > r_0$. What is the dominating contribution when $r \to \infty$? You may assume that the potential should go to zero in this limit.

Problem 6.9. A string of length ℓ is fixed at $x = 0$ and is attached to a massless ring that moves freely without friction in the transversal direction at $x = \ell$. A force F_0 is applied at the point $0 < x_0 < \ell$.

a) Determine the stationary state of the string.

b) At time $t = 0$ the force ceases to act on the string. Determine the ensuing transversal movement of the string.

You may assume that the tension in the string S and its linear density ρ_ℓ are both constant along the string and that the force F_0 is small enough for the transversal displacement to be small.

Problem 6.10. A substance is allowed to diffuse in the one-dimensional region $0 < x < \ell$. The concentration of the substance at $x = 0$ and $x = \ell$ is kept at the concentration c_0 and at the time $t = t_0 > 0$, a source starts producing the substance with a constant rate at the point $x = x_0$. At time $t = 0$, the the concentration of the substance is c_0 inside the entire region. This situation can be described by the differential equation

$$(\text{PDE}) : \ u_t(x,t) - D u_{xx}(x,t) = c_0 \xi \delta(x - x_0) \theta(t - t_0), \tag{6.268a}$$
$$(\text{BC}) : \ u(0,t) = u(\ell,t) = c_0, \tag{6.268b}$$
$$(\text{IC}) : \ u(x,0) = c_0, \tag{6.268c}$$

where $u(x,t)$ is the concentration at the point x at time t, ξ a constant, and θ is the Heaviside step function. Determine the physical dimension of the constant ξ and compute the concentration $u(x,t)$ for arbitrary $t > 0$ and $0 < x < \ell$.

Problem 6.11. A cylindrical glass contains a water solution of a substance with homogeneous concentration c_0 and depth ℓ_0. At the time $t = 0$, pure water is added on top of the solution without inducing any relevant convective currents such that the total depth is now $\ell > \ell_0$. Determine the concentration of the substance as a function of the depth and time assuming that it satisfies the diffusion equation with diffusivity D and no substance diffuses out of the glass.

Problem 6.12. The electric potential on a sphere of radius r_0 is kept at $V(r_0, \theta) =$

$V_0 \cos(2\theta)$. In addition, the interior of the sphere contains a charge density $\rho = \rho_0 j_0(\pi r/r_0)$ such that the potential fulfils

$$\text{(PDE)} : \; -\nabla^2 V(r,\theta) = \frac{\rho_0}{\varepsilon_0} j_0\left(\frac{\pi r}{r_0}\right), \tag{6.269a}$$

$$\text{(BC)} : V(r_0,\theta) = V_0 \cos(2\theta). \tag{6.269b}$$

Use the superposition principle to rewrite this as two problems, where one problem has inhomogeneous boundary conditions and a homogeneous differential equation, while the other problem has homogeneous boundary conditions and an inhomogeneous differential equation. Solve the resulting differential equations in order to to find the electric potential.

Problem 6.13. Consider heat conduction in one dimension where the initial and boundary conditions are such that the temperature $T(x,t)$ satisfies

$$\text{(PDE)} : T_t(x,t) - aT_{xx}(x,t) = \kappa_0\delta(x-x_0), \qquad (0 < x < \ell) \tag{6.270a}$$
$$\text{(BC)} : T(0,t) = T(\ell,t) = T_1, \tag{6.270b}$$
$$\text{(IC)} : T(x,0) = T_2. \tag{6.270c}$$

Solve this differential equation to find the temperature $T(x,t)$.

Problem 6.14. In an infinitely extended uranium plate of thickness h, the neutron density $n(\vec{x},t)$ satisfies the differential equation

$$\text{(PDE)} : n_t(\vec{x},t) - a\nabla^2 n(\vec{x},t) = \lambda n(\vec{x},t), \qquad (0 < x^3 < h) \tag{6.271a}$$
$$\text{(BC)} : n(\vec{x},t) = 0. \qquad (x^3 = 0, h) \tag{6.271b}$$

Determine the maximal thickness of the plate such that $n(\vec{x},t)$ does not grow exponentially with time.

Problem 6.15. A heat source is placed In the middle of a quadratic metal plate with heat diffusivity a and side length ℓ. The plate is otherwise heat isolated apart from at two adjacent sides, where it has the temperatures T_1 and T_2, respectively. The temperature $T(x^1, x^2, t)$ then satisfies the partial differential equation problem

$$\text{(PDE)} : T_t(\vec{x},t) - a\nabla^2 T(\vec{x},t) = \kappa_0\delta^{(2)}(\vec{x}-\vec{x}_0), \qquad (0 < x^1, x^2 < \ell) \tag{6.272a}$$
$$\text{(BC)} : T(\ell\vec{e}_1 + x^2\vec{e}_2, t) = T_1, \quad T(x^1\vec{e}_1 + \ell\vec{e}_2, t) = T_2,$$
$$\partial_1 T(x^2\vec{e}_2, t) = \partial_2 T(x^1\vec{e}_1, t) = 0, \tag{6.272b}$$

where $\vec{x}_0 = \ell(\vec{e}_1 + \vec{e}_2)/2$. Find the stationary temperature in the plate.

Problem 6.16. In a rocket in free fall, a metal rod of length ℓ is attached to one of the walls. At time $t = 0$, the rocket engines ignite, resulting in the rocket accelerating with an acceleration a in the longitudinal direction of the rod, see Fig. 6.20. The resulting longitudinal displacement of the rod from the initial position $u(x,t)$ satisfies the wave equation

$$\text{(PDE)} : \; u_{tt}(x,t) - c^2 u_{xx}(x,t) = 0. \tag{6.273a}$$

The boundary and initial conditions are given by

$$\text{(BC)} : u(0,t) = \frac{at^2}{2}, \quad u_x(\ell,t) = 0, \tag{6.273b}$$
$$\text{(IC)} : u(x,0) = u_t(x,0) = 0, \tag{6.273c}$$

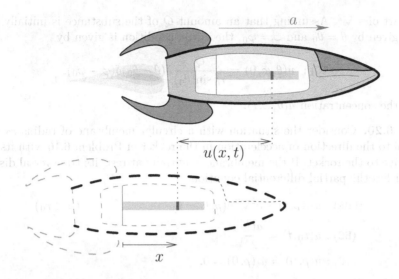

Figure 6.20 A rod inside a rocket originally at rest that starts moving with constant acceleration a at time t. The displacement of the rod is described by the differential equation to be solved in Problem 6.16.

where the first boundary condition comes from the acceleration of the rod's contact point with the rocket and the second from using that no force acts on the rod's free end. Find an expression for $u(x,t)$.

Problem 6.17. On a half-sphere with radius r_0, the flat surface is kept at temperature T_0 while the spherical surface has temperature T_1. Determine the stationary temperature in the half-sphere under the assumption that it does not contain any heat source. Your final answer may contain the integral

$$\bar{P}_\ell^m = \int_{-1}^1 P_\ell^m(\xi)d\xi \tag{6.274}$$

but all other integrals should be computed.

Problem 6.18. The wave equation for oscillations on a spherical shell of radius r_0 is given by

$$(\text{PDE}) : \ u_{tt}(\theta, \varphi, t) + \frac{c^2}{r_0^2}\hat{\Lambda}u(\theta, \varphi, t) = 0, \tag{6.275}$$

where c is the wave velocity and $\hat{\Lambda}$ is the angular part of $-\nabla^2$. Determine the resulting natural angular frequencies and eigenmodes.

Problem 6.19. A substance is allowed to diffuse on a spherical surface with radius r_0. The surface can be parametrised by the spherical coordinates θ and φ and the resulting diffusion can be described by the differential equation

$$(\text{PDE}) : \ u_t(\theta, \varphi, t) + \frac{D}{r_0^2}\hat{\Lambda}u(\theta, \varphi, t) = 0, \tag{6.276a}$$

where $u(\theta, \varphi, t)$ is the concentration of the substance, D is the diffusivity, and $\hat{\Lambda}$ is the angular part of $-\nabla^2$. Assuming that an amount Q of the substance is initially located at the point given by $\theta = \theta_0$ and $\varphi = \varphi_0$, the initial condition is given by

$$(\text{IC}): \ u(\theta, \varphi, 0) = \frac{Q}{r_0^2 \sin(\theta_0)} \delta(\theta - \theta_0) \delta(\varphi - \varphi_0). \tag{6.276b}$$

Compute the concentration $u(\theta, \varphi, t)$.

Problem 6.20. Consider the situation with a circular membrane of radius r_0 suspended orthogonal to the direction of acceleration in the rocket of Problem 6.16 with its boundary fixed relative to the rocket. If the membrane is initially at rest, its transversal displacement $u(\rho, t)$ satisfies the partial differential equation

$$(\text{PDE}): \ u_{tt}(\rho, t) - c^2 \nabla^2 u(\rho, t) = 0, \qquad (\rho < r_0) \tag{6.277a}$$

$$(\text{BC}): \ u(r_0, t) = \frac{at^2}{2}, \tag{6.277b}$$

$$(\text{IC}): \ u(\rho, 0) = u_t(\rho, 0) = 0, \tag{6.277c}$$

where ρ is the radial polar coordinate. Compute the transversal displacement $u(\rho, t)$. We have here used the rotational symmetry of the problem to deduce that $u(\rho, t)$ cannot depend on the polar angle.

Problem 6.21. One end of a string with length ℓ is kept fixed and the other is allowed to move freely in the transversal direction, but with a constant applied force F. The resulting transversal displacement $u(x, t)$ satisfies the differential equation

$$(\text{PDE}): \ u_{tt}(x, t) - c^2 u_{xx}(x, t) = 0, \qquad (0 < x < \ell) \tag{6.278a}$$

$$(\text{BC}): \ u(0, t) = 0, \quad u_x(\ell, t) = \frac{F}{S}, \tag{6.278b}$$

$$(\text{IC}): \ u(x, 0) = u_t(x, 0) = 0, \tag{6.278c}$$

where we have assumed that the string is initially at rest in a horizontal position. Find an expression for the transversal displacement of the string at times $t > 0$.

Problem 6.22. In a heat-isolated metal rod with constant cross-sectional area and length ℓ, the temperature $T(x, t)$ satisfies the heat equation

$$(\text{PDE}): \ T_t(x, t) - aT_{xx}(x, t) = 0, \tag{6.279}$$

where a is constant. For times $t < 0$, the ends of the rod are kept at the temperatures $T(0, t) = T_1$ and $T(\ell, t) = T_2$, respectively.

a) Find the stationary temperature in the rod for times $t < 0$.

b) Assuming that the rod has reached the stationary temperature and that the heat bath keeping the end at $x = \ell$ at T_2 is removed at $t = 0$, instead resulting in that end being heat-isolated, compute the resulting temperature in the rod $T(x, t)$ for $t > 0$.

Problem 6.23. Under some circumstances, acoustic oscillations in an ideal gas can be described with a velocity potential ϕ such that the velocity is given by $\vec{v} = -\nabla\phi$, which satisfies the three-dimensional wave equation. At a fixed wall, the potential satisfies the Neumann boundary condition $\vec{n} \cdot \nabla\phi = 0$. If such a gas is enclosed in a cylinder of radius r_0 and height h, determine the relation between r_0 and h such that the fundamental angular frequency is degenerate.

Problem 6.24. A spherical cavity of radius r_2 is filled with a gas such that pressure waves propagate with speed c. Find a condition that the natural angular frequencies have to satisfy if a small and heavy ball of radius $r_1 < r_2$ is suspended in the middle of the cavity. The ball is heavy enough to assume that it is not significantly displaced by the pressure waves. *Hint:* At the boundaries, the pressure can be assumed to satisfy homogeneous Neumann boundary conditions.

Problem 6.25. A sphere of radius r_0 made out of a metal alloy contains a homogeneous distribution of a radioactive material that decays with decay constant λ. As a result, there is a homogeneous heat production in the sphere declining exponentially with time. If the sphere initially has the temperature T_0, the temperature in the sphere satisfies the inhomogeneous heat equation

$$(\text{PDE}) : T_t(r,t) - a\nabla^2 T(r,t) = \kappa_0 e^{-\lambda t}, \qquad (r < r_0) \qquad (6.280a)$$

$$(\text{BC}) : T(r_0, t) = T_0, \qquad (6.280b)$$

$$(\text{IC}) : T(r, 0) = T_0, \qquad (6.280c)$$

where we have assumed that the boundary of the sphere is kept at temperature T_0 and used that the temperature is then rotationally symmetric in order to deduce that it is independent of the angular spherical coordinates. Compute the temperature $T(r,t)$ at an arbitrary radius and time.

Problem 6.26. The neutron density $n(\vec{x}, t)$ in a homogeneous uranium cylinder can be assumed to satisfy the diffusion equation with a source term proportional to the neutron density itself

$$(\text{PDE}) : n_t - a\nabla^2 n = \lambda n, \qquad (6.281)$$

where λ is the rate at which a neutron reaction with the cylinder creates another neutron. Determine the condition on the cylinder's radius r_0 and height h such that the neutron density does not grow exponentially. The neutron density can be assumed to satisfy homogeneous Dirichlet boundary condition at the surface of the cylinder.

Problem 6.27. In a homogeneous uranium ball of radius r_0, the neutron density $n(\vec{x}, t)$ satisfies the differential equation

$$(\text{PDE}) : n_t - a\nabla^2 n = \lambda n + \kappa(\vec{x}, t), \qquad (6.282)$$

as in Problem 6.26 apart from an additional source term $\kappa(\vec{x}, t)$. The neutron density is also assumed to satisfy homogeneous Dirichlet boundary conditions, indicating that neutrons reaching the surface are efficiently transported away. Determine the critical radius r_c such that the neutron density will grow exponentially if $r_0 > r_c$. In addition, assume that a point source producing neutrons at a rate K is placed in the middle of the sphere. Determine the neutron flux out of the sphere in the stationary case when $r_0 < r_c$. *Hint:* For a point source with rate K at $r = 0$, it is expected that the neutron density should diverge as $n(r) \to K/(4\pi a r)$ as $r \to 0$.

Problem 6.28. A certain type of bacteria moves randomly within a spherical cheese (without holes) of radius r_0. The bacterial concentration $u(\vec{x}, t)$ then satisfies the diffusion equation with diffusivity D. If we assume that the bacteria undergo binary fission with a rate k and that they die instantly when in contact with the air, the time evolution of their concentration is given by

$$(\text{PDE}) : u_t(\vec{x}, t) - D\nabla^2 u(\vec{x}, t) = k u(\vec{x}, t), \qquad (r < r_0) \qquad (6.283a)$$

$$(\text{BC}) : u(\vec{x}, t) = 0, \qquad (r = r_0) \qquad (6.283b)$$

$$(\text{IC}) : u(\vec{x}, t) = u_0, \qquad (r < r_0) \qquad (6.283c)$$

where the initial bacterial concentration is u_0 everywhere and r is the radial spherical coordinate. Determine the distribution of bacteria in the cheese at any given time t.

Problem 6.29. The flat ends of a homogeneous metal cylinder with height h and radius r_0 are heat isolated and its bent surface is kept at the temperature

$$(\text{BC}) : \ T(r_0, z) = T_0 \sin^2\left(\frac{\pi z}{h}\right). \qquad (6.284)$$

Compute the stationary temperature distribution inside the cylinder in the absence of heat sources in the cylinder itself.

Problem 6.30. Smoke with a temperature T_1 is being released from a cylindrical chimney of inner radius r_1, outer radius r_2, and height h. Outside of the chimney, the temperature is constant at T_0 and the bottom of the chimney is kept heat isolated. After a long time, this leads to the stationary problem

$$(\text{PDE}) : \nabla^2 T(\rho, z) = 0, \qquad (6.285a)$$

$$(\text{BC}) : T(r_1, z) = T_1, \quad T_z(\rho, 0) = 0, \quad T(r_2, z) = T(\rho, h) = T_0, \qquad (6.285b)$$

for the temperature $T(\rho, z)$ in the chimney walls, where ρ and z are cylinder coordinates and the solution does not depend on the cylinder coordinate ϕ due to the rotational symmetry of the problem. Find an expression for the temperature $T(\rho, z)$.

Problem 6.31. A homogeneous steel cylinder of radius r_0 and height h is heated to a temperature T_1. At time $t = 0$ it is lowered into an oil bath with constant temperature T_0. Assume that the surface of the cylinder immediately takes the temperature of the oil bath and determine the temperature in the cylinder as a function of position and time. Introduce any additional physical constants that are relevant to your solution and determine their physical dimension.

Problem 6.32. In Example 3.31, we considered the stationary flow of an incompressible fluid inside a cylindrical pipe subjected to a pressure gradient p_0/ℓ. The differential equation we derived was only affected by the assumption of the constant cross section and can therefore be used also in the case of a rectangular cross section. Adding no-slip conditions at the boundaries, the velocity $v(x^1, x^2)$ of the stationary flow then satisfies

$$(\text{PDE}) : \nabla^2 v = \frac{p_0}{\ell \mu}, \qquad (6.286a)$$

$$(\text{BC}) : v(0, x^2) = v(\ell_1, x^2) = v(x^1, 0) = v(x^1, \ell_2) = 0, \qquad (6.286b)$$

where ℓ_1 and ℓ_2 are the side lengths of the cross section and μ the viscosity of the fluid. Compute the velocity of the stationary flow.

Problem 6.33. A solid spherical container of radius r_0 contains an ideal gas. Initially, the entire container and its contents is travelling with velocity $v_0 \vec{e}_3$. At time $t = 0$, the container comes to a sudden halt. The resulting velocity potential ϕ such that $\vec{v} = -\nabla\phi$ of the gas satisfies the wave equation with wave velocity c. Assuming that no gas can flow out of the container, i.e., that the velocity field is parallel to the surface at the boundary, determine how the velocity field changes with time. Your result may contain an integral over the radial coordinate r.

Problem 6.34. One end of a string is attached to a mass m and allowed to move freely in the transversal direction while the other end of the string is subjected to forced oscillations with amplitude A and angular frequency ω. As long as the transversal oscillations $u(x,t)$ are small, they can then be described by the partial differential equation

$$(\text{PDE}) : u_{tt}(x,t) - c^2 u_{xx}(x,t) = 0, \qquad\qquad (0 < x < \ell) \qquad (6.287a)$$

$$(\text{BC}) : m u_{tt}(0,t) = S u_x(0,t), \quad u(\ell,t) = A\sin(\omega t), \qquad\qquad (6.287b)$$

where S is the string tension and c the wave velocity in the string. Determine the steady state solution that oscillates with the angular frequency ω.

Problem 6.35. A circular membrane of area density ρ_A and surface tension σ is clamped into a ring of radius r_0 that is undergoing forced periodic motion with amplitude A and angular frequency ω in the direction perpendicular to the membrane. At time $t = 0$ the membrane is at rest, resulting in the problem

$$(\text{PDE}) : u_{tt}(\rho,t) - c^2 \nabla^2 u(\rho,t) = 0, \qquad\qquad (\rho < r_0) \qquad (6.288a)$$

$$(\text{BC}) : u(r_0,t) = A\sin(\omega t), \qquad\qquad (6.288b)$$

$$(\text{IC}) : u(\rho,0) = u_t(\rho,0) = 0, \qquad\qquad (\rho < r_0) \qquad (6.288c)$$

for its transversal displacement $u(\rho,t)$, where $c^2 = \sigma/\rho_A$. Note that $u(\rho,t)$ does not depend on the polar coordinate ϕ due to the rotational symmetry of the problem. Determine the function $u(\rho,t)$ assuming that ω does not coincide with any of the natural angular frequencies of the membrane. Note that this is a particular realisation of Problem 3.30 with additional initial conditions

Problem 6.36. The oscillatory pressure from a sound wave of angular frequency ω provides a spatially homogeneous force density on a circular membrane with wave velocity c such that the resulting transversal oscillations satisfy the differential equation

$$(\text{PDE}) : u_{tt}(\rho,\phi,t) - c^2 \nabla^2 u(\rho,\phi,t) = f_0 \sin(\omega t), \qquad\qquad (6.289a)$$

$$(\text{BC}) : u(r_0,\phi,t) = 0, \quad |u(0,\phi,t)| < \infty. \qquad\qquad (6.289b)$$

We have here assumed that the membrane radius is r_0 and used polar coordinates. Find the amplitude of the oscillations at $r = 0$ for the steady state solution. You may assume that ω does not coincide with any of the membrane's natural angular frequencies.

Problem 6.37. Inside a spherical shell of radius r_0, the excess pressure $p(r,\theta,\varphi,t)$ satisfies the wave equation

$$(\text{PDE}) : \quad p_{tt} - c^2 \nabla^2 p = 0, \qquad\qquad (6.290a)$$

where c is the wave velocity. Assume that the shell is vibrating in such a fashion that the boundary condition is given by

$$(\text{BC}) : \quad p(r_0,\theta,\varphi,t) = p_0 \cos^3(\theta)\sin(\omega t), \qquad\qquad (6.290b)$$

where ω does not coincide with any of the natural angular frequencies of the system. Find the steady state solution for the resulting oscillations.

Problem 6.38. A string of length ℓ, linear density ρ_ℓ, and tension S is initially at rest in the equilibrium position. At time $t = 0$, a periodic force density is applied to the string such that the transversal displacement satisfies

$$\text{(PDE)}: u_{tt}(x,t) - c^2 u_{xx}(x,t) = A\cos(\omega t), \qquad (0 < x < \ell) \qquad (6.291a)$$

$$\text{(BC)}: u(0,t) = u(\ell, t) = 0, \qquad (6.291b)$$

$$\text{(IC)}: u(x,0) = u_t(x,0) = 0, \qquad (0 < x < \ell) \qquad (6.291c)$$

where $c = \sqrt{S/\rho_\ell}$ is the wave velocity. Find an expression for the resulting transversal displacement assuming that ω does not coincide with any of the natural angular frequencies of the system. How does the solution for an eigenmode change if ω does coincide with its natural angular frequency?

Problem 6.39. In the one-dimensional region $0 < x < \ell$, a physical quantity $u(x,t)$ is determined by the differential equation (see, e.g., Problem 3.24)

$$\text{(PDE)}: u_{tt}(x,t) - c^2 u_{xx}(x,t) + m^2 c^4 u(x,t) = 0, \qquad (6.292a)$$

$$\text{(BC)}: u_x(0,t) = u_x(\ell, t) = 0, \qquad (6.292b)$$

where m is a constant of appropriate physical dimensions. Determine the possible natural angular frequencies of the resulting oscillations. What is the fundamental angular frequency of the system?

Problem 6.40. Consider a long pipe with circular cross section of radius r_0. Inside the pipe, a physical quantity $u(\rho, \phi, z, t)$ satisfies the wave equation with homogeneous Dirichlet boundary conditions. Using cylinder coordinates with the z-coordinate in the pipe direction, we impose the boundary condition

$$\text{(BC)}: \quad u(\rho, \phi, 0, t) = u_0 \sin(\omega t) \qquad (6.293)$$

on one of the cylinder ends. Determine the lowest angular frequency ω such that the steady state solution is a wave propagating in the z-direction. *Hint:* Examine what happens to the steady state solution in the z-direction for low angular frequencies.

Problem 6.41. The transversal displacement of a quadratic membrane with side length ℓ satisfies the damped wave equation

$$\text{(PDE)}: \quad u_{tt} + 2k u_t - c^2 \nabla^2 u = 0 \qquad (6.294)$$

with homogeneous Dirichlet boundary conditions. Determine the number of overdamped oscillation modes if the damping constant k is three times larger than the fundamental angular frequency of the membrane. Repeat the same computation for a circular membrane of radius r_0.

Problem 6.42. For the situations described in Problem 6.41, determine the amplitudes $A_n \omega_n^2 / F_n$ and phase shifts ϕ_n for the steady states of the first three lowest natural angular frequency modes when a driving force with angular frequency ω acts homogeneously over the entire membrane. Assume that ω is five times the fundamental angular frequency of the membrane.

Problem 6.43. Assume that the excess pressure in a cylinder satisfies the damped wave equation

$$\text{(PDE)}: \quad p_{tt} + 2k p_t - c^2 p_{xx} = 0. \qquad (6.295a)$$

We can find the natural angular frequencies of the system by placing a sinusoidal tone generator at one of the cylinder ends, leading to the boundary conditions

$$(\text{BC}): \quad p_x(0,t) = 0, \quad p(\ell, t) = A\sin(\omega t), \tag{6.296}$$

where we have assumed that the pressure gradient is zero at the other end. Determine the resulting amplitude of the nth eigenmode when the driving angular frequency ω coincides with the resonant angular frequency of that eigenmode. Assume that the damping is very weak, i.e., $k \ll \omega_n$.

Problem 6.44. We have already discussed the Fourier sine transform in detail, resulting in the transform and inverse transform expressions of Eqs. (6.198). This discussion assumed a homogeneous Dirichlet boundary condition at $x = 0$. Derive the corresponding expressions for the Fourier cosine transform, which results from the same problem with the Dirichlet condition replaced by a homogeneous Neumann condition. Also show that the Fourier cosine transform may be used to describe functions that are even under the transformation $x \to -x$.

Problem 6.45. A circular membrane with radius r_0 and wave speed c is elastically bound to its equilibrium position by a force density $f = -\kappa u$, where $u(\rho, \phi, t)$ is its transversal displacement that satisfies the wave equation

$$(\text{PDE}): u_{tt} - c^2 \nabla^2 u = -\frac{\kappa}{\rho_A} u, \qquad (\rho < r_0) \tag{6.297a}$$

$$(\text{BC}): u(r_0, \phi, t) = 0, \quad |u(0, \phi, t)| < \infty. \tag{6.297b}$$

We have here assumed that the membrane is fixed at its edge $\rho = r_0$. Find the natural angular frequencies of the membrane's oscillations.

Problem 6.46. The end of a very long heat-isolated rod is subjected to a periodically varying temperature $T_0(t)$. The resulting temperature in the rod follows the heat equation

$$(\text{PDE}): T_t(x,t) - aT_{xx}(x,t) = 0, \qquad (x > 0) \tag{6.298a}$$

$$(\text{BC}): T(0,t) = T_0(t), \tag{6.298b}$$

and $T(x,t) \to 0$ when $x \to \infty$. The periodicity requirement $T_0(t + t_0) = T_0(t)$ implies that the inhomogeneity can be expanded in a Fourier series

$$T_0(t) = \sum_{n=-\infty}^{\infty} \tau_n e^{i\omega_n t}, \tag{6.299}$$

where $\omega_n = 2\pi n/t_0$. Use the superposition principle to find the solution for each term in this sum separately and thereby find the full solution to the problem. *Hint:* For the inhomogeneity proportional to $e^{i\omega_n t}$, assume that the resulting solution has the same time dependence and therefore is of the form $u_n(x,t) = X_n(x)e^{i\omega_n t}$. You should also require that the temperature is bounded as $x \to \infty$.

Problem 6.47. For a general function $f(x)$, define $\tilde{f}_c(k)$ and $\tilde{f}_s(k)$ as the Fourier cosine and sine transforms of its even and odd parts, respectively. Find the relation between the Fourier transform $\tilde{f}(k)$ and the Fourier cosine and sine transforms.

Problem 6.48. In an infinite two-dimensional plane, we have already seen how the Fourier transform basis functions $f_{\vec{k}}(\vec{x}) = e^{i\vec{k}\cdot\vec{x}}/(4\pi^2)$ may be expressed in terms of the Hankel transform basis functions $g_{n,k}(\rho, \phi) = J_n(k\rho)e^{in\phi}$. Invert this relation by finding an expression for $g_{n,k}(\rho, \phi)$ in terms of the functions $f_{\vec{k}}(\vec{x})$.

Problem 6.49. A substance is allowed to diffuse inside an infinite cylinder with radius r_0 and diffusivity a. The substance is initially contained in a thin region around $z = 0$ in cylinder coordinates and is efficiently transported away at the cylinder boundary, resulting in the problem

$$\text{(PDE)} : u_t(\rho, \phi, z, t) - a\nabla^2 u(\rho, \phi, z, t) = 0, \qquad (0 < \rho < r_0) \qquad (6.300a)$$

$$\text{(BC)} : u(r_0, \phi, z, t) = 0, \quad |u(0, \phi, z, t)| < \infty, \qquad (6.300b)$$

$$\text{(IC)} : u(\rho, \phi, z, 0) = \sigma_0 \delta(z) \qquad (6.300c)$$

for the concentration $u(\rho, \phi, z, t)$ of the substance. Determine the dimensions of the constant σ_0 and solve the partial differential equation using series and transform methods.

Problem 6.50. Consider heat conduction in an infinitely extended thin plate with thickness h. The temperature inside the plate follows the source free three-dimensional heat equation and the boundary condition is given by Newton's law of cooling with a surrounding temperature T_0. Derive a partial differential equation for the average temperature across the plate thickness and determine this temperature if the temperature at time $t = 0$ is given by

$$\text{(IC)} : T(\vec{x}, 0) = T_0 + \kappa_0 \delta(x^1)\delta(x^2), \qquad (6.301)$$

where κ_0 is a constant of appropriate dimensions and x^1 and x^2 are Cartesian coordinates along the infinite directions of the plate.

Problem 6.51. The temperature in a thin heat-isolated cylindrical shell can be described by the heat equation

$$\text{(PDE)} : T_t(x, \phi, t) - a\left[T_{xx}(x, \phi, t) + \frac{1}{r_0^2}T_{\phi\phi}(x, \phi, t)\right] = \kappa_0 \delta(x - x_0)\delta(\phi - \phi_0), \quad (6.302a)$$

where we have assumed that there is a heat source at the position given by the coordinates $x = x_0$ and $\phi = \phi_0$. Due to the cyclicity in the angle ϕ, the temperature is also 2π periodic, leading to

$$\text{(BC)} : T(x, \phi, t) = T(x, \phi + 2\pi, t). \qquad (6.302b)$$

Determine the time evolution of the temperature if the temperature at time $t = 0$ is T_0 everywhere on the cylinder. Your final expression may contain a sum and an integral of quantities that you have computed.

Problem 6.52. In a two-dimensional region given by $x^1, x^2 > 0$, a substance is allowed to diffuse with diffusivity a. At time $t = 0$, an amount Q is distributed evenly along the quarter circle $\rho = r_0$ in polar coordinates. Find the resulting concentration if no substance is allowed to flow through the boundaries $x^1 = 0$ and $x^2 = 0$. Your result may contain a single variable integral of known functions.

Problem 6.53. Consider the Sturm–Liouville operator

$$\hat{L} = -\frac{d^2}{dx^2} - \frac{1}{x_0^2}[\theta(x - a) - \theta(x + a)], \qquad (6.303)$$

where a and x_0 are constants and we are considering the operator on the entire real line. Depending on the value of the constants, this operator may have a different number of discrete eigenvalues in its spectrum. Find how the number of discrete eigenvalues depends on the ratio a/x_0.

Problem 6.54. In Example 6.21 we concluded that the operator on the form

$$\hat{L} = -\frac{d^2}{dx^2} - a\delta(x),$$
(6.304)

had one discrete eigenvalue $\lambda_0 < 0$ and computed the corresponding eigenfunction $X_0(x) = e^{-a|x|/2}$. Use the derived compatibility conditions for the eigenvalues $\lambda > 0$ to find explicit expressions for the eigenfunctions belonging to the continuous part of the spectrum and verify that these are orthogonal to $X_0(x)$.

Problem 6.55. We have already discussed an operator containing a delta distribution in Example 6.21 and Problem 6.54. Let us now consider an operator containing two delta distributions

$$\hat{L} = -\frac{d^2}{dx^2} - a[\delta(x - x_0) + \delta(x + x_0)].$$
(6.305)

Verify that $\hat{L}$ commutes with the reflection operator $\hat{R}$ such that $\hat{R}f(x) = f(-x)$. Use this result and the results of Problem 5.54 to find the requirement for $\hat{L}$ having two discrete eigenvalues.

CHAPTER 7

Green's Functions

Green's function methods are common occurrences in many areas of physics and engineering. Sometimes also referred to as an *impulse response*, they describe how a linear system behaves when responding to very particular inhomogeneities. In some sense, Green's functions may be seen as more physical than partial differential equations when it comes to describing physical situations as they will tell us directly how the system behaves in response to inhomogeneities. This being said, there is a close connection between the two and the Green's functions will be solutions to linear partial differential equations. One advantage of using Green's function methods is that the Green's function of a system may often be found by referring to known fundamental solutions to the differential equations in question by altering the solution in order to account for the particular boundary conditions. This may be accomplished by using either mirror image or series expansion techniques. Once the Green's function is known, the solution to a problem with arbitrary inhomogeneities may be found by solving an integral.

In addition to solving linear problems, Green's function methods may also be employed to solve non-linear systems where the non-linearities may be regarded as small perturbations to an otherwise linear problem. Looking at this type of perturbative expansions, we will enter the domain of *Feynman diagrams*, graphical tools that can be used to represent and keep track of the different terms in the expansions. For those who wish to study high-energy particle physics, Feynman diagrams will become an invaluable computational tool in quantum field theory.

7.1 WHAT ARE GREEN'S FUNCTIONS?

At this level of physics education, even if you have never heard the term Green's function you are very likely to have used it extensively when dealing with problems involving gravity and electrostatics at the high-school level. Let us therefore start our exposition on Green's functions with a familiar example.

Example 7.1 The *gravitational potential* of a point mass is given by

$$\Phi(\vec{x}) = -\frac{Gm}{r}, \tag{7.1}$$

where m is the mass, r the distance to the mass, and G is Newton's gravitational constant, see Fig. 7.1. Since the gravitational potential satisfies Poisson's equation

$$\nabla^2 \Phi(\vec{x}) = 4\pi G\rho(\vec{x}), \tag{7.2}$$

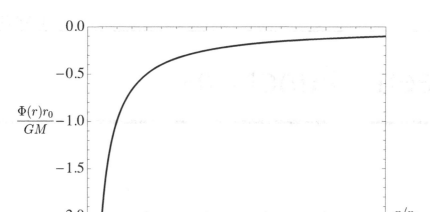

Figure 7.1 The gravitational potential $\Phi(r)$ due to a point source depends only on the distance r to the source with the well-known $1/r$ dependence. If we have several sources, the potential can be found by adding the potential from all of them.

where $\rho(\vec{x})$ is the mass density, which is an inhomogeneous linear differential equation, the general solution may be written as a sum of the contributions of all small mass elements $dm = \rho(\vec{x}) \, dV$. For a continuous mass distribution, the gravitational potential is therefore given by

$$\Phi(\vec{x}) = -G \int_V \frac{\rho(\vec{x}')}{|\vec{x} - \vec{x}'|} dV'. \tag{7.3}$$

In this case, the Green's function of the problem in question is

$$g(\vec{x}, \vec{x}') = -\frac{1}{4\pi |\vec{x} - \vec{x}'|}, \tag{7.4}$$

which we recognise from Eq. (1.260) as a function satisfying

$$\nabla^2 g(\vec{x}, \vec{x}') = \delta^{(3)}(\vec{x} - \vec{x}'), \tag{7.5}$$

where the Laplace operator acts on the unprimed coordinates. The solution to Eq. (7.2) for an arbitrary matter distribution is then the integral over $\vec{x}'$ of the Green's function multiplied by the inhomogeneity on the right-hand side. That this is the case may be checked explicitly by applying the Laplace operator to the left-hand side of Eq. (7.3)

$$\nabla^2 \Phi(\vec{x}) = 4\pi G \int_V \rho(\vec{x}')\nabla^2 g(\vec{x}, \vec{x}')dV' = 4\pi G \int_V \rho(\vec{x}')\delta^{(3)}(\vec{x} - \vec{x}')dV' = 4\pi G\rho(\vec{x}). \tag{7.6}$$

In essence, anyone who finds this example familiar already knows how to apply Green's function methods. The same approach is also valid when solving similar problems in electrostatics.

The question we now need to ask ourselves is what property allowed us to just write down the solution in the above example as an integral involving the Green's function of the problem? The answer to this question has two main ingredients. First of all, we were dealing with a linear differential operator, in this case in the form of the Laplace operator. This fact allowed us to construct the solution to a problem involving two different inhomogeneities by making a superposition of the solutions for the separate inhomogeneities, cf. Section 3.10. The second key ingredient was that we could find a function that returned a delta function when the Laplace operator was applied to it, resulting in a very easy integral when we wished to check that the integral we had written down was really a solution to our problem.

With the above in mind, let us bring these concepts to a more general setting and assume that we wish to solve an inhomogeneous linear differential equation of the form

$$(\text{DE}): \quad \hat{L}u(\vec{x}) = \rho(\vec{x}), \tag{7.7}$$

where $\hat{L}$ is a linear differential operator and $\rho(\vec{x})$ the inhomogeneity. For the time being, we do not specify whether the differential equation is ordinary or partial, as the discussion will be applicable to both cases, and assume that it is valid in some volume V with a boundary S on which we assume that the boundary conditions are linear and homogeneous. Later on we will also have a look at how to deal with inhomogeneities in the boundary conditions. With the above example in mind, we may look for a *Green's function* $G(\vec{x}, \vec{x}')$ of this problem, which we define as the solution to the differential equation

$$(\text{DE}): \quad \hat{L}G(\vec{x}, \vec{x}') = \delta^{(N)}(\vec{x} - \vec{x}'), \tag{7.8}$$

where N is the dimensionality of V, as well as the homogeneous boundary conditions. Again, as was the case for our gravity example, the operator $\hat{L}$ in this equation is taken to act on the unprimed variable $\vec{x}$.

Writing down the integral over $\vec{x}'$ with an arbitrary function $f(\vec{x}')$ multiplied by the Green's function, we find that

$$u(\vec{x}) = \int_V f(\vec{x}')G(\vec{x}, \vec{x}')dV' \tag{7.9a}$$

satisfies

$$\hat{L}u(\vec{x}) = \int_V f(\vec{x}')\hat{L}G(\vec{x}, \vec{x}')dV' = \int_V f(\vec{x}')\delta(\vec{x} - \vec{x}')dV' = f(\vec{x}), \tag{7.9b}$$

again in complete analogy to the example. By letting $f(\vec{x}) = \rho(\vec{x})$, the function $u(\vec{x})$ will now satisfy the original differential equation in Eq. (7.7). In addition, the homogeneous boundary conditions will be automatically satisfied by the requirement that the Green's function $G(\vec{x}, \vec{x}')$ satisfies the same homogeneous boundary conditions.

These considerations demonstrate the idea behind Green's functions. By finding a solution to a linear differential equation with a delta function inhomogeneity, a Green's function, we can construct the solution for any inhomogeneity as an integral involving the Green's function and the inhomogeneity itself.

7.2   GREEN'S FUNCTIONS IN ONE DIMENSION

Before applying Green's function methods to partial differential equations, it is worth looking at their application to ordinary differential equations. There are generally two types of one-dimensional problems that will concern us, problems in time with a sufficient set of initial conditions and one-dimensional problems in space, where there will generally be a set of boundary conditions at the endpoints of an interval. Differential equations in time often arise as a result of considering movement subject to Newton's equations of motion, but are also applicable to solving the ordinary differential equations we ended up with when we constructed series solutions to the heat and wave equations. In general, the problem we are interested in solving will be of the form

$$(\text{ODE}) : \hat{L}u(t) = f(t), \tag{7.10a}$$

$$(\text{IC}) : u(0) = u'(0) = \ldots = u^{(n-1)}(0) = 0, \tag{7.10b}$$

where $\hat{L}$ is a linear differential operator in the variable t, $u^{(k)}(t)$ is the k-th derivative of $u(t)$, and n is the order of $\hat{L}$. We define the Green's function of this problem as the function $G(t, t')$ satisfying

$$(\text{ODE}) : \hat{L}G(t, t') = \delta(t - t'), \tag{7.11a}$$

$$(\text{IC}) : G(0, t') = G_t(0, t') = \ldots = \left.\frac{d^n}{dt^n}G(t, t')\right|_{t=0} = 0, \tag{7.11b}$$

where $\hat{L}$ is acting on the t variable in the differential equation. In the regions where $t \neq t'$, this differential equation is homogeneous and can be solved using any method applicable to homogeneous linear differential equations. The main thing to remember here is that we need to assign different integration constants for $t < t'$ and $t > t'$, respectively, in order to satisfy the inhomogeneity at $t = t'$. We can write the solution in the form

$$G(t, t') = \theta(t' - t)G^-(t, t') + \theta(t - t')G^+(t, t'), \tag{7.12}$$

where $G^-(t, t')$ is the solution for $t < t'$, $G^+(t, t')$ the solution for $t > t'$, and θ is the Heaviside function. Both $G^\pm(t, t')$ are solutions to homogeneous differential equations and $G(t, t')$ must satisfy the homogeneous initial conditions at $t = 0 < t'$. The only possible solution for $G^-(t, t')$ is therefore $G^-(t, t') = 0$ and we are left with

$$G(t, t') = \theta(t - t')G^+(t, t'). \tag{7.13}$$

Inserting this relation into the left-hand side of Eq. (7.11a) and identifying it with the right-hand side will now provide enough conditions to fix the integration constants to completely determine $G^+(t, t')$ and we will have found the Green's function of our problem.

With the Green's function at hand, we can solve the more general problem of Eqs. (7.10) by letting

$$u(t) = \int_0^\infty f(t')G(t, t')dt' = \int_0^\infty f(t')\theta(t - t')G^+(t, t')dt' = \int_0^t f(t')G^+(t, t')dt'. \tag{7.14}$$

Applying the differential operator $\hat{L}$ to this function, we will recover the inhomogeneity $f(t)$ from Eqs. (7.10).

Example 7.2 Let us make the above discussion less abstract by considering Newton's equation of motion for a one-dimensional *harmonic oscillator* originally at rest at the equilibrium point under the influence of an external force $F(t)$

$$(\text{ODE}) : \ddot{x}(t) + \frac{k}{m}x(t) = \frac{F(t)}{m}, \tag{7.15a}$$

$$(\text{IC}) : x(0) = 0, \quad \dot{x}(0) = 0, \tag{7.15b}$$

where kx is the restoring force and m the mass of the oscillator, see Fig. 7.2. Note that

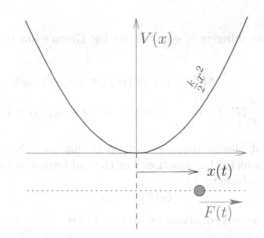

Figure 7.2 The equations of motion for a particle moving in one dimension along the dotted line while being subjected to a harmonic oscillator potential $V(x)$ and an arbitrary external time-dependent force $F(t)$ may be solved by the use of Green's function methods.

exactly this form of ordinary differential equation also appears when we solve the wave equation using series solutions. For convenience, we introduce the characteristic frequency $\omega = \sqrt{k/m}$. We now search for a Green's function for this problem that should satisfy

$$(\text{ODE}) : G_{tt}(t,t') + \omega^2 G(t,t') = \delta(t-t'), \tag{7.16a}$$

$$(\text{IC}) : G(0,t') = 0, \quad G_t(0,t') = 0. \tag{7.16b}$$

Guided by the discussion above, we directly write the solution on the form

$$G(t,t') = \theta(t-t')G^+(t,t'), \tag{7.17}$$

where $G^+(t,t')$ satisfies the homogeneous differential equation

$$\frac{\partial^2}{\partial t^2}G^+(t,t') + \omega^2 G^+(t,t') = 0, \tag{7.18a}$$

implying that

$$G^+(t,t') = A(t')\cos(\omega t) + B(t')\sin(\omega t) = C(t')\sin(\omega t + \phi_0(t')). \tag{7.18b}$$

We have here noted explicitly that the integration constants C and ϕ_0 may in general depend on the value of t'. Differentiating $G(t,t')$ with respect to t, we now obtain

$$G_t(t,t') = \delta(t-t')G^+(t',t') + \theta(t-t')\frac{\partial}{\partial t}G^+(t,t'), \tag{7.19a}$$

$$G_{tt}(t,t') = \delta'(t-t')G^+(t',t') + \delta(t-t')\frac{\partial}{\partial t}G^+(t,t') + \theta(t-t')\frac{\partial^2}{\partial t^2}G^+(t,t')$$

$$= \delta'(t-t')G^+(t',t') + \delta(t-t')\frac{\partial}{\partial t}G^+(t,t') - \omega^2 G(t,t'). \tag{7.19b}$$

Comparing this with the differential equation for the Green's function, we must satisfy the conditions

$$G^+(t', t') = C(t')\sin(\omega t' + \phi_0(t')) = 0, \tag{7.20a}$$

$$\frac{d}{dt}G^+(t, t')\bigg|_{t=t'} = \omega C(t')\cos(\omega t + \phi_0(t')) = 1. \tag{7.20b}$$

The second requirement clearly requires $C(t') \neq 0$, implying that $\sin(\omega t' + \phi_0(t')) = 0$. There are several solutions to this equation, but the end results of these are equivalent and we therefore select the solution

$$\phi_0(t') = -\omega t'. \tag{7.21a}$$

Inserting this into the second equation, we now find that

$$C(t') = \frac{1}{\omega}. \tag{7.21b}$$

Summarising these results, we now have an explicit expression for the Green's function

$$G(t, t') = \theta(t - t')\frac{1}{\omega}\sin(\omega(t - t')). \tag{7.22}$$

This is the general form of the Green's function for any harmonic oscillator and with it we can directly write down the solution for an arbitrary driving force $F(t)$ as

$$x(t) = \frac{1}{m\omega}\int_0^t F(t')\sin(\omega(t - t'))dt'. \tag{7.23}$$

We can check that this is the appropriate solution by differentiating twice with respect to t

$$\dot{x}(t) = \frac{1}{m}\int_0^t F(t')\cos(\omega(t - t'))dt', \tag{7.24a}$$

$$\ddot{x}(t) = \frac{F(t)}{m} - \frac{\omega}{m}\int_0^t F(t')\sin(\omega(t - t'))dt' = \frac{F(t)}{m} - \omega^2 x(t), \tag{7.24b}$$

where the expression for $\ddot{x}$ is the equation of motion for the driven harmonic oscillator.

It is now worth taking a step back to try and find a physical interpretation of the Green's function $G(t, t')$ for this particular problem. By definition, the Green's function is the solution to the harmonic oscillator initially at rest on which we act with a force $F(t) = m\delta(t - t')$. We may compute the impulse imparted on the oscillator between times t_1 and t_2 from the integral

$$I = \int_{t_1}^{t_2} F(t)dt = m\int_{t_1}^{t_2}\delta(t - t')dt = \begin{cases} m, & (t_1 < t' < t_2) \\ 0, & (\text{otherwise}) \end{cases}. \tag{7.25}$$

The Green's function therefore corresponds to the solution where the oscillator is given an impulse m at time $t = t'$. Note that the physical dimension of this impulse does not match our expectation of ML/T. The reason for this is that the delta function in Eq. (7.16a) does not have the same physical dimension as the inhomogeneity $F(t)/m$ in Eq. (7.15a). Instead of dimension L/T^2, the delta function has dimension $1/T$. In order for the dimensions of the differential equation defining the Green's function to match, the Green's function therefore has dimension T rather than dimension L as we expect for $x(t)$. This discrepancy will be remedied once we convolute the Green's function with the general inhomogeneity as described in Eq. (7.23). We could also have obtained a Green's function of the same physical dimension as $x(t)$ by introducing an arbitrary constant f_0 with the appropriate dimensions in front of the delta function in the Green's function differential equation. However, this constant would just be an overall factor in the Green's function and we would have to divide the final integral by it to obtain the general solution. The end result would have been the same. We will therefore be content by calling the Green's function the harmonic oscillator's response to an impulse m at time t. Since the system is linear, the general solution is the superposition of the movement resulting from all of the impulses imparted on the oscillator, i.e., the integral given by Eq. (7.23).

A different one-dimensional situation is presented to us by differential equations of spatial variables where we typically have boundary conditions at two different points in space rather than two different boundary conditions at the same coordinate as in the case of initial value problems. The problems will be of the form

$$(\text{ODE}) : \hat{L}u(x) = f(x), \tag{7.26a}$$

$$(\text{BC}) : \hat{B}_1 u(a) = \hat{B}_2 u(b) = \ldots = 0, \tag{7.26b}$$

where the $\hat{B}_i$ are linear differential operators of at least one order less than $\hat{L}$ and we have a sufficient set of independent boundary conditions to fix all of the integration constants. We will assume that we are solving the problem in the region $a < x < b$ and define the Green's function to the problem as the function satisfying

$$(\text{ODE}) : \hat{L}G(x, x') = \delta(x - x'), \tag{7.27a}$$

$$(\text{BC}) : \hat{B}_1 G(a, x') = \hat{B}_2 G(b, x') = \ldots = 0. \tag{7.27b}$$

The approach to finding the Green's function for this type of problems is exactly analogous to the way we used to find the Green's function for the initial value problems. We start by solving the homogeneous problems in the regions $a < x < x'$ and $x' < x < b$ separately. This will result in the homogeneous solutions $G^-(x, x')$ and $G^+(x, x')$, respectively, neither of which is completely determined since there are not enough boundary conditions at neither $x = a$ nor $x = b$ to do so. Both these solutions will therefore have a number of undetermined coefficients and the solution will be of the form

$$G(x, x') = \theta(x - x')G^+(x, x') + \theta(x' - x)G^-(x, x'). \tag{7.28}$$

In order to determine the remaining integration constants, we must again differentiate this expression and compare the result with the differential equation for the full Green's function.

Example 7.3 In Example 6.8, we looked at the temperature in a one-dimensional rod with a point heat source and how it changed with time. As has been mentioned on several occasions, it is often useful to find the stationary solution to a problem, either in order to find the solution for large times or to expand the solution around it. For the function $u(x, t) = T(x, t) - T_0$ in that example, the stationary solution $u_{\text{st}}(x)$ will satisfy the ordinary differential equation

$$(\text{ODE}) : u_{\text{st}}''(x) = -\frac{\kappa_0}{a}\delta(x - x_0), \tag{7.29a}$$

$$(\text{BC}) : u_{\text{st}}(0) = u_{\text{st}}(\ell) = 0. \tag{7.29b}$$

Because of the delta function in the differential equation, the stationary solution is already the Green's function of the differential operator d^2/dx^2 up to a multiplicative constant $-\kappa_0/a$, letting $G(x, x_0) = -au_{\text{st}}(x)/\kappa_0$, we find that

$$G^{\pm}(x, x_0) = A_{\pm}x + B_{\pm}. \tag{7.30a}$$

Adapting this to the boundary conditions therefore results in

$$G(x, x_0) = A_- x\theta(x_0 - x) - A_+(\ell - x)\theta(x - x_0). \tag{7.30b}$$

The second derivative of this expression is given by

$$\frac{\partial^2}{\partial x^2}G(x, x_0) = \delta(x - x_0)(A_+ - A_-) + \delta'(x - x_0)[A_+(x_0 - \ell) - A_- x_0] \tag{7.31}$$

where identification with the differential equation results in

$$A_+ = \frac{x_0}{\ell}, \quad A_- = \frac{x_0}{\ell} - 1. \tag{7.32}$$

The resulting Green's function

$$G(x, x_0) = (x_0 - \ell)\frac{x}{\ell}\theta(x_0 - x) + (x - \ell)\frac{x_0}{\ell}\theta(x - x_0) \tag{7.33}$$

may be used to find the stationary solution not only for the inhomogeneity $\kappa_0\delta(x - x_0)$ introduced in Example 6.8 but, replacing it with any inhomogeneity $\kappa(x)$, the solution will be of the form

$$u_{\text{st}}(x) = -\int_0^{\ell} G(x, x_0)\frac{\kappa(x_0)}{a}dx_0. \tag{7.34}$$

If it is possible to perform this integral analytically, it will often result in an expression that is easier to evaluate than the sum resulting from applying a series expansion method.

7.2.1 Inhomogeneous initial conditions

The problems we have been dealing with so far in this chapter have in common that they have had homogeneous initial and boundary conditions. However, Green's function methods may also be applied to situations where the initial or boundary conditions contain inhomogeneities. Let us start by considering the first order initial value problem

$$(\text{ODE}) : u'(t) + g(t)u(t) = f(t), \tag{7.35a}$$

$$(\text{IC}) : u(0) = u_0, \tag{7.35b}$$

where $g(t)$ and $f(t)$ are known functions. In order to take care of the inhomogeneity in the differential equation, we can write down the requirements on the Green's function of the homogeneous problem

$$(\text{ODE}): \frac{d}{dt}G(t,t') + g(t)G(t,t') = \delta(t-t'), \tag{7.36a}$$

$$(\text{IC}): G(0,t') = 0. \tag{7.36b}$$

Once we have found the Green's function satisfying this, the function

$$\hat{u}(t) = \int_0^t G(t,t')f(t')dt' \tag{7.37}$$

satisfies the differential equation in Eq. (7.35a), but not the initial conditions as

$$\hat{u}(0) = \int_0^0 G(t,t')f(t')dt' = 0 \tag{7.38}$$

by construction. However, we can introduce $v(t) = u(t) - \hat{u}(t)$, which will satisfy Eqs. (7.35) with $f(t) = 0$. This homogeneous differential equation is exactly the same homogeneous differential equation that $G^{\pm}(t,t')$ had to satisfy away from $t = t'$. In particular, if we let $t' = 0$, then

$$G(t,0) = \theta(t)G^+(t,0) + \theta(-t)G^-(t,0) = G^+(t,0) \tag{7.39}$$

satisfies the homogeneous differential equation for all $t > 0$. Since we are dealing with a first order differential equation, there is only one linearly independent solution and it follows that any solution can be written in the form

$$v(t) = CG^+(t,0), \tag{7.40}$$

where C is a constant that must be adapted to the initial condition by letting

$$v(0) = CG^+(0,0) = u_0 \implies C = \frac{u_0}{G^+(0,0)}. \tag{7.41}$$

Summarising, the solution to the general problem will be given by

$$u(t) = u_0 \frac{G^+(t,0)}{G^+(0,0)} + \int_0^t G(t,t')f(t')dt'. \tag{7.42}$$

It should be noted that this solution is a superposition of the contribution from the initial condition $u(0) = u_0$ and that of the inhomogeneity $f(t)$.

Example 7.4 For an object falling in a homogeneous gravitational field at a velocity low enough for turbulence to be negligible, the equation of motion for the velocity $v(t)$ can be written as

$$(\text{ODE}): \ F = mg - kv(t) = m\dot{v}(t) \iff \dot{v}(t) + \frac{k}{m}v(t) = g, \tag{7.43a}$$

see Fig. 7.3. If the object is given an initial velocity v_0 directed in the opposite direction of the gravitational field, the initial condition will be

$$(\text{IC}): \ v(0) = -v_0. \tag{7.43b}$$

The Green's function for the differential operator $d/dt + k/m$ is given by

$$G(t,t') = \theta(t-t')e^{-\frac{k}{m}(t-t')} \tag{7.44}$$

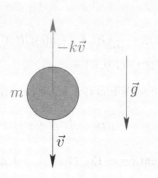

Figure 7.3 An object of mass m falling with velocity $\vec{v}$ under the influence of a gravitational field $\vec{g}$ and with air resistance modelled as a force $\vec{F} = -k\vec{v}$. Green's function methods may be used to approach this problem.

and, based on the discussion just presented, we may immediately write down the solution as

$$v(t) = -v_0 e^{-\frac{k}{m}t} + \int_0^t e^{-\frac{k}{m}(t-t')} g\, dt' = \frac{mg}{k} - e^{-\frac{k}{m}t}\left(v_0 + \frac{mg}{k}\right). \tag{7.45}$$

It is easy to check that this solution satisfies the original differential equation as well as the initial condition.

In this particular case, it might have been simpler to just rewrite the problem as a problem for $u(t) = v(t) - mg/k$, which would be homogeneous. However, this example demonstrates the possible use of the Green's function approach to find solutions also for problems with inhomogeneous initial conditions.

The approach just described apparently works well for first order differential equations. However, we must ask ourselves what to do if we encounter differential equations of higher order with inhomogeneous initial conditions. After all, for an nth order differential equation, we expect that there should be a total of n independent integration constants that must be fixed by the initial or boundary conditions and so far we have identified only one, the constant in front of $G^+(t, 0)$. Consider the initial value problem

$$\text{(ODE)} : \hat{L}u(t) = f(t), \tag{7.46a}$$

$$\text{(IC)} : u(0) = u_0, \quad u'(0) = u_1, \quad \ldots, \quad u^{(n-1)}(0) = u_{n-1}, \tag{7.46b}$$

where $u^{(k)}(t)$ is the kth derivative of $u(t)$ and $\hat{L}$ is a differential operator of order n. As for the first order equation, we can find the Green's function $G(t, t')$ for homogeneous initial conditions and use it to take care of the inhomogeneity in the differential equation. Furthermore, we know that

$$G(t, 0) = G^+(t, 0) \quad \text{and} \quad \hat{L}G^+(t, 0) = 0 \tag{7.47}$$

for $t > 0$ and so $G^+(t, 0)$ is again one of the possible independent solutions to the homogeneous problem with inhomogeneous initial conditions. We just have to find the other $n - 1$

independent solutions and may do so by noting that the function

$$G^k(t, t') = \frac{\partial^k}{\partial t'^k} G(t, t') \tag{7.48a}$$

satisfies

$$\hat{L}G^k(t, t') = \frac{\partial^k}{\partial t'^k} \hat{L}G(t, t') = \frac{\partial^k}{\partial t'^k} \delta(t - t') = (-1)^k \delta^{(k)}(t - t'), \tag{7.48b}$$

where $\delta^{(k)}$ here is the kth derivative of the delta distribution and not the k-dimensional δ distribution. Note that the derivatives in the definition of G^k are with respect to t' and that these derivatives commute with $\hat{L}$, which is a differential operator in the variable t. Since $\delta^{(k)}(t - t') = 0$ for all $t \neq t'$, this immediately implies that

$$\hat{L}G^k(t, 0) = 0 \tag{7.49}$$

for all $t > 0$ and $G^k(t, 0)$ therefore also solves the homogeneous differential equation. The functions $G^k(t, 0)$ will generally be independent for all $k < n$. We can therefore write down the general solution to the problem as

$$u(t) = \sum_{k=0}^{n-1} C_k G^k(t, 0) + \int_0^t G(t, t') f(t') dt'. \tag{7.50}$$

The constants C_k are integration constants that have to be determined by comparison with the initial conditions for $u(t)$.

Example 7.5 Returning to the harmonic oscillator of Example 7.2, we found that the Green's function was given by

$$G(t, t') = \theta(t - t') \frac{1}{\omega} \sin(\omega(t - t')). \tag{7.51}$$

If we instead of the harmonic oscillator starting at rest with zero displacement wish to solve the problem where the harmonic oscillator has an initial displacement and velocity, described by the initial conditions

$$\text{(IC)}: \quad x(0) = x_0, \quad \dot{x}(0) = v_0, \tag{7.52}$$

we may write down the general solution as

$$x(t) = C_0 \frac{1}{\omega} \sin(\omega t) - C_1 \cos(\omega t) + \frac{1}{m\omega} \int_0^t F(t') \sin(\omega(t - t')) dt', \tag{7.53}$$

where we have here used that

$$G^1(t, t') = \frac{\partial}{\partial t'} G(t, t') = \frac{1}{\omega} \frac{\partial}{\partial t'} \sin(\omega(t - t')) = -\cos(\omega(t - t')) \tag{7.54}$$

for all $t > t'$. Adapting this general solution to the initial conditions, we find that

$$C_0 = v_0 \quad \text{and} \quad C_1 = -x_0. \tag{7.55}$$

It should be noted that, while the coefficient in front of $G^1(t, 0)$ in this solution was only dependent on the initial condition $u(0)$ and the coefficient in front of $G(t, 0)$ was dependent only on the initial condition $u'(0)$, this is not generally the case. Instead, the general case will involve coefficients that depend on all of the initial conditions, giving us a set of n linear equations to solve.

7.2.2 Sturm–Liouville operators and inhomogeneities in the boundary conditions

An important special case when dealing with Green's functions in problems with conditions on the boundaries of an interval occurs when the differential operator in the problem is a Sturm–Liouville operator and we have boundary conditions that may generally be inhomogeneous

$$(\text{ODE}) : \hat{L}u(x) = -\frac{1}{w(x)}[p(x)u''(x) + p'(x)u'(x)] = f(x), \qquad (7.56a)$$

$$(\text{BC}) : \alpha_a u(a) - \beta_a u'(a) = \gamma_a, \quad \alpha_b u(b) + \beta_b u'(b) = \gamma_b. \qquad (7.56b)$$

Before we tackle the task of finding an appropriate Green's function for this problem, let us derive an identity that will be useful several times in our discussion. For any two functions $g(x)$ and $h(x)$, let us define the integral

$$\begin{aligned}
I[g,h] &= \int_a^b [g(x)\hat{L}h(x) - h(x)\hat{L}g(x)]w(x)\,dx \\
&= -\int_a^b \frac{d}{dx}[g(x)p(x)h'(x) - h(x)p(x)g'(x)]dx \\
&= p(a)[g(a)h'(a) - g'(a)h(a)] - p(b)[g(b)h'(b) - g'(b)h(b)], \qquad (7.57)
\end{aligned}$$

where we have used the explicit form of the Sturm–Liouville operator to rewrite the integrand as a total derivative that could be integrated directly. By making different choices of $g(x)$ and $h(x)$ we will be able to argue for several properties of the Green's function. To start with, we will here define the Green's function as a function satisfying

$$\hat{L}G(x,x') = -\frac{1}{w(x)}\delta(x-x'), \qquad (7.58a)$$

where the Sturm–Liouville operator acts on the unprimed variable. In addition, we will assume that the Green's function fulfils homogeneous boundary conditions of the form

$$\alpha_a G(a,x') - \beta_a G_x(a,x') = \alpha_b G(b,x') + \beta_b G_x(b,x') = 0. \qquad (7.58b)$$

The reasons for selecting these boundary conditions will become apparent shortly.

In our first application of the integral $I[g,h]$, we let $g(x) = G(x,x')$ and $h(x) = G(x,x'')$. From the properties of the Green's function, we then obtain

$$\begin{aligned}
I[G(x,x'), G(x,x'')] &= \int_a^b [G(x,x')\hat{L}G(x,x'') - G(x,x'')\hat{L}G(x,x')]w(x)\,dx \\
&= -\int_a^b [G(x,x')\delta(x-x'') - G(x,x'')\delta(x-x')]dx \\
&= -G(x'',x') + G(x',x''). \qquad (7.59)
\end{aligned}$$

At the same time, the boundary conditions on the Green's function inserted in the last expression for $I[g,h]$ in Eq. (7.57) imply that

$$I[G(x,x'), G(x,x'')] = 0 \quad \Longrightarrow \quad G(x',x'') = G(x'',x'). \qquad (7.60)$$

In other words, the Green's function defined as in Eq. (7.58a) must be *symmetric* under the exchange of the first and second argument.

Example 7.6 In Example 7.3, we had a problem of the form described here with $\hat{L} = -d^2/dx^2$. The resulting Green's function presented in Eq. (7.33) satisfies the symmetry condition

$$G(x, x_0) = (x_0 - \ell)\frac{x}{\ell}\theta(x_0 - x) + (x - \ell)\frac{x_0}{\ell}\theta(x - x_0) = G(x_0, x) \tag{7.61}$$

as expected from our discussion.

While being symmetric might be a neat feature of a Green's function, we still need to make sure that we can use the Green's function as defined above in order to solve the general differential equation with inhomogeneous boundary conditions. In order to do so, we again use the integral $I[g, h]$, this time with $g(x) = u(x)$ and $h(x) = G(x, x')$. The integral can now be simplified as

$$I[u(x), G(x, x')] = \int_a^b [u(x)\hat{L}G(x, x') - G(x, x')\hat{L}u(x)]w(x)\, dx$$

$$= \int_a^b [-u(x)\delta(x' - x) - G(x', x)f(x)w(x)]dx$$

$$= -u(x') - \int_a^b G(x', x)f(x)w(x)\, dx, \tag{7.62}$$

where we have used $\hat{L}u = f$, the symmetry of the Green's function, and the differential equation defining it. Looking at the expression for $I[g, h]$ in terms of the boundary contributions, the boundary conditions selected for the Green's function imply that

$$u(a)G_x(a, x') - u'(a)G(a, x') = \frac{1}{\beta_a}[\alpha_a u(a) - \beta_a u'(a)]G(a, x')$$

$$= \frac{\gamma_a}{\beta_a}G(a, x') = \frac{\gamma_a}{\alpha_a}G_x(a, x'), \tag{7.63}$$

where we have inserted the boundary conditions for the solution $u(x)$ at $x = a$. The boundary conditions at $x = b$ result in a similar expression and we find that

$$I[u(x), G(x, x')] = p(a)\frac{\gamma_a}{\alpha_a}G_x(a, x') - p(b)\frac{\gamma_b}{\alpha_b}G_x(b, x'). \tag{7.64}$$

Since this has to be equal to the previous expression we found for the integral we can solve for $u(x')$ and, putting all the pieces of the puzzle together, we obtain

$$u(x') = p(b)\frac{\gamma_b}{\alpha_b}G_x(b, x') - p(a)\frac{\gamma_a}{\alpha_a}G_x(a, x') - \int_a^b G(x', x)f(x)w(x)\, dx. \tag{7.65}$$

If we can find a Green's function satisfying both the required differential equation and boundary conditions, we may therefore directly write down the solution on this form regardless of the inhomogeneities in the differential equation and in the boundary conditions.

Example 7.7 Reintroducing the inhomogeneous boundary conditions from Example 6.8 such that $T(0) = T(\ell) = T_0$, we can use the Green's function from Example 7.3 to write down the solution without doing the translation to the homogeneous problem by the ansatz $u(x) = T(x) - T_0$ first. In this case, we have $p(x) = w(x) = \alpha_a = \alpha_b = 1$ and $\beta_a = \beta_b = 0$, leading to

$$p(b)\frac{1}{\alpha_b}G_x(b, x') - p(a)\frac{1}{\alpha_a}G_x(a, x') = G_x(b, x') - G_x(a, x') = 1. \qquad (7.66)$$

With $\gamma_a = \gamma_b = T_0$, we therefore recover the solution for $T(x)$ as the solution $u(x)$ for the homogeneous problem plus the constant T_0.

A word of warning is required at this point. In the discussion above, we have just assumed that a Green's function satisfying the required differential equation and boundary conditions exists. In the special case of $\alpha_a = \alpha_b = 0$, i.e., with homogeneous Neumann conditions on the Green's function, this will not be the case. We can see this by again considering the integral $I[g, h]$, but with $g(x) = 1$ and $h(x) = G(x, x')$, we find that

$$I[1, G(x, x')] = \int_a^b \hat{L}G(x, x')w(x)\,dx = -\int_a^b \delta(x - x')\,dx = -1. \qquad (7.67)$$

However, since the derivative of one is zero, the expression in terms of the boundary values would be of the form

$$g(a)h'(a) - h(a)g'(a) = G_x(a, x') - 0 = G_x(a, x'), \qquad (7.68)$$

with a similar expression holding at $x = b$. This leads to the condition

$$p(a)G_x(a, x') - p(b)G_x(b, x') = -1, \qquad (7.69)$$

which cannot be satisfied if the Green's function is required to fulfil homogeneous Neumann conditions. However, when this is the case, the inhomogeneous Neumann conditions also need to satisfy additional consistency conditions and the solution is only defined up to an arbitrary integration constant. This can be used to remedy the situation, see Problem 7.12.

7.2.3   The general structure of Green's function solutions

Before moving on to using Green's function methods for solving partial differential equations, let us have a look at the general structure of the solutions we have encountered so far. This structure will remain valid also when we deal with partial differential equations and is therefore useful for understanding how a Green's function solution is constructed.

In general, we are looking to solve a linear differential equation with linear boundary conditions, both of which may be homogeneous or inhomogeneous. Because of the linearity, the superposition of two solutions for different inhomogeneities may be added together to create the solution to the problem where both inhomogeneities are added. In particular, we are interested in problems of the form

$$(\text{DE}) : \hat{L}u(\vec{x}) = f(\vec{x}), \qquad\qquad (\vec{x} \in V) \qquad (7.70a)$$

$$(\text{BC/IC}) : \hat{B}u(\vec{x}) = g(\vec{x}), \qquad\qquad (\vec{x} \in S) \qquad (7.70b)$$

where V is the domain in which we wish to solve the differential equation, S the part of its boundary on which there is a boundary condition, and $\hat{L}$ and $\hat{B}$ are linear operators.

Furthermore, if we denote the solution to this problem by $u_{f,g}(\vec{x})$, then the solution will satisfy

$$u_{f,g}(\vec{x}) = u_{f,0}(\vec{x}) + u_{0,g}(\vec{x}). \tag{7.71}$$

In other words, the contributions from any inhomogeneous boundary conditions may be decoupled from the contribution from any inhomogeneity in the differential equation. This is precisely what we have already observed in Eqs. (7.50) and (7.65). Furthermore, the solution for an inhomogeneous differential equation and the solution for inhomogeneous boundary conditions are also linear in their respective inhomogeneities

$$u_{af_1+bf_2,0}(\vec{x}) = au_{f_1,0}(\vec{x}) + bu_{f_2,0}(\vec{x}), \tag{7.72a}$$

$$u_{0,ag_1+bg_2}(\vec{x}) = au_{0,g_1}(\vec{x}) + bu_{0,g_2}(\vec{x}), \tag{7.72b}$$

where a and b are constants. As such, the solution to the differential equation defines two linear operators

$$\hat{L}^{-1}f(\vec{x}) = u_{f,0}(\vec{x}) \quad \text{and} \quad \hat{B}^{-1}g(\vec{x}) = u_{0,g}(\vec{x}). \tag{7.73}$$

These are operators mapping the inhomogeneities to the set of functions on V, such that the differential equations with the corresponding inhomogeneities are satisfied. The operator $\hat{L}^{-1}$ is therefore an operator on functions on V to functions on V, while $\hat{B}^{-1}$ is an operator on functions on S to functions on V. As we have seen, $\hat{L}^{-1}$ will in general be given by an integral operator

$$\hat{L}^{-1}f(\vec{x}) = \int_V G(\vec{x},\vec{x}')f(\vec{x}')dV', \tag{7.74}$$

where $G(\vec{x},\vec{x}')$ is the Green's function of the problem with homogeneous boundary conditions. In the one-dimensional case, we have also seen that the operator $\hat{B}^{-1}$ is a sum over the boundary conditions at the different boundary points. In more than one dimension, this sum will turn into an integral over the boundary conditions

$$\hat{B}^{-1}g(\vec{x}) = \int_S \tilde{G}(\vec{x},\vec{x}')g(\vec{x}')dS', \tag{7.75}$$

where $\tilde{G}$ is a function satisfying the homogeneous differential equation for $\vec{x}$ in V and $\hat{B}\tilde{G}(\vec{x},\vec{x}')$ is a delta distribution on the boundary S, and therefore also be an integral operator. That these operators are integral operators should not come as a surprise, after all, $\hat{L}$ and $\hat{B}$ are both differential operators.

In some physics texts, it is common to see the notation

$$\hat{L}^{-1} = \frac{1}{\hat{L}}, \tag{7.76}$$

in particular when working with operators on a domain without boundary. Of course, this notation does not really imply that we are dividing by a differential operator, but rather an aversion to writing out the entire integral operator. In addition, it is often fine to divide by the operator after Fourier transforming it, since it will then usually turn into an algebraic expression in the Fourier transform variable, as we shall see shortly.

7.3 POISSON'S EQUATION

For most partial differential equations, things will turn out to work quite smoothly if we consider infinite regions without boundaries and instead impose regularity conditions at infinity. In addition, the Green's function solutions we can find in this fashion will turn

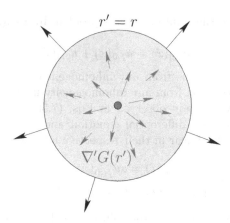

Figure 7.4 In order to find the Green's function for Poisson's equation, we can integrate Poisson's equation with a delta inhomogeneity over a spherical volume with the inhomogeneity at its center. Due to the symmetry of the problem, the magnitude of the vector field $\nabla' G(r')$ will be constant on the boundary.

out to be useful in constructing Green's functions of problems with boundaries if we apply mirror image techniques, as we shall see later. For now, let us start by considering the Green's function for Poisson's equation in $N \geq 3$ dimensions, which by definition should satisfy the differential equation

$$\text{(PDE)} : \nabla^2 G(\vec{x}, \vec{x}') = \delta^{(N)}(\vec{x} - \vec{x}'), \tag{7.77a}$$

$$\text{(BC)} : \lim_{|\vec{x}| \to \infty} G(\vec{x}, \vec{x}') = 0. \tag{7.77b}$$

For the time being, we will not consider the case of $N = 2$, since it provides an additional complication that we will deal with later. Solving the general problem involving Poisson's equation will directly provide us with the solution to several problems in gravitation and electrostatics. In fact, for $N = 3$, we already know that the solution must take the form

$$G(\vec{x}, \vec{x}') = -\frac{1}{4\pi |\vec{x} - \vec{x}'|} \tag{7.78}$$

as this is related to the potential around a point charge. Let us see if we can derive this result by solving the N-dimensional problem.

To start with, the problem of finding the Green's function is invariant under rotations around $\vec{x}'$. It follows that the Green's function can only be a function of the distance between $\vec{x}$ and $\vec{x}'$. By making the transformation $\vec{r} = \vec{x} - \vec{x}'$, we now find that

$$\nabla^2 G(r) = \delta^{(N)}(\vec{r}), \tag{7.79}$$

i.e., we have transformed our problem to an equivalent problem where the inhomogeneity is placed at the origin. Integrating this differential equation over a spherical volume of radius r, see Fig. 7.4, we obtain

$$\int_{r' < r} \nabla'^2 G(r') dV' = \oint_{r' = r} \nabla' G(r') \cdot d\vec{S}' = 1 \tag{7.80}$$

N	A_{N-1}	$G(r)$
3	4π	$-\dfrac{1}{4\pi r}$
4	$2\pi^2$	$-\dfrac{1}{4\pi^2 r^2}$
5	$\dfrac{8\pi^2}{3}$	$-\dfrac{1}{8\pi^2 r^3}$
N	$\dfrac{2\pi^{N/2}}{\Gamma(N/2)}$	$-\dfrac{\Gamma(N/2)}{(N-2)2\pi^{N/2}r^{N-1}}$

Table 7.1 The Green's function $G(r)$ for Poisson's equation and the area of an $N-1$-dimensional sphere of radius one A_{N-1} for a few selected values of the dimensionality of the space N. In the last row, $\Gamma(x)$ is the gamma function, which satisfies $\Gamma(x+1) = x\Gamma(x)$, $\Gamma(1) = 1$, and $\Gamma(1/2) = \sqrt{\pi}$.

by applying the divergence theorem. Since $G(r')$ does not depend on the angular coordinates, we find that

$$\nabla' G(r') = \vec{e}_r \partial_{r'} G(r') = \vec{e}_r G'(r'). \tag{7.81}$$

In addition, the surface element of the sphere $r' = r$ is given by $d\vec{S}' = \vec{e}_r dS'$ and we find

$$G'(r)A_{N-1}(r) = 1, \tag{7.82}$$

where we have used that $G'(r)$ is constant on the surface and $A_{N-1}(r)$ is the area of an $N-1$-dimensional sphere of radius r. The r-dependence of $A_{N-1}(r)$ is given by

$$A_{N-1}(r) = A_{N-1}(1)r^{N-1} \equiv A_{N-1}r^{N-1}, \tag{7.83}$$

i.e., the area of the sphere of radius r is equal to the area of the sphere of radius one multiplied by r^{N-1}. We now have the first order differential equation

$$G'(r) = \frac{1}{A_{N-1}r^{N-1}} \implies G(r) = -\frac{1}{(N-2)A_{N-1}r^{N-2}} + C, \tag{7.84}$$

where C is an integration constant. By the requirement that the Green's function should go to zero as $r \to \infty$, we find $C = 0$ and the sought Green's function is given by

$$G(r) = -\frac{1}{(N-2)A_{N-1}r^{N-2}}. \tag{7.85}$$

We should check that this result reduces to the one we expected for the special case of $N = 3$. The area of the two-dimensional sphere of radius one is given by $A_2 = 4\pi$ and therefore

$$G(r) = -\frac{1}{4\pi r}, \tag{7.86}$$

in agreement with our expectation. For reference, the Green's functions for a few selected values of N are shown in Table 7.1.

So what happens in the case of $N = 2$? We can argue in exactly the same way as we have done for dimensions $N \geq 3$ up to the point where we find the differential equation

$$G'(r) = \frac{1}{2\pi r} \implies G(r) = \frac{1}{2\pi}\ln(r) + C. \tag{7.87}$$

It is here impossible adjust the integration constant C in such a way that the Green's function goes to zero as $r \to \infty$ because of its logarithmic behaviour. As such, we need to assign a different criterion for finding the constant C, such as the Green's function being equal to zero at some fixed distance r_0 from the point source, leading to

$$G(r) = \frac{1}{2\pi} \ln\left(\frac{r}{r_0}\right). \tag{7.88}$$

In general, this will not cause a problem when we study different physical situations. The Green's function of Poisson's equation will represent the potential of a point source and we already know that the addition of a constant to the potential will not cause any change in the physics.

Example 7.8 Consider the electric potential $V(\vec{x})$ corresponding to a charge distribution $\rho(\vec{x})$ in two dimensions. This potential will satisfy Poisson's equation

$$\nabla^2 V(\vec{x}) = -\frac{\rho(\vec{x})}{\varepsilon_0} \tag{7.89}$$

and we can find the solution to this problem by using the Green's function

$$V(\vec{x}, r_0) = -\int G(|\vec{x} - \vec{x}'|, r_0)\frac{\rho(\vec{x}')}{\varepsilon_0} dV'. \tag{7.90}$$

We have here explicitly written out the dependence of the potential and the Green's function on the chosen parameter r_0 of the two dimensional Green's function. We can find the dependence of the potential on the choice of Green's function by comparing the potential resulting from two different choices of r_0, e.g., r_1 and r_2,

$$V(\vec{x}, r_2) - V(\vec{x}, r_1) = -\int [G(|\vec{x} - \vec{x}'|, r_2) - G(|\vec{x} - \vec{x}'|, r_1)]\frac{\rho(\vec{x}')}{\varepsilon_0} dV'$$

$$= -\frac{1}{2\pi\varepsilon_0} \ln\left(\frac{r_1}{r_2}\right)\int \rho(\vec{x}') dV' = -\frac{Q}{2\pi\varepsilon_0} \ln\left(\frac{r_1}{r_2}\right), \tag{7.91}$$

where Q is the total charge. This difference does not depend on the position $\vec{x}$ and is therefore an overall shift in the potential, which does not affect the physical electric field $\vec{E} = -\nabla V$.

7.3.1 Hadamard's method of descent

Although we have already computed the Green's function for Poisson's equation, let us take this opportunity to discuss a useful approach to relating the Green's function of a differential equation in a lower number of dimensions to the Green's function of a similar differential equation in a higher number of dimensions. This relation will be useful when finding expressions for lower dimensional Green's functions that sometimes may be more difficult to compute by other means than those in higher dimensions. This application is known as *Hadamard's method of descent*.

Let us work with an $N + M$-dimensional space which contains the vectors $\vec{x} = \vec{y} + \vec{z}$, where $\vec{y}$ belongs to an N-dimensional subspace and $\vec{z}$ to its M-dimensional orthogonal complement. We assume that we have a linear partial differential operator $\hat{L}$ of the form

$$\hat{L} = \hat{L}_1 + \hat{L}_2, \tag{7.92}$$

where $\hat{L}_1$ only contains derivatives with respect to the N coordinates in the first subspace and $\hat{L}_2$ is assumed to have at least one derivative with respect to the M remaining coordinates and otherwise does not depend on them. We further assume that we can find a Green's function $G(\vec{x}, \vec{x}')$ of $\hat{L}$ that satisfies the differential equation

$$\hat{L}G(\vec{x}, \vec{x}') = \delta^{(N+M)}(\vec{x} - \vec{x}'). \tag{7.93}$$

By using this Green's function, we can find the solution to a general differential equation

$$\hat{L}u(\vec{x}) = f(\vec{x}) \implies u(\vec{x}) = \int G(\vec{x}, \vec{x}')f(\vec{x}')dV'. \tag{7.94}$$

In particular, we can write down the solution $u(\vec{x}) = \tilde{G}(\vec{y}, \vec{y}')$ for $f(\vec{x}) = \delta^{(N)}(\vec{y}-\vec{y}')$, where $\vec{y}$ is the projection of $\vec{x}$ onto the first N dimensions. The resulting problem is translation invariant in the M-dimensional subspace and therefore satisfies

$$\hat{L}\tilde{G}(\vec{y}, \vec{y}') = \hat{L}_1\tilde{G}(\vec{y}, \vec{y}') = \delta^{(N)}(\vec{y}-\vec{y}'). \tag{7.95}$$

The solution $\tilde{G}(\vec{y}, \vec{y}')$ must therefore be the Green's function for the N-dimensional differential operator $\hat{L}_1$. We find that

$$\tilde{G}(\vec{y}, \vec{y}') = \int G(\vec{y}, \vec{y}'' + \vec{z}'')\delta^{(N)}(\vec{y}'' - \vec{y}')dV'', \tag{7.96}$$

where we can now use the delta function to perform the integral over $\vec{y}''$. This results in

$$\tilde{G}(\vec{y}, \vec{y}') = \int G(\vec{y}, \vec{y}' + \vec{z}'')dV_z'', \tag{7.97}$$

where the remaining integral is over the M-dimensional subspace only. We thus have an integral relation for the Green's function $\tilde{G}(\vec{y}, \vec{y}')$ in the N-dimensional subspace in terms of the Green's function $G(\vec{x}, \vec{x}')$ in the full $N + M$-dimensional space.

Example 7.9 Consider the four-dimensional Green's function of Poisson's equation

$$G_4(\vec{x}, \vec{x}') = -\frac{1}{4\pi^2 |\vec{x} - \vec{x}'|^2}. \tag{7.98}$$

The Laplace operator in four dimensions may be written as

$$\nabla_4^2 = \nabla_3^2 + \partial_4^2, \tag{7.99}$$

where ∇_3^2 is the three-dimensional Laplace operator with respect to the coordinates x^1, x^2, and x^3, and is therefore of the required form to apply the reasoning above. Using spherical coordinates r, θ, and φ in the three first dimensions and naming the fourth coordinate w, the three-dimensional Green's function for $\vec{y}' = 0$ is now expressed as

$$G_3(r) = \int_{-\infty}^{\infty} G_4(r\vec{e}_r, w\vec{e}_4)dw = -\frac{1}{4\pi^2} \int_{-\infty}^{\infty} \frac{dw}{r^2 + w^2}. \tag{7.100}$$

The remaining integral evaluates to π/r and we therefore find

$$G_3(r) = -\frac{1}{4\pi r} \tag{7.101}$$

as expected.

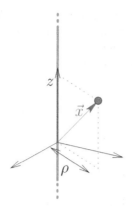

Figure 7.5 Hadamard's method can be used to find the two-dimensional Green's function of Poisson's equation, starting from the Green's function in three dimensions. We can do this by introducing an inhomogeneity that is non-zero only along the z-axis and writing down the corresponding three-dimensional solution.

Example 7.10 A slightly more involved case is trying to go from the three-dimensional Green's function for Poisson's equation to the two-dimensional one. Again, this has to do with the fact that the three-dimensional Green's function may be chosen in such a way that it goes to zero as $r \to \infty$ while the two-dimensional one may not. Following the approach from the previous example, we would formally have

$$G_2(\rho) = -\frac{1}{4\pi} \int_{-\infty}^{\infty} \frac{dz}{\sqrt{\rho^2 + z^2}}, \tag{7.102}$$

where we have introduced cylinder coordinates and put the two-dimensional inhomogeneity along the z-axis, see Fig. 7.5. As we might have suspected from our earlier findings, this integral is logarithmically divergent and we must take care of this divergence by introducing a new zero-level ρ_0 and define our new Green's function as

$$\tilde{G}_2(\rho, \rho_0) = G_2(\rho) - G_2(\rho_0), \tag{7.103}$$

which now formally involves the difference between two divergent integrals that may evaluate to anything depending on how the limits are taken. In order to find the correct two-dimensional Green's function, we need to consider the inhomogeneity $\delta^{(2)}(x^1 \vec{e}_1 + x^2 \vec{e}_2)$ as a limiting case of an inhomogeneity for which the integrals do converge. A natural choice here is given by

$$f(\vec{x}) = \delta^{(2)}(x^1 \vec{e}_1 + x^2 \vec{e}_2)[\theta(z + Z) - \theta(z - Z)] \overset{Z \to \infty}{\longrightarrow} \delta^{(2)}(x^1 \vec{e}_1 + x^2 \vec{e}_2), \tag{7.104}$$

corresponding to an inhomogeneity of finite length $2Z$ along the z-axis. We find that

$$\begin{aligned}
\tilde{G}_2(\rho, \rho_0, Z) &= -\frac{1}{4\pi} \int_{-Z}^{Z} \left(\frac{1}{\sqrt{\rho^2 + z^2}} - \frac{1}{\sqrt{\rho_0^2 + z^2}} \right) \\
&= \frac{1}{2\pi} \ln \left(\frac{\sqrt{\rho_0^2 + Z^2} + Z}{\sqrt{\rho^2 + Z^2} + Z} \frac{\rho}{\rho_0} \right).
\end{aligned} \tag{7.105}$$

Taking the limit $Z \to \infty$ results in

$$\tilde{G}_2(\rho, \rho_0) = \lim_{Z \to \infty} \tilde{G}_2(\rho, \rho_0, Z) = \frac{1}{2\pi} \ln \left(\frac{\rho}{\rho_0} \right), \tag{7.106}$$

which reproduces the previous result.

The above examples of using Hadamard's method on the Green's functions of Poisson's equation are to some extent superfluous as we already knew the answers. However, in other situations, such as for the wave equation, we will need to fall back on this method to compute Green's functions that would otherwise be difficult to find.

7.4 HEAT AND DIFFUSION

The one-dimensional inhomogeneous heat equation

$$\partial_t u(x, t) - a \partial_x^2 u(x, t) = \kappa(x, t) \tag{7.107a}$$

has a Green's function that we are almost already familiar with from Example 6.18, where we found that the solution to the homogeneous heat equation with a delta function initial condition

$$u(x, 0) = \delta(x) \tag{7.107b}$$

was given by

$$u(x, t) = \frac{1}{\sqrt{4\pi a t}} e^{-\frac{x^2}{4at}}. \tag{7.107c}$$

As a direct consequence, this $u(x, t)$ satisfies the homogeneous heat equation for all $t > 0$. Defining the function

$$G(x, t) = \theta(t) u(x, t) = \frac{\theta(t)}{\sqrt{4\pi a t}} e^{-\frac{x^2}{4at}} \tag{7.108}$$

therefore leads to

$$(\partial_t - a\partial_x^2) G(x, t) = \delta(t) u(x, t) + \theta(t) \underbrace{(\partial_t - a\partial_x^2) u(x, t)}_{=0} = \delta(t) u(x, 0) = \delta(t) \delta(x). \tag{7.109}$$

Furthermore, this function clearly satisfies $G(x, t) = 0$ for all $t < 0$ and therefore satisfies all of the necessary requirements to be the Green's function of the one-dimensional heat equation.

This result should not surprise us. After all, we have already seen that the Green's function of the ordinary differential equation

$$\partial_t u(t) - \lambda u(t) = \kappa(t), \tag{7.110}$$

to which the heat equation reduces for each Fourier mode with $\lambda = ak^2$, is also related to the corresponding initial value problem with a homogeneous differential equation, cf. Eq. (7.42).

In addition to already having the Green's function for the one-dimensional heat equation, it turns out that we can express the Green's function for the heat equation in any number of spatial dimensions N by using the one-dimensional solution. Constructing the function

$$G(\vec{x}, t) = \theta(t) \prod_{i=1}^{N} u(x^i, t) = \theta(t) \frac{1}{\sqrt{4\pi a t}^N} e^{-\frac{\vec{x}^2}{4at}} \tag{7.111}$$

and applying the N-dimensional heat operator to it, we find that

$$(\partial_t - a\nabla^2)G(\vec{x}, t) = \delta(t)\prod_{i=1}^{N} u(x^i, t) + \theta(t)(\partial_t - a\nabla^2)\prod_{i=1}^{N} u(x^i, t). \tag{7.112}$$

Using the delta function, the first term in this expression can be rewritten as

$$\delta(t)\prod_{i=1}^{N} u(x^i, t) = \delta(t)\prod_{i=1}^{N} u(x^i, 0) = \delta(t)\prod_{i=1}^{N} \delta(x^i) = \delta(t)\delta^{(N)}(\vec{x}) \tag{7.113a}$$

and for the second term we find that

$$(\partial_t - a\nabla^2)\prod_{i=1}^{N} u(x^i, t) = \sum_{j=1}^{N} \underbrace{[(\partial_t - a\partial_j^2)u(x^j, t)]}_{=0}\prod_{i\neq j} u(x^i, t) = 0. \tag{7.113b}$$

Consequently, it holds that

$$(\partial_t - a\nabla^2)G(\vec{x}, t) = \delta(t)\delta^{(N)}(\vec{x}) \tag{7.114}$$

and the function $G(\vec{x}, t)$ is therefore the Green's function of the N-dimensional heat equation.

The general solution to an initial value problem of the type

$$\text{(PDE)} : u_t(\vec{x}, t) - a\nabla^2 u(\vec{x}, t) = \kappa(\vec{x}, t), \tag{7.115a}$$
$$\text{(IC)} : u(\vec{x}, t) = u_0(\vec{x}), \tag{7.115b}$$

can now be written down directly by using the Green's function

$$u(\vec{x}, t) = \int G(\vec{x} - \vec{x}', t)u_0(\vec{x}')dV' + \int_{t'=0}^{t}\int G(\vec{x} - \vec{x}', t - t')\kappa(\vec{x}', t')\,dV'dt'. \tag{7.116}$$

Just as in the one-dimensional case, the first term in this expression takes care of the inhomogeneous initial condition whereas the second makes sure that the inhomogeneous differential equation is satisfied.

Naturally, we could also have found the N-dimensional Green's function for the heat equation by other means. By performing an N-dimensional Fourier transform of the defining differential equation, we would have found

$$(\partial_t + a\vec{k}^2)\tilde{G}(\vec{k}, t) = \delta(t) \tag{7.117}$$

for the Fourier transform $\tilde{G}(\vec{k}, t)$. This is the definition of the Green's function for an ordinary differential equation involving the linear differential operator $\partial_t + a\vec{k}^2$, which we have already discussed. Taking the inverse Fourier transform will give the result in Eq. (7.111).

Example 7.11 In a homogeneous three-dimensional medium with heat diffusivity a initially at temperature T_0, a heat source is introduced in the origin at $t = 0$ such that the temperature in the medium satisfies the partial differential equation

$$\text{(PDE)} : T_t(\vec{x}, t) - a\nabla^2 T(\vec{x}, t) = \kappa_0\delta^{(3)}(\vec{x}), \tag{7.118a}$$
$$\text{(IC)} : T(\vec{x}, 0) = T_0, \tag{7.118b}$$

for all times $t > 0$. By direct insertion into the integral expression for the solution in Eq. (7.116), we find that the contribution from the inhomogeneous initial condition is given by

$$T^{(\text{IC})}(\vec{x}, t) = T_0 \frac{1}{\sqrt{4\pi a t}^3} \int e^{-\vec{x}^2/4at} dV = T_0, \tag{7.119}$$

which is hardly surprising. If there were no heat source, the medium would keep its homogeneous temperature. In a similar fashion, the contribution from the heat source is given by

$$T^{(\text{PDE})}(\vec{x}, t) = \int_0^t \frac{\kappa_0}{\sqrt{4\pi a t'}^3} e^{-\vec{x}^2/4at'} dt', \tag{7.120a}$$

where we have applied the transformation $t' \to t - t'$ relative to the usual expression. This integral evaluates to

$$T^{(\text{PDE})}(\vec{x}, t) = \frac{\kappa_0}{4\pi a r} \operatorname{erfc}\left(\frac{r}{\sqrt{4at}}\right), \tag{7.120b}$$

where erfc is the complementary error function

$$\operatorname{erfc}(\xi) = \frac{2}{\sqrt{\pi}} \int_\xi^\infty e^{-\zeta^2} d\zeta \tag{7.121}$$

and $r^2 = \vec{x}^2$. We can here note that, for a fixed $\vec{x}$, this solution satisfies

$$\lim_{t \to \infty} T^{(\text{PDE})}(\vec{x}, t) = \frac{\kappa_0}{4\pi a r}. \tag{7.122}$$

This corresponds to the stationary solution, which is the solution to Poisson's equation for a point source. As such, we should expect to recover the Green's function for Poisson's equation in three dimensions up to a factor $-\kappa_0/a$, which is precisely what we have done. This is yet another example of Hadamard's method of descent, where we have used the Green's function of the heat equation to find the Green's function of Poisson's equation. In the other limit, as $t \to 0$, we find that $T^{(\text{PDE})}(\vec{x}, t) \to 0$ whenever $\vec{x} \neq 0$, which satisfies the initial condition. To summarise, we have found that the solution to our problem is given by

$$T(\vec{x}, t) = T^{(\text{IC})}(\vec{x}, t) + T^{(\text{PDE})}(\vec{x}, t) = T_0 + \frac{\kappa_0}{4\pi a r} \operatorname{erfc}\left(\frac{r}{\sqrt{4at}}\right). \tag{7.123}$$

7.5 WAVE PROPAGATION

Unlike in the case of the heat equation, the Green's function of the wave equation in N dimensions will not just be a product of the Green's functions of the one-dimensional wave equations. The underlying reason for why the heat equation displays this feature was the fact that it only contains a first derivative with respect to time and therefore splits into several terms where the time derivative only acts upon one of the factors. In the case of the wave equation, the second derivative with respect to time acting on such a product would lead to cross terms where the first derivative of two different factors would appear. Because of this, we need to seek an alternative way of finding the Green's functions for the wave equation and will do so in one, three, and two dimensions. The reason we will do the three-dimensional case before the two-dimensional is that the two-dimensional one is rather complicated to do from scratch and it is simpler to apply Hadamard's method of descent.

7.5.1 One-dimensional wave propagation

Beginning with the case of the one-dimensional wave equation subjected to an external source

$$u_{tt}(x,t) - c^2 u_{xx}(x,t) = f(x,t), \tag{7.124}$$

we define our Green's function as the function satisfying

$$(\text{PDE}) : G_{tt}(x,t) - c^2 G_{xx}(x,t) = \delta(x)\delta(t), \tag{7.125a}$$

$$(\text{IC}) : G(x, t < 0) = 0. \tag{7.125b}$$

By Fourier transform, we now find

$$\hat{G}_{tt}(k,t) + c^2 k^2 \hat{G}(k,t) = \delta(t) \tag{7.126}$$

as the ordinary differential equation that the Fourier modes $\hat{G}(k,t)$ need to satisfy. This is a differential equation that we have already solved as it is the Green's function for a harmonic oscillator. Taking the result directly from Eq. (7.22), we can immediately write down the solution

$$\hat{G}(k,t) = \frac{1}{ck} \sin(ckt)\theta(t). \tag{7.127}$$

Taking the inverse Fourier transform of this expression, we find that

$$G(x,t) = \frac{\theta(t)}{2\pi c} \int_{-\infty}^{\infty} \frac{e^{ik(x+ct)} - e^{ik(x-ct)}}{2ik} dk, \tag{7.128a}$$

which satisfies

$$G_x(x,t) = \frac{\theta(t)}{4\pi c} \int_{-\infty}^{\infty} (e^{ik(x+ct)} - e^{ik(x-ct)}) dk = \frac{\theta(t)}{2c} [\delta(x+ct) - \delta(x-ct)]. \tag{7.128b}$$

Integrating while keeping the initial conditions in mind now results in

$$G(x,t) = \frac{\theta(t)}{2c} [\theta(x+ct) - \theta(x-ct)]. \tag{7.128c}$$

The behaviour of this Green's function is shown in Fig. 7.6.

Any initial conditions can be taken care of in precisely the same fashion as for the one-dimensional harmonic oscillator and the solution to the general inhomogeneous wave equation in one spatial dimension is given by

$$u(x,t) = \int_{-\infty}^{\infty} [G(x-x',t)v_0(x') + G_t(x-x',t)u_0(x')]dx'$$

$$+ \int_{t'=0}^{t} \int_{-\infty}^{\infty} G(x-x',t-t')f(x',t')dx'dt', \tag{7.129a}$$

where

$$G_t(x,t) = \frac{\theta(t)}{2} [\delta(x+ct) + \delta(x-ct)] \tag{7.129b}$$

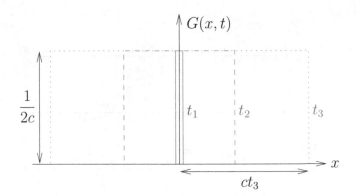

Figure 7.6 The behaviour of the Green's function of the one-dimensional wave equation for $t_1 <$ $t_2 < t_3$. The wave front travels away from the origin at the wave velocity c and once it has passed the function is constant at a value of $1/2c$.

and we have imposed the initial conditions

$$u(x, 0) = u_0(x) \quad \text{and} \quad u_t(x, 0) = v_0(x). \tag{7.129c}$$

For the special case of the homogeneous wave equation and $t > 0$, we can use the particular form of the Green's function and its derivative to simplify this expression as

$$u(x, t) = \frac{1}{2}[u_0(x + ct) + u_0(x - ct)] + \frac{1}{2c} \int_{x-ct}^{x+ct} v_0(x')dx'. \tag{7.130}$$

This particular form of the one-dimensional solution is known as *d'Alembert's formula* and it is easy to explicitly check that it satisfies the correct differential equation and initial conditions.

An alternative way of arriving at d'Alembert's formula is to consider that the homogeneous wave equation can be written on the form

$$u_{tt}(x, t) - c^2 u_{xx}(x, t) = (\partial_t + c\partial_x)(\partial_t - c\partial_x)u(x, t) = 0. \tag{7.131}$$

This implies that any function on the form

$$u(x, t) = f_-(x - ct) + f_+(x + ct) \tag{7.132}$$

will be a solution since $(\partial_t \pm c\partial_x)f(x \mp ct) = 0$ for regardless of the function f. Adapting this to the initial conditions, we find that

$$u_0(x) = u(x, 0) = f_-(x) + f_+(x), \tag{7.133a}$$
$$v_0(x) = u_t(x, 0) = c[f'_+(x) - f'_-(x)]. \tag{7.133b}$$

Integrating the last equation, we find that

$$f_+(x) - f_-(x) = \frac{1}{c} \int_{x_0}^{x} v_0(x')dx', \tag{7.134}$$

where the lower integration boundary x_0 provides an arbitrary integration constant. Solving for $f_\pm(x)$ gives

$$f_\pm(x) = \frac{1}{2}u_0(x) \pm \frac{1}{2c} \int_{x_0}^{x} v_0(x')dx'. \tag{7.135}$$

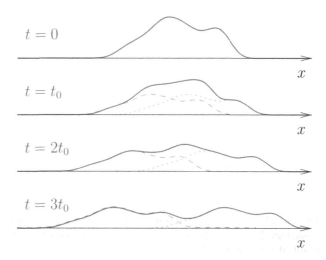

Figure 7.7 The time evolution of a string that is released from rest having the shape shown in the upper diagram. The resulting wave on the string is a superposition of two waves with the same shape and half the amplitude of the original shape, one travelling to the left (dashed curve) and one travelling to the right (dotted curve).

Inserted into Eq. (7.132), this implies that

$$u(x,t) = \frac{1}{2}\left[u_0(x-ct) + u_0(x+ct) + \frac{1}{c}\int_{x-ct}^{x+ct} v_0(x')dx'\right], \tag{7.136}$$

which is exactly the same expression we found by applying Green's function methods.

Example 7.12 Consider an infinite string with tension S and linear density ρ_ℓ. The transversal deviation $u(x,t)$ of this string from the equilibrium position then satisfies the wave equation with $c^2 = S/\rho_\ell$. If the string is released from rest in the position

$$u(x,0) = u_0(x) \tag{7.137}$$

at time $t = 0$, we find that

$$u(x,t) = \frac{1}{2}\left[u_0(x+ct) + u_0(x-ct)\right]. \tag{7.138}$$

Both of the terms in this solution have the same shape as the original, but half the amplitude, and one of them is moving to lower x while the other is moving to larger x, see Fig. 7.7.

7.5.2   Three-dimensional wave propagation

In three spatial dimensions, the Green's function of the wave equation needs to satisfy

$$\text{(PDE)} : G_{tt}(\vec{x},t) - c^2\nabla^2 G(\vec{x},t) = \delta^{(3)}(\vec{x})\delta(t), \tag{7.139a}$$
$$\text{(IC)} : G(\vec{x},t<0) = 0. \tag{7.139b}$$

As for the one-dimensional wave equation, we take the Fourier transform of the differential equation and obtain

$$\hat{G}_{tt}(\vec{k}, t) + c^2 \vec{k}^2 \hat{G}(\vec{k}, t) = \delta(t) \tag{7.140}$$

and consequently

$$\hat{G}(\vec{k}, t) = \frac{\theta(t)}{ck} \sin(ckt), \tag{7.141}$$

where $k = |\vec{k}|$. The inverse Fourier transform is given by

$$G(\vec{x}, t) = \frac{\theta(t)}{(2\pi)^3} \int \frac{1}{ck} \sin(ckt) e^{i\vec{k} \cdot \vec{x}} d\mathcal{K}, \tag{7.142}$$

which we can rewrite in terms of spherical coordinates in $\mathcal{K}$, taking the angle θ_k as the angle between $\vec{x}$ and $\vec{k}$, to obtain

$$G(\vec{x}, t) = \frac{\theta(t)}{(2\pi)^2 c} \int_{k=0}^{\infty} \int_{\theta_k=0}^{\pi} k \sin(ckt) e^{ikr \cos(\theta_k)} \sin(\theta_k) d\theta_k dk$$

$$= \frac{\theta(t)}{(2\pi)^2 cri} \int_{k=0}^{\infty} \sin(ckt)[e^{ikr} - e^{-ikr}] dk = -\frac{\theta(t)}{(2\pi)^2 cr} \int_{-\infty}^{\infty} [e^{ik(r+ct)} - e^{ik(r-ct)}] dk$$

$$= -\frac{\theta(t)}{(2\pi)cr}[\delta(r + ct) - \delta(r - ct)], \tag{7.143}$$

where $r = |\vec{x}|$. From the Heaviside function $\theta(t)$, we know that the Green's function is non-zero only for $t > 0$. In addition, $r > 0$ and therefore $r + ct > 0$, implying that the first delta function in this equation is zero for all $t > 0$ and the Green's function reduces to

$$G(\vec{x}, t) = \frac{\theta(t)}{4\pi cr} \delta(r - ct). \tag{7.144}$$

This expression has several properties that have straightforward physical interpretations. First of all, it is a spherically expanding delta pulse that expands at the wave velocity c. This is something which is peculiar for the three dimensional wave propagation and that did not occur in one dimension. We will discuss this in more detail after looking at the two-dimensional wave equation. Apart from this, the amplitude of the wave decreases as $1/r$, just as the potential outside a spherically symmetric source distribution when considering Poisson's equation. This is in good agreement with the conservation of energy carried by the wave, as the energy is generally proportional to the square of the amplitude. As the wave expands, the total area of the wave front grows as r^2, but is compensated by the decrease in amplitude, leading to conservation of the energy in the wave.

When it comes to the wave being an expanding delta function and therefore singular, it should be remembered that this is a wave generated by a delta impulse in the wave equation. An actual physical source will generally not have this property. In addition, one effect that we have not, and will not, look at is dispersion effects. In many physical situations, the wave velocity c will depend on the wave vector $\vec{k}$. When this occurs, the Fourier modes will no longer conspire to form the delta function in the radial direction, but rather an expanding and dispersing wave front.

Example 7.13 Pressure waves are being created with a frequency ω by a point generator in an extended medium such that the pressure satisfies the wave equation

$$(\text{PDE}): \quad p_{tt}(\vec{x}, t) - c^2 \nabla^2 p(\vec{x}, t) = A\delta^{(3)}(\vec{x}) \sin(\omega t). \tag{7.145}$$

Assuming that the generator is the only relevant source and that any waves from the initial conditions have long since passed, the pressure at a point $\vec{x}$ will be given by

$$p(\vec{x}, t) = \frac{A}{4\pi c} \int_{-\infty}^{t} \int \frac{1}{|\vec{x} - \vec{x}'|} \delta(|\vec{x} - \vec{x}'| - c(t - t')) \delta^{(3)}(\vec{x}') \sin(\omega t') dV' dt'$$

$$= \frac{A}{4\pi c r} \int_{-\infty}^{t} \delta(r - c(t - t')) \sin(\omega t') dt' = \frac{A}{4\pi c^2 r} \sin\left(\omega\left(t - \frac{r}{c}\right)\right), \qquad (7.146)$$

where $r = |\vec{x}|$. Due to the oscillating source at $\vec{x} = 0$, the amplitude of the pressure is constant and proportional to $1/r$. Since it takes the wave a time r/c to travel from the source to $\vec{x}$, the oscillatory behaviour of the solution is phase shifted with respect to the source by an amount $\omega r/c$.

7.5.3 Two-dimensional wave propagation

The Green's function for the two-dimensional wave equation has the same Fourier transform as its one- and three-dimensional counterparts. However, inverting the Fourier transform becomes slightly more involved than in the three-dimensional case due to a different power of k appearing after going to polar coordinates. Instead, the simplest way forward is to use Hadamard's method of descent and we find the two-dimensional Green's function by writing down the three-dimensional problem

$$G_{tt}^2(\rho, t) - c^2 \nabla^2 G^2(\rho, t) = \delta(t)\delta(x^1)\delta(x^2), \qquad (7.147)$$

where ρ is the radial cylinder coordinate. Using the three-dimensional Green's function, we now immediately find

$$G^2(\rho, t) = \frac{\theta(t)}{4\pi c} \int_{-\infty}^{\infty} \frac{1}{\sqrt{\rho^2 + z^2}} \delta(\sqrt{\rho^2 + z^2} - ct) dz$$

$$= \frac{\theta(t)}{2\pi c} \int_{0}^{\infty} \frac{ct\delta(z - \sqrt{c^2 t^2 - \rho^2})}{\sqrt{c^2 t^2 - \rho^2}} \frac{dz}{\sqrt{\rho^2 + z^2}} = \frac{\theta(t)}{2\pi c} \frac{\theta(ct - \rho)}{\sqrt{c^2 t^2 - \rho^2}}$$

$$= \frac{1}{2\pi c} \frac{\theta(ct - \rho)}{\sqrt{c^2 t^2 - \rho^2}}, \qquad (7.148)$$

where, in the last step, we have used that $\theta(t) = 1$ for any value of t for which $\theta(ct - \rho)$ is non-zero. Just as for the one- and three-dimensional Green's function, we find that the resulting wave has a wave front moving at the wave speed c. However, each of the Green's functions in one, two, and three dimensions display certain features that are unique for each dimensionality.

7.5.4 Physics discussion

Let us discuss the differences and similarities between the wave equation Green's functions in different numbers of spatial dimensions. For reference, we show the behaviour of the Green's functions in Fig. 7.8. As can be seen from this figure, the one-dimensional Green's function has a wave front that travels at speed c. After the wave front has passed, the solution remains at the same level as the wave front. As such, applying a delta impulse to a

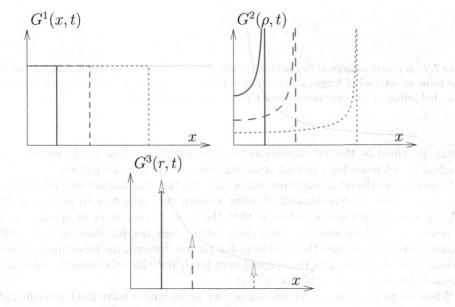

Figure 7.8 The qualitative behaviour of the Green's function of the wave equation in one (left), two (right), and three (bottom) dimensions for different times with the solid curve representing a time t_0, the dashed curve a time $2t_0$, and the dotted curve a time $4t_0$. The thin gray line represents the amplitude of the wave as a function of the position of the wave front. Note that the arrows in the three-dimensional case represent delta distributions, the amplitudes in the corresponding figure represent the multipliers of these distributions.

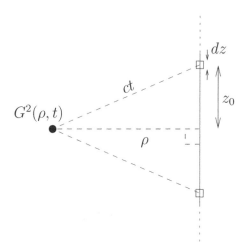

Figure 7.9 A one-dimensional line-inhomogeneity in three dimensions. By looking at the contributions from an interval of length dz around the points $z = z_0$ and $z = -z_0$, we can reason our way to the behaviour of the two-dimensional Green's function $G^2(\rho, t)$.

system described by the one-dimensional wave equation will lead to the system eventually changing its reference level to that resulting from the passing wave front.

In contrast to the situation in one dimension, the two dimensional solution has a singular wave front, since the denominator contains a factor that goes to zero as $ct \to \rho$. Common with the one-dimensional situation is that the wave front travels at speed c away from the source in all directions and that the solution does not immediately fall back to the original reference level once the wave front has passed. Instead, the two-dimensional solution gradually decreases towards the original zero level, but this will formally take an infinite time and the decay goes as $1/t$ for $ct \gg \rho$.

When we go to three spatial dimensions, we again have a wave front travelling at speed c. As in the two-dimensional case, but to an even larger extent, this wave front is singular in nature as it is described by a delta function. The particular property of the three-dimensional Green's function is that, once the wave front has passed, the solution immediately returns to the original reference level and the wave therefore has a non-zero amplitude only at a distance $\rho = ct$ from the source. In general, we see that the Green's function becomes more and more singular and concentrated to $\rho = ct$ as the number of spatial dimensions increases. This behaviour can be physically understood from considering the one- and two-dimensional cases as special cases resulting from particular inhomogeneities according to Hadamard's method.

Let us start by considering the two-dimensional Green's function resulting from a one-dimensional line-inhomogeneity at time $t = 0$ in three-dimensions as depicted in Fig. 7.9. The two-dimensional Green's function is now given by the integral in Eq. (7.148). If we consider the contribution to the Green's function resulting from the inhomogeneity at $z = z_0$, its distance to the point at $z = 0$ and an arbitrary ρ is given by

$$r^2 = z_0^2 + \rho^2. \tag{7.149}$$

Since the Green's function of the three-dimensional wave equation is non-zero only when

$r = ct$, the contribution from $z = z_0$ is therefore only going to give a contribution when

$$c^2 t^2 = z_0^2 + \rho^2. \tag{7.150}$$

For any time $t < \rho/c$, none of the contributions will have reached a distance ρ away from the inhomogeneity and this is the reason for the appearance of the Heaviside function in the two-dimensional Green's function. As different contributions to the inhomogeneity are at different distances from the point of interest, their effects will continue to arrive also for times later than this. However, as the inhomogeneities come from further and further away, we will generally expect a decrease in the two-dimensional Green's function amplitude with time. To start with, a general decrease proportional to the distance $\sqrt{z_0^2 + \rho^2} = ct$ will appear from the three-dimensional Green's function, this factor is apparent in the first two lines of Eq. (7.148). In addition, in a certain time interval between t and $t + dt$, the point of interest is affected by the inhomogeneities between z_0 and $z_0 + dz$, where dz is given by

$$c^2 t\, dt \simeq z_0\, dz \quad \Longrightarrow \quad \frac{dz}{dt} = \frac{c^2 t}{\sqrt{c^2 t^2 - \rho^2}}, \tag{7.151}$$

indicating that this should also be a factor in the two-dimensional Green's function. Putting these pieces together, the two-dimensional Green's function should be given by

$$G^2(\rho, t) = \frac{\theta(t)}{4\pi c^3 t} \frac{2c^2 t}{\sqrt{c^2 t^2 - \rho^2}} \theta(ct - \rho), \tag{7.152a}$$

where the first factor comes from the three-dimensional Green's function, the middle factor is proportional to the total source affecting the point of interest at time t, and the last factor is the Heaviside function describing the necessity of the wave front of the closest source to arrive. Note that there is an additional factor of c in the denominator of the three-dimensional Green's function appearing from the integration over time of the delta function. The two in the middle factor arises due to the source being symmetric and having a contribution from $z = -z_0$ as well as from $z = z_0$. Simplifying this expression, we arrive at

$$G^2(\rho, t) = \frac{1}{2\pi c} \frac{\theta(ct - \rho)}{\sqrt{c^2 t^2 - \rho^2}}, \tag{7.152b}$$

which is exactly the two-dimensional Green's function found in Eq. (7.148). Taking the three factors together therefore accounts exactly for the behaviour of this Green's function.

The above argument can also be applied to the one-dimensional Green's function. At a given time t, the one-dimensional Green's function at $z > 0$ is given by the inhomogeneity at a circular area of radius $\rho = \sqrt{c^2 t^2 - z^2}$, see Fig. 7.10, where we have placed the inhomogeneity at $z = 0$ and used polar coordinates in its plane. The area contributing to the Green's function between times t and $t + dt$ is given by

$$dA = 2\pi\rho\, d\rho = 2\pi c^2 t\, dt \tag{7.153}$$

by arguments exactly equivalent to those used for the two-dimensional Green's function with the roles of z and ρ reversed. Multiplying the factors together, we find

$$G^1(z > 0, t) = \frac{1}{4\pi c^3 t} 2\pi c^2 t\, \theta(ct - z) = \frac{1}{2c}\theta(ct - z). \tag{7.154a}$$

By reflection symmetry in the plane $z = 0$, we must therefore have

$$G^1(z, t) = \frac{1}{2c}\theta(ct - |z|) \tag{7.154b}$$

for arbitrary z. This may also be rewritten on the exact same form as Eq. (7.128c).

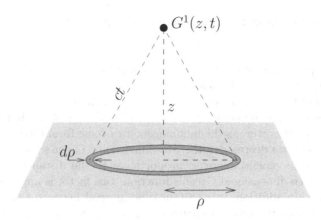

Figure 7.10 The same as Fig. 7.9, but for the case of using a two-dimensional planar inhomogeneity in order to find the Green's function of the one-dimensional wave equation.

7.6   PROBLEMS WITH A BOUNDARY

Up to this point, we have only discussed the Green's function solutions in several dimensions for problems where the region of interest has been all of space. However, just as for the case of one-dimensional Green's function solutions, we will generally encounter problems where the region of interest has one or more boundaries as we discussed briefly in Section 7.2.3. The general Green's function will be defined as having a delta function inhomogeneity in the differential equation and homogeneous boundary conditions

$$\text{(PDE)}: \hat{L}G(\vec{x}, \vec{x}') = \delta(\vec{x} - \vec{x}'), \qquad\qquad (\vec{x} \in V) \qquad\qquad (7.155a)$$

$$\text{(BC)}: \hat{B}G(\vec{x}, \vec{x}') = 0. \qquad\qquad (\vec{x} \in S) \qquad\qquad (7.155b)$$

Solving this differential equation may generally be difficult, but there are some tricks we can use in order to find Green's functions for bounded domains using the tools we have already developed.

7.6.1   Inhomogeneous boundary conditions

As you may have noted above, we will generally demand that the Green's function satisfies homogeneous boundary conditions. We have already seen that this is a reasonable requirement in the case of a one-dimensional problem involving a Sturm–Liouville operator, but let us briefly discuss also the cases of Poisson's equation and the heat equation. For Poisson's equation, imagine that we wish to solve the general problem

$$\text{(PDE)}: \nabla^2 u(\vec{x}) = f(\vec{x}), \qquad\qquad (\vec{x} \in V) \qquad\qquad (7.156a)$$

$$\text{(BC)}: \alpha(\vec{x})u(\vec{x}) + \beta(\vec{x})\vec{n} \cdot \nabla u(\vec{x}) = g(\vec{x}), \qquad (\vec{x} \in S) \qquad\qquad (7.156b)$$

where V is some finite volume and S its boundary. We will work with the case $\alpha(\vec{x}) \neq 0$ as the case $\alpha(\vec{x}) = 0$ runs into exactly the same problems as those encountered for pure Neumann boundary conditions in Section 7.2.2. By a reasoning similar to that presented in that section, we can also deduce that the Green's function for Poisson's equation is symmetric as long as it satisfies homogeneous boundary conditions and we will therefore work under the assumption that

$$G(\vec{x}, \vec{x}') = G(\vec{x}', \vec{x}), \tag{7.157}$$

see Problem 7.14. Requiring that the Green's function satisfies

$$\nabla^2 G(\vec{x}, \vec{x}') = \delta^{(N)}(\vec{x} - \vec{x}'), \tag{7.158}$$

we can rewrite any function $u(\vec{x}')$ according to

$$u(\vec{x}') = \int_V \delta^{(N)}(\vec{x} - \vec{x}')u(\vec{x})dV = \int_V u(\vec{x})\nabla^2 G(\vec{x}, \vec{x}')dV. \tag{7.159}$$

Applying the divergence theorem to this expression twice now results in the relation

$$u(\vec{x}') = \int_V G(\vec{x}, \vec{x}')\nabla^2 u(\vec{x})dV + \oint_S [u(\vec{x})\nabla G(\vec{x}, \vec{x}') - G(\vec{x}, \vec{x}')\nabla u(\vec{x})] \cdot d\vec{S}. \tag{7.160}$$

While this is true for any function $u(\vec{x}')$, this expression is of particular interest whenever $u(\vec{x}')$ satisfies Eqs. (7.156) as it then takes the form

$$u(\vec{x}') = \int_V G(\vec{x}, \vec{x}')f(\vec{x})dV + \oint_S \frac{g(\vec{x})}{\alpha(\vec{x})}\nabla G(\vec{x}, \vec{x}') \cdot d\vec{S}$$
$$- \oint_S \frac{\vec{n} \cdot \nabla u(\vec{x})}{\alpha(\vec{x})}[\beta(\vec{x})\vec{n} \cdot \nabla G(\vec{x}, \vec{x}') + \alpha(\vec{x})G(\vec{x}, \vec{x}')]dS, \tag{7.161}$$

which is true regardless of what boundary conditions we choose to apply to the Green's function. However, a wise choice is to let the Green's function satisfy the homogeneous boundary condition

$$\beta(\vec{x})\vec{n} \cdot \nabla G(\vec{x}, \vec{x}') + \alpha(\vec{x})G(\vec{x}, \vec{x}') = 0 \tag{7.162}$$

as the entire last line of the preceding expression then vanishes and we are left with

$$u(\vec{x}') = \int_V G(\vec{x}, \vec{x}')f(\vec{x})dV + \oint_S \frac{g(\vec{x})}{\alpha(\vec{x})}\nabla G(\vec{x}, \vec{x}') \cdot d\vec{S}, \tag{7.163}$$

where the dependence of the right-hand side on $u(\vec{x})$ has been removed. Selecting the homogeneous boundary conditions of Eq. (7.162) for the Green's function, we have therefore found an expression for the solution to Eqs. (7.156) in terms of the Green's function and the inhomogeneities of the differential equation and boundary conditions. While we here assumed a finite volume V, the argumentation above holds true also for the case of an infinite volume with boundaries as long as the functions involved fall off sufficiently fast as we approach infinity.

For the heat equation, we can follow a similar approach. Starting from the problem

$$\text{(PDE)} : u_t(\vec{x}, t) - a\nabla^2 u(\vec{x}, t) = \kappa(\vec{x}, t), \qquad (\vec{x} \in V) \tag{7.164a}$$
$$\text{(BC)} : u(\vec{x}, t) = f(\vec{x}, t), \qquad (\vec{x} \in S) \tag{7.164b}$$
$$\text{(IC)} : u(\vec{x}, 0) = g(\vec{x}), \qquad (\vec{x} \in V) \tag{7.164c}$$

we consider the relation

$$u(\vec{x}', t') = \int_{t=0}^{\infty} \int_V \delta(t' - t)\delta^{(N)}(\vec{x}' - \vec{x})u(\vec{x}, t)dV\, dt$$
$$= \int_{t'=0}^{\infty} \int_V [\partial_{t'} G(\vec{x}', \vec{x}, t' - t) - a\nabla^2 G(\vec{x}', \vec{x}, t' - t)]u(\vec{x}, t)dV\, dt$$
$$= -\int_{t'=0}^{\infty} \int_V [\partial_t G(\vec{x}', \vec{x}, t' - t) + a\nabla^2 G(\vec{x}', \vec{x}, t' - t)]u(\vec{x}, t)dV\, dt. \tag{7.165}$$

Performing the partial integrations requiring that $G(\vec{x}', \vec{x}, t) = 0$ for all $t < 0$ and that $u(\vec{x}, t)$ satisfies the heat equation, we find that

$$u(\vec{x}', t') = \int_{t=0}^{\infty} \int_{V} G(\vec{x}', \vec{x}, t' - t)\kappa(\vec{x}, t)dV\,dt + \int_{V} G(\vec{x}', \vec{x}, t')u(\vec{x}, 0)dV$$
$$+ a \oint_{S} [G(\vec{x}', \vec{x}, t' - t)\nabla u(\vec{x}, t) - u(\vec{x}, t)\nabla G(\vec{x}', \vec{x}, t' - t)] \cdot d\vec{S}\,dt. \tag{7.166}$$

Applying the boundary and initial conditions from Eqs. (7.164) and choosing homogeneous Dirichlet boundary conditions

$$G(\vec{x}', \vec{x}, t) = 0 \tag{7.167}$$

on S, we obtain the final expression

$$u(\vec{x}', t') = \int_{t=0}^{\infty} \int_{V} G(\vec{x}', \vec{x}, t' - t)\kappa(\vec{x}, t)dV\,dt + \int_{V} G(\vec{x}', \vec{x}, t')g(\vec{x})dV$$
$$- a \int_{t=0}^{\infty} \oint_{S} f(\vec{x}, t)\nabla G(\vec{x}', \vec{x}, t' - t) \cdot d\vec{S}\,dt. \tag{7.168}$$

We here recognise the two first terms from Eq. (7.116), which described the solution to the heat equation in an infinite domain without boundaries. The last term describes the influence of the boundary condition.

7.6.2 Method of images

It should be observed that the differential equation for the Green's function only needs to be satisfied within the volume V. The *method of images* makes use of this fact and attempts to reformulate a given problem with boundaries by considering a problem in a larger domain that has the same solution as the original problem within V. Let us start by considering the Green's function for the one-dimensional heat equation in the region $x > 0$ subjected to a homogeneous boundary condition at $x = 0$

$$(\text{PDE}) : \partial_t G(x, x', t) - a\partial_x^2 G(x, x', t) = \delta(t)\delta(x - x'), \tag{7.169a}$$
$$(\text{BC}) : G(0, x', t) = 0, \tag{7.169b}$$
$$(\text{IC}) : G(x, x', t < 0) = 0. \tag{7.169c}$$

After seeing how to use a mirror image to solve this problem, we will generalise the result to more complicated situations. Based on the given boundary condition, we will approach this situation by first taking the Fourier sine transform of the Green's function, this is given by

$$\tilde{G}_s(k, x', t) = \int_0^{\infty} G(x, x', t)\sin(kx)dx = \frac{1}{2i}\int_0^{\infty} G(x, x', t)(e^{ikx} - e^{-ikx})dx$$
$$= -\frac{1}{2i}\int_{-\infty}^{\infty} \bar{G}(x, x', t)e^{-ikx}dx, \tag{7.170}$$

where we have extended the Green's function to negative values of x by defining

$$\bar{G}(x, x', t) = \begin{cases} G(x, x', t), & (x > 0) \\ -G(-x, x', t), & (x < 0) \end{cases}. \tag{7.171}$$

This anti-symmetric extension is compatible with our boundary conditions as all anti-symmetric functions must be equal to zero at $x = 0$. The extended Green's function $\tilde{G}(x, x', t)$ will therefore automatically be equal to the sought Green's function $G(x, x', t)$ in the region $x > 0$ and satisfy the required boundary condition at $x = 0$. Looking at the Fourier sine transform of $G(x, x', t)$, we can see that it is directly given by the regular Fourier transform of $\tilde{G}(x, x', t)$ as

$$\tilde{G}_s(k, x', t) = -\frac{1}{2i}\tilde{\tilde{G}}(k, x', t). \tag{7.172}$$

Taking the Fourier sine transform of the partial differential equation, we now obtain

$$\partial_t \tilde{G}_s(k, x', t) + ak^2 \tilde{G}_s(k, x', t) = \frac{1}{2i}\delta(t)(e^{ikx'} - e^{-ikx'}). \tag{7.173}$$

Rewriting this in terms of $\tilde{\tilde{G}}(k, x', t)$, we obtain

$$\partial_t \tilde{\tilde{G}}(k, x', t) + ak^2 \tilde{\tilde{G}}(k, x', t) = \delta(t)(e^{-ikx'} - e^{ikx'}). \tag{7.174}$$

We could go right ahead and solve this ordinary differential equation, but it is much more illuminating to take its inverse Fourier transform, which leads to

$$\partial_t \bar{G}(x, x', t) - a\partial_x^2 \bar{G}(x, x') = \delta(t)[\delta(x - x') - \delta(x + x')] \tag{7.175}$$

as an extended problem for $-\infty < x < \infty$. The inhomogeneity on the right-hand side is now the sum of two different contributions, the original one at $x = x'$ and an additional one with the same strength, but opposite sign, at $x = -x' < 0$. As $\delta(x + x') = 0$ for all $x > 0$, the additional contribution does not affect the differential equation in this region and the extended function $\bar{G}(x, x', t)$ still satisfies the original differential equation there. In addition, we also know that this extended function will automatically satisfy the required boundary conditions and can therefore conclude that solving this new problem will give us the Green's function for our problem as

$$G(x, x', t) = \bar{G}(x, x', t) \tag{7.176}$$

for all $x > 0$. In addition, we already know the solution to the new problem as it must be the superposition of the solutions for the sink and source terms and we can directly write it down as

$$G(x, x', t) = G_0(x, x', t) - G_0(x, -x', t) = \frac{1}{\sqrt{4\pi at}}\left(e^{-(x-x')^2/4at} - e^{-(x+x')^2/4at}\right), \tag{7.177}$$

where $G_0(x, x', t)$ is the Green's function for the heat equation in the full range $-\infty < x < \infty$.

So what just happened? In the end, we could find the Green's function of a problem in a restricted space by rewriting the problem as a problem extended to the full infinite domain in which we already knew the Green's function. The final Green's function was a linear combination of the Green's function for an inhomogeneity placed at the same position as the original inhomogeneity and a mirror image inhomogeneity placed outside of the region of interest.

The underlying reason why the addition of the mirror image provides the solution is twofold. First of all, the introduction of any new inhomogeneities outside of the region of interest is not going to affect the differential equation inside it. We can therefore take the

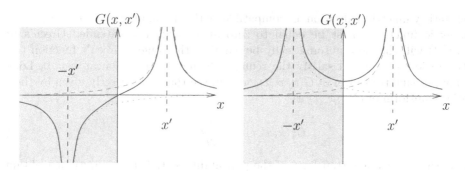

Figure 7.11 The figure shows the odd (left figure) and even (right figure) mirror image approaches to solving problems with Dirichlet and Neumann boundary conditions, respectively. The dashed curves show the Green's function for the actual inhomogeneity at $x = x'$, while the dotted curves show the Green's function for the mirror image at $x = -x'$. The thick curves show the sum of the two, which satisfies the required boundary conditions and have the correct inhomogeneity for the differential equation in the target region. Placing the mirror image in the gray shaded region does not change the inhomogeneity in the region of interest (not shaded).

Green's function of the differential operator in the full space, which will have an inhomogeneity in the correct position, and add to this any Green's function in the full space with an inhomogeneity that is outside of the actual region of interest and we will still have a function that satisfies the differential equation, we just have to find the correct combination to do so while also satisfying the boundary conditions. Second, the addition of the Green's function corresponding to a contribution of the opposite sign at the mirror point makes the new Green's function anti-symmetric. Since anti-symmetric functions automatically satisfy the homogeneous Dirichlet boundary condition, this new Green's function will be the appropriate one.

We could also make the very same argument for a homogeneous Neumann boundary condition at $x = 0$. The first line of the argumentation, that we can add any inhomogeneity outside of the region of interest will still be true. However, we cannot satisfy the Neumann condition by making the Green's function anti-symmetric, but instead we must make it symmetric in order to have a zero derivative at $x = 0$. The Green's function for this situation would therefore instead be given by

$$G(x, x', t) = G_0(x, x', t) + G_0(x, -x', t). \tag{7.178}$$

The odd and even mirror image approaches to problems with Dirichlet and Neumann boundary conditions are illustrated in Fig. 7.11.

The mirror image approach may be applied to any situation where we have homogeneous boundary conditions on a flat surface also in a lager number of spatial dimensions. A requirement for this to work is of course that we can find the Green's function for the differential operator in the extended domain resulting from the mirroring process. Consider the search for a Green's function in N spatial dimensions of either Poisson's equation, the heat equation, or the wave equation in the half-infinite space $x^1 > 0$. Constructing the anti-symmetrised Green's function, with the possible time-dependence suppressed,

$$G_a(\vec{x}, \vec{x}') = G_0(|\vec{x} - \vec{x}'|) - G_0(|\vec{x} - \hat{R}_1\vec{x}'|), \tag{7.179}$$

where we have introduced the linear reflection operator $\hat{R}_1$ that satisfies

$$\hat{R}_1 \vec{e}_i = \begin{cases} -\vec{e}_1, & (i = 1) \\ \vec{e}_i, & (i \neq 1) \end{cases} \tag{7.180}$$

and therefore reflects the vector $\vec{x}'$ in the plane $x'^1 = 0$. In addition, we have used that the Green's function in the full space only depends on the distance between $\vec{x}$ and $\vec{x}'$. Whenever $x^1 = 0$, we find that

$$G_a(\vec{x}, \vec{x}') = 0 \tag{7.181}$$

due to the distance from $\vec{x}$ to $\vec{x}'$ being the same as the distance from $\vec{x}$ to $\hat{R}_1\vec{x}'$, which follows from

$$(\vec{x} - \hat{R}_1\vec{x}')^2 = (-x'^1)^2 + \sum_{i=2}^{N}(x^i - x'^i)^2 = (x'^1)^2 + \sum_{i=2}^{N}(x^i - x'^i)^2 = (\vec{x} - \vec{x}')^2. \tag{7.182}$$

In other words, the anti-symmetrised Green's function $G_a(\vec{x}, \vec{x}')$ is the Green's function for the problem involving homogeneous Dirichlet conditions on $x^1 = 0$. In the same fashion, the symmetrised Green's function

$$G_s(\vec{x}, \vec{x}') = G_0(|\vec{x} - \vec{x}'|) + G_0(|\vec{x} - \hat{R}_1\vec{x}'|) \tag{7.183}$$

will be the Green's function that satisfies homogeneous Neumann conditions on $x^1 = 0$.

Example 7.14 Let us consider heat diffusion in a three-dimensional space described by $x^1 > 0$ with the boundary at $x^1 = 0$ being heat isolated. In addition, we assume that there is a space-dependent heat production that is constant in time such that the stationary temperature is described by Poisson's equation

$$\text{(PDE)}: \nabla^2 T(\vec{x}) = -\kappa(\vec{x}), \qquad\qquad (x^1 > 0) \tag{7.184}$$

$$\text{(BC)}: \partial_1 T(\vec{x}) = 0. \qquad\qquad (x^1 = 0) \tag{7.185}$$

We can construct the Green's function for this problem by performing an even mirroring in the plane $x^1 = 0$ and using the Green's function for Poisson's equation in three dimensions without boundary

$$G(\vec{x}, \vec{x}') = -\frac{1}{4\pi}\left(\frac{1}{|\vec{x} - \vec{x}'|} + \frac{1}{|\vec{x} - \hat{R}_1\vec{x}'|}\right). \tag{7.186}$$

Using this Green's function, we can immediately write down the stationary temperature distribution as

$$T(\vec{x}) = \frac{1}{4\pi}\int_{x'^1 > 0}\left(\frac{1}{|\vec{x} - \vec{x}'|} + \frac{1}{|\vec{x} - \hat{R}_1\vec{x}'|}\right)\kappa(\vec{x}')dV' = \frac{1}{4\pi}\int_{\mathbb{R}^3}\frac{\bar{\kappa}(\vec{x}')}{|\vec{x} - \vec{x}'|}dV', \tag{7.187}$$

where we have rewritten the second term in the original integral as an integral over $x'^1 < 0$ and defined the function

$$\bar{\kappa}(\vec{x}) = \begin{cases} \kappa(\vec{x}), & (x^1 > 0) \\ \kappa(\hat{R}_1\vec{x}), & (x^1 < 0) \end{cases}, \tag{7.188}$$

which is even with respect to the reflection $\hat{R}_1$. Thus, the solution to the problem with the boundary is the same as the solution found when extending the problem to all of space by making an even expansion of the source term. Naturally, this is directly related to the fact that we can find the Green's function by performing an even extension.

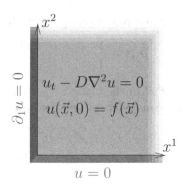

$$\partial_1 u = 0$$

$$u_t - D\nabla^2 u = 0$$
$$u(\vec{x}, 0) = f(\vec{x})$$

$$u = 0$$

Figure 7.12 We are searching for a Green's function that can be used to solve the diffusion equation in the region $x^1, x^2 > 0$ with homogeneous Dirichlet boundary conditions on the boundary $x^2 = 0$ and homogeneous Neumann boundary conditions on the boundary $x^1 = 0$.

7.6.2.1 Multiple mirrors

In some situations, the region of interest will be such that one image is not sufficient to account for the boundary conditions. This may occur in several different ways. The first of these is a rather straightforward generalisation of the case with one mirror image that we have just seen. Consider a domain which is such that several of the coordinates are restricted to be positive, e.g., $x^i > 0$ if $i \leq k$, where k is some integer smaller than or equal to the number of dimensions N, and we assume that we have homogeneous Dirichlet or Neumann conditions on the boundaries. In this situation, we can start by extending the Green's function to the full range $-\infty < x^k < \infty$ by performing the appropriate extension in $x^k = 0$

$$G_1(\vec{x}, \vec{x}') = G_0(\vec{x}, \vec{x}') + s_k G_0(\vec{x}, \hat{R}_k \vec{x}'), \tag{7.189}$$

where $G_0(\vec{x}, \vec{x}')$ is the Green's function for the differential operator in the entire N-dimensional space and $s_k = \pm 1$ depending on the boundary conditions at $x^k = 0$. This Green's function satisfies the differential equation in the region of interest as well as the boundary conditions at $x^k = 0$ by construction. We now have a problem that has one boundary less than the original problem and we can repeat the procedure for x^{k-1} to reduce it further. Doing this type of mirroring for all of the x^i with $i \leq k$, we obtain a Green's function

$$G_k(\vec{x}, \vec{x}') = \sum_\alpha s_1^{\alpha_1} s_2^{\alpha_2} \ldots s_k^{\alpha_k} G(\vec{x}, \hat{R}_1^{\alpha_1} \hat{R}_2^{\alpha_2} \ldots \hat{R}_k^{\alpha_k} \vec{x}'), \tag{7.190}$$

where α is a multi-index with k entries ranging from zero to one. This Green's function will have 2^k terms and satisfy all of the boundary conditions as well as the original differential equation in the region of interest.

Example 7.15 A substance is allowed to diffuse on a two-dimensional surface described by coordinates $x^1 > 0$ and $x^2 > 0$ with diffusivity D. Furthermore, the conditions are such that no substance may flow through the boundary at $x^1 = 0$ and that any substance reaching $x^2 = 0$ is efficiently transported away, see Fig. 7.12. This situation may be described by the differential equation

$$(\text{PDE}) : u_t(\vec{x}, t) - D\nabla^2 u(\vec{x}, t) = 0, \qquad (x^1, x^2, t > 0) \tag{7.191a}$$

$$(\text{BC}): \partial_1 u(\vec{x}, t) = 0, \qquad\qquad (x^1 = 0) \qquad\qquad (7.191\text{b})$$

$$u(\vec{x}, t) = 0, \qquad\qquad (x^2 = 0) \qquad\qquad (7.191\text{c})$$

$$(\text{IC}): u(\vec{x}, 0) = f(\vec{x}), \qquad\qquad (x^1, \ x^2 > 0) \qquad\qquad (7.191\text{d})$$

where $f(\vec{x})$ describes the initial distribution of the substance. Even though there is no inhomogeneity in the differential equation itself, we can use the Green's function derived with an impulse in the differential equation in order to express the solution as an integral over the initial condition as described earlier.

The Green's function we are looking for should satisfy the homogeneous boundary conditions and in order to accomplish this we make an odd extension in the x^2 direction, resulting in the Green's function

$$G_1(\vec{x}, \vec{x}', t) = G_0(\vec{x} - \vec{x}', t) - G_0(\vec{x} - \hat{R}_2 \vec{x}', t), \qquad\qquad (7.192)$$

where $G_0(\vec{x}, t)$ is the Green's function of the diffusion equation in two infinite spatial dimensions. By construction, this Green's function evaluates to zero whenever $x^2 = 0$. In order to also satisfy the boundary condition in the x^1 direction, we now make an even extension of $G_1(\vec{x}, \vec{x}', t)$ to the region $x^1 < 0$ by defining

$$
\begin{aligned}
G_2(\vec{x}, \vec{x}', t) &= G_1(\vec{x}, \vec{x}', t) + G_1(\vec{x}, \hat{R}_1 \vec{x}', t) \\
&= G_0(\vec{x} - \vec{x}', t) - G_0(\vec{x} - \hat{R}_2 \vec{x}', t) + G_0(\vec{x} - \hat{R}_1 \vec{x}', t) - G_0(\vec{x} - \hat{R}_1 \hat{R}_2 \vec{x}', t) \\
&= \frac{\theta(t)}{4\pi D t} \left[e^{-\frac{(\vec{x} - \vec{x}')^2}{4Dt}} - e^{-\frac{(\vec{x} - \hat{R}_2 \vec{x}')^2}{4Dt}} + e^{-\frac{(\vec{x} - \hat{R}_1 \vec{x}')^2}{4Dt}} - e^{-\frac{(\vec{x} - \hat{R}_1 \hat{R}_2 \vec{x}')^2}{4Dt}} \right], \quad (7.193)
\end{aligned}
$$

where we have inserted the explicit expression for $G_0(\vec{x} - \vec{x}', t)$ from Eq. (7.111) in the last step. Applying the diffusion operator $\partial_t - D\nabla^2$ to this Green's function, we find that

$$
\begin{aligned}
(\partial_t - D\nabla^2) G_2(\vec{x}, \vec{x}', t) = \delta(t) [&\delta(\vec{x} - \vec{x}') - \delta(\vec{x} - \hat{R}_2 \vec{x}') \\
&+ \delta(\vec{x} - \hat{R}_1 \vec{x}') - \delta(\vec{x} - \hat{R}_1 \hat{R}_2 \vec{x}')].
\end{aligned}
\qquad (7.194)
$$

Among the delta functions in this expression, only the first one is non-zero in the region $x^1, \ x^2 > 0$ and this Green's function therefore satisfies both the differential equation as well as the necessary boundary conditions. The solution may now be written directly in the form

$$u(\vec{x}, t) = \int_{x'^1, x'^2 > 0} G_2(\vec{x}, \vec{x}', t) f(\vec{x}') dx'^1 dx'^2. \qquad\qquad (7.195)$$

The mirroring procedure in this example is illustrated in Fig. 7.13.

In some situations, we may also need to place several image charges even in problems with boundaries only in one direction. In particular, this will occur if the region of interest is finite and extending the problem by mirroring it in one of the boundaries will result in a new problem in a domain that is twice as large. To be specific, let us consider the one-dimensional heat equation with homogeneous Dirichlet conditions for which the Green's function will satisfy

$$(\text{PDE}): G_t(x, x', t) - a G_{xx}(x, x', t) = \delta(t)\delta(x - x'), \qquad\qquad (7.196\text{a})$$

$$(\text{BC}): G(0, x', t) = G(\ell, x', t) = 0. \qquad\qquad (7.196\text{b})$$

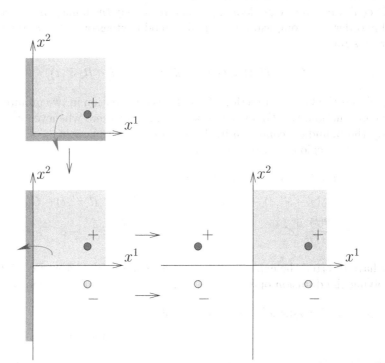

Figure 7.13 An illustration of the mirroring procedure used in Example 7.15. Starting from the original domain (light shaded region) and inhomogeneity, an odd mirror image is positioned at the mirror point with respect to the x^1-axis followed by an even mirror image at the mirror points with respect to the x^2-axis results in a problem that has the same solution in the original domain and that automatically satisfies the required boundary conditions.

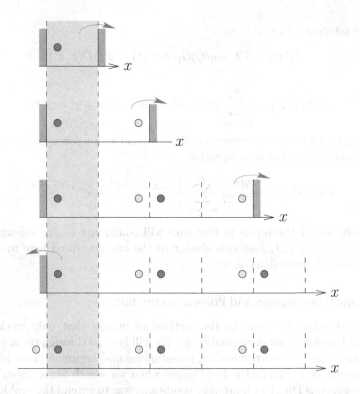

Figure 7.14 For a one-dimensional finite region with homogeneous Dirichlet boundary conditions, we may find the Green's function of the heat equation by repeatedly using odd extensions. For every step, the inhomogeneity in the original domain (shaded) remains the same while the boundary conditions are automatically satisfied due to the imposed symmetry conditions. We end up with an infinite set of periodically placed mirror images.

By extending the Green's function anti-symmetrically around $x = \ell$, see Fig. 7.14, we find that the solution to the problem

$$(\text{PDE}) : G_t^1(x, x', t) - aG_{xx}^1(x, x', t) = \delta(t)[\delta(x - x') - \delta(x - 2\ell + x')], \qquad (7.197a)$$

$$(\text{BC}) : G^1(0, x', t) = G(2\ell, x', t) = 0, \qquad (7.197b)$$

where the differential equation holds for $0 < x < 2\ell$, must coincide with the solution to Eqs. (7.196) in the region $0 < x < \ell$. Repeating the procedure to extend the solution to $0 < x < 4\ell$, and continuing on in the same fashion to larger and larger intervals, the problem is found to be equivalent to

$$(\text{PDE}) : G_t^\infty(x, x', t) - aG_{xx}^\infty(x, x', t) = \delta(t)D(x, x') \qquad (7.198a)$$

$$(\text{BC}) : G^\infty(0, x', t) = 0, \qquad (7.198b)$$

where the function $D(x, x')$ in the inhomogeneity is given by an infinite sum of delta functions

$$D(x, x') = \sum_{k=0}^{\infty} [\delta(x - x' - 2k\ell) - \delta(x + x' - 2(k+1)\ell)]. \qquad (7.198c)$$

Finally, we can perform an extension that is odd with respect to $x = 0$ and obtain a problem on the entire interval $-\infty < x < \infty$

$$(\text{PDE}): \ (\partial_t - a\partial_x^2)\bar{G}_t(x, x', t) = \delta(t)\bar{D}(x, x'), \qquad (7.199\text{a})$$

where

$$\bar{D}(x, x') = \sum_{k=-\infty}^{\infty} [\delta(x - x' - 2k\ell) - \delta(x + x' - 2k\ell)]. \qquad (7.199\text{b})$$

Since this is a problem on the entire real line, we can write down the solution in terms of the Green's function for the heat equation

$$\bar{G}(x, x', t) = \frac{\theta(t)}{\sqrt{4\pi at}} \sum_{k=-\infty}^{\infty} \left[e^{-\frac{(x-x'-2k\ell)^2}{4at}} - e^{-\frac{(x+x'-2k\ell)^2}{4at}} \right]. \qquad (7.200)$$

Note that only one of the terms in this sum will contribute to the inhomogeneity in the original domain $0 < x < \ell$. Methods similar to the one described here may be applied to problems in several dimensions as well.

7.6.3 Spherical boundaries and Poisson's equation

There is a trick when it comes to the method of images that only works for Poisson's equation with Dirichlet boundary conditions. We will here go through the argument for three dimensions and leave the corresponding two-dimensional argumentation as Problem 7.37. The entire argument in the method of images when we search for a Green's function that satisfies homogeneous Dirichlet boundary conditions was to extend the problem to a problem without boundaries by placing images of the inhomogeneity at strategic locations outside the region of interest in order to automatically satisfy the boundary conditions. In the case of flat mirror surfaces, the image inhomogeneities turned out to be of the same magnitude and displaced by the same distance from the surface. It now makes sense to ask the question whether we can encounter situations where the image charges are of different magnitude or displaced by a different distance from the boundary surface. We can investigate this by considering two charges of different magnitude and opposite sign Q and $-q$ placed as shown in Fig. 7.15. If we can find a surface on which the resulting potential is zero, we can consider Poisson's equation either within it or outside of it by using the method of images. In the case when the charge Q is within the region of interest, the charge $-q$ will be the corresponding mirror charge and vice versa.

Based on the argumentation above, we wish to find out where the differential equation

$$(\text{PDE}): \ \nabla^2 V(\vec{x}) = -[Q\delta(\vec{x} - d\vec{e}_1) - q\delta(\vec{x})] \qquad (7.201)$$

results in $V(\vec{x}) = 0$. Before we do this, we start by concluding that there must exist a surface where this holds. In particular, close to the positive inhomogeneity Q, the solution will tend to positive infinity, while close to the negative inhomogeneity $-q$, the solution will tend to negative infinity. In between, the function is continuous and must therefore have a level surface where $V(\vec{x}) = 0$. Since we know the Green's function of Poisson's equation in three dimensions, we can immediately write down the solution for $V(\vec{x})$ in the form

$$V(\vec{x}) = \frac{q}{4\pi} \left(\frac{\xi}{|\vec{x} - d\vec{e}_1|} - \frac{1}{|\vec{x}|} \right) \equiv \frac{q}{4\pi} \left(\frac{\xi}{r_1} - \frac{1}{r_2} \right), \qquad (7.202)$$

where $\xi = Q/q$. This is zero only if $r_1^2 = \xi^2 r_2^2$, which can be explicitly written out as

$$\xi^2(z^2 + \rho^2) = (z - d)^2 + \rho^2. \qquad (7.203\text{a})$$

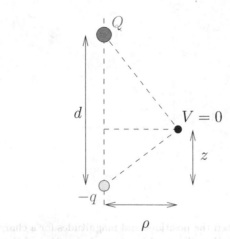

Figure 7.15 For two charges Q and $-q$ of different magnitude and sign we can look for the set of points for which the overall potential is equal to zero. This will turn out to be a spherical surface and we can use the result for using mirror image techniques when solving problems with spherical symmetry.

After some rewriting, this expression becomes

$$\left(z + \frac{d}{\xi^2 - 1}\right)^2 + \rho^2 = \frac{\xi^2 d^2}{(\xi^2 - 1)^2}, \tag{7.203b}$$

which is the equation for a sphere of radius $R = \xi d / |\xi^2 - 1|$ centred on $z = -d/(\xi^2 - 1)$ and $\rho = 0$. For the sake of the argument, we can assume that $\xi > 1$ and perform a translation along the z-axis such that the sphere is centred at the origin. We can also eliminate the distance d from the problem in favour of the distance

$$R_1 = \frac{d}{\xi^2 - 1}, \tag{7.204}$$

which is the distance of the charge $-q$ from the new origin. With this in mind, the distance of the charge Q from the origin will be

$$R_2(R, R_1) = d + \frac{d}{\xi^2 - 1} = \frac{R^2}{R_1} \tag{7.205}$$

as shown in Fig. 7.16. In addition, by comparing the expressions for R and R_1, we can compute the ratio ξ as

$$\xi = \frac{R}{R_1}. \tag{7.206}$$

Thus, if we have a situation where we are looking to find a Green's function to Poisson's equation with Dirichlet boundary conditions on a sphere

$$\text{(PDE)}: \ -\nabla^2 G(\vec{x}, \vec{x}') = \delta^{(3)}(\vec{x} - \vec{x}'), \qquad (|\vec{x}| < R) \tag{7.207a}$$

$$\text{(BC)}: \ G(\vec{x}, \vec{x}') = 0, \qquad (|\vec{x}| = R) \tag{7.207b}$$

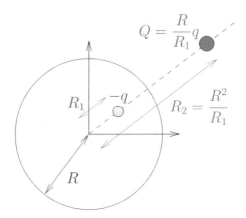

Figure 7.16 The relations between the positions and magnitudes for a charge and its mirror charge when using a mirror image such that the resulting Green's function of Poisson's equation satisfies homogeneous Dirichlet boundary conditions on the sphere with radius R.

we may extend this problem to a problem in the entire three-dimensional space by introducing a mirror inhomogeneity of magnitude $-\xi = -R/|\vec{x}'|$ at the mirror point

$$\hat{R}_R\vec{x}' = \frac{R^2}{|\vec{x}'|^2}\vec{x}' = \xi^2\vec{x}'. \tag{7.208}$$

By construction, the solution $\tilde{G}(\vec{x}, \vec{x}')$ to the extended problem

$$(\text{PDE}): \ -\nabla^2\tilde{G}(\vec{x}, \vec{x}') = \delta^{(3)}(\vec{x} - \vec{x}') - \frac{R}{|\vec{x}'|}\delta^{(3)}(\vec{x} - \hat{R}_R\vec{x}') \tag{7.209}$$

will be equal to the original Green's function in the region $|\vec{x}| < R$ and we can therefore directly write down the Green's function on the form

$$G(\vec{x}, \vec{x}') = \frac{1}{4\pi}\left(\frac{1}{|\vec{x} - \vec{x}'|} - \frac{R}{r'}\frac{1}{|\vec{x} - \hat{R}_R\vec{x}'|}\right) = \frac{1}{4\pi}\left(\frac{1}{|\vec{x} - \vec{x}'|} - \frac{Rr'}{|\vec{x}r'^2 - R^2\vec{x}'|}\right), \tag{7.210}$$

where $r' = |\vec{x}'|$. By construction, this Green's function is automatically equal to zero on the sphere $\vec{x}^2 = R^2$. The exact same argument may be applied to a situation where we are studying Poisson's equation in the region outside of a spherical boundary with Dirichlet conditions. We can also use this type of image when dealing with multiple images due to several different boundaries.

Example 7.16 Consider the electric potential inside a half-sphere with a spherical coordinate system in place such that $r < R$ and $\theta < \pi/2$, see Fig. 7.17. Let us assume that there are no charges inside the half-sphere and that the spherical surface is being kept at potential V_0, while the plane $\theta = \pi/2$ is kept grounded with potential zero. By the rotational symmetry of the problem about the x^3-axis, the potential can only be a function of the spherical coordinates r and θ and is independent of φ. This can be mathematically described by the differential equation

$$(\text{PDE}): \ -\nabla^2 V(r, \theta) = 0, \tag{7.211a}$$

$$(\text{BC}): \ V(R, \theta) = V_0, \quad V(r, \pi/2) = 0. \tag{7.211b}$$

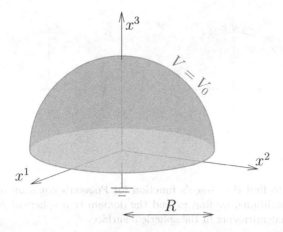

Figure 7.17 We are looking for the Green's function for Poisson's equation in a half-sphere of radius R with Dirichlet boundary conditions. Once found, we will apply it to the problem where the spherical surface is being kept at a potential V_0 while the plane $x^3 = 0$ is grounded.

The Green's function of this problem is the function that satisfies Poisson's equation with homogeneous boundary conditions

$$\text{(PDE)} : \ -\nabla^2 G(\vec{x}, \vec{x}') = \delta(\vec{x} - \vec{x}'), \tag{7.212a}$$

$$\text{(BC)} : G(\vec{x}, \vec{x}') = 0. \qquad\qquad (r = R \text{ or } \theta = \pi/2) \tag{7.212b}$$

The final solution for $V(r, \theta)$ will then be given by Eq. (7.163) as

$$V(\vec{x}) = -V_0 \int_{r=R, \theta < \pi/2} \nabla' G(\vec{x}, \vec{x}') \cdot d\vec{S}'. \tag{7.213}$$

In order to find the Green's function, we first extend the problem to a full sphere $r < R$ by making an odd extension to the domain $\theta > \pi/2$, see Fig. 7.18, resulting in the problem

$$\text{(PDE)} : \ -\nabla^2 G^1(\vec{x}, \vec{x}') = \delta(\vec{x} - \vec{x}') - \delta(\vec{x} - \hat{R}_3 \vec{x}'), \tag{7.214a}$$

$$\text{(BC)} : G^1(\vec{x}, \vec{x}') = 0, \qquad\qquad (|\vec{x}| = R) \tag{7.214b}$$

where $\hat{R}_3$ is the reflection operator in the plane $\theta = \pi/2$ (or, equivalently, $x^3 = 0$). This new problem is a problem inside a sphere of radius R and we can extend it to the full three-dimensional space it by introducing the spherical mirror images of both of the inhomogeneities

$$\text{(PDE)} : \ -\nabla^2 G^2(\vec{x}, \vec{x}') = \delta(\vec{x} - \vec{x}') - \delta(\vec{x} - \hat{R}_3 \vec{x}')$$

$$- \frac{R}{r'} \delta(\vec{x} - \hat{R}_R \vec{x}') + \frac{R}{r'} \delta(\vec{x} - \hat{R}_R \hat{R}_3 \vec{x}'). \tag{7.215}$$

The solution to this problem is directly given by

$$G^2(\vec{x}, \vec{x}') = \frac{1}{4\pi} \left(\frac{1}{|\vec{x} - \vec{x}'|} - \frac{1}{|\vec{x} - \hat{R}_3 \vec{x}'|} - \frac{Rr'}{|\vec{x}r'^2 - R^2\vec{x}'|} + \frac{Rr'}{|\vec{x}r'^2 - R^2\hat{R}_3\vec{x}'|} \right). \tag{7.216}$$

Inserted into Eq. (7.213), this gives the solution to the problem in integral form.

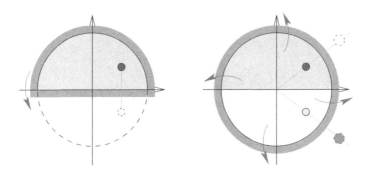

Figure 7.18 In order to find the Green's function for Poisson's equation on the half-sphere with Dirichlet boundary conditions, we first extend the domain to a spherical domain by mirroring in the plane and after that mirroring in the spherical surface.

7.6.4 Series expansions

In some situations, we will encounter problems where the domain and differential operator are such that the solution to the problem may be expressed in the eigenfunctions of Sturm–Liouville operators. We have discussed these cases extensively in Chapter 6 and it therefore makes sense to ask the question of whether or not we can apply that framework to find the Green's function of a given problem. The answer is that not only can we do this, but the resulting Green's function takes a very particular and illuminating form.

Let us start by assuming that we are searching for a Green's function satisfying

$$\text{(PDE)}: \quad \hat{L}G(\vec{x}, \vec{x}') = -\frac{\delta^{(N)}(\vec{x} - \vec{x}')}{w(\vec{x})}, \quad (\vec{x} \in V) \tag{7.217}$$

where the operator $\hat{L}$ is a linear differential operator for which there exists a complete orthonormal eigenbasis of functions $u_n(\vec{x})$ under the inner product with weight function $w(\vec{x})$. We take the eigenvalue of $u_n(\vec{x})$ to be λ_n such that $\hat{L}u_n(\vec{x}) = \lambda_n u_n(\vec{x})$. Expanding the Green's function $G(\vec{x}, \vec{x}')$ in terms of the eigenbasis we know that

$$G(\vec{x}, \vec{x}') = \sum_n G_n(\vec{x}') u_n(\vec{x}), \tag{7.218}$$

where $G_n(\vec{x}')$ is the expansion coefficient of $u_n(\vec{x})$ for a fixed $\vec{x}'$. Inserting this expression into the differential equation for the Green's function, we find

$$\sum_n G_n(\vec{x}') \lambda_n u_n(\vec{x}) = -\frac{\delta(\vec{x} - \vec{x}')}{w(\vec{x})}. \tag{7.219}$$

Since the inner product is of the form

$$\langle f, g \rangle = \int_V f(\vec{x}) g(\vec{x}) w(\vec{x}) \, dV \tag{7.220}$$

we can take the inner product between Eq. (7.219) and $u_m(\vec{x})$ to obtain

$$G_m(\vec{x}') \lambda_m = -\int_V \frac{\delta(\vec{x} - \vec{x}')}{w(\vec{x})} u_m(\vec{x}) w(\vec{x}) \, dV = -u_m(\vec{x}'). \tag{7.221}$$

Solving for the expansion coefficients, we can therefore express the Green's function as

$$G(\vec{x}, \vec{x}') = -\sum_n \frac{1}{\lambda_n} u_n(\vec{x}') u_n(\vec{x}). \tag{7.222}$$

Note that the symmetry $G(\vec{x}, \vec{x}') = G(\vec{x}', \vec{x})$ is obvious from this expression. We should not be very surprised by the form of this equation. If we take a general problem

$$(\text{PDE}): \hat{L}u(\vec{x}) = f(\vec{x}) \tag{7.223}$$

subject to the same homogeneous boundary conditions, the solution will be given by

$$u(\vec{x}) = -\int_V G(\vec{x}, \vec{x}') f(\vec{x}') w(\vec{x}') dV' = \sum_n \frac{u_n(\vec{x})}{\lambda_n} \int_V u_n(\vec{x}') f(\vec{x}') w(\vec{x}') \, dV'$$

$$= \sum_n \frac{u_n(\vec{x})}{\lambda_n} \langle u_n, f \rangle. \tag{7.224}$$

This result is exactly the same as that obtained when expanding the functions $u(\vec{x})$ and $f(\vec{x})$ as sums over the eigenfunction basis. Writing

$$u(\vec{x}) = \sum_n U_n u_n(\vec{x}) \tag{7.225}$$

and taking the inner product between Eq. (7.223) and $u_m(\vec{x})$, we would find

$$\lambda_n U_n = \langle u_n, f \rangle \implies u(\vec{x}) = \sum_n \frac{u_n(\vec{x})}{\lambda_n} \langle u_n, f \rangle. \tag{7.226}$$

The solution of the differential equation by series expansion of the Green's function and the solution found by directly using series expansions of the solution and inhomogeneities are therefore equivalent.

Example 7.17 Let us look at an example where we can find an exact solution relatively easy just in order to exemplify the approach described above. If we wish to find the stationary temperature in a long heat-isolated rod with constant heat diffusivity a where the ends at $x = 0$ and $x = \ell$ are held at temperatures zero and T_0, respectively, this can be described by the second order differential equation

$$(\text{ODE}): -T''(x) = 0, \tag{7.227a}$$

$$(\text{BC}): T(0) = 0, \quad T(\ell) = T_0. \tag{7.227b}$$

The normalised eigenfunctions of the differential operator $-d^2/dx^2$ under the inner product

$$\langle f, g \rangle = \frac{2}{\ell} \int_0^\ell f(x) g(x) \, dx \tag{7.228}$$

are given by

$$u_n(x) = \sin(k_n x), \tag{7.229}$$

where $k_n = \pi n / \ell$. With $w(x) = 2/\ell$, we can immediately write down the corresponding Green's function as

$$G(x, x') = -\sum_{n=1}^\infty \frac{2}{\ell k_n^2} \sin(k_n x') \sin(k_n x) \tag{7.230a}$$

and its derivative with respect to the first argument is therefore

$$G_x(x, x') = -\sum_{n=1}^{\infty} \frac{2}{\ell k_n} \sin(k_n x') \cos(k_n x). \tag{7.230b}$$

According to the discussion in Section 7.2.2, and by using the fact that the differential equation as well as the boundary condition at $x = 0$ are homogeneous, we can now write down the solution for the temperature $T(x)$ directly as

$$T(x') = T_0 G_x(\ell, x') = -\sum_{n=1}^{\infty} \frac{2T_0}{\pi n}(-1)^n \sin(k_n x'). \tag{7.231}$$

Naturally, this is exactly the Fourier series expansion of the function

$$T(x) = T_0 \frac{x}{\ell}, \tag{7.232}$$

which is the solution to the given problem.

7.7   PERTURBATION THEORY

We have so far dealt with differential equations that are linear or linearisable around some stationary or steady state solution. However, physics is a descriptive science and it is not always possible to describe nature using only linear equations. Unfortunately, non-linear equations are often difficult to solve and must sometimes be dealt with numerically. As we shall see in this section, the Green's function formalism can help us go beyond the purely linear regime to solve non-linear differential equations perturbatively. Although it will not give us an exact solution, including just the first terms of a perturbative expansion is going to give us the leading non-linear effects as long as they are relatively small. Perturbation theory and Green's functions plays a leading role in the study of high-energy particle physics and we will end this section by introducing the concept of Feynman diagrams, which are graphical representations of terms in a perturbation series.

The idea behind *perturbation theory* is to consider a differential equation of the form

$$(\text{PDE}): \hat{L}u(\vec{x}) = \lambda f(u(\vec{x}), \vec{x}), \tag{7.233}$$

where $\hat{L}$ is a linear differential operator and the function $f(u(\vec{x}), \vec{x})$ in general is a non-linear function of $u(\vec{x})$. In addition, a condition for perturbation theory to work is that the non-linearities may be considered as small in comparison to the linear terms, which may be emphasised by considering λ to be small. With this assumption, we can write down the solution $u(\vec{x})$ as a power series in the small parameter λ according to

$$u(\vec{x}) = \sum_{n=0}^{\infty} \lambda^n u_n(\vec{x}). \tag{7.234}$$

By computing the first few terms of this sum and neglecting the rest with the argument that λ^n becomes smaller and smaller with increasing n, we can find an approximate solution to our non-linear problem.

In order to find the different functions $u_n(\vec{x})$, we start by inserting the expression for

$u(\vec{x})$ into the non-linear right-hand side of the differential equation and expand the result around $u_0(\vec{x})$

$$\lambda f(u(\vec{x}), \vec{x}) = \lambda \sum_{k=0}^{\infty} \frac{1}{k!} \left.\frac{\partial^k f}{\partial u^n}\right|_{u=u_0(\vec{x})} [u(\vec{x}) - u_0(\vec{x})]^k. \tag{7.235}$$

Collecting terms of the same order in λ, we find that $[u(\vec{x}) - u_0(\vec{x})]^k$ is at least of order λ^k and the leading term on the right-hand side is therefore at least of order λ, resulting from the $k = 0$ term due to the λ multiplying $f(u(\vec{x}), \vec{x})$. We now identify the terms of the same order in λ on both sides of the differential equation. In particular, for the term proportional to λ^0, we will find

$$\hat{L}u_0(\vec{x}) = 0. \tag{7.236}$$

This is a linear problem that may be solved using the framework we have already discussed and so we can find $u_0(\vec{x})$. From the term proportional to λ, we obtain the relation

$$\hat{L}u_1(\vec{x}) = f(u_0(\vec{x}), \vec{x}). \tag{7.237}$$

Here, the right-hand side is a known function, since $u_0(\vec{x})$ is known. If we can find the Green's function $G(\vec{x}, \vec{x}')$ for the linear differential operator $\hat{L}$, we can therefore write down the solution for $u_1(\vec{x})$ on the form

$$u_1(\vec{x}) = \int_V G(\vec{x}, \vec{x}') f(u_0(\vec{x}'), \vec{x}') \, dV'. \tag{7.238}$$

In the same fashion, the term proportional to λ^2 will result in a linear differential equation for $u_2(\vec{x})$ where the inhomogeneity will be expressed in terms of the lower order solutions $u_0(\vec{x})$ and $u_1(\vec{x})$. We can continue in this fashion also to higher orders of perturbation theory and for the term proportional to λ^n, we will obtain an inhomogeneous differential equation for $u_n(\vec{x})$ with an inhomogeneity that only depends on the expansion terms up to $u_{n-1}(\vec{x})$, which we can compute from the lower orders.

Example 7.18 Consider an *anharmonic oscillator* described by the equation of motion

$$(\text{ODE}): \quad \ddot{x}(t) + \omega^2 x(t) = \lambda x(t)^2 \tag{7.239a}$$

and assume that we release it from rest at X_0 at time $t = 0$, which can be described by the initial conditions

$$(\text{IC}): \quad x(0) = X_0, \quad \dot{x}(0) = 0. \tag{7.239b}$$

An anharmonic oscillator generally results when corrections to the quadratic behaviour of a potential around the potential minimum have to be taken into account, see Fig. 7.19. If we assume that the anharmonic term is small, we can write the solution as a power series in λ

$$x(t) = x_0(t) + \lambda x_1(t) + \lambda^2 x_2(t) + \dots. \tag{7.240}$$

Inserted into the differential equation, we find that

$$[\ddot{x}_0(t) + \omega^2 x_0(t)] + \lambda[\ddot{x}_1(t) + \omega^2 x_1(t)] + \mathcal{O}(\lambda^2) = \lambda x_0(t)^2 + \mathcal{O}(\lambda^2). \tag{7.241}$$

Identifying the terms independent of λ on both sides results in

$$\ddot{x}_0(t) + \omega^2 x_0(t) = 0. \tag{7.242}$$

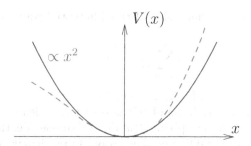

Figure 7.19 The potential of an anharmonic oscillator (dashed curve) does not follow the x^2 behaviour of a harmonic oscillator (solid curve) around its minimum. For this to play a significant role in the motion, the amplitude of the oscillations needs to be large enough for the anharmonic terms to be noticeable.

Since the initial conditions are independent of λ, we find that

$$x_0(t) = X_0 \cos(\omega t). \tag{7.243}$$

For the first correction term $x_1(t)$, the terms proportional to λ in the differential equation result in

$$\ddot{x}_1(t) + \omega^2 x_1(t) = x_0(t)^2 = X_0^2 \cos^2(\omega t). \tag{7.244}$$

From our previous discussion on Green's functions, we know that the Green's function to the differential operator $\partial_t^2 + \omega^2$ is given by

$$G(t) = \frac{\theta(t)}{\omega} \sin(\omega t), \tag{7.245}$$

see Eq. (7.22), and we can therefore write down the integral expression

$$x_1(t) = \int_0^\infty G(t - t') x_0(t')^2 dt' = \frac{X_0^2}{\omega} \int_0^t \sin(\omega(t - t')) \cos^2(\omega t') \, dt'$$

$$= \frac{2X_0^2}{3\omega^2} \sin^2\left(\frac{\omega t}{2}\right) [2 + \cos(\omega t)]. \tag{7.246}$$

The solution $x(t)$ is therefore given by

$$x(t) \simeq X_0 \cos(\omega t) + \lambda \frac{2X_0^2}{3\omega^2} \sin^2\left(\frac{\omega t}{2}\right) [2 + \cos(\omega t)] \tag{7.247}$$

to first order in λ. This solution is shown together with the solution $x_0(t)$ to the corresponding harmonic oscillator as well as the numerical solution and the solution including the second order corrections in Fig. 7.20. Note that, by construction, the perturbation solution $x_1(t)$ satisfies homogeneous initial conditions, which therefore do not interfere with the function $x_0(t)$ satisfying the same initial conditions as $x(t)$.

Some properties of perturbation theory will become clearer if we work with a specific

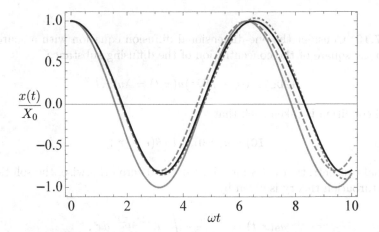

Figure 7.20 The numerical solution (black) to the anharmonic oscillator of Example 7.18 along with the approximative solutions (grey) to zeroth (solid line), first (dashed line), and second (dotted line) order in perturbation theory for $\lambda = 0.3\omega^2/X_0$. Note that the first order perturbation is not able to account for the phase shift of the extreme values, while this is corrected by the second order perturbation.

$f(u(\vec{x}), \vec{x})$ that is a monomial in $u(\vec{x})$. For definiteness, let us take

$$f(u(\vec{x}), \vec{x}) = u(\vec{x})^2, \tag{7.248}$$

as was the case in the preceding example. Performing the expansion of this monomial in λ, we find that

$$u(\vec{x})^2 = u_0(\vec{x})^2 + \lambda 2 u_0(\vec{x}) u_1(\vec{x}) + \lambda^2 [2 u_0(\vec{x}) u_2(\vec{x}) + u_1(\vec{x})^2] + \mathcal{O}(\lambda^3). \tag{7.249}$$

The first of these terms is the term that appears as the inhomogeneity in the differential equation for $u_1(\vec{x})$ and we can immediately write down

$$u_1(\vec{x}) = \int G(\vec{x}, \vec{x}') u_0(\vec{x}')^2 dV'. \tag{7.250}$$

When it comes to the second term, the solutions for $u_0(\vec{x})$ and $u_1(\vec{x})$ are going to appear in the inhomogeneity in the differential equation for $u_2(\vec{x})$

$$\hat{L} u_2(\vec{x}) = 2 u_0(\vec{x}) u_1(\vec{x}), \tag{7.251}$$

resulting in the solution

$$u_2(\vec{x}) = 2 \int G(\vec{x}, \vec{x}') u_0(\vec{x}') u_1(\vec{x}') dV'$$
$$= 2 \int_{\vec{x}'} \int_{\vec{x}''} G(\vec{x}, \vec{x}') G(\vec{x}', \vec{x}'') u_0(\vec{x}') u_0(\vec{x}'')^2 dV' dV'', \tag{7.252}$$

where we have been able to express $u_2(\vec{x})$ only in terms of the Green's function $G(\vec{x}, \vec{x}')$ and the solution $u_0(\vec{x})$ to the linear problem by inserting the expression already found for $u_1(\vec{x})$. In fact, for any differential equation of this type, it will be possible to express the perturbation order λ^n using only the Green's function and $u_0(\vec{x})$ and n integrals.

Example 7.19 Consider the one-dimensional diffusion equation with a source term proportional to the square of the concentration of the diffusing substance

$$(\text{PDE}): (\partial_t - D\nabla^2)u(x,t) = \lambda u(x,t)^2. \tag{7.253}$$

If the initial condition is taken such that

$$(\text{IC}): u(x,0) = U_0\theta(\ell - |x|), \tag{7.254}$$

i.e., the initial concentration is U_0 for all $|x| < \ell$ and zero otherwise, the solution to zeroth order in perturbation theory is given by

$$u_0(x,t) = \frac{U_0}{\sqrt{4\pi Dt}} \int_{-\ell}^{\ell} e^{-\frac{(x-x')^2}{4Dt}} dx', \tag{7.255}$$

where we have used that the Green's function is given by

$$G(x-x',t) = \frac{\theta(t)}{\sqrt{4\pi Dt}} e^{-\frac{(x-x')^2}{4Dt}}. \tag{7.256}$$

The differential equation for the first order correction has an inhomogeneity $\lambda u_0(x,t)^2$ and is therefore given by

$$u_1(x,t) = \int_{t'=0}^{t} \int_{x'=-\infty}^{\infty} G(x-x',t-t')u_0(x',t')^2 dx' \, dt'$$

$$= \int_{t'=0}^{t} \int_{x'=-\infty}^{\infty} \frac{u_0(x',t')^2}{\sqrt{4\pi D(t-t')}} e^{-\frac{(x-x')^2}{4D(t-t')}} dx' \, dt'. \tag{7.257}$$

In the same way, with the inhomogeneity $\lambda^2 2u_0(x,t)u_1(x,t)$, the second order correction can be written down as

$$u_2(x,t) = \int_{t'=0}^{t} \int_{x'=-\infty}^{\infty} \int_{t''=0}^{t'} \int_{x''=-\infty}^{\infty} \frac{u_0(x',t')u_0(x'',t'')^2}{4\pi D\sqrt{(t-t')(t'-t'')}}$$

$$\times e^{-\frac{(x-x')^2}{4D(t-t')}} e^{-\frac{(x'-x'')^2}{4D(t'-t'')}} dx'' \, dt'' \, dx' \, dt'. \tag{7.258}$$

The higher order corrections can be written down by using similar arguments.

7.7.1 Feynman diagrams

For the non-linearities and at the orders in perturbation theory we have dealt with so far, there has only appeared a single term in the inhomogeneity. In general, the inhomogeneity will contain several terms. As an example, the differential equation for $u_3(\vec{x})$ will be of the form

$$\hat{L}u_3(\vec{x}) = 2u_0(\vec{x})u_2(\vec{x}) + u_1(\vec{x})^2 \tag{7.259}$$

in the case when $f(u(\vec{x}),\vec{x}) = u(\vec{x})^2$. Of course, due to the linearity of the left-hand side in $u_3(\vec{x})$, we can still construct it in terms of the basic building blocks we have found, the

Green's function $G(\vec{x}, \vec{x}')$, the zeroth order solution $u_0(\vec{x})$, and a number of integrals. At higher orders in perturbation theory, or for more involved inhomogeneities, the number of terms can become very large. In order to keep track of all the terms, there is a nice pictorial representation of the different terms in the perturbative expansion called *Feynman diagrams*. In order to construct this representation, let us first consider the zeroth order term $u_0(\vec{x})$. We will represent this term as

$$u_0(\vec{x}) = \quad \times\!\!-\!\!-\!\!\bullet^{\,\vec{x}} , \tag{7.260}$$

where the cross represents any inhomogeneity in the zeroth order differential equation or in the boundary conditions, the line represents its propagation to the point $\vec{x}$ through the Green's function, and the point $\vec{x}$ is represented by the black dot. In the same fashion, we can represent $u_0(\vec{x})^2$ as

$$u_0(\vec{x})^2 = \quad \times\!\!\!\!\!\diagdown\!\!\!\!\bullet^{\,\vec{x}} . \tag{7.261}$$

The value of this diagram is given by the multiplication of the lines entering the point $\vec{x}$. We have two lines, each representing one power of $u_0(\vec{x})$ as given by the inhomogeneities propagated by the Green's function.

If we look at the product of two factors of $u_0(\vec{x})$, we note that it is exactly the inhomogeneity appearing in Eq. (7.250) for the first order perturbative correction $u_1(\vec{x})$ for the problem with $f(u, \vec{x}) = u^2$, which is therefore pictorially given by

$$u_1(\vec{x}) = \int G(\vec{x}, \vec{x}') \; \times\!\!\!\!\!\diagdown\!\!\!\!\bullet^{\,\vec{x}'} \, dV'. \tag{7.262}$$

Of course, this looks rather ugly and we would like to include the Green's function as well as the integral over $\vec{x}'$ in the pictorial framework. The Green's function is a function of two different points $\vec{x}$ and $\vec{x}'$ and we can therefore represent it as a line between the them

$$G(\vec{x}, \vec{x}') = \quad \bullet^{\,\vec{x}'}\!\!-\!\!-\!\!\bullet^{\,\vec{x}} . \tag{7.263}$$

At the same time, we can represent the integral over $\vec{x}'$ by using a different kind of point, for example a white circle instead of a black one

$$\int \ldots dV' = \quad \circ^{\,\vec{x}'} . \tag{7.264}$$

With these ingredients, the pictorial representation of $u_1(\vec{x})$ becomes

$$u_1(\vec{x}) = \quad \times\!\!\!\!\!\diagdown\!\!\!\!\circ\!\!-\!\!-\!\!\bullet^{\,\vec{x}} . \tag{7.265}$$

In this diagram, we have also suppressed the $\vec{x}'$ coordinate as it is a dummy coordinate that is anyway integrated over.

So we have found a way of graphically representing the integrals that are going to appear when we do perturbation theory, but, apart from being a more compact form than writing out all of the Green's functions and integrals, how does this help us to actually perform any computations and keep track of the different terms that are going to appear? To this point, we have only worried about finding a pictorial representation of a mathematical object, but we need to find a set of rules describing what terms we are allowed to write down for a given differential equation. This set of rules will be called the *Feynman rules* for the differential equation and let us discuss how we can find them in the special case of the differential equation

$$\hat{L}u(\vec{x}) = \lambda u(\vec{x})^2. \tag{7.266}$$

When we consider the λ^n term in the perturbative expansion of the solution we will obtain

$$\hat{L}u_n(\vec{x}) = \sum_{k=0}^{n-1} u_k(\vec{x})u_{n-1-k}(\vec{x}). \tag{7.267}$$

Solving for $u_n(\vec{x})$, we will therefore use a Green's function $G(\vec{x}, \vec{x}')$ and sum over all of the possible inhomogeneities that contain two functions $u_k(\vec{x})$ and $u_{n-1-k}(\vec{x})$ such that the sum of their orders in perturbation theory equals $n - 1$. These two functions will be represented by two separate Green's function lines going into the vertex at $\vec{x}'$, which is integrated over. In addition, the propagation of the inhomogeneity by the Green's function will be represented by a line going out of the vertex. Since no inhomogeneity contains a product of three or more functions, there will be no vertices with more than two incoming Green's function lines. Therefore, the only vertex which we will be allowed is the vertex

with two lines coming in from the left and one exiting to the right. Since each order of perturbation theory is adding a Green's function integral and $u_0(\vec{x})$ does not contain any vertices that are integrated over, the λ^n term in perturbation theory will be given by all diagrams containing n vertices.

Looking at second order in perturbation theory, we can write $u_2(\vec{x})$ as

$$u_2(\vec{x}) = \quad + \quad , \tag{7.268}$$

where the diagrams result from the two contributing terms in Eq. (7.267). However, in this case the inhomogeneities are $u_0(\vec{x})u_1(\vec{x})$ and $u_1(\vec{x})u_0(\vec{x})$, respectively, and are therefore equivalent. Indeed, it is easy to see that both of the diagrams in the above expression will be mathematically equivalent. This is a general feature at any order in perturbation theory and topologically equivalent diagrams, i.e., diagrams that can be transformed into each other by moving the vertices and edges around, will always be equal to each other. As such, we only need to write down one of these diagrams and multiply them by a symmetry factor. This symmetry factor can be deduced for each vertex as the number of possible inequivalent ways the incoming legs can be attached. In the example above, there are two possible cases for the right vertex, since we can choose to attach the $u_1(\vec{x})$ as the upper or as the lower leg. For the left vertex, which connects two incoming $u_0(\vec{x})$, the symmetry factor is equal to one as the diagram will be the same regardless of which $u_0(\vec{x})$ is connected to which line. Ultimately, this symmetry factor comes from the binomial theorem as the number of possible ways to select one term from each factor in $u(\vec{x})^2$ and combine them to a term on the form $u_k(\vec{x})u_{n-1-k}(\vec{x})$. Thus, taking symmetry into account, we may write

$$u_2(\vec{x}) = 2 \times \quad . \tag{7.269}$$

The computation of the correction at order λ^n in perturbation theory can now be performed by following just a few basic rules:

1. Deduce the Green's function of the linear operator $\hat{L}$.

2. Deduce the allowed vertex or vertices from the non-linearity (if there are several terms in the inhomogeneity $f(u, \vec{x})$, there may be several different vertices).

3. Draw all possible topologically inequivalent Feynman diagrams with n internal vertices and multiply them with the correct symmetry factor.

4. Compute the sum of all of the diagrams computed in the previous step.

Since we will generally be interested in the solution up to a certain accuracy, we want to compute all the contributions in perturbation theory up to some given order in λ. In this case, we can compute $u(\vec{x})$ to nth order by replacing the third step in the process above by drawing all possible topologically inequivalent Feynman diagrams with *at most* n internal vertices and multiplying them with the correct symmetry factor. In this case, each diagram should also be multiplied by λ^k, where k is the number of vertices in the diagram, in order to account for the order in perturbation theory at which the diagram enters.

Example 7.20 Let us write down the third order contribution to $u(\vec{x})$ in the case when the differential equation is that given in Eq. (7.266). We are already aware of the allowed vertex based on our previous discussion and we just need to draw all of the possible topologically inequivalent diagrams with three vertices. There are two such diagrams, and we obtain

$$u_3(\vec{x}) = S_1 \times \quad\text{(diagram)}\quad + S_2 \times \quad\text{(diagram)}, \tag{7.270a}$$

where the S_i are the symmetry factors of their respective diagrams. For the first diagram, there is a symmetry factor of two for the right-most and the middle internal vertices as the legs connecting to them from the left are not equivalent and we therefore obtain $S_1 = 2^2 = 4$. In the case of the second diagram, the right vertex has two equivalent legs connecting from the left, each representing one factor of $u_1(\vec{x})$, while the vertices to the left also have equivalent legs, representing factors of $u_0(\vec{x})$. Because of this, the symmetry factor of the right diagram is given by $S_2 = 1$ and therefore

$$u_3(\vec{x}) = 4 \times \quad\text{(diagram)}\quad + \quad\text{(diagram)}. \tag{7.270b}$$

Collecting all terms up to third order in perturbation theory, the solution $u(\vec{x})$ is now given by

$$u(\vec{x}) = \quad\text{(diagram)}\quad + \lambda \times \quad\text{(diagram)}\quad + 2\lambda^2 \times \quad\text{(diagram)}$$

$$+ 4\lambda^3 \times \quad\text{(diagram)}\quad + \lambda^3 \times \quad\text{(diagram)}\quad + \mathcal{O}(\lambda^4). \tag{7.271}$$

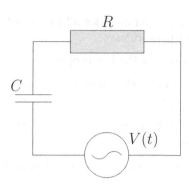

Figure 7.21 An RC circuit with a varying driving voltage $V(t)$. Finding the behaviour of this circuit using Green's function methods is the subject of Problem 7.3.

7.8 PROBLEMS

Problem 7.1. Consider the Green's function of a one-dimensional linear differential operator $\hat{L}$ of order n with constant coefficients where the initial conditions are given by letting all of the derivatives of order $n-1$ or lower be equal to zero at $t = 0$. This Green's function must satisfy the differential equation

$$(\text{ODE}): \quad \hat{L}G(t,t') = \delta(t - t') \tag{7.272}$$

for all $t, t' > 0$.

a) Extend the Green's function domain to the entire real line by defining $G(t,t') = 0$ whenever $t < t'$. Show that the resulting Green's function still satisfies the same differential equation for all t.

b) By making the translations $t \to t + s$ and $t' \to t' + s$, verify that the Greens function can be written in terms of a function of only one argument $G(t,t') = G(t - t')$.

Problem 7.2. The equation of motion for the damped harmonic oscillator is given by

$$m\ddot{x}(t) + 2\alpha\dot{x}(t) + kx(t) = F(t). \tag{7.273}$$

Find the Green's function for this equation. You may assume that the oscillator is underdamped, i.e, that $\alpha^2 < mk$.

Problem 7.3. An RC circuit consists of a single resistor of resistance R, a capacitor with capacitance C, and a variable voltage source $V(t)$, see Fig. 7.21. The voltage across the resistance is given by Ohm's law $V_R = RI = R\,dQ/dt$ and the voltage across the capacitor by $V_C = Q/C$, where Q is the charge stored in the capacitor.

a) Find a differential equation for the charge Q by using the fact that the net change in the potential when going around the circuit should be zero.

b) Construct the Green's function of this differential equation and write down the general solution for an arbitrary initial condition $Q(0) = Q_0$ and driving voltage $V(t)$.

c) Using the result from (b), find the charge Q as a function of time when there is no driving voltage $V(t) = 0$.

d) In the same fashion, find the solution when there is a driving voltage $V(t) = V_0 \sin(\omega t)$ and the capacitor is initially uncharged $Q_0 = 0$.

Problem 7.4. For an RCL circuit, an inductance L is added to the circuit in Problem 7.3. The voltage across the inductance is given by $V_L = L\, d^2Q/dt^2$. In this case, it is not sufficient to know the charge Q at $t = 0$ and an additional initial condition $\dot{Q}(0) = I_0$ is necessary. Show that the differential equation for $Q(t)$ is now that of a damped harmonic oscillator, find its Green's function in the overdamped case ($R^2C > 4L$), and write down the general solution for arbitrary Q_0, I_0, and $V(t)$. What is the resonant frequency of the oscillator? Discuss what happens in the limit where $L \to 0$ and compare with the results of Problem 7.3.

Problem 7.5. Green's function methods can also be used to solve coupled linear differential equations. In nuclear medicine, radioisotopes are often administered to patients and the concentration in different parts of the body can be modelled by coupled linear differential equations. Consider the rudimentary model where a patient is administered a radioisotope with a mean-life of $\tau = 1/\lambda_0$ at a rate $r(t)$. The isotope is modelled to be instantly taken up by the patient's blood stream from which it can either leave the body with decay constant λ_1 or transfer to the patient's thyroid with decay constant λ_T. From the thyroid, the isotope leaves the body with decay constant λ_2 and transfers to the blood with decay constant λ_B. The number of atoms in the blood N_B and thyroid N_T then follow the coupled differential equations

$$\dot{N}_B = r(t) + \lambda_B N_T - \tilde{\lambda}_B N_B, \tag{7.274a}$$
$$\dot{N}_T = \lambda_T N_B - \tilde{\lambda}_T N_T, \tag{7.274b}$$

where we have defined $\tilde{\lambda}_B = \lambda_0 + \lambda_1 + \lambda_T$ and $\tilde{\lambda}_T = \lambda_0 + \lambda_2 + \lambda_B$.

a) Write this linear differential equation on the matrix form

$$\dot{N} + \Lambda N = R(t), \tag{7.275}$$

where $N = (N_B, N_T)^T$ and find an explicit expressions for the 2×2 matrix Λ and the column vector $R(t)$.

b) Define the Green's function matrix

$$G(t) = \begin{pmatrix} G_{BB} & G_{TB} \\ G_{BT} & G_{TT} \end{pmatrix} \tag{7.276}$$

of the system as the time-dependent matrix such that

$$\dot{G} + \Lambda G = \delta(t) \begin{pmatrix} 1 & 0 \\ 0 & 1 \end{pmatrix}. \tag{7.277}$$

Find an explicit expression for this Green's function matrix with the initial condition that $G = 0$ for $t < 0$.

c) Show that the original differential equation is solved by

$$N(t) = \int_0^\infty G(t - t')R(t')dt'. \tag{7.278}$$

d) Given that the rate of radioactive decays in the thyroid is $\Gamma = \lambda_0 N_T$, compute the number of atoms of a total administered dose of N_0 atoms that end up decaying in the thyroid.

Problem 7.6. Define and find the Green's function of the differential equation

$$(\text{ODE}) : u'(t) + g(t)u(t) = h(t), \tag{7.279a}$$

$$(\text{IC}) : u(0) = u_0, \tag{7.279b}$$

where $g(t)$ is an arbitrary function.

Problem 7.7. Consider the two linear differential operators $\hat{L}_1$ and $\hat{L}_2$ with corresponding Green's functions $G_1(t,t')$ and $G_2(t,t')$, respectively. Assume that the Green's functions satisfy $G_i(t,t') = 0$ for $t < t'$.

a) Show that the Green's function composition

$$G(t,t') = \int_{-\infty}^{\infty} G_2(t,t'')G_1(t'',t')dt'' \tag{7.280}$$

is the Green's function of the operator $\hat{L} = \hat{L}_1 \hat{L}_2$ that satisfies $G(t,t') = 0$ for $t < t'$.

b) For the case of $\hat{L}_1 = \hat{L}_2 = d/dt$, compute the Green's function of $\hat{L}_1 \hat{L}_2 = d^2/dt^2$ in the manner described above. Verify that your result is the appropriate Green's function.

Problem 7.8. Consider the function $F(x) = f(x)\theta(x)$ and find expressions for the first and second derivatives $F'(x)$ and $F''(x)$. Use your result to find the conditions that must be imposed on the functions $G^+(x,x')$ and $G^-(x,x')$ in order for $G(x,x') = \theta(x-x')G^+(x,x') + \theta(x'-x)G^-(x,x')$ to be a solution to the differential equation

$$\partial_x^2 G(x,x') + h(x)\partial_x G(x,x') + w(x)G(x,x') = \delta(x-x'). \tag{7.281}$$

Problem 7.9. For the differential equation

$$(\text{ODE}) : -u''(x) = \kappa(x) \tag{7.282}$$

in the region $0 < x < \ell$, consistency in the Neumann boundary conditions require that

$$u'(\ell) = u'(0) - \int_0^\ell \kappa(x)dx. \tag{7.283}$$

We can still solve this problem using Green's function methods up to an arbitrary constant. Perform the following steps:

a) Write down the differential equation that the Green's function $G(x,x')$ must satisfy and the corresponding consistency condition.

b) Assume that $G_x(0,x') = h_0$ and $G(0,x) = g_0$ and find an explicit expression for the Green's function. Verify that your result satisfies the consistency condition.

c) Make the ansatz

$$u(x) = v(x) + \int_0^\ell G(x,x')\kappa(x')dx' \tag{7.284}$$

and find the differential equation and boundary conditions that $v(x)$ must satisfy. Solve the resulting homogeneous differential equation. Note that you should get one arbitrary integration constant that cannot be fixed by the Neumann boundary conditions. Verify that the solution is unique up to an arbitrary shift $u(x) \to u(x) + D$, where D is a constant.

Problem 7.10. Consider the differential equation

$$(\partial_t + \gamma)^2 x(t) = f(t), \qquad (t > 0)$$

where γ is a constant of appropriate dimension. Assume the initial conditions

$$x(0) = x_0, \quad \dot{x}(0) = v_0.$$

a) Compute the Green's function for this system and use it to write down the solution for arbitrary x_0, v_0, and $f(t)$.

b) Find the solution to the problem $x(0) = 0$, $\dot{x}(0) = 0$, $f(t) = f_0 e^{i\gamma t}$, using your result from (a).

Problem 7.11. The position of a particle moving in one dimension is described by the coordinate $x(t)$. If the particle motion is damped and in addition subjected to a time-dependent force $F(t)$, the motion satisfies the differential equation

$$\ddot{x}(t) + \frac{k}{m}\dot{x}(t) = \frac{F(t)}{m}, \tag{7.285}$$

where k is the damping constant and m the particle's mass. Find the Green's function of this differential equation and write down the general solution for the case where $x(0) = \dot{x}(0) = 0$.

Problem 7.12. For the Sturm–Liouville problem

$$(\text{ODE}) : \hat{L}u(x) = -\frac{1}{w(x)}[p(x)u''(x) + p'(x)u'(x)] = f(x), \tag{7.286a}$$

$$(\text{BC}) : u'(a) = \gamma_a, \quad u'(b) = \gamma_b, \tag{7.286b}$$

show that the constants γ_a and γ_b must satisfy the consistency condition

$$p(a)\gamma_a - p(b)\gamma_b = \int_a^b f(x)w(x)dx \tag{7.287}$$

in order for the problem to be solvable. Verify that the condition on the Green's function boundary conditions given in Eq. (7.69) are sufficient for

$$u(x) = -\int_a^b G(x, x')f(x')w(x')dx' \tag{7.288}$$

to automatically satisfy the consistency condition.

Problem 7.13. Find the Green's function $G(x, x')$ of the one-dimensional Sturm–Liouville operator $\hat{L} = -d^2/dx^2$ on the interval $0 < x < \ell$ with the boundary conditions

$$G(0, x') = 0 \quad \text{and} \quad G_x(\ell, x') = 0. \tag{7.289}$$

Use your result to find the stationary solution to the heat equation

$$(\text{PDE}) : T_t(x, t) - aT_{xx}(x, t) = \kappa_0, \tag{7.290a}$$

$$(\text{BC}) : T(0, t) = 0, \quad T_x(\ell, t) = 0, \tag{7.290b}$$

describing one-dimensional heat conduction with a constant source where one boundary is kept at a temperature $T = 0$ and the other is heat isolated.

Problem 7.14. In an arbitrary number of dimensions, use Green's second identity to show that the Green's function of Poisson's equation, which satisfies

$$\text{(PDE)}: \quad \nabla^2 G(\vec{x}, \vec{x}') = \delta^{(N)}(\vec{x} - \vec{x}'), \tag{7.291a}$$

with the homogeneous boundary condition

$$\text{(BC)}: \quad \alpha(\vec{x})G(\vec{x}) + \beta(\vec{x})\vec{n} \cdot \nabla u(\vec{x}) = 0 \tag{7.291b}$$

is symmetric, i.e., that $G(\vec{x}, \vec{x}') = G(\vec{x}', \vec{x})$.

Problem 7.15. Consider a very long heat isolated rod that can be approximated as infinite in both directions. At the time $t = 0$, the rod has the constant temperature T_0 everywhere. For times $t > 0$, a point-like heat source at $x = 0$ produces heat in a time-dependent fashion such that the temperature $T(x, t)$ satisfies

$$\text{(PDE)}: \quad T_t(x, t) - a T_{xx}(x, t) = \frac{\alpha}{\sqrt{t}} \delta(x). \tag{7.292}$$

Write down an integral expression for the temperature at an arbitrary position at any time. In particular, evaluate the temperature at $x = 0$ explicitly by performing the integrals.

Problem 7.16. A string of length ℓ with linear density ρ_ℓ and tension S is hanging in a gravitational field g. One end of the string is kept fixed while the other is allowed to move transversally under a linear restoring force. The transversal displacement $u(x, t)$ of the string then satisfies the differential equation

$$\text{(PDE)}: u_{tt} - c^2 u_{xx} = -g, \qquad\qquad (0 < x < \ell) \tag{7.293a}$$
$$\text{(BC)}: u(0, t) = 0, \quad S u_x(\ell, t) + k u(\ell, t) = 0, \tag{7.293b}$$

where k is the spring constant of the restoring force. Find the Green's function for the stationary state equation and use it to find the stationary state. Verify that you get the same result as you would get if you solved the stationary state differential equation directly.

Problem 7.17. An infinite string with tension S and linear density ρ_ℓ is at rest at times $t < 0$. At $t = 0$, the string is given a transversal impulse density $p_0(x)$ and the resulting transversal oscillations are described by

$$\text{(PDE)}: \quad u_{tt}(x, t) - \frac{S}{\rho_\ell} u_{xx}(x, t) = \delta(t) \frac{p_0(x)}{\rho_\ell}. \tag{7.294}$$

Show the quantity

$$K(t) = \int_{-\infty}^{\infty} \rho_\ell u(x, t) dx \tag{7.295a}$$

grows linearly with time for times $t > 0$, resulting in its time derivative

$$P(t) = K'(t) = \int_{-\infty}^{\infty} \rho_\ell u_t(x, t) dx, \tag{7.295b}$$

i.e., the total transversal momentum of the string, being constant.

Problem 7.18. The results of Problem 7.17 can be generalised to the wave equation in more than one dimension. Starting from a state with $u(\vec{x}, t) = 0$ for all times $t < 0$ and the differential equation

$$(\text{PDE}): \ u_{tt}(\vec{x}, t) - c^2 \nabla^2 u(\vec{x}, t) = \delta(t) v_0(\vec{x}) \tag{7.296}$$

in two and three dimensions, show that

$$X(t) = \int u(\vec{x}, t) dV \tag{7.297a}$$

grows linearly with time and therefore

$$V(t) = X'(t) \int u_t(\vec{x}, t) dV \tag{7.297b}$$

is a constant.

Problem 7.19. We wish to study the sourced diffusion of a substance on the disc $x^2 + y^2 \leq R^2$. We assume that substance escapes the disc as it reaches the border, leading to the problem

$$(\text{PDE}): (\partial_t - D\nabla^2) u(\vec{x}, t) = g(\vec{x}, t), \tag{7.298a}$$

$$(\text{BC}): \ u(\vec{x}, t)|_{\rho=R} = 0, \tag{7.298b}$$

$$(\text{IC}): \ u(\vec{x}, 0) = f(\vec{x}). \tag{7.298c}$$

Find an expression for the Green's function of this problem and write down the general solution.

Problem 7.20. A total electric charge Q is evenly spread out over a spherical shell of radius R. The electrostatic potential $V(\vec{x})$ due to this charge distribution satisfies the partial differential equation

$$(\text{PDE}): \ -\nabla^2 V(\vec{x}) = \frac{Q}{4\pi\varepsilon_0 R^2} \delta(R - r), \tag{7.299a}$$

$$(\text{BC}): \ \lim_{r \to \infty} V(\vec{x}) = 0, \tag{7.299b}$$

in spherical coordinates with the origin at the shell's center. Compute the electric potential for an arbitrary value of the radius r using Green's function methods.

Problem 7.21. A total mass m of a substance is released at time $t = 0$ at the coordinates $x^1 = x^2 = 0$ and $x^3 = x_0^3$ and is allowed to diffuse within the region $x^3 > 0$. Any substance that reaches the boundary at $x^3 = 0$ is immediately adsorbed and the concentration $u(\vec{x}, t)$ can therefore be assumed to satisfy a homogeneous Dirichlet boundary condition at this boundary. Let $\sigma(x^1, x^2, t)$ be the total area density of mass of the substance that has been adsorbed at the coordinates x^1 and x^2 after time t and compute the limit

$$\Sigma(x^1, x^2) = \lim_{t \to \infty} \sigma(x^1, x^2, t). \tag{7.300}$$

Problem 7.22. The inhomogeneous Helmholtz equation is given by

$$\nabla^2 u(\vec{x}) + k^2 u(\vec{x}) = f(\vec{x}), \tag{7.301}$$

where $k^2 > 0$. Write down the partial differential equation that the Green's function of the Helmholtz equation has to satisfy and find a Green's function in the one- and three-dimensional cases. For the three-dimensional case, verify that you recover the Green's function of the Poisson equation in the limit $k^2 \to 0$.

Problem 7.23. For a diffusing substance that decays with decay constant $\lambda > 0$, the concentration $u(\vec{x}, t)$ follows the differential equation

$$(\text{PDE}) : u_t - a\nabla^2 u + \lambda u = \kappa(\vec{x}, t), \tag{7.302a}$$

where $\kappa(\vec{x}, t)$ is a source term. Find the Green's function of this problem in an arbitrary number of spatial dimensions.

Problem 7.24. A long heat isolated rod originally at temperature T_0 at $t = 0$ is subjected to a temperature variation $T_1 \sin(\omega t) + T_0$ at one end. The resulting temperature in the rod satisfies the heat equation

$$(\text{PDE}) : T_t(x, t) - aT_{xx}(x, t) = 0, \tag{7.303a}$$

$$(\text{BC}) : T(0, t) = T_1 \sin(\omega t) + T_0, \quad \lim_{x \to \infty} T(x, t) = T_0, \tag{7.303b}$$

$$(\text{IC}) : T(x, 0) = T_0. \tag{7.303c}$$

Express the temperature $T(x, t)$ in terms of an integral.

Problem 7.25. A half-infinite three-dimensional region given by $x^3 > 0$ is filled with a material of heat conductivity a and the boundary at $x^3 = 0$ is heat isolated. The stationary temperature $T(\vec{x})$ then satisfies the differential equation

$$(\text{PDE}) : \nabla^2 T(\vec{x}) = -\kappa(\vec{x}), \quad (x^3 > 0) \tag{7.304a}$$

$$(\text{BC}) : \partial_3 T(\vec{x}) = 0, \quad (x^3 = 0) \tag{7.304b}$$

where $\kappa(\vec{x})$ is a stationary source term. Write down the solution to this problem in terms of an integral and solve the integral for the source term

$$\kappa(\vec{x}) = \begin{cases} \kappa_0, & (r < R) \\ 0, & (r \geq R) \end{cases}, \tag{7.305}$$

where r is the radius in spherical coordinates.

Problem 7.26. In order to determine the diffusivity of silver in copper, a thin film of a radioactive silver isotope is placed on one end of a long copper rod with cross-sectional area A. Let $u(x, t)$ be the concentration of the silver isotope and assume that it satisfies the one-dimensional diffusion equation with a sink proportional to $u(x, t)$ due to the decays of the isotope. At the time $t = 0$ of the application of the silver film on the rod, the initial condition for the diffusion is given by

$$(\text{IC}) : u(x, 0) = \begin{cases} \frac{m}{\delta A}, & (x < \delta) \\ 0, & (x \geq \delta) \end{cases}, \tag{7.306}$$

where m is the total mass of the silver isotope and δ the thickness of the film. Determine $u(x, t)$, in particular in the case where δ can be considered small, and estimate the diffusivity if the concentration after a time t_0 at the distance ℓ_2 from the rod's end is 10 % of the concentration a distance $\ell_1 < \ell_2$ from the end at the same time.

Problem 7.27. Hadamard's method of descent can also be used on a finite region if the solution to the higher-dimensional problem is independent of the additional coordinate.

a) Find the Green's function of the Laplace operator in two dimensions on a square of side length ℓ with homogeneous Dirichlet boundary conditions in the x^1 direction and homogeneous Neumann boundary conditions in the x^2 direction.

b) Verify that the solution to the two-dimensional problem with the inhomogeneity given by $\delta(x^1 - x'^1)$ is the Green's function of the one-dimensional operator ∂_1^2 with homogeneous Dirichlet boundary conditions on an interval with length ℓ.

Problem 7.28. In the discussion around Eq. (7.117), it was claimed that the Green's function of the N-dimensional heat equation can be found by performing a Fourier transform, solving the ordinary differential equation for the Fourier modes, and then performing the inverse Fourier transform. Perform this computation and verify that you recover the correct Green's function.

Problem 7.29. Verify that Hadamard's method of descent applied to the Green's function of the N-dimensional heat equation can be used to find the Green's function of the heat equation in any lower-dimensional space.

Problem 7.30. Explicitly verify that d'Alembert's formula, see Eq. (7.130), is a solution to the one-dimensional wave equation with the appropriate initial conditions.

Problem 7.31. In the case of the one-dimensional wave equation in a half-infinite region $x > 0$, use mirror image techniques to find the appropriate Green's functions in the case of homogeneous Dirichlet and Neumann boundary conditions at $x = 0$, respectively.

Problem 7.32. Consider an infinite string with wave speed c for the transversal displacement. At time $t = 0$, the string satisfies the initial conditions

$$(\text{IC}) : \quad u(x,0) = u_0 e^{-x^2/a^2} \quad \text{and} \quad u_t(x,0) = -\alpha\delta(x). \tag{7.307}$$

Find the physical dimension of the constant α and determine the resulting transversal displacement $u(x,t)$ by using the Green's function of the one-dimensional wave equation. Determine the shape of the string when $ct \gg x$ and $ct \gg a$.

Problem 7.33. Find the solution to the one-dimensional wave equation with an oscillating point source

$$(\text{PDE}) : \quad u_{tt}(x,t) - c^2 u_{xx}(x,t) = A\delta(x)\sin(\omega t) \tag{7.308}$$

when the initial conditions at $t = 0$ are homogeneous.

Problem 7.34. For a quantity $u(x,t)$ that satisfies the one-dimensional wave-equation, consider the initial conditions

$$(\text{IC}) : \quad u(x,0) = 0 \quad \text{and} \quad u_t(x,0) = v_0\sin(kx), \tag{7.309}$$

where k is a constant. Show that the resulting solution is a standing wave of the form $u(x,t) = X(x)T(t)$ by using Green's function methods and determine the function $T(t)$. Determine where the amplitude is equal to zero.

Problem 7.35. Use an expansion in terms of the eigenfunctions of the Laplace operator in the three-dimensional region given by $r < r_0$ with homogeneous Dirichlet boundary conditions to write down a series that represents the Green's function of the problem

$$(\text{PDE}) : \quad -\nabla^2 u(\vec{x}) = f(\vec{x}), \qquad (r < r_0) \tag{7.310a}$$
$$(\text{BC}) : \quad u(\vec{x}) = 0, \qquad (r = r_0) \tag{7.310b}$$

where r is the radial spherical coordinate.

Problem 7.36. In Problem 7.11, we considered the damped motion of a particle in one dimension under the influence of an external force $F(t)$. In a model for *Brownian motion*, a particle subjected to collisions with the smaller molecules in a suspending medium is under the influence of a random external force with zero expectation value that is also uncorrelated between different times. This is described through the expectation values

$$\langle F(t) \rangle = 0 \quad \text{and} \quad \langle F(t)F(t') \rangle = F_0^2 \delta(t - t'). \tag{7.311}$$

a) Using your solution from Problem 7.11 and assuming that $x(0) = \dot{x}(0) = 0$, write down the solution $x(t)$ for the position of the particle.

b) Since $F(t)$ is a stochastic function, $x(t)$ will be a stochastic variable. Compute the expectation value $\langle x(t) \rangle$ and the variance $\langle x(t)^2 \rangle - \langle x(t) \rangle^2$ of the position as a function of t.

c) For large t, the probability distribution of $x(t)$ will be Gaussian, i.e., its probability density function will be given by

$$p(x, t) = \frac{1}{\sqrt{2\pi\sigma^2(t)}} e^{-\frac{(x - \mu(t))^2}{2\sigma^2(t)}}, \tag{7.312}$$

where $\mu(t)$ is the expectation value and $\sigma^2(t)$ the variance at time t. Identifying with your result from (b), find expressions for μ and σ^2.

d) By definition, the probability density function gives the probability density of the particle moving a distance x in time t. If we have several particles undergoing the same type of motion, their concentration will have the form $u(x, t) = N_0 p(x, t)$, where N_0 is the number of particles. Show that this distribution satisfies the diffusion equation with initial condition $u(x, 0) = N_0 \delta(x)$. Find an expression for the diffusivity in terms of the parameters of the original differential equation for $x(t)$.

Problem 7.37. Let us consider the method of mirror images for Poisson's equation in a circular domain in two dimensions.

a) For two point sources with unit magnitude and opposite sign that are separated by a distance d in two dimensions, find the set of points for which the solution takes a particular value u_0. Show that the expressions are mathematically equivalent to those obtained for the surface of zero potential in the three-dimensional case and, therefore, that the region where the solution is equal to u_0 in the two-dimensional case is a circle.

b) Shifting the entire solution by $-u_0$, we will obtain a new solution to Poisson's equation that is identically zero for the set of points considered in (a). Write down an expression for the new solution.

c) Use your result to write down the Green's function for Poisson's equation inside a circular domain with homogeneous Dirichlet boundary conditions by applying the method of images.

Problem 7.38. A half-cylindrical shell is kept at a potential of $V = 0$ on the cylindrical surface and closed off by a flat surface with potential $V = V_0$. The inside of the shell is free from any charges. This results in the following two-dimensional problem

$$\begin{align}
(\text{PDE}) : \nabla^2 V &= 0, & (\rho < r_0) \tag{7.313a} \\
(\text{BC}) : V(\vec{x}) &= 0, & (\rho = r_0) \tag{7.313b} \\
V(\vec{x}) &= V_0. & (x^1 = 0) \tag{7.313c}
\end{align}$$

Write down the formal solution to this problem using a Green's function. Your answer may contain an integral and derivatives of known functions, i.e., you should write down the Green's function explicitly.

Problem 7.39. We wish to study the sourced diffusion of a substance on the disc of radius r_0. We assume that substance escapes the disc as it reaches the border, leading to the problem

$$
\begin{align}
&(\text{PDE}) : (\partial_t - a\nabla^2)u(\vec{x}, t) = g(\vec{x}, t), & (\rho < r_0) \tag{7.314a} \\
&(\text{BC}) : u(\vec{x}, t) = 0, & (\rho = r_0) \tag{7.314b} \\
&(\text{IC}) : u(\vec{x}, 0) = f(\vec{x}). & (\rho < r_0) \tag{7.314c}
\end{align}
$$

Find an expression for the Green's function of this problem in terms of a series expansion and write down the general solution.

Problem 7.40. A two-dimensional region Ω is described by $\rho > 0$ and $0 < \phi < \pi/4$ in polar coordinates. Determine the Green's function of Poisson's equation on Ω that satisfies homogeneous Dirichlet boundary conditions, i.e., solve the differential equation

$$
\begin{align}
&(\text{PDE}) : \nabla^2 G(\vec{x}, \vec{x}') = \delta^{(2)}(\vec{x} - \vec{x}'), & (\vec{x} \in \Omega) \tag{7.315a} \\
&(\text{BC}) : G(\vec{x}, \vec{x}') = 0, & (\vec{x} \in \partial\Omega) \tag{7.315b}
\end{align}
$$

where $\partial\Omega$ is the boundary of Ω.

Problem 7.41. Consider the four-dimensional Laplace operator $\nabla_4^2 = \nabla_3^2 + \partial_w^2$, where the dimension corresponding to the coordinate w is cyclic, i.e., any function $f(\vec{x}, w)$ satisfies $f(\vec{x}, w) = f(\vec{x}, w + L)$, where L is the size of the fourth dimension.

a) Find the Green's function of this operator by solving the differential equation

$$
\nabla_4^2 G(\vec{x}, w) = \sum_{n=-\infty}^{\infty} \delta(w - nL)\delta^{(3)}(\vec{x}) \tag{7.316}
$$

in an infinite four-dimensional space.

b) Use your result from (a) to find an expression for the Green's function $G(\vec{x}, 0)$.

c) When $r \gg L$, where r is the radial spherical coordinate in three dimensions, we expect to recover the Green's function of the three-dimensional Laplace operator. Starting from your result from (b), show that this is the case and that the corresponding charge is proportional to $1/L$. Also verify that $G(\vec{x}, 0)$ behaves as the Green's function of the four-dimensional Laplace operator when $r \ll L$.

Hint: You may find the identity

$$
\sum_{n=-\infty}^{\infty} \frac{1}{n^2 + k^2} = \frac{\pi}{k} \coth(\pi k) \tag{7.317}
$$

useful.

Problem 7.42. A substance is allowed to diffuse in the two-dimensional region $x^1, x^2 > 0$ with diffusivity a. At the time $t = 0$, a total mass M of the substance is evenly distributed in a quarter circle with radius r_0 and it is assumed that no substance can diffuse through the boundaries at $x^1 = 0$ and $x^2 = 0$, respectively. This can be modelled by the differential equation

$$(\text{PDE}) : u_t(x^1, x^2, t) - a\nabla^2 u(x^1, x^2, t) = 0, \qquad (x^1, x^2, t > 0) \qquad (7.318a)$$

$$(\text{BC}) : u_1(0, x^2, t) = u_2(x^1, 0, t) = 0, \qquad (7.318b)$$

$$(\text{IC}) : u(x^1, x^2, 0) = \xi_0 \delta(\rho - r_0), \qquad (x^1, x^2 > 0) \qquad (7.318c)$$

where ρ is the radial polar coordinate and $u(x^1, x^2, t)$ is the concentration of the substance at the point with coordinates x^1 and x^2 at time t. Determine the physical dimension of the constant ξ_0 and relate it to the total amount M of the substance and then solve the differential equation.

Problem 7.43. The potential $\phi(\vec{x})$ satisfies the Laplace equation in the half-infinite region $x^3 > 0$ and additionally assumes the boundary values

$$(\text{BC}) : \phi(\vec{x}) = \begin{cases} V_0, & (r < r_0) \\ 0, & (r \geq r_0) \end{cases}. \qquad (x^3 = 0) \qquad (7.319)$$

In addition $\phi(\vec{x}) \to 0$ when $r \to \infty$. Use Green's function methods in order to compute the potential along the positive x^3-axis.

Problem 7.44. Find the stationary temperature $T_{\text{st}}(x^1, x^2)$ in the half-plane $x^2 > 0$ if the temperature on the x^1-axis is given by

$$(\text{BC}) : T(x^1, 0, t) = \begin{cases} T_0, & (|x| < \ell) \\ 0, & (|x| \geq \ell) \end{cases}. \qquad (7.320)$$

The temperature $T(x^1, x^2, t)$ may be assumed to satisfy the heat-equation.

Problem 7.45. A thin half-infinite rod without heat-isolation is losing heat proportionally to the temperature difference to the environment. If its end is additionally held at the environmental temperature T_0, the temperature $T(x, t)$ in the rod will satisfy the differential equation

$$(\text{PDE}) : T_t(x, t) - aT_{xx}(x, t) = -\alpha[T(x, t) - T_0], \qquad (x > 0) \qquad (7.321a)$$

$$(\text{BC}) : T(0, t) = T_0, \qquad (7.321b)$$

$$(\text{IC}) : T(x, 0) = T_0 + T_1 e^{-\mu x}, \qquad (7.321c)$$

where we have also assumed that the temperature difference relative to the environment at $t = 0$ is given by $T_1 e^{-\mu x}$. Compute the temperature in the rod for an arbitrary position x and time t.

Problem 7.46. In Example 7.4, we discussed the behaviour of an object falling under the influence of gravitation and a force proportional to the velocity v. For larger velocities, the force is no longer proportional to the velocity but receives a contribution proportional to v^2, resulting in the differential equation

$$(\text{PDE}) : m\dot{v} = mg - kv - \lambda v^2, \qquad (7.322)$$

where λ is a constant of suitable physical dimension. For small λ, use perturbation theory to find the effect of this new contribution on the terminal velocity of the object to first order in λ. Verify your result by finding an exact expression for the terminal velocity and Taylor expanding it around $\lambda = 0$. You may assume that the initial velocity is $v(0) = 0$.

Problem 7.47. While the equation of motion for the harmonic oscillator

$$\ddot{x}(t) + \omega^2 x(t) = 0 \tag{7.323}$$

is linear, it may still be treated by perturbation theory, assuming that ω^2 is a small number or, more precisely, that ωt is small.

a) Considering the linear differential operator d^2/dt^2, find the Green's function that satisfies $G(t) = \theta(t)g(t)$.

b) With ω^2 being a small number, write the general solution as an expansion in ω^2 according to

$$x(t) = \sum_{n=0}^{\infty} \omega^{2n} x_n(t). \tag{7.324}$$

Inserting this ansatz into the harmonic oscillator equation of motion, find the differential equations that must be satisfied for each power of ω^2.

c) Solve the differential equation for the leading order ($n = 0$) contribution with the initial conditions $x(0) = x_0$ and $\dot{x}(0) = v_0$.

d) Using the Green's function found in (a), compute the solution to all orders, i.e., find $x_n(t)$ for all n. Verify that your solution is equivalent to the analytical solution

$$x(t) = x_0 \cos(\omega t) + \frac{v_0}{\omega} \sin(\omega t). \tag{7.325}$$

Problem 7.48. For small values of α, the damping term for the damped harmonic oscillator may be considered as a perturbation

$$(\text{ODE}): \ m\ddot{x}(t) + kx(t) = -2\alpha\dot{x}. \tag{7.326a}$$

Assume that the oscillator is subjected to the initial conditions

$$(\text{IC}): \ x(0) = 0 \quad \text{and} \quad \dot{x}(0) = v_0 \tag{7.326b}$$

and that the solution can be written as a series in α

$$x(t) = \sum_{n=0}^{\infty} \alpha^n x_n(t). \tag{7.327}$$

Find the differential equation each $x_n(t)$ must satisfy. Construct the exact solution for the given initial conditions using your result from Problem 7.2, find the first order correction $x_1(t)$ by expanding it to first order in α, and verify that the $x_1(t)$ you find satisfies the corresponding differential equation.

Problem 7.49. The non-linear differential equation

$$(\text{ODE}): \ \dot{u}(t) = \mu u(t) - \lambda u(t)^2 \tag{7.328}$$

can be solved exactly by integration of $\dot{u}/(\mu u - \lambda u^2) = 1$ with respect to time. Considering λ to be a small parameter, find the first order correction to the zeroth order approximation $u(t) = u(0)e^{\mu t}$ and verify that it coincides with the first order contribution in λ in the Taylor expansion of the exact solution.

Problem 7.50. Consider the non-linear differential equation

$$\hat{L}u(\vec{x}) = \lambda u(\vec{x})^3 \qquad (7.329)$$

where $\hat{L}$ is a linear differential operator. Derive the Feynman rules of this differential equation and write down the solution in terms of Feynman diagrams to third order in perturbation theory.

CHAPTER 8

Variational Calculus

Many physical concepts may be formulated as *variational principles*, which are conditions on the functions describing a physical system such that some given quantity is stationary, i.e., small changes in the functions do not change the quantity. Examples of such principles are Fermat's principle of light propagation, the principle that a system out of which energy dissipates will eventually end up in the state with lowest energy, and Hamilton's principle in classical mechanics. In this chapter, we will eventually introduce and work with all of these principles as we study the language in which they are formulated. In order to do this, we will work with *functionals*, a concept that we briefly encountered in Chapter 5, and discuss how to find their extreme values.

8.1 FUNCTIONALS

When we were discussing different function spaces and distributions, we mentioned functionals as mappings from a function space to the real or the complex numbers. In that discussion, the functionals we were considering were all linear such that

$$F[a_1\varphi_1 + a_2\varphi_2] = a_1 F[\varphi_1] + a_2 F[\varphi_2], \tag{8.1}$$

where F is the functional and the φ_i are functions. In this chapter, we will consider more general functionals that do not necessarily satisfy the requirement of being linear. Our task will be to develop the calculus of variations, which is based on finding the stationary functions of a given functional, i.e., the functions for which the change of the functional is zero when the function is varied. The number of different physical properties that may be described using functionals is large and although not all of them are in integral form, many of them can be expressed in terms of integrals of the form

$$F[\varphi] = \int f(x, \varphi(x), \varphi'(x), \ldots)dx, \tag{8.2}$$

where $f(x, \varphi(x), \varphi'(x), \ldots)$ is some function of x, $\varphi(x)$, and its derivatives. Let us start by giving just a couple of examples of functionals that are of physical relevance.

Example 8.1 Consider a general curve $\vec{x}(t)$ in two dimensions as shown in Fig. 8.1, where t is a curve parameter. The curve may be parametrised by the two coordinate functions $x^1(t)$ and $x^2(t)$. We can always choose the curve parameter such that $\vec{x}(0)$ and $\vec{x}(1)$ are two

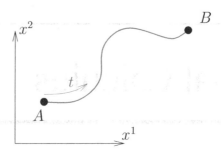

Figure 8.1 A general curve in two dimensions may be described by giving its x^1 and x^2 coordinates as functions of a curve parameter t, which may be chosen such that $\vec{x}(0) = \vec{x}_A$ and $\vec{x}(1) = \vec{x}_B$ are the position vectors of the curve endpoints A and B, respectively. The length of the curve may be described as a functional $\ell[x^1, x^2]$.

given points A and B on the curve and we can compute the distance $\ell[x^1, x^2]$ between A and B along the curve by evaluating the integral

$$\ell[x^1, x^2] = \int_0^1 \sqrt{\dot{x}^1(t)^2 + \dot{x}^2(t)^2}\, dt. \tag{8.3}$$

This is a functional mapping two functions $x^1(t)$ and $x^2(t)$ to the real numbers. Using this expression, we can compute the length of any curve as long as we are given these two functions.

Example 8.2 Given a rope of linear density ρ_ℓ in a homogeneous gravitational field $\vec{g} = -g\vec{e}_3$, see Fig. 8.2, we can express its gravitational potential energy as a functional. For a small part of the rope of mass dm, the gravitational potential is given by

$$dV = gh(x)\, dm = gh(x)\rho_\ell d\ell, \tag{8.4}$$

where $d\ell$ is the length of the part and $h(x)$ is the vertical distance from the reference level. As long as the part is small enough, its length is given by

$$d\ell = \sqrt{dx^2 + dh^2} = \sqrt{1 + h'(x)^2}\, dx. \tag{8.5}$$

Summing the contributions to the gravitational potential from all parts of the rope, we find that the potential energy is given by the functional

$$V[h] = \int_0^{x_0} \rho_\ell gh(x)\sqrt{1 + h'(x)^2}\, dx. \tag{8.6}$$

Note that x_0 is the horizontal distance between the rope endpoints and not the length of the rope, which is given as a functional by summing the lengths of all of the parts as

$$L[h] = \int d\ell = \int_0^{x_0} \sqrt{1 + h'(x)^2}\, dx. \tag{8.7}$$

Note that this length is the same expression as that found in Example 8.1, but using the horizontal coordinate x as the curve parameter.

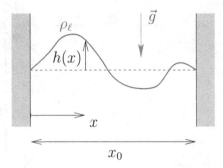

Figure 8.2 The potential energy of a rope of a fixed linear density ρ_ℓ in a gravitational field may be described as a functional by summing the contributions from each small part of the rope.

Both of the examples above have the common property of being non-linear functionals and are of the form given in Eq. (8.2). In a similar fashion to these examples, functionals can often be used to describe other physical quantities of interest.

8.2   FUNCTIONAL OPTIMISATION

A large number of physics problems deal with the optimisation of different functions and functionals, i.e., the search for extreme values. While we are already familiar with how this works for functions in single- and multi-variable calculus, we need to develop a framework for finding the extrema of functionals. This may seem like a daunting task, but it turns out to be very similar to the approach used in ordinary calculus.

Let us start by making the observation that the functional $F[\varphi]$ is a mapping from a functional space to the real numbers. As such, we can always make a small change in the argument of the functional and we define

$$F(\varepsilon) = F[\varphi + \varepsilon\eta], \qquad (8.8)$$

where η is some function in the function space and ε is a small number such that $\varepsilon\eta$ is a small change in the function φ. If we keep the functions φ and η fixed, then $F(\varepsilon)$ is a function of ε and we can find its stationary points in the usual fashion in single-variable calculus, by differentiating it with respect to the argument and setting the derivative equal to zero. Naturally, the exact form of this derivative is generally going to depend on the form of the functional $F[\varphi]$. If we can find a function φ such that $F(0)$ is stationary regardless of the function η, then it corresponds to a stationary value of the functional $F[\varphi]$. In particular, if $F(\varepsilon)$ has a minimum at $\varepsilon = 0$ regardless of η, then φ corresponds to a minimum of the functional.

Example 8.3   Consider the problem of finding the shortest path between two points in a two-dimensional Euclidean space, see Fig. 8.3. We can introduce coordinates such that point A corresponds to the origin and point B corresponds to $x = x_0$ and $y = y_0$. Based on our intuition, the shortest path should be given by the straight line

$$y(x) = y_0 \frac{x}{x_0} \equiv kx \qquad (8.9)$$

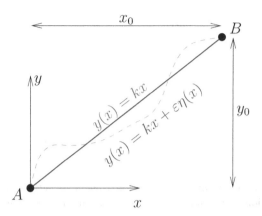

Figure 8.3 There are several possible curves starting at A and ending at B. Here we show the straight line described by the function $y(x) = kx$, with $k = y_0/x_0$, and an arbitrary small variation $y(x) = kx + \varepsilon\eta(x)$. We wish to show that the straight line gives a stationary value for the curve length regardless of the variation $\eta(x)$.

between the points, a fact we can use the argumentation above to verify. In general, the length of any curve $y(x)$ between A and B can be written as

$$L[y] = \int_0^{x_0} \sqrt{1 + y'(x)^2}\, dx, \tag{8.10}$$

where we require that $y(0) = 0$ and $y(x_0) = y_0$ in order for the curve to be a curve between the given points. We now define the function

$$L(\varepsilon) = L[kx + \varepsilon\eta] = \int_0^{x_0} \sqrt{1 + [k + \varepsilon\eta'(x)]^2}\, dx, \tag{8.11}$$

which gives the length of the curve $y(x) = kx + \varepsilon\eta(x)$. Differentiating with respect to ε, we find that

$$L'(\varepsilon) = \int_0^{x_0} \frac{[k + \varepsilon\eta'(x)]\eta'(x)}{\sqrt{1 + [k + \varepsilon\eta'(x)]^2}}\, dx, \tag{8.12a}$$

which for $\varepsilon = 0$ evaluates to

$$L'(0) = \frac{k}{\sqrt{1 + k^2}} \int_0^{x_0} \eta'(x)dx = \frac{k}{\sqrt{1 + k^2}}[\eta(x_0) - \eta(0)] = 0 \tag{8.12b}$$

as $\eta(x_0) = \eta(0) = 0$ must hold for the curve $kx + \varepsilon\eta(x)$ to start at A and end at B. From this follows that $y(x) = kx$ is a stationary function for the curve length between A and B. Computing the second derivative $L''(0)$, we can also verify that this curve is a minimum, see Problem 8.1.

The example above shows that the straight line between two points indeed gives a stationary function for the path length. However, we already had a pretty good idea about what the solution should be and it does not really bring us any closer to a general method for finding the stationary functions.

8.2.1 Euler–Lagrange equations

In order to find the stationary functions, let us for now specialise to functionals acting on functions of one variable x in the interval $a < x < b$ of the form

$$F[\varphi] = \int_a^b \mathcal{L}(\varphi(x), \varphi'(x), x)dx, \tag{8.13}$$

where $\mathcal{L}(\varphi, \varphi', x)$ is some function of three variables. For the time being, let us assume that the function φ is required to take a particular value at the boundaries, as in Example 8.3, where the endpoint values were necessary for the curve to start and end at the correct points. As before, we introduce the function

$$F(\varepsilon) = F[\varphi + \varepsilon\eta] = \int_a^b \mathcal{L}(\varphi(x) + \varepsilon\eta(x), \varphi'(x) + \varepsilon\eta'(x), x)dx \tag{8.14}$$

and take the derivative with respect to ε, leading to

$$
\begin{aligned}
F'(\varepsilon) &= \int_a^b \frac{d}{d\varepsilon}\mathcal{L}(\varphi(x) + \varepsilon\eta(x), \varphi'(x) + \varepsilon\eta'(x), x)dx \\
&= \int_a^b \left[\frac{\partial\mathcal{L}}{\partial\varphi}\eta(x) + \frac{\partial\mathcal{L}}{\partial\varphi'}\eta'(x)\right]dx,
\end{aligned} \tag{8.15a}
$$

where $\partial\mathcal{L}/\partial\varphi$ is the derivative of $\mathcal{L}$ with respect to its first argument and $\partial\mathcal{L}/\partial\varphi'$ is the derivative with respect to its second argument. Using partial integration on the second term, we find that

$$
\begin{aligned}
F'(\varepsilon) &= \eta(b)\left.\frac{\partial\mathcal{L}}{\partial\varphi'}\right|_{x=b} - \eta(a)\left.\frac{\partial\mathcal{L}}{\partial\varphi'}\right|_{x=a} + \int_a^b\left[\frac{\partial\mathcal{L}}{\partial\varphi} - \frac{d}{dx}\frac{\partial\mathcal{L}}{\partial\varphi'}\right]\eta(x)dx \\
&= \int_a^b\left[\frac{\partial\mathcal{L}}{\partial\varphi} - \frac{d}{dx}\frac{\partial\mathcal{L}}{\partial\varphi'}\right]\eta(x)dx,
\end{aligned} \tag{8.15b}
$$

where we have used that the *variation* $\varepsilon\eta(x)$ must vanish at the endpoints. In order for φ to be a stationary function of $F[\varphi]$, this derivative must vanish when evaluated at $\varepsilon = 0$ for all variations $\eta(x)$. For this to hold, the other factor in the integrand must be identically equal to zero and $\varphi(x)$ therefore must satisfy the differential equation

$$\frac{\partial\mathcal{L}}{\partial\varphi} - \frac{d}{dx}\frac{\partial\mathcal{L}}{\partial\varphi'} = 0. \tag{8.16}$$

Again, it needs to be stressed that the partial derivatives with respect to φ and φ' are the partial derivatives of $\mathcal{L}$ with respect to the first and second arguments, respectively. In both cases, they should be evaluated for $\varepsilon = 0$. This equation is known as the *Euler–Lagrange equation*.

Example 8.4 Again consider the problem of finding the shortest path between two points studied in Example 8.3. Since the length of a curve $y(x)$ is given by

$$L[y] = \int_0^{x_0} \sqrt{1 + y'(x)^2}\, dx, \tag{8.17a}$$

we find that

$$\mathcal{L}(y, y', x) = \sqrt{1 + y'^2}, \tag{8.17b}$$

which is only a function of y'. Computing the partial derivatives now leads to

$$\frac{\partial \mathcal{L}}{\partial y} = 0 \quad \text{and} \quad \frac{d}{dx}\frac{\partial \mathcal{L}}{\partial y'} = \frac{d}{dx}\frac{y'}{\mathcal{L}}. \tag{8.18}$$

Insertion into the Euler–Lagrange equation results in the differential equation

$$y''(x) = 0 \tag{8.19a}$$

with the general solution

$$y(x) = kx + m. \tag{8.19b}$$

Using the same boundary conditions as in Example 8.3, we find that $m = 0$ and $k = y_0/x_0$, which is exactly the straight line from A to B that we earlier could only verify as being the solution after taking it as an ansatz.

It is common that the small number ε is considered part of the variation η. Naturally, if this is done, we would instead consider the limit $\eta \to 0$ rather than $\varepsilon \to 0$. The number ε is usually reintroduced when considering very particular shapes of the perturbation in order to give a well defined limit.

In some situations, the functional will depend not only on one, but on several different functions $\varphi_i(x)$. When this occurs, the functions should be varied independently by assigning different variations to each of the functions

$$\varphi_i(x) \to \varphi_i(x) + \varepsilon \eta_i(x). \tag{8.20}$$

In particular, we can consider variations $\eta_i(x) = 0$ for all $i \neq j$, where j is some fixed value. This results in the Euler–Lagrange equation for φ_j

$$\frac{\partial \mathcal{L}}{\partial \varphi_j} - \frac{d}{dx}\frac{\partial \mathcal{L}}{\partial \varphi_j'} = 0. \tag{8.21}$$

Repeating this for all possible j results in one Euler–Lagrange equation for each function $\varphi_j(x)$.

8.2.1.1 Natural boundary conditions

In the derivation of the Euler-Lagrange equation, we considered the endpoint values of the function $\varphi(x)$ to be fixed such that the partial integration could be carried out without caring about the boundary conditions. As we shall see, this is a rather constraining situation and many times it will not be the case that the endpoint values can be fixed by some underlying principle. In fact, keeping the general boundary term will actually provide us with a boundary condition that the stationary functions of the functional must satisfy. Since these boundary conditions are not imposed, this type of boundary condition is called a *natural boundary condition*. Looking back at Eq. (8.15b), we found that

$$F'(\varepsilon) = \eta(b) \left.\frac{\partial \mathcal{L}}{\partial \varphi'}\right|_{x=b} - \eta(a) \left.\frac{\partial \mathcal{L}}{\partial \varphi'}\right|_{x=a} + \int_a^b \left[\frac{\partial \mathcal{L}}{\partial \varphi} - \frac{d}{dx}\frac{\partial \mathcal{L}}{\partial \varphi'}\right] \eta(x) dx. \qquad (8.22)$$

We argued that in order for φ to be a local stationary function of the functional, $F'(0)$ necessarily had to be equal to zero regardless of the variation $\eta(x)$. In particular, this must hold for all perturbations such that $\eta(a) = \eta(b) = 0$, which again leads us back to the Euler–Lagrange equation. As such, the absence of the fixed boundary conditions really does not affect the derivation of the Euler–Lagrange equation, but will instead impose additional constraints on $\varphi(x)$ in order for $F'(0)$ to be identically equal zero for all perturbations.

If we assume that $\varphi(x)$ does satisfy the Euler–Lagrange equation, the requirement that $F'(0) = 0$ reduces to the identity

$$\eta(b) \left.\frac{\partial \mathcal{L}}{\partial \varphi'}\right|_{x=b} - \eta(a) \left.\frac{\partial \mathcal{L}}{\partial \varphi'}\right|_{x=a} = 0, \qquad (8.23)$$

where all the relevant quantities are evaluated for $\varepsilon = 0$. In the case where the endpoints were fixed, this requirement was trivially fulfilled. However, in the case where the endpoints are not fixed, the perturbations $\eta(a)$ and $\eta(b)$ may take any independent values and for the equation to hold for any perturbation, we must therefore have

$$\left.\frac{\partial \mathcal{L}}{\partial \varphi'}\right|_{x=b} = \left.\frac{\partial \mathcal{L}}{\partial \varphi'}\right|_{x=a} = 0. \qquad (8.24)$$

These requirements are equations in terms of the value of $\varphi(x)$ and its derivative at the boundary points and therefore constitute the sought natural boundary conditions. Naturally, we may also encounter the situation where $\varphi(x)$ is fixed at one of the boundaries and free at the other. In such a situation, only the free boundary will be subjected to the natural boundary condition.

Example 8.5 Consider a string with tension S in a gravitational field. Working under the assumption that the deviation of the string is mainly in the vertical direction, the mass of the string between x and $x + dx$ is given by

$$dm = \rho_\ell dx, \qquad (8.25)$$

where ρ_ℓ is the linear density of the string when it has a length ℓ (due to stretching, the actual linear density will vary). The potential energy of the string now has two contributions, one coming from the gravitational potential and one resulting from stretching the string. The first of these is given by

$$V_g[y] = \int_0^\ell gy(x)\rho_\ell dx \qquad (8.26)$$

while the second is proportional to the length of the string and the string tension

$$V_t[y] = \int_0^\ell S\sqrt{1 + y'(x)^2}\, dx. \qquad (8.27)$$

We here assume that the deviation from $y = 0$ will not be large enough to significantly affect the tension S. The total potential energy is now given by

$$V[y] = \int_0^\ell \underbrace{\left[gy(x)\rho_\ell + S\sqrt{1 + y'(x)^2} \right]}_{\equiv \mathcal{L}} dx. \tag{8.28}$$

In order to minimise the potential energy, we perform the derivatives needed for the Euler–Lagrange equation

$$\frac{\partial \mathcal{L}}{\partial y} = g\rho_\ell \quad \text{and} \quad \frac{d}{dx}\frac{\partial \mathcal{L}}{\partial y'} = S\frac{d}{dx}\frac{y'}{\sqrt{1 + y'^2}} = \frac{Sy''}{(1 + y'^2)^{3/2}}, \tag{8.29}$$

resulting in the differential equation

$$Sy''(x) = g\rho_\ell \sqrt{1 + y'(x)^2}^3, \tag{8.30}$$

which may be linearised to

$$Sy''(x) = g\rho_\ell \tag{8.31}$$

for deviations such that $|y'(x)| \ll 1$. If we fix the string at $x = 0$ but allow its end at $x = \ell$ to move freely in the vertical direction, the boundary condition at $x = \ell$ will be a natural boundary condition given by

$$\left. \frac{\partial \mathcal{L}}{\partial y'} \right|_{x=\ell} = \frac{Sy'(\ell)}{\sqrt{1 + y'(\ell)^2}} = 0 \quad \Longrightarrow \quad y'(\ell) = 0. \tag{8.32}$$

Note that whereas the fixed boundary at $x = 0$ has the boundary condition $y(0) = y_0$, the condition at $x = \ell$ is exactly the same boundary condition that we found from force considerations on the free end of an oscillating string in Example 3.14.

8.2.2 Higher order derivatives

So far we have only considered functionals in integral form where the integrand's dependence on the function was restricted to an expression in terms of the function itself and its first derivative. In general, the integrand may also depend on higher derivatives and we will have a functional of the form

$$F[\varphi] = \int_a^b \mathcal{L}(\varphi(x), \varphi'(x), \varphi''(x), \ldots, x) \, dx. \tag{8.33}$$

In such a situation, we can still follow the same procedure as above but the derivative of $F(\varepsilon)$ will be given by

$$F'(\varepsilon) = \sum_{k=0}^n \int_a^b \frac{\partial \mathcal{L}}{\partial y^{(k)}} \eta^{(k)}(x) \, dx, \tag{8.34}$$

where $y^{(k)}$ and $\eta^{(k)}$ are the kth derivatives of $y(x)$ and $\eta(x)$, respectively, and n is the highest order of the derivatives on which $\mathcal{L}$ depends. Partially integrating the kth term k

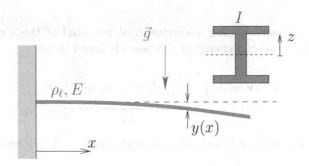

Figure 8.4 The potential energy of a beam with its own weight as a load may be described by a functional including second derivatives of the deflection $y(x)$. It also depends on the Young's modulus E of the beam material and on the area moment of inertia of the beam cross section. A common beam cross section, that of a so-called *I-beam*, is shown in the upper right. This type of cross section gives a large area moment of inertia while requiring a low amount of material.

times now results in

$$F'(\varepsilon) = \int_a^b \eta(x) \sum_{k=0}^n (-1)^k \frac{d^k}{dx^k} \frac{\partial \mathcal{L}}{\partial y^{(k)}} \, dx, \tag{8.35}$$

where we have ignored the boundary terms that are assumed to vanish either due to constraints on the variation $\eta(x)$ or due to imposed natural boundary conditions. As before, in order for $\varphi(x)$ to be a stationary function of the functional $F[\varphi]$, we must have $F'(0) = 0$ independent of the variation $\eta(x)$, implying that

$$\sum_{k=0}^n (-1)^k \frac{d^k}{dx^k} \frac{\partial \mathcal{L}}{\partial y^{(k)}} = 0. \tag{8.36}$$

As a special case, when $n = 2$, we find that

$$\frac{\partial \mathcal{L}}{\partial y} - \frac{d}{dx} \frac{\partial \mathcal{L}}{\partial y'} + \frac{d^2}{dx^2} \frac{\partial \mathcal{L}}{\partial y''} = 0. \tag{8.37}$$

Example 8.6 In solid mechanics, the elastic bending of a beam, see Fig. 8.4, gives rise to a potential energy per unit length that is inversely proportional to the radius of curvature squared. In the limit of small vertical deformations, the potential energy due to the bending is therefore given by

$$V_b[y] = \frac{1}{2} \int_0^\ell EI y''(x)^2 dx, \tag{8.38}$$

where E is *Young's modulus* of the beam material and I is the *area moment of inertia* for the cross section of the beam in the direction of bending, given by

$$I = \int_S z^2 dS - \frac{1}{A} \left(\int_S z \, dS \right)^2 \tag{8.39}$$

with z being the vertical coordinate, S the cross sectional surface of the beam, and A the total area of this surface. Adding a gravitational potential of the same form as that in Eq. (8.26), the total potential energy of the beam is found to be

$$V[y] = \int_a^b \underbrace{\left[\frac{EI}{2}y''(x)^2 + \rho_\ell g y(x)\right]}_{\equiv \mathcal{L}} dx. \tag{8.40}$$

Performing the derivatives of $\mathcal{L}$ with respect to $y(x)$ and its derivatives, we find that

$$\frac{\partial \mathcal{L}}{\partial y} = \rho_\ell g \quad \text{and} \quad \frac{d^2}{dx^2}\frac{\partial \mathcal{L}}{\partial y''} = \frac{d^2}{dx^2}EIy'' = EIy''''(x). \tag{8.41}$$

In order to minimise the potential energy $V[y]$, the beam displacement must therefore satisfy the fourth order differential equation

$$EIy''''(x) = -\rho_\ell g. \tag{8.42}$$

Note that, in order to solve this differential equation, we need four different boundary conditions. These can come either from imposing conditions on the displacements at the endpoints or from natural boundary conditions, see Problem 8.12.

A relevant comment with regards to the appearance of higher order derivatives is that some functionals with different integrands may result in the same Euler–Lagrange equation. For example, if we consider the functionals

$$F_1[\varphi] = \int_a^b \varphi(x)\varphi''(x)\,dx \quad \text{and} \quad F_2[\varphi] = -\int_a^b \varphi'(x)^2 dx \tag{8.43}$$

the functions minimising them must satisfy the same differential equation

$$\varphi''(x) = 0. \tag{8.44}$$

The underlying reason for this is that the difference between the functionals is given by

$$F_1[\varphi] - F_2[\varphi] = \int_a^b [\varphi(x)\varphi''(x) + \varphi'(x)^2]\,dx = \int_a^b \frac{d}{dx}[\varphi(x)\varphi'(x)]\,dx$$
$$= \varphi(b)\varphi'(b) - \varphi(a)\varphi'(a), \tag{8.45}$$

which is independent of any variation of φ apart from variations at the boundary. The same argument can be applied to any functional for which the integrand may be written as a *total derivative*

$$\mathcal{L} = \frac{d\mathcal{I}}{dx}, \tag{8.46}$$

where $\mathcal{I}$ is some function of $\varphi(x)$ and its derivatives. Any total derivative can therefore be added to the integrand without affecting the resulting Euler–Lagrange equation.

8.2.3 Comparison to finite spaces

Just as we can introduce the concept of derivatives of functions, we can introduce the notion of *functional derivatives* when dealing with functionals. The functional derivative of the functional $F[\varphi]$ with respect to the function $\varphi(x)$ is defined as the function $\delta F / \delta \varphi(x)$ satisfying

$$\left. \frac{d}{d\varepsilon} F[\varphi + \varepsilon \eta] \right|_{\varepsilon=0} = \int \frac{\delta F}{\delta \varphi(x)} \eta(x) \, dx \tag{8.47}$$

for all variations $\eta(x)$. In particular, we can use the variation $\eta(x) = \delta_{x_0}(x) = \delta(x - x_0)$ to formally obtain

$$\frac{\delta F}{\delta \varphi(x_0)} = \left. \frac{d}{d\varepsilon} F[\varphi + \varepsilon \delta_{x_0}] \right|_{\varepsilon=0} . \tag{8.48}$$

Going through the same exercise as we did when deriving the Euler–Lagrange equation, we find that the functional derivative of a functional on the form given in Eq. (8.33) will be given by

$$\frac{\delta F}{\delta \varphi(x)} = \sum_{k=0}^{n} (-1)^k \frac{d^k}{dx^k} \frac{\partial \mathcal{L}}{\partial y^{(k)}}, \tag{8.49}$$

implying that the Euler–Lagrange equation may be written as

$$\frac{\delta F}{\delta \varphi(x)} = 0. \tag{8.50}$$

This looks surprisingly similar to the requirement of finding local stationary points of functions, for which all partial derivatives, and therefore the gradient, must be equal to zero and thus

$$\nabla \phi(\vec{x}) = 0 \tag{8.51}$$

holds for any local stationary points of $\phi(\vec{x})$.

The analogue between the functional derivative and the gradient goes deeper than the mere appearance described above. Consider the situation where we discretise the functional

$$F[\varphi] = \int_0^\ell \mathcal{L}(\varphi, \varphi', x) \, dx \tag{8.52}$$

by introducing a grid $x_k = k\ell/N \equiv k\Delta$, where N is some large number. The functional is now approximated by

$$F[\varphi] \simeq \Delta \sum_{k=1}^{N-1} \mathcal{L}(\varphi(x_k), \varphi'(x_k), x_k), \tag{8.53}$$

where the original integral has been replaced by a sum. Note that we have only included internal points x_k in this sum, i.e., no boundary points, in order to be able to use the symmetric discrete approximation of the derivative. Introducing the notation $\varphi_k = \varphi(x_k)$ this approximation is given by the finite difference

$$\varphi'(x_k) \simeq \frac{\varphi_{k+1} - \varphi_{k-1}}{2\Delta}. \tag{8.54}$$

In addition we also introduce the notation

$$f_k = f(\varphi_k, \tfrac{\varphi_{k+1} - \varphi_{k-1}}{2\Delta}) \tag{8.55}$$

as a short-hand for any function $f(\varphi, \varphi')$ evaluated at x_k approximated using the finite difference approximation for the derivative $\varphi'(x_k)$. The functional $F[\varphi]$ is now approximated by

$$F[\varphi] \simeq F(\vec{\varphi}) = \Delta \sum_{k=1}^{N-1} \mathcal{L}_k \tag{8.56}$$

which is no longer a functional but a function of the $N + 1$-dimensional space spanned by vectors of the form $\vec{\varphi} = \varphi_k \vec{e}_k$. As usual, the stationary points of $F(\vec{\varphi})$ are found by taking its gradient and equating it to zero. In particular, taking the partial derivative with respect to φ_m results in

$$
\begin{aligned}
\frac{\partial F(\vec{\varphi})}{\partial \varphi_m} &= \Delta \sum_{k=1}^{N-1} \left[\frac{\partial \mathcal{L}_k}{\partial \varphi} \frac{\partial \varphi_k}{\partial \varphi_m} + \frac{1}{2\Delta} \frac{\partial \mathcal{L}_k}{\partial \varphi'} \frac{\partial (\varphi_{k+1} - \varphi_{k-1})}{\partial \varphi_m} \right] \\
&= \Delta \sum_{k=1}^{N-1} \left[\frac{\partial \mathcal{L}_k}{\partial \varphi} \delta_{mk} + \frac{1}{2\Delta} \frac{\partial \mathcal{L}_k}{\partial \varphi'} \varphi_m (\delta_{m,k+1} - \delta_{m,k-1}) \right] \\
&= \Delta \left[\frac{\partial \mathcal{L}_m}{\partial \varphi} - \frac{1}{2\Delta} \left(\frac{\partial \mathcal{L}_{m+1}}{\partial \varphi'} - \frac{\partial \mathcal{L}_{m-1}}{\partial \varphi'} \right) \right].
\end{aligned}
\tag{8.57}
$$

Comparing with our results from the functional minimisation, we can see that this is nothing but the discrete approximation of the functional derivative at x_m. Equating all partial derivatives to zero therefore results in a discretised version of the Euler–Lagrange equation, which may be thought of as the continuum limit as $N \to \infty$.

8.3 CONSTANTS OF MOTION

There are few things in physics that are as important as *constants of motion* and conservation laws. In the setting when we consider a functional $F[\varphi]$, constants of motion will generally be expressions in terms of x, $\varphi(x)$, and its derivatives, that for an arbitrary function $\varphi(x)$ will depend on x, but are constant for any stationary functions of the functional. Such constants of motion often take the form of *first integrals*, which are differential equations of one order lower than the Euler–Lagrange equation and for which the value of the constant of motion is a parameter representing an integration constant.

Example 8.7 The perhaps most iconic constant of motion is the total energy in a conservative system. Consider the trajectory $\vec{x}(t)$ of a particle of mass m moving in three dimensions with a potential $V(\vec{x})$. The equations of motion for the particle are given by Newton's second law

$$
m\ddot{\vec{x}} = -\nabla V(\vec{x})
\tag{8.58}
$$

and the total energy of the particle is given by

$$
E = \frac{m}{2} \dot{\vec{x}}^2 + V(\vec{x}).
\tag{8.59}
$$

Using the equations of motion results in

$$
\frac{dE}{dt} = m\dot{\vec{x}} \cdot \ddot{\vec{x}} + \dot{\vec{x}} \cdot \nabla V = \dot{\vec{x}} \cdot (m\ddot{\vec{x}} + \nabla V) = 0.
\tag{8.60}
$$

Naturally, this is nothing other than a statement about the conservation of the total energy of the conservative system. However, the expression for the total energy in Eq. (8.59) is only constant for particle trajectories $\vec{x}(t)$ that satisfy Newton's second law. We note that if the total energy is known, then Eq. (8.59) is a first order differential equation for $\vec{x}(t)$, while Newton's second law is of second order.

Assuming a functional of the form given in Eq. (8.13), there are a few special cases in which there are easily accessible constants of motion. A more systematic approach to finding the constants of motion for a given functional is supplied by *Noether's theorem*, which relates constants of motion to continuous symmetries of the functional and is one of the more important contributions in the mathematical development of theoretical physics. We postpone the treatment of constants of motion in the framework of Noether's theorem to Section 10.2.4 and instead quench our curiosity by considering two very important special cases.

8.3.1 Integrand independent of the function

We have already seen a few situations where the actual integrand $\mathcal{L}$ does not depend explicitly on the function $\varphi(x)$ itself, but only on its derivative and the independent variable x

$$\mathcal{L} = \mathcal{L}(\varphi'(x), x). \tag{8.61}$$

This is equivalent to finding that the partial derivative of the integrand with respect to the function is given by

$$\frac{\partial \mathcal{L}}{\partial \varphi} = 0 \tag{8.62}$$

and consequently the Euler–Lagrange equation reduces to

$$\frac{d}{dx} \frac{\partial \mathcal{L}}{\partial \varphi'} = 0 \quad \Longrightarrow \quad \frac{\partial \mathcal{L}}{\partial \varphi'} = C, \tag{8.63}$$

where C is an integration constant and hence $\partial \mathcal{L}/\partial \varphi'$ is a first integral of the equations of motion.

Example 8.8 Consider again the minimisation of the distance between the points A and B as described in Example 8.3. The functional describing the length of a curve between these two points resulted in an integrand

$$\mathcal{L} = \sqrt{1 + y'(x)^2} \tag{8.64}$$

that is not explicitly dependent on $y(x)$. By the discussion we have just had, this means that

$$\frac{\partial \mathcal{L}}{\partial y'} = \frac{y'(x)}{\sqrt{1 + y'(x)^2}} = C \quad \Longrightarrow \quad y'(x) = \frac{C}{\sqrt{1 - C^2}} = k, \tag{8.65}$$

where the constants C and k can be used interchangeably. In this case, we find that the derivative $y'(x)$ is a constant of motion and will therefore be the same along the entire curve whenever the curve is a stationary function of the curve length. Solving the resulting first order differential equation directly gives

$$y(x) = kx + m. \tag{8.66}$$

In other words, we again find that such curves must be straight lines.

8.3.2 Integrand independent of the variable

Another special case of importance appears when the integrand does not depend explicitly on the variable x, but only on the function $\varphi(x)$ and its derivative

$$\mathcal{L} = \mathcal{L}(\varphi(x), \varphi'(x)). \tag{8.67}$$

By differentiating the integrand with respect to x, we find that

$$\frac{d\mathcal{L}}{dx} = \frac{\partial \mathcal{L}}{\partial x} + \frac{\partial \mathcal{L}}{\partial \varphi}\varphi'(x) + \frac{\partial \mathcal{L}}{\partial \varphi'}\varphi''(x) = \frac{\partial \mathcal{L}}{\partial \varphi}\varphi'(x) + \frac{\partial \mathcal{L}}{\partial \varphi'}\varphi''(x) \tag{8.68}$$

since $\partial \mathcal{L}/\partial x = 0$. If $\varphi(x)$ is a solution to the Euler–Lagrange equation, we can rewrite this as

$$\frac{d\mathcal{L}}{dx} = \varphi'(x)\frac{d}{dx}\frac{\partial \mathcal{L}}{\partial \varphi'} + \varphi''(x)\frac{\partial \mathcal{L}}{\partial \varphi'} = \frac{d}{dx}\left[\varphi'(x)\frac{\partial \mathcal{L}}{\partial \varphi'}\right] \tag{8.69}$$

and it follows that

$$\frac{d}{dx}\left[\varphi'(x)\frac{\partial \mathcal{L}}{\partial \varphi'} - \mathcal{L}\right] = 0 \quad \implies \quad \varphi'(x)\frac{\partial \mathcal{L}}{\partial \varphi'} - \mathcal{L} = C, \tag{8.70}$$

where C is an integration constant and the expression on the left-hand side is therefore a constant of motion. This relation is known as the *Beltrami identity*. Note that if the integrand $\mathcal{L}$ depends on several functions $\varphi_i(x)$, then the corresponding argumentation leads to the conserved quantity

$$\varphi_i'(x)\frac{\partial \mathcal{L}}{\partial \varphi_i'} - \mathcal{L} = C, \tag{8.71}$$

where a sum over i is implicit in the first term. In other words, we will only have one Beltrami identity involving all of the functions φ_i, not one Beltrami identity per function.

Example 8.9 Coincidentally, the path length minimisation of Example 8.3 also satisfies the criterion required for this special case, since the integrand is not only independent of $y(x)$, but also independent of x. We now find that

$$y'(x)\frac{\partial \mathcal{L}}{\partial y'} - \mathcal{L} = \frac{y'(x)^2}{\sqrt{1 + y'(x)^2}} - \sqrt{1 + y'(x)^2} = C, \tag{8.72}$$

from which we can deduce

$$y'(x)^2 = \frac{1}{C^2} - 1. \tag{8.73}$$

Again, this indicates that $y'(x)$ is a constant of motion. It should be noted that the constant of motion due to the independence of x being equivalent to that due to the independence of $\varphi(x)$ is a coincidence in this special case and not a general feature. Indeed, there are many cases where this will no longer be true and, in particular, there are also many cases where the integrand is independent only of one of the two.

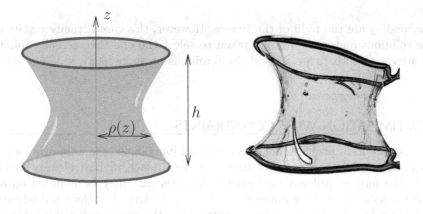

Figure 8.5 The potential energy of a soap film suspended between two rings will be directly proportional to its surface area. A smaller radius $\rho(z)$ implies less area per z coordinate, but at the same time a changing $\rho(z)$ leads to a larger area. These two effects are competing to give the soap film its shape. The right image shows the experimental result.

Example 8.10 The potential energy of soap film due to the surface tension S can be written as

$$V_T = 2SA, \tag{8.74}$$

where A is the total area of the film. The factor of two arises due to the film having two sides, thus implying a contribution from each side of the film. If other effects such as gravitation are ignored, which is often a good approximation, the shape of a soap film suspended between two circular frames that lie in planes parallel to the x^1-x^2-plane and are centred on the x^3-axis, see Fig. 8.5, will therefore take the shape that minimises the surface area A, which may be expressed as the functional

$$A[\rho] = \int_0^h 2\pi\rho(z)\sqrt{1 + \rho'(z)^2}\, dz. \tag{8.75}$$

Here, $\rho(z)$ is the radial coordinate in cylinder coordinates, which can be described as a function of the z-coordinate. The rotational symmetry of the problem implies that this radius does not depend on the angular coordinate.

Searching for the function $\rho(z)$ that minimises the surface area, the Euler–Lagrange equation becomes rather cumbersome, but we can use the fact that the integrand does not depend explicitly on z and apply the Beltrami identity, leading to

$$C = \frac{\rho(z)}{\sqrt{1 + \rho'(z)^2}}. \tag{8.76}$$

This differential equation is separable and performing the integration we end up with the relation

$$\rho(z) = C \cosh\left(\frac{z - z_0}{C}\right), \tag{8.77}$$

where z_0 is an integration constant. The constants C and z_0 need to be fixed by satisfying the boundary conditions

$$\rho(0) = r_1 \quad \text{and} \quad \rho(h) = r_2 \tag{8.78}$$

where r_1 and r_2 are the radii of the frames. However, this case actually results in several possible solutions and care must be taken to select the one that actually minimises the surface area as the others are, at best, local minima or, more likely, saddle points.

8.4   OPTIMISATION WITH CONSTRAINTS

In many optimisation problems for functions and functionals, the arguments are not arbitrary but subjected to constraints of different types. In fact, we have already seen examples of this in the form of different boundary conditions that may be imposed on a solution and that the solutions behave differently if those boundary conditions are relaxed and the boundaries are free. When considering constraints in the optimisation of functions, we generally wish to find the extreme values of some function $f(\vec{x})$ subjected to one or more constraints on the coordinates x^i of the form

$$g(\vec{x}) = g_0. \tag{8.79}$$

Such constraints restrict the possible extrema $\vec{x}$ to lie on a particular level surface of $g(\vec{x})$, given by the constant g_0. As long as the constraints are independent, every constraint will result in constraining the argument $\vec{x}$ to a space of one dimension lower than if the constraint was not present. Therefore, if we are looking for the extrema of a function in N dimensions and we have k constraints, we are generally searching for the extrema within a $N - k$-dimensional subspace. In addition, it should be noted that the function $f(\vec{x})$ does not need to have local or global extrema in the full N-dimensional space in order for extrema to exist within the subspace, the constrained problem is a problem in its own right and generally very different from the problem of finding extrema in the full space.

Example 8.11 Consider the case of the function

$$f(\vec{x}) = x^1 + x^2 \tag{8.80}$$

in $\mathbb{R}^2$. Clearly, this function does not have any local extrema since the gradient is given by

$$\nabla f(\vec{x}) = \vec{e}_1 + \vec{e}_2 \neq 0 \tag{8.81}$$

everywhere. In addition, the function is also not bounded from below or above. However, if we impose the constraint

$$g(\vec{x}) = \vec{x}^2 = R^2, \tag{8.82}$$

i.e., if we restrict the argument $\vec{x}$ to be on the circle of radius R, then the function $f(\vec{x})$ has two extreme values at $\vec{x}_\pm = \pm R(\vec{e}_1 + \vec{e}_2)/\sqrt{2}$, corresponding to a maximum and a minimum with the function values $f(\vec{x}_\pm) = \pm\sqrt{2}R$ as illustrated in Fig. 8.6. Although there are several ways of finding these extreme values, we will use this example to demonstrate the application of Lagrange multipliers to function optimisation, a method we will later carry over by analogue to the case of functional optimisation.

Just as we can look for stationary points of functions whose arguments are subject to constraints, we can look for stationary functions of functionals whose arguments are constrained. The basic idea is the same as in the function case, we assume that we have some

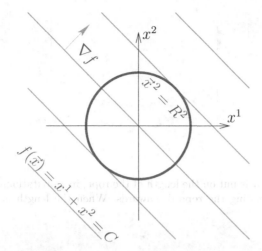

Figure 8.6 A graphical representation of the problem of finding the extreme values of the function $f(\vec{x}) = x^1 + x^2$ subjected to the constraint $\vec{x}^2 = R^2$. The light lines represent the level surfaces of $f(x)$ while the dark circle is the constraint curve on which we wish to find the extreme values. The gradient ∇f indicates the direction in which $f(\vec{x})$ grows.

functional $F[\varphi]$ for which we wish to find the stationary functions under some constraints on the solution $\varphi(x)$. There are two different types of constraints that we will consider, the first of which are *isoperimetric constraints* of the form

$$G[\varphi] = G_0, \tag{8.83}$$

where $G[\varphi]$ is some functional and G_0 a constant. This is the exact analogue of Eq. (8.79) in the case of functional minimisation. The nomenclature here comes from the fact that constraints of this type often correspond to having a variational problem in which some physical quantity, such as the length of a bounding curve, is kept fixed.

The second type of constraints that we will consider when finding extreme values of functionals is *holonomic constraints*. These constraints arise when we wish to find the stationary functions of a functional that takes several functions $\varphi_i(x)$ as an argument and there is some relation among these that must be satisfied for all x

$$g(\varphi_1(x), \varphi_2(x), \ldots, x) = g_0. \tag{8.84}$$

In many cases, these constraints can be satisfied by introducing a smaller set of functions as relations of this form will generally make one function dependent on the others. However, in other cases it will be simpler to find a different approach and there will also be some physical insights that can be gained by doing so.

Example 8.12 We have earlier expressed the gravitational potential of a rope in Example 8.2 as

$$V[h] = \int_a^b \rho_\ell g h(x) \sqrt{1 + h'(x)^2} \, dx. \tag{8.85}$$

Just as the function $f(\vec{x})$ in Example 8.11, this functional does not have any extrema unless

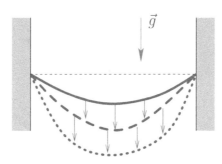

Figure 8.7 If no constraint is put on the length of the rope, its gravitational potential can be made arbitrarily small by extending the rope downwards. When the length is constrained, this is no longer possible.

we impose additional constraints on the allowed stationary functions $h(x)$. In particular, we can always lower the potential by just extending the rope downwards, see Fig. 8.7.

If we have an actual rope that is inelastic such that its length ℓ is constant and we attach its ends at given heights at $h(a) = h_a$ and $h(b) = h_b$, then we find the isoperimetric requirement that the rope length $L[h] = \ell$, where the functional $L[h]$ is given by

$$L[h] = \int_a^b \sqrt{1 + h'(x)^2}\, dx. \tag{8.86}$$

This constraint will be sufficient to guarantee that the rope has a minimal (and a maximal) potential energy given that its endpoints are fixed.

Example 8.13 Let us imagine that we again want to find the shortest path between two points A and B, but this time we are restricted to move on a two-dimensional spherical surface $\vec{x}^2 = R^2$ embedded in a three dimensional space. Using a parametrisation such that $\vec{x}(0) = \vec{x}_A$ and $\vec{x}(1) = \vec{x}_B$, the length of an arbitrary path is given by

$$L[\vec{x}] = \int_A^B ds = \int_0^1 \sqrt{\dot{\vec{x}}^2}\, dt = \int_0^1 \sqrt{\dot{r}(t)^2 + r(t)^2\dot{\theta}(t)^2 + r(t)^2 \sin^2(\theta(t))\dot{\varphi}(t)^2}\, dt, \tag{8.87}$$

where the last expression is written in spherical coordinates. The requirement that the path should be restricted to the two-dimensional sphere of radius R can be expressed as the holonomic constraint

$$g(\vec{x}(t)) = \vec{x}(t)^2 = R^2 \tag{8.88}$$

or, in spherical coordinates, $r(t) = R$.

8.4.1 Lagrange multipliers

The method we will use to find extrema of functionals under different constraints is based on the method of *Lagrange multipliers*. In order to make the analogy to finding the extrema

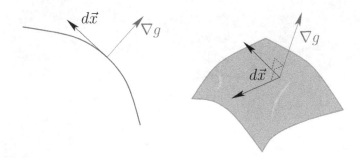

Figure 8.8 In order to be within a given level surface of the function $g(\vec{x})$, and so to satisfy the given constraint $g(\vec{x}) = g_0$, any variation $d\vec{x}$ must necessarily be orthogonal to the gradient $\nabla g(\vec{x})$ as the change in $g(\vec{x})$ under the variation is given by $d\vec{x} \cdot \nabla g(\vec{x})$. These examples illustrate this for constraints in two and three dimensions, respectively.

of functions with constraints, we will first review how Lagrange multipliers arise in regular multi-variable calculus and introduce the application to functionals by analogy. Let us start by considering the problem of finding the minima of the function $f(\vec{x})$ under the constraint

$$g(\vec{x}) = g_0. \tag{8.89}$$

This constraint represents an $N - 1$-dimensional level surface in the N-dimensional space. For any small deviations $d\vec{x}$, the function $g(\vec{x})$ will change as

$$g(\vec{x} + d\vec{x}) \simeq g(\vec{x}) + d\vec{x} \cdot \nabla g(\vec{x}), \tag{8.90}$$

where we have neglected terms of second order in the deviation and higher. If the deviation is within the level surface, then $g(\vec{x} + d\vec{x}) = g(\vec{x}) = g_0$ and we find that

$$d\vec{x} \cdot \nabla g(\vec{x}) = 0 \tag{8.91}$$

or, in other words, the gradient $\nabla g(\vec{x})$ is orthogonal to all perturbations $d\vec{x}$ within the level surface, see Fig. 8.8.

In the same fashion as we expressed the change in the function $g(\vec{x})$ under small perturbations, we can express the change in the function $f(\vec{x})$ as

$$f(\vec{x} + d\vec{x}) - f(\vec{x}) \simeq d\vec{x} \cdot \nabla f(\vec{x}). \tag{8.92}$$

In order for the point $\vec{x}$ to be an extremum of $f(\vec{x})$ under the given constraints, this change needs to be equal to zero as long as $d\vec{x}$ is within the level surface determined by $g(\vec{x})$, implying that

$$d\vec{x} \cdot \nabla f(\vec{x}) = 0 \tag{8.93}$$

for such perturbations. If this is not the case, small deviations within the level surface will lead to changes in $f(\vec{x})$, which will therefore not be an extreme value. This can be satisfied only if $\nabla f(\vec{x})$ is parallel to $\nabla g(\vec{x})$, i.e., if $\nabla f(\vec{x})$ is also a normal of the constraint level surface. This can generally be written as

$$\nabla f(\vec{x}) = \lambda \nabla g(\vec{x}), \tag{8.94a}$$

which results in

$$d\vec{x} \cdot \nabla f(\vec{x}) = \lambda \, d\vec{x} \cdot \nabla g(\vec{x}) = 0 \tag{8.94b}$$

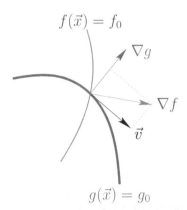

$$f(\vec{x}) = f_0$$

$$\nabla g$$

$$\nabla f$$

$$\vec{v}$$

$$g(\vec{x}) = g_0$$

Figure 8.9 The gradient $\nabla f(\vec{x})$ can always be decomposed into a part parallel to $\nabla g(\vec{x})$ and a part $\vec{v}$ orthogonal to $\nabla g(\vec{x})$. As long as $\vec{v} \neq 0$, the variation $d\vec{x} = \vec{v}\,dx$ is allowed and results in a change in the function $f(\vec{x})$, since $\vec{v}\,dx$ is not a variation within the level surface $f(\vec{x}) = f_0$. In order for $\vec{x}$ to be a stationary point, we must therefore have $\nabla f(\vec{x}) = \lambda \nabla g(\vec{x})$ for some constant λ.

for all small perturbations within the level surface. If $\nabla f(\vec{x})$ would not be parallel to $\nabla g(\vec{x})$, then we could always write

$$\nabla f(\vec{x}) = \lambda \nabla g(\vec{x}) + \vec{v}(\vec{x}), \tag{8.95}$$

where $\vec{v}(\vec{x})$ is a non-zero vector orthogonal to $\nabla g(\vec{x})$. Taking a perturbation $d\vec{x} = \vec{v}(\vec{x})\,dx$ would then lead to

$$d\vec{x} \cdot \nabla g(\vec{x}) = 0 \quad \text{while} \quad d\vec{x} \cdot \nabla f(\vec{x}) = \vec{v}(\vec{x})^2 dx \neq 0, \tag{8.96}$$

implying that $\vec{x}$ is not an extremum. This is illustrated in Fig. 8.9.

Using the fact that $\nabla f(\vec{x})$ needs to be parallel to $\nabla g(\vec{x})$, we can write down the condition for an extremum of $f(\vec{x})$ under the given constraint as

$$\nabla f(\vec{x}) - \lambda \nabla g(\vec{x}) = 0 \quad \Longrightarrow \quad \nabla h_\lambda(\vec{x}) \equiv \nabla[f(\vec{x}) - \lambda g(\vec{x})] = 0, \tag{8.97}$$

where we have defined the new function $h_\lambda(\vec{x}) = f(\vec{x}) - \lambda g(\vec{x})$. This is nothing but the requirement for a stationary point of $h_\lambda(\vec{x})$ and so the problem of finding points where $\nabla f(\vec{x})$ is parallel to $\nabla g(\vec{x})$ is equivalent to finding the global stationary points of the function $h_\lambda(\vec{x})$. In general, the locations of these extreme values will depend on λ and in order to solve our original problem, we need to fix λ in such a way that the solution actually satisfies the constraint $g(\vec{x}) = g_0$.

Example 8.14 Let us apply the above framework to the problem described in Example 8.11. Constructing the function

$$h_\lambda(\vec{x}) = f(\vec{x}) - \lambda g(\vec{x}) = x^1 + x^2 - \lambda \vec{x}^2, \tag{8.98}$$

we find that it has a stationary point when

$$\nabla h_\lambda(\vec{x}) = \vec{e}_1 + \vec{e}_2 - 2\lambda \vec{x} = 0 \quad \Longrightarrow \quad \vec{x} = \frac{1}{2\lambda}(\vec{e}_1 + \vec{e}_2). \tag{8.99}$$

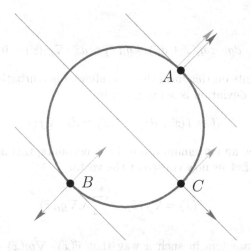

Figure 8.10 The same as Fig. 8.6 but without the coordinate axes and with the gradient vectors $\nabla f(\vec{x})$ (light) and $\nabla g(\vec{x})$ (dark) shown at the global maximum A, the global minimum B, and at C, which is not a stationary point. The gradients are proportional at both A and B while at C the allowed variation is parallel to $\nabla f(\vec{x})$.

This solution implies that

$$g(\vec{x}) = \vec{x}^2 = \frac{1}{2\lambda^2}, \tag{8.100a}$$

which satisfies the constraint $g(\vec{x}) = g_0$ only if

$$\lambda = \pm\frac{1}{\sqrt{2}R}. \tag{8.100b}$$

The stationary points of $f(\vec{x})$ subject to the given constraint are therefore given by

$$\vec{x}_\pm = \pm\frac{R}{\sqrt{2}}(\vec{e}_1 + \vec{e}_2), \tag{8.101}$$

corresponding to the extreme values $f(\vec{x}_\pm) = \pm\sqrt{2}R$ as expected. This is illustrated in Fig. 8.10.

8.4.1.1 Several constraints

It sometimes happens that we need to find the extrema of a function $f(\vec{x})$ that is subject to more than one constraint. Let us therefore consider the changes that occur when we have several constraints that need to be satisfied and that are of the form

$$g_i(\vec{x}) = g_{i,0}, \tag{8.102}$$

where $1 \leq i \leq k$ and k is the number of constraints. Just as we did when we had only one constraint, we can write down the necessary conditions for $\vec{x} + d\vec{x}$ to satisfy the ith constraint if $\vec{x}$ does as

$$dg_i \equiv g_i(\vec{x} + d\vec{x}) - g_i(\vec{x}) \simeq d\vec{x} \cdot \nabla g_i(\vec{x}) = 0 \tag{8.103}$$

and we have k constraints on this form that the allowed perturbations $d\vec{x}$ must satisfy. The change in $f(\vec{x})$ for any deviation is again given by

$$df \equiv f(\vec{x} + d\vec{x}) - f(\vec{x}) \simeq d\vec{x} \cdot \nabla f(\vec{x}) \tag{8.104}$$

and in order for $\vec{x}$ to be an extremum given the constraints, this must be equal to zero for all allowed deviations. Let us now construct the vector

$$\vec{v}(\vec{x}) = \nabla f(\vec{x}) - \sum_{i=1}^{k} \lambda_i \nabla g_i(\vec{x}) \tag{8.105}$$

and adjust the λ_i parameters in such a way that $\vec{v}(\vec{x}) \cdot \nabla g_i(\vec{x}) = 0$, implying that the variation $d\vec{x} = \vec{v}(\vec{x}) \, dx$ satisfies all of the constraints given by Eq. (8.103). Just as when we had only one constraint, this leads to

$$df = \vec{v}(\vec{x}) \cdot \nabla f(\vec{x}) \, dx = \vec{v}(\vec{x})^2 dx, \tag{8.106}$$

which is zero only if $\vec{v}(\vec{x}) = 0$. It follows that if $\vec{x}$ is a stationary point of $f(\vec{x})$, then we must have

$$\nabla h_\lambda(\vec{x}) = 0, \quad \text{where} \quad h_\lambda(\vec{x}) = f(\vec{x}) - \sum_{i=1}^{k} \lambda_i g_i(\vec{x}). \tag{8.107}$$

As in the case with only one constraint, we can find the extreme values of $f(\vec{x})$ under several constraints by constructing $h_\lambda(\vec{x})$ in this manner, finding its stationary points, and then adjusting the parameters λ_i such that all of the constraints are satisfied.

8.4.2 Isoperimetric constraints

When we deal with finding the stationary functions of functionals under *isoperimetric constraints* of the form

$$G[\varphi] = G_0, \tag{8.108}$$

the situation is essentially analogous to that encountered when finding the stationary points of functions, replacing the gradient with the functional derivative and having the inner product as the integral of two functions. In order to vary the function $\varphi(x)$ within the constraints, we can only allow variations $\delta\varphi(x)$ such that

$$G[\varphi + \delta\varphi] = G[\varphi] = G_0. \tag{8.109}$$

To leading order in $\delta\varphi(x)$, we find that

$$\delta G \equiv G[\varphi + \delta\varphi] - G[\varphi] = \int_a^b \frac{\delta G}{\delta\varphi(x)} \delta\varphi(x) \, dx = 0, \tag{8.110}$$

which is the direct analogue of Eq. (8.91). In order to find the stationary functions of the functional $F[\varphi]$ under the given isoperimetric constraint, it is necessary that

$$\delta F = F[\varphi + \delta\varphi] - F[\varphi] = \int_a^b \frac{\delta F}{\delta\varphi(x)} \delta\varphi(x) \, dx = 0 \tag{8.111}$$

for all allowed variations $\delta\varphi(x)$. Just as for the gradients in Eq. (8.95), we can define

$$v(x) = \frac{\delta F}{\delta\varphi(x)} - \lambda\frac{\delta G}{\delta\varphi(x)} \tag{8.112a}$$

and adjust the constant λ such that

$$\int_a^b \frac{\delta G}{\delta\varphi(x)} v(x)\, dx = 0. \tag{8.112b}$$

Taking a variation $\delta\varphi(x) = \varepsilon v(x)$, we can now express the change in the functional $F[\varphi]$ as

$$\delta F = \varepsilon \int_a^b \left[\lambda\frac{\delta G}{\delta\varphi(x)} + v(x) \right] v(x)\, dx = \varepsilon \int_a^b v(x)^2 dx, \tag{8.113}$$

which is zero only if $v(x) = 0$. As a result $\varphi(x)$ is only a stationary function of $F[\varphi]$ under the constraint if

$$\frac{\delta F}{\delta\varphi(x)} - \lambda\frac{\delta G}{\delta\varphi(x)} = \frac{\delta\,(F - \lambda G)}{\delta\varphi(x)} = 0 \tag{8.114}$$

for some λ or, in other words, if $\delta F/\delta\varphi(x)$ is directly proportional to $\delta G/\delta\varphi(x)$. Just as for the stationary points of the constrained functions, this condition is exactly equivalent to finding the stationary functions of the new functional

$$H_\lambda[\varphi] = F[\varphi] - \lambda G[\varphi]. \tag{8.115}$$

This does not actually incorporate the constraint $G[\varphi] = G_0$ and again we must adjust the value of the parameter λ in such a way that the constraint is satisfied in order to find the extreme values of $F[\varphi]$ subject to the constraint.

Example 8.15 Let us attempt to solve the problem of minimising the potential energy of a rope of a fixed length hanging between two points as described in Example 8.12. We start by constructing the new functional

$$H_\lambda[h] = V[h] - \lambda L[h] = \int_a^b \underbrace{\sqrt{1 + h'(x)^2}\left[\rho\ell g h(x) - \lambda\right]}_{\equiv \mathcal{L}(h, h')} dx, \tag{8.116}$$

which is what we will need to find the stationary functions of in order to find the stationary functions of our actual constrained problem. The integrand $\mathcal{L}(h, h')$ here does not depend explicitly on x and so the Beltrami identity must be satisfied and we have a first integral given by

$$\lambda - \rho\ell g h(x) = C\sqrt{1 + h'(x)^2}. \tag{8.117}$$

This first order differential equation is separable and we can integrate it to find the solution

$$h(x) = \frac{C}{\rho\ell g} \cosh\left(\frac{\rho\ell g}{C}(x - x_0) \right) + \frac{\lambda}{\rho\ell g}, \tag{8.118}$$

where x_0 is an integration constant. The requirement that the rope length should be ℓ is now of the form

$$L[h] = \int_a^b \sqrt{1 + h'(x)^2}\, dx = \int_a^b \cosh\left(\frac{\rho\ell g}{C}(x - x_0) \right) dx$$
$$= \frac{C}{\rho\ell g} \left[\sinh\left(\frac{\rho\ell g}{C}(b - x_0) \right) - \sinh\left(\frac{\rho\ell g}{C}(a - x_0) \right) \right] = \ell. \tag{8.119}$$

Figure 8.11   A chain hanging under its own weight and fixed at each of its two ends. An idealised chain would form a perfect catenary. Actual chains, such as the one shown here, are rather well described by this approximation.

Although this equation does not depend explicitly on λ, it does fix the values of λ as well as the constants C and x_0 together with the two boundary conditions that will be given by the heights at which the ends of the rope are fixed.

The form of the solution in Eq. (8.118) is known as a *catenary* and, as shown in this example, describes the shape of an idealised rope, chain, or wire, hanging under its own weight and supported at its ends, see Fig. 8.11.

Just as we can subject a function to several constraints, we can also subject a functional to several isoperimetric constraints. The way to handle this is also exactly analogous to the function case and we can do this by finding the extrema of the new functional

$$H_\lambda[\varphi] = F[\varphi] - \sum_{i=1}^{k} \lambda_i G_i[\varphi], \tag{8.120}$$

where the k constraints are given by $G_i[\varphi] = G_{i,0}$. Adapting the constants λ_i to the constraints will then ensure that the found stationary functions actually satisfy the constraints.

8.4.3 Holonomic constraints

When considering *holonomic constraints*, we again need to figure out what the different allowed variations are. In general, we will be working with a functional depending on a number of different functions $\varphi_i(x)$, which we will commonly denote as $\vec{\varphi}(x)$. A holonomic constraint given by

$$g(\vec{\varphi}(x), x) = g_0 \tag{8.121}$$

implies that

$$g(\vec{\varphi}(x) + \delta\vec{\varphi}(x), x) - g_0 \simeq \sum_i \frac{\partial g}{\partial \varphi_i} \delta\varphi_i(x) = 0 \tag{8.122}$$

for all variations around $\vec{\varphi}(x)$ that satisfy it. In particular, we may look at variations such that $\delta\varphi_i(x) = 0$ unless $i = 1$ or 2, which leads to the relation

$$\frac{\partial g}{\partial \varphi_1} \delta\varphi_1(x) = -\frac{\partial g}{\partial \varphi_2} \delta\varphi_2(x). \tag{8.123}$$

Looking at the variation of a functional $F[\vec{\varphi}]$ under such variations, we find that

$$\delta F = \int_a^b \left[\frac{\delta F}{\delta\varphi_1(x)} \delta\varphi_1(x) + \frac{\delta F}{\delta\varphi_2(x)} \delta\varphi_2(x) \right] dx = 0 \qquad (8.124a)$$

if $\vec{\varphi}(x)$ is an extremum of $F[\vec{\varphi}]$. Using the constraint on the allowed variation, this can be rewritten as

$$\int_a^b \left[-\frac{\delta F}{\delta\varphi_1(x)} \left(\frac{\partial g}{\partial\varphi_1} \right)^{-1} + \frac{\delta F}{\delta\varphi_2(x)} \left(\frac{\partial g}{\partial\varphi_2} \right)^{-1} \right] \frac{\partial g}{\partial\varphi_2} \delta\varphi_2(x) dx = 0. \qquad (8.124b)$$

In order for this relation to hold regardless of the variation $\delta\varphi_2(x)$, we can conclude that

$$\frac{\delta F}{\delta\varphi_1(x)} \left(\frac{\partial g}{\partial\varphi_1} \right)^{-1} = \frac{\delta F}{\delta\varphi_2(x)} \left(\frac{\partial g}{\partial\varphi_2} \right)^{-1} = \lambda(x), \qquad (8.125)$$

where $\lambda(x)$ is some function that is independent of the actual variation. Naturally, there is nothing special with the choice of $\delta\varphi_1(x)$ and $\delta\varphi_2(x)$ in the variation and selecting any other pair will lead to the corresponding result for that pair. This leads to the fact that

$$\frac{\delta F}{\delta\varphi_i(x)} = \lambda(x) \frac{\partial g}{\partial\varphi_i} \quad \Longleftrightarrow \quad \frac{\delta F}{\delta\varphi_i(x)} - \lambda(x) \frac{\partial g}{\partial\varphi_i} = 0 \qquad (8.126)$$

regardless of the value of i. This is exactly the Euler–Lagrange equation resulting from the variation $\delta\varphi_i(x)$ of the functional

$$H[\vec{\varphi}, \lambda] = F[\vec{\varphi}] - \int_a^b \lambda(x)[g(\vec{\varphi}(x)) - C]dx, \qquad (8.127)$$

where C is an integration constant. We can also find the Euler–Lagrange equation corresponding to the variation of $\lambda(x)$, which is trivially found to be

$$\frac{\delta H}{\delta\lambda(x)} = C - g(\vec{\varphi}(x)) = 0. \qquad (8.128)$$

Letting $C = g_0$, this reproduces the original holonomic constraint and therefore the requirements that $\vec{\varphi}(x)$ has to satisfy in order for it to be a stationary function of $F[\vec{\varphi}]$ subjected to the holonomic constraint are equivalent the Euler–Lagrange equations for $H[\vec{\varphi}, \lambda]$.

It should be noted that holonomic constraints may be implemented by using coordinates such that the constraint restricts the solutions to a coordinate surface. In these cases, the function $\lambda(x)$ will drop out of the solution for the remaining coordinates as will be discussed in the next section. However, the $\lambda(x)$ that will result from applying the above procedure will sometimes have a relevant physical interpretation, see, e.g., Problem 8.39, and it may therefore be relevant to compute it.

8.5   CHOICE OF VARIABLES

When we deal with functionals that depend on several functions, the problem at hand may be simpler if it can be reformulated in a different way by using a different set of functions. Typically, such reformulations are in the form of changes of variables, where we define a new set of functions

$$\phi_i(x) = \Phi_i(\vec{\varphi}(x), x), \qquad (8.129a)$$

where the functions $\Phi_i(\vec{\varphi}(x), x)$ are invertible for each value of x, i.e., there are also functions $\Phi_i^{-1}(\vec{\phi}(x), x)$ such that

$$\varphi_i(x) = \Phi_i^{-1}(\vec{\phi}(x), x). \tag{8.129b}$$

For convenience, we will refer to these functions using ϕ_i and φ_i and not explicitly write out the transformations Φ_i.

Given a functional of the form

$$F[\vec{\varphi}] = \int_a^b \mathcal{L}(\vec{\varphi}(x), \vec{\varphi}'(x), x) dx, \tag{8.130}$$

we can make a transformation such that

$$\mathcal{L}'(\vec{\phi}(x), \vec{\phi}'(x), x) = \mathcal{L}(\vec{\varphi}(\vec{\phi}, x), \tfrac{d}{dx}\vec{\varphi}(\vec{\phi}, x), x). \tag{8.131}$$

Note that, with $\vec{\varphi}$ being a function of $\vec{\phi}(x)$ and x, the derivative $d\vec{\varphi}/dx$ is generally a function of $\vec{\phi}(x)$, $\vec{\phi}'(x)$, and x, given by

$$\frac{d\varphi_i}{dx} = \frac{\partial \varphi_i}{\partial \phi_j}\phi_j'(x) + \frac{\partial \varphi_i}{\partial x}. \tag{8.132}$$

This defines a new functional, now of the $\vec{\phi}(x)$, as

$$\tilde{F}[\vec{\phi}] = \int_a^b \mathcal{L}'(\vec{\phi}(x), \vec{\phi}'(x), x) dx = F[\vec{\varphi}], \tag{8.133}$$

where $\vec{\phi}(x)$ and $\vec{\varphi}(x)$ are related through the transformation $\vec{\varphi} = \vec{\varphi}(\vec{\phi}, x)$. The Euler–Lagrange equations for the functional $\tilde{F}[\vec{\phi}]$ are given by

$$\frac{\partial \mathcal{L}'}{\partial \phi_i} - \frac{d}{dx}\frac{\partial \mathcal{L}'}{\partial \phi_i'} = 0. \tag{8.134}$$

Applying the chain rule and using $\mathcal{L}' = \mathcal{L}$, we find that

$$\frac{\partial \mathcal{L}'}{\partial \phi_i} = \frac{\partial \mathcal{L}}{\partial \varphi_j}\frac{\partial \varphi_j}{\partial \phi_i} + \frac{\partial \mathcal{L}}{\partial \varphi_j'}\frac{\partial \varphi_j'}{\partial \phi_i} = \frac{\partial \mathcal{L}}{\partial \varphi_j}\frac{\partial \varphi_j}{\partial \phi_i} + \frac{\partial \mathcal{L}}{\partial \varphi_j'}\left(\frac{\partial^2 \varphi_j}{\partial \phi_i \partial \phi_k}\phi_k' + \frac{\partial^2 \varphi_j}{\partial \phi_i \partial x}\right). \tag{8.135a}$$

Similarly, we can express the second term in Eq. (8.134) as

$$\frac{d}{dx}\frac{\partial \mathcal{L}'}{\partial \phi_i'} = \frac{d}{dx}\frac{\partial \mathcal{L}}{\partial \varphi_j'}\frac{\partial \varphi_j'}{\partial \phi_i'} = \frac{d}{dx}\frac{\partial \mathcal{L}}{\partial \varphi_j'}\frac{\partial \varphi_j}{\partial \phi_i} = \frac{\partial \varphi_j}{\partial \phi_i}\frac{d}{dx}\frac{\partial \mathcal{L}}{\partial \varphi_j'} + \frac{\partial \mathcal{L}}{\partial \varphi_j'}\frac{d}{dx}\frac{\partial \varphi_j}{\partial \phi_i}$$
$$= \frac{\partial \varphi_j}{\partial \phi_i}\frac{d}{dx}\frac{\partial \mathcal{L}}{\partial \varphi_j'} + \frac{\partial \mathcal{L}}{\partial \varphi_j'}\left(\frac{\partial^2 \varphi_j}{\partial \phi_k \partial \phi_i}\phi_k' + \frac{\partial^2 \varphi_j}{\partial x \partial \phi_i}\right). \tag{8.135b}$$

Finally, collecting our results, we arrive at the relation

$$\frac{\partial \varphi_j}{\partial \phi_i}\left(\frac{\partial \mathcal{L}}{\partial \varphi_j} - \frac{d}{dx}\frac{\partial \mathcal{L}}{\partial \varphi_j'}\right) = 0. \tag{8.136a}$$

Since the transformation is invertible, this directly implies that

$$\frac{\partial \mathcal{L}}{\partial \varphi_j} - \frac{d}{dx}\frac{\partial \mathcal{L}}{\partial \varphi_j'} = 0 \tag{8.136b}$$

and therefore the Euler–Lagrange equations for the functional $F[\vec{\varphi}]$ must be satisfied if the Euler–Lagrange equations for $\tilde{F}[\vec{\phi}]$ are. This should not come as a surprise. After all, the functionals are really the same functional with the only difference being how the functions they depend on are represented.

Example 8.16 One way of handling the holonomic constraint when minimising the path along the surface of a sphere, see Example 8.13, is to directly express the path in spherical coordinates such that

$$x^1(t) = r(t)\sin(\theta(t))\cos(\varphi(t)), \tag{8.137a}$$

$$x^2(t) = r(t)\sin(\theta(t))\cos(\varphi(t)), \tag{8.137b}$$

$$x^3(t) = r(t)\cos(\theta(t)). \tag{8.137c}$$

In the original Cartesian coordinate system, the holonomic constraints result in four Euler–Lagrange equations, three for the functions $x^i(t)$ and one for the Lagrange multiplier $\lambda(t)$, which are rather complicated. However, in the spherical coordinate system, the holonomic constraint is of the form $r(t) = R$, directly implying that $r(t)$ is equal to the radius. Making the change of variables in the integrand, we find that

$$L[r, \theta, \varphi] = \int_0^1 \underbrace{\sqrt{\dot{r}^2 + r^2\dot{\theta}^2 + r^2\sin^2(\theta)\dot{\varphi}^2}}_{\equiv \mathcal{L}}\, dt. \tag{8.138}$$

From our discussion on holonomic constraints, we know that the path length will be minimised whenever the Euler–Lagrange equations

$$\frac{\partial \mathcal{L}}{\partial r} - \frac{d}{dt}\frac{\partial \mathcal{L}}{\partial \dot{r}} = \lambda(t)\frac{\partial r}{\partial r} = \lambda(t), \tag{8.139a}$$

$$\frac{\partial \mathcal{L}}{\partial \theta} - \frac{d}{dt}\frac{\partial \mathcal{L}}{\partial \dot{\theta}} = \lambda(t)\frac{\partial r}{\partial \theta} = 0, \tag{8.139b}$$

$$\frac{\partial \mathcal{L}}{\partial \varphi} - \frac{d}{dt}\frac{\partial \mathcal{L}}{\partial \dot{\varphi}} = \lambda(t)\frac{\partial r}{\partial \varphi} = 0, \tag{8.139c}$$

$$r(t) = R, \tag{8.139d}$$

are satisfied. The zeros on the right-hand side of the Euler–Lagrange equations for the angular coordinates are a direct result of the constraint not being affected by these coordinates. As such, these equations are just the regular Euler–Lagrange equations that we would find from inserting the constraint $r(t) = R$ into the original functional. Doing so we end up with two coupled differential equations for two unknown functions and this is all the information required to solve the problem. The path length is therefore given by

$$L[\theta, \varphi] = R\int_0^1 \sqrt{\dot{\theta}^2 + \sin^2(\theta)\dot{\varphi}^2}\, dt, \tag{8.140}$$

which can be minimised without reference to the Lagrange multiplier $\lambda(t)$. By selecting a set of coordinates such that the holonomic constraint is a coordinate level surface, we can therefore just ignore the Lagrange multiplier and solve the problem directly. Should the form of $\lambda(t)$ be of interest, it can be read off directly from the Euler–Lagrange equation resulting from radial variations.

The above example illustrates a general idea that can be used to handle holonomic constraints. Imagine a setting in which a functional depends on N functions subject to a holonomic constraint. Choosing variables φ_i such that the functional is given by

$$F[\vec{\varphi}] = \int \mathcal{L}(\vec{\varphi}, \vec{\varphi}', x)\, dx \tag{8.141}$$

and the holonomic constraint takes the form

$$\varphi_N(x) = C, \tag{8.142}$$

where C is a constant, we can apply the approach developed in Section 8.4.3 and find the stationary functions of

$$H[\vec{\varphi}, \lambda] = F[\vec{\varphi}] - \int \lambda(x)[\varphi_N(x) - C]dx. \tag{8.143}$$

The Euler–Lagrange equations for $H[\vec{\varphi}, \lambda]$ now take the form

$$\frac{\partial \mathcal{L}}{\partial \varphi_i} - \frac{d}{dx}\frac{\partial \mathcal{L}}{\partial \varphi_i'} = 0, \qquad (i < N) \tag{8.144a}$$

$$\frac{\partial \mathcal{L}}{\partial \varphi_N} - \frac{d}{dx}\frac{\partial \mathcal{L}}{\partial \varphi_N'} = \lambda(x), \tag{8.144b}$$

$$\varphi_N(x) = C. \tag{8.144c}$$

The first of these equations with the third inserted is now a set of $N-1$ differential equations that are equivalent to the Euler–Lagrange equations that would have resulted from inserting the third equation into the functional from the beginning and considered it a functional of the remaining $N-1$ functions φ_i for $i < N$. The second relation can be used to express the function $\lambda(x)$ and the third is the holonomic constraint itself. Based on this argumentation, we do not need to go through the trouble of introducing the functional H, we can instead solve a problem involving holonomic constraints by choosing variables that parametrise the constraint surface and find the stationary functions of the functional resulting from inserting the holonomic constraint into the original one.

8.6 FUNCTIONALS AND HIGHER-DIMENSIONAL SPACES

In the discussion so far, all of our functionals have been maps from functions that depend only on one variable to the real numbers. This is a rather restrictive set of functions, but luckily the extension to general functions that depend on a larger number of parameters is rather straightforward. Let us consider functionals of the form

$$F[\varphi] = \int_V \mathcal{L}(\varphi(\vec{x}), \nabla\varphi(\vec{x}), \vec{x})dV, \tag{8.145}$$

where $\vec{x}$ is a vector in N dimensions and the integral is taken over some N-dimensional volume V.

Example 8.17 A square membrane with tension σ and side length ℓ is subjected to a homogeneous force density $f(\vec{x})$, see Fig. 8.12. For small deviations from the flat state $f(\vec{x}) = 0$, we can write the potential energy of the membrane as a sum of the contribution

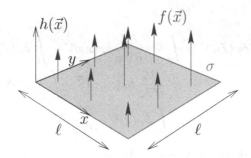

Figure 8.12 The potential energy of a membrane with a tension σ subjected to an external transversal force density $f(\vec{x})$ can be described as a sum of the potential energies due to the elastic deformation of the membrane and that due to displacement relative to the external force. The result is a functional taking the transversal displacement function $h(\vec{x})$ as argument.

from the tension and the contribution from the external force density. The contribution from the tension is proportional to the total area of the membrane and therefore

$$V_T[h] = \sigma \int \sqrt{1 + (\nabla h(\vec{x}))^2}\, dS, \tag{8.146}$$

where $\vec{x}$ is a two-dimensional vector defining the position on the membrane and $h(\vec{x})$ is the transversal deviation of the membrane at $\vec{x}$. At the same time, the potential energy due to the external force density is given by

$$V_f[h] = - \int f(\vec{x}) h(\vec{x})\, dS. \tag{8.147}$$

If we wish to minimise the total potential energy, we must therefore try to find the minimum of the sum $V[h] = V_T[h] + V_f[h]$.

The approach to the general scenario with a number of dimensions larger than one is very similar to what we have already seen. We start by considering a variation $\delta\varphi(\vec{x})$ of the function $\varphi(\vec{x})$. To linear order in the variation, the change in the functional $F[\varphi]$ can be written as

$$\delta F = F[\varphi + \delta\varphi] - F[\varphi] \simeq \int_V \left[\frac{\partial \mathcal{L}}{\partial \varphi} \delta\varphi(\vec{x}) + \frac{\partial \mathcal{L}}{\partial \nabla\varphi} \cdot \nabla\delta\varphi(\vec{x}) \right] dV, \tag{8.148a}$$

where we have introduced the notation

$$\frac{\partial \mathcal{L}}{\partial \nabla\varphi} = \vec{e}_i \frac{\partial \mathcal{L}}{\partial(\partial_i\varphi)} \tag{8.148b}$$

as the multi-dimensional analogue of $\partial \mathcal{L}/\partial\varphi'$. Assuming the variation $\delta\varphi(\vec{x})$ vanishes on the boundary S, we can use the divergence theorem to rewrite the second term of δF according to

$$\int_V \frac{\partial \mathcal{L}}{\partial \nabla \varphi} \cdot \nabla \delta \varphi(\vec{x}) dV = \int_V \nabla \cdot \frac{\partial \mathcal{L}}{\partial \nabla \varphi} \delta \varphi(x) dV - \int_V \delta \varphi(\vec{x}) \nabla \cdot \frac{\partial \mathcal{L}}{\partial \nabla \varphi} dV$$

$$= \oint_S \frac{\partial \mathcal{L}}{\partial \nabla \varphi} \delta \varphi(x) \cdot d\vec{S} - \int_V \delta \varphi(\vec{x}) \nabla \cdot \frac{\partial \mathcal{L}}{\partial \nabla \varphi} dV$$

$$= - \int_V \delta \varphi(\vec{x}) \nabla \cdot \frac{\partial \mathcal{L}}{\partial \nabla \varphi} dV. \tag{8.149}$$

This allows us to write down the variation δF as

$$\delta F = \int_V \left(\frac{\partial \mathcal{L}}{\partial \varphi} - \nabla \cdot \frac{\partial \mathcal{L}}{\partial \nabla \varphi} \right) \delta \varphi(\vec{x}) dV \tag{8.150}$$

or, in terms of a functional derivative,

$$\frac{\delta F}{\delta \varphi(\vec{x})} = \frac{\partial \mathcal{L}}{\partial \varphi} - \nabla \cdot \frac{\partial \mathcal{L}}{\partial \nabla \varphi}. \tag{8.151}$$

In order for $\varphi(\vec{x})$ to be an extremum of $F[\varphi]$, the change δF must vanish for all allowed variations $\delta \varphi(\vec{x})$ and therefore the functional derivative must be equal to zero, just as in the case with functions in one dimension. Written explicitly in terms of the coordinate derivatives, we therefore have the multi-dimensional Euler–Lagrange equation

$$\frac{\partial \mathcal{L}}{\partial \varphi} - \frac{\partial}{\partial x^i} \frac{\partial \mathcal{L}}{\partial (\partial_i \varphi)} = 0. \tag{8.152}$$

A word of warning regarding the notation is in order here. The partial derivatives of $\mathcal{L}$ in this expression are partial derivatives of $\mathcal{L}$ with respect to its corresponding arguments as given explicitly in Eq. (8.145). However, the partial derivative $\partial/\partial x^i$ is a partial derivative treating $\partial \mathcal{L}/\partial(\partial_i \varphi)$ as a function of the coordinates x^i only. With this in mind, this term could also be written as

$$\frac{\partial}{\partial x^i} \frac{\partial \mathcal{L}}{\partial (\partial_i \varphi)} = \frac{\partial^2 \mathcal{L}}{\partial \varphi \partial (\partial_i \varphi)} (\partial_i \varphi) + \frac{\partial^2 \mathcal{L}}{\partial (\partial_k \varphi) \partial (\partial_i \varphi)} (\partial_i \partial_k \varphi) + \frac{\partial^2 \mathcal{L}}{\partial x^i \partial (\partial_i \varphi)}, \tag{8.153}$$

where all derivatives of $\mathcal{L}$ are to be treated as partial derivatives with respect to its full set of arguments. However, this formula is rather cumbersome and it is usually simpler to just remember that the derivative $\partial/\partial x^i$ also acts on any functions of x^i and perform the derivative explicitly. The corresponding relation in the one-dimensional case would have been

$$\frac{d}{dx} \frac{\partial \mathcal{L}}{\partial \varphi'} = \frac{\partial^2 \mathcal{L}}{\partial \varphi \partial \varphi'} \varphi' + \frac{\partial^2 \mathcal{L}}{\partial \varphi'^2} \varphi'' + \frac{\partial^2 \mathcal{L}}{\partial x \partial \varphi'}. \tag{8.154}$$

The reason we did not encounter this difficulty before was that there was no ambiguity in writing d/dx while we really cannot write d/dx^i in a meaningful way in the multi-dimensional case.

Example 8.18 Consider the square membrane from Example 8.17 with a fixed boundary such that the boundary conditions are given by

$$\text{(BC)}: h(0, x^2) = h(\ell, x^2) = h(x^1, 0) = h(x^1, \ell) = 0. \tag{8.155}$$

The total potential energy of the membrane is given by

$$V[h] = \int \underbrace{\left[\sigma\sqrt{1 + (\nabla h(\vec{x}))^2} - f(\vec{x})h(\vec{x})\right]}_{\equiv \mathcal{L}} dx^1 dx^2 \tag{8.156}$$

and we therefore find that

$$\frac{\partial \mathcal{L}}{\partial h} = -f(\vec{x}), \tag{8.157a}$$

$$\frac{\partial \mathcal{L}}{\partial \nabla h} = \sigma\frac{\nabla h}{\sqrt{1 + (\nabla h)^2}}, \tag{8.157b}$$

$$\nabla \cdot \frac{\partial \mathcal{L}}{\partial \nabla h} = \sigma\left[\frac{\nabla^2 h}{\sqrt{1 + (\nabla h)^2}} - \frac{\nabla h \cdot [(\nabla h) \cdot \nabla]\nabla h}{\sqrt{1 + (\nabla h)^2}^3}\right] \simeq \sigma\nabla^2 h, \tag{8.157c}$$

where we have linearised the problem in the last step, assuming that the deviation $h(\vec{x})$ and its derivative are small. The Euler–Lagrange equation for $h(\vec{x})$ is therefore given by

$$(\text{PDE}): \quad \nabla^2 h(\vec{x}) = -\frac{f(\vec{x})}{\sigma}, \tag{8.158}$$

i.e., it is Poisson's equation. The resulting problem may be solved using the methods described in the previous chapters.

It should also be noted that we could have saved ourselves some of the pain we suffered when doing the derivatives by expanding the expression for the potential energy to second order in $h(\vec{x})$ before performing the derivatives instead of doing so at the very end.

8.6.1   Conservation laws

In the one-dimensional setting we encountered a number of constants of motion, quantities that were the same everywhere as long as the Euler–Lagrange equations were satisfied. In the multi-dimensional case, it is no longer clear what the generalisation of this statement is and, as it will turn out, we are no longer talking about quantities that take the same value everywhere.

Let us start by examining the case in which the integrand $\mathcal{L}(\varphi, \nabla\varphi, \vec{x}) = \mathcal{L}(\nabla\varphi, \vec{x})$ does not explicitly depend on the function $\varphi(\vec{x})$ itself, but only on its derivatives. In the one-dimensional setting, this led us to the constant of motion

$$C = \frac{\partial \mathcal{L}}{\partial \varphi'}. \tag{8.159}$$

When going to the multi-dimensional case, the Euler–Lagrange equation is given by

$$0 = \nabla \cdot \frac{\partial \mathcal{L}}{\partial \nabla\varphi} \equiv \nabla \cdot \vec{J}, \tag{8.160}$$

where we have introduced the *conserved current*

$$\vec{J} = \frac{\partial \mathcal{L}}{\partial \nabla\varphi}. \tag{8.161}$$

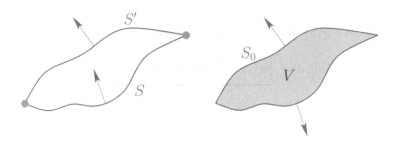

Figure 8.13   Two different $N-1$-dimensional surfaces S' and S with the same boundary (represented by the points) for the case of $N = 2$. By taking the difference of the flux integrals of the current $\vec{J}$ over these surfaces, we obtain the closed surface integral over S_0, which can be rewritten as a volume integral of $\nabla \cdot \vec{J} = 0$ over the volume V.

While this certainly allows for the possibility that $\vec{J}$ is constant, it is no longer a necessity. Instead, when working with a problem in N dimensions, we can consider an $N-1$-dimensional region S, which has an $N-2$-dimensional boundary. In the two-dimensional case, this corresponds to taking a curve which has two endpoints, while in the three-dimensional case it corresponds to taking a surface whose boundary is a one-dimensional curve, see Fig. 8.13. If we continuously deform the surface into a new surface S' while keeping the boundary fixed, we can write the difference in the flux of $\vec{J}$ through the surfaces as

$$\Delta\Phi = \int_{S'} \vec{J} \cdot d\vec{S} - \int_{S} \vec{J} \cdot d\vec{S} = \oint_{S_0} \vec{J} \cdot d\vec{S}, \tag{8.162}$$

where S_0 is the boundary of the volume enclosed by the surfaces S and S', i.e., it is the union of the surfaces S and S' where the normal direction of S has been inverted. Using the divergence theorem we now find

$$\Delta\Phi = \int_{V} \nabla \cdot \vec{J} \, dV = 0 \tag{8.163}$$

or, in other words, the fluxes of the conserved current $\vec{J}$ through the surfaces are the same. This has several profound physical consequences, among them generalised conservation laws that allow for a non-constant current $\vec{J}$.

Example 8.19   Consider the situation where we are working in a four-dimensional setting and one of the coordinates is time t and the others are the spatial coordinates x^i. The current $\vec{J}$ is now generally of the form

$$\vec{J} = \vec{e}_0 \rho + \vec{\jmath}, \tag{8.164}$$

where $\vec{e}_0$ is the basis vector in the time direction and $\vec{\jmath}$ is the spatial part of the conserved current $\vec{J}$. Separating the time derivative from the spatial derivatives in the current conservation law $\nabla \cdot \vec{J} = 0$, we find that

$$\frac{\partial \rho}{\partial t} + \nabla \cdot \vec{\jmath} = 0. \tag{8.165}$$

This is an equation we have seen before! It is the source free *continuity equation* for a

Figure 8.14 The equivalent to Fig. 8.13 for $N = 1$ with the single dimension being the t-axis. The zero-dimensional surfaces are now single points with no boundary. The corresponding argumentation results in that the value of the current, now with only one component, at these points must be the same and it is therefore a constant of motion.

quantity with density ρ and a current density $\vec{j}$. Therefore, if a current is conserved in this setting, it implies that there is a concentration that follows the continuity equation with no sources or sinks. Naturally, this leads us to more generally consider the case when the integrand $\mathcal{L}$ does depend on φ and we instead end up with the relation

$$\frac{\partial \rho}{\partial t} + \nabla \cdot \vec{j} = \frac{\partial \mathcal{L}}{\partial \varphi}, \tag{8.166}$$

which is the continuity equation with a source term $\partial \mathcal{L}/\partial \varphi$.

We can now go back and compare with the one-dimensional setting. In this situation, an $N - 1$-dimensional surface is just a single point which has no boundary and therefore the continuous deformation consists of moving the point, see Fig. 8.14. At the same time, the surface integral reduces to evaluating the current $J = \partial \mathcal{L}/\partial \varphi'$ at the point and the conservation law then directly states that the value of J is the same at all points, just as we concluded earlier.

The current we just considered was the multi-dimensional equivalent of the constant of motion implied by the integrand $\mathcal{L}$ not depending explicitly on the function itself. This is not the only type of conserved current which will appear and we may ask ourselves the question if there is also a current corresponding to the Beltrami identity. As we shall now see, the answer to this question is that there is not only one, but several such currents, each corresponding to an integrand that does not depend explicitly on a particular coordinate.

Consider the situation where the integrand $\mathcal{L}$ does not depend explicitly on the coordinate x^i. In this situation, we can differentiate $\mathcal{L}$ with respect to x^i (remember that $\mathcal{L}$ may still depend on x^i implicitly through its dependence on $\varphi(\vec{x})$) and we obtain

$$\frac{\partial}{\partial x^i} \mathcal{L} = \frac{\partial \mathcal{L}}{\partial \varphi} \partial_i \varphi + \frac{\partial \mathcal{L}}{\partial(\partial_k \varphi)} \partial_i \partial_k \varphi. \tag{8.167a}$$

If the Euler–Lagrange equation for φ is satisfied, we can use it to replace $\partial \mathcal{L}/\partial \varphi$, just as we did when deriving the Beltrami identity, and find

$$\frac{\partial}{\partial x^i} \mathcal{L} = (\partial_i \varphi) \frac{\partial}{\partial x^k} \frac{\partial \mathcal{L}}{\partial(\partial_k \varphi)} + \frac{\partial \mathcal{L}}{\partial(\partial_k \varphi)} \partial_k \partial_i \varphi = \frac{\partial}{\partial x^k} \left[\frac{\partial \mathcal{L}}{\partial(\partial_k \varphi)} \partial_i \varphi \right]. \tag{8.167b}$$

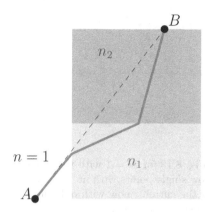

Figure 8.15 If the index of refraction is not constant, the straight path, here shown as a dashed line, may not be the fastest way for light to propagate from A to B. Instead, the alternative path shown as a solid curve will be faster and will be the path along which we can send a signal between the points according to Fermat's principle.

We can now use $\partial/\partial x^i = \delta_i^k \partial/\partial x^k$ on the left-hand side and rearrange this to

$$0 = \frac{\partial}{\partial x^k} \left[\frac{\partial \mathcal{L}}{\partial(\partial_k \varphi)} \partial_i \varphi - \delta_i^k \mathcal{L} \right] \equiv \partial_k T^k{}_i. \tag{8.168}$$

If $\mathcal{L}$ does not depend explicitly on x^i for some fixed i, then $T^k{}_i$ is therefore a conserved current.

8.7 BASIC VARIATIONAL PRINCIPLES IN PHYSICS

There are several variational principles in modern physics. Many times, they are equivalent to other formulations of the same theory, but offer new insights and simplify further development and computations. We have already seen that variational calculus is of use when we need to find the lowest energy state in a system and we shall now discuss two of the more prominent variational principles in classical physics.

8.7.1 Fermat's principle

Within the area of optics, the propagation of light is often described using the *Huygens–Fresnel principle*, which treats the propagation of a wave by considering every point the wave reaches as the new source of a spherical wave. In fact, this is well in line with our earlier treatment using Green's functions as, given the state of a wave at a particular time t, we can find the entire future propagation of the wave by propagating this state with the Green's function of the wave equation, which is spherically symmetric. Wave rays are then the lines that are orthogonal to the wave fronts.

An alternative formulation for how light behaves is offered by *Fermat's principle*, which states that light rays propagate along the path that makes the travel time stationary. The time taken for a light ray to propagate along a path Γ in a medium with variable index of refraction, see Fig. 8.15, is given by

$$T[\vec{x}] = \int_\Gamma dt = \int \frac{ds}{v} = \int_0^1 \frac{n(\vec{x})}{c} \sqrt{\vec{x}'^2}\, d\theta, \tag{8.169}$$

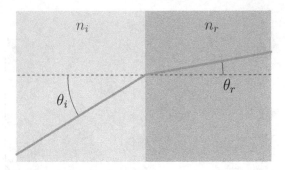

Figure 8.16 When passing between media of different refractive index, a light ray is refracted. The relation between the incident angle and the refraction angle is given by Snell's law, which may be derived based on Fermat's principle.

where θ is a curve parameter, here assumed to go from zero to one, c is the speed of light in vacuum, $n(\vec{x})$ the variable index of refraction, v the speed of light in the medium, and $\vec{x} = \vec{x}(\theta)$ describes the path of the light ray. Equivalently, we can work with the *optical length* of a path, defined by

$$L[\vec{x}] = \int n(\vec{x})ds = \int_0^1 n(\vec{x})\sqrt{\vec{x}'^2}\, d\theta = cT[\vec{x}], \qquad (8.170)$$

as it only differs from $T[\vec{x}]$ by the constant c. Finding the stationary paths of the optical length is a general variational problem for which we can apply all of the framework developed so far in this chapter.

Example 8.20 *Snell's law* (or the *law of refraction*) states that the direction of a refracted light ray is related to its incident direction as

$$n_i \sin(\theta_i) = n_r \sin(\theta_r), \qquad (8.171)$$

where n_i and n_r are the refractive indices in the medium before and after the refraction, respectively, θ_i is the angle of the incoming light ray relative to the normal of the refractive surface, and θ_r is the angle of the refracted light ray relative to the normal, see Fig. 8.16.

Assuming that we wish to send a light signal from point A in a medium with refractive index n_i to point B in the medium with refractive index n_r, we need to find the path for which the optical length takes a stationary value. Since the refractive index in each medium is constant, we may write the total optical length as a sum of the optical length of the path in each medium and find that

$$L[y] = L_1[y] + L_2[y] = n_i\ell_i[y] + n_r\ell_r[y], \qquad (8.172)$$

where $\ell_i[y]$ and $\ell_r[y]$ are the functionals giving the length of the paths in each medium, respectively. Within each medium, the Euler–Lagrange equations are just the same as that for finding the extrema of the path length $\ell_i[y]$ and the paths will therefore be straight lines within the media. However, there are still several possibilities depending on where the path crosses from one medium to the next, see Fig. 8.17, and we can compute the total optical path as a function of the location of this point

$$L[y] = L(y_c) = n_i\sqrt{x_i^2 + y_c^2} + n_r\sqrt{x_r^2 + (y_0 - y_c)^2}. \qquad (8.173)$$

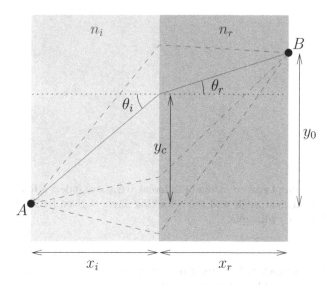

Figure 8.17 Finding the stationary optical length between the points A and B, we know that the straight lines within each medium satisfy the Euler–Lagrange equations. We can find the stationary path by checking all possibilities for the distance y_c.

This is a function of one variable only and we can easily find its stationary points from

$$L'(y_c) = \frac{n_i y_c}{\sqrt{x_i^2 + y_c^2}} - \frac{n_r(y_0 - y_c)}{\sqrt{x_r^2 + (y_0 - y_c)^2}} = n_i \sin(\theta_i) - n_r \sin(\theta_r) = 0, \qquad (8.174)$$

which proves that the stationary path for the optical length indeed satisfies Snell's law.

Example 8.21 We can use variational calculus to express Snell's law in a more general form. If we let a light ray propagate in a medium with a variable refractive index that depends only on the x-coordinate, then the optical length is given by

$$L[y] = \int \underbrace{n(x)\sqrt{1 + y'(x)^2}}_{\equiv \mathcal{L}} \, dx, \qquad (8.175)$$

where the integrand $\mathcal{L}$ does not depend explicitly on the function $y(x)$, but only on its derivative and x. Because of this, we have a constant of motion given by

$$C = \frac{\partial \mathcal{L}}{\partial y'} = \frac{n(x)y'(x)}{\sqrt{1 + y'(x)^2}} = n(x) \sin(\theta(x)), \qquad (8.176)$$

where $\theta(x)$ is the angle that the path makes with the x-axis at x. For any two points x_1 and x_2 on the path, it must therefore hold that

$$n(x_1) \sin(\theta(x_1)) = n(x_2) \sin(\theta(x_2)), \qquad (8.177)$$

which is Snell's law for the case when the index of refraction may vary in any manner as a function of x. Note that if $n(x_1) > n(x_2)$, then a requirement for this equation to have a solution is that

$$\sin(\theta(x_1)) \leq \frac{n(x_2)}{n(x_1)}, \tag{8.178}$$

if the light ray is sent from x_1, any larger angle $\theta(x_1)$ will result in a reflection of the ray before reaching x_2.

8.7.2 Hamilton's principle

In the context of classical mechanics, we are already very familiar with Newton's formulation in terms of forces and *Newton's laws*:

1. In an inertial frame, a body moves with a constant velocity unless acted upon by a force.

2. If a force $\vec{F}$ acts on a body of mass m, its acceleration is given by $\vec{a} = \vec{F}/m$. More generally, for a system with variable mass, this is often expressed in terms of the change in momentum according to $\vec{F} = d\vec{p}/dt$.

3. If a body A exerts a force $\vec{F}$ on a body B, then B exerts a force $-\vec{F}$ on A.

As it turns out, as long as we deal only with conservative internal forces and external potentials that do not depend on velocities, this may be reformulated into a variational principle.

Let us first define a few objects that will be useful in describing the evolution of a system with a general set $q^i(t)$ of degrees of freedom, starting with the *Lagrangian* $\mathcal{L}(\vec{q}, \dot{\vec{q}}, t)$, which in classical mechanics takes the form

$$\mathcal{L}(\vec{q}, \dot{\vec{q}}, t) = T - V, \tag{8.179}$$

where T is the kinetic energy of the system and V its potential energy. We also define the *action*

$$S[\vec{q}] = \int_{t_0}^{t_1} \mathcal{L}(\vec{q}, \dot{\vec{q}}, t) dt \tag{8.180}$$

as the integral of the Lagrangian between the time t_0, at which we know the state of the system, to the time t_1, at which we wish to know the state of the system. *Hamilton's principle* (or the *principle of stationary action*) states that the system's evolution will be described by functions that are stationary functions of the action. As such, the equations of motion for the system will be the Euler–Lagrange equations for the action. In more compact terms, the physical solutions $\vec{q}(t)$ are such that

$$\delta S = 0. \tag{8.181}$$

Hamilton's principle is also sometimes referred to as the principle of *least* action. However, this is generally a misnomer as the stationary function may also be a maximum or a saddle point. To separate the two approaches to mechanics, although they are often equivalent, we will refer to them as *Newtonian* and *Lagrangian mechanics*, respectively. It is also of interest to note that there is a third formulation of classical mechanics, known as *Hamiltonian mechanics*. Although Hamilton's principle appears in Lagrangian mechanics, the two should generally not be confused with each other.

At face value, Hamilton's principle may look unfamiliar and even a bit strange. Let us therefore verify that we can recover Newton's laws by the following series of examples.

Example 8.22 First, let us consider a particle of mass m moving in three spatial dimensions. If this particle is not subjected to any forces, the potential V may be set to zero while the kinetic energy is given by

$$T = m\frac{\vec{v}^2}{2} = m\frac{\dot{\vec{x}}^2}{2}. \tag{8.182}$$

With a potential of zero, the Lagrangian $\mathcal{L}$ is equal to the kinetic energy, resulting in an action

$$S[\vec{x}] = \frac{m}{2}\int \dot{\vec{x}}^2 dt. \tag{8.183}$$

Since the Lagrangian does not depend explicitly on $\vec{x}$, we find that

$$\vec{p} = \frac{\partial \mathcal{L}}{\partial \dot{\vec{x}}} = m\dot{\vec{x}} = m\vec{v} \tag{8.184}$$

is a constant of motion. Thus, if there are no forces acting on the particle, its velocity will be constant. This is Newton's first law.

Example 8.23 Let us now lift the restriction of no forces acting upon the particle and instead consider a particle moving in a force field such that $\vec{F}(\vec{x}, t) = -\nabla V$, where the potential energy $V(\vec{x}, t)$ is independent of the velocity of the particle. With the kinetic energy given by the same expression as before, the Lagrangian is now of the form

$$\mathcal{L} = \frac{m}{2}\dot{\vec{x}}^2 - V(\vec{x}, t). \tag{8.185}$$

From this expression, we can easily find the Euler–Lagrange equations

$$\frac{\partial \mathcal{L}}{\partial \vec{x}} - \frac{d}{dt}\frac{\partial \mathcal{L}}{\partial \dot{\vec{x}}} = -\nabla V(\vec{x}, t) - \frac{d}{dt}m\dot{\vec{x}} = -\nabla V - \dot{\vec{p}} = 0, \tag{8.186}$$

where again $\vec{p} = m\dot{\vec{x}}$ is the momentum of the particle. Inserting the expression for the force in terms of the potential, we end up with

$$\dot{\vec{p}} = \vec{F}, \tag{8.187}$$

which is Newton's second law.

Example 8.24 Finally, let us consider the situation where we are looking at a system of two particles with masses m_1 and m_2 that has a potential energy

$$V(\vec{x}_1, \vec{x}_2) = V(\vec{x}_2 - \vec{x}_1) = V(\vec{X}), \tag{8.188}$$

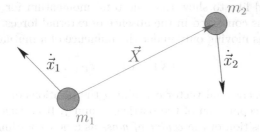

Figure 8.18 Two particles with masses m_1 and m_2, respectively, moving freely apart from being subjected to a force due to a potential $V(\vec{X})$ depending on the separation $\vec{X} = \vec{x}_2 - \vec{x}_1$ between them. Hamilton's principle will imply that the forces on the particles are equal in magnitude and opposite in direction.

which depends on the separation $\vec{X} = \vec{x}_2 - \vec{x}_1$ between the particles, see Fig. 8.18. In addition, each of the particles contributes with its own kinetic energy, leading to the Lagrangian

$$\mathcal{L} = \frac{1}{2}(m_1 \dot{\vec{x}}_1^2 + m_2 \dot{\vec{x}}_2^2) - V(\vec{X}). \tag{8.189}$$

The Euler–Lagrange equations for the first particle are now given by

$$\frac{\partial \mathcal{L}}{\partial \vec{x}_1} - \frac{d}{dt}\frac{\partial \mathcal{L}}{\partial \dot{\vec{x}}_1} = -\nabla_1 V(\vec{X}) - \frac{d}{dt}m\dot{\vec{x}}_1 = 0, \tag{8.190}$$

where ∇_1 is the gradient with respect to the coordinates $\vec{x}_1$. Using the chain rule, this may be rewritten as

$$\vec{F}_1 = \dot{\vec{p}}_1 = -\nabla_1 V(\vec{X}) = -\vec{e}_i \frac{\partial X^k}{\partial x_1^i}\frac{\partial V}{\partial X^k} = \vec{e}_i \frac{\partial V}{\partial X^i}, \tag{8.191}$$

where we have used the fact that $\partial X^k/\partial x_1^i = -\delta_{ik}$. The exact same computation holds for the second particle with the only difference being that $\partial X^k/\partial x_2^i = \delta_{ik}$, which leads to

$$\vec{F}_2 = \dot{\vec{p}}_2 = -\vec{e}_i \frac{\partial V}{\partial X^i} = -\vec{F}_1. \tag{8.192}$$

The mutual potential $V(\vec{X})$ therefore leads to the appearance of a force $\vec{F}_2$ acting on the second particle which is the negative of the force $\vec{F}_1$ acting on the first, showing that Newton's third law also holds in Lagrangian mechanics, at least for this type of systems.

8.7.2.1 Constants of motion

Our earlier discussion on constants of motion may have been illuminating in itself and provided us with sharp tools for facing some variational problems that might otherwise have been more difficult. However, it is now about to bear fruit in a whole different context as we shall see how they relate to a large number of conservation laws that we are used to from Newtonian mechanics. As a small appetiser, we have already seen that the momentum of a single particle is conserved unless there are any external forces acting on it in Example 8.22.

In general, we would like to show that the total momentum for a system containing any number of particles is conserved in the absence of external forces. Let us therefore study a system of N particles moving only under the influence of a mutual potential

$$V(\vec{X}) = V(\vec{x}_{12}, \vec{x}_{13}, \ldots),\tag{8.193}$$

where $\vec{X}$ is a $3N$-dimensional vector containing the positions of all of the N particles and $\vec{x}_{ij} = \vec{x}_i - \vec{x}_j$ is the separation of the particles i and j. If particle i has mass m_i, then we can introduce the position of the *center of mass* as a new coordinate

$$\vec{x}_{\mathrm{cm}} = \frac{\sum_i m_i \vec{x}_i}{\sum_i m_i} \equiv \frac{\sum_i m_i \vec{x}_i}{M},\tag{8.194}$$

where we have introduced $M = \sum_i m_i$ as the total mass of the system. In order to describe the system fully, the remaining coordinates may be given in terms of the particles' displacement from the center of mass. Note that only $N - 1$ of those displacements will be independent and we will therefore still have a total of $3N$ coordinates describing the system. Since the potential only depends on the separation of the particles, we find that

$$\frac{\partial V}{\partial x_{\mathrm{cm}}^k} = 0.\tag{8.195}$$

In addition, the kinetic energy of the system can be written as

$$T = \sum_i \frac{m_i}{2} \dot{\vec{x}}_i^2 = \sum_i \frac{m_i}{2} (\dot{\vec{x}}_{\mathrm{cm}} + \dot{\vec{d}}_i)^2,\tag{8.196}$$

where $\vec{d}_i$ is the displacement of particle i from the center of mass. Expanding the square leads to

$$T = \sum_i \frac{m_i}{2} \dot{\vec{x}}_{\mathrm{cm}}^2 + \dot{\vec{x}}_{\mathrm{cm}} \cdot \frac{d}{dt} \sum_i m_i \vec{d}_i + \sum_i \frac{m_i}{2} \dot{\vec{d}}_i^2 = \frac{M}{2} \dot{\vec{x}}_{\mathrm{cm}}^2 + \sum_i \frac{m_i}{2} \dot{\vec{d}}_i^2,\tag{8.197}$$

where we have used that

$$\vec{x}_{\mathrm{cm}} = \frac{1}{M} \sum_i m_i \vec{x}_i = \frac{1}{M} \sum_i m_i (\vec{x}_{\mathrm{cm}} + \vec{d}_i) = \vec{x}_{\mathrm{cm}} + \frac{1}{M} \sum_i m_i \vec{d}_i,\tag{8.198}$$

implies

$$\sum_i m_i \vec{d}_i = 0.\tag{8.199}$$

We can now write down the Lagrangian in this coordinate system as

$$\mathcal{L} = \frac{M}{2} \dot{\vec{x}}_{\mathrm{cm}}^2 - V(\vec{D}) + \sum_i \frac{m_i}{2} \dot{\vec{d}}_i^2,\tag{8.200}$$

where $\vec{D}$ is a $3N$-dimensional vector containing the displacements $\vec{d}_i$. It is possible to express the potential in this fashion since

$$\vec{x}_{ij} = \vec{x}_i - \vec{x}_j = (\vec{x}_{\mathrm{cm}} + \vec{d}_i) - (\vec{x}_{\mathrm{cm}} + \vec{d}_j) = \vec{d}_i - \vec{d}_j.\tag{8.201}$$

We can get rid of the additional three degrees of freedom introduced by using N displacement vectors $\vec{d}_i$ either by directly using Eq. (8.199) to express one of the $\vec{d}_i$ in terms of the others or by using it as a holonomic constraint.

Once we have done the above exercise of rewriting the Lagrangian, it is clear that the Lagrangian does not depend explicitly on the center of mass coordinates, but only their derivatives through the kinetic term. It therefore follows that the *total momentum*

$$\vec{P} = \frac{\partial \mathcal{L}}{\partial \dot{\vec{x}}_{\mathrm{cm}}} = M\dot{\vec{x}}_{\mathrm{cm}} = \sum_i m_i \dot{\vec{x}}_i = \sum_i \vec{p}_i \qquad (8.202)$$

is a constant of motion.

In addition to the constant of motion resulting from the Lagrangian not being explicitly dependent on the center of mass position, it is also clear that the given Lagrangian does not depend explicitly on the time t. Since this is true also in the original coordinates and the discussion is slightly simpler using the original coordinates, we use the Lagrangian

$$\mathcal{L} = \sum_i \frac{m_i}{2} \dot{\vec{x}}_i^2 - V(\vec{X}). \qquad (8.203)$$

The Beltrami identity now implies that

$$\sum_i \dot{\vec{x}}_i \cdot \frac{\partial \mathcal{L}}{\partial \dot{\vec{x}}_i} - \mathcal{L} = \sum_i \frac{m_i}{2} \dot{\vec{x}}_i^2 + V(\vec{X}) = T + V = E, \qquad (8.204)$$

where E is a constant of motion. Being the sum of the kinetic and potential energies of the system, we can directly identify this constant with its *total energy*.

In a similar fashion to what we have described here, if the Lagrangian can be shown to be independent of any single coordinate, there is a constant of motion that corresponds to this independence. This may be used to show that several different quantities, such as the angular momentum, are invariant under certain conditions, see Problem 8.37.

8.8   MODELLING WITH VARIATIONAL CALCULUS

As we have seen in this chapter, variational principles generally lead to differential equations that need to be solved. This is good news! We spent the major part of Chapter 3 using different lines of argumentation to do just that and admittedly some of the derivations were rather lengthy and required us to go down to the level of force diagrams for small parts of an object. Variational calculus provides us with an additional tool that can be used in model building to derive certain differential equations from basic variational principles by just writing down the functional for which we wish to find the stationary functions and applying the Euler–Lagrange equations.

In addition to allowing us to find the differential equations describing a system in a more direct fashion, the variational framework may also provide us with additional insights into how the system behaves. This will generally result from considering the different constants of motion and conserved currents that arise from determining whether the functional contains an integrand that is not explicitly dependent on some of its arguments.

Example 8.25 As an example of a situation where we can derive a differential equation directly from variational arguments is given by the vibrating string with a tension S and linear density ρ_ℓ discussed in Section 3.5.1. After a rather lengthy argument using the free body diagram of a small part of the string, we found that the transversal deviation $u(x,t)$ of the string should satisfy the wave equation

$$u_{tt}(x,t) - c^2 u_{xx}(x,t) = \frac{f}{\rho_\ell}, \qquad (8.205)$$

where $c^2 = S/\rho_\ell$ is the wave velocity. Let us see how we could treat this derivation by applying variational methods.

According to Hamilton's principle, the system will evolve in such a way that the action is stationary. In order to find out what this implies, we need to find the Lagrangian of the system, which is written in terms of the kinetic and potential energies. Since we have already treated the potential energy of the elastic string in Example 8.5, we can just write down the resulting potential energy as

$$V[u] = \int_a^b \left[\frac{S}{2} u_x(x,t)^2 + f(x,t)u(x,t) \right] dx, \tag{8.206}$$

where we have replaced the gravitational force from Example 8.5 by a general time dependent force density $f(x,t)$ and kept only the leading non-constant term in the potential due to the extension of the string. Looking at the kinetic energy T, it is a sum of the contributions of all small masses dm along the string, which have kinetic energies $dT = u_t(x,t)^2 dm/2$. Using that $dm = \rho_\ell dx$, we therefore find

$$T = \int dT = \int_a^b \frac{\rho_\ell}{2} u_t(x,t)^2 dx. \tag{8.207}$$

The action is now given by the functional

$$\mathcal{S}[u] = \int_{t_0}^{t_1} (T-V)dt = \int_{t_0}^{t_1} \int_a^b \underbrace{\left[\frac{\rho_\ell}{2} u_t(x,t)^2 - \frac{S}{2} u_x(x,t)^2 - f(x,t)u(x,t) \right]}_{\equiv \mathcal{L}} dx\, dt, \tag{8.208}$$

where $\mathcal{L}$ is not the Lagrangian, since it is not the kinetic energy minus the potential, but the *Lagrangian density*, i.e., the density of kinetic energy minus the density of potential energy. However, it is often common to also refer to the Lagrangian density just as the Lagrangian although not technically accurate.

Having found an expression for the action functional, we can find the equations of motion for the system by requiring that $\delta \mathcal{S} = 0$ or, in other terms, that the Lagrangian density satisfies the Euler–Lagrange equation

$$\frac{\partial \mathcal{L}}{\partial u} - \partial_t \frac{\partial \mathcal{L}}{\partial u_t} - \partial_x \frac{\partial \mathcal{L}}{\partial u_x} = 0. \tag{8.209}$$

Performing the derivatives of the Lagrangian density, we find that

$$\frac{\partial \mathcal{L}}{\partial u} = -f(x,t), \tag{8.210a}$$

$$\partial_t \frac{\partial \mathcal{L}}{\partial u_t} = \rho_\ell u_{tt}(x,t), \tag{8.210b}$$

$$\partial_x \frac{\partial \mathcal{L}}{\partial u_x} = -S u_{xx}(x,t), \tag{8.210c}$$

leading to the differential equation

$$\rho_\ell u_{tt}(x,t) - S u_{xx}(x,t) = f(x,t). \tag{8.210d}$$

Dividing this result by the linear density ρ_ℓ gives us the result in Eq. (8.205).

Example 8.26 Let us see what we can say about a vibrating string like the one in the previous example based on the particular expression for the Lagrangian density

$$\mathcal{L} = \frac{1}{2}[\rho_\ell u_t(x,t)^2 - S u_x(x,t)^2], \tag{8.211}$$

where we have assumed that there is no external force acting on it and therefore $f(x,t) = 0$. This Lagrangian density has no explicit dependence on the coordinates x and t, nor does it depend on the transversal displacement $u(x,t)$ itself, but only on its derivatives.

Starting with the Lagrangian density not being explicitly dependent on the function $u(x,t)$ itself, we find that the corresponding conserved current is given by

$$J^t = \frac{\partial \mathcal{L}}{\partial u_t} = \rho_\ell u_t(x,t), \quad J^x = \frac{\partial \mathcal{L}}{\partial u_x} = -S u_x. \tag{8.212}$$

What physical quantity does this correspond to? Having in mind the well-known formula $p = mv$, the momentum of a small part dx of the string in the transversal direction must be given by

$$dp_T = u_t(x,t)dm = u_t(x,t)\rho_\ell dx. \tag{8.213}$$

As a result, we can interpret the time component J^t of the current as the momentum density in the transversal direction. By our discussion in Example 8.19, the corresponding current in the continuity equation is the spatial component J^x. This is exactly what we expect! For a given x, the expression for the spatial component is nothing but the transversal force from the string to the left of x on the string to the right of x as we argued already in Section 3.5.1. We have not only recovered the continuity equation for the transversal momentum but also found the correct expression for the corresponding force. We can find another relation of interest by considering the integral

$$\mathcal{I} = \int_{t_0}^{t_1} \int_a^b [\partial_t J^t + \partial_x J^x] dx\, dt = 0, \tag{8.214}$$

which holds due to the conserved current, see Fig. 8.19. As advertised when we discussed conserved currents in general, this can be rewritten as

$$\mathcal{I} = \int_a^b [J^t(x,t_1) - J^t(x,t_0)]dx + \int_{t_0}^{t_1} [J^x(b,t) - J^x(a,t)]dt = 0 \tag{8.215}$$

by using the divergence theorem. For the spatial integrals we find that

$$\int_a^b J^t(x,\tau)dx = \int_a^b \rho_\ell u_t(x,\tau)dx = p_T(\tau) \tag{8.216}$$

is the total transversal momentum of the string at time τ. We therefore obtain the relation

$$p_T(t_1) - p_T(t_0) = S \int_{t_0}^{t_1} [u_x(b,t) - u_x(a,t)]dt, \tag{8.217}$$

where the right-hand side describes the transversal momentum flow into the string at the endpoints. In particular, we note that if the string endpoints are allowed to move freely, then the natural boundary condition at $x = a$ and $x = b$ will be given by

$$\frac{\partial \mathcal{L}}{\partial u_x} = J^x = -S u_x = 0 \implies u_x = 0 \tag{8.218}$$

and the total transversal momentum in the string will be conserved. This is no longer true if we fix the string at the endpoints, see Fig. 8.20.

Since the Lagrangian density is also not explicitly dependent on the coordinates x and t, there are two conserved currents related to this which can be deduced through the generalisation of the Beltrami identity. In particular, taking the current from Eq. (8.168) corresponding to the Lagrangian density not depending explicitly on the time coordinate t, we find that the density

$$T^t{}_t = u_t(x,t)\frac{\partial \mathcal{L}}{\partial u_t} - \mathcal{L} = \frac{1}{2}\left[\rho_\ell u_t(x,t)^2 + S u_x(x,t)^2\right] \qquad (8.219a)$$

is the sum of the kinetic and potential energy densities and therefore *total energy density* and that the corresponding current is

$$T^x{}_t = u_t(x,t)\frac{\partial \mathcal{L}}{\partial u_x} = -S u_t(x,t)u_x(x,t). \qquad (8.219b)$$

Again, this is just what we should expect! We already know that $-S u_x(x,t)$ is the force acting from the string left of x on the string to the right of x. Furthermore, the movement of the point of application of this force parallel to the force direction is given by the velocity $u_t(x,t)$. The work done on the part of the string to the right of x per time unit is therefore given by the force multiplied by $u_t(x,t)$, which gives exactly the above expression for $T^x{}_t$ and is therefore the power dissipated from the left to right, i.e., the energy current density.

Finally, there is the current related to the Lagrangian density not explicitly depending on the x coordinate. The corresponding time component of the conserved current is

$$T^t{}_x = u_x(x,t)\frac{\partial \mathcal{L}}{\partial u_t} = \rho_\ell u_t(x,t)u_x(x,t) \qquad (8.220a)$$

which should be interpreted as a density of some sort and the corresponding spatial current is given by

$$T^x{}_x = u_x(x,t)\frac{\partial \mathcal{L}}{\partial u_x} - \mathcal{L} = -\frac{1}{2}\left[\rho_\ell u_t(x,t)^2 + S u_x(x,t)^2\right]. \qquad (8.220b)$$

As it turns out, the density corresponds to a *longitudinal* momentum density carried by the transversal wave, or more precisely, the negative of the longitudinal momentum density. Treating this properly in terms of the string motion requires looking at the longitudinal motion of the string as well as the transversal, a fact we have been ignoring throughout. This is therefore out of the scope for this text.

8.9 VARIATIONAL METHODS IN EIGENVALUE PROBLEMS

Variational methods also have some interesting applications to Sturm–Liouville problems. Let us start by considering a functional of the form

$$I[\varphi] = \left\langle \varphi, \hat{L}\varphi \right\rangle = \int_a^b w(x)\varphi(x)\hat{L}\varphi(x)dx, \qquad (8.221)$$

where $\hat{L}$ is a Sturm–Liouville operator and see if variational methods can give us some insights when we look for eigenfunctions of $\hat{L}$. Since eigenfunctions are only defined up

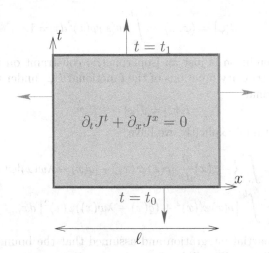

Figure 8.19 Integrating a conserved current over the spatial and temporal boundary surfaces will give the result zero due to the divergence of the current being identically zero. The current integral over the temporal boundary surfaces (darker boundary) relate to an extensive quantity of the system at that time, while the current integral over the spatial boundary surfaces (lighter boundary) give the total flow of the quantity out of the system. If this flow is zero, the extensive quantity must be the same at $t = t_0$ and $t = t_1$ and is therefore conserved in the system.

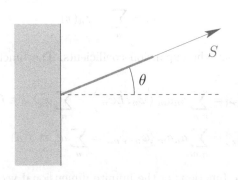

Figure 8.20 For a string with a fixed endpoint, the force due to the string tension S at the end-point will generally have a component in the transverse direction unless $\theta = 0$. This means that momentum in the transverse direction will be transferred between the string and the wall, thereby resulting in the transverse momentum in the string not being constant in time.

to a multiplicative constant, let us look for normalised eigenfunctions in particular, i.e., eigenfunctions that satisfy the normalisation condition

$$J[\varphi] = \langle \varphi, \varphi \rangle = \int_a^b w(x)\varphi(x)^2 dx = 1. \tag{8.222}$$

This normalisation condition is just an isoperimetric constraint on the function $\varphi(x)$ and in order to find the stationary functions of the functional $I[\varphi]$ under this constraint, we can define the new functional

$$K[\varphi] = I[\varphi] - \lambda J[\varphi]. \tag{8.223}$$

Writing out this functional explicitly, we have

$$K[\varphi] = \int_a^b \left\{ -\varphi(x)\frac{d}{dx}\left[p(x)\varphi'(x)\right] + [q(x) - \lambda w(x)]\varphi(x)^2 \right\} dx$$

$$= \int_a^b \left[p(x)\varphi'(x)^2 + (q(x) - \lambda w(x))\varphi(x)^2 \right] dx, \tag{8.224}$$

where we have used partial integration and assumed that the boundary terms vanish due to the boundary conditions. The Euler–Lagrange equation corresponding to this functional is given by

$$\frac{1}{2}\frac{\delta K}{\delta \varphi(x)} = -\frac{d}{dx}\left[p(x)\varphi'(x)\right] + [q(x) - \lambda w(x)]\varphi(x) = 0, \tag{8.225a}$$

which after division by $w(x)$ becomes the Sturm–Liouville eigenvalue equation

$$\hat{L}\varphi(x) = \lambda \varphi(x). \tag{8.225b}$$

As a direct consequence of this, any normalised eigenfunction of the Sturm–Liouville operator $\hat{L}$ is a local stationary function of $I[\varphi]$ in the set of normalised functions.

Let us see if we can understand this result from a different perspective by expressing a general function $\varphi(x)$ as a linear combination of the normalised eigenfunctions $\varphi_n(x)$ such that $\hat{L}\varphi_n(x) = \lambda_n \varphi_n(x)$. This linear combination will be of the form

$$\varphi(x) = \sum_n a_n \varphi_n(x), \tag{8.226}$$

where the constants a_n are the expansion coefficients. The functionals $I[\varphi]$ and $J[\varphi]$ can now be written as

$$I[\varphi] = \sum_{n,m} a_n a_m \left\langle \varphi_n, \hat{L}\varphi_m \right\rangle = \sum_n a_n^2 \lambda_n \equiv I(\vec{a}), \tag{8.227a}$$

$$J[\varphi] = \sum_{n,m} a_n a_m \langle \varphi_n, \varphi_m \rangle = \sum_n a_n^2 \equiv J(\vec{a}), \tag{8.227b}$$

where $I(\vec{a})$ and $J(\vec{a})$ are functions of the infinite dimensional vector $\vec{a}$ with the expansion coefficients as the vector components. The corresponding function $K(\vec{a}) = I(\vec{a}) - \lambda J(\vec{a})$ is therefore given by

$$K(\vec{a}) = \sum_n a_n^2 (\lambda_n - \lambda). \tag{8.228}$$

As a direct consequence, we find that the extrema of $K(\vec{a})$ must satisfy

$$\frac{\partial K}{\partial a_m} = 2a_m(\lambda_m - \lambda) = 0 \tag{8.229}$$

for all m. The only ways of accommodating this for each m are either $a_m = 0$ or $\lambda = \lambda_m$. Since the eigenvalues are distinct, we therefore find that the extrema correspond to functions indexed by the integer k such that $a_m = 0$ for all $m \neq k$. With the additional normalisation condition, we must also have $a_k = 1$, indicating that the stationary functions of the functional $I[\varphi]$ constrained to $J[\varphi] = 1$ are the normalised eigenfunctions $\varphi_n(x)$.

We can also go one step further and ask what type of stationary function each eigenfunction corresponds to. For the eigenfunction $\varphi_k(x)$, let us consider a variation

$$\varphi(x) = N(\vec{\varepsilon}) \left[\varphi_k(x) + \sum_{n \neq k} \varepsilon_n \varphi_n(x) \right] \tag{8.230}$$

such that the parameters ε_n may be considered small and $N(\vec{\varepsilon})$ is a normalisation constant given by

$$J[\varphi] = N(\vec{\varepsilon})^2 \left(1 + \sum_{n \neq k} \varepsilon_n^2 \right) = 1 \implies N(\vec{\varepsilon})^2 \simeq 1 - \sum_{n \neq k} \varepsilon_n^2. \tag{8.231}$$

Inserting this into the expression for $I[\varphi]$ leads to

$$I[\varphi] = N(\vec{\varepsilon})^2 \left[\lambda_k + \sum_{n \neq k} \lambda_n \varepsilon_n^2 \right] = \lambda_k + \sum_{n \neq k} (\lambda_n - \lambda_k)\varepsilon_n^2 + \mathcal{O}(\varepsilon^4). \tag{8.232}$$

If we order the eigenfunctions such that $\lambda_1 < \lambda_2 < \lambda_3 < \ldots$, then a variation such that $\varepsilon_n = 0$ for all $n > k$ will lead to a decrease in $I[\varphi]$, while a variation such that $\varepsilon_n = 0$ for all $n < k$ will lead to an increase. Since there is an infinite set of eigenfunctions and eigenvalues, it follows that $\varphi_k(x)$ is a saddle point for all k except for the eigenfunction with the lowest eigenvalue, for which it is a minimum.

8.9.1 The Ritz method

Based on the above, we can use variational methods in order to find an approximation to the lowest eigenvalue of a Sturm–Liouville operator by guessing how the corresponding eigenfunction will behave. This is mainly important in the cases where the Sturm–Liouville problem is difficult, or even impossible, to solve analytically and the better the guess we can make, the closer to the actual eigenvalue we get. Given a guess $\tilde{\varphi}_1(x)$, it will generally be a linear combination of all of the eigenfunctions $\varphi_n(x)$ such that

$$\tilde{\varphi}_1(x) = k \left[\varphi_1(x) + \sum_{n > 1} \varepsilon_n \varphi_n(x) \right]. \tag{8.233}$$

Note that we do not need to actually know the eigenfunctions to make this argument, it is sufficient to know that they exist. We have also extracted the coefficient of $\varphi_1(x)$ from the entire sum, which is just a matter of normalisation. If we have made a good guess, then all of the ε_n will be small and we will find that the *Rayleigh quotient*

$$R[\varphi] = \frac{I[\varphi]}{J[\varphi]} = \frac{\langle \varphi, \hat{L}\varphi \rangle}{\langle \varphi, \varphi \rangle} \tag{8.234a}$$

is given by

$$R[\varphi] \simeq \lambda_1 + \sum_{n > 1} (\lambda_n - \lambda_1)\varepsilon_n^2 \geq \lambda_1 \tag{8.234b}$$

to second order in ε_n. Naturally, this is just the same expression as in Eq. (8.232) with $k = 1$ and minimising the Rayleigh quotient is therefore equivalent to minimising $I[\varphi]$ under the criteria that $J[\varphi] = 1$. This is hardly surprising as $R[\varphi]$ is independent of the normalisation of φ and $R[\varphi] = I[\varphi]$ when $J[\varphi] = 1$. If we have made a good enough guess, then the ε_n should be small numbers and we can use the approximation

$$\bar{\lambda}_1 = R[\varphi] = \lambda_1 + \mathcal{O}(\varepsilon^2). \tag{8.235}$$

Note that, e if we make an error of $\mathcal{O}(\varepsilon)$ in our guess of the eigenfunction, the leading correction in this approximation is of $\mathcal{O}(\varepsilon^2)$.

The method described above for finding an approximation of the eigenfunction corresponding to the lowest eigenvalue is known as the *Ritz method* and there is a very straightforward way in which we can improve our estimate. Our chances of finding a good approximation will increase significantly if we can make several guesses at once. In particular, we can indeed make an infinite number of guesses at once by considering a one-parameter family of functions $\tilde{\varphi}_1^\kappa(x)$, where κ is a continuous parameter that we can vary. The Rayleigh quotient will now be given by a function of κ

$$R(\kappa) = R[\tilde{\varphi}_1^\kappa], \tag{8.236}$$

but still satisfy $R(\kappa) \geq \lambda_1$. If we can compute this function, it is only a matter of minimising it to find the best approximation of λ_1. Not only will this generally result in a better approximation than pure guessing, but it will also produce the best approximation of the actual eigenfunction $\varphi_1(x)$ within the one-parameter family. Note that this method works not only in one, but also in a higher number of dimensions.

Example 8.27 Consider the problem of finding the lowest eigenvalue of the Laplace operator in a circular domain with Dirichlet boundary conditions

$$(\text{PDE}) : -\nabla^2 \varphi(\vec{x}) = \lambda \varphi(\vec{x}), \tag{8.237a}$$

$$(\text{BC}) : \varphi(\vec{x}) = 0. \qquad\qquad (\vec{x}^2 = 1) \tag{8.237b}$$

This is a problem we have seen before and one that we know is relevant, for example in order to find the dominant mode in describing a diffusion problem in such a circular domain. It is also a problem that we can solve exactly and the lowest eigenfunction is

$$\varphi_1(\vec{x}) = J_0(\alpha_{01}\rho), \tag{8.238}$$

where α_{01} is the first zero of the Bessel function $J_0(x)$ and ρ is the radial coordinate in polar coordinates with the center of the circular domain as the origin. The corresponding eigenvalue is given by α_{01}^2.

Let us for a blissful moment imagine that we did not know anything about Bessel functions but still wanted to know the lowest eigenvalue for this problem. We could start by guessing that the corresponding eigenfunction is approximately given by

$$\tilde{\varphi}_1^\kappa(\vec{x}) = 1 - \rho^\kappa. \tag{8.239}$$

For all $\kappa > 0$, this function is non-trivial and satisfies the given boundary conditions. We can now use the explicit form of this function to compute

$$I[\tilde{\varphi}_1^\kappa] = -\int_0^1 \rho(1-\rho^\kappa)\left(\frac{\partial^2}{\partial\rho^2} + \frac{1}{\rho}\frac{\partial}{\partial\rho}\right)(1-\rho^\kappa)d\rho = \frac{\kappa}{2}, \tag{8.240a}$$

$$J[\tilde{\varphi}_1^\kappa] = \int_0^1 \rho(1-\rho^\kappa)^2 d\rho = \frac{\kappa^2}{4+6\kappa+2\kappa^2}, \tag{8.240b}$$

leading to the Rayleigh quotient

$$R(\kappa) = \frac{I[\tilde{\varphi}_1^\kappa]}{J[\tilde{\varphi}_1^\kappa]} = \frac{2}{\kappa} + 3 + \kappa. \tag{8.241}$$

This function takes its minimum value when

$$R'(\kappa) = -\frac{2}{\kappa^2} + 1 = 0 \implies \kappa = \sqrt{2},\ R(\sqrt{2}) = 3 + 2\sqrt{2}. \tag{8.242}$$

Our upper bound for the lowest eigenvalue is therefore given by

$$\lambda_1 \leq \bar{\lambda}_1 = 3 + 2\sqrt{2} \simeq 5.828. \tag{8.243}$$

Going back to the world where we do know about Bessel functions and their zeros, we find that this is reasonably close to the true value $\alpha_{01}^2 \simeq 5.784$, which can be found from tables of Bessel function zeros.

If a given guessed eigenfunction can be guaranteed to be orthogonal to $\varphi_1(x)$ by some means, we can go on to find a similar bound for the second eigenvalue λ_2 by making that guess. However, there is a different extension of Ritz method which is generally more effective and that requires only the diagonalisation of a matrix.

8.9.2 The Rayleigh-Ritz method

Instead of considering only one guessed eigenfunction as in Ritz method, we can always consider a finite subspace of our entire function space, spanned by N functions $\tilde{\varphi}_n(\vec{x})$ that we assume to be linearly independent. The action of the Sturm–Liouville operator $\hat{L}$ on this finite dimensional space can be approximated in terms of an $N \times N$ matrix L with the entries

$$L_{nm} = \left\langle \tilde{\varphi}_n, \hat{L}\tilde{\varphi}_m \right\rangle. \tag{8.244}$$

At the same time, we may also define another $N \times N$ matrix J by its entries

$$J_{nm} = \langle \tilde{\varphi}_n, \tilde{\varphi}_m \rangle. \tag{8.245}$$

If the $\tilde{\varphi}_n(\vec{x})$ have been chosen in such a way that they are orthonormal, then $J_{nm} = \delta_{nm}$, but this is not a requirement for the following method to work. Due to the symmetric property of the inner product, the matrix J is by definition symmetric and we can therefore find a new basis

$$\bar{\varphi}_n(\vec{x}) = P_{nm}\tilde{\varphi}_m(\vec{x}) \tag{8.246}$$

such that

$$\bar{J}_{nm} = \langle \bar{\varphi}_n, \bar{\varphi}_m \rangle = \delta_{nm} \tag{8.247}$$

by diagonalising and rescaling the original basis. We can also write down the matrix $\bar{L}$ with the entries

$$\bar{L}_{nm} = \left\langle \bar{\varphi}_n, \hat{L}\bar{\varphi}_m \right\rangle. \tag{8.248}$$

However, the simplest way of finding these elements will be to do the same matrix transformation as was performed in order to diagonalise and normalise J. As long as we pick functions $\tilde{\varphi}_n$ such that they are mainly composed of N eigenfunctions φ_m, the eigenvalues and eigenvectors of $\hat{L}$ restricted to the finite dimensional space may be taken as approximations of the eigenvalues and eigenfunctions in the full space. This method, known as the *Rayleigh-Ritz method*, works best when the functions $\tilde{\varphi}_n(\vec{x})$ are chosen in such a fashion that they are actual linear combinations of N eigenfunctions of $\hat{L}$. The method will then return those exact eigenfunctions and eigenvalues.

Example 8.28 Let us consider the example of the operator $\hat{L} = -d^2/dx^2$ on the interval $0 < x < \ell$ with homogeneous Dirichlet boundary conditions $\varphi(0) = \varphi(\ell) = 0$. In this case, we have already seen several times that the normalised eigenfunctions are of the form

$$\varphi_n(x) = \sqrt{\frac{2}{\ell}} \sin(k_n x) \tag{8.249}$$

for $k_n = \pi n/\ell$ with corresponding eigenvalues $\lambda_n = k_n^2$. Let us therefore investigate if we can get anywhere close to this by using the Rayleigh-Ritz method. We aim for the two lowest eigenvalues and choose a two-dimensional subspace of the functional space, given by

$$\tilde{\varphi}_1(x) = \sqrt{\frac{30}{\ell}} \frac{x}{\ell} \left(1 - \frac{x}{\ell}\right), \tag{8.250a}$$

$$\tilde{\varphi}_2(x) = \sqrt{\frac{840}{\ell}} \frac{x}{\ell} \left(1 - \frac{x}{\ell}\right) \left(\frac{1}{2} - \frac{x}{\ell}\right), \tag{8.250b}$$

where the prefactors have been chosen in such a way that these functions are already normalised. Furthermore, since $\tilde{\varphi}_1(x)$ is symmetric around $x = \ell/2$ and $\tilde{\varphi}_2(x)$ is anti-symmetric around the same point, the functions are also orthogonal. These functions are shown in Fig. 8.21 along with the actual first three eigenfunctions and the function $\tilde{\varphi}_3$, which we will define in Eq. (8.252).

Because $\tilde{\varphi}_1(x)$ and $\tilde{\varphi}_2(x)$ are orthonormal, we do not need to diagonalise the 2×2 matrix J, since it already is diagonal. The elements of the matrix $\bar{L}$ can be found according to

$$\bar{L}_{11} = -\int_0^\ell \tilde{\varphi}_1(x)\tilde{\varphi}_1''(x) = \frac{10}{\ell^2}, \tag{8.251a}$$

$$\bar{L}_{22} = -\int_0^\ell \tilde{\varphi}_2(x)\tilde{\varphi}_2''(x) = \frac{42}{\ell^2}, \tag{8.251b}$$

$$\bar{L}_{12} = \bar{L}_{21} = -\int_0^\ell \tilde{\varphi}_1(x)\tilde{\varphi}_2''(x) = 0. \tag{8.251c}$$

The resulting matrix

$$\bar{L} = \frac{1}{\ell^2} \begin{pmatrix} 10 & 0 \\ 0 & 42 \end{pmatrix} \tag{8.251d}$$

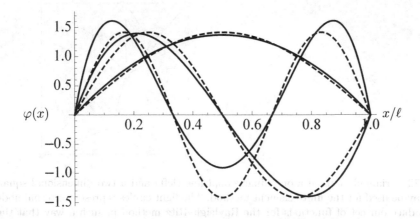

Figure 8.21 The three functions defined in Eqs. (8.250) and (8.252) used in our example of the Rayleigh–Ritz method (solid curves) along with the three actual eigenfunctions of the problem with the smallest eigenvalues (dashed curves).

clearly has the eigenvalues $\bar{\lambda}_1 = 10/\ell^2$ and $\bar{\lambda}_2 = 42/\ell^2$. These approximations should be compared to the true lowest eigenvalues $\lambda_1 = \pi^2/\ell^2 \simeq 9.8696/\ell^2$ and $\lambda_2 = 4\pi^2/\ell^2 \simeq 39.5/\ell^2$ of the Sturm–Liouville operator $\hat{L}$. If we extend our set of functions to also include the function

$$\tilde{\varphi}_3(x) = \sqrt{\frac{17010}{\ell}} \frac{x}{\ell} \left(1 - \frac{x}{\ell}\right) \left(\frac{1}{3} - \frac{x}{\ell}\right) \left(\frac{2}{3} - \frac{x}{\ell}\right), \tag{8.252}$$

the approximation for the first eigenvalue improves to the significantly more accurate value $\bar{\lambda}_1 \simeq 9.8698/\ell^2$.

8.9.2.1 Finite element method

A particular choice of the functions $\tilde{\varphi}_n(\vec{x})$ in the Rayleigh-Ritz method is very useful in for finding approximations of the eigenvalues and eigenfunctions numerically. In the *finite element method* (often abbreviated *FEM*), we start by selecting a number of points $\vec{x}_n$ within the domain of study and then perform a triangulation of the domain based on this set of points as vertices. A triangulation is a division of the domain into a set of simplices, which in one dimension is just the interval between two of the points, in two dimensions it is a set of triangles and in three dimensions a set of irregular tetrahedrons, see Fig. 8.22. In general, a simplex in N dimensions is uniquely defined by its $N+1$ corners. Mathematically, a point within a simplex can be written as

$$\vec{x} = \sum_{k=1}^{N+1} \xi_k \vec{x}_{s(k)}, \tag{8.253a}$$

where $\vec{x}_{s(k)}$ is the kth corner of the simplex and the numbers ξ_k satisfy

$$\sum_{k=1}^{N+1} \xi_k = 1 \quad \text{and} \quad \xi_k \geq 0. \tag{8.253b}$$

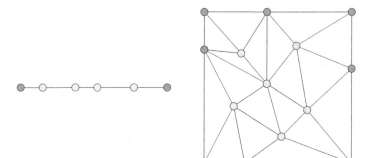

Figure 8.22 Triangulations of a one dimensional line (left) and a two dimensional square (right), which may be used for the finite element method. The light circles represent internal nodes and we will introduce our set of functions for the Rayleigh–Ritz method in such a way that there is one function per internal node. In general, the more nodes that are introduced, the more accurate the result will be.

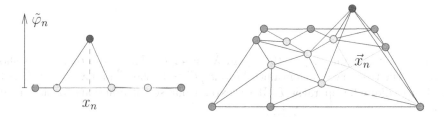

Figure 8.23 The functions $\tilde{\varphi}_n(\vec{x})$ corresponding to the dark internal nodes for the triangulations presented in Fig. 8.22. The functions are taken to have the value one at the dark node, zero at all other nodes, and to be linear within each simplex.

For our set of functions $\tilde{\varphi}_n(\vec{x})$, we pick functions such that

$$\tilde{\varphi}_n(\vec{x}_m) = \delta_{nm} \tag{8.254}$$

and such that $\tilde{\varphi}_n(\vec{x})$ is linear within each simplex, see Fig. 8.23.

Now consider the problem of finding the eigenvalues of the Laplace operator on some arbitrary domain with homogeneous Dirichlet boundary conditions. We can then rewrite the functional $I[\varphi]$ as

$$I[\varphi] = -\int_V \varphi(\vec{x})\nabla^2\varphi(\vec{x})dV = \int_V [\nabla\varphi(\vec{x})]^2 dV \tag{8.255}$$

by using Green's first identity. The reason for doing this is that our selected set of functions has derivatives that are piecewise smooth and behave much nicer than the second derivatives. Because of the way $\tilde{\varphi}_n(\vec{x})$ is defined, it is going to have a vanishing gradient everywhere but in the simplices in which $\vec{x}_n$ is a corner. As such, most of the matrix elements L_{nm} and J_{nm} will vanish as well, generally leading to very sparse matrices that are well suited for numerical diagonalisation.

8.10   PROBLEMS

Problem 8.1. In Example 8.3, we showed that the straight line between two points is a stationary function of the distance functional

$$L[y] = \int_0^{x_0} \sqrt{1 + y'(x)^2}\, dx. \tag{8.256}$$

We did this by introducing the function

$$L(\varepsilon) = L[kx + \varepsilon\eta] \tag{8.257}$$

and verifying that $L'(0) = 0$ for all $\eta(x)$. Show that the straight line is not only a stationary function, but indeed a minimum, by computing $L''(\varepsilon)$ and evaluating it for $\varepsilon = 0$. If it is a minimum then we must have $L''(0) \geq 0$ regardless of the function $\eta(x)$.

Problem 8.2. That the straight line is a minimum for the path length has been shown in Example 8.3 and Problem 8.1 by checking that any variation around the straight line leads to a longer path.

a) If we had not been able to guess this solution, we would have needed to solve a set of differential equations. Find this set by writing down the Euler–Lagrange equations corresponding to the path length functional

$$L[x, y] = \int_0^1 \sqrt{\dot{x}(t)^2 + \dot{y}(t)^2}\, dt. \tag{8.258}$$

Give an interpretation of the result you find.

b) By changing coordinates according to

$$x(t) = \rho(t)\cos(\phi(t)) \quad \text{and} \quad y(t) = \rho(t)\sin(\phi(t)), \tag{8.259}$$

write down the differential equation that a straight line must satisfy in polar coordinates when parametrised by the curve length.

Problem 8.3. A particle moves in a non-conservative time-dependent force field $\vec{F}(\vec{x}, t)$. Write down a functional of the path $\vec{x}(t)$ that describes the total work done on the particle when moving along the path from $\vec{x}(0)$ to $\vec{x}(t_0)$.

Problem 8.4. A fluid is flowing with a stationary velocity field $\vec{v}(\vec{x})$ and has a density given by $\rho(\vec{x})$. Write down functionals that describe

a) the total kinetic energy in a volume V and

b) the net mass flow per unit time through a surface S.

Problem 8.5. Write down the Euler–Lagrange equations for the functionals

a) $F_1[\phi] = \int_a^b \sqrt{\phi'(x) + 1}\phi(x)dx,$

b) $F_2[\phi] = \int_a^b \dfrac{x\phi(x)}{\phi'(x)^2}dx,$

c) $F_3[\phi] = \int_a^b \phi(x)^3 \phi'(x)dx,$

where $\phi(x)$ is a function of x. Discuss the meaning of your result in (c).

Problem 8.6. When studying special relativity, it will turn out that space-time has a very different geometry compared to Euclidean space. As a result, the *proper time* between $t = 0$ and $t = t_1$ of an observer whose motion is described by the function $x(t)$ is given by

$$\tau[x] = \int_0^{t_1} \sqrt{1 - \frac{\dot{x}(t)^2}{c^2}}\, dt, \tag{8.260}$$

where c is the speed of light.

a) Assume that $\Delta x = |x(t_1) - x(0)| < ct_1$ and find the Euler–Lagrange equations describing a stationary proper time.

b) Show that the straight line $x(t) = \Delta x\, t/t_1 + x(0)$ satisfies the Euler–Lagrange equations and that it is a maximum for the proper time.

Problem 8.7. The functional

$$F[\phi_1, \phi_2] = \int_0^{x_0} [x_0 \phi_1'(x) + \phi_2(x)][x_0 \phi_2'(x) + 2\phi_1(x)]dx \tag{8.261}$$

depends on the two functions $\phi_1(x)$ and $\phi_2(x)$. Find both sets of Euler–Lagrange equations for this functional corresponding to the variation with respect to the different functions. Determine the solution to the resulting differential equations.

Problem 8.8. Determine the natural boundary conditions that the argument $\phi(x)$ of the functionals defined in Problem 8.5 must satisfy in order to provide a stationary function of the functionals if it is not fixed at the endpoints.

Problem 8.9. The length of a curve $y(x)$ between the lines $x = 0$ and $x = x_0$ is given by the functional

$$\ell[y] = \int_0^{x_0} \sqrt{1 + y'(x)^2}\, dx. \tag{8.262}$$

Assuming that we place no requirement on the boundary values $y(0)$ and $y(x_0)$ of the curve and find the resulting natural boundary conditions. Verify that the straight lines with constant $y(x) = y_0$ minimise $\ell[y]$.

Problem 8.10. Just as Snell's law was derived in Example 8.20, we may use a similar argument to deduce the law of reflection. Consider the situation shown in Fig. 8.24 where we wish to find the stationary functions of the optical length between the points A and B given by $(x, y) = (-x_0, y_0)$ and $(x, y) = (x_0, y_0)$, respectively.

a) Start by considering reflection in the surface $y = 0$ and show that the total path is minimal when the path touches the surface at $x = 0$.

b) Now consider reflection in a curved surface $y = kx^2$. Show that the path touching the surface at $x = 0$ is still an extremum of the path length. In addition, find the values of k for which that path is not a minimum.

Problem 8.11. Consider the functional

$$F[\varphi] = \int_a^b \mathcal{L}(\varphi''(x), x)dx, \tag{8.263}$$

where the integrand does not depend explicitly on the function $\varphi(x)$, but only on its second derivative. Use the same kind of reasoning as we did around Eq. (8.63) to show that there exist two constants of motion related to this form of the integrand.

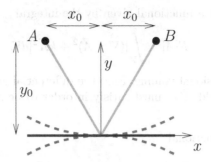

Figure 8.24 The situation considered in Problem 8.10. The reflective surface considered in (a) is shown as a solid line, while two of the surfaces in (b) are shown dashed. Depending on the curvature of the surface at the point of reflection, the path taken (light gray curve) by the light may correspond to different types of a stationary point.

Problem 8.12. When discussing the bending of a beam in solid mechanics in Example 8.6, we only derived the differential equation describing the transversal deviation under the assumption that the boundary terms would vanish when performing the partial integrations. Perform the partial integrations explicitly and deduce the form of these boundary terms. Discuss the restrictions on the variation $\eta(x)$ that will guarantee that this is the case. In the situation where these restrictions are not imposed, what are the natural boundary conditions that will ensure that the change in potential energy is equal to zero regardless of the variation $\eta(x)$? What is the physical interpretation of the natural boundary conditions?

Problem 8.13. Consider a string under tension with some stiffness. In such a string, the potential energy due to transversal displacement between the coordinates x and $x + dx$ can be written as

$$dV = \frac{1}{2}\left[Su_x(x,t)^2 + \mu u_{xx}(x,t)^2\right] dx. \tag{8.264}$$

Find the equation of motion for this string using Hamilton's principle. Also find the natural boundary conditions of the system.

Problem 8.14. A string that can be approximated to have zero stiffness, i.e., $\mu = 0$ in Problem 8.13, has ends at $x = 0$ and $x = \ell$ that are allowed to move in the transversal direction subject to a linear restoring force from springs with spring constant k. Write down a functional that describes the total potential energy of the system and use variational arguments to find the boundary conditions resulting from requiring the potential energy to be minimal.

Problem 8.15. We showed that the natural boundary conditions imposed on a stationary function of a functional in one dimension are

$$\frac{\partial \mathcal{L}}{\partial \dot{\varphi}} = 0, \tag{8.265}$$

where $\mathcal{L}$ is the integrand of the functional. Use the same kind of reasoning as we did in one dimension to deduce the generalisation of the natural boundary conditions to an arbitrary number of dimensions.

Problem 8.16. Consider a functional given by the integral

$$F[\vec{A}] = \int_V [(\nabla \times \vec{A})^2 + k\vec{A}^2] dV, \tag{8.266}$$

where V is a three-dimensional volume. Find the differential equations and boundary conditions that the vector field $\vec{A}(\vec{x})$ must satisfy in order to be a stationary function of this functional.

Problem 8.17. For a functional

$$F[\varphi_1, \varphi_2] = \int \mathcal{L}(\varphi_1, \varphi_2, \varphi_1', \varphi_2') dx, \tag{8.267}$$

where the integrand does not depend explicitly on the integration variable x, show that the same argumentation as that leading to the Beltrami identity for functionals depending only on one function leads to

$$\varphi_1'(x) \frac{\partial \mathcal{L}}{\partial \varphi_1'} + \varphi_2'(x) \frac{\partial \mathcal{L}}{\partial \varphi_2'} - \mathcal{L} = C \tag{8.268}$$

being a constant of motion, i.e., verify Eq. (8.71) in the case where the functional depends on two functions.

Problem 8.18. Show that solving Poisson's equation in a volume V with the inhomogeneity $\rho(\vec{x})$ is equivalent to finding the stationary functions of the functional

$$F[u] = \int_V \left[\frac{1}{2} (\nabla u(\vec{x})) \cdot (\nabla u(\vec{x})) - u(\vec{x})\rho(\vec{x}) \right] dV. \tag{8.269}$$

Also find the natural boundary conditions on the boundary ∂V of V that will be implied if the value of $u(\vec{x})$ on the boundary is not known.

Problem 8.19. One functional where derivatives appear at an order higher than one and where the function it depends on is a function on a higher-dimensional space is

$$F[\phi] = \frac{1}{2} \int_V [\nabla^2 \phi(\vec{x})]^2 dV. \tag{8.270}$$

Find the variation of this functional with respect to its argument ϕ and write down the partial differential equation that needs to be satisfied in order to find its stationary functions.

Problem 8.20. The problem of finding the shape of a curve of fixed length ℓ that encloses the largest area is known as *Dido's problem*, see Fig. 8.25.

a) The area of any two dimensional region S is given by

$$A = \int_S dA. \tag{8.271}$$

Use the the divergence theorem to rewrite this as an integral over the boundary curve of S. You may parametrise this curve with some curve parameter t and after solving this part you should have an expression for the area enclosed by the curve as a functional $A[x^1, x^2]$. *Hint:* You may want to use the fact that $\nabla \cdot \vec{x} = N$ in N dimensions.

b) Write down an expression for the total length of the curve in terms of a functional $L[x^1, x^2]$.

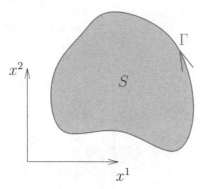

Figure 8.25 Dido's problem is the problem of finding the region S with the largest possible area A that is bounded by a curve Γ of a fixed length ℓ.

c) Use the method of Lagrange multipliers to solve Dido's problem.

Problem 8.21. A function $y(x) > 0$ that satisfies $y(0) = y(x_0) = 0$ is rotated around the x-axis. Find the maximal value of the area $A[y]$ of the rotated shape under the constraint that the arc-length $L[y]$ of the function $y(x)$ between $x = 0$ and $x = x_0$ should be given by $L[y] = \ell$.

Problem 8.22. A cylinder of radius R is partially filled with water and is rotated at an angular velocity ω. The stationary state occurs when the water is rotating along with the cylinder and will minimise the total potential energy of the water in the rotating frame. The potential energy of a small water element of volume dV in this frame will be given by

$$d\Phi = \rho \left(gz - \frac{r^2\omega^2}{2} \right) dV, \tag{8.272}$$

where ρ is the water density, and we have used cylinder coordinates with r as the radial distance rather than ρ in order to avoid mixing it up with the density. Find an expression for the total potential energy as a functional of the water depth $h(r)$, see Fig. 8.26, and solve for $h(r)$ under the constraint that the total volume of water in the cylinder is fixed.

Problem 8.23. Assume a particle starts at rest and is then constrained to move without friction along a curve $y(x)$ under the influence of a gravitational field in the negative y-direction. Show that the curve $y(x)$ that minimises the time taken to reach $y(\ell) = -y_0$ starting from $y(0) = 0$, see Fig. 8.27, is a cycloid, i.e., it can be parametrised as

$$x(s) = k[s - \sin(s)], \quad y(s) = -k[1 - \cos(s)], \tag{8.273}$$

where k is a constant. The problem of finding this curve is known as the *brachistochrone problem*.

Problem 8.24. Instead of moving in a homogeneous gravitational field as in the brachistochrone problem, consider a particle of charge q moving in an electric field $\vec{E} = E_0 y \vec{e}_2$, again constrained to move along some curve $y(x)$. Starting at $y(0) = 0$ and initially being at rest, find the curve that minimises the time taken to reach the line $x = \ell$. *Note:* Although $y = 0$ is a stationary point, it is unstable. The solution you will find may therefore be interpreted as the limiting case as the particle is only slightly offset from $y = 0$ at $t = 0$.

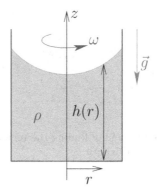

Figure 8.26 In a rotating cylinder partially filled with water, the water surface will eventually settle in such a way that the potential energy in the rotating system is minimised. This potential energy will be a functional with contributions from the gravitational field $\vec{g}$ and from the rotation. There will also be an isoperimetric constraint based on the fixed volume of water in the cylinder.

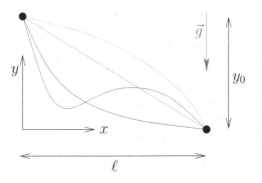

Figure 8.27 The brachistochrone problem is based on finding the fastest path between two points for a particle affected only by a homogeneous gravitational field. Taking x as a curve parameter, the time taken is a functional of $y(x)$.

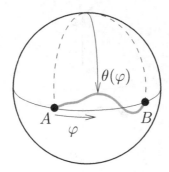

Figure 8.28 Looking for the shortest path between points A and B on a sphere, a coordinate system may be introduced such that both points lie on the equator $\theta = \pi/2$. A curve between the points may then be described by the function $\theta(\varphi)$ and its length in terms of a functional taking this function as its argument.

Problem 8.25. Consider a particle that moves in a rotationally symmetric potential such that its velocity is given by

$$v(\rho) = v_0 \frac{\rho^2}{R^2} \tag{8.274}$$

in polar coordinates, where v_0 and R are constants. Find the constraint curve $\rho(\phi)$ that minimises the time for the particle to move from the point given by $\phi = 0$ and $\rho = R$ to the line $\phi = \pi/4$.

Problem 8.26. In general relativity, the motion of a particle moving radially outside a spherically symmetric mass distribution is described by the curve $r(t)$ that makes the functional

$$S[r] = \int \left[c^2 \phi(r) - \frac{\dot{r}^2}{\phi(r)} \right] dt \tag{8.275}$$

take a stationary value. The function $\phi(r)$ is given by

$$\phi(r) = 1 - \frac{2GM}{c^2 r}, \tag{8.276}$$

where G is Newton's gravitational constant, M the total mass, and c the speed of light in vacuum. Derive a second order differential equation describing how r depends on t and show that it is equivalent to Newton's law of gravitation when $|\dot{r}| \ll c$ and $GM \ll c^2 r$.

Problem 8.27. We essentially solved the problem of finding the shape that minimises the area, and therefore the energy, of a soap film suspended between two concentric rings in Example 8.10. Consider the situation where the rings are placed at $z = \pm h$ and are both of radius $\rho(\pm h) = r_0$. By adapting the constants z_0 and C found in the example to these boundary conditions, show that there are one, two, or zero solutions, depending on the relation between h and r_0. In the case when there are two solutions, only one of these solutions will correspond to a minimum area. Find out which one by comparing the resulting areas.

Problem 8.28. Looking to minimise the distance between two points A and B on a sphere of radius R, see Fig. 8.28, we may introduce a spherical coordinate system such that $\theta = \pi/2$ for both A and B. In addition, we may select the φ coordinate in such a way that

$$\varphi(A) = 0 \quad \text{and} \quad \varphi(B) = \varphi_0. \tag{8.277}$$

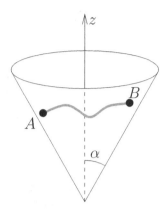

Figure 8.29 We are looking for the shortest path between points A and B on the cone with the opening angle 2α. In order to apply the resulting holonomic constraint, we can write the cylinder coordinate ρ as a function of z.

a) Using the coordinate φ as the curve parameter, verify that the curve

$$\theta(\varphi) = \frac{\pi}{2} \tag{8.278}$$

is a stationary function of the curve length functional.

b) The curve described in (a) is a *great circle*. If allowed to continue to larger values of φ than φ_0, it will return to A as $\varphi \to 2\pi$. In the same fashion, we could have used the coordinate $\varphi(B) = -2\pi + \varphi_0$ and let $-2\pi + \varphi_0 < \varphi < 0$ in order to describe a curve between A and B. Verify that the curve defined by Eq. (8.278) is still a stationary function for the curve length when we approach B in this way.

c) You have just found that there are at least two curves that are stationary functions for the curve length between A and B. Determine the nature of these two extrema if $0 < \varphi_0 < \pi$, i.e., determine whether they are minima, maxima, or saddle points.

Problem 8.29. Looking at the problem of finding the shortest distance between two points on a sphere, see Problem 8.28, find the differential equations that a general curve (not necessarily at $\theta = \pi/2$) must satisfy in order to minimise the distance when using the angle φ as the curve parameter. Once this is done, use geometry to parametrise the curve where the sphere intersects a plane through its center. Verify that this curve satisfies the differential equations.

Problem 8.30. A two-dimensional conical surface may be described using the holonomic constraint $\rho(z) = \tan(\alpha)z$ in cylinder coordinates, where 2α is the opening angle of the cone, see Fig. 8.29. Determine the general form of the curves describing the shortest path between two given points on the surface.

Problem 8.31. Consider a circular disc of a material with a refractive index that is inversely proportional to the distance from the center of the disc $n = k/\rho$ up to the radius $\rho = k$. Introducing polar coordinates, we wish to send a light signal from the point $\rho = \rho_0 < k$, $\phi = 0$, to the point $\rho = \rho_1$, $\phi = \phi_1$. In which direction must we direct the signal at the first point for it to reach the second?

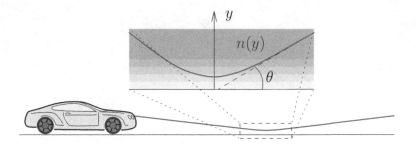

Figure 8.30 On a sunny day, inferior mirages appear at roads due to a layer of heated air that reflects light just above the road, in this case from the car's headlights. In Problem 8.33, we look for an expression for the minimal angle θ for which the road will be visible instead of the mirage.

Problem 8.32. On the paraboloid $z = k\rho^2/2$ in cylinder coordinates, a path may be parametrised by the polar coordinate ϕ and uniquely defined by the function $\rho(\phi)$. Write down a functional describing the path length $L[\rho]$ and find a first order differential equation that $\rho(\phi)$ has to satisfy in order to provide a stationary value for the distance between two points.

Problem 8.33. An *inferior mirage* is a phenomenon that occurs when there is a layer of air below eye-level which is much hotter than the air at eye-level. When viewed at a small angle, the phenomenon manifests itself as if the layer of hot air is reflective. Most people will be familiar with this phenomenon in terms of the optical effects occurring on a road on a sunny summer's day, see Fig. 8.30, when the asphalt is heated up by the Sun and in turn heats up a layer of air above it. Ultimately, the mirage is due to hot air being less dense and therefore having a lower index of refraction. Viewing the index of refraction as a function of the height above the road, find an expression for the minimal angle θ for which the road will be seen instead of the mirage if the index of refraction is n at eye-level and n_0 just above the asphalt.

Problem 8.34. A different type of mirage is the *superior mirage*, which may occur when there is a thermal inversion, i.e., when the air temperature increases with altitude. In order to illustrate this, consider a situation where the index of refraction varies as $n = n_0(1 - ky)$, where y is the distance above the surface and $k > 0$. If a superior mirage is observed from ground level at $x = 0$ in the direction $y'(0) = \alpha_0$, what is the actual distance to the object being seen, which is also assumed to be at ground level? You may ignore any effects due to the curvature of the Earth.

Problem 8.35. A *gradient-index fibre* is an optical fibre where the index of refraction decreases with the distance from the optical axis. For the purpose of this problem, assume that the index of refraction is of the form $n(\rho) = n_0(1 - k^2\rho^2)$. This index of refraction will allow helical optical paths, i.e., paths for which ρ is constant and the path moves around the optical axis at a constant rate $d\varphi/dz = \omega$. Determine what ω is necessary for such a path as a function of the constant radius ρ.

Problem 8.36. For a vibrating string that is subjected to a restoring force density proportional to the transversal displacement, there is an additional contribution to the potential energy that for a small string element between x and $x + dx$ is given by

$$dV_f = \frac{1}{2}ku(x, t)^2 dx, \qquad (8.279)$$

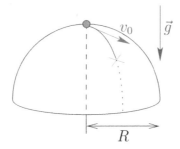

Figure 8.31 A particle moving from the top of a sphere of radius R in a gravitational field $\vec{g}$ with initial speed v_0. At some point, the particle will leave the sphere if there is no constraining force which can hold it down.

where k is a constant of appropriate physical dimension. Use this together with the kinetic energy due to string movement and the potential energy due to the stretching of the string in order to find a partial differential equation describing the transversal motion of the string.

Problem 8.37. A particle of mass m is moving in a *central potential*, i.e., in a potential that only depends on the distance from its center. Due to the symmetry of the problem, this will imply that the particle can only move within a two-dimensional plane. Introduce polar coordinates ρ and ϕ in this plane and use Hamilton's principle to show that the angular momentum

$$L = m\rho^2\dot{\phi} \tag{8.280}$$

is a constant of motion.

Problem 8.38. Consider a collection of particles that has two contributions to the potential energy. The first contribution only depends on the separation of the particles

$$V(\vec{X}) = V_i(\vec{x}_{12}, \vec{x}_{13}, \ldots), \tag{8.281}$$

where $\vec{x}_{ij} = \vec{x}_i - \vec{x}_j$, and the second is an external potential energy $V_e(\vec{X})$ that satisfies

$$\frac{\partial V_e}{\partial x_i^1} = 0 \tag{8.282}$$

for all i. Show that the total momentum of the entire system in the $\vec{e}_1$ direction is a constant of motion.

Problem 8.39. A particle is allowed to slide without friction starting with velocity v_0 on the top of a spherical surface, see Fig. 8.31.

a) Assuming the particle is constrained to the surface, use the holonomic constraint $r = R$ in spherical coordinates to write down the action for the particle. Use the fact that the Lagrangian is not explicitly time-dependent to derive a constant of motion.

b) Write down the problem using the holonomic constraint and a Lagrange multiplier function $\lambda(t)$. Write down the resulting Euler–Lagrange equation for the motion of the particle in the radial direction. Based on your result, what is your interpretation of the function $\lambda(t)$?

c) Using your results from (a) and (b) and with the assumption that the constraining force can only be directed in the positive r direction, at what position will the particle fall off the sphere?

Problem 8.40. The one-dimensional harmonic oscillator can be implemented as a single particle of mass m moving in one direction with a potential energy $V(x) = kx^2/2$.

a) Find the equation of motion for this harmonic oscillator by requiring that the action is stationary.

b) Write down the equation of motion for a single particle of mass m constrained to move on the parabola $y(x) = \kappa x^2/2$ in the presence of a gravitational potential $V(y) = mgy$. Verify that this situation does not correspond to a harmonic oscillator, but that the harmonic oscillator behaviour is recovered for small oscillations.

Problem 8.41. Consider a particle of mass m that is restricted to moving on a surface defined by the function $z = f(\rho)$ in cylinder coordinates.

a) Write down the Lagrangian for this particle in cylinder coordinates. Use the holonomic constraint in order to remove one of the degrees of freedom.

b) Your Lagrangian should not depend explicitly on either the time t or on the function $\phi(t)$. Use these facts to find two constants of motion for the particle. What is the physical interpretation of these constants of motion?

You may assume that there is no potential energy for the particle.

Problem 8.42. A homogeneous cylinder is allowed to roll without slipping on an inclined plane making an angle α with the horizontal. Use Hamilton's principle to find the acceleration of the cylinder in the direction along the plane.

Problem 8.43. When discussing the conserved currents of the vibrating string in Example 8.26, we considered the implications of performing an integral of the divergence of the current J^a over the full domain by use of the divergence theorem. Repeat these considerations for the current $T^a{}_t$, see Eqs. (8.219), and show that the result gives you the difference in the total energy of the string at different times. In addition, verify that the total energy in the string is conserved both in the case when the ends of the string are fixed and when the string is allowed to move freely.

Problem 8.44. Consider a string with tension S and length ℓ that is fixed at its endpoints. In the case when the string's linear density $\rho_\ell(x)$ is not constant, but depends on the coordinate x, show that the energy due to the transversal movement is still conserved, but that the longitudinal momentum of the string is not. *Note:* The longitudinal momentum density and current are the components of $T^a{}_x$.

Problem 8.45. The Lagrangian density corresponding to the transversal movement of an elastic beam is given by

$$\mathcal{L} = \frac{1}{2}[\rho_\ell u_t(x,t)^2 - EI u_{xx}(x,t)^2].$$

Since this Lagrangian density does not depend explicitly on $u(x,t)$, but only on its derivatives, there exists a corresponding conserved current J^a such that $\partial_a J^a = 0$. Find this current and use it to interpret the physical meaning of the third derivative $u_{xxx}(x,t)$.

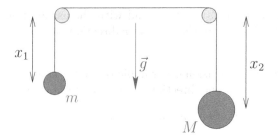

Figure 8.32 The Atwood machine consists of two masses m and M that are connected by an inextensible massless string via frictionless pulleys. We can describe the motion of the masses in a gravitational field $\vec{g}$ by using Hamilton's principle and the holonomic constraint $x_1 + x_2 = \ell$.

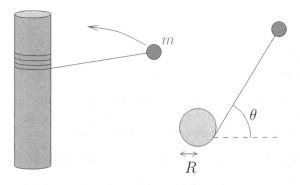

Figure 8.33 A string with a mass m tied to one end and the other end winding around a pole of radius R. The setting can be parametrised using the coordinate θ and analysed based on Hamilton's principle.

Problem 8.46. The *Atwood machine* consists of a weight M and a counter-weight m hanging by the same fixed-length string over friction- and mass-less pulleys in a homogeneous gravitational field g, see Fig. 8.32.

a) The fixed length of the string leads to the holonomic constraint $x_1(t) + x_2(t) = \ell_0$. Write down the action of the system for arbitrary $x_1(t)$ and $x_2(t)$ and use the Lagrange multiplier method for holonomic constraints to find the equations of motion of the system. Make sure that your result has the correct limits when $m \ll M$ and $m = M$, respectively. What is the interpretation of the Lagrange multiplier function $\lambda(t)$ that you have introduced?

b) Implement the holonomic constraint from (a) already at the Lagrangian level. This should result in a system with only one degree of freedom instead of three. Write down the Euler–Lagrange equation for this new system and verify that it is the same as those found in (a) as long as the holonomic constraint is satisfied.

Problem 8.47. Consider a string that is tied to a mass m in one end and with the other end free to wind around a pole of radius R. As long as the string is taut, the situation may be modelled by a single coordinate θ, see Fig. 8.33. Using Hamilton's principle, write down the equation of motion for the system. Can you find any constants of motion? Solve the equation of motion and determine the time taken for the mass to hit the rod if the initial length of the free string is ℓ and $\dot{\theta}(0) = \omega$.

Problem 8.48. Consider the two-dimensional domain defined by $R/2 < \rho < R$ and $0 < \phi < \pi/2$ in polar coordinates. Assuming homogeneous Dirichlet conditions on the boundary, find an approximation of the lowest eigenvalue of $-\nabla^2$ on this domain by guessing an approximate form of the corresponding eigenfunction.

Problem 8.49. We already know that the lowest eigenvalue function to the operator $-\nabla^2$ in the domain bounded by a sphere of radius R with homogeneous Dirichlet boundary conditions is given by

$$u(\vec{x}) = j_0(\pi r/R) = \frac{R\sin(\pi r/R)}{\pi r} \tag{8.283}$$

and that the corresponding eigenvalue is $\lambda = \pi^2/R^2$. Use the Ritz method to find an approximation of this result using the trial function $\tilde{u}^\kappa(r) = R^\kappa - r^\kappa$. Compare your result with the exact result.

CHAPTER 9

Calculus on Manifolds

In Chapters 1 and 2, we covered the analysis of vectors and tensors in Cartesian coordinate systems as well as in general curvilinear ones. However, in all of our exposition we always assumed that the underlying space was Euclidean so that every position could be uniquely defined by a position vector. While this assumption is often reasonable and very useful in the description of physical processes, there are several cases when it turns out to be too constraining. In particular, there are many situations in both classical mechanics as well as in general relativity where a physical system must be described using a more advanced mathematical framework that allows configurations that are not always possible to describe using a Euclidean space.

Example 9.1 A *spherical pendulum* consists of a mass m that is constrained to move freely on a sphere of fixed radius R, see Fig. 9.1. For example, this may be achieved either by fixing the mass on the end of a rod of length R with the other end allowed to rotate freely. While this system is embedded in a three-dimensional space, it really is a two-dimensional system. Any possible position of the mass may be referred to by using only two coordinates, e.g., the spherical coordinates θ and φ, and the system is described by giving a point on a sphere, which is not a Euclidean space. In order to uniquely define how the pendulum moves, it is therefore sufficient to state how θ and φ change with time.

As should be clear from the example above, where we might want to describe the movement of the spherical pendulum by a velocity vector, we will often want to describe movement or other quantities that we are used to describing as vector quantities in terms of some similar objects on the sphere. Since the spherical pendulum has a natural embedding in three dimensions, we may be tempted to apply our full three-dimensional thinking to it, which will result in the appearance of holonomic constraints such as the ones encountered in Chapter 8. However, in many other applications where the space involved is not Euclidean, there is no such natural embedding and it therefore makes sense to look for a description of this type of situation that does not rely on it, a description that only deals with the properties within the space we are describing.

9.1 MANIFOLDS

The first order of business in working with more general spaces is to find a framework with which we may describe them. When we chose curvilinear coordinates in a Euclidean space, we could base our description on coordinate functions on an underlying Cartesian

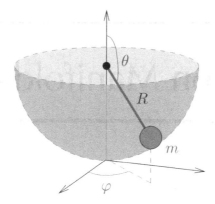

Figure 9.1 A pendulum of mass m and length m is allowed to move freely at a fixed distance R from a central point. The configuration of this system may be described by giving the two angles θ and φ and so the configuration space is two-dimensional.

coordinate system. This is no longer the case in a more general space, but we still want to be able to create a description in terms of coordinates.

Let us consider a space that we may call M. Since it is not always going to be possible to define a set of global coordinates valid in the entire space, we will be content with requiring that there exists a one-to-one mapping between a set of N local coordinates y^a and a part of the space. The number of coordinates required to achieve this mapping is called the *dimension* of the space. This set of local coordinates is restricted to an open set U in $\mathbb{R}^N$ and this together with the identification of each point in U with a point in the space M is called a *chart*.

Example 9.2 Let us make an attempt to describe the surface of the Earth. One option for doing so would be to describe it using a three-dimensional Euclidean space with a Cartesian or spherical coordinate system imposed on it. The drawback of this is that one of the coordinates, e.g., the radial coordinate in spherical coordinates, would be redundant, in particular if we are only moving on a small part of the surface. Instead, we can draw a map of our surroundings on a two-dimensional piece of paper. If we are reasonably good cartographers, each point on the map will correspond to one single point on the Earth's surface. We can therefore use a coordinate system with two coordinates to describe a position on the map and therefore also a position on the Earth's surface. This is the basic idea behind a chart.

Looking at larger regions, we could also use different projections of the Earth's surface onto a two-dimensional plane in order to describe them. These also constitute charts and may also be used as such, see Fig. 9.2.

There is no guarantee that a single chart will cover all points in M, the only requirement is that the chart works in some part of the space. In order to describe the entire space, we may therefore need to use different charts at different positions. In order to have a complete description of M, it is necessary to have a complete collection of charts such that each point in M is covered by at least one chart. Such a collection of charts is called an *atlas* of M and if we can describe all of M like this it is a *manifold*.

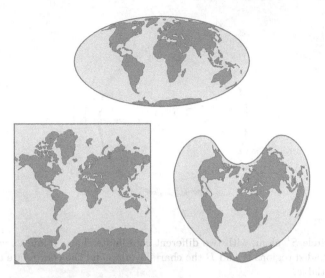

Figure 9.2   Three different possible projections of the Earth's surface onto a two-dimensional plane: The Mollweide (top), Mercator (left), and Bonne (right) projections. Each of the projections is a chart of the Earth's surface.

Example 9.3   None of the charts we showed in Example 9.2 covered the entire Earth surface (some of them covered *almost* all of it, but were singular in at least one point, which would have to be removed, or could not cover the entire surface while remaining a one-to-one map based on an open set). In fact, it is impossible to cover a sphere just using a single chart and our atlas will need at least two charts. In connection to the Earth's surface, an atlas is just a collection of two-dimensional maps of different parts of the Earth's surface. Before the existence of online map services accessible via the internet, these collections were often called a *World atlas*, printed on paper and bound into a book stored on a library shelf. Some maps may partially cover the same region, i.e., New York can be on several maps, but may only appear once in each map. In order for our atlas to be complete, each point on the Earth's surface must be on at least one map.

As mentioned in the example above, there is no requirement that a particular point in a manifold M is described in only one coordinate chart. On the contrary, unless there is one chart covering the entire manifold, there must necessarily exist points that can be referred to using several different charts and so there will be regions of M where the charts overlap and where points can be described using either one set of coordinates y^a or another $y'^{a'}$. Since each coordinate chart is a one-to-one mapping, wherever such overlaps occur we can always express one set of coordinates as functions of the other

$$y^a = \phi^a(y') \quad \text{or} \quad y'^{a'} = \varphi^{a'}(y). \tag{9.1}$$

Since this notation is a bit cumbersome, we will use the coordinate names and infer from context whether they are taken to mean the coordinate itself or the coordinate expressed as a function of another set of coordinates. Although we have been rather vague with specifying particular conditions on functions throughout this text, it is here of particular importance to point out a very relevant assumption: In the remainder of this text, the

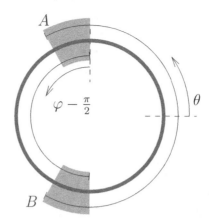

Figure 9.3 The circle S^1 along with two different coordinate charts θ and φ, which together form an atlas. In the shaded regions A and B the charts overlap and the coordinate transformations are infinitely differentiable.

coordinate transformations will all be assumed to be smooth, i.e., infinitely differentiable. The manifolds we will encounter in physics will generally be of this type.

Example 9.4 Consider the one dimensional circle S^1, which is a one-dimensional manifold, and let us impose two coordinate systems on parts of the circle according to Fig. 9.3. The ranges of the coordinates θ and φ are given by

$$-\frac{2\pi}{3} < \theta < \frac{2\pi}{3} \quad \text{and} \quad \frac{\pi}{2} < \varphi < \frac{3\pi}{2}, \tag{9.2}$$

respectively. The coordinate transformation on the overlap A is given by

$$\theta = \varphi, \tag{9.3a}$$

while that on the overlap B is of the form

$$\theta = \varphi - 2\pi \quad \Longleftrightarrow \quad \varphi = \theta + 2\pi. \tag{9.3b}$$

These coordinate transformations are linear functions, which are clearly infinitely differentiable.

9.2 FORMALISATION OF VECTORS

In order to generalise our concept of vectors and tensors to manifolds, let us take a step back and think about which properties of vectors we are actually using and try to understand whether or not we can formalise these in a way that does not inherently depend on the space being Euclidean. The vectors we have become accustomed to generally appear to be of two different types, vectors that describe displacements or quantities related to displacements and vectors that describe how scalar fields change depending on the position. These are fundamentally different concepts and we should not be surprised to find that they will be described by different types of vectors in the general setting, *tangent vectors* and *dual vectors*, respectively.

Example 9.5 Vectors related to displacements are relatively straightforward and the very first mention of a vector in Chapter 1 was concerned with velocities and accelerations, quantities related to how a displacement changes with time. Let us take displacements on the surface of the Earth as an example here. Introducing any coordinate system, such as using the longitude and latitude as coordinates, the movement of an object may be described by giving these as functions of time t. It is worth noting here that any two coordinates y^a that uniquely define a position on the Earth's surface will do. The velocity $v^a(t)$, i.e., the displacement per time unit, may now be specified by giving the derivatives

$$v^a(t) = \frac{dy^a}{dt} \equiv \dot{y}^a(t). \tag{9.4}$$

It is tempting to immediately also attempt to define acceleration as $a^a(t) = \ddot{y}^a(t)$, but this definition comes with some problems that we have already encountered when dealing with the derivatives of vectors in curvilinear coordinate systems in Chapter 2. We will deal with these complications in a while.

Example 9.6 The second type of vectors that we have already encountered is vectors describing changes in scalar quantities, such as the temperature gradient and the force related to a potential field. Choosing to consider the altitude h as a function of the position on the Earth's surface, the gradient

$$w_a = \frac{\partial h}{\partial y^a} \equiv \partial_a h \tag{9.5}$$

describes how the altitude changes depending on the chosen coordinates y^a.

Given these examples of the two different types of vectors, it is relatively straightforward to put the two concepts together. If we wish to describe how a scalar quantity φ, which depends on the position, varies with time along a path described by the coordinate functions $y^a(t)$, we are essentially interested in the time derivative of $\varphi(y(t))$, which can be expressed in terms of the velocity and the gradient by applying the chain rule

$$\frac{d\varphi(y(t))}{dt} = \frac{\partial \varphi}{\partial y^a} \frac{dy^a}{dt} = (\partial_a \varphi) v^a. \tag{9.6}$$

Naturally, this directly reminds us of the form of the inner product encountered in Euclidean spaces, but we need to be wary; we still have not shown that the gradient $\partial_a \varphi$, describing the change in a scalar quantity, and the velocity v^a, describing a displacement, are vectors of the same type. Indeed, we will find that this is generally not the case and that in order for a true inner product to be defined, we will need to define a metric tensor describing distances in the space. However, the rate of change $d\varphi/dt$ is well defined even without reference to a metric as it only requires knowledge on how φ changes with the coordinates and how the coordinates change with time.

Example 9.7 Given Examples 9.5 and 9.6, we may describe how the altitude h changes with time when we follow a path $y^a(t)$ along the Earth's surface. The change is given by

$$\frac{dh}{dt} = \frac{\partial h}{\partial y^a}\frac{dy^a}{dt} = w_a v^a. \tag{9.7}$$

Note how this naturally splits the rate of change in the altitude into two separate objects, the altitude gradient $w_a = \partial_a h$, which only depends on how the altitude changes as a function of position, and the velocity $v^a = \dot{y}^a$, which describes the movement on the surface.

With the above considerations in mind, let us see how we can use our insights about the two different types of vectors in order to define vectors in a general non-Euclidean space.

9.2.1 Tangent vectors

Starting with the vectors describing displacements, we have seen that it is natural to define the components of a velocity vector as

$$v^a = \frac{dy^a}{dt} \tag{9.8}$$

along some curve described by the coordinate functions $y^a(t)$. Naturally, these components will generally depend on the chosen coordinate system. Choosing a different set of coordinates $y'^{a'}$ and applying the chain rule, we find that the components of the velocity vector should transform as

$$v^{a'} = \frac{dy'^{a'}}{dt} = \frac{\partial y'^{a'}}{\partial y^a}\frac{dy^a}{dt} = \frac{\partial y'^{a'}}{\partial y^a}v^a. \tag{9.9}$$

Looking back at Eq. (1.153), this is exactly the transformation property we found for the contravariant vector components when we dealt with vector analysis in curvilinear coordinates. We now recall that the corresponding basis vectors were given by

$$\vec{E}_a = \frac{\partial \vec{x}}{\partial y^a} = \partial_a \vec{x}, \tag{9.10}$$

where $\vec{x}$ is the position vector. The appearance of the position vector poses a direct problem in generalising this to a basis which we can formally use when describing vectors in curved spaces. Since the base space is not necessarily a vector space on its own, the position vector generally does not exist. The solution to this problem may be as shocking as it is ingenious. We can just drop the position vector from the definition of the vector basis and define the *tangent vector basis* as

$$\vec{E}_a \equiv \partial_a. \tag{9.11}$$

The first questions that pop into our minds are: Does this makes sense at all? Was the appearance of the position vector irrelevant in defining the directions? If the vector basis is a set of partial derivatives, does this mean that a vector is a derivative? The answers to those questions are yes, yes, and yes, but let us elaborate on why this is so.

First of all, what does it mean for a vector to be a derivative? Given the tangent vector basis as the partial derivatives with respect to the coordinates, a general vector X (note that we will here skip the vector arrows and generally use capital letters for tangent vectors) may be written as

$$X = X^a \vec{E}_a = X^a \partial_a. \tag{9.12}$$

Looking back to the case of vectors in Euclidean spaces, this directly reminds us of the directional derivative defined by

$$\frac{d}{d\vec{n}} = \vec{n} \cdot \nabla = n^a \partial_a, \tag{9.13}$$

which does have a one-to-one correspondence with the vectors $\vec{n}$, since $d\vec{x}/d\vec{n} = \vec{n}$. It is also easy to see that the set of directional derivatives forms a vector space as there are natural ways of defining vector addition and multiplication by a scalar

$$X + Y = (X^a + Y^a)\partial_a \quad \text{and} \quad kX = kX^a\partial_a. \tag{9.14}$$

We therefore generalise the concept of a tangent vector to manifolds by defining it as a directional derivative. As a directional derivative, the action of a vector on a scalar function φ is given by

$$X\varphi = X^a \partial_a \varphi. \tag{9.15}$$

As might be expected, when the vector X is the velocity, as in the examples above, this is again the rate of change in φ.

So what about the role of the position vector $\vec{x}$ that we used in our definition of the tangent vector basis in a Euclidean space? Its role in the Euclidean case was to provide us with a reference in terms of the Cartesian basis vectors $\vec{e}_i$ as

$$\vec{E}_a = \frac{\partial \vec{x}}{\partial y^a} = \frac{\partial \vec{x}}{\partial x^i}\frac{\partial x^i}{\partial y^a} = \vec{e}_i \frac{\partial x^i}{\partial y^a}. \tag{9.16}$$

In the case of a more general base space, we do not have an underlying Euclidean space and the concept of a Cartesian coordinate system with its corresponding vector basis is not relevant. Instead, a displacement direction is now defined by a directional derivative in the direction of the displacement.

Finally, let us verify that the definition makes sense in terms of coordinate transformations. Changing coordinates from y^a to $y'^{a'}$, we change our vector basis from ∂_a to $\partial_{a'}$. A general vector X, i.e., a general directional derivative, can be expressed as

$$X = X^a \partial_a = X^{a'} \partial_{a'}, \tag{9.17}$$

where X^a and $X^{a'}$ are the vector components in the different bases. Using the chain rule to express ∂_a in terms of $\partial_{a'}$, we find that

$$X^a \partial_a = X^a \frac{\partial y'^{a'}}{\partial y^a} \partial_{a'} \tag{9.18a}$$

implying that

$$X^{a'} = \frac{\partial y'^{a'}}{\partial y^a} X^a, \tag{9.18b}$$

which is the expected transformation rule for the contravariant vector components and precisely what we would expect from a displacement vector as argued in relation to Eq. (9.9).

Example 9.8 A curve γ on a manifold M is a set of points on M that may be parametrised using a single continuous variable t. It may be described locally taking a coordinate system and writing down how the coordinates depend on the curve parameter t. Just as in the

Euclidean case, if the curve is differentiable, we may define a tangent vector of the curve at the point $\gamma(t)$. Since tangent vectors are derivatives, we can define the tangent vector of the curve to be the vector $\dot{\gamma}$ such that

$$\dot{\gamma}f = \frac{df(\gamma(t))}{dt} = \frac{dy^a}{dt}\partial_a f, \tag{9.19}$$

where f is a scalar field, in any set of coordinates. In other terms, the tangent vector has the components dy^a/dt, where t is the curve parameter.

9.2.1.1 Vector fields

Just as in the case when we studied vectors in curvilinear coordinate systems in a Euclidean space, there is really no obvious way of comparing vectors at different points in a general space by looking at the vector components. In the Euclidean space, this could be solved by expressing the vector basis in terms of the Cartesian basis, but in a general space this is no longer possible as we do not have access to a Cartesian basis. As a direct consequence of this, there is no longer any possibility of directly comparing vectors at different points in the base space. When we see the vectors as directional derivatives, this is not very strange. As a vector is a directional derivative of a function, the direction is only well defined at the particular point where we take the derivative. It is not possible to add or compare directions at different points in the space in a meaningful way. Because of this, a tangent vector is inherently associated with a particular point p in the space M. The vector space that this vector belongs to is the set of all vectors at p, which is called the *tangent vector space* at p and denoted $T_p M$.

An important concept when we discussed vector analysis was the concept of scalar and vector fields. When it comes to scalar fields, it generalises directly to a more general base space as a function assigning a scalar value to each point in the base space. When we introduced vector fields, we defined them as the assignment of a vector to each point in space. Naturally, when we talk about vector fields in more general spaces, the vector assigned to the point p must be an element of the tangent vector space $T_p M$ at p in order for the assignment of the vector to p to make sense.

Example 9.9 Consider the two-dimensional sphere S^2 with spherical coordinates θ and φ. An example of a vector field on this sphere is

$$X(\theta, \varphi) = \sin(\theta)\partial_\theta, \tag{9.20a}$$

which has the vector components

$$X^\theta = \sin(\theta) \quad \text{and} \quad X^\varphi = 0. \tag{9.20b}$$

Although both vector components are given as functions of the coordinates, no inherent comparison can be made between vectors at different points, see Fig. 9.4.

In some situations it is useful to talk about the *flow* of a tangent vector field. For each point p in the manifold M, the flow $\phi_X(p, t)$ of the vector field X is a function of a single

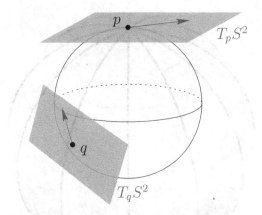

Figure 9.4 Tangent vectors at different points on the two-dimensional sphere S^2 belong to different tangent spaces. Here we show tangent vectors at the points p and q with their respective tangent spaces $T_p S^2$ and $T_q S^2$. There is no inherent way of adding or subtracting these as they belong to different vector spaces and we may only do this with tangent vectors in the same tangent space.

variable t to M itself, i.e., a curve, such that for every t, the tangent vector of the curve is equal to $X(\phi_X(p,t))$ and $\phi_X(p,0) = p$. In other words, we require that in any coordinate system

$$\frac{dy^a(\phi_X(p,t))}{dt} = X^a(\phi_X(p,t)).$$ (9.21)

For a fixed starting point p, the curve $\phi_X(p,t)$ is called a *flow line* of X. This is a coupled system of N ordinary differential equations for N variables, the coordinates as a function of the curve parameter t, and is solvable given N initial conditions set by specifying the point p, i.e., by giving its N coordinates.

Example 9.10 Let us get back to our favourite example of the two-dimensional sphere S^2 with a coordinate system given by the spherical coordinates θ and φ. Looking at the vector field defined by the coordinate basis vector ∂_θ, we find that Eq. (9.21) takes the form

$$\frac{d\theta}{dt} = 1 \quad \text{and} \quad \frac{d\varphi}{dt} = 0.$$ (9.22)

Therefore, the flow of ∂_θ starting at the point with coordinates θ_0 and φ_0 is given by

$$\theta(t) = \theta_0 + t, \quad \varphi(t) = \varphi_0.$$ (9.23)

In other words, the flow of the coordinate basis vector ∂_a is just the coordinate line on which all of the other coordinates are kept constant, see Fig. 9.5. Naturally, this holds also for any other coordinate basis vector field.

9.2.2   Dual vectors

The above treatment of tangent vectors revealed that they are directly related to displacements and their components transform as contravariant vector components under coordinate

Figure 9.5 The θ and φ coordinate lines on the sphere shown in black and red, respectively. The θ coordinate lines correspond to keeping φ fixed and varying θ and vice versa. The coordinate lines are also the flow lines of the coordinate basis vectors ∂_θ and ∂_φ, respectively.

transformations. It should therefore come as no surprise that dual vectors will turn out to be related to gradients with components that transform covariantly.

Let us start by formally defining the *dual space* T_p^*M at the point p in the manifold M as the set of all linear mappings from the tangent vector space T_pM to the real numbers. The physical meaning of this definition is not yet very clear, but we can make some sort of sense of it by noting that in the Euclidean setting, we found that

$$\vec{v} \cdot \vec{w} = v^a w_a, \tag{9.24}$$

where the v^a were the contravariant components of $\vec{v}$ and the w_a were the covariant components of $\vec{w}$. Since this product is linear in v^a, it is therefore a linear map from vectors with contravariant components to the real numbers. The covariant components in the dual basis therefore define just the type of linear transformation we are looking for.

The first thing to note about the dual space is that it is a vector space in itself under the natural definitions of addition and multiplication by a scalar

$$(\omega + \xi)(X) = \omega(X) + \xi(X), \tag{9.25a}$$
$$(a\omega)(X) = a\omega(X), \tag{9.25b}$$

where X is any tangent vector. Since the dual vectors are linear mappings, we can write down the action of a dual vector ω on an arbitrary tangent vector as

$$\omega(X) = \omega(X^a \partial_a) = X^a \omega(\partial_a) \equiv X^a \omega_a, \tag{9.26}$$

where the dual vector components $\omega_a = \omega(\partial_a)$ are given by the action of the dual vector on the tangent vector coordinate basis. For any given coordinate system, it is therefore sufficient to specify the components ω_a in order to fully define the dual vector, and the dual space therefore has the same dimension as the tangent vector space.

We can now examine how the coefficients ω_a transform from one set of coordinates to another. Taking a different set of coordinates $y'^{a'}$, we find that

$$\omega_{a'} = \omega(\partial_{a'}) = \omega\left(\frac{\partial y^a}{\partial y'^{a'}}\partial_a\right) = \frac{\partial y^a}{\partial y'^{a'}}\omega(\partial_a) = \frac{\partial y^a}{\partial y'^{a'}}\omega_a. \tag{9.27}$$

Just as the tangent vector components transformed as contravariant vector components, we have now found that the dual vector components transform as covariant vector components. While this is satisfying in its own right, we still have only found the transformation of the dual vector components. When discussing the dual basis in the Euclidean setting, these components belonged to a set of basis vectors $\vec{E}^a$ such that the vector could be written on the form $\vec{w} = w_a\vec{E}^a$. A property of the dual basis $\vec{E}^a$ was that

$$\vec{E}^a \cdot \vec{E}_b = \delta^a_b. \tag{9.28}$$

To generalise this relation to the dual space, we wish to find out what sort of objects the dual basis E^a that satisfies

$$E^a(\partial_b) = \delta^a_b \tag{9.29}$$

consists of and how we can construct them. Once we have performed this construction, we will be able to write any dual vector on the form

$$\omega = \omega_a E^a \tag{9.30a}$$

since this will lead to

$$\omega(X) = \omega_a E^a(X^b\partial_b) = \omega_a X^b E^a(\partial_b) = \omega_a X^b \delta^a_b = \omega_a X^a \tag{9.30b}$$

as expected.

9.2.2.1 Differentials as dual vectors

In the case of the dual basis in curvilinear coordinates in Euclidean space, we defined it in terms of the gradient of the coordinate functions y^a as

$$\vec{E}^a = \nabla y^a. \tag{9.31}$$

In turn, the gradient ∇f for any function f was defined in terms of an underlying Cartesian coordinate system such that $\nabla f = \vec{e}_i\partial_i f$. In a manifold, there is no longer any possible reference to a Cartesian coordinate system and so the gradient cannot be defined in this manner, just as we could not define the position vector $\vec{x}$ when dealing with the definition of the tangent vector basis. Instead, the only thing which may be salvaged from this is the definition of the dual basis in terms of small changes of the coordinates y^a. Let us therefore see if we can find a description making use of infinitesimal changes in different functions.

For any scalar field f, we can define a dual vector field df according to

$$df(X) = Xf = X^a\partial_a f, \tag{9.32}$$

i.e., its components are given by $df_a = \partial_a f$. It is rather straightforward to show that the components df_a have the proper transformation properties under coordinate transformations (see Problem 9.5) and they exactly correspond to the components of the gradient that we found in the Euclidean case. This dual vector is called the *differential* of f and the fact that it is closely related to the gradient provides us with exactly the property we need in order to find the dual basis. Since the dual basis in the Euclidean setting was related to the gradient of the coordinate functions, we consider the differentials dy^a of the coordinate functions, which have the property

$$dy^a(\partial_b) = \frac{\partial y^a}{\partial y^b} = \delta_b^a. \tag{9.33}$$

This is precisely the property we expected from the dual basis and we therefore identify $E^a = dy^a$. In particular, this lets us write any dual vector on the form

$$\omega = \omega_a dy^a. \tag{9.34}$$

Furthermore, we also find the familiar relation

$$df = (\partial_a f)dy^a, \tag{9.35}$$

which is a statement on how the general scalar field f varies due to small changes in the coordinates.

Example 9.11 Looking for a graphical representation of tangent vectors and dual vectors, we consider the case of the Earth's surface. Being associated with a direction of displacement, we can represent a tangent vector X with an arrow on a map, see Fig. 9.6. If we use a chart where objects are represented by larger regions on the map, the arrow will grow accordingly, representing an increase of the tangent vector components. On the other hand, a dual vector ω is related to a rate of change and the closest graphical interpretation is in terms of level curves, with the dual vector components being proportional to the density of the level curves. As such, using the chart where objects are represented by larger regions on the map, the density of the level curves, and therefore also the dual vector components, decrease accordingly. The number of level curves representing the dual vector ω crossed by the arrow representing a tangent vector X is kept constant and should be interpreted as $\omega(X) = \omega_a X^a$.

9.2.3 Tensors

With the concepts of tangent vectors and dual vectors in place, we are ready to follow through with the construction of tensors in exactly the same fashion as was done in Chapter 2. We define a *type* (n, m) *tensor* at the point p in a manifold M as a linear combination of tensor products of n tangent vectors and m dual vectors belonging to $T_p M$ and $T_p^* M$, respectively. We find that a coordinate basis for the resulting vector space is given by

$$e_{a_1 \ldots a_n}^{b_1 \ldots b_m} = \bigotimes_{k=1}^{n} \partial_{a_k} \bigotimes_{\ell=1}^{m} dy^{b_\ell} \tag{9.36a}$$

and consequently a general tensor of type (n, m) can be written as

$$T = T_{b_1 \ldots b_m}^{a_1 \ldots a_n} e_{a_1 \ldots a_n}^{b_1 \ldots b_m}, \tag{9.36b}$$

where the coefficients $T_{b_1 \ldots b_m}^{a_1 \ldots a_n}$ are the *tensor components* in the given coordinate system. Since the tangent vector and dual bases transform just as they did in Chapter 2 under

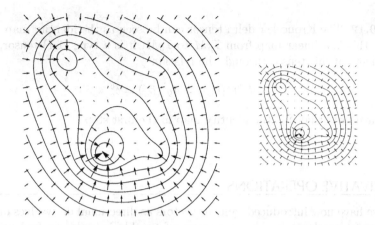

Figure 9.6 A tangent vector may be represented by an arrow with a certain length. Changing the scaling of the chart changes the arrow length accordingly. A dual vector may then be represented by level curves and the contraction between a tangent vector and a dual vector is given by the number of level curves the arrow crosses. When the chart is scaled, the density of the level curves changes in precisely the correct fashion in order to preserve this number.

coordinate transformations, the tensor components also follow the transformation rules we are already accustomed to from working with curvilinear coordinates in Euclidean space

$$T^{a'_1 \ldots a'_n}_{b'_1 \ldots b'_m} = T^{a_1 \ldots a_n}_{b_1 \ldots b_m} \left(\prod_{k=1}^{n} \frac{\partial y'^{a'_k}}{\partial y^{a_k}} \right) \left(\prod_{\ell=1}^{m} \frac{\partial y^{b_\ell}}{\partial y'^{b'_\ell}} \right), \tag{9.37}$$

just as stated in Eq. (2.18).

In addition to the coordinate transformations, all of our discussion regarding tensor algebra in Section 2.2 also still holds. It should now become clear why contractions are only allowed between contravariant and covariant indices, there just is no natural inner product between two tangent vectors, nor between two dual vectors, and our introduction of the contraction explicitly required one to exist. As in Chapter 2, we will usually not write out the tensor basis explicitly and refer to the tensor components $T^{a_1 \ldots a_n}_{b_1 \ldots b_m}$ and the tensor T interchangeably. In these cases, the assumption will always be that the tensor components are written in the coordinate basis of some coordinate system on the manifold.

As a tensor of type (n, m) includes a tensor product of n tangent vectors and m dual vectors and we have introduced dual vectors as linear mappings from $T_p M$ to the real numbers, a tensor of this type may also be naturally seen as a linear mapping from $(T_p M)^{\otimes m}$ to $(T_p M)^{\otimes n}$, where $V^{\otimes k}$ represents the tensor product of k copies of V. If ω is a tensor of this rank, this linear mapping may be defined as

$$\omega(X_1, \ldots, X_m) = \omega^{a_1 \ldots a_n}_{b_1 \ldots b_m} \left(\prod_{\ell=1}^{m} dy^{b_\ell}(X_\ell) \right) e_{a_1 \ldots a_n}. \tag{9.38}$$

In fact, this is the most general form of linear maps from $(T_p M)^{\otimes m}$ to $(T_p M)^{\otimes n}$ and an alternative way of defining general mixed tensors. As such, we will sometimes specify a tensor by defining this linear transformation.

Example 9.12 The Kronecker delta tensor can be seen as the identity map $\delta(X) = X$ on T_pM. Since this is a linear map from T_pM to T_pM, it is a type (1,1) tensor. We can find the components of this tensor through the relation

$$dy^b(\delta(\partial_a)) = dy^b(\partial_a) = \delta_a^b \tag{9.39}$$

where we have used that δ is the identity map in the first step.

9.3 DERIVATIVE OPERATIONS

Although we have now introduced tangent vectors as directional derivatives of scalar fields, we will have to go through a large amount of trouble in order to find something that resembles a derivative of a vector or tensor field. The underlying reason for this is the same as when we considered derivatives of vector fields in curvilinear coordinates on a Euclidean space. Apart from differentiating the vector field components, we will also need to know how the derivative acts upon the vector bases, an issue which is not yet very clear when it comes to vectors on manifolds.

9.3.1 The Lie bracket

The first thing we might attempt in trying to define a derivative of a vector field Y is to apply a directional derivative $X = X^a\partial_a$ to it, resulting in

$$XY = X^a\partial_a Y^b\partial_b = X^a(\partial_a Y^b)\partial_b + X^a Y^b\partial_a\partial_b. \tag{9.40}$$

This is a differential operator of order two and therefore not a vector field and our attempt was therefore in vain. However, the first term is a first order derivative, although it does not display the correct transformation properties under coordinate transformations, see Problem 9.8. The second term, while being a second order derivative, is symmetric in X and Y. As a result, the *Lie bracket*

$$[X, Y] = XY - YX = X^a(\partial_a Y^b)\partial_b - Y^a(\partial_a X^b)\partial_b \tag{9.41}$$

is a first order derivative of Y, sometimes referred to as the *Lie derivative* $\mathcal{L}_X Y$. The parts of the components $X^a(\partial_a Y^b) - Y^a(\partial_a X^b)$ that do not display the correct transformation properties exactly cancel during coordinate transformations and the Lie bracket is therefore a tangent vector. In addition, the Lie bracket has several properties that we would expect from a derivative of a vector field. In particular, it satisfies the product rule

$$\mathcal{L}_X fY = [X, fY] = X(f)Y + f[X, Y] = df(X)Y + f\mathcal{L}_X Y, \tag{9.42}$$

where f is a scalar field.

So what is this object we have created? Does it have a geometrical interpretation? In order to answer this, let us consider the flows of the vector fields X and Y starting from a point p with coordinates y_0^a. Following the flow line of X for a short parameter change dt leaves us at the point p_X with coordinates

$$y_X^a = y^a(\phi_X(p, dt)) = y_0^a + X^a(p)dt + \frac{dt^2}{2}X^b\partial_b X^a. \tag{9.43}$$

Here and in what follows in this paragraph, we will keep only terms of quadratic order or lower in small parameters such as dt. Continuing from this point by following the flow of Y for a parameter change ds we end up at the point p_{YX} with coordinates

$$y_{YX}^a = y_0^a + X^a dt + \frac{dt^2}{2} X^b \partial_b X^a + Y^a ds + \frac{ds^2}{2} Y^b \partial_b Y^a, \tag{9.44}$$

where X^a and its derivatives are evaluated at p and Y^a and its derivatives at p_X. If we instead wish to use the components Y^a at p as well, we can do so in terms of the expansion

$$Y^a(p_X) = Y^a(p) + X^b(p)\partial_b Y^a(p)\, dt, \tag{9.45}$$

which leads to

$$y_{YX}^a = y_0^a + X^a dt + Y^a ds + \frac{1}{2}\left(X^b \partial_b X^a dt^2 + Y^b \partial_b Y^a ds^2\right) + X^b \partial_b Y^a ds\, dt. \tag{9.46}$$

In the same fashion as we here followed the flow of X first and the flow of Y second, we could just as well have done it in the other order and ended up at the point p_{XY} with coordinates

$$y_{XY}^a = y_0^a + X^a dt + Y^a ds + \frac{1}{2}\left(X^b \partial_b X^a dt^2 + Y^b \partial_b Y^a ds^2\right) + Y^b \partial_b X^a ds\, dt. \tag{9.47}$$

The points p_{YX} and p_{XY} have the same coordinates only if

$$y_{YX}^a - y_{XY}^a = dt\, ds\, (X^b \partial_b Y^a - Y^b \partial_b X^a) = dt\, ds\, [X, Y]^a = 0. \tag{9.48}$$

In other words, following the flow of X and then the flow of Y, we end up at the same point as if we first follow the flow of Y and then the flow of X only if the Lie bracket vanishes. If this is the case, we say that the flows *commute* and the Lie bracket is a measure of whether this is the case or not, see Fig. 9.7.

Example 9.13 The property of the Lie bracket relating to whether flows commute or not can be used to determine whether it is possible to use the flows of vector fields to construct a local coordinate system. If the Lie bracket vanishes, we can use the parameters t and s of the different flows above in order to uniquely refer to a point close to p. On the other hand, given a coordinate system, the coordinate basis vectors always commute

$$[\partial_a, \partial_b] = \partial_a \partial_b - \partial_b \partial_a = 0 \tag{9.49}$$

and so their flows commute. As we have seen, using the coordinates θ and φ on the sphere, the flows of ∂_θ and ∂_φ are the coordinate lines of constant θ and φ, respectively, and the flows commute, see Fig. 9.8.

9.3.2 Affine connections

While the Lie bracket $[X, Y]$ works as a derivative of Y, it lacks some of the properties we have gotten used to from the directional derivative in Euclidean space. Most strikingly, the covariant derivative introduced in Chapter 2 satisfies the relation

$$\nabla_{f\vec{v}} T = f \nabla_{\vec{v}} T \tag{9.50a}$$

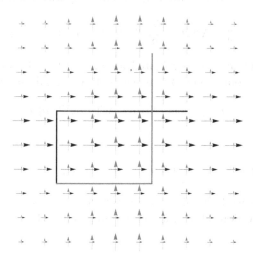

Figure 9.7 Two vector fields $X = \frac{1}{1+(x^2)^2}\partial_1$ and $Y = \frac{1}{1+(x^1)^2}\partial_2$, shown in black and gray, respectively, with non-commuting flows in $\mathbb{R}^2$. The curves represent the flows with the black curve first taking the flow along Y and then along X, while the gray curve starts at the same point, but follows the flow along X first and then Y with the same increases in the parameter values along the curves. The curves not ending at the same point is an indication of the non-commutativity of the flows.

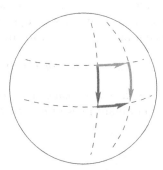

Figure 9.8 The flows of the tangent vector basis fields ∂_θ and ∂_φ on the sphere commute, as do all coordinate bases. The dark curve first follows the ∂_θ direction and then the ∂_φ direction, while the light reverses the order. Both curves start and end at the same point.

for an arbitrary tensor field T and scalar field f, while the Lie derivative we just introduced instead gives

$$\mathcal{L}_{fX}Y = [fX, Y] = f[X, Y] - df(Y)X = f\mathcal{L}_X Y - df(Y)X, \qquad (9.50b)$$

introducing an additional term $-df(Y)X$. We therefore want to find a more direct generalisation and start by recalling some of the properties of the covariant derivative.

When we considered vector and tensor fields in general coordinate systems on a Euclidean space, we defined the covariant derivative $\nabla_{\vec{n}} = n^a \nabla_a$ as a generalisation of the directional derivative to arbitrary tensor fields. Our introduction of the covariant derivative relied upon the partial derivatives of the tangent vector and dual bases in order to define the Christoffel symbols according to

$$\partial_a \vec{E}_b = \Gamma^c_{ab} \vec{E}_c. \qquad (9.51)$$

In a general manifold, there are several reasons why these derivatives no longer make sense. To start with, the basis vectors were defined in relation to an underlying Cartesian coordinate system with a position vector, which no longer exists in the manifold case. More importantly, the directional derivative itself was implicitly defined as the limit

$$n^a \partial_a \vec{E}_b(\vec{x}) = \lim_{\varepsilon \to 0} \frac{\vec{E}_b(\vec{x} + \varepsilon \vec{n}) - \vec{E}_b(\vec{x})}{\varepsilon}. \qquad (9.52)$$

This definition assumes that there exists an intrinsic way of subtracting a vector at the point $\vec{x}$ from a vector at the point $\vec{x} + \varepsilon \vec{n}$. In a Euclidean space, this is not problematic as all vectors may be traced back to their expression in the underlying Cartesian vector basis, but in a general manifold this is no longer the case. Instead, tangent vectors at different points p and q in the manifold belong to different vector spaces $T_p M$ and $T_q M$, respectively. As a consequence of this, it is impossible to define the derivative of the tangent vector basis in the same fashion.

So how do we recover from this setback? A large portion of the utility of vector analysis came from the possibility of differentiating vector fields and we have seemingly just lost it. The way out is to define an operator ∇_X, which has the important properties expected from a directional derivative operator in the direction determined by the tangent vector X. In particular we want this operator to map a tangent vector field to another tangent vector field, to satisfy the product rule for derivatives

$$\nabla_X(fY) = df(X)Y + f\nabla_X Y \qquad (9.53a)$$

for the product of a scalar and a vector field, and to be linear in both the direction and in the tangent vector field it acts upon, i.e.,

$$\nabla_{f_1 X_1 + f_2 X_2} Y = f_1 \nabla_{X_1} Y + f_2 \nabla_{X_2} Y, \qquad (9.53b)$$
$$\nabla_X(a_1 Y_1 + a_2 Y_2) = a_1 \nabla_X Y_1 + a_2 \nabla_X Y_2, \qquad (9.53c)$$

where a_1 and a_2 are constants and f_1 and f_2 are scalar functions. An *affine connection* introduces an operator that satisfies these conditions and thereby generalises the directional derivative concept to manifolds. Because of the linearity in the direction X, we find that

$$\nabla_X Y = \nabla_{X^a \partial_a} Y = v^a \nabla_a Y, \qquad (9.54)$$

where we have defined $\nabla_a \equiv \nabla_{\partial_a}$. In order to know how ∇_X acts on Y, it is therefore enough to know the action of the operators ∇_a and the components of X. In fact, in the remainder of this text, we are going to interchangeably refer to ∇_a and ∇_X as an affine connection. Furthermore, it is also customary to use the notation $\nabla_a Y^b$ for the coefficient of the basis vector ∂_b in $\nabla_a Y$, i.e., $\nabla_a Y^b = dy^b(\nabla_a Y)$. This is particularly true when we are not writing out the vector basis explicitly and in accordance with the notation introduced in Eq. (2.74).

Developing the affine connection further, we can expand the vector field it acts on in the same coordinate basis to obtain

$$\nabla_a Y = \nabla_a(Y^b \partial_b) = (\partial_a Y^b)\partial_b + Y^b \nabla_a \partial_b \tag{9.55}$$

by use of the product rule. While the first term is straightforward to evaluate, it only contains the partial derivatives of the components Y^b and the tangent vector basis, the second term contains the object $\nabla_a \partial_b$, which we do not know how to evaluate. Since the tangent vectors have ∂_a as their coordinate basis, this is nothing other than the generalisation of the term $\partial_a \vec{E}_b$ that appeared when we treated Euclidean spaces in curvilinear coordinates. All that we currently know about these objects is that they are tangent vectors and it must therefore be possible to write them on the form

$$\nabla_a \partial_b = \Gamma_{ab}^c \partial_c, \tag{9.56}$$

since any vector may be expressed as a linear combination of the basis vectors. The components Γ_{ab}^c are the *connection coefficients* of ∇_a and depend only on the chosen coordinates. There is no unique way of assigning the connection coefficients and, as a result, there are going to be many different possible affine connections on a given manifold. As we shall discover later, it will be possible to define a unique affine connection with a particular set of properties if our manifold is endowed with a metric, but for now we leave the ambiguity in having several possible generalisations of the directional derivative to manifolds and instead focus on how the connection coefficients must behave under coordinate transformations. We will return to the issue of defining different connections later.

9.3.2.1 Coordinate transformations

If we know the expression for the connection coefficients in one coordinate system, it is relatively straightforward to express them in any other coordinate system. This may be done by expressing the basis vectors of the new coordinate system in terms of those of the old by applying the chain rule according to

$$\nabla_{a'}\partial_{b'} = \Gamma_{a'b'}^{c'}\partial_{c'} = \frac{\partial y^a}{\partial y'^{a'}} \nabla_a \left(\frac{\partial y^b}{\partial y'^{b'}} \partial_b\right). \tag{9.57}$$

Use of the product rule now results in

$$\Gamma_{a'b'}^{c'}\partial_{c'} = \frac{\partial y^a}{\partial y'^{a'}} \left[\frac{\partial y^b}{\partial y'^{b'}} \nabla_a \partial_b + \left(\frac{\partial}{\partial y^a} \frac{\partial y^b}{\partial y'^{b'}}\right)\partial_b\right] = \left[\frac{\partial y^a}{\partial y'^{a'}} \frac{\partial y^b}{\partial y'^{b'}} \Gamma_{ab}^c + \frac{\partial^2 y^c}{\partial y'^{a'} \partial y'^{b'}}\right]\partial_c$$

$$= \left[\frac{\partial y^a}{\partial y'^{a'}} \frac{\partial y^b}{\partial y'^{b'}} \Gamma_{ab}^c + \frac{\partial^2 y^c}{\partial y'^{a'} \partial y'^{b'}}\right] \frac{\partial y'^{c'}}{\partial y^c} \partial_{c'} \tag{9.58a}$$

or, in other words,

$$\Gamma_{a'b'}^{c'} = \frac{\partial y^a}{\partial y'^{a'}} \frac{\partial y^b}{\partial y'^{b'}} \frac{\partial y'^{c'}}{\partial y^c} \Gamma_{ab}^c + \frac{\partial y'^{c'}}{\partial y^c} \frac{\partial^2 y^c}{\partial y'^{a'} \partial y'^{b'}}. \tag{9.58b}$$

In fact, these are the very same transformation rules as those of the Christoffel symbols that you have found in Problem 2.17. Clearly this is no coincidence as the Christoffel symbols encoded the partial derivatives of the tangent vector basis in the definition of the covariant derivative that we encountered in Chapter 2.

9.3.2.2 Affine connections and tensor fields

We can extend the action of the affine connection to a general tensor field in a fashion very similar to how we earlier extended the covariant derivative to general tensor fields in the Euclidean setting. In order to do this, we require that the action of the operator ∇_X satisfies the conditions

$$\nabla_X(T_1 \otimes T_2) = (\nabla_X T_1) \otimes T_2 + T_1 \otimes (\nabla_X T_2), \tag{9.59a}$$

$$\nabla_X(\omega(Y)) = (\nabla_X \omega)(Y) + \omega(\nabla_X Y), \tag{9.59b}$$

where T_1 and T_2 are arbitrary tensors, X and Y are tangent vector fields, and ω is a dual vector field. The first of these conditions is the generalised product rule for derivatives and the second is the product rule applied to a contraction of a tangent vector field and a dual vector field. Starting with the second condition, $\omega(Y)$ is given by the scalar field $\omega_a Y^a$ in any given set of coordinates. The action of ∇_X on a scalar field f is just that of the directional derivative defined by Xf, resulting in

$$\nabla_X(\omega(Y)) = \nabla_X(\omega_a Y^a) = X^b \partial_b(\omega_a Y^b). \tag{9.60}$$

On the other hand, the right-hand side of Eq. (9.59b) can be written as

$$\begin{aligned}(\nabla_X \omega)(Y) + \omega(\nabla_X Y) &= X^a[(\nabla_a \omega)(Y) + \omega(\nabla_a Y)] \\ &= X^a[(\partial_a \omega_b)Y^b + \omega_b(\nabla_a dy^b)(Y) + \omega_b(\partial_a Y^b + Y^c \Gamma^b_{ac})] \\ &= X^a \partial_a(\omega_b Y^b) + X^a \omega_b Y^c[C^b_{ac} + \Gamma^b_{ac}], \end{aligned} \tag{9.61}$$

where we have defined the coefficients C^b_{ac} according to

$$\nabla_a dy^b = C^b_{ac} dy^c. \tag{9.62}$$

Comparing the different expressions for $\nabla_X(\omega(Y))$, we now find that

$$C^b_{ac} = -\Gamma^b_{ac} \tag{9.63a}$$

must hold since the fields X, Y, and ω are arbitrary, or in other words

$$\nabla_a dy^b = -\Gamma^b_{ac} dy^c. \tag{9.63b}$$

This relation should be compared to Eq. (2.80b), where we found a similar relation for the partial derivative of the dual basis in a Euclidean space. The general action of the ∇_X on an arbitrary tensor field follows directly from applying the product rule, see Problem 9.14.

Just as we introduced the covariant derivative of tensor components in Chapter 2, we will often not write out the tangent vector, dual, or tensor basis explicitly when using the affine connection. Thus, when we write expressions of the form $\nabla_a X^b$, this should be interpreted as the component in the ∂_b direction of the vector field resulting from applying the ∇_a to the vector field X, i.e.,

$$\nabla_a X \equiv (\nabla_a X^b) \partial_b. \tag{9.64}$$

In the same spirit, we will from now on also only use $\partial_a X^b$ when we refer to the partial derivative of the vector field component X^b. With this notation, we find that

$$\nabla_a X^b = \partial_a X^b + \Gamma^b_{ac} X^c, \tag{9.65a}$$

$$\nabla_a \omega_b = \partial_a \omega_b - \Gamma^c_{ab} \omega_c \tag{9.65b}$$

along with similar relations for arbitrary tensor fields. Note that this notation is consistent with that already introduced for the covariant derivative in a Euclidean space.

9.3.3 Parallel fields

As we have just discussed, an affine connection provides us with a way of differentiating vectors as well as more general tensors. In particular, the relations

$$\nabla_a \partial_b = \Gamma^c_{ab} \partial_c \quad \text{and} \quad \nabla_a dy^b = -\Gamma^b_{ac} dy^c \tag{9.66a}$$

are the direct generalisations of the identities

$$\partial_a \vec{E}_b = \Gamma^c_{ab} \vec{E}_c \quad \text{and} \quad \partial_a \vec{E}^b = -\Gamma^b_{ac} \vec{E}^c \tag{9.66b}$$

that we encountered in curvilinear coordinates on a Euclidean space. As such, we may interpret the connection coefficients as a way of describing how the tangent vector and dual bases change in different directions.

If we know how the bases change, we also get an idea about how a general field varies with the position. In a Euclidean space, there was a natural way of determining whether a field was constant or not by referring back to the Cartesian basis in order to compare vectors at different points by using it. For a constant vector field $\vec{v}$, we then find the requirement that

$$\partial_a \vec{v} = 0 \tag{9.67}$$

for all a. In a general manifold, this is no longer possible and this was our reason to introduce the affine connection in the first place. Since talking about a constant field no longer has any clear meaning due to the tangent spaces at different points being different, we instead generalise the concept of a constant field by defining a field T as a *parallel tensor field* if it satisfies

$$\nabla_X T = 0 \tag{9.68}$$

for all vector fields X. This requirement generally results in an overdetermined system of partial differential equations and it is by no means guaranteed that a given affine connection allows the system to have non-trivial solutions.

Example 9.14 In a Euclidean space with Cartesian coordinates x^i, the connection coefficients all vanish. Because of this, the requirement for a vector field Y to be parallel reduces to

$$\nabla_i Y^j = \partial_i Y^j + \Gamma^j_{ik} Y^k = \partial_i Y^j = 0. \tag{9.69}$$

This is the same requirement as that for the Cartesian vector components deduced from $\partial_i \vec{v} = 0$ and therefore precisely what we would expect. The resulting differential equations are solved by letting all vector components be constant in the Cartesian coordinate system. Transferring to a curvilinear coordinate system, the connection coefficients would become non-zero, but the equations would still be solvable.

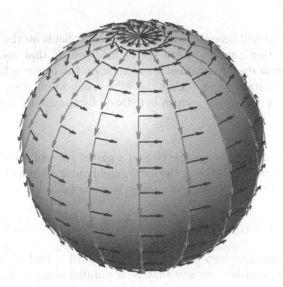

Figure 9.9 The vector fields $X = \partial_\theta$ (light) and $Y = \frac{1}{\sin(\theta)}\partial_\varphi$ (dark) on the sphere. These fields are defined as parallel when introducing an affine connection in Example 9.15.

Example 9.15 The affine connection may be defined by specifying N linearly independent vector fields that are assumed to be parallel. The connection coefficients Γ^c_{ab} can then be computed by applying ∇_a to these vector fields and identifying the coefficients. Let us consider the following two vector fields on the two-dimensional sphere S^2

$$X = \partial_\theta \quad \text{and} \quad Y = \frac{1}{\sin(\theta)}\partial_\varphi, \tag{9.70}$$

which are linearly independent everywhere where they are well defined, i.e., everywhere but at $\theta = 0$ and $\theta = \pi$. When we later introduce a metric on S^2, these vector fields will turn out to be the vectors of unit length in the θ and φ directions, respectively, see Fig. 9.9. The affine connection defined by assuming these vector fields to be parallel will therefore correspond to keeping the vector length and angle with the coordinate lines constant during parallel transport.

The connection coefficients are now given by the assumption that X and Y are parallel, leading to

$$0 = \nabla_\theta X = \nabla_\theta \partial_\theta = \Gamma^a_{\theta\theta}\partial_a, \tag{9.71a}$$

$$0 = \nabla_\varphi X = \nabla_\varphi \partial_\theta = \Gamma^a_{\varphi\theta}\partial_a, \tag{9.71b}$$

$$0 = \nabla_\theta Y = \left(\partial_\theta \frac{1}{\sin(\theta)}\right)\partial_\varphi + \frac{1}{\sin(\theta)}\Gamma^a_{\theta\varphi}\partial_a = \frac{1}{\sin(\theta)}[-\cot(\theta)\partial_\varphi + \Gamma^a_{\theta\varphi}\partial_a], \tag{9.71c}$$

$$0 = \nabla_\varphi Y = \frac{1}{\sin(\theta)}\nabla_\varphi \partial_\varphi = \frac{1}{\sin(\theta)}\Gamma^a_{\varphi\varphi}\partial_a. \tag{9.71d}$$

Direct identification of the connection coefficients now results in the only non-zero connection coefficient being $\Gamma^\varphi_{\theta\varphi} = \cot(\theta)$.

It should be noted that these vector fields are not global as both of them are singular at the poles. However, we will here consider them as vector fields on the sphere with the poles removed. In general, there are no global vector fields on S^2 that are non-zero everywhere and so a global affine connection cannot be defined on the entire sphere in this fashion. In addition, another important point is that this affine connection is not the affine connection that is usually used on the sphere, which is induced by the metric tensor we will introduce in a while.

While it may not always be possible to find a global parallel vector field, we can always consider how a vector field behaves in the local neighbourhood of a curve γ. The directional derivative of the vector field X along the curve is given by

$$\nabla_{\dot\gamma} X = \dot{y}^a \nabla_a X = \dot{y}^a (\partial_a X^b + \Gamma^b_{ac} X^c)\partial_b = \left(\frac{dX^b}{dt} + \Gamma^b_{ac} \dot{y}^a X^c \right)\partial_b, \qquad (9.72)$$

where t is the curve parameter, $\dot\gamma$ is the tangent vector of γ, and $\dot{y}^a$ its components. If this directional derivative vanishes, we say that X is parallel along γ. Unlike the case of fully parallel fields, the requirement

$$\nabla_{\dot\gamma} X = 0 \qquad (9.73)$$

in an N-dimensional manifold is a more limited set of N ordinary differential equations for any given curve γ and for any given initial conditions there will be a solution. Taking a curve in the manifold M such that $\gamma(0) = p$ and $\gamma(1) = q$ and a vector X in $T_p M$, the vector obtained in $T_q M$ by requiring that X is parallel along γ is called the *parallel transport* of X from p to q along γ.

A special case of a parallel transported vector field occurs when the tangent vector $\dot\gamma$ itself is parallel transported along the curve γ itself, leading to

$$\nabla_{\dot\gamma}\dot\gamma = 0 \implies \dot{y}^a \partial_a \dot{y}^b + \Gamma^b_{ac}\dot{y}^a \dot{y}^c = \ddot{y}^b + \Gamma^b_{ac}\dot{y}^a\dot{y}^c = 0. \qquad (9.74)$$

A curve satisfying this condition is called a *geodesic* and the condition itself is called the *geodesic equation*. The interpretation of the geodesic equation is that the tangent vector keeps pointing in the same direction as defined by the affine connection.

Example 9.16 Going back to a Euclidean space with a set of Cartesian coordinates x^i, the connection coefficients vanish leading to the geodesic equation

$$\ddot{x}^i = 0, \qquad (9.75)$$

indicating that the tangent vector components $\dot{x}^i$ are constant. This means that the geodesics in Euclidean space are just straight lines. On other manifolds, geodesics may be considered as a generalisation of straight lines, see Fig. 9.10.

Example 9.17 Let us now consider the affine connection on the sphere defined in Example 9.15. The geodesic equations for this connection are given by

$$\ddot\theta = -\Gamma^\theta_{ab}\dot{y}^a\dot{y}^b = 0, \qquad (9.76a)$$

$$\ddot\varphi = -\Gamma^\varphi_{ab}\dot{y}^a\dot{y}^b = -\cot(\theta)\dot\theta\dot\varphi. \qquad (9.76b)$$

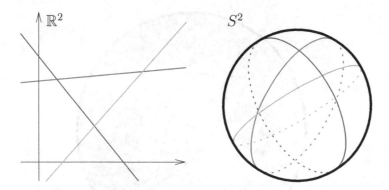

Figure 9.10 Examples of geodesics in the two-dimensional plane $\mathbb{R}^2$ and on the two-dimensional sphere S^2. In $\mathbb{R}^2$ the geodesics correspond to straight lines, while on the two-dimensional sphere, we shall later find that a natural choice for the affine connection (see Section 9.4.3) leads to the geodesics being great circles.

The first of these implies that $\dot\theta = v_\theta$ is a constant and therefore

$$\theta(t) = v_\theta t + \theta_0. \tag{9.77}$$

Inserting this into the second geodesic equation and integrating yields

$$\dot\varphi = \frac{A}{\sin(v_\theta t + \theta_0)}, \tag{9.78}$$

where A is an integration constant. Performing another integration for $\theta < \pi/2$ then results in

$$\varphi(t) = \frac{A}{v_\theta} \ln\left(\frac{\sin(v_\theta t + \theta_0)}{1 + \cos(v_\theta t + \theta_0)}\right) + B. \tag{9.79}$$

Note that we in total have four constants characterising the geodesics. These can be determined by specifying where the geodesic starts at $t = 0$ and its tangent vector at that point. The geodesics for this affine connection are illustrated in Fig. 9.11, where we also note that they correspond to curves that always have a given bearing, i.e., go in a given compass direction. The corresponding affine connection on the surface of the Earth would tell us that the direction in which a compass needle points is parallel transported as we travel around the world.

9.3.4 Torsion

Let us now examine the relation between the affine connection and the Lie bracket. When acting on a scalar field, the operator ∇_X acts in precisely the same way as the vector field X since the scalar field does not have a basis in need of a connection in order to define the directional derivative. We can therefore write

$$[X, Y]f = XYf - YXf = X(\nabla_Y f) - Y(\nabla_X f) = (\nabla_X \nabla_Y - \nabla_Y \nabla_X)f, \tag{9.80}$$

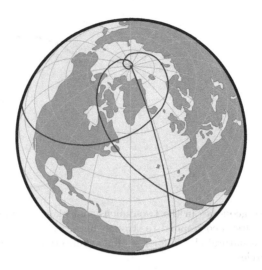

Figure 9.11 The geodesics of the connection defined in Example 9.15 correspond to the curves with a constant bearing. Here exemplified on the Earth with three different geodesics. Note how the geodesics always intersect the coordinate lines at the same angles.

i.e., we can express the Lie bracket in terms of the commutator of ∇_X and ∇_Y acting on a scalar field. On the other hand, the expression $\nabla_X \nabla_Y f$ may also be rewritten as

$$\nabla_X \nabla_Y f = \nabla_X(Y^a \partial_a f) = (\partial_a f)\nabla_X Y^a + Y^a \nabla_X(\nabla_a f) \tag{9.81}$$

in a local coordinate system, where we now consider the expression $\partial_a f = \nabla_a f$ as a component of the differential df. Using this relation, the expression for the Lie bracket's action on f can be written as

$$[X,Y]f = (\nabla_X Y - \nabla_Y X)f + X^a Y^b(\nabla_a \nabla_b - \nabla_b \nabla_a)f. \tag{9.82}$$

It should here be noted that the commutator $\nabla_a \nabla_b - \nabla_b \nabla_a$ acting on f is not necessarily zero. Instead, we can rewrite the above relation as a vector equation

$$T(X,Y) \equiv \nabla_X Y - \nabla_Y X - [X,Y] = -X^a Y^a(\nabla_a \nabla_b - \nabla_b \nabla_a), \tag{9.83}$$

where we have introduced the vector field $T(X,Y)$, which generally is a function of the vector fields X and Y. We can easily verify (see Problem 9.16) that $T(X,Y)$ is linear in both arguments

$$T(h_1 X_1 + h_2 X_2, Y) = h_1 T(X_1, Y) + h_2 T(X_2, Y), \tag{9.84a}$$
$$T(X, h_1 Y_1 + h_2 Y_2) = h_1 T(X, Y_2) + h_2 T(X, Y_2), \tag{9.84b}$$

where h_1 and h_2 are scalar fields. It follows that the mapping $T(X,Y)$ defines a type $(1,2)$ tensor field with components given by

$$T^c_{ab}\partial_c f = -(\nabla_a \nabla_b - \nabla_b \nabla_a)f. \tag{9.85}$$

This tensor is called the *torsion tensor* and is only dependent on the affine connection defining it. In any given coordinate system, we find that

$$\nabla_a \nabla_b f = \nabla_a \partial_b f = \partial_a \partial_b f - \Gamma^c_{ab}\partial_c f. \tag{9.86}$$

Since the partial derivatives commute, this leads to the relation

$$T^c_{ab} = \Gamma^c_{ab} - \Gamma^c_{ba} = 2\Gamma^c_{[ab]}, \tag{9.87}$$

i.e., the components of the torsion are given by the anti-symmetrised connection coefficients. It directly follows that an affine connection is torsion free if its connection coefficients are symmetric in the lower indices or, equivalently, if

$$\nabla_a \partial_b = \nabla_b \partial_a. \tag{9.88}$$

As for the Lie bracket, we can find a geometrical interpretation of the torsion by considering a set of different curves in the manifold M. When considering the Lie bracket, we looked at the flows of two vector fields X and Y, but for the torsion there is a more natural choice as it is intimately related to the affine connection and therefore to the notion of parallel transport. Let us therefore start at a point p and consider the two vectors X and Y in T_pM. Each of these vectors defines a unique geodesic and the geodesic equation becomes

$$\ddot{y}^a + \Gamma^a_{bc}\dot{y}^b\dot{y}^c = 0, \tag{9.89}$$

where we have $\dot{y}^a(0) = X^a$ for the geodesic defined by X and $\dot{y}^a(0) = Y^a$ for that defined by Y. Focusing on the former of these and using a slight abuse of notation, we define $X^a(t)$ to be the tangent vector $\dot{y}^a(t)$, which is parallel transported along the geodesic, and $Y^a(t)$ to be the parallel transport of Y^a along the same geodesic, implying that

$$\dot{Y}^a(t) + \Gamma^a_{bc}\dot{y}^bY^c(t) = 0. \tag{9.90}$$

For small values dt of the geodesic parameter, we can expand both the coordinates y^a and the vector Y^a in terms of dt according to

$$y^a(dt) \simeq y^a_0 + dt\,\dot{y}^a(0) + \frac{dt^2}{2}\ddot{y}^a(0) = y^a_0 + dt\,X^a - \frac{dt^2}{2}\Gamma^a_{bc}X^bX^c, \tag{9.91a}$$

$$Y^a(dt) \simeq Y^a + dt\,\dot{Y}^a(0) = Y^a - dt\,\Gamma^a_{bc}X^bY^c, \tag{9.91b}$$

where terms have been kept to second order in dt in the first expression and to first order in the last. The vector $Y^a(dt)$ now defines a unique geodesic $y^a(dt, s)$ starting at $y^a(dt)$, whose parameter we can call s. Its geodesic equation is given by

$$\frac{\partial^2 y^a(dt, s)}{\partial s^2} + \Gamma^a_{bc}\frac{\partial y^b(dt, s)}{\partial s}\frac{\partial y^c(dt, s)}{\partial s} = 0, \tag{9.92}$$

with $y^a(dt, 0) = y^a(dt)$ and $\partial y^a(dt, 0)/\partial s = Y^a(dt)$. Letting the geodesic parameter take the small value ds, we can expand $y^a(dt, ds)$ to combined second order in dt and ds and find that

$$y^a(dt, ds) \simeq y^a(dt) + ds\,Y^a(dt) - \frac{ds^2}{2}\Gamma^a_{bc}Y^b(dt)Y^c(dt)$$

$$\simeq y^a_0 + dt\,X^a - \frac{dt^2}{2}\Gamma^a_{bc}X^bX^c + ds[Y^a - dt\,\Gamma^a_{bc}X^bY^c] - \frac{ds^2}{2}\Gamma^a_{bc}Y^bY^c$$

$$\simeq y^a_0 + dt\,X^a + ds\,Y^a - \frac{1}{2}\Gamma^a_{bc}\left(X^bX^cdt^2 + Y^bY^cds^2 + 2X^bY^cdt\,ds\right). \tag{9.93}$$

Defining this point as p_{YX} and the point obtained by first following the geodesic defined by Y and then the geodesic defined by the parallel transported X as p_{XY}, calling their coordinates y^a_{YX} and y^a_{XY}, respectively, results in

$$y^a_{YX} - y^a_{XY} = dt\,ds\,\Gamma^a_{bc}(Y^b X^c - X^b Y^c) = dt\,ds\,T^a_{bc}Y^b X^c. \tag{9.94}$$

In other terms, just as the Lie bracket measures the failure of the flows of two vector fields to commute, the torsion measures the failure of following the flows of the parallel transported vector fields to commute.

Example 9.18 We can exemplify this reasoning by looking at the affine connection on the sphere defined in Example 9.15 and taking the vector fields $X = \partial_\theta$ and $Y = (1/\sin(\theta))\partial_\varphi$ at a point given by the coordinates $\theta = \theta_0$ and $\varphi = \varphi_0$. Comparing with the results of Example 9.17, the geodesic in the X-direction is given by

$$\theta(t) = t + \theta_0, \quad \varphi(t) = \varphi_0. \tag{9.95}$$

Furthermore, since the fields X and Y are parallel with respect to the given connection, we already know that the parallel transport of Y along this geodesic will be given by

$$Y(t) = Y(\theta(t), \varphi(t)) = \frac{1}{\sin(t + \theta_0)}\partial_\varphi. \tag{9.96}$$

The geodesic in the $Y(t)$ direction now has the form

$$\theta(t, s) = t + \theta_0, \quad \varphi(t, s) = \frac{s}{\sin(t + \theta_0)} + \varphi_0. \tag{9.97}$$

Going along the geodesics in the opposite order, but for the same values of the geodesic parameters s and t, leads to

$$\tilde{\theta}(s, t) = t + \theta_0, \quad \tilde{\varphi}(s, t) = \frac{s}{\sin(\theta_0)} + \varphi_0 \tag{9.98}$$

or, in other terms,

$$v^\theta(t, s) = \theta(t, s) - \tilde{\theta}(s, t) = 0, \tag{9.99a}$$

$$v^\varphi(t, s) = \varphi(t, s) - \tilde{\varphi}(s, t) = s\left(\frac{1}{\sin(t + \theta_0)} - \frac{1}{\sin(\theta_0)}\right). \tag{9.99b}$$

Taking the derivatives of these functions with respect to t and s and evaluating at $t = s = 0$ results in

$$\left.\frac{\partial^2 v^\theta}{\partial t \partial s}\right|_{t=s=0} = 0 = T^\theta_{ab}X^a Y^b, \tag{9.100a}$$

$$\left.\frac{\partial^2 v^\varphi}{\partial t \partial s}\right|_{t=s=0} = -\frac{\cot(\theta_0)}{\sin(\theta_0)} = T^\varphi_{ab}Y^a X^b = T^\varphi_{\varphi\theta}\frac{1}{\sin(\theta_0)}, \tag{9.100b}$$

where we have used $T^\varphi_{\varphi\theta} = \Gamma^\varphi_{\varphi\theta} - \Gamma^\varphi_{\theta\varphi} = -\cot(\theta)$. This is exactly what we expect from Eq. (9.94).

Remembering our treatment of the covariant derivative in curvilinear coordinates in a

Euclidean space, we remember that one of the properties of the corresponding Christoffel symbols was their symmetry $\Gamma^c_{ab} = \Gamma^c_{ba}$ stemming from the relation $\partial_a \vec{E}_b = \partial_b \vec{E}_a$. From this we can deduce that this covariant derivative defined a torsion free connection. In fact, in most physical situations, we will deal with affine connections such that the torsion vanishes, in many cases by assumption. For the remainder of this text, the manifolds will be assumed to be torsion free unless non-zero torsion is mentioned explicitly.

9.3.5 Curvature

When considering the geometric interpretation of the Lie bracket and the torsion we have been dealing with different curves that did not form a closed loop. However, there is also a great deal of geometry involved when parallel transporting an arbitrary vector X along a closed loop or, equivalently, from a starting point to an endpoint along different paths. Just as we arrived at the expression for the torsion by considering the commutator of ∇_X and ∇_Y acting on a scalar field and then removing the commutator $[X, Y]$, let us now consider the corresponding procedure for a vector field Z and look at the expression

$$R(X, Y)Z = (\nabla_X \nabla_Y - \nabla_Y \nabla_X - \nabla_{[X,Y]})Z. \tag{9.101}$$

The natural extension of the Lie bracket acting on scalar fields to an operator on vector fields is replacing $[X, Y]$ by $\nabla_{[X,Y]}$ as $[X, Y]f = \nabla_{[X,Y]}f$. Because of the definition of the expression, $R(X, Y)Z$ is clearly a vector field, but we would like to know how it behaves when we change the fields X, Y, and Z. Replacing X by fX, we find that

$$\begin{aligned}
R(fX, Y)Z &= (\nabla_{fX} \nabla_Y - \nabla_Y \nabla_{fX} - \nabla_{[fX,Y]})Z \\
&= (f\nabla_X \nabla_Y - \nabla_Y f\nabla_X - f\nabla_{[X,Y]} + Y(f)\nabla_X)Z \\
&= (f\nabla_X \nabla_Y - Y(f)\nabla_X - f\nabla_Y \nabla_X - f\nabla_{[X,Y]} + Y(f)\nabla_X)Z \\
&= fR(X, Y)Z, \tag{9.102a}
\end{aligned}$$

where we have used the product rule properties of the various differential operators. It follows that $R(X, Y)Z$ is linear in X and a similar computation holds for Y. For the dependence when changing Z to fZ we have

$$\begin{aligned}
R(X, Y)fZ &= (\nabla_X \nabla_Y - \nabla_Y \nabla_X - \nabla_{[X,Y]})fZ \\
&= \nabla_X (Yf)Z + \nabla_X f\nabla_Y Z - \nabla_Y (Xf)Z - \nabla_Y f\nabla_X Z - ([X,Y]f)Z - f\nabla_{[X,Y]}Z \\
&= (XYf)Z + (Yf)\nabla_X Z + (Xf)\nabla_Y Z + f\nabla_X \nabla_Y Z - (YXf)Z \\
&\quad - (Xf)\nabla_Y Z - (Yf)\nabla_X Z - f\nabla_Y \nabla_X Z - ([X,Y]f)Z - f\nabla_{[X,Y]}Z \\
&= fR(X, Y)Z \tag{9.102b}
\end{aligned}$$

and hence $R(X, Y)Z$ is linear in Z as well and only depends on its value at the given point. Since $R(X, Y)Z$ is a tangent vector field, it defines a linear map from three tangent vector fields to one. It therefore follows that it defines a type $(1, 3)$ tensor field. This tensor is called the *curvature tensor* of the affine connection (alternatively the *Riemann tensor* or *Riemann curvature tensor*) and its components may be found as

$$\begin{aligned}
R^d_{cab}\partial_d &\equiv R(\partial_a, \partial_b)\partial_c = \nabla_a \nabla_b \partial_c - \nabla_b \nabla_a \partial_c \\
&= \nabla_a \Gamma^e_{bc}\partial_e - \nabla_b \Gamma^e_{ac}\partial_e = (\partial_a \Gamma^d_{bc} + \Gamma^e_{bc}\Gamma^d_{ae} - \partial_b \Gamma^d_{ac} - \Gamma^e_{ac}\Gamma^d_{be})\partial_d. \tag{9.103a}
\end{aligned}$$

Carefully note the positioning of the indices here, the indices of the ∂_a and ∂_b in $R(\partial_a, \partial_b)\partial_c$ are placed in the end of the list of covariant indices! Identifying the components from this expression we find that

$$R^d_{cab} = \partial_a \Gamma^d_{bc} - \partial_b \Gamma^d_{ac} + \Gamma^e_{bc} \Gamma^d_{ae} - \Gamma^e_{ac} \Gamma^d_{be}. \tag{9.103b}$$

It should be clear that these components are anti-symmetric in the last two indices due to the anti-symmetry $R(X,Y)Z = -R(Y,X)Z$.

Just like the torsion, the curvature has a rather intriguing geometrical interpretation. If we consider the parallel transport of a vector X around a closed loop starting and ending at the same point p, we will end up with a new vector X^* at that point. Since X and X^* are in the same vector space T_pM, they can be directly compared without any reference to the affine connection other than through the parallel transport equations around the curve, which take the form

$$\frac{dX^a}{dt} + \Gamma^a_{bc} \dot{y}^b X^c = 0. \tag{9.104}$$

This is an ordinary linear differential equation for the components X^a. It follows that the components of X^* are linearly dependent on the initial conditions, given by the components of X, and we may write this as

$$X^* = P_\gamma(X), \tag{9.105}$$

where P_γ is a type (1,1) tensor at p, which depends on the curve γ and the affine connection.

Example 9.19 If we study a Euclidean space with a set of Cartesian coordinates x^i, the connection coefficients all vanish and the parallel transport equation gives us

$$\frac{dX^a}{dt} = 0 \quad \Longrightarrow \quad X^{*a} = X^a \tag{9.106}$$

regardless of the curve γ. In this case, we therefore find that $P_\gamma(X) = X$ is the identity map and therefore the Kronecker delta tensor δ, see Fig. 9.12. At the same time, the curvature tensor also vanishes due to the connection coefficients all vanishing. A manifold where the curvature tensor vanishes everywhere is called *flat* and any closed curve that may be continuously deformed to a point in such a space will result in $P_\gamma = \delta$.

Example 9.20 We will soon introduce the concept of a metric and when doing so an affine connection may be defined in such a way that when parallel transporting a vector along a geodesic, its length and angle to the geodesic remain constant. With the natural metric on the sphere S^2 the geodesics will turn out to be great circles. Parallel transporting a vector around the curve γ in Fig. 9.13, composed by three different geodesics, one along the equator and the other two going from the equator to one of the poles, we find that the final vector has been rotated by an angle α relative to the original one. The angle α will be equal to the angle between the geodesics that meet at the pole.

In general, if the vector X^* is not equal to X, i.e., the tensor P_γ is not the Kronecker delta, for curves γ that can be continuously deformed to a point, then the manifold is said to be *curved*. This has an intricate relation to the curvature tensor. Let us study the parallel transport of X along two different paths between two points. Technically, this will be the same as parallel transporting one of the resulting vectors around the loop to the final position, but the computations we will have to do with this approach are more

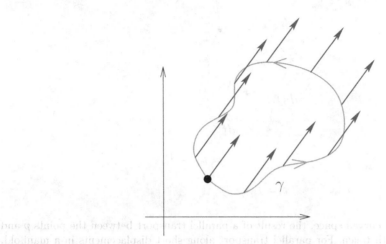

Figure 9.12 For any curve γ starting and ending at the same point in a Euclidean space, the parallel transport of a vector around the curve gives back the same vector. The map P_γ, which is a map from the tangent space at the point to itself, is therefore the identity map. This will generally no longer be the case in a curved space.

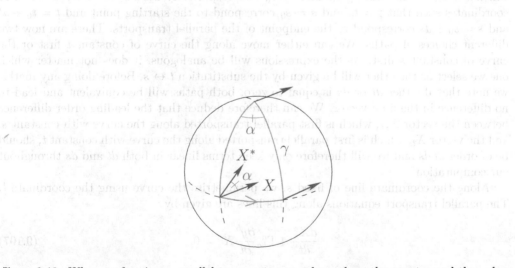

Figure 9.13 When performing a parallel transport on a sphere along the equator and then along the θ coordinate lines to a pole, with a torsion free affine connection based on the metric (see Section 9.4.3), the angle α between the initial and final vectors will be given by the angle between the coordinate lines at the pole.

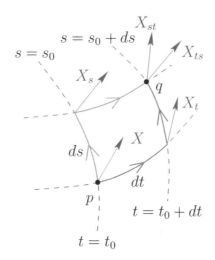

Figure 9.14 In a general curved space, the result of a parallel transport between the points p and q will depend on the path taken. For parallel transport along short displacements in a manifold, the difference between paths will be given by the curvature tensor R^a_{cbd}.

straightforward. We will assume that we are studying parallel transport in a region so small that the curves may be parametrised by the two parameters t and s that can be taken as two of the local coordinates and consider the curves defined by the coordinate lines for which $s = s_0$, $s = s_0 + ds$, $t = t_0$, and $t = t_0 + dt$, respectively, see Fig. 9.14. We select these coordinates such that $t = t_0$ and $s = s_0$ correspond to the starting point and $t = t_0 + dt$ and $s = s_0 + ds$ correspond to the endpoint of the parallel transports. There are now two different choices of paths. We can either move along the curve of constant t first or the curve of constant s first. As the expressions will be analogous, it does not matter which one we select as the other will be given by the substitution $t \leftrightarrow s$. Before doing any maths, we note that if either dt or ds is equal to zero, both paths will be equivalent and lead to no difference in the final vector. We can therefore deduce that the leading order difference between the vector X_{st}, which is first parallel transported along the curve with constant s, and the vector X_{ts} which is first parallel transported along the curve with constant t, should be of order $dt\,ds$ and we will therefore only keep terms linear in both dt and ds throughout our computation.

Along the coordinate line of fixed s, we parametrise the curve using the coordinate t. The parallel transport equations along this lines are given by

$$\frac{dX^a}{dt} + \Gamma^a_{bc} \frac{\partial y^b}{\partial t} X^c = 0, \tag{9.107}$$

where $y^b = y^b(t, s)$ are the coordinates. By expanding the parallel transport equation using $X^a(t_0, s_0) = X^a_0$ as the initial condition, we find that, to first order in dt,

$$X^a_t \equiv X^a(t_0 + dt, s) \simeq X^a_0 + dt \left.\frac{dX^a}{dt}\right|_{t=t_0} = X^a_0 - dt\,\Gamma^a_{bc}(t_0, s_0)\frac{\partial y^b}{\partial t} X^c_0. \tag{9.108}$$

In the same fashion, expanding the parallel transport along the curve of constant $t = t_0 + dt$

we find that

$$X_{st}^a \simeq X_t^a + ds \left. \frac{dX^a}{ds} \right|_{s=s_0} = X_t^a - ds\, \Gamma_{bc}^a(t_0 + dt, s_0) \frac{\partial y^b}{\partial s} X_t^c. \tag{9.109}$$

Inserting our expression for X_t into this and expanding

$$\Gamma_{bc}^a(t_0 + dt, s_0) \simeq \Gamma_{bc}^a(t_0, s_0) + dt \frac{\partial y^d}{\partial t} \partial_d \Gamma_{bc}^a \tag{9.110}$$

to first order in dt, again keeping only terms linear in dt as well as in ds, i.e., including also the $dt\, ds$ term, we end up with

$$X_{st}^a \simeq X_0^a - \Gamma_{bc}^a \left(\frac{\partial y^b}{\partial t} dt + \frac{\partial y^b}{\partial s} ds \right) X_0^c - dt\, ds\, \frac{\partial y^b}{\partial t} \frac{\partial y^d}{\partial s} X_0^c (\partial_b \Gamma_{dc}^a - \Gamma_{de}^a \Gamma_{bc}^e). \tag{9.111}$$

Obtaining the corresponding expression for X_{ts}^a by substituting $t \leftrightarrow s$ results in the difference

$$X_{ts}^a - X_{st}^a = dt\, ds\, \frac{\partial y^b}{\partial t} \frac{\partial y^d}{\partial s} X_0^c (\partial_b \Gamma_{dc}^a - \partial_d \Gamma_{bc}^a + \Gamma_{be}^a \Gamma_{dc}^e - \Gamma_{de}^a \Gamma_{bc}^e)$$

$$= dt\, ds\, \frac{\partial y^b}{\partial t} \frac{\partial y^d}{\partial s} X_0^c R_{cbd}^a. \tag{9.112}$$

We have therefore found that the change in the vector component X after parallel transport around a small closed loop in the directions $Y = (\partial y^b/\partial t)\partial_b$ and $Z = (\partial y^d/\partial s)\partial_d$ is directly proportional to $R(Y, Z)X$.

9.4 METRIC TENSOR

In Chapter 2 we discussed the metric tensor with components defined by $g_{ab} = \vec{E}_a \cdot \vec{E}_b$. In a general manifold, there is no longer a natural choice of an inner product between tangent vectors and we will instead need to work in the opposite direction. The metric tensor in an arbitrary coordinate system in the Euclidean setting displayed two important properties, symmetry and positive definiteness. We therefore define a *metric* (or *Riemannian metric*) on a manifold M to be a type $(0, 2)$ tensor field g for which

$$g(X, Y) = g(Y, X), \tag{9.113a}$$
$$g(X, X) \geq 0, \tag{9.113b}$$
$$g(X, X) = 0 \implies X = 0, \tag{9.113c}$$

where X and Y are vectors in the same tangent space $T_p M$. The last two conditions are related to the positive definiteness of the metric, but are in some cases replaced by the less strict condition that the metric is non-degenerate, i.e., if

$$g(X, Y) = 0 \tag{9.114}$$

for all Y, then $X = 0$. A metric satisfying this condition rather than positive definiteness is usually called a *pseudo metric* (or *pseudo-Riemannian metric*). However, this construction is most common in special and general relativity, where the tensor is usually still referred to as a metric.

Example 9.21 We have already seen the example of the standard metric in a Euclidean N-dimensional space, defined by the Cartesian inner product. However, this is not the only possible metric on $\mathbb{R}^N$ and there are many possible choices that will all be perfectly valid metrics. In two dimensions, another possible choice would be given by

$$g_{11} = e^{-(x^1)^2}, \quad g_{22} = e^{-(x^2)^2}, \quad g_{12} = g_{21} = 0. \tag{9.115}$$

The geometry implied by this metric will be very different from that given by the standard one.

Just as in Chapter 2, we can also introduce the inverse metric tensor as the type $(2,0)$ tensor with components g^{ab} such that

$$g_{ab} g^{bc} = \delta_a^c. \tag{9.116}$$

Together, the metric tensor and its inverse define a natural one-to-one correspondence between tangent vectors and dual vectors such that for a tangent vector X^a, there exists a unique dual vector X_a such that

$$X_a = g_{ab} X^b \quad \text{and} \quad X^a = g^{ab} X_b. \tag{9.117}$$

Thanks to the metric tensor, we have now recovered the relation between tangent vectors and dual vectors that was previously lost by not being able to refer back to an underlying Cartesian coordinate system. As a type $(0,2)$ tensor, the metric is a linear map from tangent vectors to dual vectors, while the inverse metric as a type $(2,0)$ tensor is a linear map from dual vectors to tangent vectors. In a similar fashion, when there exists a metric tensor, we can lower and raise the indices of tensors of arbitrary type and although a tensor such as the curvature may be naturally considered as a type $(1,3)$ tensor, it then generally suffices to refer to it as a rank four tensor.

9.4.1 Inner products

Before the introduction of a metric, there is no clear definition of an inner product on the tangent space $T_p M$. Indeed, an inner product between two vectors should be a bilinear positive definite mapping to the real numbers, but this is exactly what the metric provides. In a manifold equipped with a metric tensor, we therefore define the *inner product* between two tangent vectors X and Y to be

$$X \cdot Y = g(X, Y) = g_{ab} X^a Y^b = X_a Y^a = X^a Y_a. \tag{9.118}$$

Since the metric defines a unique relation between tangent vectors and dual vectors, it also removes the need to separate them on a notational level and we will therefore also use the notation $\omega \cdot X$, where ω is a dual vector and X a tangent vector, to denote the contraction

$$\omega \cdot X = \omega(X) = \omega_a X^a. \tag{9.119}$$

With an inner product in place, we can start introducing some of the concepts we used in the earlier chapters. In particular, we define the *norm* of a vector X as

$$|X| \equiv \sqrt{X \cdot X} = \sqrt{g(X, X)} = \sqrt{g_{ab} X^a X^b} \tag{9.120}$$

and the angle α between the vectors X and Y through the relation

$$X \cdot Y = g_{ab} X^a Y^b \equiv |X| \, |Y| \cos(\alpha). \tag{9.121}$$

Example 9.22 As we shall see later in this chapter, there is a natural metric on the two-dimensional sphere arising from its embedding into a three-dimensional Euclidean space, given by

$$g_{\theta\theta} = 1, \quad g_{\varphi\varphi} = \sin^2(\theta), \quad \text{and} \quad g_{\theta\varphi} = g_{\varphi\theta} = 0 \tag{9.122}$$

in the usual spherical coordinates. With this metric, we find that

$$|\partial_\theta| = \sqrt{g_{\theta\theta}} = 1, \quad |\partial_\varphi| = \sqrt{g_{\varphi\varphi}} = \sin(\theta), \quad g(\partial_\theta, \partial_\varphi) = 0. \tag{9.123}$$

Thus, the norms of the coordinate basis vectors are one and $\sin(\theta)$, respectively, and the angle between them is $\pi/2$, i.e., they are orthogonal

9.4.2 Length of curves

With the inner product in place, we now have a tool for defining the norm of any given tangent vector. Looking back at the Euclidean case, we found the distance between two nearby points whose coordinates differed by dy^a to be given by

$$ds^2 = g_{ab}dy^a dy^b. \tag{9.124}$$

Lifting this definition directly to the general manifold, the expression

$$|\dot\gamma| = \sqrt{g(\dot\gamma, \dot\gamma)} \tag{9.125}$$

for a curve γ is the norm of the curve tangent $\dot\gamma$, i.e., the distance covered per unit increase in the curve parameter. Integrating this norm over the curve parameter, we find the total length of the curve

$$s_\gamma = \int_\gamma ds = \int_\gamma |\dot\gamma|\, dt = \int_\gamma \sqrt{g(\dot\gamma, \dot\gamma)}\, dt = \int_\gamma \sqrt{g_{ab}\dot y^a \dot y^b}\, dt, \tag{9.126}$$

where t is the curve parameter.

Example 9.23 Consider a curve of constant θ on the sphere. This is a closed curve that may be parametrised as

$$\theta = \theta_0, \quad \varphi = 2\pi t, \tag{9.127}$$

where $0 < t < 1$. The tangent vector to the curve is given by

$$\dot\gamma = \frac{dy^a}{dt}\partial_a = 2\pi\partial_\varphi. \tag{9.128}$$

This implies that

$$|\dot\gamma| = \sqrt{g_{\varphi\varphi}(2\pi)^2} = 2\pi\sin(\theta_0), \tag{9.129}$$

where we are using the metric introduced in Example 9.22. The total length of the closed curve is therefore given by

$$s_\gamma = \int_0^1 |\dot\gamma|\, dt = \int_0^1 2\pi\sin(\theta_0)dt = 2\pi\sin(\theta_0). \tag{9.130}$$

This is perfectly compatible with our intuition that the curve should be shorter when it is close to one of the poles, see Fig. 9.15.

Figure 9.15 We expect the curves on the sphere of constant θ coordinate to be shorter closer to the pole $\theta = 0$. The definition of the curve length s_γ in terms of the metric introduced in Example 9.22 gives us this expected result.

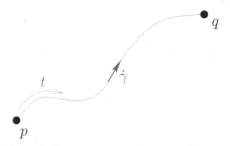

Figure 9.16 For a curve γ from p to q parametrised by the curve parameter t, we compute its length by integrating $ds = \sqrt{g(\dot\gamma, \dot\gamma)}\, dt$ along the curve.

We now have an expression in place to describe the length of a curve parametrised by t, given by the coordinate functions $y^a(t)$. If we fix the curve endpoints p and q, see Fig. 9.16, we can consider the expression

$$s[\gamma] = s_\gamma = \int \underbrace{\sqrt{g_{ab}\dot y^a \dot y^b}}_{\equiv \mathcal{L}}\, dt \tag{9.131}$$

as a functional of the coordinate functions and apply the machinery developed in Chapter 8 in order to find its stationary curves, i.e., the curves that provide a stationary curve length between the given endpoints. The Euler–Lagrange equation for the variation of $y^a(t)$ is given by

$$\frac{\partial \mathcal{L}}{\partial y^a} - \frac{d}{dt}\frac{\partial \mathcal{L}}{\partial \dot y^a} = 0, \tag{9.132a}$$

where the individual terms are

$$\frac{\partial \mathcal{L}}{\partial y^a} = \frac{1}{\mathcal{L}}\dot y^b \dot y^c \partial_a g_{bc}, \tag{9.132b}$$

$$\frac{d}{dt}\frac{\partial \mathcal{L}}{\partial \dot y^a} = \frac{d}{dt}\left(\frac{1}{\mathcal{L}}g_{bc}\dot y^c \delta_a^b\right) = \frac{1}{\mathcal{L}}\left(-\frac{1}{\mathcal{L}}\frac{d\mathcal{L}}{dt}g_{ac}\dot y^c + g_{ac}\ddot y^c + \dot y^b \dot y^c \partial_b g_{ac}\right). \tag{9.132c}$$

Inserting this into the Euler–Lagrange equation and multiplying by $\mathcal{L}g^{da}$ results in the equation

$$\ddot{y}^d + \dot{y}^b \dot{y}^c g^{da} \left(\partial_b g_{ac} - \frac{1}{2} \partial_a g_{bc} \right) = \dot{y}^d \frac{d}{dt} \ln(\mathcal{L}). \tag{9.133}$$

Any curve satisfying this differential equation is going to give a stationary value for the curve length $s[\gamma]$.

Example 9.24 Let us go back to the case of Euclidean space with Cartesian coordinates. In this situation, the metric is diagonal, its entries constant, and the Euler–Lagrange equations to find the stationary curves of the path length between two points are reduced to

$$\ddot{x}^i = \dot{x}^i \frac{d}{dt} \ln \left(|\dot{\vec{x}}| \right) \quad \Longleftrightarrow \quad \ddot{\vec{x}} = \dot{\vec{x}} \frac{d}{dt} \ln \left(|\dot{\vec{x}}| \right). \tag{9.134}$$

In other words, we find that the derivative of the tangent vector $\dot{\vec{x}}$ of the curve is directly proportional to $\dot{\vec{x}}$ itself. This is hardly surprising as we already know that the curves that give a stationary curve length are going to be straight lines. The right-hand side arises due to an ambiguity in the parametrisation of the curve. In particular, we can write the equation of a straight line with a constant tangent vector as

$$\vec{x} = \vec{a}s + \vec{x}_0, \tag{9.135}$$

where s is the curve parameter and $\vec{a}$ the constant tangent vector. This line clearly satisfies the required Euler–Lagrange equations since both sides of the equations vanish. We could now reparametrise this line by introducing a new curve parameter t and expressing the old curve parameter in terms of it as a function $s = s(t)$. We find that

$$\frac{d\vec{x}}{dt} = \vec{a} \frac{ds}{dt}, \quad \frac{d^2 \vec{x}}{dt^2} = \vec{a} \frac{d}{dt} \frac{ds}{dt} = \frac{d\vec{x}}{dt} \frac{1}{ds/dt} \frac{d}{dt} \frac{ds}{dt} = \frac{d\vec{x}}{dt} \frac{d}{dt} \ln \left(\left| \frac{d\vec{x}}{dt} \right| \right), \tag{9.136}$$

which is exactly the form of Eq. (9.134). The interpretation of this result is that, since the re parametrisation of the curve does not affect its length, it is sufficient to require that the change in the tangent vector is proportional to the tangent vector itself. However, it is always possible to change the curve parametrisation in such a way that the tangent vector has constant norm and for this situation the right-hand side vanishes.

Coming back to the case of the general manifold M, we can also reparametrise the curve in such a way that the right-hand side vanishes, i.e., the norm of the tangent vector is constant. Such a parametrisation of a curve is a parametrisation in terms of the curve length and is called an *affine parametrisation*. We are left with the relation

$$\ddot{y}^d + \dot{y}^b \dot{y}^c g^{da} \left(\partial_b g_{ac} - \frac{1}{2} \partial_a g_{bc} \right) = 0. \tag{9.137}$$

This looks very similar to the geodesic equation

$$\ddot{y}^d + \Gamma^d_{bc} \dot{y}^b \dot{y}^c = 0. \tag{9.138}$$

In fact, it *is* the geodesic equation if we have a connection such that

$$\Gamma^d_{bc} \dot{y}^b \dot{y}^c = \dot{y}^b \dot{y}^c g^{da} \left(\partial_b g_{ac} - \frac{1}{2} \partial_a g_{bc} \right), \tag{9.139}$$

which should not be very surprising. After all, in Example 9.24 we saw that the straight line with a constant norm tangent vector was that where the tangent vector was constant. As this relation contains the factor $\dot{y}^b \dot{y}^c$, there are several possible connections that satisfy this relation, as only the symmetric part $\Gamma^c_{\{ab\}}$ of the Christoffel symbols contribute to the left-hand side.

It should be noted that Eq. (9.137) can be obtained directly by variation of the functional

$$L[\gamma] = \int_\gamma \mathcal{L}^2 dt = \int_\gamma g_{ab} \dot{y}^a \dot{y}^b dt \tag{9.140}$$

rather than of $s[\gamma]$. The Euler–Lagrange equations for $L[\gamma]$ are given by

$$\frac{\partial \mathcal{L}^2}{\partial y^a} - \frac{d}{dt} \frac{\partial \mathcal{L}^2}{\partial \dot{y}^a} = \dot{y}^b \dot{y}^c \partial_a g_{bc} - 2\frac{d}{dt} g_{ac} \dot{y}^c = \dot{y}^b \dot{y}^c (\partial_a g_{bc} - 2\partial_b g_{ac}) - 2g_{ac} \ddot{y}^c = 0. \tag{9.141}$$

Multiplying this by $-g^{da}/2$ results in Eq. (9.137). The curves that give stationary values for $L[\gamma]$ are therefore precisely the curves that also give stationary values of $s[\gamma]$ and have a parametrisation such that the tangent vector has constant norm. It is also often easier to vary $L[\gamma]$ than $s[\gamma]$ and we will therefore use this for all practical computations.

9.4.3   The Levi-Civita connection

For any manifold M with a metric tensor g, an affine connection is said to be *metric compatible* if the metric tensor is parallel, i.e., if

$$\nabla_X g = 0 \tag{9.142}$$

for all vectors X. The reason for this is that if we take two vectors X and Y parallel transported along a curve γ, then their inner product $g(X, Y)$ satisfies

$$\frac{d}{dt} g(X, Y) = \nabla_{\dot{\gamma}} g(X, Y) = (\nabla_{\dot{\gamma}} g)(X, Y) + g(\nabla_{\dot{\gamma}} X, Y) + g(X, \nabla_{\dot{\gamma}} Y) = 0, \tag{9.143}$$

where the first term is zero due to the metric compatibility and the last two due to X and Y being parallel along γ and hence $\nabla_{\dot{\gamma}} X = \nabla_{\dot{\gamma}} Y = 0$. Thus, a metric compatible connection preserves the inner product between parallel transported vectors and therefore also the vector norms as well as the angles between parallel transported vectors.

In any given coordinate system, a metric compatible connection satisfies

$$\nabla_c g_{ab} = \partial_c g_{ab} - \Gamma^d_{ca} g_{db} - \Gamma^d_{cb} g_{ad} = 0. \tag{9.144}$$

This expression is exactly equivalent to Eq. (2.84) that we found in Chapter 2 by differentiating $g_{ab} = \vec{E}_a \cdot \vec{E}_b$ with respect to y^c. If we furthermore require that our connection is torsion free, we can follow exactly the same steps as those leading up to Eq. (2.86) and we find that there is a unique metric compatible torsion free affine connection given by the connection coefficients

$$\Gamma^c_{ab} = \frac{1}{2} g^{cd} (\partial_a g_{db} + \partial_b g_{da} - \partial_d g_{ab}). \tag{9.145}$$

This connection is known as the *Levi-Civita connection* and, just as we did in Chapter 2, we will refer to its connection coefficients as the *Christoffel symbols*.

With the Levi-Civita connection in place, we can write down its geodesic equations and find that

$$\ddot{y}^c + \Gamma^c_{ab}\dot{y}^a\dot{y}^b = \ddot{y}^c + \dot{y}^a\dot{y}^b\frac{1}{2}g^{cd}(\partial_a g_{db} + \partial_b g_{da} - \partial_d g_{ab})$$

$$= \ddot{y}^c + \dot{y}^a\dot{y}^b g^{cd}\left(\partial_a g_{db} - \frac{1}{2}\partial_d g_{ab}\right) = 0. \tag{9.146}$$

This is exactly the condition given in Eq. (9.137) for a curve that gives a stationary value for the curve length $s[\gamma]$. Thus, for the Levi-Civita connection, the geodesics are the stationary curves of the curve length. Inversely, the Christoffel symbols of the Levi-Civita connection can be directly inferred from the Euler–Lagrange equations resulting from variation of the functional $L[\gamma]$ defined in Eq. (9.140).

Example 9.25 Let us compute the Christoffel symbols on a sphere with the metric given in Example 9.22. The functional $L[\gamma]$ is given by

$$L[\gamma] = \int g_{ab}\dot{y}^a\dot{y}^b dt = \int (\dot{\theta}^2 + \sin^2(\theta)\dot{\varphi}^2)dt. \tag{9.147}$$

The Euler–Lagrange equation due to variations of θ are now given by

$$\sin(\theta)\cos(\theta)\dot{\varphi}^2 - \ddot{\theta} = 0, \tag{9.148}$$

which we identify with the geodesic equation

$$\ddot{\theta} + \Gamma^\theta_{ab}\dot{y}^a\dot{y}^b = 0, \tag{9.149}$$

indicating that the only non-zero Christoffel symbol Γ^θ_{ab} is given by $\Gamma^\theta_{\varphi\varphi} = -\sin(\theta)\cos(\theta)$. In a similar fashion, the variation with respect to φ results in

$$\frac{d}{dt}\sin^2(\theta)\dot{\varphi} = \sin^2(\theta)\ddot{\varphi} + 2\sin(\theta)\cos(\theta)\dot{\theta}\dot{\varphi} = 0. \tag{9.150}$$

Dividing by $\sin^2(\theta)$ and identifying with

$$\ddot{\varphi} + \Gamma^\varphi_{ab}\dot{y}^a\dot{y}^b = \ddot{\varphi} + \Gamma^\varphi_{\theta\theta}\dot{\theta}^2 + \Gamma^\varphi_{\varphi\varphi}\dot{\varphi}^2 + 2\Gamma^\varphi_{\theta\varphi}\dot{\theta}\dot{\varphi} = 0, \tag{9.151}$$

where we have used the fact that the Levi-Civita connection is assumed to be torsion free, we arrive at the conclusion that the only non-zero Christoffel symbols of the form Γ^φ_{ab} are $\Gamma^\varphi_{\theta\varphi} = \Gamma^\varphi_{\varphi\theta} = \cot(\theta)$. In summary, the non-zero Christoffel symbols on the sphere are given by

$$\Gamma^\theta_{\varphi\varphi} = -\sin(\theta)\cos(\theta) = -\frac{1}{2}\sin(2\theta) \quad \text{and} \quad \Gamma^\varphi_{\theta\varphi} = \Gamma^\varphi_{\varphi\theta} = \cot(\theta). \tag{9.152}$$

Of course, we could also have computed these Christoffel symbols starting directly from Eq. (9.145). However, this is generally more tedious and easier to get wrong than the approach shown in this example.

9.4.4 Curvature revisited

With the Levi-Civita connection in place, we can return to the concept of curvature and discuss the implications for the curvature tensor. It turns out that it is sometimes easier to work with the type $(0, 4)$ tensor

$$R_{dcab} \equiv g_{de} R^e_{cab} \tag{9.153}$$

rather than the curvature tensor itself. In addition, some of the symmetries of the curvature tensor will be more apparent when it is written on this form. Of course, these are completely equivalent as the type $(1, 3)$ tensor may be recovered by raising the first index of R_{dcab}.

In order to find the symmetries of R_{dcab}, we first need to derive a few helpful relations. We start by defining the Christoffel symbols with all indices down as

$$\Gamma_{abc} = g_{ad} \Gamma^d_{bc}. \tag{9.154}$$

These are sometimes referred to as Christoffel symbols *of the first kind*, while the Γ^a_{bc} that we are used to are then referred to as Christoffel symbols of the *second kind*. Due to the symmetry in the indices b and c of the left-hand side, it should be clear that $\Gamma_{abc} = \Gamma_{acb}$. We will also need the relation

$$0 = \partial_a \delta^b_c = \partial_a g^{bd} g_{cd} = g^{bd} \partial_a g_{cd} + g_{cd} \partial_a g^{bd} \quad \Longleftrightarrow \quad g^{bd} \partial_a g_{cd} = -g_{cd} \partial_a g^{bd}. \tag{9.155}$$

We now set out to express R_{dcab} in such a way that its symmetries become apparent, starting by rewriting the derivatives of the Christoffel symbols in the expression for the curvature tensor in terms of the metric

$$\begin{aligned}
R_{dcab} &= g_{de} (\partial_a \Gamma^e_{bc} - \partial_b \Gamma^e_{ac} + \Gamma^f_{bc} \Gamma^e_{af} - \Gamma^f_{ac} \Gamma^e_{bf}) \\
&= g_{de} [\partial_a (g^{ef} \Gamma_{fbc}) - \partial_b (g^{ef} \Gamma_{fac})] + \Gamma^f_{bc} \Gamma_{daf} - \Gamma^f_{ac} \Gamma_{dbf} \\
&= \partial_a \Gamma_{dbc} - \partial_b \Gamma_{dac} + g_{de} \Gamma_{fbc} \partial_a g^{ef} - g_{de} \Gamma_{fac} \partial_b g^{ef} + \Gamma^f_{bc} \Gamma_{daf} - \Gamma^f_{ac} \Gamma_{dbf} \\
&= \frac{1}{2} (\partial_a \partial_b g_{cd} + \partial_a \partial_c g_{bd} - \partial_a \partial_d g_{bc} - \partial_b \partial_a g_{cd} - \partial_b \partial_c g_{ad} + \partial_b \partial_d g_{ac}) \\
&\quad - \Gamma^f_{bc} (\partial_a g_{df} - \Gamma_{daf}) + \Gamma^f_{ac} (\partial_b g_{df} - \Gamma_{dbf}). \tag{9.156a}
\end{aligned}$$

Noting that Eq. (9.144) can be written on the form

$$\partial_a g_{df} = \Gamma_{fad} + \Gamma_{daf} \tag{9.156b}$$

this result can be rewritten as

$$R_{dcab} = \frac{1}{2} (\partial_a \partial_c g_{bd} - \partial_a \partial_d g_{bc} - \partial_b \partial_c g_{ad} + \partial_b \partial_d g_{ac}) - \Gamma^f_{bc} \Gamma_{fad} + \Gamma^f_{ac} \Gamma_{fbd}. \tag{9.156c}$$

It is now evident that the curvature tensor has the symmetries

$$R_{dcab} = -R_{cdab} = -R_{dcba} = R_{abdc}. \tag{9.157}$$

These symmetries are important when considering the possible contractions of the curvature tensor. In particular, the contractions of the first two indices as well as that of the last two both vanish due to the anti-symmetries

$$g^{dc} R_{dcab} = 0 \quad \text{and} \quad g^{ab} R_{dcab} = 0. \tag{9.158}$$

The remaining four possibilities are all related as

$$R_{cb} = R^a_{cab} = g^{ad} R_{dcab} = -g^{ad} R_{cdab} = -g^{ad} R_{dcba} = g^{ad} R_{cdba} \tag{9.159}$$

and results in a single type $(0,2)$ tensor called the *Ricci tensor*, which is symmetric

$$R_{cb} = g^{ad} R_{dcab} = g^{ad} R_{abdc} = R_{bc}. \tag{9.160}$$

Finally, the *Ricci scalar*

$$R = g^{bc} R_{bc} \tag{9.161}$$

is defined as the trace of the Ricci tensor. These concepts will turn out to be central in the formulation of general relativity.

Example 9.26 Computing the components of the curvature tensor on the two-dimensional sphere with the metric introduced in Example 9.22, its symmetries imply that the only independent non-zero component is

$$R_{\theta\varphi\theta\varphi} = -R_{\varphi\theta\theta\varphi} = -R_{\theta\varphi\varphi\theta} = R_{\varphi\theta\varphi\theta}. \tag{9.162}$$

Using the non-zero components $g_{\theta\theta} = 1$ and $g_{\varphi\varphi} = \sin^2(\theta)$ of the metric together with the corresponding Christoffel symbols of the Levi-Civita connection, computed in Example 9.25, we find that

$$
\begin{aligned}
R_{\theta\varphi\theta\varphi} &= \frac{1}{2}(\partial_\theta\partial_\varphi g_{\theta\varphi} - \partial_\theta^2 g_{\varphi\varphi} - \partial_\varphi^2 g_{\theta\theta} + \partial_\varphi\partial_\theta g_{\varphi\theta}) - \Gamma^a_{\varphi\varphi}\Gamma_{a\theta\theta} + \Gamma^a_{\varphi\theta}\Gamma_{a\varphi\theta} \\
&= -\frac{1}{2}\partial_\theta^2\sin^2(\theta) + \Gamma^\varphi_{\varphi\theta}\Gamma_{\varphi\varphi\theta} = -\partial_\theta\sin(\theta)\cos(\theta) + (\Gamma^\varphi_{\varphi\theta})^2 g_{\varphi\varphi} \\
&= -\cos^2(\theta) + \sin^2(\theta) + \cot^2(\theta)\sin^2(\theta) = \sin^2(\theta).
\end{aligned}
\tag{9.163}
$$

The components of the Ricci tensor are now given by

$$R_{\theta\theta} = R^a_{\theta a\theta} = g^{\theta\theta} R_{\theta\theta\theta\theta} + g^{\varphi\varphi} R_{\varphi\theta\varphi\theta} = \frac{1}{\sin^2(\theta)}\sin^2(\theta) = 1 = g_{\theta\theta}, \tag{9.164a}$$

$$R_{\varphi\varphi} = R^a_{\varphi a\varphi} = g^{\theta\theta} R_{\theta\varphi\theta\varphi} + g^{\varphi\varphi} R_{\varphi\varphi\varphi\varphi} = \sin^2(\theta) = g_{\varphi\varphi}, \tag{9.164b}$$

$$R_{\theta\varphi} = R^a_{\theta a\varphi} = g^{\theta\theta} R_{\theta\theta\theta\varphi} + g^{\varphi\varphi} R_{\varphi\theta\varphi\varphi} = 0 = g_{\theta\varphi}. \tag{9.164c}$$

In other words, we find that the Ricci tensor is equal to the metric tensor $R_{ab} = g_{ab}$ and consequently the Ricci scalar is given by

$$R = R_{ab} g^{ab} = g_{ab} g^{ab} = \delta^a_a = 2. \tag{9.165}$$

In general, a space where the Ricci tensor is proportional to the metric tensor is called an *Einstein space*.

9.5 INTEGRATION ON MANIFOLDS

A large part of Chapter 1 was devoted to defining integrals over lines, volumes, and surfaces. We also spent some time in Chapter 2 on generalising the expression of these integrals to general coordinate systems. With the previous parts of this chapter fresh in our memory, we are now ready to start a discussion on how to generalise the integral concept to manifolds. This will be done in the language of *differential forms* and we will therefore start by introducing these and looking at some of their properties.

9.5.1 Differential forms

Looking at the expression in Eq. (2.165), it is clear that the volume element has the property that it depends on a set of tangent vectors and is completely anti-symmetric in these. The volume element was generally given by

$$dV = \vec{E}_1 \cdot (\vec{E}_2 \times \vec{E}_3) dy^1 dy^2 dy^3 = \frac{\partial \vec{x}}{\partial y^1} \cdot \left(\frac{\partial \vec{x}}{\partial y^2} \times \frac{\partial \vec{x}}{\partial y^3} \right) dy^1 dy^2 dy^3, \qquad (9.166)$$

where it should be clear that the vector triple product is completely anti-symmetric in the three tangent vectors $\partial \vec{x}/\partial y^a$. Indeed, also the line element

$$d\vec{x} = \frac{d\vec{x}}{dt} dt \qquad (9.167)$$

contains a tangent vector and a differential and, since there are no other tangent vectors, is completely anti-symmetric in all possible exchanges of tangent vectors. All in all, these properties of volume, surface, and line elements all gently hint towards the relevance of completely anti-symmetric multi-linear maps from tangent vectors to the real numbers. By definition, when such a map takes p tangent vector arguments, it is a completely anti-symmetric tensor of type $(0, p)$ and we call such tensors p-*forms*. The set of 1-forms is just the cotangent space as all type $(0, 1)$ tensors are completely antisymmetric. It is also common to talk about scalar fields as 0-forms as they are completely anti-symmetric type $(0, 0)$ tensors.

Just as a general type $(0, p)$ tensor can be written in the basis $e^{a_1 \cdots a_p} = dy^{a_1} \otimes \ldots \otimes dy^{a_p}$, a basis for the set of p-forms can be constructed through the *wedge product* of p 1-forms defined as

$$dy^{a_1} \wedge \ldots \wedge dy^{a_p} = \sum_{\sigma \in S_p} \text{sgn}(\sigma) dy^{a_{\sigma(1)}} \otimes \ldots \otimes dy^{a_{\sigma(p)}}, \qquad (9.168)$$

where S_p is the symmetric group for p elements, i.e., the set of all possible permutations of those elements (see Section 4.3.3). It should be clear that this product is anti-symmetric in all indices a_k and so if any two indices are the same, the wedge product will be zero. Because of this, the set of non-zero basis elements are found by selecting p different indices, which in an N-dimensional manifold can be done in

$$n_p^N = \binom{N}{p} = \frac{N!}{(N-p)! p!} \qquad (9.169)$$

different ways. Due to the anti-symmetry, any ordering of the indices will result in linearly dependent forms and the set of p-forms on an N-dimensional manifold therefore has dimension n_p^N. Since it is impossible to select more than N different indices, any p-form with $p > N$ is going to be equal to zero.

Example 9.27 Let us consider a three-dimensional manifold with coordinates y^1, y^2, and y^3. From our discussion above, we will only have non-zero p-forms for $p = 0, 1, 2$, and 3 and the numbers of basis p-forms are going to be given by

$$n_0^3 = 1, \quad n_1^3 = 3, \quad n_2^3 = 3, \quad \text{and} \quad n_3^3 = 1, \qquad (9.170)$$

respectively. The different bases can be written down explicitly as

$$\text{0-forms}: \quad 1, \tag{9.171a}$$

$$\text{1-forms}: \quad dy^1, \ dy^2, \ dy^3, \tag{9.171b}$$

$$\text{2-forms}: \quad dy^1 \wedge dy^2 = -dy^2 \wedge dy^1 = dy^1 \otimes dy^2 - dy^2 \otimes dy^1,$$
$$dy^2 \wedge dy^3 = -dy^3 \wedge dy^2 = dy^2 \otimes dy^3 - dy^3 \otimes dy^2,$$
$$dy^3 \wedge dy^1 = -dy^1 \wedge dy^3 = dy^3 \otimes dy^1 - dy^1 \otimes dy^3, \tag{9.171c}$$

$$\text{3-forms}: \quad dy^1 \wedge dy^2 \wedge dy^3, \tag{9.171d}$$

where we have written out the expansion of the 2-forms explicitly for clarity.

When writing a p-form ω in terms of its completely anti-symmetric components $\omega_{a_1 \ldots a_p}$ of the type $(0, p)$ tensor, we know that

$$\omega = \omega_{a_1 \ldots a_p} dy^{a_1} \otimes \ldots \otimes dy^{a_p} = \omega_{a_1 \ldots a_p} dy^{[a_1} \otimes \ldots \otimes dy^{a_p]}, \tag{9.172}$$

where we have used the fact that $\omega_{a_1 \ldots a_p}$ is completely anti-symmetric in order to anti-symmetrise the basis. We now note that

$$dy^{[a_1} \otimes \ldots \otimes dy^{a_p]} = \frac{1}{p!} \sum_{\sigma \in S_p} \text{sgn}(\sigma) dy^{a_{\sigma(1)}} \otimes \ldots \otimes dy^{a_{\sigma(p)}} = \frac{1}{p!} dy^{a_1} \wedge \ldots \wedge dy^{a_p}. \tag{9.173}$$

In other words, we find that

$$\omega = \frac{1}{p!} \omega_{a_1 \ldots a_p} dy^{a_1} \wedge \ldots \wedge dy^{a_p}, \tag{9.174}$$

where it must be noted that for $\omega_{a_1 \ldots a_p}$ to be the components of the form in the $dy^{a_1} \otimes \ldots \otimes dy^{a_p}$ basis, they must be completely anti-symmetric.

Example 9.28 Consider the 2-form $\omega = dy^1 \wedge dy^2$ in a two dimensional manifold with coordinates y^1 and y^2. Writing out the wedge product in terms of the tensor product of 1-forms, we find that

$$\omega = dy^1 \otimes dy^2 - dy^2 \otimes dy^1. \tag{9.175}$$

In other words, ω is a type $(0, 2)$ tensor with the components

$$\omega_{11} = \omega_{22} = 0, \quad \omega_{12} = -\omega_{21} = 1. \tag{9.176}$$

However, when we sum ω_{ab} with the wedge product $dy^a \wedge dy^b$, we obtain

$$\omega_{ab} dy^a \wedge dy^b = \omega_{12} dy^1 \wedge dy^2 + \omega_{21} dy^2 \wedge dy^1 = dy^1 \wedge dy^2 - dy^2 \wedge dy^1$$
$$= 2 dy^1 \wedge dy^2 = 2\omega. \tag{9.177}$$

Consequently, we find that we must divide $\omega_{ab} dy^a \wedge dy^b$ by $2! = 2$ in order to recover ω.

9.5.1.1 The exterior derivative

There is an important derivative operator on a p-form that returns a $p+1$-form. As we will soon discover through examples, it is a generalisation of the gradient, the divergence, and the curl all in one. We define the *exterior derivative* $d\omega$ of a p-form ω as

$$d\omega = \frac{1}{p!}(\partial_{a_1}\omega_{a_2...a_{p+1}})\,dy^{a_1} \wedge dy^{a_2} \wedge \ldots \wedge dy^{a_{p+1}}. \tag{9.178}$$

Note that even though we have used the partial derivative of the tensor components $\omega_{a_2...a_{p+1}}$, this expression still transforms as a tensor under coordinate transformations due to the anti-symmetrisation implied by the wedge product.

Example 9.29 Let us look at the different possible exterior derivatives in three dimensions. Starting from the exterior derivative of a 0-form, we should obtain a 1-form as a result. As a 0-form is equivalent to a scalar field f, we find that its exterior derivative is given by

$$df = (\partial_a f)dy^a = (\partial_1 f)dy^1 + (\partial_2 f)dy^2 + (\partial_3 f)dy^3, \tag{9.179}$$

which is nothing but the definition of the differential of f as a type $(0,1)$ tensor. Since we have already used df to denote the differential earlier, we can here see the consistency in using df to also denote the exterior derivative, at least when applied to 0-forms. We note that there is a close connection between the differential of a 0-form and the gradient of a scalar field that we encountered in Chapter 1, since the gradient of f has components $\partial_i f$ in a Cartesian coordinate system. The *gradient* of a scalar field f in a manifold with a metric is usually defined as the tangent vector field with components $g^{ab}\partial_a f$, i.e., the tangent vector that the metric maps to the differential df.

Moving on to the exterior derivative of a 1-form $\omega = \omega_1 dy^1 + \omega_2 dy^2 + \omega_3 dy^3$, we find that

$$\begin{aligned}
d\omega &= (\partial_2\omega_1)dy^2 \wedge dy^1 + (\partial_3\omega_1)dy^3 \wedge dy^1 + (\partial_1\omega_2)dy^1 \wedge dy^2 \\
&\quad + (\partial_3\omega_2)dy^3 \wedge dy^2 + (\partial_1\omega_3)dy^1 \wedge dy^3 + (\partial_2\omega_3)dy^2 \wedge dy^3 \\
&= (\partial_2\omega_3 - \partial_3\omega_2)dy^2 \wedge dy^3 + (\partial_3\omega_1 - \partial_1\omega_3)dy^3 \wedge dy^1 \\
&\quad + (\partial_1\omega_2 - \partial_2\omega_1)dy^1 \wedge dy^2.
\end{aligned} \tag{9.180}$$

The observant reader will here notice that the prefactors of all the wedge products are exactly the components of the *curl* of a vector field with components ω_1, ω_2, and ω_3.

Finally, we look at the exterior derivative of the most general 2-form

$$\omega = \omega_1 dy^2 \wedge dy^3 + \omega_2 dy^3 \wedge dy^1 + \omega_3 dy^1 \wedge dy^2. \tag{9.181}$$

The definition of the exterior derivative now leads to

$$d\omega = (\partial_1\omega_1 + \partial_2\omega_2 + \partial_3\omega_3)dy^1 \wedge dy^2 \wedge dy^3, \tag{9.182}$$

where we have used that $dy^2 \wedge dy^3 \wedge dy^1 = dy^3 \wedge dy^1 \wedge dy^2 = dy^1 \wedge dy^2 \wedge dy^3$. As we are now used to finding expressions similar to the ones discussed in Chapter 1, we merely acknowledge that the pre-factor of $dy^1 \wedge dy^2 \wedge dy^3$ in this expression is very reminiscent of the *divergence* of a vector field with components ω_1, ω_2, and ω_3.

9.5.2 Integration of differential forms

We shall now put our differential forms to work and we start by examining how they relate to line, area, and volume elements. In general, if we have a set of p tangent vectors X_i, we look for the properties we expect from the p-dimensional volume of the infinitesimal parallelogram spanned by the displacements $dy_i^a = X_i^a dt_i$ (no sum over i). First of all, if the vectors X_i are not linearly independent, we expect the p-dimensional volume to be zero, since the parallelogram spanned by the displacements will now be at most $p-1$-dimensional. Furthermore, we expect that the volume is going to be linear in all of the X_i, i.e., if we increase the length of one of the sides, the p-dimensional volume will increase accordingly. Both of these properties are directly satisfied by letting the p-dimensional *volume element* of the parallelogram be given by

$$dV_p = \omega(X_1, \ldots, X_p) dt_1 \ldots dt_p, \tag{9.183}$$

where ω is a p-form. In addition to satisfying the requirements we have already mentioned, this p-volume is directed and the volume will alternate between being positive and negative depending on whether the vectors X_i form a right- or left-handed set.

Example 9.30 Consider the case of a three-dimensional Euclidean space in Cartesian coordinates. The standard volume spanned by the vectors $\vec{v}_1$, $\vec{v}_2$, and $\vec{v}_3$ is given by the triple product

$$V = \vec{v}_1 \cdot (\vec{v}_2 \times \vec{v}_3) = \varepsilon_{ijk} v_1^i v_2^j v_3^k. \tag{9.184}$$

At the same time, we note that

$$\eta = \frac{1}{3!}\varepsilon_{ijk} dx^i \wedge dx^j \wedge dx^k = dx^1 \wedge dx^2 \wedge dx^3 \tag{9.185}$$

is a 3-form with precisely the property that

$$\eta(\vec{v}_1, \vec{v}_2, \vec{v}_3) = V \tag{9.186}$$

when the $\vec{v}_i$ are considered as tangent vectors. Generalising to arbitrary curvilinear coordinates, we must replace the permutation symbol by the completely anti-symmetric tensor $\eta_{abc} = \sqrt{g}\,\varepsilon_{abc}$, where g is the metric determinant. Our 3-form then becomes

$$\eta = \frac{1}{3!}\sqrt{g}\,\varepsilon_{abc} dy^a \wedge dy^b \wedge dy^c = \sqrt{g}\, dy^1 \wedge dy^2 \wedge dy^3 \tag{9.187}$$

and the infinitesimal volume spanned by the tangent vectors $X_1 dt_1$, $X_2 dt_2$, and $X_3 dt_3$ is found to be

$$dV_3 = \eta(X_1 dt_1, X_2 dt_2, X_3 dt_3) = \sqrt{g}\,\varepsilon_{abc} X_1^a X_2^b X_3^c dt_1 dt_2 dt_3. \tag{9.188}$$

This should be compared with what we found in Eq. (2.166).

We consider a p-dimensional volume S, parametrised by the variables t_i, in an N-dimensional manifold with local coordinates y^a. Of course, we will generally have $N \geq p$ and based on viewing a p-form ω taking p tangent vectors as an argument as a p-volume element, possibly multiplied by some function, we define the *integral* of ω over S as

$$\int_S \omega \equiv \int_{S^*} \omega\,(\dot{\gamma}_1, \ldots, \dot{\gamma}_p)\, dt_1 \ldots dt_p, \tag{9.189}$$

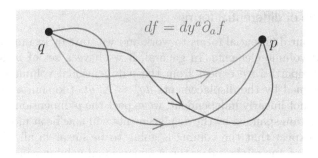

Figure 9.17 The integral of a one-form df which is the exterior derivative of a function f along a curve γ from q to p is given by the difference of the function values at the endpoints $f(p) - f(q)$ and does not depend on the curve γ itself, only the positions of its endpoints.

where $\dot{\gamma}_i$ is the tangent vector with components $\partial y^a / \partial t_i$ and S^* is the range of parameter values t_i that map to S.

Example 9.31 Going back to the integration of the three-form $dx^1 \wedge dx^2 \wedge dx^3$ in Cartesian coordinates over a volume V in a Euclidean space, we may use the coordinates x^1, x^2, and x^3 themselves as parameters. Naturally, this directly implies that $V^* = V$ and we find that, if $\rho = \rho(\vec{x})$ is a scalar field, then

$$\int_V \rho\omega = \int_{V^*} \rho(\vec{x})(dx^1 \wedge dx^2 \wedge dx^3)(\vec{e}_1, \vec{e}_2, \vec{e}_3)dx^1 dx^2 dx^3 = \int_V \rho(\vec{x})\, dx^1 dx^2 dx^3. \quad (9.190)$$

If $\rho(\vec{x})$ describes a density of an extensive property, this integral therefore evaluates to the total of that property enclosed in the volume V.

Example 9.32 Consider the one-form $df = (\partial_a f)dy^a$, i.e., the differential of the scalar field f, in a two-dimensional manifold M. The integral of this one-form along a curve γ going from point q to point p, see Fig. 9.17, is given by

$$\int_\gamma df = \int_0^1 df(\dot{\gamma})dt, \quad (9.191)$$

where we have assumed that the curve parameter t is such that $\gamma(0) = q$ and $\gamma(1) = p$. Inserting the definition of $df(\dot{\gamma})$, we find that

$$\int_\gamma df = \int_0^1 (\partial_a f)\frac{dy^a}{dt}dt = \int_0^1 \frac{df(\gamma(t))}{dt}dt = f(p) - f(q). \quad (9.192)$$

In other words, integrating the differential of f along a curve gives us the difference of the value of f at the curve endpoints. This is a result which is familiar from when we first considered line integrals.

Without a metric, there is no notion of distances on an N-dimensional manifold and so there is also no intrinsic concept of what constitutes an actual volume element corresponding to a physical volume. However, when a metric g is introduced, there will be a particular N-form η that fulfils exactly this purpose. Since we are looking for an N-form in an N-dimensional manifold, it must be proportional to

$$\varepsilon = dy^1 \wedge \ldots \wedge dy^N \qquad (9.193)$$

everywhere in a general coordinate system and we should find that

$$\eta = \tilde{f}\varepsilon, \qquad (9.194)$$

where $\tilde{f}$ is a function of the coordinates. It should here be noted that

$$\varepsilon = dy^1 \wedge \ldots \wedge dy^N \neq \varepsilon' = dy'^1 \wedge \ldots \wedge dy'^N \qquad (9.195)$$

and ε is therefore a tensor density rather than an actual tensor. In order for η to be a differential form, and hence a tensor, the function $\tilde{f}$ must be a scalar density of the appropriate weight. As was argued already in Section 2.5.2, the square root of the metric determinant g has the correct weight and we therefore assume that

$$\tilde{f} = f\sqrt{g}, \qquad (9.196)$$

where f is now a scalar field. To find an appropriate value of f, we can go to a coordinate system where the metric tensor is diagonal at the point p. In this system, the tangent vector basis satisfies

$$g_{ab} = g(\partial_a, \partial_b) = \delta_{ab}h_a^2, \qquad \text{(no sum)} \qquad (9.197)$$

where the scale factor h_a is the norm of ∂_a. Since the metric is diagonal, the metric determinant is given by the product of the diagonal elements

$$g = \prod_{a=1}^{N} h_a^2 \implies \sqrt{g} = \prod_{a=1}^{N} h_a. \qquad (9.198)$$

At the same time, since the vectors $X_a = dt\,\partial_a$ are orthogonal, they span a small cuboid with side lengths $h_a dt$, which should have the volume

$$dV = dt^N \prod_{a=1}^{N} h_a = dt^N \sqrt{g}. \qquad (9.199)$$

By construction, we want this to be equal to $dt^N \eta(\partial_1, \ldots, \partial_N)$ and we conclude that

$$dt^N \sqrt{g} = dt^N \eta(\partial_1, \ldots, \partial_N) = dt^N f\sqrt{g}\,\varepsilon_{12\ldots N} = dt^N f\sqrt{g}. \qquad (9.200)$$

Direct identification now gives us $f = 1$ and since f is a scalar field, taking the same value in any coordinate system, this will be the case regardless of the coordinate system we select. We conclude that the N-form corresponding to a physical volume is given by

$$\eta = \sqrt{g}\,dy^1 \wedge \ldots \wedge dy^N \qquad (9.201)$$

in any coordinate system with coordinates y^a.

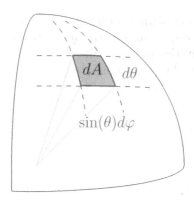

Figure 9.18 The area element on the sphere spanned by changing θ by $d\theta$ and φ by $d\varphi$ is given by $dA = \sin(\theta) d\theta \, d\varphi$. This may be deduced by considering the small area spanned by the orthogonal vectors ∂_θ and ∂_φ.

Example 9.33 Looking for the physical volume element on the two-dimensional unit sphere S^2, we use the coordinates θ and φ as usual. In this coordinate system, the metric tensor is already diagonal and we know that $h_\theta = 1$ and $h_\varphi = \sin(\theta)$. The two-volume, i.e., area, spanned by the tangent vectors $dt \, \partial_\theta$ and $dt \, \partial_\varphi$ should therefore be $\sin(\theta) \, dt^2$, see Fig. 9.18. Writing down the area two-form, we find that

$$\eta = \tilde{f} \, d\theta \wedge d\varphi, \tag{9.202}$$

which should satisfy

$$\eta(\partial_\theta, \partial_\varphi) = \tilde{f} \left(\frac{\partial \theta}{\partial \theta} \frac{\partial \varphi}{\partial \varphi} - \frac{\partial \varphi}{\partial \theta} \frac{\partial \theta}{\partial \varphi} \right) = \tilde{f} = \sin(\theta). \tag{9.203}$$

The area two-form is therefore given by

$$\eta = \sin(\theta) \, d\theta \wedge d\varphi. \tag{9.204}$$

In passing we also note that $\sqrt{g} = \sin(\theta)$ in this coordinate system, in accordance to what we expect from the discussion above.

9.5.3 Stokes' theorem

Some of the most important results of Chapter 1 were the integral theorems, i.e., the divergence and curl theorems, relating integrals over closed surfaces and loops to integrals over the volumes and surfaces they bound. In the language of integration of differential forms, it will turn out that these are special cases of a more general relation known as *Stokes' theorem*, not to be confused with the fact that the curl theorem is also sometimes referred to as Stokes' theorem. The theorem states that for any p-form ω and $p+1$-volume V, it holds that

$$\int_V d\omega = \oint_{\partial V} \omega, \tag{9.205}$$

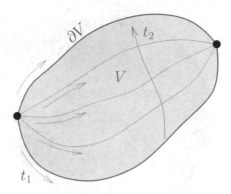

Figure 9.19 A two-dimensional example of the parametrisation used for the derivation of Stokes' theorem. The parameter t_1 is used as the curve parameter in a family of curves going from one point on the boundary to another while the parameter t_2 parametrises this family of curves, which sweeps the integration volume V. The boundary ∂V of V corresponds to the extreme values of the parameter t_2.

where ∂V is the closed p-dimensional boundary of V. In order to show that this is the case, let us consider a $p+1$-volume, which we can parametrise using the $p+1$ parameters t_i for $i = 1, 2, \ldots, p+1$ assigned such that $0 \leq t_{p+1} \leq 1$ and the remaining coordinates for a fixed t_{p+1} take values in the range V_p^* and parametrise a p-dimensional surface with a t_{p+1}-independent boundary, see Fig. 9.19. This implies that whenever the remaining parameters take values on the boundary of V_p^*, we have $\partial y^a/\partial t_1 = 0$. Although this is a bit restrictive on the type of volumes we can consider, we can always use volumes of this kind to build an arbitrary volume.

Example 9.34 To give a concrete example of a volume parametrised in this fashion, let us consider the the volume V given by $r \leq 1$ in spherical coordinates in $\mathbb{R}^3$. In order to parametrise this volume we use the three variables t_1, t_2, and t_3 such that

$$x^1 = t_1, \quad x^2 = t_2, \quad \text{and} \quad x^3 = (2t_3 - 1)\sqrt{1 - t_1^2 - t_2^2}. \tag{9.206}$$

With this choice, V_2^* is the region $t_1^2 + t_2^2 \leq 1$ and we note that $t_3 = 1$ corresponds to the upper half-sphere of radius one and $t_3 = 0$ to the lower one, see Fig. 9.20.

With ω being a p-form, we can construct the integral

$$\int_V d\omega = \sum_{\sigma \in S_{p+1}} \frac{\mathrm{sgn}(\sigma)}{p!} \int_{V^*} \partial_{a_{p+1}} \omega_{a_1 \ldots a_p} \frac{\partial y^{a_{\sigma(1)}}}{\partial t_1} \cdots \frac{\partial y^{a_{\sigma(p+1)}}}{\partial t_{p+1}} dt_1 \ldots dt_{p+1}$$

$$= \sum_{\sigma \in S_{p+1}} \frac{\mathrm{sgn}(\sigma)}{p!} \underbrace{\int_{V^*} \partial_{a_{p+1}} \omega_{a_1 \ldots a_p} \frac{\partial y^{a_1}}{\partial t_{\sigma(1)}} \cdots \frac{\partial y^{a_{p+1}}}{\partial t_{\sigma(p+1)}} dt_1 \ldots dt_{p+1}}_{\equiv \mathcal{I}(\sigma, \omega)}. \tag{9.207}$$

Figure 9.20 A cross section of the example parametrisation from Example 9.34 used for the derivation of Stokes' theorem in the case of three dimensions with the volume being a ball V. The surfaces inside the sphere correspond to fixed values of t_3 and the dark circle the region where the mapping does not depend on t_3, i.e., the boundary of the family of surfaces. The outer surface corresponds to the boundary ∂V of V, which is a sphere.

Looking closer at the integral $\mathcal{I}(\sigma, \omega)$, we find that

$$
\begin{aligned}
\mathcal{I}(\sigma,\omega) &= \int_{V^*} \frac{\partial y^{a_{p+1}}}{\partial t_{\sigma(p+1)}} (\partial_{a_{p+1}}\omega_{a_1\ldots a_p}) \frac{\partial y^{a_1}}{\partial t_{\sigma(1)}} \cdots \frac{\partial y^{a_p}}{\partial t_{\sigma(p)}} dt_1 \ldots dt_{p+1} \\
&= \int_{V^*} \frac{\partial \omega_{a_1\ldots a_p}}{\partial t_{\sigma(p+1)}} \frac{\partial y^{a_1}}{\partial t_{\sigma(1)}} \cdots \frac{\partial y^{a_p}}{\partial t_{\sigma(p)}} dt_1 \ldots dt_{p+1} \\
&= \int_{t_{\sigma(p+1)}=t^+_{\sigma(p+1)}} \omega_{a_1\ldots a_p} \frac{\partial y^{a_1}}{\partial t_{\sigma(1)}} \cdots \frac{\partial y^{a_p}}{\partial t_{\sigma(p)}} dt_1 \ldots \cancel{dt_{\sigma(p+1)}} \ldots dt_{p+1} \\
&\quad - \int_{t_{\sigma(p+1)}=t^-_{\sigma(p+1)}} \omega_{a_1\ldots a_p} \frac{\partial y^{a_1}}{\partial t_{\sigma(1)}} \cdots \frac{\partial y^{a_p}}{\partial t_{\sigma(p)}} dt_1 \ldots \cancel{dt_{\sigma(p+1)}} \ldots dt_{p+1} \\
&\quad + \int_{V^*} \omega_{a_1\ldots a_p} \frac{\partial}{\partial t_{\sigma(p+1)}} \left(\frac{\partial y^{a_1}}{\partial t_{\sigma(1)}} \cdots \frac{\partial y^{a_p}}{\partial t_{\sigma(p)}} \right) dt_1 \ldots dt_{p+1},
\end{aligned}
\tag{9.208}
$$

where $t_{\sigma(p+1)\pm}$ are the largest and smallest possible values of $t_{\sigma(p+1)}$ given the values of the other parameters and the first two integrals are taken with respect to those. In order to obtain the result in the last step we have performed a partial integration in $t_{\sigma(p+1)}$. While this expression looks quite bulky, most of the terms are going to either cancel or be equal to zero. We start by noting that if $\sigma(p+1) \neq p+1$, then $t_{\sigma(p+1)} = t^{\pm}_{\sigma(p+1)}$ implies that the leftover integration domain is on the boundary of V^*_p. Consequently, there is going to be a k such that $1 \leq k \leq p$ and $\sigma(k) = p+1$ for which $\partial y^{a_k}/\partial t_{p+1} = \partial y^{a_k}/\partial t_{\sigma(k)} = 0$ and the integrals in the first two lines of the final expression therefore evaluate to zero. On the other hand, if $\sigma(p+1) = p+1$, we find that $t^+_{p+1} = 1$ and $t^-_{p+1} = 0$ and that the remaining integration domain is V^*_p. Looking at the integrand in the final line, it is always going to contain a term of the form

$$
\frac{\partial}{\partial t_{\sigma(p+1)}} \frac{\partial y^{a_k}}{\partial t_{\sigma(k)}} = \frac{\partial^2 y^{a_k}}{\partial t_{\sigma(k)} \partial t_{\sigma(p+1)}},
\tag{9.209}
$$

which is symmetric with respect to $p+1$ and k. However, the sum over S_{p+1} is anti-symmetric

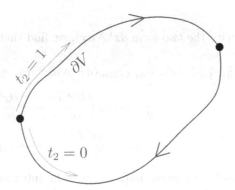

Figure 9.21 In order to obtain the integral over the closed directed boundary ∂V, the integral for the curve $t_2 = 1$ must be taken in the positive t_1 direction, but the integral for the curve $t_2 = 0$ must be taken in the negative t_1 direction. This is precisely the type of integral that results from the argumentation leading to Stokes' theorem.

under exchange of $p+1$ and k and these terms therefore cancel out in the sum. We are left with the result

$$\int_V d\omega = \sum_{\tilde{\sigma} \in S_p} \mathrm{sgn}(\tilde{\sigma}) \left(\int_{V_p^*} \omega_{a_1 \dots a_p} \frac{\partial y^{a_1}}{\partial t_{\tilde{\sigma}(1)}} \dots \frac{\partial y^{a_p}}{\partial t_{\tilde{\sigma}(p)}} dt_1 \dots dt_p \Big|_{t_{p+1}=1} \right.$$
$$\left. - \int_{V_p^*} \omega_{a_1 \dots a_p} \frac{\partial y^{a_1}}{\partial t_{\tilde{\sigma}(1)}} \dots \frac{\partial y^{a_p}}{\partial t_{\tilde{\sigma}(p)}} dt_1 \dots dt_p \Big|_{t_{p+1}=0} \right). \tag{9.210}$$

The surfaces parametrised by V_p^* are nothing but the bounding surfaces of V and the minus sign tells us that the the surfaces have opposite orientation, essentially meaning that both of these surfaces taken together form the bounding surface ∂V, see Fig. 9.21, and therefore

$$\int_V d\omega = \oint_{\partial V} \omega, \tag{9.211}$$

which is Stokes' theorem for integration of differential forms.

Example 9.35 All of the integral theorems that we encountered in Chapter 1 are special cases of Stokes' theorem. Let us recover them one by one! For $p = 0$, we have a zero-form f, i.e., a scalar, and V is a one-dimensional curve γ. We have already seen the result in Example 9.32

$$\int_\gamma df = f(p) - f(q), \tag{9.212}$$

where q is the starting point and p the endpoint of γ and thus its boundary. Restricting ourselves to a three-dimensional Euclidean space in Cartesian coordinates, we can consider a one-form $\omega = \omega_i dx^i$ and integrate its exterior derivative over a two-dimensional surface S.

Looking at any integral with the two-form $dx^i \wedge dx^j$, we find that

$$(dx^i \wedge dx^j)(\dot{\gamma}_t, \dot{\gamma}_s) dt\, ds = \frac{1}{2} \varepsilon_{ijk} \varepsilon_{k\ell m} (dx^\ell \wedge dx^m)(\dot{\gamma}_t, \dot{\gamma}_s) dt\, ds$$

$$= \frac{1}{2} \varepsilon_{ijk} \varepsilon_{k\ell m} \left(\frac{\partial x^\ell}{\partial t} \frac{\partial x^m}{\partial s} - \frac{\partial x^\ell}{\partial s} \frac{\partial x^m}{\partial t} \right) dt\, ds$$

$$= \varepsilon_{ijk} \varepsilon_{k\ell m} \frac{\partial x^\ell}{\partial t} \frac{\partial x^m}{\partial s} dt\, ds = \varepsilon_{ijk} dS_k, \tag{9.213}$$

where we have used t and s as curve parameters. The integral over the surface S can therefore be rewritten according to

$$\int_S d\omega = \int_S (\partial_i \omega_j) dx^i \wedge dx^j = \int_S (\partial_i \omega_j) \varepsilon_{ijk} dS_k = \int_S (\nabla \times \vec{\omega}) \cdot d\vec{S}. \tag{9.214a}$$

At the same time, we find that

$$\oint_{\partial S} \omega = \oint_{\partial S} \omega_i dx^i = \oint_{\partial S} \vec{\omega} \cdot d\vec{x} \tag{9.214b}$$

and therefore Stokes' theorem implies

$$\int_S (\nabla \times \vec{\omega}) \cdot d\vec{S} = \int_S d\omega = \oint_{\partial S} \omega = \oint_{\partial S} \vec{\omega} \cdot d\vec{x}, \tag{9.214c}$$

i.e., the *curl theorem*. Moving on to the two-form

$$\omega = \frac{1}{2} \omega_i \varepsilon_{ijk} dx^j \wedge dx^k, \tag{9.215}$$

the integral of its exterior derivative over the volume V is given by

$$\int_V d\omega = \frac{1}{2} \int_V (\partial_\ell \omega_i) \varepsilon_{ijk} dx^\ell \wedge dx^j \wedge dx^k = \frac{1}{2} \int_V (\partial_\ell \omega_i) \underbrace{\varepsilon_{ijk} \varepsilon_{\ell jk}}_{=2\delta_{i\ell}} dx^1 dx^2 dx^3$$

$$= \int_V (\partial_i \omega_i) dV = \int_V (\nabla \cdot \vec{\omega}) dV, \tag{9.216a}$$

while the integral of ω itself over the boundary surface ∂V becomes

$$\oint_{\partial V} \omega = \frac{1}{2} \oint_{\partial V} \omega_i \varepsilon_{ijk} dx^j \wedge dx^k = \frac{1}{2} \oint_{\partial V} \omega_i \underbrace{\varepsilon_{ijk} \varepsilon_{jk\ell}}_{=2\delta_{i\ell}} dS_\ell$$

$$= \oint_{\partial V} \omega_i dS_i = \oint_{\partial V} \vec{\omega} \cdot d\vec{S}. \tag{9.216b}$$

From Stokes' theorem we now obtain

$$\int_V (\nabla \cdot \vec{\omega}) dV = \int_V d\omega = \oint_{\partial V} \omega = \oint_{\partial V} \vec{\omega} \cdot d\vec{S}, \tag{9.216c}$$

which is nothing other than the *divergence theorem*.

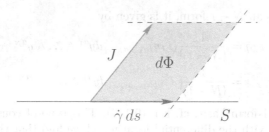

Figure 9.22 The two-dimensional case of a flux through a one-dimensional surface S, i.e., a curve. The differential flux $d\Phi$ of the tangent vector J through the surface is proportional to the volume spanned by the tangent vector $\dot{\gamma}\,ds$ and J.

9.5.4 The continuity equation revisited

A basic ingredient used to model physical systems in Chapter 3 was the *continuity equation*, derived from equating the change of an extensive property within a volume to the production minus the flux out of the volume. We are now equipped to perform the corresponding considerations in a general manifold of arbitrary dimension N.

9.5.4.1 Flux

Since we will deal with the flux out of the volume, we first need to examine the generalisation of the flux integral. In the Euclidean setting, the flux of the vector field $\vec{j}$ through the $N-1$-dimensional surface S was given by

$$\Phi = \int_S \vec{j} \cdot d\vec{S} \tag{9.217}$$

and so we expect our flux integral to reduce to this when the special case is considered. Let us assume that the flux is described by some tangent vector field J determining the direction and magnitude of the flux. If S is parametrised with the $N-1$ parameters s_i for $i = 1, 2, \ldots, N-1$, then the the flux through a small surface element should be equal to the volume spanned by J together with the tangent vectors $\dot{\gamma}_i ds_i$ (no sum), see Fig. 9.22. Knowing the physical volume N-form, we therefore find that the flux is given by

$$d\Phi = \eta(J, \dot{\gamma}_1, \ldots, \dot{\gamma}_{N-1})ds_1 \ldots ds_{N-1}. \tag{9.218}$$

Summing up the small contributions to the flux to obtain the total flux, we find that

$$\Phi = \int_S \eta(J, \dot{\gamma}_1, \ldots, \dot{\gamma}_{N-1})ds_1 \ldots ds_{N-1}. \tag{9.219}$$

This is exactly the result we would obtain from integrating the $N-1$ form $i_J\eta$ given by

$$i_J\eta(X_1, \ldots, X_{N-1}) \equiv \eta(J, X_1, \ldots, X_{N-1}) \tag{9.220a}$$

whose components are

$$i_J\eta(\partial_{a_1}, \ldots, \partial_{a_{N-1}}) = J^b \eta_{ba_1 \ldots a_{N-1}}. \tag{9.220b}$$

Note that since $i_J \eta$ is an $N - 1$-form, it is given by

$$i_J \eta = \frac{1}{(N-1)!}(i_J \eta)_{a_1 \ldots a_{N-1}} dy^{a_1} \wedge \ldots \wedge y^{a_{N-1}}$$

$$= \frac{1}{(N-1)!} J^b \eta_{ba_1 \ldots a_{N-1}} dy^{a_1} \wedge \ldots \wedge y^{a_{N-1}} \tag{9.221}$$

in terms of the $N - 1$-form basis, cf. Eq. (9.174). This type of construction is called the *interior product* of J with the differential form η and we find that the flux integral can be written as

$$\Phi = \int_S i_J \eta, \tag{9.222}$$

where η is the physical volume form.

Example 9.36 Let us make sure that the above definition of the flux integral gives back the expression we have previously used in a three-dimensional Euclidean space with Cartesian coordinates. In this case, the physical volume element is just $\eta_{ijk} = \varepsilon_{ijk}$, which means that

$$\int_S i_J \eta = \frac{1}{2} \int_S J^i \varepsilon_{ijk} dx^j \wedge dx^k = \frac{1}{2} \int_S J^i \underbrace{\varepsilon_{ijk}\varepsilon_{jkm}}_{=2\delta_{im}} dS_m = \int_S J^i dS_i = \int_S \vec{J} \cdot d\vec{S} \tag{9.223}$$

as expected.

In addition to reducing to the expected formula in the Euclidean setting, we also note that the flux is going to be zero if J is linearly dependent on the tangent vectors $\dot{\gamma}_i$ of the surface S, i.e., if the J is parallel to the surface.

9.5.4.2 Production, concentration, and continuity

Just as we did when deriving the continuity equation in Chapter 3, we assume that the production of the extensive property within a volume V is given by a source density κ and that the extensive property Q itself is related to an intensive property q, which we will refer to as the concentration. We can express Q and its production K in V as the integrals

$$Q = \int_V q\eta \quad \text{and} \quad K = \int_V \kappa\eta, \tag{9.224}$$

respectively. We can now go through exactly the same motions as we originally did when deriving the continuity equation. The time derivative of Q is going to be given by

$$\frac{dQ}{dt} = \int_V \frac{\partial q}{\partial t}\eta \tag{9.225}$$

as long as we select a volume V that does not change with time. The continuity assumption is now given by $dQ/dt = K - \Phi$, leading to

$$\int_V \frac{\partial q}{\partial t}\eta - \int_V \kappa\eta + \oint_{\partial V} i_J \eta = \int_V \left(\frac{\partial q}{\partial t}\eta - \kappa\eta + di_J \eta\right) = 0, \tag{9.226}$$

where we have used Stokes' theorem for the flux integral. For this to hold for any volume V, we obtain the continuity requirement

$$\frac{\partial q}{\partial t}\eta + di_J \eta = \kappa\eta. \tag{9.227}$$

This may be simplified to get rid of the volume form η. We start by expressing the exterior derivative of $i_J\eta$ as

$$
\begin{aligned}
di_J\eta &= \frac{1}{(N-1)!}\partial_{a_1}\left(J^b\sqrt{g}\,\varepsilon_{ba_2\ldots a_N}\right)dy^{a_1}\wedge\ldots\wedge dy^{a_N}\\
&= \frac{1}{(N-1)!}\varepsilon_{ba_2\ldots a_N}\varepsilon^{a_1\ldots a_N}\left(\sqrt{g}\,\partial_{a_1}J^b + J^b\partial_{a_1}\sqrt{g}\right)dy^1\wedge\ldots\wedge dy^N\\
&= \delta_b^{a_1}\left(\sqrt{g}\,\partial_{a_1}J^b + J^b\partial_{a_1}\sqrt{g}\right)dy^1\wedge\ldots\wedge dy^N\\
&= \left(\sqrt{g}\,\partial_b J^b + J^b\partial_b\sqrt{g}\right)dy^1\wedge\ldots\wedge dy^N\\
&= (\partial_b J^b)\eta + J^b[\partial_b\ln(\sqrt{g})]\eta.
\end{aligned}
\tag{9.228}
$$

Using the relation $\Gamma^b_{ab} = \partial_a\ln(\sqrt{g})$, see Problem 2.18, we can rewrite this as

$$
di_J\eta = (\partial_b J^b + \Gamma^b_{ab}J^a)\eta = (\nabla_b J^b)\eta.
\tag{9.229}
$$

Inserted into Eq. (9.227), we find the continuity equation

$$
\frac{\partial q}{\partial t} + \nabla_b J^b = \kappa,
\tag{9.230}
$$

which is of exactly the same form as the continuity equation we are familiar with, where the divergence $\nabla\cdot\vec{\jmath}$ has been replaced by the generalised divergence $\nabla_b J^b$.

Example 9.37 Taking the case of diffusion into consideration, *Fick's law* may be generalised to

$$
J^a = -\lambda g^{ab}\partial_b q.
\tag{9.231}
$$

This leads us to the divergence of the current

$$
\nabla_a J^a = -\lambda g^{ab}\nabla_a\partial_b q = -\lambda g^{ab}(\partial_a\partial_b - \Gamma^c_{ab}\partial_c)q \equiv -\lambda\nabla^2 q,
\tag{9.232}
$$

where we have introduced the *generalised Laplace operator*

$$
\nabla^2 = g^{ab}(\partial_a\partial_b - \Gamma^c_{ab}\partial_c)
\tag{9.233}
$$

that was also mentioned in Eq. (2.101).

9.6 EMBEDDINGS

Although a manifold, such as the sphere, is well defined without considering it as a subspace of a higher-dimensional space, it is often of interest to do so. In particular, when embedding a manifold in a higher-dimensional space it will turn out that some properties, such as the metric tensor, may be inherited from it. Before we can discuss this properly, we need to go through some of the underlying concepts. We start by looking at a map f from a manifold M to another manifold N. For any given point p in M, the map f induces a natural map f_*, called the *pushforward*, between the tangent spaces T_pM and $T_{f(p)}N$. As a tangent vector is a directional derivative, the pushforward should bring a directional derivative of functions on M to a directional derivative of functions on N. Taking a vector V in T_pM, we define the pushforward by its action on a function φ on N as

$$
(f_*V)\varphi = V(\varphi\circ f),
\tag{9.234}
$$

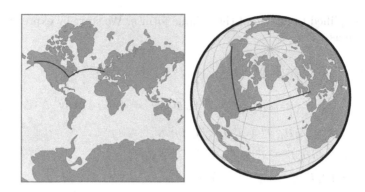

Figure 9.23 The projection of the geodesics going from New York to Paris and from New York to Anchorhage, respectively, onto the two-dimensional chart of the Earth's surface given by the Mercator projection and the corresponding three-dimensional paths in $\mathbb{R}^3$ shown together with the natural embedding of the Earth's surface.

where $\varphi \circ f$ is the composite function of φ and f such that

$$(\varphi \circ f)(p) = \varphi(f(p)). \tag{9.235}$$

Since φ is a function on N and f is a function from M to N, this composition is a function on M, which we may act upon with the vector V. If we introduce coordinates y^a on M and z^μ on N, the chain rule results in

$$(f_*V)^\mu \partial_\mu \varphi = V^a \partial_a (\varphi \circ f) = V^a \frac{\partial z^\mu}{\partial y^a} \partial_\mu \varphi, \tag{9.236}$$

where $\partial z^\mu / \partial y^a$ are the partial derivatives of the coordinates z^μ with respect to y^a given the function f. We can here identify the components of f_*V as

$$(f_*V)^\mu = V^a \frac{\partial z^\mu}{\partial y^a}. \tag{9.237}$$

Example 9.38 Since this definition of the pushforward is rather abstract, let us see how it works in practice when considering the tangent vector of a curve γ in M. By definition, γ is a function taking a single parameter t and mapping it to M. If we have a map f from M to N, there is a corresponding curve $\tilde{\gamma} = f \circ \gamma$, which is the composition of f and γ and therefore maps t to N instead, see Fig. 9.23. With the coordinates y^a on M and z^μ on N, the tangent vector of γ has the components

$$\dot{\gamma}^a = \frac{dy^a}{dt}, \tag{9.238a}$$

while the tangent vector of $\tilde{\gamma}$ has the components

$$\dot{\tilde{\gamma}}^\mu = \frac{dz^\mu}{dt}. \tag{9.238b}$$

Using the relation we found for the pushforward, we find that

$$(f_*\dot{\gamma})^\mu = \dot{\gamma}^a \frac{\partial z^\mu}{\partial y^a} = \frac{dy^a}{dt} \frac{\partial z^\mu}{\partial y^a} = \frac{dz^\mu}{dt} = \dot{\tilde{\gamma}}^\mu. \tag{9.239}$$

In other words, the pushforward of $\dot{\gamma}$ is just the tangent vector of $\tilde{\gamma}$.

Example 9.39 Consider the natural map from the sphere S^2 to $\mathbb{R}^3$ given by

$$x^1 = R\sin(\theta)\cos(\varphi), \quad x^2 = R\sin(\theta)\sin(\varphi), \quad \text{and} \quad x^3 = R\cos(\theta), \tag{9.240}$$

where R is a constant, θ and φ the coordinates on the sphere, and the x^i are Cartesian coordinates on $\mathbb{R}^3$. Taking a curve γ in S^2 given by

$$\theta(t) = t + \theta_0, \quad \varphi(t) = \varphi_0, \tag{9.241}$$

where θ_0 and φ_0 are constants, the corresponding tangent vector $\dot{\gamma}$ is given by

$$\dot{\gamma} = \dot{\gamma}^a \partial_a = \partial_\theta, \tag{9.242}$$

or in other terms $\dot{\gamma}^\theta = 1$ and $\dot{\gamma}^\varphi = 0$. It follows that the pushforward of this vector is given by

$$\dot{\tilde{\gamma}} = \frac{\partial x^i}{\partial \theta}\partial_i = R[\cos(\theta)\cos(\varphi)\partial_1 + \cos(\theta)\sin(\varphi)\partial_2 - \sin(\theta)\partial_3]. \tag{9.243}$$

Identifying this with the basis vectors of $\mathbb{R}^3$ expressed in spherical coordinates, see Eq. (1.209), we find that

$$\dot{\tilde{\gamma}} = R\vec{e}_\theta = \vec{E}_\theta. \tag{9.244}$$

With the function f inducing a map from $T_p M$ to $T_{f(p)} N$, we can also construct a map from dual vectors on N to dual vectors on M. For a dual vector ω in $T_{f(p)}^* N$, we define the *pullback* $f^*\omega$ by its action on any tangent vector X in $T_p M$ according to

$$f^*\omega(X) = \omega(f_* X). \tag{9.245}$$

This is well defined as the pushforward $f_* X$ is a tangent vector in $T_{f(p)} N$. For the components of $f^*\omega$ in the coordinate systems y^a defined above, we find that

$$(f^*\omega)_a = f^*\omega(\partial_a) = \omega(f_* \partial_a) = \frac{\partial z^\mu}{\partial y^a}\omega_\mu. \tag{9.246}$$

Thus, the pullback transforms with exactly the same coefficients $\frac{\partial z^\mu}{\partial y^a}$ as the pushforward, but with the other index contracted.

Although both the pushforward and pullback will always be defined, they may not be invertible. However, if one of them is invertible, so is the other. Any map f that induces an invertible pushforward and pullback is called an *immersion* of M into N. If, in addition, f itself is invertible, then it is an *embedding* of M in N. For any embedding, the image $f(M)$ in N is called a *submanifold* of N.

Example 9.40 Let us look at one of the most straightforward examples of a map f that is *not* an immersion. We consider the map from the real line $\mathbb{R}$ to $\mathbb{R}^2$ with global coordinate systems t and x^1, x^2, respectively, such that

$$x^1(t) = t^2 \quad \text{and} \quad x^2(t) = t^3. \tag{9.247}$$

For any tangent X vector at $t = 0$, this implies that the pushforward is given by

$$f_* X = X^t \frac{dx^1}{dt} \partial_1 + X^t \frac{dx^2}{dt} \partial_2 = 0. \tag{9.248}$$

Since any vector is mapped to the zero vector by the pushforward, it is clearly not invertible, even though f itself is.

Example 9.41 We now look for an example where we have a map f that is an immersion, but not an embedding. We do so by looking at a map from the circle S^1, parametrised by an angle θ, to $\mathbb{R}^2$, again with the coordinates x^1 and x^2, defined by

$$x^1(\theta) = \cos(\theta) \quad \text{and} \quad x^2(\theta) = \sin(2\theta). \tag{9.249}$$

The pushforward of a vector X in $T_\theta S^1$ with the component X^θ is given by

$$f_* X = X^\theta \left(\frac{dx^1}{d\theta} \partial_1 + \frac{dx^2}{d\theta} \partial_2 \right) = X^\theta (-\sin(\theta)\partial_1 + 2\cos(2\theta)\partial_2). \tag{9.250}$$

Since $\sin(\theta) = 0$ only when θ is an integer multiple of π and $\cos(2\pi n) = 1$ when n is an integer, the pushforward is always invertible. The function f itself is not invertible since the points $\theta = \pi/2$ and $\theta = 3\pi/2$ both map to

$$x^1 = \cos\left(\frac{\pi}{2}\right) = \cos\left(\frac{3\pi}{2}\right) = 0 \quad \text{and} \quad x^2 = \sin(\pi) = \sin(3\pi) = 0. \tag{9.251}$$

Note that while $\theta = \pi/2$ and $\theta = 3\pi/2$ map to the same point $x^1 = x^2 = 0$ in $\mathbb{R}^2$, the pushforward of the respective tangent spaces are linearly independent. Because of this, f is an immersion, but not an embedding of the circle S^1 in $\mathbb{R}^2$. However, it is perfectly possible to embed the circle in the plane, the most common embedding being given by

$$x^1(\theta) = R\cos(\theta) \quad \text{and} \quad x^2 = R\sin(\theta), \tag{9.252}$$

which maps the circle to the set of points in $\mathbb{R}^2$ which are a distance R away from the origin, see Fig. 9.24. In the same way, the map from the sphere to $\mathbb{R}^3$ discussed in Example 9.39 is also an embedding.

The pushforward and pullback may be generalised to higher order contravariant and covariant tensors, respectively. For example, the pullback of a type $(0,2)$ tensor ω to M is given by

$$f^* \omega(X, Y) = \omega(f_* X, f_* Y). \tag{9.253}$$

In particular, this works very well if we have a metric g on N and wish to select a suitable metric on M as well. If f is an embedding, the pullback tensor $f^* g$ will be a metric on M called the *induced metric*.

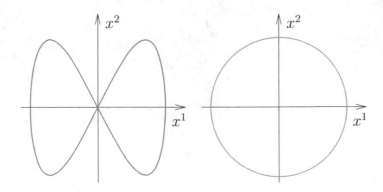

Figure 9.24 The images of the two mappings from the circle S^1 to $\mathbb{R}^2$ discussed in Example 9.41. The left figure shows an immersion that is not an embedding due to two points on the circle mapping to the origin of $\mathbb{R}^2$. The right figure shows the canonical embedding, where S^1 is mapped to a circle of constant radius.

We have already seen that the metric tensor is directly related to the path length of curves through the relation

$$s_\gamma = \int_\gamma ds = \int_0^1 \sqrt{g(\dot\gamma, \dot\gamma)}\, dt, \tag{9.254}$$

where we have assumed the curve parameter to go from 0 to 1. So how does this relate to the induced metric? Consider the curve γ in M with the induced metric f^*g, where g is a metric on N. The curve length of γ in M is now given by

$$s_\gamma = \int_0^1 \sqrt{f^*g(\dot\gamma, \dot\gamma)}\, dt = \int_0^1 \sqrt{g(f_*\dot\gamma, f_*\dot\gamma)}\, dt = \int_0^1 \sqrt{g(\dot{\tilde\gamma}, \dot{\tilde\gamma})}\, dt = s_{\tilde\gamma}, \tag{9.255}$$

where $\tilde\gamma$ is the curve $f \circ \gamma$ in N. The path length of a curve in M can therefore be found by computing the length of the corresponding path in N induced by the map f. Considering that the line element ds can be written in terms of the metric tensor components

$$ds^2 = g_{ab}dy^a dy^b = g_{\mu\nu}\frac{\partial z^\mu}{\partial y^a}\frac{\partial z^\nu}{\partial y^b}dy^a dy^b = g_{\mu\nu}dz^\mu dz^\nu, \tag{9.256}$$

where we have used g_{ab} to denote the components of f^*g and $g_{\mu\nu}$ for the components of the original metric g, we find that the easiest way of computing the components of the induced metric is to take the line element and restrict it to displacements within the submanifold.

Example 9.42 Consider the embedding of the sphere into $\mathbb{R}^3$ discussed in Example 9.39. The metric tensor on $\mathbb{R}^3$ is given by

$$g_{11} = g_{22} = g_{33} = 1 \tag{9.257a}$$

and the off-diagonal terms all being equal to zero. Consequently, the line element is given by

$$ds^2 = g_{ij}dx^i dx^j = (dx^1)^2 + (dx^2)^2 + (dx^3)^2. \tag{9.257b}$$

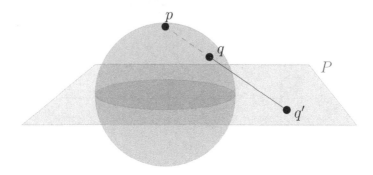

Figure 9.25 The stereographic projection based on the point p is constructed by projecting a point q to the the point q' in the plane P by where a straight line through p and q passes through P. The stereographic projection is a chart that covers all of the sphere S^2 except the point p itself.

Expressing the differentials dx^i in terms of the differentials $d\theta$ and $d\varphi$, we find that

$$dx^1 = R[\cos(\theta)\cos(\varphi)d\theta - \sin(\theta)\sin(\varphi)d\varphi], \tag{9.258a}$$
$$dx^2 = R[\cos(\theta)\sin(\varphi)d\theta + \sin(\theta)\cos(\varphi)d\varphi], \tag{9.258b}$$
$$dx^3 = -R\sin(\theta)d\theta. \tag{9.258c}$$

Squaring and inserting into the expression for the line element now results in

$$ds^2 = R^2[d\theta^2 + \sin^2(\theta)d\varphi^2] \tag{9.259}$$

from which we can directly identify the components

$$g_{\theta\theta} = R^2, \quad g_{\varphi\varphi} = R^2\sin^2(\theta), \quad \text{and} \quad g_{\theta\varphi} = g_{\varphi\theta} = 0 \tag{9.260}$$

of the induced metric. If we use $R = 1$, i.e., if we consider the unit sphere in $\mathbb{R}^3$, this exactly corresponds to the metric which was first introduced in Example 9.22.

9.7 PROBLEMS

Problem 9.1. We can construct a set of coordinates on the two-dimensional sphere covering everything but a single point through the *stereographic projection*. It is produced by selecting a point p on the sphere and the plane P through the sphere's center which is orthogonal to the radius going to p. For any other point q on the sphere, a straight line through p and q will cross P at a single point q' which is unique for each q, see Fig. 9.25. We can therefore use the coordinates x^1 and x^2 in P as a coordinate system for the sphere. Derive the set of coordinate transformations going from the usual coordinates on the sphere θ and φ to x^1 and x^2.

Problem 9.2. The set of possible positions for a double pendulum, see Fig. 2.10, can be described as a two-dimensional manifold. Construct an atlas for this manifold and explicitly write down the coordinate transformations between the different charts.

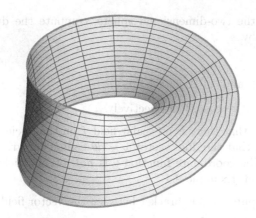

Figure 9.26   The Möbius strip is a non-orientable two-dimensional manifold for which Problem 9.3 asks for the construction of an atlas.

Problem 9.3. The *Möbius strip*, see Fig. 9.26, is a two-dimensional example of a non-orientable manifold. Construct an atlas for the Möbius strip and explicitly write down the coordinate transformations in the regions where the charts overlap.

Problem 9.4. Consider the curve given by

$$\theta(t) = \frac{\pi}{2} + \cos(t) \quad \text{and} \quad \varphi(t) = t \tag{9.261}$$

in the regular coordinates on a two-dimensional sphere. Compute the components of the tangent vector of this curve at an arbitrary point along the curve in the given coordinates as well as in the stereographic coordinates introduced in Problem 9.1.

Problem 9.5. Show that the components $df_a = \partial_a f$ of the differential df transform as you would expect the components of a dual vector to transform under arbitrary coordinate transformations.

Problem 9.6. Verify that the maps $\omega + \xi$ and $a\omega$ defined in Eqs. (9.25) are dual vectors, i.e., that they are linear maps from the tangent vector space $T_p M$ to real numbers.

Problem 9.7. On the two-dimensional sphere, consider the vector fields X and Y with the components

$$X^\theta = 0, \ X^\varphi = 1, \quad \text{and} \quad Y^\theta = \sin(\varphi), \ Y^\varphi = \cot(\theta)\cos(\varphi), \tag{9.262}$$

respectively. Find the scalar fields resulting from letting these vector fields act on the scalar fields

$$f_1 = \cos(\theta), \quad f_2 = \sin(\varphi). \tag{9.263}$$

Problem 9.8. Determine how the term $X^a(\partial_a Y^b)\partial_b$ appearing in the expression for XY in Eq. (9.40) transforms under coordinate transformations and use the result to argue why it cannot be a component of a tangent vector Z. Having done so, determine the transformation properties of the Lie bracket $[X, Y] = XY - YX$ and argue that it is a tangent vector.

Problem 9.9. Compute the flow lines of the vector field X on the two-dimensional sphere as defined in Problem 9.7 starting at an arbitrary point $\theta(0) = \theta_0$ and $\varphi(0) = \varphi_0$. Also write down the differential equations that need to be satisfied for the flow lines of the vector field Y.

Problem 9.10. On the two-dimensional sphere, compute the dual vector df when the scalar field f is given by

   a) $f = \sin(\theta)$,

   b) $f = \cos(\theta)\sin(\varphi)$, and

   c) $f = \cos(2\theta)[\sin(2\varphi) - \cos(\varphi)]$, respectively.

Problem 9.11. Show that there is no scalar field f on the circle for which df is non-zero everywhere and, thus, that the dual vector $\omega = d\theta$ cannot be written as the differential of a scalar field. *Note:* The coordinate function θ is *not* a scalar field on the full circle, only on a particular chart, cf. Example 9.4.

Problem 9.12. Compute the Lie bracket between the vector fields X and Y given by:

   a) $X = \partial_\theta$ and $Y = [1/\sin(\theta)]\partial_\varphi$

   b) $X = [1/(1 + (x^2)^2)]\partial_1$ and $Y = [1/(1 + (x^1)^2)]\partial_2$

Problem 9.13. Given N vector fields of the form $X_i = f_i(y^i)\partial_i$ (no sum), on an N-dimensional manifold, show that all of them Lie commute with each other, i.e., $[X_i, X_j] = 0$.

Problem 9.14. Use the product rule

$$\nabla_X(T_1 \otimes T_2) = (\nabla_X T_1) \otimes T_2 + T_1 \otimes (\nabla_X T_2), \tag{9.264}$$

where T_1 and T_2 are arbitrary tensors to deduce the components of the result of ∇_X acting on a type (n, m) tensor.

Problem 9.15. In this chapter, we have seen two different connections on the two-dimensional sphere, one which was based on a particular set of vector fields being parallel and one which was the Levi-Civita connection based on the natural embedding of the sphere in a three-dimensional Euclidean space. In order to illustrate that vector fields that are parallel with respect to one affine connection are not necessarily parallel with respect to others, compute $\nabla_\theta X$, $\nabla_\theta Y$, $\nabla_\varphi X$, and $\nabla_\varphi Y$, where the connection is the Levi-Civita connection with the non-zero Christoffel symbols

$$\Gamma^\theta{}_{\varphi\varphi} = -\sin(\theta)\cos(\theta) = -\frac{1}{2}\sin(2\theta) \quad \text{and} \quad \Gamma^\varphi{}_{\theta\varphi} = \Gamma^\varphi{}_{\varphi\theta} = \cot(\theta) \tag{9.265}$$

and the vector fields X and Y are given by

$$X = \partial_\theta \quad \text{and} \quad Y = \frac{1}{\sin(\theta)}\partial_\varphi, \tag{9.266}$$

respectively.

Problem 9.16. Verify that Eqs. (9.84) hold and use them to conclude that the torsion $T(X, Y)$ only depends on the components of X and Y and not on their derivatives.

Problem 9.17. In an N-dimensional manifold M with an affine connection, a map $f_p(X)$ from T_pM to M can be defined in such a way that $f_p(X)$ is the point given by following the geodesic with the tangent vector X at p while increasing the geodesic parameter by one. Given a set of N linearly independent vectors X_i at the point p in an N-dimensional manifold M, verify that $f_p(X)$ defines a local coordinate system on M by mapping the N coordinates t^i to M according to

$$\phi(t^1, \ldots, t^N) = f_p(t^i X_i). \tag{9.267}$$

In other words, show that this map is locally invertible.

Problem 9.18. For an arbitrary coordinate change, deduce the transformation rules of the curvature tensor components specified in Eq. (9.103b). By doing so, verify that they transform in exactly the way you would expect a type $(1,3)$ tensor to transform under coordinate changes.

Problem 9.19. For a general affine connection that may not be torsion free, write down the condition on the deviation $\tilde{\Gamma}^c_{ab}$ of the connection coefficients from the Christoffel symbols

$$\tilde{\Gamma}^c_{ab} = \Gamma^c_{ab} - \frac{1}{2}g^{cd}\left(\partial_a g_{db} + \partial_b g_{ad} - \partial_d g_{ab}\right) \tag{9.268}$$

that must be satisfied in order for the connection to be metric compatible.

Problem 9.20. In Example 9.15, we introduced an affine connection on the sphere for which the only non-zero connection coefficient was $\Gamma^\varphi_{\theta\varphi} = \cot(\theta)$. Show that this connection is compatible with the standard metric

$$g_{\theta\theta} = 1, \quad g_{\varphi\varphi} = \sin^2(\theta), \quad \text{and} \quad g_{\theta\varphi} = g_{\varphi\theta} = 0 \tag{9.269}$$

given by the embedding of the sphere in $\mathbb{R}^3$.

Problem 9.21. In the punctured three-dimensional space for which $r > 0$ and $\sin(\theta) \neq 0$ in spherical coordinates, define a connection such that the vector fields

$$X_1 = \partial_r, \quad X_2 = \frac{1}{r}\partial_\theta, \quad \text{and} \quad X_3 = \frac{1}{r\sin(\theta)}\partial_\varphi \tag{9.270}$$

are all parallel. Show that this connection is compatible with the standard Euclidean metric and compute the connection coefficients and the components of the torsion tensor in spherical coordinates.

Problem 9.22. When an N-dimensional manifold is endowed with a metric tensor, we can define a *vielbein* as a set of N tangent vectors X_i such that

$$g(X_i, X_j) = \delta_{ij}. \tag{9.271}$$

When working in a specific set of coordinates, the vielbein vector fields may be expressed through their components e^a_i

$$X_i = e^a_i \partial_a. \tag{9.272}$$

Show that any vector $Y = Y^a \partial_a$ can be written as a linear combination $Y = Y^i X_i$ and express the numbers Y^i in terms of the coordinate components Y^a and the vielbein components e^a_i.

Problem 9.23. Use the vielbein from Problem 9.22 in order to construct a local coordinate system at the point p based on the procedure in Problem 9.17. Using the Levi-Civita connection, show that the first order partial derivatives of the metric components vanish at p in this coordinate system and that the Christoffel symbols are therefore all zero at p. These coordinates are called *normal coordinates* at p.

Problem 9.24. Verify that the Lie bracket satisfies the *Jacobi identity*

a) $[X,[Y,Z]] + [Y,[Z,X]] + [Z,[X,Y]] = 0$.

In addition to its symmetries, the curvature tensor has some additional properties. In particular, show that

b) $R(X,Y)Z + R(Y,Z)X + R(Z,X)Y = 0$ and

c) $(\nabla_X R)(Y,Z)W + (\nabla_Y R)(Z,X)W + (\nabla_Z R)(X,Y)W = 0$

in the case of a torsion free connection. These relations are known as the first and second *Bianchi identities*, respectively.

d) Write down the relations between the components of the curvature tensor that correspond to the Bianchi identities.

Problem 9.25. Determine the number of independent components of the curvature tensor using the following line of argumentation:

a) Determine the number of independent components that are allowed by the symmetries of the curvature tensor.

b) The component form of the first Bianchi identity introduced in Problem 9.24(d) can be used to further constrain the number of independent components. For each choice of the indices in the first Bianchi identity which is not trivially satisfied by the curvature tensor symmetries, an additional constraint is implied and the number of independent components decreases by one. Determine the number of such additional constraints.

c) Find the total number of independent constraints by comparing your results from (a) and (b).

Problem 9.26. The length of a curve γ in a manifold with a metric is given by

$$s_\gamma = \int_0^1 \sqrt{g(\dot\gamma, \dot\gamma)}\, dt, \tag{9.273}$$

where we have assumed that the curve parameter goes from zero to one. Show that this curve length is independent of the parametrisation of the curve, i.e., if we select a different curve parameter s such that $s = s(t)$ is a monotonically increasing function, then

$$s_\gamma = \int_{s(0)}^{s(1)} \sqrt{g(\gamma', \gamma')}\, ds, \tag{9.274}$$

where γ' is the tangent vector of the curve when parametrised by s.

Problem 9.27. Show that the inverse metric tensor g^{ab} is a parallel tensor field for any metric compatible connection and that the Kronecker delta δ_a^b is always a parallel tensor field regardless of the connection.

Problem 9.28. Using the standard coordinates θ and φ on the unit sphere with the standard metric, write down the generalised Laplace operator and show that it corresponds exactly to the operator $-\hat\Lambda$ appearing in the Laplace operator in spherical coordinates on $\mathbb{R}^3$

$$\nabla^2 = \frac{1}{r^2}\partial_r r^2 \partial_r - \frac{1}{r^2}\hat\Lambda. \tag{9.275}$$

Problem 9.29. In a manifold M with a metric g, consider a set of geodesics $\gamma_X(t)$ of length s_0 originating at the point p such that the unit tangent vectors X at p vary continuously with the parameter t. The resulting curve endpoints of the geodesics form a curve $\tilde\gamma_{s_0}(t)$ in M. Show that this curve is orthogonal to the geodesics, i.e., that

$$g(\dot\gamma_X(s_0), \dot{\tilde\gamma}_{s_0}) = 0 \tag{9.276}$$

for all t.

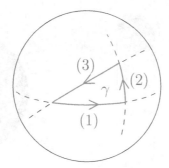

Figure 9.27 In Problem 9.32 we consider the parallel transport of a vector X around the loop γ formed by the three curves (1), (2), and (3).

Problem 9.30. The scalar curvature S at a point p in a two-dimensional manifold with a metric g can be computed as

$$S = -6 \, \frac{d^2}{d\rho^2} \left(\frac{\ell_p(\rho)}{2\pi\rho} \right) \bigg|_{\rho=0}, \tag{9.277}$$

where $\ell_p(\rho)$ is the circumference of a circle of radius ρ around p, which is defined as the set of points that are a distance ρ from p (defined as in Problem 9.29). Compute the scalar curvature of a unit sphere and verify that it is equal to the Ricci scalar.

Problem 9.31. Find the expressions for the components of the metric tensor on the two-dimensional sphere in stereographic coordinates, see Problem 9.1, and use it to derive the Christoffel symbols of the Levi-Civita connection in those coordinates.

Problem 9.32. On the two-dimensional sphere with the standard metric and the Levi-Civita connection, consider the parallel transport of the vector $X = X^\theta \partial_\theta + X^\varphi \partial_\varphi$ around the curve γ given by the three segments

$$(1): \quad \theta(t) = \frac{\pi}{2}, \quad \varphi(t) = t\varphi_0, \qquad\qquad (0 \leq t \leq 1) \tag{9.278a}$$

$$(2): \quad \theta(t) = \frac{\pi}{2} - t\theta_0, \quad \varphi(t) = \varphi_0, \qquad (0 \leq t \leq 1) \tag{9.278b}$$

$$(3): \quad \theta(t) = \frac{\pi}{2} - (1-t)\theta_0, \quad \varphi(t) = \varphi_0(1-t), \qquad (0 \leq t \leq 1) \tag{9.278c}$$

in that order, see Fig. 9.27. Find the angle between the original vector X and the vector that has been parallel transported once around γ. How does your result change if you instead use the connection defined in Example 9.15?

Problem 9.33. In Problem 8.29, you should have found a parametrisation of a general great circle on the sphere. Check whether or not this parametrisation leads to a tangent vector of constant length. If it does not, find a parametrisation of the great circle such that the length of the tangent vector is constant.

Problem 9.34. In the case of the Levi-Civita connection, a *Killing vector field* is a vector field X that satisfies the relation

$$g(\nabla_Y X, Z) + g(Y, \nabla_Z X) = 0 \tag{9.279}$$

for all other vector fields Y and Z.

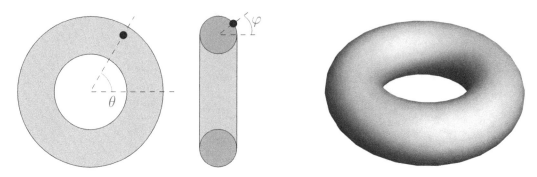

Figure 9.28 The coordinates θ and φ on the torus, here shown embedded in $\mathbb{R}^3$ as described in Problem 9.36.

a) Show that the Killing vector field satisfies

$$\nabla_a X_b + \nabla_b X_a = 0 \tag{9.280}$$

in any coordinate system and that this is also a sufficient condition for the vector field X to be a Killing vector field.

b) Show that the Lie bracket of any two Killing vector fields is also a Killing vector field.

c) For any geodesic γ with an affine parameter t, show that

$$Q = g(X, \dot{\gamma}), \tag{9.281}$$

where $\dot{\gamma}$ is the tangent vector of γ, is constant along the geodesic.

Problem 9.35. Combining Problems 9.33 and 9.34, verify that the vector field

$$X = \partial_\varphi \tag{9.282}$$

is a Killing vector field and compute the conserved quantity $Q = g(X, \dot{\gamma})$ for the vector field X when γ is a great circle parametrised by its curve length.

Problem 9.36. A two-dimensional *torus*, denoted T^2, is a manifold that may be parametrised using two cyclic angular coordinates θ and φ, see Fig. 9.28. For two constants $R > \rho > 0$, we can define an embedding of the torus into $\mathbb{R}^3$ as

$$x^1 = [R + \rho\sin(\varphi)]\cos(\theta), \tag{9.283a}$$
$$x^2 = [R + \rho\sin(\varphi)]\sin(\theta), \tag{9.283b}$$
$$x^3 = \rho\cos(\varphi). \tag{9.283c}$$

Based on this embedding, compute the following:

a) The components g_{ab} of the induced metric on the torus resulting from the standard Euclidean metric in $\mathbb{R}^3$.

b) The Christoffel symbols of the Levi-Civita connection.

c) The independent non-zero component $R_{\theta\varphi\theta\varphi}$ of the curvature tensor.

d) The components of the Ricci tensor R_{ab} and the Ricci scalar R. Verify that R is not constant and that it has a different sign on different points in the manifold.

In addition, argue for why the requirement $\rho < R$ is necessary for the mapping from T^2 to $\mathbb{R}^3$ to be an embedding. *Note:* If you are worried about having a proper atlas, we can construct charts for the torus in much the same way as we created charts for the circle in Example 9.4. In fact, the circle is a one-dimensional torus!

Problem 9.37. In an arbitrary coordinate system with coordinates y^a on a four-dimensional manifold, write down the coordinate basis for all p-forms for $0 \leq p \leq 4$.

Problem 9.38. Write down an explicit expression for the components of the exterior product $\omega \wedge \xi$, where ω is a p-form and ξ an r-form, in terms of the components of ω and ξ. In particular, find how $\omega \wedge \xi$ is related to $\xi \wedge \omega$ and use your result to discuss the requirements for $\omega^2 = \omega \wedge \omega$ to be non-zero.

Problem 9.39. We defined the exterior derivative of a p-form as

$$d\omega = \frac{1}{p!} (\partial_{a_1} \omega_{a_2 \ldots a_{p+1}}) \, dy^{a_1} \wedge \ldots \wedge dy^{a_{p+1}}. \tag{9.284}$$

Normally we could not use the partial derivative ∂_{a_1} on the tensor components and obtain a new tensor, but instead we would need to use the operator ∇_{a_1}. Show that the anti-symmetry of $dy^{a_1} \wedge \ldots \wedge dy^{a_{p+1}}$ solves this problem and that using the partial derivative is equivalent to using ∇_{a_1}.

Problem 9.40. Show that acting twice with the exterior derivative on any p-form ω gives zero, i.e., that $d^2\omega = 0$.

Problem 9.41. Compute the components of the area form η on the torus defined in Problem 9.36. Also use your results from Problem 9.36 to find an expression for the generalised Laplace operator on the torus.

Problem 9.42. Working in the Euclidean space $\mathbb{R}^2$ and considering the integral of the one-form $\omega = P \, dx^1 + Q \, dx^2$ around a closed loop γ, show that Stokes' theorem

$$\oint_\gamma \omega = \int_S d\omega, \tag{9.285}$$

where S is the area bounded by γ, reduces to Green's formula in the plane.

Problem 9.43. We wish to study heat conduction in a thin isolated paraboloid that is embedded in $\mathbb{R}^3$ with the relation $x^3 = k[(x^1)^2 + (x^2)^2]$. Generalise Fourier's law to the case of heat conduction on a manifold and derive an expression for the resulting partial differential equation in polar coordinates ρ and ϕ.

Problem 9.44. For a stationary flow $J(\theta, \phi)$ on the sphere, the flow at the curves $\phi = 0$ and $\phi = \pi/2$ are given by

$$J(\theta, 0) = \sin(\theta)\partial_\theta + \partial_\varphi \quad \text{and} \quad J(\theta, \pi/2) = -\sin(\theta)\partial_\theta + \frac{\sin(2\theta)}{\sin(\theta)}\partial_\varphi, \tag{9.286}$$

respectively. Compute the total source in the region $0 < \varphi < \pi/2$.

Problem 9.45. Consider the two-dimensional plane with the line element

$$ds^2 = d\rho^2 + R_0^2 \sinh^2\left(\frac{\rho}{R_0}\right) d\phi^2 \tag{9.287}$$

in polar coordinates, where R_0 is a constant. Determine the value of the Ricci scalar for this space and write down the geodesic equations for the Levi-Civita connection. This is an example of a *hyperbolic space*.

Problem 9.46. Verify that the standard mapping of the sphere S^2 into $\mathbb{R}^3$

$$x^1 = R\sin(\theta)\cos(\varphi), \tag{9.288a}$$
$$x^2 = R\sin(\theta)\sin(\varphi), \tag{9.288b}$$
$$x^3 = R\cos(\theta) \tag{9.288c}$$

is an embedding, i.e., that both the mapping itself and its corresponding pushforward are invertible.

Problem 9.47. The punctured plane, i.e., a two-dimensional plane with a single point removed, can be mapped to $\mathbb{R}^3$ with the embedding

$$x^1 = r_0\cos(\varphi), \quad x^2 = r_0\sin(\varphi), \quad \text{and} \quad x^3 = r_0\ln\left(\frac{\rho}{r_0}\right), \tag{9.289}$$

where ρ and φ are polar coordinates on the plane centred on the removed point.

  a) Compute the induced metric in the coordinates ρ and φ on the punctured plane.

  b) Compute the Christoffel symbols of the corresponding Levi-Civita connection.

  c) Verify that the metric is flat, i.e., that the curvature tensor vanishes.

Problem 9.48. Consider the embedding of the punctured two-dimensional plane as a cone in a three-dimensional Euclidean space given by

$$x^1 = \rho\cos(\varphi), \quad x^2 = \rho\sin(\varphi), \quad x^3 = k\rho. \tag{9.290}$$

Write down an expression for the physical area form corresponding to the induced metric tensor.

Problem 9.49. For the cone submanifold discussed in Problem 9.48, show that a parallel transport of a vector around a closed loop enclosing the cone apex at $\rho = 0$ results in a net rotation and compute the corresponding rotation angle, see Fig. 9.29. Also verify that the manifold itself is flat and, consequently, any parallel transport around a closed loop that does not enclose the apex returns the the same vector.

Problem 9.50. Using your result from Problem 9.48, find the area on the surface of the cone that is enclosed by the closed curve

$$\rho(t) = r_0 + r_1\cos(t), \quad \varphi(t) = t, \tag{9.291}$$

where $0 < t < 2\pi$ and $r_1 < r_0$.

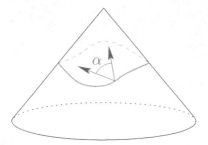

Figure 9.29 When parallel transporting a vector around a closed loop that encloses the cone apex, the resulting vector will be turned an angle α in relation to the original vector.

Problem 9.51. An alternative chart for the two-dimensional sphere embedded in $\mathbb{R}^3$ that covers all but one point on the sphere has coordinates t and s given by the relations

$$x^1 = t\frac{R}{\rho}\sin\left(\frac{\rho}{R}\right), \quad x^2 = s\frac{R}{\rho}\sin\left(\frac{\rho}{R}\right), \quad x^3 = R\cos\left(\frac{\rho}{R}\right), \tag{9.292}$$

where $\rho = \sqrt{t^2 + s^2} < R$.

a) Verify that including $\rho = R$ in the chart is not possible as all t and s satisfying this condition would map to the same point.

b) Compute the components of the metric tensor in the t and s coordinates.

Problem 9.52. For a two-dimensional submanifold M of $\mathbb{R}^3$ given by the hyperboloid

$$(x^1)^2 - (x^2)^2 - (x^3)^2 = R^2, \tag{9.293}$$

show that the pullback of the type $(0, 2)$ tensor

$$\omega = -dx^1 \otimes dx^1 + dx^2 \otimes dx^2 + dx^3 \otimes dx^3 \tag{9.294}$$

to the M is a metric although ω is not. Compute the components of this metric after introducing polar coordinates in the x^2-x^3-plane. *Note:* The tensor ω is a pseudo metric on $\mathbb{R}^3$.

Problem 9.53. The two dimensional submanifold of $\mathbb{R}^3$ defined by

$$z^2 = 4r_s(\rho - r_s) \tag{9.295}$$

in cylinder coordinates, see Fig. 9.30, is called *Flamm's paraboloid*.

a) Using the cylinder coordinates ρ and ϕ as coordinates on a chart for $z > 0$, find the components of the metric tensor on this surface.

b) Write down the geodesic equations and determine whether the coordinate lines of constant ρ or constant ϕ are geodesics.

Problem 9.54. Consider an embedding f of the manifold M_1 in the manifold M_2 and a p-form ω on M_2. Furthermore, let K be a p-dimensional subspace of M_1 and $f(K)$ its image under the embedding. Prove the integral relation

$$\int_{f(K)} \omega = \int_K f^*\omega, \tag{9.296}$$

i.e., that the integral of ω over the image of K is equal to the integral of the pullback $f^*\omega$ over K.

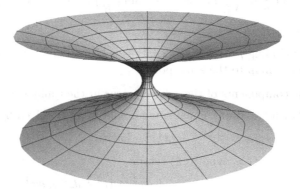

Figure 9.30 The submanifold of $\mathbb{R}^3$ called Flamm's paraboloid. It is sometimes used to illustrate the spatial geometry of a Schwarzschild black hole.

CHAPTER 10

Classical Mechanics and Field Theory

Throughout this book, we have introduced many mathematical concepts that have applications in a wide variety of advanced physics subjects, including quantum mechanics and general relativity. However, we have also seen a large number of examples where these new concepts have been applied to more familiar physics. In many cases, these examples have been taken from the field of classical mechanics. In order to wrap things up, we will therefore take some time to discuss the application of the introduced concepts to classical mechanics in more detail. Starting with a quick look at Newton's formulation, we will then treat both the Lagrangian and Hamiltonian formulations and conclude with a short introduction to field theory.

10.1 NEWTONIAN MECHANICS

As most students at this level will be familiar with, Newton's formulation of classical mechanics is based upon *Newton's laws of mechanics*:

1. A body upon which no force is acting remains at rest or in rectilinear motion.

2. The net force on a body is equal to the rate of change in its momentum

$$\vec{F} = \frac{d\vec{p}}{dt}. \tag{10.1}$$

3. The forces with which two bodies affect each other are equal in magnitude and opposite in direction

$$\vec{F}_{12} = -\vec{F}_{21}, \tag{10.2}$$

where $\vec{F}_{ij}$ is the force on body j from body i.

These axioms are assumed to hold in any *inertial frame* and different inertial frames are related through the *Galilei transformations* (see Section 4.6.2.3). Within this formulation there is an inherent assumption of an underlying homogeneous Euclidean space as well as a universal time, assumptions that would later be overthrown by the arrival of special relativity. However, Newton's mechanics is a very good approximation for most cases that we encounter in daily life and of which we have an intuitive understanding. We have already used Newton's mechanics extensively in examples and problems, but let us remind ourselves of some basic concepts in rigid body mechanics, where the dynamics of a rigid body are considered.

Example 10.1 For an object with mass m, the momentum is given by

$$\vec{p} = m\vec{v} = m\dot{\vec{x}}, \tag{10.3}$$

where $\vec{x}$ is its center of mass position. As a result of the second law, $\vec{x}$ satisfies the differential equation

$$m\ddot{\vec{x}} = \vec{F} \tag{10.4}$$

when the mass m is constant. If the net force $\vec{F} = 0$, the solution to this differential equation results in $\vec{x} = \vec{p}t/m + \vec{x}_0$, where $\vec{p}$ is the constant momentum and $\vec{x}_0$ the initial position. This describes a rectilinear motion with velocity $\vec{v} = \vec{p}/m$, in accordance with the first law.

Example 10.2 Apart from providing a way of relating how two objects affect each other, the third law ensures the conservation of overall momentum as long as all systems are considered. If the only forces on two objects are those acting between them, the rate of change of the total momentum $\vec{p} = \vec{p}_1 + \vec{p}_2$ is given by

$$\frac{d\vec{p}}{dt} = \frac{d\vec{p}_1}{dt} + \frac{d\vec{p}_2}{dt} = \vec{F}_{21} + \vec{F}_{12} = 0, \tag{10.5}$$

where the last step is provided by the third law. However, it should be noted that this conservation in general will require us to consider an isolated system, which is something we might not always be interested in doing.

10.1.1 Motion of a rigid body

In order to consider the motion of an extended *rigid body*, not only its position but also its orientation in space is of relevance. We will here consider a rigid body initially occupying a volume V_0, see Fig. 10.1. For any given point in the body, we will assume that its position at time t is given by $\vec{x}(\vec{\xi}, t)$, where $\vec{\xi}$ is its position at $t = 0$. In order for the body to be rigid, the separation vector between any two points in the body can only change by a rotation that generally depends on the time t, i.e.,

$$\vec{x}(\vec{\xi}_2, t) - \vec{x}(\vec{\xi}_1, t) = \hat{R}(t)(\vec{\xi}_2 - \vec{\xi}_1), \tag{10.6}$$

where $\hat{R}(t)$ is the time-dependent rotation operator. Selecting a particular point $\vec{\xi}_0$ in the rigid body, its motion can be described by the function

$$\vec{x}_0(t) = \vec{x}(\vec{\xi}_0, t). \tag{10.7}$$

If we know the movement of this point and the rotation $\hat{R}(t)$, we can find the position of an arbitrary point in the rigid body through the relation

$$\vec{x}(\vec{\xi}, t) = \vec{x}_0(t) + \hat{R}(t)(\vec{\xi} - \vec{\xi}_0) \tag{10.8}$$

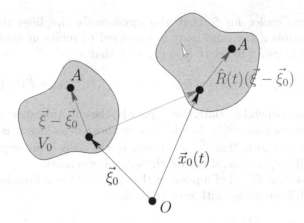

Figure 10.1 The motion of a point A in a rigid body relative to a fixed origin O if originally at the position $\vec{\xi}$ can be described by the motion $\vec{x}_0(t)$ of a reference point originally at $\vec{\xi}_0$ and a rotation $\hat{R}(t)$ of the separation vector between A and the reference point.

that follows directly from Eq. (10.6) with $\vec{\xi}_1 = \vec{\xi}_0$ and $\vec{\xi}_2 = \vec{\xi}$. Consequently, the motion of the rigid body can be uniquely described by specifying the motion of a single reference point $\vec{x}_0(t)$ and the time-dependent rotation $\hat{R}(t)$. The choice of the point $\vec{\xi}_0$ is arbitrary, but selecting it to be the *center of mass* $\vec{\xi}_{cm}$ defined by

$$\vec{\xi}_{cm} = \frac{1}{M} \int_{V_0} \vec{\xi}\rho_0(\vec{\xi})dV_0, \tag{10.9a}$$

where $\rho_0(\vec{\xi})$ is the mass density distribution of the rigid body at time $t = 0$ and the total mass M is given by

$$M = \int_{V_0} \rho_0(\vec{\xi})dV_0, \tag{10.9b}$$

is often preferable from a practical point of view.

Example 10.3 The motion of the rigid body generally results in a time-dependent density $\rho(\vec{x}, t)$. Since the motion is rigid, it follows that

$$\rho(\vec{x}(\vec{\xi}, t), t) = \rho_0(\vec{\xi}), \tag{10.10}$$

i.e., the density at the point $\vec{x}(\vec{\xi}, t)$ is the same as that at the point $\vec{\xi}$ at time $t = 0$. This relation can be used in order to show that the center of mass motion $\vec{x}_{cm}(t)$ defined by

$$\vec{x}_{cm}(t) = \frac{1}{M} \int_{V(t)} \vec{x}\rho(\vec{x}, t)dV, \tag{10.11}$$

where $V(t)$ is the volume occupied by the rigid body at time t coincides with the motion of the original center of mass, defined by $\vec{x}(\vec{\xi}_{cm}, t)$, see Problem 10.2.

An alternative choice for $\vec{\xi}_0$ that also significantly simplifies the expressions we will encounter is possible if the rigid body is restricted to rotate around a fixed point $\vec{\xi}_{\text{fix}}$. In this situation, it is useful to select $\vec{\xi}_0 = \vec{\xi}_{\text{fix}}$ such that

$$\vec{x}(\vec{\xi}_0, t) = \vec{x}(\vec{\xi}_{\text{fix}}, t) = \vec{\xi}_{\text{fix}} \quad \Longrightarrow \quad \vec{x}(\vec{\xi}, t) = \vec{\xi}_{\text{fix}} + \hat{R}(t)(\vec{\xi} - \vec{\xi}_{\text{fix}}), \tag{10.12}$$

where $\vec{\xi}_{\text{fix}}$ does not depend on time. Consequently, the only time dependence in this situation is encoded in the rotation $\hat{R}(t)$. In addition, in these situations it is often useful to select a coordinate system such that $\vec{\xi}_{\text{fix}} = 0$, which will reduce the expressions even further. However, we will keep a general $\vec{\xi}_{\text{fix}}$ for the sake of generality.

Given the position $\vec{x}(\vec{\xi}, t)$ of a point in the rigid body as a function of time, we can find its velocity by differentiating with respect to time

$$\vec{v}(\vec{\xi}, t) \equiv \frac{\partial \vec{x}}{\partial t} = \partial_t [\vec{x}_0(t) + \hat{R}(t)(\vec{\xi} - \vec{\xi}_0)] = \vec{v}_0(t) + \dot{\hat{R}}(t)(\vec{\xi} - \vec{\xi}_0), \tag{10.13}$$

where $\vec{v}_0(t) = \dot{\vec{x}}_0(t)$ is the velocity of the reference point $\vec{\xi}_0$. For the second term, we note that

$$\hat{R}(t + dt) - \hat{R}(t) \simeq [\hat{R}_{\hat{n}}^{\omega\,dt} - 1]\hat{R}(t), \tag{10.14}$$

where $\vec{n}\omega = \vec{\omega}(t)$ is the angular velocity at time t and $R_{\hat{n}}^{\theta}$ is the rotation introduced in Section 4.4.1. Since $\omega\,dt$ is a small angle, we can use the small angle expansion of Eq. (4.38) and find

$$\dot{\hat{R}}(t)\vec{a}\,dt \equiv [\hat{R}(t + dt) - \hat{R}(t)]\vec{a} \simeq \vec{\omega} \times \hat{R}(t)\vec{a}\,dt \tag{10.15}$$

for any vector $\vec{a}$. In particular, this implies that

$$\vec{v}(\vec{\xi}, t) = \vec{v}_0(t) + \vec{\omega}(t) \times \hat{R}(t)(\vec{\xi} - \vec{\xi}_0) = \vec{v}_0(t) + \vec{\omega}(t) \times [\vec{x}(\vec{\xi}, t) - \vec{x}_0(t)]. \tag{10.16}$$

The velocity of an arbitrary point can therefore be determined in terms of the velocity $\vec{v}_0(t)$ of the reference point, the instantaneous angular velocity $\vec{\omega}(t)$, and the instantaneous displacement $\vec{x}(\vec{\xi}, t) - \vec{x}_0(t)$ from the reference point.

Example 10.4 The expression for the velocity of an arbitrary point in the rigid body becomes particularly simple in the case when the body is rotating around a fixed point $\vec{\xi}_{\text{fix}}$ that we use as reference. In this case $\vec{v}_0(t) = 0$ and consequently

$$\vec{v}(\vec{\xi}, t) = \vec{\omega}(t) \times [\vec{x}(\xi, t) - \vec{\xi}_{\text{fix}}]. \tag{10.17}$$

10.1.2 Dynamics of a rigid body

So far we have only discussed the kinematics of how a rigid body may move. In order to find out how external forces affect the body, we now move on to studying its dynamics using Newton's laws. The momentum of a small volume dV_0 around the point defined by $\vec{\xi}$ in the rigid body is given by $d\vec{p} = dm\,\vec{v}(\vec{\xi}, t) = \rho_0(\vec{\xi})\vec{v}(\vec{\xi}, t)dV_0$. Consequently, the total *momentum* of the body at time t is given by

$$\vec{p}(t) = \int d\vec{p} = \int_{V_0} \rho_0(\vec{\xi})\vec{v}(\vec{\xi}, t)dV_0. \tag{10.18a}$$

As the only time dependence in this expression arises from the velocity $\vec{v}(\vec{\xi}, t)$, we find that the time derivative of the total momentum, i.e., the force on the rigid body, is

$$\vec{F}(t) = \dot{\vec{p}}(t) = \int_{V_0} \rho_0(\vec{\xi}) \frac{\partial \vec{v}}{\partial t} dV_0 \equiv \int_{V_0} \rho_0(\vec{\xi}) \vec{a}(\vec{\xi}, t) dV_0, \tag{10.18b}$$

where $\vec{a}(\vec{\xi}, t)$ is the acceleration of the point in the rigid body defined by $\vec{\xi}$ at time t. Looking at a small volume element around $\vec{\xi}$, the forces acting on it at time t must satisfy

$$\vec{f}(\vec{\xi}, t) = \rho_0(\vec{\xi}) \vec{a}(\vec{\xi}, t) \tag{10.19}$$

according to Newton's second law, where $\vec{f}(\vec{\xi}, t)$ is the force density. The force density can be split into two contributions

$$\vec{f}(\vec{\xi}, t) = \vec{f}_{\text{int}}(\vec{\xi}, t) + \vec{f}_{\text{ext}}(\vec{\xi}, t), \tag{10.20}$$

where the first contribution is due to internal forces in the rigid body and the second is due to external forces. Because of Newton's third law, it must hold that

$$\int_{V_0} \vec{f}_{\text{int}}(\vec{\xi}, t) dV_0 = 0 \implies \vec{F}(t) = \int_{V_0} \vec{f}_{\text{ext}}(\vec{\xi}, t) dV_0 \tag{10.21}$$

as any force from one part of the rigid body on another is equal in magnitude but opposite in direction to the force of the second part from the first. We therefore conclude that the total force on the rigid body is the sum of the external forces on all its parts.

In order to find out what this implies for the movement of the rigid body, let us again look at the total momentum $\vec{p}(t)$, but now inserting our explicit expression for $\vec{v}(\vec{\xi}, t)$, which results in

$$\vec{p}(t) = \int_{V_0} \rho_0(\vec{\xi})[\vec{v}_0(t) + \vec{\omega}(t) \times (\vec{x}(\vec{\xi}, t) - \vec{x}_0(t))] dV_0$$

$$= M[\vec{v}_0(t) - \vec{\omega}(t) \times \vec{x}_0(t)] + \vec{\omega}(t) \times \int_{V_0} \rho_0(\vec{\xi}) \vec{x}(\vec{\xi}, t) dV_0, \tag{10.22}$$

where we have used that neither $\vec{v}_0(t)$, $\vec{x}_0(t)$, or $\vec{\omega}(t)$ depend on the integration variable $\vec{\xi}$. Applying the definition of the center of mass to the last term, we finally find that

$$\vec{p}(t) = M[\vec{v}_0(t) + \vec{\omega}(t) \times (\vec{x}_{\text{cm}}(t) - \vec{x}_0(t))]. \tag{10.23}$$

Taking the time derivative of this expression now leads to

$$\vec{F}(t) = M[\vec{a}_0(t) + \vec{\alpha}(t) \times (\vec{x}_{\text{cm}}(t) - \vec{x}_0(t)) + \vec{\omega}(t) \times \{\vec{\omega}(t) \times (\vec{x}_{\text{cm}}(t) - \vec{x}_0)(t)\}], \tag{10.24}$$

where $\vec{\alpha}(t) = \dot{\vec{\omega}}(t)$ is the angular acceleration of the rotation.

While these are a rather cumbersome expression, the expression for the momentum simplifies to

$$\vec{p}(t) = M\vec{v}_{\text{cm}}(t) \tag{10.25a}$$

when $\vec{\xi}_0 = \vec{\xi}_{\text{cm}}$ and to

$$\vec{p}(t) = M\vec{\omega}(t) \times \hat{R}(t)(\vec{\xi}_{\text{cm}} - \vec{\xi}_{\text{fix}}) \tag{10.25b}$$

when $\vec{\xi}_0$ is a fixed point $\vec{\xi}_{\text{fix}}$ of the motion. In particular, by differentiating the first of these relations, we find that

$$\vec{F}(t) = \dot{\vec{p}}(t) = M\dot{\vec{v}}_{\text{cm}}(t) \equiv M\vec{a}_{\text{cm}}(t), \tag{10.26}$$

where $\vec{a}_{\text{cm}}(t)$ is the center of mass acceleration. In other words, Newton's second law takes the familiar form $\vec{F} = m\vec{a}$ for a rigid body with fixed mass $m = M$ when the acceleration is taken to be the center of mass acceleration.

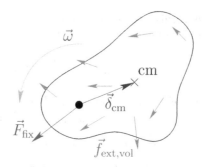

Figure 10.2 From the angular velocity $\vec{\omega}$ and the external volume forces $\vec{f}_{\text{vol,ext}}$ on a rigid object rotating around a fixed point we can find the force $\vec{F}_{\text{fix}}$ that needs to be applied at that fixed point.

Example 10.5 For a rigid body with a fixed point $\vec{\xi}_{\text{fix}}$, we can use the above reasoning in order to derive a condition on the external force at the fixed point, see Fig. 10.2. If we define the force $\vec{F}_{\text{fix}}(t)$ at the fixed point through the relation

$$\vec{f}_{\text{ext}}(\vec{\xi},t) = \vec{f}_{\text{ext,vol}}(\vec{\xi},t) + \vec{F}_{\text{fix}}(t)\delta^{(3)}(\vec{\xi}-\vec{\xi}_0), \tag{10.27}$$

where the first term describes all other external forces, we find that

$$\vec{F}(t) = \vec{F}_{\text{fix}}(t) + \int_{V_0} \vec{f}_{\text{ext,vol}}(\vec{\xi},t)dV_0$$
$$= M\vec{\alpha}(t) \times \vec{\delta}_{\text{cm}}(t) + M\vec{\omega}(t) \times [\vec{\omega}(t) \times \vec{\delta}_{\text{cm}}(t)], \tag{10.28}$$

where $\vec{\delta}_{\text{cm}}(t) = \hat{R}(t)(\vec{\xi}_{\text{cm}} - \vec{\xi}_0)$. Solving for $\vec{F}_{\text{fix}}(t)$ now results in

$$\vec{F}_{\text{fix}}(t) = M\vec{\alpha}(t) \times \vec{\delta}_{\text{cm}}(t) + M\vec{\omega}(t) \times [\vec{\omega}(t) \times \vec{\delta}_{\text{cm}}(t)] - \vec{F}_{\text{ext,vol}}(t), \tag{10.29}$$

where $\vec{F}_{\text{ext,vol}}(t)$ is the volume integral of $\vec{f}_{\text{ext,vol}}(\vec{\xi},t)$. We will soon see how the angular acceleration $\vec{\alpha}(t)$ can be related to the external force density. However, we can already here note that if there is no angular acceleration and no net external force apart from the force at $\vec{\xi}_0$, then

$$\vec{F}_{\text{fix}}(t) = M\vec{\omega}(t) \times [\vec{\omega}(t) \times \vec{\delta}_{\text{cm}}(t)], \tag{10.30}$$

which is just the regular expression for the centripetal force of an object with mass M moving in a circle with angular velocity $\vec{\omega}(t)$ offset from the center of the circle by the displacement $\vec{\delta}_{\text{cm}}(t)$, see Fig. 10.3. It should be noted that there can be a non-zero angular acceleration even when the external force density is equal to zero.

Studying the change in the momentum of the entire rigid body essentially resulted in a differential equation for the movement of its center of mass. In order to know the complete motion of the body we also need to find out how the rotation evolves with time based on the forces that act upon it. We can do so by studying the *angular momentum* $\vec{L}(t)$ of the

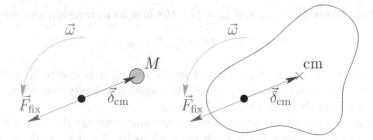

Figure 10.3 If the net external force density apart from $\vec{F}_{\text{fix}}$ is zero and the angular acceleration $\vec{\alpha}$ is zero, then the force at the fixed point will be the same as the force needed to keep a particle placed in the center of mass with the same total mass as the rigid body rotating about the fixed point. Note that the external force density $\vec{f}_{\text{ext,vol}} = 0$ does not necessarily imply zero angular acceleration.

body relative to the point $\vec{x}_0(t)$, which is given by

$$\vec{L}(t) = \int_{V_0} [\vec{x}(\vec{\xi},t) - \vec{x}_0(t)] \times \rho_0(\vec{\xi})\vec{v}(\vec{\xi},t)dV_0. \tag{10.31}$$

Inserting the expressions for $\vec{x}(\vec{\xi},t)$, $\vec{v}(\vec{\xi},t)$, and suppressing the $\vec{\xi}$ and t dependence of all the quantities for brevity, we find that

$$\vec{L} = \int_{V_0} (\vec{x} - \vec{x}_0) \times \rho_0[\vec{v}_0 + \vec{\omega} \times (\vec{x} - \vec{x}_0)]dV_0$$

$$= -\vec{v}_0 \times \int_{V_0} (\vec{x} - \vec{x}_0)\rho_0 dV_0 + \int_{V_0} (\vec{x} - \vec{x}_0) \times [\vec{\omega} \times (\vec{x} - \vec{x}_0)]dV_0$$

$$= M\vec{\delta}_{\text{cm}} \times \vec{v}_0 + \int_V \rho(\vec{x} - \vec{x}_0) \times [\vec{\omega} \times (\vec{x} - \vec{x}_0)]dV, \tag{10.32}$$

where in the last step we have used the definition of the center of mass and changed integration variables in the remaining integral to $\vec{x}$ instead of $\vec{\xi}$. Writing down the ith component of the last term of this relation in Cartesian coordinate tensor notation, we find that

$$L_i - M(\vec{\delta}_{\text{cm}} \times \vec{v}_0)_i = \int_V \rho\varepsilon_{ijk}(x^j - x_0^j)\varepsilon_{k\ell m}\omega^\ell(x^m - x_0^m)dV$$

$$= \omega^\ell \int_V \rho(\delta_{i\ell}\delta_{jm} - \delta_{im}\delta_{j\ell})(x^j - x_0^j)(x^m - x_0^m)dV$$

$$= \omega^\ell \int_V \rho[\delta_{i\ell}(x^j - x_0^j)(x^j - x_0^j) - (x^i - x_0^i)(x^\ell - x_0^\ell)]dV$$

$$\equiv I_{i\ell}\omega^\ell, \tag{10.33}$$

where we have introduced the *moment of inertia tensor*

$$I_{i\ell} = \int_V \rho[\delta_{i\ell}(x^j - x_0^j)(x^j - x_0^j) - (x^i - x_0^i)(x^\ell - x_0^\ell)]dV \tag{10.34}$$

relative to $\vec{x}_0$, see also Section 2.6.3.1. Summarising, this leads to

$$L^i = M\varepsilon_{ijk}\delta_{\text{cm}}^j v_0^k + I_{ij}\omega^j. \tag{10.35}$$

For both of the choices $\vec{\xi}_0 = \vec{\xi}_{cm}$ and $\vec{\xi}_0 = \vec{\xi}_{fix}$, the first term vanishes and we find that

$$L^i = I_{ij}\omega^i. \tag{10.36}$$

In the first case this is due to $\vec{\delta}_{cm} = 0$ and in the second it is due to $\vec{v}_0 = 0$. This is the relation between the angular momentum and the angular velocity that we used when first introducing tensors in the beginning of Chapter 2.

As we did in the case of the momentum, we can now compute the time derivative of the angular momentum and insert the explicit expressions for $\vec{x}$ and $\vec{v}$, resulting in

$$\dot{\vec{L}} = \int_V (\vec{x} - \vec{x}_0) \times \rho\vec{a}dV + \int_V (\vec{v} - \vec{v}_0) \times \rho\vec{v}dV$$
$$= \int_V (\vec{x} - \vec{x}_0) \times \vec{f}_{ext}dV - \vec{v}_0 \times \int_V \rho\vec{v}dV = \vec{\tau} - \vec{v}_0 \times \vec{p}, \tag{10.37}$$

where we have introduced the *torque*

$$\vec{\tau} = \int_V (\vec{x} - \vec{x}_0) \times \vec{f}_{ext}dV \tag{10.38}$$

relative to $\vec{x}_0$. Again, this expression takes a particularly simple form in the cases $\vec{\xi}_0 = \vec{\xi}_{cm}$ and $\vec{\xi}_0 = \vec{\xi}_{fix}$. In the former case, $\vec{v}_{cm} \times \vec{p} = 0$ as $\vec{p} = M\vec{v}_{cm}$, while in the latter $\vec{v}_0 = 0$, for both cases resulting in

$$\dot{\vec{L}} = \vec{\tau} \tag{10.39a}$$

or, in terms of the Cartesian components,

$$\dot{L}^i = \frac{d}{dt}I_{ij}\omega^j = \dot{I}_{ij}\omega^j + I_{ij}\dot{\omega}^j = \tau^i. \tag{10.39b}$$

This is reminiscent of Newton's second law, replacing the momentum $\vec{p}$ by the angular momentum $\vec{L}$ and the force $\vec{F}$ by the torque $\vec{\tau}$. Using the results from Problem 2.45, we find that the differential equation governing the rotation is given by

$$I_{ij}\alpha^j = \tau^i - \dot{I}_{ij}\omega^j = \tau^i - \varepsilon_{ijk}\omega^j I_{k\ell}\omega^\ell, \tag{10.40}$$

where $\vec{\alpha}$ is the angular acceleration.

Example 10.6 Returning to the rigid body rotating around a fixed point $\vec{\xi}_{fix}$ of Example 10.5, we note that the torque $\vec{\tau}$ is given by

$$\vec{\tau} = \int_V (\vec{x} - \vec{\xi}_{fix}) \times [\vec{f}_{ext,vol} + \delta^{(3)}(\vec{x} - \vec{\xi}_{fix})\vec{F}_{fix}]dV = \int_V (\vec{x} - \vec{\xi}_{fix}) \times \vec{f}_{ext,vol}dV \tag{10.41}$$

and therefore is independent of the force $\vec{F}_{fix}$, since $\vec{x} - \vec{\xi}_{fix} = 0$ at its point of application. As a consequence, the differential equation for the rotation can be solved without knowledge of $\vec{F}_{fix}$ and the resulting angular acceleration can be inserted into Eq. (10.29) in order to find the force $\vec{F}_{fix}$.

The moment of inertia tensor does not only show up in the expression for the angular momentum. Indeed, if we wish to express the total kinetic energy of the rigid body, we note that the kinetic energy of a small part of the body is given by $dT = \rho \vec{v}^2 dV/2$. As a result the total kinetic energy is

$$
\begin{aligned}
T = \int dT &= \int_V \frac{\rho}{2} \vec{v}^2 dV = \frac{1}{2} \int_V \rho [\vec{v}_0 + \vec{\omega} \times (\vec{x} - \vec{x}_0)]^2 dV \\
&= \frac{1}{2} \int_V \rho [\vec{v}_0^2 + 2\vec{v}_0 \cdot [\vec{\omega} \times (\vec{x} - \vec{x}_0)] + [\vec{\omega} \times (\vec{x} - \vec{x}_0)]^2] dV \\
&= \frac{M\vec{v}_0^2}{2} + M\vec{v}_0 \cdot [\vec{\omega} \times (\vec{x}_{\text{cm}} - \vec{x}_0)] + \frac{1}{2} \vec{\omega} \cdot \int_V \rho(\vec{x} - \vec{x}_0) \times [\vec{\omega} \times (\vec{x} - \vec{x}_0)] dV \\
&= \frac{M\vec{v}_0^2}{2} + M\vec{v}_0 \cdot [\vec{\omega} \times (\vec{x}_{\text{cm}} - \vec{x}_0)] + \frac{1}{2} \omega^i I_{ij} \omega^j.
\end{aligned}
\tag{10.42}
$$

For the case $\vec{\xi}_0 = \vec{\xi}_{\text{cm}}$, the middle term is zero and the total kinetic energy is given by

$$
T = \frac{M\vec{v}_{\text{cm}}^2}{2} + \frac{1}{2} \omega^i I_{ij}^{\text{cm}} \omega^j.
\tag{10.43a}
$$

In the case with a fixed point, both the first term and the middle term vanish due to $\vec{v}_0 = 0$ and we obtain

$$
T = \frac{1}{2} \omega^i I_{ij}^{\text{fix}} \omega^j.
\tag{10.43b}
$$

This is an example of a case in classical mechanics where the kinetic energy is quadratic in the variables describing the movement of the system, here given by $\vec{v}_0$ and $\vec{\omega}$, and also generally depends on the configuration of the system, here through the dependence of the moment of inertia tensor on how the rigid body is oriented.

10.1.3 Dynamics in non-inertial frames

Newton's three laws are invariant under the Galilei transformations, discussed in Section 4.6.2, that relate different inertial frames to each other. However, for some applications, it happens that there is a *non-inertial frame*, see Fig. 10.4, that is more natural to use as a reference. For a non-inertial frame, we will generally want to describe a process using the coordinates y^a based on an origin that is undergoing an arbitrary motion $\vec{x}_0(t)$ in an inertial frame as well as with a set of basis vectors $\vec{e}_a(t)$ that depend on time. Let us consider the motion of an object with the position vector $\vec{x}(t)$ in an inertial frame. For any given time, we must then have

$$
\vec{x}(t) = \vec{x}_0(t) + \vec{y}(t),
\tag{10.44a}
$$

where

$$
\vec{y}(t) = y^a(t)\vec{e}_a(t)
\tag{10.44b}
$$

describes the motion of the object relative to the origin of the non-inertial frame using the coordinates $y^a(t)$ corresponding to the basis $\vec{e}_a(t)$ of that frame. By differentiating this relation with respect to time, we find that

$$
\vec{v}(t) = \dot{\vec{x}}(t) = \dot{\vec{x}}_0(t) + \dot{\vec{y}}(t) = \vec{v}_0(t) + \dot{\vec{y}}(t),
\tag{10.45}
$$

where $\vec{v}(t)$ is the velocity of the object in the inertial frame, $\vec{v}_0(t)$ is the velocity of the origin of the non-inertial reference frame, and

$$
\dot{\vec{y}}(t) = \dot{y}^a(t)\vec{e}_a(t) + y^a(t)\dot{\vec{e}}_a(t).
\tag{10.46}
$$

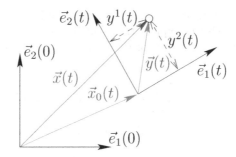

Figure 10.4 The relation between the non-inertial frame and the inertial frame that coincides with the non-inertial frame at $t = 0$. The origin of the non-inertial frame moves according to the function $\vec{x}_0(t)$ and the basis vectors generally depends on t as well. The coordinates $y^a(t)$ used in the non-inertial frames are the coordinates with respect to the changing basis vectors.

We now introduce the vector

$$\overset{\circ}{\vec{y}}(t) = \dot{y}^a(t)\vec{e}_a(t) \tag{10.47}$$

that has the time derivative of the non-inertial frame coordinates y^a as its components. This lets us write

$$\dot{\vec{y}} = \overset{\circ}{\vec{y}} + \vec{\omega} \times \vec{y}, \tag{10.48}$$

where we have suppressed, and will continue to suppress, the time-dependence of the vectors for brevity and used that $\dot{\vec{e}}_a = \vec{\omega} \times \vec{e}_a$, where $\vec{\omega}$ is the angular velocity with which the basis vectors of the non-inertial frame rotate. With this notation, we find that the velocity in the non-inertial frame can be written as

$$\overset{\circ}{\vec{y}} = \vec{v} - \vec{v}_0 - \vec{\omega} \times \vec{y}. \tag{10.49}$$

Example 10.7 There are two basic examples of non-inertial reference frames that we will have a closer look at in this example. The first example is a uniformly accelerated non-rotating frame, for which

$$\vec{x}_0 = \frac{\vec{a}_0 t^2}{2} \quad \text{and} \quad \vec{\omega} = 0. \tag{10.50}$$

For this frame we find that $\vec{v}_0 = \vec{a}_0 t$ and therefore

$$\overset{\circ}{\vec{y}} = \vec{v} - \vec{a}_0 t. \tag{10.51}$$

If the acceleration of an object in the inertial frame is $\vec{a}_0$, it will therefore move with constant velocity in the accelerated frame. As an example of this, consider an object moving under the influence of a homogeneous gravitational field $\vec{g}$. The velocity of this object in an inertial frame is given by

$$\vec{v} = \vec{v}(0) + \vec{g}t, \tag{10.52}$$

where $\vec{v}(0)$ is the velocity of the object at time $t = 0$. If we go to a frame with acceleration $\vec{g}$, the velocity of the object will be given by

$$\overset{\circ}{\vec{y}} = \vec{v}(0) + \vec{g}t - \vec{g}t = \vec{v}(0) \tag{10.53}$$

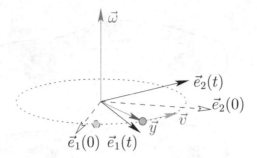

Figure 10.5 A point with fixed $y^a(t)$ coordinates in a system that rotates about its origin with angular velocity $\vec{\omega}$ will move with a velocity $\vec{v} = \vec{\omega} \times \vec{y}$ in an inertial frame. If $\vec{\omega}$ is constant, the point will move in a circle around the axis of rotation.

and therefore be constant. In particular, this holds for a frame where the position of the object itself is used as the origin, for which we would also have $\vec{v}(0) = 0$ and therefore $\mathring{\vec{y}} = 0$.

The second basic example of a non-inertial frame is a frame for which $\vec{x}_0 = 0$ and $\vec{\omega}$ is a constant vector, i.e., a frame that rotates around a fixed axis with a constant angular velocity. For this frame, we find that

$$\mathring{\vec{y}} = \vec{v} - \vec{\omega} \times \vec{y}. \tag{10.54}$$

In particular, we find that if an object is at rest relative to this frame, i.e., if $\mathring{\vec{y}} = 0$, then

$$\vec{v} = \vec{\omega} \times \vec{y}. \tag{10.55}$$

This is the familiar expression for the velocity of an object rotating about the origin with angular velocity $\vec{\omega}$, see Fig. 10.5.

In the same fashion as we introduced the velocity $\mathring{\vec{y}}$ in the non-inertial frame, we can introduce the acceleration in the non-inertial frame as

$$\mathring{\mathring{\vec{y}}} = \ddot{y}^a \vec{e}_a. \tag{10.56}$$

By taking the second derivative of $\vec{y}$ with respect to time, we find that

$$\ddot{\vec{y}} = \ddot{y}^a \vec{e}_a + 2\dot{y}^a \dot{\vec{e}}_a + y^a \ddot{\vec{e}}_a = \mathring{\mathring{\vec{y}}} + 2\vec{\omega} \times \mathring{\vec{y}} + \vec{\alpha} \times \vec{y} + \vec{\omega} \times (\vec{\omega} \times \vec{y}), \tag{10.57}$$

where $\vec{\alpha} = \dot{\vec{\omega}}$ is the angular acceleration. For an object moving with acceleration $\ddot{\vec{x}} = \vec{a}$ in the inertial frame, we obtain the relation

$$\vec{a} = \vec{a}_0 + \ddot{\vec{y}}, \tag{10.58}$$

where $\vec{a}_0 = \ddot{\vec{x}}_0$ is the acceleration of the origin of the non-inertial frame. Solving for $\mathring{\mathring{\vec{y}}}$ results in

$$\mathring{\mathring{\vec{y}}} = \vec{a} - \vec{a}_0 - 2\vec{\omega} \times \mathring{\vec{y}} - \vec{\alpha} \times \vec{y} - \vec{\omega} \times (\vec{\omega} \times \vec{y}). \tag{10.59}$$

If the object is moving under the influence of a force $\vec{F}$ in the inertial frame, its acceleration

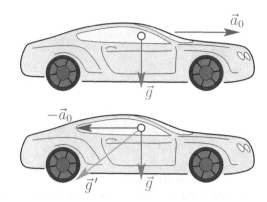

Figure 10.6 In an inertial system (upper figure) a car is accelerating with acceleration $\vec{a}_0$. In the presence of gravitational field $\vec{g}$, an object will be subject to gravitational acceleration in the inertial frame. In an accelerating frame fixed to the car (lower figure), the object will be subjected to an effective gravitational field $\vec{g}' = \vec{g} - \vec{a}_0$.

will be given by $\vec{a} = \vec{F}/m$, where m is the mass of the object. In the non-inertial frame, we would therefore find that

$$m\overset{\circ\circ}{\vec{y}} = \vec{F} - m\vec{a}_0 - 2m\vec{\omega} \times \overset{\circ}{\vec{y}} - m\vec{\alpha} \times \vec{y} - m\vec{\omega} \times (\vec{\omega} \times \vec{y}). \tag{10.60}$$

If attempting to interpret this in terms of Newton's second law in the non-inertial frame, it will appear as if the force acting on the object is given by

$$\vec{F}_{\text{acc}} = \vec{F} + \vec{F}_{\text{inertial}}, \tag{10.61a}$$

where

$$\vec{F}_{\text{inertial}} = -m[\vec{a}_0 + 2\vec{\omega} \times \overset{\circ}{\vec{y}} + \vec{\alpha} \times \vec{y} + \vec{\omega} \times (\vec{\omega} \times \vec{y})] \tag{10.61b}$$

is referred to as an *inertial force* (also commonly called *fictitious force* or *pseudo force*).

Example 10.8 Returning to the first example presented in Example 10.7, the inertial force on an object in the uniformly accelerated frame is given by

$$\vec{F}_{\text{inertial}} = -m\vec{a}_0, \tag{10.62}$$

since $\vec{\omega} = \vec{\alpha} = 0$. As an example of such a situation, consider the reference system of an accelerating car, see Fig. 10.6. If an object is subjected to a gravitational force $m\vec{g}$ and an additional external force $\vec{F}$, the total force on the object in the frame of the car will be given by

$$\vec{F}_{\text{acc}} = m(\vec{g} - \vec{a}_0) + \vec{F}. \tag{10.63}$$

This situation is completely equivalent to movement in an inertial frame where the gravitational field is instead given by

$$\vec{g}' = \vec{g} - \vec{a}_0. \tag{10.64}$$

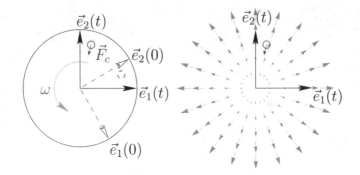

Figure 10.7 A mass rotating around the origin of an inertial frame (left) requires a centripetal force $\vec{F}_c$ in order to maintain the circular motion. The centrifugal field in a coordinate system rotating with the same angular velocity ω is shown to the right. In such a system, the mass is at rest and the centrifugal acceleration exactly cancels the acceleration due to $\vec{F}_c$.

Example 10.9 Looking at the second example of Example 10.7, the inertial force on an object in a frame rotating with constant angular velocity $\vec{\omega}$ is found to be

$$\vec{F}_{\text{inertial}} = -2m\vec{\omega} \times \mathring{\vec{y}} - m\vec{\omega} \times (\vec{\omega} \times \vec{y}). \tag{10.65}$$

The first of these terms is known as the *Coriolis force* while the second is referred to as the *centrifugal force*. The centrifugal force is always directed away from the axis of rotation and depends only on the position of an object in the rotating frame and the angular velocity $\vec{\omega}$, see Fig. 10.7. On the other hand, the Coriolis force requires the object to be moving relative to the non-inertial frame and results in a force orthogonal to both the direction of motion $\mathring{\vec{y}}$ and the angular velocity $\vec{\omega}$.

Considering two persons sitting still on a carousel, the centrifugal force will be apparent for them both as the centrifugal force will need to be countered by a centripetal force in order for them not to fly off. In an inertial frame, this centripetal force will instead keep the persons moving in a circle along with the carousel. The Coriolis force does not become apparent until the persons try to move on the carousel or attempt to throw a ball between each other, see Fig. 10.8.

10.2 LAGRANGIAN MECHANICS

As we have already briefly discussed in Chapter 8, *Lagrangian mechanics* is a reformulation of classical mechanics based on *Hamilton's principle* (also often referred to as the *principle of stationary action*), a variational principle that states that a physical system will evolve according to the stationary functions of the *action* functional

$$S = \int \mathcal{L} \, dt, \tag{10.66}$$

where $\mathcal{L}$ is the *Lagrangian* of the system. For many cases that do not involve dissipative forces, Lagrangian mechanics is equivalent to Newtonian mechanics if the Lagrangian is

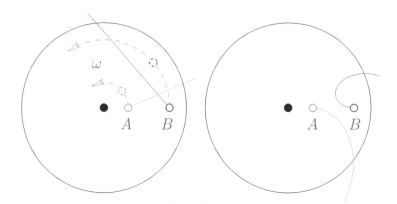

Figure 10.8 Two persons A and B sitting on a carousel spinning with angular velocity ω simultaneously throw a ball in the other's direction as seen in the rotating frame. The left figure shows the trajectories of the balls in an inertial frame with the dashed positions of A and B being the positions of the persons at the time the ball from A reaches the radius at which B is located. The right figure shows the corresponding trajectories in the rotating frame. In this frame, the initial deviation from a straight line is due to the Coriolis force while, for later times, both the centrifugal and Coriolis forces bend the trajectory. The Coriolis force is always perpendicular to the velocity and therefore does not change the speed of the balls, while the centrifugal force in general affects both speed and direction.

chosen to be

$$\mathcal{L} = T - V, \tag{10.67}$$

where T is the kinetic energy and V the potential energy of the system. However, in a more general setting, the Lagrangian may take a different form and then defines the physical system. In this sense, specifying Hamilton's principle along with the Lagrangian is the equivalent of stating Newton's laws and modelling the forces between objects in Newtonian mechanics.

10.2.1 Configuration space

We just mentioned that the action $\mathcal{S}$ is a functional, but we did not specify the functions on which it depends. Given a physical system, we will assume that its configuration at a given time t can be described by a set of numbers $q(a, t)$, called *generalised coordinates*, where a is a label that belongs to some indexing set. In general this set may not be finite, or even countable, but if it is we will instead use the notation $q^a(t)$. The set of all $q(a, t)$ that describe a physically possible configuration for a given t is called the *configuration space* of the system. The functions that the action depends upon are functions of time to the configuration space, i.e., paths in configuration space, and generally the Lagrangian is a function of the generalised coordinates $q(a, t)$, their time derivatives $\dot{q}(a, t)$, and time t.

Example 10.10 In the case of a particle moving in three dimensions, the configuration space consists of the three coordinates x^1, x^2, and x^3 that can take any real value and the corresponding index set can be taken to be $\{1, 2, 3\}$. The Lagrangian, given by the difference of the kinetic and potential energies, is a function of $\vec{x}$, $\dot{\vec{x}}$, and t and can be written as

$$\mathcal{L}(\vec{x}, \dot{\vec{x}}, t) = T - V = \frac{m}{2}\dot{\vec{x}}^2 - V(\vec{x}, t), \tag{10.68}$$

where the potential is assumed to be a function of the system configuration and time and m is the mass of the particle.

Example 10.11 A situation for which the configuration space has an uncountable index set is given by a string of length ℓ that undergoes transversal oscillations. The configuration of the system is then the shape of the string, which may be described by a function $u(x)$ whose value is the transversal displacement from the equilibrium position, where the index set is the interval $0 < x < \ell$. As discussed in Example 8.25, the Lagrangian of this system is given by

$$\mathcal{L} = \frac{1}{2}\int_0^\ell [\rho_\ell u_t(x,t)^2 - S u_x(x,t)^2]dx, \tag{10.69}$$

where ρ_ℓ is the linear density of the string and S its tension.

The choice of the numbers $q(a, t)$ that define the configuration space is generally not unique and the configuration space can be described in many different equivalent ways. Selecting a different representation of the configuration space in terms of the numbers $Q(b, t)$ amounts to a *coordinate change*. Such a coordinate change should always be invertible for any t, i.e., we must be able to write the $Q(b, t)$ as functions of the $q(a, t)$ and vice versa. That such coordinate changes does not change the physical solution was discussed in Section 8.5.

Example 10.12 For the particle moving in three dimensions described in Example 10.10, the configuration space may be described in any coordinate system. For example, we may use the spherical coordinates r, θ, and φ instead of the Cartesian coordinates x^1, x^2, and x^3. The configuration space is then restricted to the ranges

$$r \geq 0, \quad 0 \leq \theta \leq \pi, \quad \text{and} \quad 0 \leq \varphi < 2\pi. \tag{10.70}$$

Example 10.13 In the case of Example 10.11 with homogeneous Dirichlet boundary conditions, the function $u(x)$ is completely determined by its Fourier series

$$u(x) = \sum_{n=1}^{\infty} u_n \sin\left(\frac{\pi n x}{\ell}\right). \tag{10.71}$$

Instead of using the function $u(x)$ to describe the configuration space, we could therefore decide to use the Fourier coefficients u_n instead. The index set will then no longer be the interval $0 < x < \ell$, but instead the countable set of positive integers $\mathbb{N}_+ = \{1, 2, 3, \ldots\}$.

Inserting the Fourier series into the definition of the Lagrangian, we find that it is now given by

$$\mathcal{L} = \ell \sum_{n=1}^{\infty} \left[\rho_\ell \dot{u}_n(t)^2 - S \frac{\pi^2 n^2}{\ell^2} u_n(t)^2 \right]. \tag{10.72}$$

The configuration space in terms of the Fourier coefficients with this Lagrangian is equivalent to the configuration space in terms of the functions $u(x, t)$ and the Lagrangian of Eq. (10.69).

Once the configuration space and the action have been identified, we can find out whether a particular path $q(a, t)$ in configuration space provides a stationary value for the action using the variational principles discussed in Chapter 8. In general, a path provides a stationary action if the variation $\delta \mathcal{S}$ of the action around that path is equal to zero. Such a path describes a physical time evolution of the system and it is clear that not every possible path in configuration space is going to fulfil this condition. A path that provides a stationary action is generally referred to as being *on-shell*, whereas a path that does not is referred to as *off-shell*. Comparing to Newtonian mechanics, a path is on-shell if it satisfies Newton's equations and off-shell if it does not.

10.2.2 Finite number of degrees of freedom

The evolution of a system that has a finite index set can be described using only a finite number N of functions $q^a(t)$, where the index set may be taken to just be the numbers 1 through N. In this situation, the action is a functional that depends on N functions and a path in configuration space is on-shell if it satisfies the Euler–Lagrange equations

$$\frac{\partial \mathcal{L}}{\partial q^a} - \frac{d}{dt} \frac{\partial \mathcal{L}}{\partial \dot{q}^a} = 0, \tag{10.73}$$

where we have assumed that $\mathcal{L}$ does not depend on higher order derivatives of q^a. In many situations, the generalised coordinates q^a will define the spatial position of an object or an angle describing its orientation in space.

Example 10.14 Consider a rigid body in two dimensions, see Fig. 10.9. The configuration space for such a body can be described by the spatial coordinates x^1 and x^2 of its center of mass relative to its position at time $t = 0$ and its rotation angle θ relative to some reference orientation. As discussed earlier, the kinetic energy of this body is given by

$$T = \frac{M}{2} \dot{\vec{x}}^2 + \frac{I}{2} \dot{\theta}^2, \tag{10.74}$$

where $\vec{x} = x^1 \vec{e}_1 + x^2 \vec{e}_2$, M is the mass of the body, and I its momentum of inertia with respect to the center of mass. With the Lagrangian given by $\mathcal{L} = T - V(\theta, \vec{x})$, the Euler–Lagrange equations result in

$$m\ddot{\vec{x}} = -\nabla V, \tag{10.75a}$$

$$I\ddot{\theta} = -\partial_\theta V, \tag{10.75b}$$

where we have defined $\nabla = \vec{e}_1 \partial_1 + \vec{e}_2 \partial_2$ to act on the position coordinates only. For a conservative force, we have earlier seen that the force $\vec{F}$ can be written in terms of the

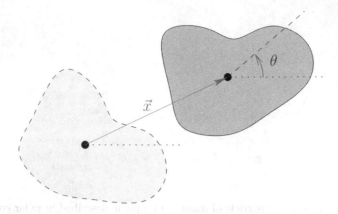

Figure 10.9   The configuration space of a rigid body in two dimension can be constructed using the location of its center of mass $\vec{x}$ and its rotation angle θ relative to those at time $t = 0$.

potential as $\vec{F} = -\nabla V$. Consequently, the first of the Euler–Lagrange equations is just Newton's second law applied to the rigid body. In the same fashion, we can identify $-\partial_\theta V$ with the torque τ relative to the center of mass. This results in the relation

$$I\ddot{\theta} = \tau, \tag{10.76}$$

which in two dimensions, where the moment of inertia is constant, is nothing but the equivalent of Eqs. (10.39).

As illustrated by the example above, the equations of motion resulting from generalised coordinates that are positions will be of the same form as Newton's second law, relating acceleration to the applied force. We had also already noted that the relation $\dot{\vec{L}} = \vec{\tau}$ between the angular momentum and the torque was very reminiscent of Newton's second law and the Lagrangian approach makes it clear that this stems from being derived in the same way, but with the generalised coordinate being an angle instead of a position. In general, if the kinetic energy is quadratic in $\dot{q}^a$, then the Lagrangian is of the form

$$\mathcal{L} = \frac{1}{2} M_{ab} \dot{q}^a \dot{q}^b - V, \tag{10.77}$$

where M_{ab} and V are functions of the generalised coordinates and time, i.e., they should be independent of $\dot{q}^a$. The equation of motion now becomes

$$\frac{dp_a}{dt} = -\partial_a V + \frac{1}{2} \dot{q}^b \dot{q}^c \partial_a M_{bc}, \tag{10.78a}$$

where we have introduced the *canonical momentum*

$$p_a = \frac{\partial \mathcal{L}}{\partial \dot{q}^a} = M_{ab} \dot{q}^b. \tag{10.78b}$$

This is the equation of motion corresponding to the variation of the action with respect to the generalised coordinate q^a. The functions M_{ab} together form the *generalised inertia tensor* that was briefly discussed in Section 2.6.3.

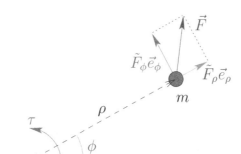

Figure 10.10 The motion of a particle of mass m in a plane described in polar coordinates reduce to an equation describing the radial acceleration and one describing the change in the angular momentum as the torque τ. The torque only depends on the tangential component $\tilde{F}_\phi$ of the force $\vec{F}$.

Example 10.15 Let us look at the two-dimensional motion of a particle under the influence of a force $\vec{F} = -\nabla V$ in polar coordinates. The kinetic energy is

$$T = \frac{m}{2}\dot{\vec{x}}^2 = \frac{1}{2}m\delta_{ij}\dot{x}^i\dot{x}^j = \frac{1}{2}mg_{ab}\dot{y}^a\dot{y}^b \tag{10.79}$$

and so the generalised inertia tensor in this case is

$$M_{ab} = mg_{ab}. \tag{10.80}$$

In polar coordinates, this implies that

$$M_{\rho\rho} = m, \quad M_{\phi\phi} = m\rho^2, \quad \text{and} \quad M_{\rho\phi} = M_{\phi\rho} = 0. \tag{10.81}$$

The canonical momenta become

$$p_\rho = m\dot{\rho} \quad \text{and} \quad p_\phi = m\rho^2\dot{\phi}, \tag{10.82}$$

respectively, where p_ϕ can be identified with the angular momentum L with respect to the origin $\rho = 0$. The equations of motion are now given by

$$m\ddot{\rho} = -\partial_\rho V + m\rho\dot{\phi}^2 = \tilde{F}_\rho + m\rho\dot{\phi}^2, \tag{10.83a}$$

$$\dot{L} = -\partial_\phi V = \rho\tilde{F}_\phi = \tau, \tag{10.83b}$$

where $\tilde{F}_a$ are the physical components of the force relative to the orthonormal set of basis vectors $\vec{e}_\rho$ and $\vec{e}_\phi$, and τ is the torque, see Fig. 10.10. The second term in the equation of motion for the ρ-coordinate arises from the coordinate system being curvilinear.

10.2.3   Non-inertial frames in Lagrangian mechanics

In our discussion on Newtonian mechanics, we spent a lot of effort on finding the proper equations of motion in a non-inertial coordinate frame. The corresponding procedure is rather straightforward in Lagrangian mechanics. As we have already discussed, the action is invariant under changes of generalised coordinates, even changes that depend on time t. When going to a non-inertial frame, we introduce its coordinates in the same fashion as we did before, by requiring that

$$\vec{x} = \vec{x}_0 + \vec{y} \tag{10.84}$$

and using time-dependent basis vectors $\vec{e}_a$. For a point object of mass m, the kinetic energy is given by

$$T = \frac{m}{2}\dot{\vec{x}}^2 = \frac{m}{2}[\vec{v}_0^2 + 2\vec{v}_0 \cdot (\mathring{\vec{y}} + \vec{\omega} \times \vec{y}) + \mathring{\vec{y}}^2 + 2\mathring{\vec{y}} \cdot (\vec{\omega} \times \vec{y}) + \vec{\omega} \cdot (\vec{y} \times (\vec{\omega} \times \vec{y}))]. \tag{10.85}$$

The components of the vectors $\vec{v}_0$ and $\vec{\omega}$ depend only on the time t and the first term will therefore not affect the equations of motion. With the Lagrangian $\mathcal{L} = T - V$, this now results in

$$\frac{\partial \mathcal{L}}{\partial y^a} = -\partial_a V - m\varepsilon_{abc}\omega^b[v_0^c + \dot{y}^c + \varepsilon_{cde}\omega^d y^e], \tag{10.86a}$$

$$\frac{d}{dt}\frac{\partial \mathcal{L}}{\partial \dot{y}^a} = m(\dot{v}_0^a + \ddot{y}^a + \varepsilon_{abc}\dot{\omega}^b y^c + \varepsilon_{abc}\omega^b \dot{y}^c) \tag{10.86b}$$

and multiplying both of these by $\vec{e}_a$ gives us the relation

$$m\mathring{\mathring{\vec{y}}} = \vec{F} - m\vec{a}_0 - 2m\vec{\omega} \times \mathring{\vec{y}} - m\vec{\alpha} \times \vec{y} - m\vec{\omega} \times (\vec{\omega} \times \vec{y}), \tag{10.87}$$

where we have used the fact that $\vec{a}_0 \equiv \dot{\vec{v}}_0 = \mathring{\vec{v}}_0 + \vec{\omega} \times \vec{v}_0$ and that $\vec{\alpha} \equiv \dot{\vec{\omega}} = \mathring{\vec{\omega}} + \vec{\omega} \times \vec{\omega} = \mathring{\vec{\omega}}$. This is precisely the same relation as that presented in Eq. (10.60). The corresponding conclusions for the inertial forces follow in the same fashion as in the Newtonian case.

Example 10.16 The change of variables to a rotating system in Lagrangian mechanics are well illustrated by considering the motion of an object in two dimensions. We define the coordinates y^1 and y^2 in the rotating system by the time-dependent coordinate transformation

$$x^1 = y^1 \cos(\omega t) - y^2 \sin(\omega t) \quad \text{and} \quad x^2 = y^1 \sin(\omega t) + y^2 \cos(\omega t), \tag{10.88}$$

see Fig. 10.11. The resulting time derivatives of the inertial coordinates x^1 and x^2 are given by

$$\dot{x}^1 = (\dot{y}^1 - \omega y^2)\cos(\omega t) - (\dot{y}^2 + \omega y^1)\sin(\omega t), \tag{10.89a}$$

$$\dot{x}^2 = (\dot{y}^1 - \omega y^2)\sin(\omega t) + (\dot{y}^2 + \omega y^1)\cos(\omega t), \tag{10.89b}$$

directly leading to

$$\dot{\vec{x}}^2 = (\dot{y}^1)^2 + (\dot{y}^2)^2 - 2\omega(\dot{y}^1 y^2 - \dot{y}^2 y^1) + \omega^2[(y^1)^2 + (y^2)^2]. \tag{10.90}$$

With the kinetic energy given by $T = m\dot{\vec{x}}^2/2$, the Euler–Lagrange equations imply that

$$m\ddot{y}^1 = 2m\omega\dot{y}^2 + m\omega^2 y^1, \tag{10.91a}$$

$$m\ddot{y}^2 = -2m\omega\dot{y}^1 + m\omega^2 y^2 \tag{10.91b}$$

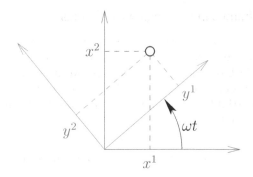

Figure 10.11 A graphical representation of the relation between the coordinates y^1 and y^2 of a point in a rotating coordinate system to the coordinates x^1 and x^2 in an inertial frame. As the angle ωt increases with time, the y^a coordinates for the same point in space will change.

in the absence of a potential energy V. The terms on the right-hand side exactly correspond to the inertial Coriolis and centrifugal forces, respectively. The same result can be found by restricting the expressions for the three dimensional motion to a plane, upon which the angular velocity $\vec{\omega}$ is constrained to be equal to $\vec{\omega} = \omega \vec{n}$, where $\vec{n}$ is a unit vector perpendicular to the plane.

10.2.4 Noether's theorem

Throughout this book, we have several times referred to *Noether's theorem* as one of the more important results in classical physics and it is time that we discuss what it is and how it is applied. We will start by discussing the theorem for a finite number of degrees of freedom and return to a field theory discussion of the theorem later in this chapter. When we discussed constants of motion in Section 8.3, we found that if the Lagrangian $\mathcal{L}$ does not depend explicitly on the generalised coordinate q^a, then the corresponding canonical momentum is conserved, i.e,

$$\frac{dp_a}{dt} = \frac{d}{dt}\frac{\partial \mathcal{L}}{\partial \dot{q}^a} = 0. \tag{10.92}$$

We also found that if $\mathcal{L}$ does not depend explicitly on the time t, then we have a corresponding conserved quantity $\mathcal{H}$ given by

$$\mathcal{H} = \dot{q}^a \frac{\partial \mathcal{L}}{\partial \dot{q}^a} - \mathcal{L}, \tag{10.93}$$

i.e., $d\mathcal{H}/dt = 0$. These statements will both turn out to be special cases of Noether's theorem.

In order to state Noether's theorem, we start by defining a continuous one-parameter transformation of time and the configuration space such that

$$t \to \tau(t, s) \quad \text{and} \quad q^a \to Q^a(q, s) \tag{10.94}$$

such that $\tau(t, 0) = t$ and $Q^a(q, 0) = q^a$. In order to find how the generator of this transformation acts on time and configuration space, we use an infinitesimal parameter $s = \varepsilon$ and define

$$\tau(t, \varepsilon) = t + \varepsilon\,\delta t \quad \text{and} \quad Q^a(q, \varepsilon) = q^a + \varepsilon\,\delta q^a, \tag{10.95}$$

where we ignore terms quadratic in ε or higher. For our purposes here, we will also restrict ourselves to the cases where $\delta t = 0$ or 1. Since infinitesimal symmetry transformations involving time can be rewritten in such a way that $\delta t = 1$ by normalisation of the continuous parameter s, this poses no real restriction on the types of symmetries that we consider. Noether's theorem deals with transformations such that

$$\varepsilon\,\delta\mathcal{L} \equiv \mathcal{L}(Q(q,\varepsilon), \dot{Q}(q,\varepsilon), \tau(t,\varepsilon)) - \mathcal{L}(q, \dot{q}, t) = \varepsilon\frac{dF}{dt}, \tag{10.96}$$

where F is some function that may depend on t, q^a and the derivatives of q^a. Transformations of this sort are called *quasi-symmetries* or, when $F = 0$, *symmetries*, of the Lagrangian. The statement of Noether's theorem is that if there exists a quasi-symmetry of the Lagrangian, then there exists a corresponding constant of motion, i.e., a function J that is generally an expression in terms of t, q^a, and the derivatives of q^a that does not change with time for any on-shell solution to the equations of motion.

In order to show that Noether's theorem holds, let us start by examining the quantity $\delta\mathcal{L}$, which to leading order in ε is given by

$$\delta\mathcal{L} = \frac{\partial\mathcal{L}}{\partial q^a}\delta q^a + \frac{\partial\mathcal{L}}{\partial\dot{q}^a}\delta\dot{q}^a + \frac{\partial\mathcal{L}}{\partial t}\delta t. \tag{10.97}$$

The partial derivative of $\mathcal{L}$ with respect to t is part of the total derivative of $\mathcal{L}$ with respect to t and can therefore be written as

$$\frac{\partial\mathcal{L}}{\partial t} = \frac{d\mathcal{L}}{dt} - \frac{\partial\mathcal{L}}{\partial q^a}\dot{q}^a - \frac{\partial\mathcal{L}}{\partial\dot{q}^a}\ddot{q}^a. \tag{10.98}$$

For any on-shell solution, the Euler–Lagrange equations of Eq. (10.73) must hold and we therefore find that

$$\delta\mathcal{L} = \delta q^a\frac{d}{dt}\frac{\partial\mathcal{L}}{\partial\dot{q}^a} + \frac{\partial\mathcal{L}}{\partial\dot{q}^a}\frac{d}{dt}\delta q^a + \left(\frac{d\mathcal{L}}{dt} - \dot{q}^a\frac{d}{dt}\frac{\partial\mathcal{L}}{\partial\dot{q}^a} - \frac{\partial\mathcal{L}}{\partial\dot{q}^a}\frac{d\dot{q}^a}{dt}\right)\delta t$$

$$= \frac{d}{dt}\left(p_a\delta q^a - \mathcal{H}\,\delta t\right), \tag{10.99}$$

where $p_a = \partial\mathcal{L}/\partial\dot{q}^a$ is the canonical momentum corresponding to q^a and

$$\mathcal{H} = p_a\dot{q}^a - \mathcal{L} \tag{10.100}$$

is the *Hamiltonian* function of the system. If the transformation is a quasi-symmetry of the Lagrangian, it follows that

$$\frac{d}{dt}\left(\mathcal{H}\,\delta t - p_a\delta q^a + F\right) = 0 \tag{10.101a}$$

and therefore that

$$J = \mathcal{H}\,\delta t - p_a\delta q^a + F \tag{10.101b}$$

is a constant of motion.

Example 10.17 As a first example of an application of Noether's theorem, let us see how the conservation of momentum and energy follow from the Lagrangian being independent of the generalised coordinate q^a and time t, respectively. We start by looking at the transformation for which

$$\delta t = 0 \quad \text{and} \quad \delta q^a = -\delta_b^a \tag{10.102}$$

for some fixed b and find that

$$\delta\mathcal{L} = \frac{\partial\mathcal{L}}{\partial q^a}\delta q^a + \frac{\partial\mathcal{L}}{\partial\dot{q}^a}\delta\dot{q}^a = -\frac{\partial\mathcal{L}}{\partial q^a}\delta_b^a = -\frac{\partial\mathcal{L}}{\partial q^b} = 0 \qquad (10.103)$$

if the Lagrangian is assumed to not depend explicitly on q^b. In this case, we can pick $F = 0$ and it follows that there exists a corresponding conserved quantity

$$J = p_a\delta_b^a = p_b, \qquad (10.104)$$

i.e., the canonical momentum p_b is a constant of motion. In the same fashion, if the Lagrangian is assumed not to depend explicitly on t, we can study the transformation given by

$$\delta t = 1 \quad \text{and} \quad \delta q^a = 0. \qquad (10.105)$$

Under this transformation

$$\delta\mathcal{L} = \frac{\partial\mathcal{L}}{\partial t} = 0, \qquad (10.106)$$

due to our assumption on the Lagrangian. Again, we may pick $F = 0$ and thus

$$J = \mathcal{H}\,\delta t = \mathcal{H} \qquad (10.107)$$

is a constant of motion.

Example 10.18 In Problem 8.37, it was shown that the angular momentum

$$L = m\rho^2\dot{\phi} \qquad (10.108)$$

is a constant of motion for a particle of mass m moving in a two-dimensional central potential $V(\rho)$ by looking at the behaviour of the Lagrangian in polar coordinates. If we stick to Cartesian coordinates x^1 and x^2, the potential can be seen as a function of $\vec{x}^2$ and we find that

$$\mathcal{L} = \frac{m}{2}\dot{\vec{x}}^2 - V(\vec{x}^2). \qquad (10.109)$$

This Lagrangian is invariant under the rotation

$$x^1 \to x^1\cos(s) + x^2\sin(s), \quad x^2 \to -x^1\sin(s) + x^2\cos(s), \qquad (10.110a)$$

which corresponds to

$$\delta t = 0, \quad \delta x^1 = x^2, \quad \text{and} \quad \delta x^2 = -x^1 \qquad (10.110b)$$

for small transformations. The resulting constant of motion is therefore

$$J = -x^2\frac{\partial\mathcal{L}}{\partial\dot{x}^1} + x^1\frac{\partial\mathcal{L}}{\partial\dot{x}^2} = m(x^1\dot{x}^2 - x^2\dot{x}^1). \qquad (10.111)$$

We can verify that this is the angular momentum from Problem 8.37 by again introducing polar coordinates, leading to

$$\dot{x}^1 = -\rho\sin(\phi)\dot{\phi} + \dot{\rho}\cos(\phi), \quad \dot{x}^2 = \rho\cos(\phi)\dot{\phi} + \dot{\rho}\sin(\phi) \qquad (10.112a)$$

and therefore

$$x^1\dot{x}^2 - x^2\dot{x}^1 = \rho^2\dot{\phi}. \qquad (10.112b)$$

Example 10.19 Some transformations that are quasi-symmetries of the Lagrangian result in the same conserved quantity as a symmetry of the Lagrangian. Consider again the case where the Lagrangian does not depend explicitly on time and take the infinitesimal transformation to be given by

$$\delta t = 0, \quad \text{and} \quad \delta q^a = -\dot{q}^a. \tag{10.113}$$

With this transformation, we find that

$$\delta \mathcal{L} = -\frac{\partial \mathcal{L}}{\partial q^a} \dot{q}^a - \frac{\partial \mathcal{L}}{\partial \dot{q}^a} \ddot{q}^a = -\frac{d\mathcal{L}}{dt}, \tag{10.114}$$

since $\partial \mathcal{L}/\partial t = 0$. In this case, we can pick $F = -\mathcal{L}$ and it follows that the corresponding constant of motion is

$$J = -\frac{\partial \mathcal{L}}{\partial \dot{q}^a} \delta q^a + F = p_a \dot{q}^a - \mathcal{L} = \mathcal{H}. \tag{10.115}$$

Thus, using this transformation is merely an alternative way to deduce that the Hamiltonian is a constant of motion if the Lagrangian does not depend explicitly on time.

10.2.5   Effective potentials

In many situations of physical interest, it is possible to use external constraints or constants of motion in order to rewrite a problem in terms of an *effective potential*. Let us first consider the situation where we have N generalised coordinates Q^A and impose a holonomic constraint $f(Q, t) = 0$. We furthermore assume that the Lagrangian takes the form

$$\mathcal{L} = \frac{1}{2} M_{AB} \dot{Q}^A \dot{Q}^B - V(Q, t). \tag{10.116}$$

Based on the holonomic constraint, the coordinates Q^A can be written as functions $Q^A(q, t)$ of $N - 1$ coordinates q^a and we find that

$$\dot{Q}^A = \partial_t Q^A + (\partial_a Q^A) \dot{q}^a. \tag{10.117}$$

Inserting this into the Lagrangian, we find the *effective Lagrangian*

$$\mathcal{L}_{\text{eff}} = \frac{1}{2} M_{AB} \left[(\partial_a Q^A)(\partial_b Q^B) \dot{q}^a \dot{q}^b + 2(\partial_t Q^A)(\partial_b Q^B) \dot{q}^b + (\partial_t Q^A)(\partial_t Q^B) \right] - V, \tag{10.118}$$

where it should be noted that both $\partial_t Q^A$ and V are now functions of q^a and t. If we can find coordinates q^a such that

$$M_{AB}(\partial_t Q^A)(\partial_b Q^B) = 0 \tag{10.119}$$

for all b, this simplifies to

$$\mathcal{L} = \frac{1}{2} M_{ab} \dot{q}^a \dot{q}^b - V_{\text{eff}}(q, t), \tag{10.120a}$$

where we have introduced the *effective inertia tensor*

$$M_{ab} = (\partial_a Q^A)(\partial_b Q^B) M_{AB} \tag{10.120b}$$

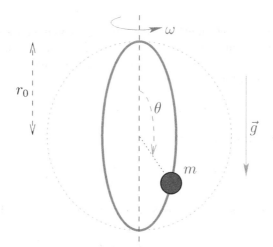

Figure 10.12 A bead of mass m moving freely on a ring with radius r_0 that rotates with angular frequency ω in a gravitational field $\vec{g}$. The motion of the bead can be described as motion on a sphere parametrised by the angles θ and φ with the constraint $\varphi = \omega t$.

and the *effective potential*

$$V_{\text{eff}}(q,t) = V(Q(q,t),t) - (\partial_t Q^A)(\partial_t Q^B) M_{AB}. \tag{10.120c}$$

We have then reduced the problem with a potential $V(Q,t)$ and a holonomic constraint $f(Q,t) = 0$ to a lower-dimensional problem with an effective potential $V_{\text{eff}}(q,t)$.

Example 10.20 Consider a bead of mass m moving on a rotating circular loop in a gravitational field parallel to the rotational axis, see Fig. 10.12. This situation may be described as the bead moving on the surface of a sphere with radius r_0 subjected to the holonomic constraint

$$f(\theta, \varphi, t) = \varphi - \omega t = 0, \tag{10.121}$$

where ω is the angular velocity of the rotation. The Lagrangian for this motion is given by

$$\mathcal{L} = \frac{m}{2} g_{AB} \dot{Q}^A \dot{Q}^B - V(Q) = \frac{mr_0^2}{2} (\dot{\theta}^2 + \sin^2(\theta) \dot{\varphi}^2) - mgr_0 \cos(\theta). \tag{10.122}$$

Using the holonomic constraint, we can express the coordinates on the sphere using only a single coordinate θ and time t as

$$\theta(\theta, t) = \theta \quad \text{and} \quad \varphi(\theta, t) = \omega t. \tag{10.123}$$

Note that this is a slight abuse of notation as we have used θ both as one of the coordinates Q^A as well as the coordinate q, but this really does not result in any confusion in this case since $\theta(\theta, t)$ is equal to θ and $\varphi(\theta, t)$ is independent of θ. We find that, with this choice of coordinates,

$$g_{AB}(\partial_t Q^A)(\partial_\theta Q^B) = r_0^2 (\partial_t \theta)(\partial_\theta \theta) + r_0^2 \sin^2(\theta)(\partial_t \omega t)(\partial_\theta \omega t) = 0 \tag{10.124}$$

and we therefore satisfy the requirement of Eq. (10.119). We could also see this by direct insertion of the holonomic constraint into the Lagrangian, upon which we find the effective Lagrangian

$$\mathcal{L}_{\text{eff}} = \frac{mr_0^2}{2}\dot{\theta}^2 + \frac{mr_0^2\omega^2}{2}\sin^2(\theta) - mgr_0\cos(\theta). \tag{10.125}$$

From this expression, we can identify the effective potential

$$V_{\text{eff}} = mr_0\left(g\cos(\theta) - \frac{r_0\omega^2}{2}\sin^2(\theta)\right). \tag{10.126}$$

Instead of using the Lagrangian with the two-dimensional configuration space, we can therefore study the motion of the bead based only on this effective potential with just one degree of freedom.

The equation of motion for θ is given by

$$mr_0^2\ddot{\theta} = -V_{\text{eff}}'(\theta) \tag{10.127}$$

and therefore the possible equilibrium points of the system are given by the stationary points of the effective potential. Differentiating the potential, we find that

$$V_{\text{eff}}'(\theta) = -mr_0\sin(\theta)\left[g + r_0\omega^2\cos(\theta)\right] \tag{10.128}$$

and it is therefore clear that the points $\theta = 0$ and $\theta = \pi$ are always stationary due to the factor $\sin(\theta)$. In addition, if $r_0\omega^2 > g$, then we have two additional stationary points that satisfy

$$\cos(\theta) = -\frac{g}{r_0\omega^2}, \tag{10.129}$$

see Fig. 10.13. From the shape of the potential, we conclude that for $r_0\omega^2 < g$, the point $\theta = \pi$ is a stable equilibrium for the system while the point $\theta = 0$ is unstable. In particular, this is compatible with our expectation from the case when the angular velocity $\omega = 0$. However, as the angular velocity is increased such that $r_0\omega^2 > g$, $\theta = 0$ remains unstable whereas $\theta = \pi$ now also becomes an unstable stationary point. Instead, the new stationary points given by Eq. (10.129) are both stable. We also note that when $r_0\omega^2 \gg g$, these stable points approach $\cos(\theta) = 0$, corresponding to $\theta = \pi/2$ and $\theta = 3\pi/2$. This is also in accordance to our intuition. In the rotating frame, the centrifugal force drives the bead to be as far away from the rotational axis as possible and when the gravitational field is weak, this will be the dominating effect.

There are some conclusions that can be drawn directly from looking at the effective potential. We first note that the time-translation symmetry of the effective Lagrangian indicates that

$$\mathcal{H} = \dot{\theta}\frac{\partial\mathcal{L}_{\text{eff}}}{\partial\dot{\theta}} - \mathcal{L} = \frac{mr_0^2}{2}\dot{\theta}^2 + V_{\text{eff}} = E \tag{10.130}$$

is a constant of motion that may be interpreted as the total energy in the rotating frame. Since $\dot{\theta}^2 \geq 0$, we find that

$$V_{\text{eff}} \leq E, \tag{10.131}$$

indicating that the effective potential at any time must be smaller than the constant E. The maximum of the effective potential is at the point $\theta = 0$ and is given by

$$V_{\text{eff,max}} = V_{\text{eff}}(0) = mr_0 g. \tag{10.132}$$

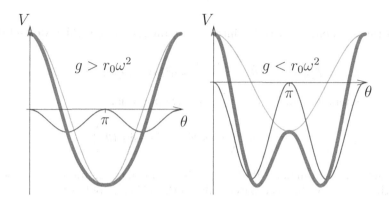

Figure 10.13 The effective potential for the bead on the rotating ring for the two cases of $g > r_0\omega^2$ (left) and $g < r_0\omega^2$ (right), respectively. The thick curve shows the effective potential V_{eff} while the thin light and dark curves show the gravitational and rotational contributions, respectively. In the former case there is one global minimum at $\theta = \pi$ and in the latter two global minima develop on either side of this point, which itself becomes a local maximum.

Thus, if $E < mr_0g$, then the bead will never be able to reach $\theta = 0$ and instead oscillate between the extremal points given by

$$V_{\text{eff}}(\theta_{\text{extr}}) = E, \tag{10.133}$$

where $\dot\theta = 0$. On the other hand, if $E > mr_0g$, then $\dot\theta^2 > 0$ and the bead will never change direction and therefore go around the loop indefinitely. In the case when $r_0\omega^2 < g$, these are the only two possibilities, but when $r_0\omega^2 > g$, the former case can be further split into two different cases. For this case, the local maximum of the effective potential at $\theta = \pi$ is given by

$$V_{\text{eff}}(\pi) = -mr_0g. \tag{10.134}$$

It follows that if $E < -mr_0g$, then the bead can never pass this point and will be constrained to move on one side of the ring. However, if $-mgr_0 < E < mgr_0$, then $\dot\theta$ cannot change sign anywhere except at the extremal points on either side of $\theta = \pi$ and will therefore oscillate between those, see Fig. 10.14.

A general comment about the period of motion in a potential is in order as we shall use this in the next section. If a one-dimensional motion has a constant of motion of the form

$$E = \frac{M(q)}{2}\dot q^2 + V(q) \tag{10.135}$$

with $M(q) > 0$ and $V(q)$ is such that the motion is constrained to be between two extremal points given by

$$V(q_{\text{extr}}) = E, \tag{10.136}$$

then $\dot q^2 > 0$ everywhere in between the extremal points and will not change sign in this region. The time τ to go from one extremal point q_1 to the other q_2 can be found by solving

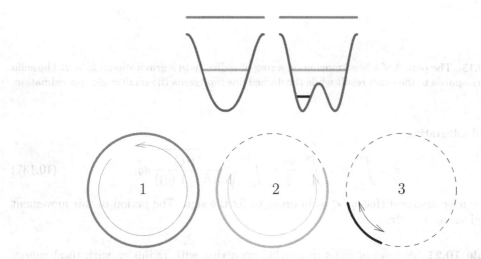

Figure 10.14 Depending on the total energy of the system and the relation between g and $r_0\omega^2$, the motion of the bead will fall into one of three categories. In the first category (1), the energy is large enough for the bead to pass the global maximum of the effective potential and the bead will go around the ring indefinitely. If the energy is lower than the maximum of the effective potential the bead will oscillate between two points on the ring. If the bead has enough energy to pass $\theta = \pi$, i.e., the lower part of the ring, it will oscillate between two points at the same height on opposite sides of the ring (2). The third case (3), when the bead cannot pass $\theta = 0$, only occurs when $g < r_0\omega^2$ and the bead will then oscillate between two points on the same side of the ring. For reference, we also show the effective potentials in both cases with examples of the energy levels corresponding to each case, see also Fig. 10.13. In principle, there are also special cases on the boundary between the cases, where the bead will asymptotically move towards a local maximum of the effective potential.

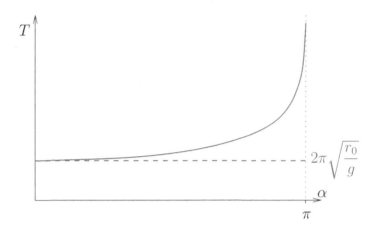

Figure 10.15 The period of a bead moving on a ring of radius r_0 in a gravitational field g. The solid curve corresponds to the exact result while the dashed line represents the small angle approximation.

for $\dot q$ and integrating

$$\tau = \int_{t_1}^{t_2} dt = \int_{q_1}^{q_2} \frac{dq}{\dot q} = \int_{q_1}^{q_2} \sqrt{\frac{M(q)}{2(E - V(q))}}\, dq, \tag{10.137}$$

where we have assumed that $q_1 < q_2$ in order to fix the sign. The period of this movement is then given by $T = 2\tau$.

Example 10.21 A bead of mass m moving on a ring with radius r_0 with total energy $E < mgr_0$ in a gravitational field has turning points at the angles θ_0 that solve the equation

$$E = mgr_0 \cos(\theta_0). \tag{10.138}$$

This implies that

$$\theta_0 - \pi = \pm \arccos\left(-\frac{E}{mgr_0}\right) \equiv \alpha \tag{10.139}$$

and the period of the resulting oscillations is given by

$$T = \int_{-\alpha}^{\alpha} \sqrt{\frac{2mr_0^2}{mgr_0[\cos(s) - \cos(\alpha)]}}\, ds = \sqrt{\frac{8r_0}{g}} \int_0^{\alpha} \frac{ds}{\sqrt{\cos(s) - \cos(\alpha)}}. \tag{10.140}$$

For small α, the integral in the last expression is approximately equal to $\pi/\sqrt2$ and we find that $T = 2\pi\sqrt{r_0/g}$. For larger α, the integral can still be computed numerically and the result is shown in Fig. 10.15 as a function of α.

Another situation where a problem may be rewritten in terms of an effective one for fewer generalised coordinates arises in some cases where the Lagrangian does not depend explicitly on time, indicating that $\mathcal{H}$ is a constant of motion, and there is another constant of motion on top of this. Assume that we have a Lagrangian of the form

$$\mathcal{L} = \frac{1}{2} M_{ab}\dot q^a \dot q^b - V(q). \tag{10.141}$$

The Lagrangian not depending explicitly on time results in the conserved quantity

$$\mathcal{H} = \frac{1}{2} M_{ab} \dot{q}^a \dot{q}^b + V(q) = E, \tag{10.142}$$

which may replace one of the equations of motion. Let us furthermore assume that M_{ab} and $V(q)$ do not depend explicitly on the coordinate $q^c = Q$ for some fixed c. With this assumption, it follows that we have a constant of motion

$$J = \frac{\partial \mathcal{L}}{\partial \dot{q}^c} = M_{ca} \dot{q}^a = \mu \dot{Q} + \lambda_i \dot{q}^i, \tag{10.143}$$

where we have denoted $M_{cc} = \mu$ (no sum) and $M_{ci} = \lambda_i$ and i can take all values except c. Squaring J, we find that

$$\frac{J^2}{\mu} = \mu \dot{Q}^2 + 2\lambda_i \dot{Q} \dot{q}^i + \frac{\lambda_i \lambda_j}{\mu} \dot{q}^i \dot{q}^j. \tag{10.144}$$

At the same time, we can rewrite the first term in the expression for $\mathcal{H}$ according to

$$\frac{1}{2} M_{ab} \dot{q}^a \dot{q}^b = \frac{1}{2} \left(\mu \dot{Q}^2 + 2\lambda_i \dot{Q} \dot{q}^i + M_{ij} \dot{q}^i \dot{q}^j \right) = \frac{1}{2} \underbrace{\left(M_{ij} - \frac{\lambda_i \lambda_j}{\mu} \right)}_{= m_{ij}} \dot{q}^i \dot{q}^j + \frac{J^2}{2\mu}, \tag{10.145}$$

which is completely independent of Q and its time derivative $\dot{Q}$. Since the potential V was also assumed to be independent of Q, the new integrated equation of motion becomes

$$E = \frac{1}{2} m_{ij} \dot{q}^i \dot{q}^j + V_{\text{eff}}(q), \tag{10.146a}$$

where we have introduced the effective potential

$$V_{\text{eff}}(q) = V(q) + \frac{J^2}{2\mu}. \tag{10.146b}$$

as well as the effective inertia m_{ij}. Based on this, one might imagine that the equations of motion would be equivalent to those of the effective Lagrangian

$$\mathcal{L}_{0,\text{eff}} = \frac{1}{2} m_{ij} \dot{q}^i \dot{q}^j - V_{\text{eff}}. \tag{10.147a}$$

However, this is not necessarily true as any effective Lagrangian of the form

$$\mathcal{L}_{\text{eff}} = \mathcal{L}_{0,\text{eff}} + \dot{q}^i \kappa_i(q) \tag{10.147b}$$

will lead to the same constant of motion E although having different equations of motion unless $\kappa_i(q) = \partial_i K(q)$ for some function $K(q)$, which would make the additional term a total derivative. Deriving the equations of motion based on $\mathcal{L}_{\text{eff}}$ and identifying them with the equations of motion for the full problem, we find that the correct equations of motion for the q^i are obtained when

$$\kappa_i(q) = \frac{J \lambda_i}{\mu}. \tag{10.148}$$

The effective Lagrangian giving rise to the correct equations of motion for the q^i is therefore on the form

$$\mathcal{L}_{\text{eff}} = \frac{1}{2} m_{ij} \dot{q}^i \dot{q}^j + \dot{q}^i \frac{J \lambda_i}{\mu} - V_{\text{eff}}. \tag{10.149}$$

Note that, in the case of a one-dimensional effective problem, the additional term $\dot{q}\kappa(q)$ will always be a total derivative and therefore not affect the equations of motion.

A word of warning is in order. We obtained this result by using the fact that J is a constant of motion. This is only true for an on-shell solution to the equations of motion and we therefore *cannot* insert this result directly into the Lagrangian as the Lagrangian is necessary to derive the equations of motion in the first place. However, inserting it into the equation of motion is perfectly fine as the equation of motion is only valid on-shell anyway.

Example 10.22 Consider a particle of mass m moving freely on the surface of a sphere with radius r_0 under the influence of an external gravitational field, see Fig. 10.16. This situation is equivalent to that of Example 10.20 with the holonomic constraint removed and we can write the Lagrangian as

$$\mathcal{L} = \frac{mr_0^2}{2}[\dot{\theta}^2 + \sin^2(\theta)\dot{\varphi}^2] - mgr_0\cos(\theta). \tag{10.150}$$

As this Lagrangian does not depend explicitly on time t nor on the angle φ, we have two corresponding constants of motion

$$E = \frac{mr_0^2}{2}[\dot{\theta}^2 + \sin^2(\theta)\dot{\varphi}^2] + mgr_0\cos(\theta) \quad \text{and} \quad L = mr_0^2\sin^2(\theta)\dot{\varphi}. \tag{10.151}$$

The first of these constants correspond to the total energy of the particle and the second to the angular momentum about the vertical axis. Solving for $\dot{\varphi}$ in terms of L and inserting the result into the expression for the energy E, we find that

$$E = \frac{mr_0^2}{2}\dot{\theta}^2 + \frac{L^2}{2mr_0^2\sin^2(\theta)} + mgr_0\cos(\theta), \tag{10.152}$$

indicating that effective potential of the problem with the φ coordinate removed is given by

$$V_{\text{eff}}(\theta) = \frac{L^2}{2mr_0^2\sin^2(\theta)} + mgr_0\cos(\theta). \tag{10.153}$$

The shape of this effective potential is shown in Fig. 10.17 and since the potential becomes infinite for $\theta = 0$ and $\theta = \pi$ if $L \neq 0$, the particle can never pass through these points. Instead, for a given energy E and angular momentum L, we find that the particle will move between the two angles θ for which

$$\frac{L^2}{2mr_0^2\sin^2(\theta)} + mgr_0\cos(\theta) = E \tag{10.154}$$

and therefore $\dot{\theta} = 0$.

10.3 CENTRAL POTENTIALS AND PLANAR MOTION

An important problem that will apply to several different settings is that of a particle of mass m moving in a *central potential* such that

$$\mathcal{L} = \frac{m}{2}\dot{\vec{x}}^2 - V(r), \tag{10.155}$$

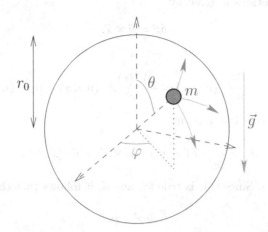

Figure 10.16 A particle with mass m allowed to move freely on the surface of a sphere of radius r_0. The particle is furthermore subjected to a gravitational field $\vec{g}$. The configuration space is described by the spherical coordinates θ and φ.

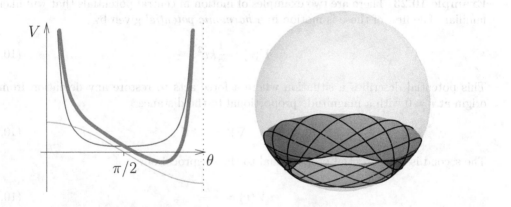

Figure 10.17 To the left, the shape of the effective potential for the particle moving freely on the sphere is shown as a thick curve along with its gravitational (light curve) and angular momentum (dark curve) contributions. The right figure shows a typical region on the sphere to which the motion is constrained for a particular energy along with an actual trajectory for the same values of E and L.

where $r = |\vec{x}|$ is the particle's distance from the origin. This Lagrangian is completely rotationally invariant as can be seen from making an arbitrary rotation around the vector $\vec{n}$, which for small rotations is given by

$$\delta\vec{x} = \vec{n} \times \vec{x}. \tag{10.156}$$

This results in

$$\delta\mathcal{L} = \frac{\partial\mathcal{L}}{\partial\vec{x}} \cdot (\vec{n} \times \vec{x}) + \frac{\partial\mathcal{L}}{\partial\dot{\vec{x}}} \cdot (\vec{n} \times \dot{\vec{x}}) = -\frac{V'(r)}{r}\vec{x} \cdot (\vec{n} \times \vec{x}) + m\dot{\vec{x}} \cdot (\vec{n} \times \dot{\vec{x}}) = 0 \tag{10.157}$$

and by Noether's theorem the quantity

$$J_{\vec{n}} = \frac{\partial\mathcal{L}}{\partial\dot{\vec{x}}} \cdot \delta\vec{x} = m\dot{\vec{x}} \cdot (\vec{n} \times \vec{x}) = \vec{n} \cdot (\vec{x} \times m\dot{\vec{x}}) \tag{10.158}$$

is a constant of motion. Since this is true for any $\vec{n}$, it follows that the angular momentum vector

$$\vec{L} = \vec{x} \times m\dot{\vec{x}} \tag{10.159}$$

is a constant of motion. By selecting coordinates such that $\vec{L} = L\vec{e}_3$, we find that

$$\vec{x} \cdot \vec{L} = Lx^3 = \vec{x} \cdot (\vec{x} \times m\dot{\vec{x}}) = 0, \tag{10.160}$$

indicating that $x^3 = 0$ and the motion occurs completely in the x^1-x^2-plane. Because of this, we can neglect the x^3 direction completely and it is further convenient to use polar coordinates in the plane of motion so that

$$\mathcal{L} = \frac{m}{2}(\dot{\rho}^2 + \rho^2\dot{\phi}^2) - V(\rho). \tag{10.161}$$

Example 10.23 There are two examples of motion in central potentials that will likely be familiar. The first of these is motion in a *harmonic potential* given by

$$V(r) = \frac{1}{2}kr^2. \tag{10.162}$$

This potential describes a situation where a force acts to restore any deviation from the origin at $r = 0$ with a magnitude proportional to the distance

$$\vec{F} = -\nabla V = -k\vec{x}. \tag{10.163}$$

The second is the potential proportional to the reciprocal of r

$$V(r) = -\frac{k}{r}. \tag{10.164}$$

This potential is encountered both in *Newton's law of gravitation* with $k = Gm_1m_2$ and in the electrostatic *Coulomb law* with $k = -Q_1Q_2/4\pi\varepsilon_0$. The corresponding force for this type of potential is given by

$$\vec{F} = -\nabla V = -k\frac{\vec{x}}{r^3} = -k\frac{\vec{e}_r}{r^2}. \tag{10.165}$$

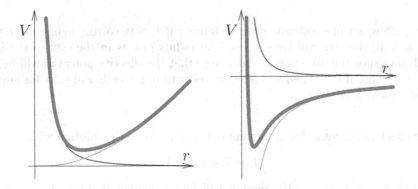

Figure 10.18 The effective potentials for motion in a central harmonic potential (left) and in a Kepler potential (right) are shown as thick curves. The individual contributions from the potential $V(r)$ itself and the angular momentum barrier are shown as light and dark curves, respectively.

With the restriction to planar motion, we already know that $L = m(x^1\dot{x}^2 - x^2\dot{x}^1)$ is a conserved quantity. We can also see this directly from the two-dimensional Lagrangian, which does not depend explicitly on the angle ϕ and therefore has the constant of motion

$$L = \frac{\partial \mathcal{L}}{\partial \dot{\phi}} = m\rho^2 \dot{\phi}. \tag{10.166}$$

Using the results of Section 10.2.5, we find that the problem can be rewritten as an effective one-dimensional problem for ρ with the effective Lagrangian

$$\mathcal{L}_{\text{eff}} = \frac{m}{2}\dot{\rho}^2 - V_{\text{eff}}(\rho), \tag{10.167a}$$

where we have introduced the effective potential

$$V_{\text{eff}}(\rho) = \frac{L^2}{2m\rho^2} + V(\rho) \tag{10.167b}$$

in which the first term is sometimes referred to as the *angular momentum barrier*. The total energy of the system is given by

$$E = \frac{m}{2}\dot{\rho}^2 + V_{\text{eff}}(\rho) \tag{10.168}$$

and is also a constant of motion.

Example 10.24 For both of the cases discussed in Example 10.23, the effective potentials for $L \neq 0$ are shown in Fig. 10.18. As $V_{\text{eff}} \to \infty$ as $\rho \to 0$ for both of these cases, there will be a minimal radius $\rho_{\text{min}} > 0$ for any value of E. In the case of the harmonic potential, we also find that $V_{\text{eff}} \to \infty$ as $\rho \to \infty$ and thus there will also exist a maximal radius ρ_{max} for any energy E. The minimal and maximal radii are the solutions to the equation

$$V_{\text{eff}}(\rho) = \frac{L^2}{2m\rho^2} + \frac{1}{2}k\rho^2 = E \implies \rho^2 = \frac{E}{k} \pm \sqrt{\frac{E^2}{k^2} - \frac{L^2}{mk}}. \tag{10.169}$$

For the $1/r$ potential and a positive k, i.e., an attractive potential, we find that $V_{\text{eff}} \to 0$

as $\rho \to \infty$. Thus, ρ will eventually grow to infinity if $E > 0$, corresponding to an unbound state. If $E < 0$, then we will have a maximal radius just as in the case of the harmonic potential, corresponding to a bound state. Note that the effective potential will be negative for all $\rho > L^2/2mk$ if $k > 0$ and so bound states with negative E will exist for any value of the angular momentum L.

If the effective potential has a minimum at $\rho = \rho_0$, then the solution with

$$E = V_{\text{eff}}(\rho_0) \equiv V_0 \tag{10.170}$$

will only allow $\dot{\rho} = 0$ and so the motion will be completely circular with radius ρ_0. For energies slightly larger than this value, only small deviations from $\rho = \rho_0$ will be allowed and we can use perturbation theory to find the linear effects. In order to do this, we assume that $\rho = \rho_0 + x$, where x is assumed to be small at all times. The effective potential now takes the form

$$V_{\text{eff}}(\rho) \simeq V_0 + \frac{1}{2}V_{\text{eff}}''(\rho_0)x^2 \equiv V_0 + \frac{1}{2}V_0''x^2 \tag{10.171}$$

to second order in x. If the energy is assumed to be

$$E = V_0 + \varepsilon, \tag{10.172}$$

then the turning points are given by

$$x_{\pm} \simeq \pm\sqrt{\frac{2\varepsilon}{V_0''}}. \tag{10.173}$$

The period of small oscillations around ρ_0 is therefore found through the expression

$$T_\rho = 2\int_{x_-}^{x_+} \sqrt{\frac{m}{2\varepsilon - V_0''x^2}}\, dx = 2\pi\sqrt{\frac{m}{V_0''}}. \tag{10.174}$$

This should not come as a surprise as the equation of motion for x to leading order is the same as that for a harmonic oscillator with angular frequency $\omega = \sqrt{V_0''/m}$. The period T_ρ is the period of small oscillations in the ρ coordinate, which we can compare with the time T_ϕ taken for the angular coordinate to increase by 2π. To leading order, we find that

$$\dot{\phi} = \frac{L}{m\rho_0^2} \implies T_\phi = \int_0^{2\pi} \frac{d\phi}{\dot{\phi}} = 2\pi\frac{m\rho_0^2}{L}. \tag{10.175}$$

Generally, there is nothing requiring the periods T_ρ and T_ϕ to be the same and if they are not, the point at which the radius ρ takes its smallest value will not be the same for successive orbits. This phenomenon is known as *orbital precession* and the angle between successive closest approaches is approximately given by

$$\Delta\phi = 2\pi\frac{T_\phi - T_\rho}{T_\phi} = 2\pi\left(1 - \frac{L}{\sqrt{m\rho_0^4 V_0''}}\right), \tag{10.176}$$

see Fig. 10.19. Note that we have here only computed the orbital precession to leading order in deviations from the fully circular orbit. In order to conclude that there is no orbital precession at all, we would have to verify that the higher order corrections also vanish or use exact methods.

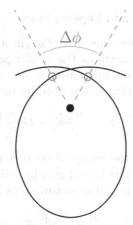

Figure 10.19 If the period of radial oscillations does not match the orbital period, the point of closest approach to the center of the potential will precess. The angle $\Delta\phi$ between successive closest approaches can be found by comparing the periods.

Example 10.25 Consider a central potential of the form

$$V(r) = \frac{k}{\alpha}r^\alpha, \tag{10.177}$$

where α is a fixed number. The effective potential for the ρ coordinate is given by

$$V_{\text{eff}}(\rho) = \frac{L^2}{2m\rho^2} + \frac{k}{\alpha}\rho^\alpha \tag{10.178}$$

and its minimum is found for the ρ_0 that satisfies

$$V_{\text{eff}}'(\rho_0) = -\frac{L^2}{m\rho_0^3} + k\rho_0^{\alpha-1} = 0 \implies \rho_0 = \left(\frac{L^2}{mk}\right)^{\frac{1}{\alpha+2}}. \tag{10.179}$$

A second differentiation now results in

$$m\rho_0^4 V_0'' = 3L^2 + mk(\alpha-1)\rho_0^{\alpha+2} = (\alpha+2)L^2. \tag{10.180}$$

It follows that

$$\Delta\phi = 2\pi\left(1 - \frac{1}{\sqrt{\alpha+2}}\right) \tag{10.181}$$

and, in particular, $\Delta\phi = \pi$ for the special case $\alpha = 2$ of the harmonic potential. Thus, for the harmonic potential, the motion reaches the smallest ρ value twice per orbit. While we here have only shown this to leading order, it is a result that holds to all orders and is rather straightforward to derive by solving the problem exactly. It follows that the orbit is closed, i.e., the orbit returns to the same position with the same velocity on every turn. The same is true when $\alpha = -1$, which results in $\Delta\phi = 0$. This result is also true to all orders and we will discuss this case in detail below.

10.3.1   The two-body problem and Kepler's laws

One of the more important applications of motion in a central potential is the *two-body problem* in a $1/r$ potential, also known as the *Kepler problem*. We here assume that two masses m_1 and m_2 are moving freely apart from a mutual potential given by by Newton's gravitational potential. This situation is described by the Lagrangian

$$\mathcal{L} = \frac{m_1}{2}\dot{\vec{x}}_1^2 + \frac{m_2}{2}\dot{\vec{x}}_2^2 + G\frac{m_1 m_2}{|\vec{x}_2 - \vec{x}_1|} \tag{10.182}$$

and has six degrees of freedom in the two position vectors $\vec{x}_1$ and $\vec{x}_2$ of the corresponding masses. It is preferable to rewrite this Lagrangian in a different set of coordinates by using the center of mass position and separation vector, given by

$$\vec{X} = \frac{m_1\vec{x}_1 + m_2\vec{x}_2}{m_1 + m_2} = \frac{\mu}{m_2}\vec{x}_1 + \frac{\mu}{m_1}\vec{x}_2 \quad \text{and} \quad \vec{x} = \vec{x}_2 - \vec{x}_1, \tag{10.183}$$

respectively, where

$$\mu = \frac{m_1 m_2}{m_1 + m_2} = \frac{m_1 m_2}{M} \tag{10.184a}$$

is the *reduced mass* of the system and $M = m_1 + m_2$ the total mass. For future reference, we note that the inversion of this transformation is given by

$$\vec{x}_1 = \vec{X} - \frac{\mu}{m_1}\vec{x} \quad \text{and} \quad \vec{x}_2 = \vec{X} + \frac{\mu}{m_2}\vec{x}. \tag{10.184b}$$

In this coordinate system, the Lagrangian takes the simpler form

$$\mathcal{L} = \frac{M}{2}\dot{\vec{X}}^2 + \frac{\mu}{2}\dot{\vec{x}} + G\frac{m_1 m_2}{|\vec{x}|}, \tag{10.185}$$

which does not have any cross terms involving both $\vec{X}$ and $\vec{x}$ and therefore can be treated as two separate problems, one for $\vec{X}$ and the other for $\vec{x}$. Since the Lagrangian does not depend explicitly on the center of mass coordinates $\vec{X}$, the center of mass momentum

$$\vec{P} = M\dot{\vec{X}} \tag{10.186}$$

is a constant of motion and therefore the center of mass moves at constant velocity

$$\vec{v}_0 = \frac{\vec{P}}{M}. \tag{10.187}$$

The problem for the separation vector is a central potential problem with mass μ and potential

$$V(r) = -G\frac{m_1 m_2}{r} = -G\frac{\mu M}{r}. \tag{10.188}$$

Let us therefore examine the properties of the orbits of this type of potential in more detail to see if we can find some familiar results.

   We start by looking at the conserved energy E, which is given by

$$E = \frac{\mu}{2}\dot{\rho}^2 + \frac{L^2}{2\mu\rho^2} - G\frac{\mu M}{\rho}. \tag{10.189}$$

In order to determine the spatial shape of the orbit, we would like to find ρ as a function of

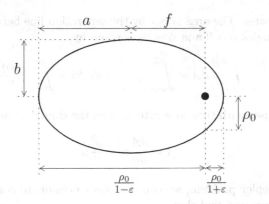

Figure 10.20 Some different parameters that may be used in describing an ellipse. The semi-major axis a, the semi-minor axis b, the semi-latus rectum ρ_0, and the focal distance f. The eccentricity ε is defined as the quotient f/a. The black dot represents one of the foci of the ellipse.

the angle ϕ rather than time t. Luckily, ϕ is a monotonic function of t and we can therefore make the substitution

$$\dot{\rho} = \frac{d\rho}{dt} = \frac{d\rho}{d\phi}\frac{d\phi}{dt} = \frac{d\rho}{d\phi}\dot{\phi} = \rho'\frac{L}{\mu\rho^2}, \tag{10.190}$$

where we have introduced the notation $\rho' = d\rho/d\phi$ and used the expression for the angular momentum in the last step. Substituting this into the expression for the energy and solving for ρ'^2 results in

$$\rho'^2 = \frac{2\mu}{L^2}\left(E\rho^4 + G\mu M\rho^3\right) - \rho^2. \tag{10.191}$$

While this differential equation might look unfamiliar, consider an ellipse of eccentricity $\varepsilon < 1$ and one of its focal points in the origin. Such an ellipse is described by the curve

$$\rho(\phi) = \frac{\rho_0}{1 + \varepsilon\cos(\phi)}, \tag{10.192}$$

where the parameter ρ_0 determines the size of the ellipse, see Fig. 10.20. Differentiating this expression and substituting in the expression for $\rho(\phi)$ results in

$$\rho'^2 = \frac{\varepsilon^2 - 1}{\rho_0^2}\rho^4 + \frac{2}{\rho_0}\rho^3 - \rho^2. \tag{10.193}$$

Identification with the differential equation for the orbit, we find that

$$\rho_0 = \frac{L^2}{G\mu^2 M} \quad \text{and} \quad \varepsilon^2 = 1 + \frac{2L^2 E}{G^2\mu^3 M^2}. \tag{10.194}$$

Note that the condition $\varepsilon < 1$ is satisfied as long as we have a bound state, i.e., as long as $E < 0$. We have thus showed that the shape of the orbit is an ellipse with a focal point at $\rho = 0$ and determined its orbital parameters ρ_0 and ε in terms of the constants of motion. The fact that the orbit is an ellipse is *Kepler's first law*. For unbound states with $E > 0$, the bodies no longer orbit each other and the relative trajectory takes a different shape, see Problem 10.33.

Having determined that the orbits of the Kepler problem are ellipses, let us find a geometrical implication of the conservation of angular momentum. Consider a general curve

$\rho(\phi)$ in polar coordinates. The area swept by the separation line between the origin and the curve between the angles $\phi = 0$ and $\phi = \phi_0$ is given by

$$A(\phi_0) = \int_{\phi=0}^{\phi_0} \int_{r=0}^{\rho(\phi)} dA = \int_{\phi=0}^{\phi_0} \int_{r=0}^{\rho(\phi)} r \, dr \, d\phi = \int_0^{\phi_0} \frac{\rho(\phi)^2}{2} d\phi. \tag{10.195}$$

If the angle ϕ_0 increases with time at a rate $\dot{\phi}$, then the time derivative of the swept area is

$$\dot{A} = \frac{dA}{d\phi}\dot{\phi} = \frac{\rho^2}{2}\dot{\phi}. \tag{10.196}$$

In the case of the Kepler problem, we can use this to eliminate $\dot{\phi}$ from the expression for the angular momentum and find that

$$\dot{A} = \frac{L}{2\mu}. \tag{10.197}$$

The area swept per time unit in the Kepler problem is therefore constant, which is *Kepler's second law*, and equal to $L/2\mu$.

Finally, we look at the semi-major axis a of an ellipse, see Fig. 10.20, which is given by

$$a = \frac{\rho_0}{2}\left(\frac{1}{1-\varepsilon} + \frac{1}{1+\varepsilon}\right) = \frac{\rho_0}{1-\varepsilon^2}. \tag{10.198}$$

At the same time, the semi-minor axis b satisfies the relation

$$b^2 = (1 - \varepsilon^2)a^2 \tag{10.199}$$

and the squared area of the ellipse is therefore given by

$$A^2 = \pi^2 a^2 b^2 = \pi^2 (1 - \varepsilon^2)a^4 = \pi^2 a^3 \rho_0 = \pi^2 a^3 \frac{L^2}{G\mu^2 M}. \tag{10.200}$$

Since the area swept per time is constant, we can relate the area to the orbital period through the relation

$$T = \frac{A}{\dot{A}} \implies T^2 = \frac{A^2}{\dot{A}^2} = \frac{4\pi^2}{GM}a^3, \tag{10.201}$$

where we have used the expressions for $\dot{A}$ and A from Eqs. (10.196) and (10.200), respectively. We therefore find that, regardless of the orbital parameters E and L, the square of the orbital period is proportional to the cube of the semi-major axis, which is the statement of *Kepler's third law*. In addition, if one of the bodies is much heavier than the other, i.e., if $M \simeq m_1 \gg m_2$, then the proportionality constant does not depend significantly on the lesser mass. In particular, our solar system is dominated by the mass of the Sun, and all the planets are significantly lighter. This means that all planets will have approximately the same proportionality constant in the relation between the orbital period and the semi-major axis, which is what led Kepler to draw the conclusion of the third law in the first place. We have now derived all of Kepler's laws directly from Newton's theory of gravitation. It should be pointed out that Kepler did not have this luxury as Newton had not yet presented his theory and worked out the implications. Kepler based his conclusions solely on experimental observations. An argument for Kepler's third law for the case of a circular orbit can also be made using dimensional analysis only, see Problem 3.43.

So what about orbital precession? As we have seen, Newton's theory predicts that the orbits are closed ellipses and therefore that there should not be any orbital precession in the case of the pure two-body problem. In reality, the solar system is not a two-body system, but all constituents affect each other gravitationally, leading to corrections to the first order approximation where the Sun is the only gravitational attractor. Taking the disturbances of other bodies into account will generally lead to orbital precession. However, even in doing so, the orbit of Mercury shows a precession that is not compatible with Newton's theory. This observation was one of the earliest experimental evidence for general relativity, which matches the experimental data perfectly.

10.3.2   The restricted three-body problem

Having solved the two-body problem, the next natural question is whether we can solve the general three-body problem as well. Ultimately, this turns out to be a situation that cannot be solved by exact methods although some particular solutions do exist. We will therefore constrain our discussion to the special case of the *restricted three-body problem*, which assumes that the three masses satisfy $m_1 \geq m_2 \gg m_3$ so that the motion of the two heavier bodies is not significantly affected by the third and therefore can be approximated with the solution to the two-body problem. We furthermore assume that all of the motion occurs in the plane and that the orbit of the two heavier masses have an eccentricity $\varepsilon = 0$, i.e., their orbits are circular. We therefore find that

$$E = -\frac{G^2 \mu^3 M^2}{L^2} = V_{\text{eff}}(\rho_0) \tag{10.202}$$

and that the constant orbital angular velocity is given by

$$\omega^2 = \frac{4\pi^2}{T^2} = \frac{GM}{\rho_0^3}, \tag{10.203}$$

in accordance with Kepler's third law.

When looking at the motion of the third body in the plane, it is preferable to do so in the co-rotating frame where the other two bodies are at rest, which rotates with angular frequency ω. We also restrict ourselves to look at the case where the center of mass is stationary at the origin as the case when the center of mass is moving can be obtained by performing a Galilei boost of the resulting solution. The Lagrangian for the third body in the rotating frame is given by

$$\mathcal{L}_3 = \frac{m_3}{2}\dot{\vec{y}}_3^2 - m_3\omega(\dot{y}_3^1 y_3^2 - \dot{y}_3^2 y_3^1) + \frac{m_3}{2}\omega^2 \vec{y}_3^2 - V_{31}(r_{13}) - V_{32}(r_{23}), \tag{10.204}$$

where V_{3i} is the gravitational potential of the interaction between the ith and third bodies and r_{i3} the distance between them

$$V_{3i} = -\frac{Gm_i m_3}{r_{i3}}. \tag{10.205}$$

Furthermore, we are free to select coordinates in the rotating frame in such a way that the separation between the first two bodies lies entirely along the $\vec{e}_1$ direction. With the solution to the two-body problem, this means that the positions of the first two masses are given by (cf. Eq. (10.184b))

$$\vec{y}_1 = -\frac{\mu\rho_0}{m_1}\vec{e}_1 \quad \text{and} \quad \vec{y}_2 = \frac{\mu\rho_0}{m_2}\vec{e}_1, \tag{10.206}$$

respectively, and that

$$r_{i3} = \sqrt{(y_3^1 - y_i^1)^2 + (y_3^2)^2}. \tag{10.207}$$

Defining the effective potential

$$V_{3\text{eff}} = -\frac{m_3}{2}\omega^2\vec{y}^2 + V_{31}(r_{13}) + V_{32}(r_{23}), \tag{10.208}$$

we find that the equations of motion are given by

$$m_3\ddot{y}_3^1 = 2m_3\omega\dot{y}_3^2 - \frac{\partial V_{3\text{eff}}}{\partial y_3^1}, \tag{10.209a}$$

$$m_3\ddot{y}_3^2 = -2m_3\omega\dot{y}_3^1 - \frac{\partial V_{3\text{eff}}}{\partial y_3^2}. \tag{10.209b}$$

As the Coriolis force depends linearly on the velocity of the body, it does not affect the stationary points of the problem for which the coordinates being constant is a valid solution to the equations of motion. Instead, such a solution will still be given by the stationary points of the effective potential $V_{3\text{eff}}$, i.e., the points where its gradient is equal to zero. Imposing this condition results in

$$\omega^2 y_3^1 = \frac{Gm_1}{r_{13}^3}\left(y_3^1 + \frac{\mu\rho_0}{m_1}\right) + \frac{Gm_2}{r_{23}^3}\left(y_3^1 - \frac{\mu\rho_0}{m_2}\right), \tag{10.210a}$$

$$\omega^2 y_3^2 = \frac{Gm_1}{r_{13}^3}y_3^2 + \frac{Gm_2}{r_{23}^3}y_3^2. \tag{10.210b}$$

The latter of these equations is trivially solved by $y_3^2 = 0$, which is also intuitive since the effective potential is symmetric about the y_3^1-axis, so let us start by looking for solutions to the first one where this is the case. In Fig. 10.21, the behaviour of the left- and right-hand sides of the first equation for this situation is shown. In particular, regardless of the masses m_1 and m_2, the right-hand side will always be monotonically decreasing and go from zero to $-\infty$ to the left of the first mass, from ∞ to $-\infty$ in the region between the two masses, and from ∞ to zero to the right of the second mass. As a result, the straight line given by $\omega^2 y_3^1$ must intersect the graph for the right-hand side once in each of these regions, each providing one stationary solution to the equations of motion. These points are known as *Lagrange points* of the m_1-m_2-system and are labelled $L1$, $L2$, and $L3$ as shown in the figure. If a small body is placed at one of those points, the force of gravity of the two other bodies and the centrifugal force exactly cancel out and the smaller body will rotate along with the other two at a fixed position in the rotating frame.

Let us examine whether there are any Lagrange points away from $y_3^2 = 0$. In this situation, Eq. (10.210b) can be divided by y_3^2 and we find that

$$\frac{Gm_1}{r_{13}^3} = \omega^2 - \frac{Gm_2}{r_{23}^3}. \tag{10.211}$$

Inserting this into Eq. (10.210a) now yields

$$\omega^2 = \frac{GM}{r_{23}^3} \tag{10.212}$$

and comparison with Eq. (10.203) directly gives $r_{23} = \rho_0$, which also implies that $r_{13} = \rho_0$ and thus

$$r_{13} = r_{12} = \rho_0. \tag{10.213}$$

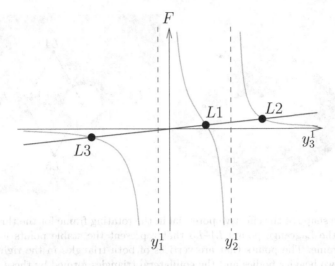

Figure 10.21 The behaviour of the left- and right-hand sides of Eq. (10.210a) for $y_3^2 = 0$, representing the centrifugal (dark) and gravitational (light) forces in the co-rotating frame. When the two-contributions are equal, the net force on a third body at rest at that point is zero and the point is stable. Such points are Lagrange points of the system and there are three such points with $y_3^2 = 0$, called $L1$, $L2$, and $L3$.

Since ρ_0 is the distance between the first and second masses, this implies that any of the stationary points we seek along with the positions of the first two masses form the vertices of an equilateral triangle. There are two such points in the plane, labelled $L4$ and $L5$, see Fig. 10.22, where we show the shape of the effective potential $V_{3\text{eff}}$ along with the positions of all five Lagrange points.

10.4   HAMILTONIAN MECHANICS

Apart from the Newtonian and Lagrangian formulations of classical mechanics, there is a third formulation that is also equivalent in many situations. The idea behind *Hamiltonian mechanics* is to treat the generalised coordinates and their corresponding canonical momenta on a more equal footing and while problems are not necessarily easier to solve in this formulation, it provides a deeper understanding for how classical mechanics work. In addition, many of the concepts of Hamiltonian mechanics will be recognisable in quantum mechanics, which might otherwise appear as a set of ad hoc assumptions.

10.4.1   Phase space

In Lagrangian mechanics, we were faced with a Lagrangian that generally depended on the generalised coordinates q^a and their time derivatives $\dot{q}^a$, leading N second order differential equations in the case of N generalised coordinates. Such a differential requires $2N$ initial conditions in order for the solution to be completely specified. For a function $x(t)$ that is determined by a second order differential equation with initial conditions $x(0) = x_0$ and $\dot{x}(0) = v_0$, it is possible to introduce an additional auxiliary variable

$$v(t) = \dot{x}(t) \tag{10.214}$$

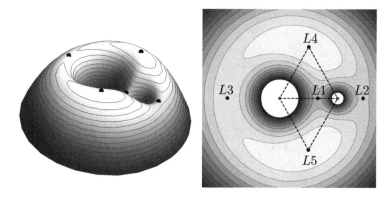

Figure 10.22 The shape of the effective potential in the rotating frame for the three-body problem. Also shown are the Lagrange points $L1$–$L5$ that represent the stable points for the third body in the rotating frame. The points that are vertices of both triangles in the right figure show the positions of the two heavier bodies and the equilateral triangles formed by these positions and the Lagrange points $L4$ and $L5$ are also drawn explicitly.

in order to rewrite the second order differential equation as a first order one, replacing $\ddot{x}(t)$ with $\dot{v}(t)$. The new differential equation along with Eq. (10.214) will then form a set of two coupled first order differential equations and will therefore also require two initial conditions, just as the original second order one, given by $x(0) = x_0$ and $v(0) = v_0$.

Example 10.26 For the harmonic oscillator, the equations of motion are given by

$$m\ddot{x}(t) + kx(t) = 0. \tag{10.215}$$

Introducing $v(t) = \dot{x}(t)$, this second order differential equation is equivalent to the set

$$m\dot{v}(t) + kx(t) = 0, \tag{10.216a}$$
$$\dot{x}(t) - v(t) = 0 \tag{10.216b}$$

of first order differential equations. In this particular case, the differential equations are linear and may be written as

$$\begin{pmatrix} \dot{v}(t) \\ \dot{x}(t) \end{pmatrix} = \underbrace{\begin{pmatrix} 0 & -\frac{k}{m} \\ 1 & 0 \end{pmatrix}}_{=\Omega} \underbrace{\begin{pmatrix} v(t) \\ x(t) \end{pmatrix}}_{=X} \iff \dot{X} = \Omega X. \tag{10.217}$$

This differential equation has the solution $X = \exp(\Omega t)X(0)$. The eigenvalues λ of the matrix Ω are given by the characteristic equation

$$\lambda^2 + \frac{k}{m} = 0 \implies \lambda = \pm i\sqrt{\frac{k}{m}} \equiv \pm i\omega. \tag{10.218}$$

Since the eigenvalues are purely imaginary, the resulting solution will describe periodic oscillations with the characteristic frequency ω of the harmonic oscillator.

The above discussion and example demonstrates the possibility of rewriting a second order differential equation in terms of two first order ones. However, it is not always the case that the most convenient choice for the auxiliary variable is just the time derivative of the original one. In fact, as we shall see shortly, in classical mechanics with a set of generalised coordinates q^a, it will be natural to use the canonical momenta

$$p_a = \frac{\partial \mathcal{L}}{\partial \dot{q}^a} \tag{10.219}$$

instead. The set of N generalised coordinates along with the N corresponding canonical momenta is known as the *phase space* of a system and describes not only its configuration, but also its current state of motion.

Example 10.27 In the case of the harmonic oscillator, the Lagrangian is given by

$$\mathcal{L} = \frac{m}{2}\dot{x}^2 - \frac{k}{2}x^2. \tag{10.220}$$

The canonical momentum is therefore given by

$$p = \frac{\partial \mathcal{L}}{\partial \dot{x}} = m\dot{x} \tag{10.221}$$

and the corresponding equations of motion in phase space are given by this definition and

$$\dot{p} + kx = 0. \tag{10.222}$$

This second equation is just Newton's second law for the linear restoring force $-kx$.

Viewing q^a and p_a as coordinates on the phase space, the time derivative of the coordinates will be uniquely determined by a function of the coordinates themselves. This means that if we draw a diagram of phase space, we can represent the evolution from every point in phase space by a vector with components $\dot{q}^a$ and $\dot{p}_a$ at that point, given by the $2N$ differential equations. The time evolution of the system will be the flow lines of the resulting vector field. Since these lines are the flow lines of a vector field, they will never cross each other.

Example 10.28 The vector field with components $\dot{x}$ and $\dot{p}$ in phase space in the case of the harmonic oscillator is shown in Fig. 10.23. The corresponding flow lines are closed circles, illuminating the fact that the harmonic oscillator solutions are periodic.

Example 10.29 Let us consider a bead of mass m moving on a ring in a gravitational field of strength g. Using the angle θ between the bead's position and the position on top of the ring as the generalised coordinate, the Lagrangian is given by

$$\mathcal{L} = \frac{mr_0^2}{2}\dot{\theta}^2 - mgr_0\cos(\theta). \tag{10.223}$$

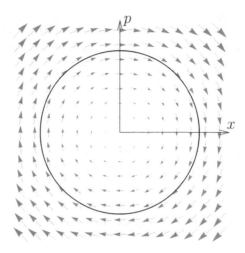

Figure 10.23 The phase space of the harmonic oscillator and the vector field with components $\dot{x}$ and $\dot{p}$. The black circle represents one of the flow lines of the vector field.

The corresponding canonical momentum now takes the form

$$p = \frac{\partial \mathcal{L}}{\partial \dot{\theta}} = mr_0^2 \dot{\theta}, \tag{10.224}$$

which can be interpreted as the angular momentum about the ring's center. This results in the equations of motion

$$\dot{\theta} = \frac{p}{mr_0^2} \quad \text{and} \quad \dot{p} = mgr_0 \sin(\theta) \tag{10.225}$$

and the corresponding phase space flow is shown in Fig. 10.24.

10.4.2   The Hamiltonian

We have already encountered the *Hamiltonian*

$$\mathcal{H} = p_a \dot{q}^a - \mathcal{L} \tag{10.226}$$

as a quantity that is conserved whenever the Lagrangian $\mathcal{L}$ is invariant under time translations. In the Hamiltonian formalism, it will take a much more prominent role. In general, the differential of the Hamiltonian will be given by

$$d\mathcal{H} = \dot{q}^a dp_a + p_a d\dot{q}^a - \frac{\partial \mathcal{L}}{\partial q^a} dq^a - \frac{\partial \mathcal{L}}{\partial \dot{q}^a} d\dot{q}^a - \frac{\partial \mathcal{L}}{\partial t} dt = \dot{q}^a dp_a - \dot{p}_a dq^a - \frac{\partial \mathcal{L}}{\partial t} dt, \tag{10.227}$$

where the second step follows from the definition of the canonical momentum and the equations of motion and therefore holds for on-shell solutions. When we dealt with the Hamiltonian earlier, we viewed the canonical momenta p_a as functions of the generalised coordinates and their time derivatives. However, we can use the definition of the canonical

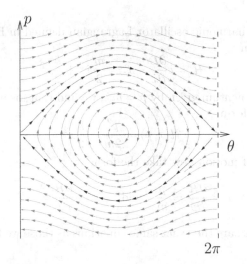

Figure 10.24 The phase space flow for a bead moving on a ring in a gravitational field with the generalised coordinate θ on the horizontal axis and its corresponding canonical momentum p on the vertical one. Note that the $\theta = 0$ and $\theta = 2\pi$ correspond to the same point in phase space and flowing out on the right edge means flowing in the left edge. Marked in black are the flows corresponding to the solutions separating the families of solutions where the bead motion oscillates around the bottom of the ring and where the bead continues around the loop indefinitely. These flows move asymptotically toward the unstable fixed point at $\theta = 0$.

momenta to express $\dot{q}^a$, and therefore also the Hamiltonian, as a function of q^a, p_a, and t and as such, we find that

$$d\mathcal{H} = \frac{\partial \mathcal{H}}{\partial q^a} dq^a + \frac{\partial \mathcal{H}}{\partial p_a} dp^a + \frac{\partial \mathcal{H}}{\partial t} dt. \tag{10.228}$$

Comparing the two expressions for $d\mathcal{H}$ now gives us the relations

$$\dot{q}^a = \frac{\partial \mathcal{H}}{\partial p_a}, \quad \dot{p}_a = -\frac{\partial \mathcal{H}}{\partial q^a}, \quad \text{and} \quad \frac{\partial \mathcal{H}}{\partial t} = -\frac{\partial \mathcal{L}}{\partial t}. \tag{10.229}$$

The first two of these equations are *Hamilton's equations of motion* and are precisely the first order differential equations in phase space that we need to solve in order to determine the time evolution of the system. If we have an expression for the Hamiltonian, we can therefore find the corresponding equations of motion by taking its partial derivatives with respect to the phase space coordinates q^a and p_a.

It should be noted that just as the configuration space and the Lagrangian defines a physical model in Lagrangian mechanics, the phase space and the Hamiltonian along with Hamilton's equations of motion define the model in Hamiltonian mechanics. The above argumentation serves only to conclude that with the Hamiltonian defined according to Eq. (10.226) with respect to the Lagrangian, known as the *Legendre transform* of the Lagrangian with respect to $\dot{q}^a$, then Hamilton's equations are equivalent to the equations of motion based on Lagrangian mechanics.

Example 10.30 The harmonic oscillator Lagrangian defined in Eq. (10.220) has the corresponding Hamiltonian

$$\mathcal{H} = \dot{x}\frac{\partial \mathcal{L}}{\partial \dot{x}} - \mathcal{L} = \frac{m}{2}\dot{x}^2 + \frac{k}{2}x^2. \tag{10.230}$$

With the canonical momentum given by $p = m\dot{x}$, we find that, as a function of x and p, the Hamiltonian takes the form

$$\mathcal{H}(x,p) = \frac{p^2}{2m} + \frac{k}{2}x^2. \tag{10.231}$$

Hamilton's equations of motion now take the form

$$\dot{x} = \frac{\partial \mathcal{H}}{\partial p} = \frac{p}{m} \quad \text{and} \quad \dot{p} = -\frac{\partial \mathcal{H}}{\partial x} = -kx, \tag{10.232}$$

which, hardly surprising, are the exact same equations as those we found in Examples 10.26 and 10.27.

Example 10.31 Consider a particle moving in a central potential $V(r)$. According to our previous discussion on central potentials, this motion occurs in a plane in which we can introduce polar coordinates ρ and ϕ upon which the Lagrangian takes the form

$$\mathcal{L} = \frac{m}{2}(\dot{\rho}^2 + \rho^2\dot{\phi}^2) - V(\rho). \tag{10.233}$$

The canonical momenta corresponding to the ρ and ϕ coordinates are given by

$$p_\rho = \frac{\partial \mathcal{L}}{\partial \dot{\rho}} = m\dot{\rho} \quad \text{and} \quad p_\phi = \frac{\partial \mathcal{L}}{\partial \dot{\phi}} = m\rho^2\dot{\phi}, \tag{10.234}$$

respectively, resulting in the Hamiltonian

$$\mathcal{H} = \frac{p_\rho^2}{2m} + \frac{p_\phi^2}{2m\rho^2} + V(\rho). \tag{10.235}$$

The equations of motion for the angular coordinate ϕ are now given by

$$\dot{\phi} = \frac{\partial \mathcal{H}}{\partial p_\phi} = \frac{p_\phi}{m\rho^2} \quad \text{and} \quad \dot{p}_\phi = -\frac{\partial \mathcal{H}}{\partial \phi} = 0. \tag{10.236}$$

These are just the expressions of the angular momentum p_ϕ in terms of the angular velocity $\dot{\phi}$ and the conservation of angular momentum, respectively. We also find the equations of motion for the radial coordinate to be

$$\dot{\rho} = \frac{\partial \mathcal{H}}{\partial p_\rho} = \frac{p_\rho}{m} \quad \text{and} \quad \dot{p}_\rho = -\frac{\partial \mathcal{H}}{\partial \rho} = \frac{p_\phi^2}{m\rho^3} - V'(\rho), \tag{10.237}$$

which coincides with the equations of motion derived in Lagrangian mechanics.

10.4.3   Poisson brackets

Hamilton's equation can be written in a rather abstract manner with the help of a mathematical construct called the *Poisson bracket*. Given two functions f and g of the phase space coordinates q^a and p_a, we define the Poisson bracket between those functions as

$$\{f,g\} = \frac{\partial f}{\partial q^a}\frac{\partial g}{\partial p_a} - \frac{\partial g}{\partial q^a}\frac{\partial f}{\partial p_a}. \tag{10.238}$$

This construction has a few important properties that are good to keep in mind when doing computations:

1. *Anti-symmetry*: From the definition of the Poisson bracket, it follows directly that $\{f,g\} = -\{g,f\}$.

2. *Linearity*: Due to the derivatives being linear, it is always true that

$$\{\alpha_1 f_1 + \alpha_2 f_2, g\} = \alpha_1\{f_1,g\} + \alpha_2\{f_2,g\} \tag{10.239}$$

   for any constants α_i. The fact that the Poisson bracket is anti-symmetric also means that a similar relation holds for the second argument.

3. *Leibniz rule*: Due to the Leibniz rule for partial derivatives, the Poisson bracket automatically satisfies the Leibniz rule

$$\{fg,h\} = f\{g,h\} + \{f,h\}g. \tag{10.240}$$

4. *Jacobi identity*: The Poisson bracket satisfies the relation

$$\{f,\{g,h\}\} + \{g,\{h,f\}\} + \{h,\{f,g\}\} = 0. \tag{10.241}$$

   This relation also follows from directly inserting the definition of the Poisson bracket, upon which all terms cancel pairwise.

Example 10.32   The most basic functions of the phase space coordinates, apart from constant functions, are just the coordinates themselves. Taking the Poisson bracket between different coordinates, we find the *canonical commutation relations*

$$\{q^a,q^b\} = \frac{\partial q^a}{\partial q^c}\frac{\partial q^b}{\partial p_c} - \frac{\partial q^b}{\partial q^c}\frac{\partial q^a}{\partial p_c} = 0, \tag{10.242a}$$

$$\{p_a,p_b\} = \frac{\partial p_a}{\partial q^c}\frac{\partial p_b}{\partial p_c} - \frac{\partial p_b}{\partial q^c}\frac{\partial p_a}{\partial p_c} = 0, \tag{10.242b}$$

$$\{q^a,p_b\} = \frac{\partial q^a}{\partial q^c}\frac{\partial p_b}{\partial p_c} - \frac{\partial p_b}{\partial q^c}\frac{\partial q^a}{\partial p_c} = \delta^a_c\delta^c_b = \delta^a_b. \tag{10.242c}$$

In words, all of the Poisson brackets between the phase space coordinates are zero except for that between a generalised coordinate q^a and its corresponding canonical momentum p_a.

Taking the Poisson bracket between the Hamiltonian and the phase space coordinates q^a and p_a, we find that

$$\{q^a, \mathcal{H}\} = \frac{\partial q^a}{\partial q^b}\frac{\partial \mathcal{H}}{\partial p_b} - \frac{\partial \mathcal{H}}{\partial q^b}\frac{\partial q^a}{\partial p_b} = \delta_b^a \frac{\partial \mathcal{H}}{\partial p_b} = \frac{\partial \mathcal{H}}{\partial p_a}, \qquad (10.243a)$$

$$\{p_a, \mathcal{H}\} = \frac{\partial p_a}{\partial q^b}\frac{\partial \mathcal{H}}{\partial p_b} - \frac{\partial \mathcal{H}}{\partial q^b}\frac{\partial p_a}{\partial p_b} = -\frac{\partial \mathcal{H}}{\partial q^b}\delta_a^b = -\frac{\partial \mathcal{H}}{\partial q^a}. \qquad (10.243b)$$

With the Poisson bracket, Hamilton's equations of motion therefore take the form

$$\dot{q}^a = \{q^a, \mathcal{H}\} \quad \text{and} \quad \dot{p}_a = \{p_a, \mathcal{H}\}. \qquad (10.244)$$

So what about other functions of the phase space coordinates? Let us consider an arbitrary function f that generally depends on q^a, p_a, and t and a Hamiltonian $\mathcal{H}$. Using the definition of the Poisson bracket, we find that

$$\{f, \mathcal{H}\} = \frac{\partial f}{\partial q^a}\frac{\partial \mathcal{H}}{\partial p_a} - \frac{\partial \mathcal{H}}{\partial q^a}\frac{\partial f}{\partial p_a} = \frac{\partial f}{\partial q^a}\dot{q}^a + \dot{p}_a \frac{\partial f}{\partial p_a}. \qquad (10.245)$$

We can also look at the total derivative of f with respect to time, which is given by

$$\frac{df}{dt} = \frac{\partial f}{\partial q^a}\dot{q}^a + \frac{\partial f}{\partial p_a}\dot{p}_a + \frac{\partial f}{\partial t} = \{f, \mathcal{H}\} + \frac{\partial f}{\partial t}. \qquad (10.246)$$

In particular, if the function f does not depend explicitly on time, then

$$\dot{f} \equiv \frac{df}{dt} = \{f, \mathcal{H}\}. \qquad (10.247)$$

If, in addition, f *Poisson commutes* with the Hamiltonian, i.e., if

$$\{f, \mathcal{H}\} = 0, \qquad (10.248)$$

then the function f is a constant of motion. We have encountered constants of motion earlier in this chapter when we discussed Noether's theorem in Lagrangian mechanics. We will discuss the corresponding statements in Hamiltonian mechanics once we have developed the framework a bit further.

Example 10.33 The Hamiltonian itself is a function of q^a and p_a and we find that

$$\dot{\mathcal{H}} = \{\mathcal{H}, \mathcal{H}\} + \frac{\partial \mathcal{H}}{\partial t} = \frac{\partial \mathcal{H}}{\partial t}, \qquad (10.249)$$

where we have used that the anti-symmetry of the Poisson bracket implies that $\{\mathcal{H}, \mathcal{H}\} = 0$. It follows that the Hamiltonian is a constant of motion if it does not depend explicitly on t. This sounds familiar. In the framework of Lagrangian mechanics, we found that the Hamiltonian was a constant of motion if the Lagrangian did not depend explicitly on time. These statements are equivalent as we have already established the relationship

$$\frac{\partial \mathcal{H}}{\partial t} = -\frac{\partial \mathcal{L}}{\partial t}. \qquad (10.250)$$

Hence, if the Lagrangian does not depend explicitly on time, neither does the Hamiltonian and vice versa.

Example 10.34 Consider a particle of mass m moving in a potential $V(\vec{x})$. The corresponding Hamiltonian is given by

$$\mathcal{H} = \frac{\vec{p}^2}{2m} + V(\vec{x})$$

(10.251)

and the angular momentum relative to the origin $\vec{x} = 0$ is given by

$$\vec{L} = \vec{x} \times \vec{p} \quad \Longleftrightarrow \quad L_i = \varepsilon_{ijk} x^j p^k.$$

(10.252)

It follows that

$$\dot{L}_i = \{L_i, \mathcal{H}\} = \varepsilon_{ijk}(\{x^j, \mathcal{H}\} p^k + x^j \{p^k, \mathcal{H}\}) = \varepsilon_{ijk}\left(\frac{p^j}{m}p^k - x^j \partial_k V\right)$$

$$= -\varepsilon_{ijk} x^j \partial_k V.$$

(10.253a)

or, in other words,

$$\dot{\vec{L}} = \vec{x} \times \vec{F},$$

(10.253b)

where $\vec{F} = -\nabla V$ is the force acting on the particle. We recognise $\vec{x} \times \vec{F}$ as the torque $\vec{\tau}$ on the particle relative to the origin.

Given two functions f_1 and f_2 of the phase space coordinates, we may ask the question of how their Poisson bracket evolves. Just as any function depending only on the phase space coordinates, the Poisson bracket $\{f_1, f_2\}$ will evolve according to

$$\frac{d}{dt}\{f_1, f_2\} = \{\{f_1, f_2\}, \mathcal{H}\}.$$

(10.254)

Using the Jacobi identity for the Poisson bracket, we can rewrite this as

$$\{\{f_1, f_2\}, \mathcal{H}\} = -\{\{f_2, \mathcal{H}\}, f_1\} - \{\{\mathcal{H}, f_1\}, f_2\} = -\{\dot{f}_2, f_1\} + \{\dot{f}_1, f_2\}.$$

(10.255)

In particular, if both f_1 and f_2 are constants of motion, then so is their Poisson bracket $\{f_1, f_2\}$. Assuming the Poisson bracket is non-zero and independent from the functions themselves, we can therefore extend the number of known constants of motion by including the Poisson bracket $\{f_1, f_2\}$.

Example 10.35 The angular momentum vector $\vec{L} = \vec{x} \times \vec{p}$ has three components as described in Example 10.34. Taking the Poisson bracket between two of those components using a Cartesian basis, we find that

$$\{L_i, L_j\} = \varepsilon_{ik\ell}\varepsilon_{jmn}\{x^k p^\ell, x^m p^n\} = \varepsilon_{ik\ell}\varepsilon_{jmn}(p^\ell x^m \delta_{kn} - x^k p^n \delta_{\ell m})$$

$$= (\delta_{im}\delta_{\ell j} - \delta_{ij}\delta_{\ell m})(p^\ell x^m - x^\ell p^m) = x^i p^j - x^j p^i.$$

(10.256)

We also note that a general property of the angular momentum components is

$$\varepsilon_{ijk} L_k = \varepsilon_{ijk}\varepsilon_{k\ell m} x^\ell p^m = x^i p^j - x^j p^i$$

(10.257)

and therefore

$$\{L_i, L_j\} = \varepsilon_{ijk}L_k. \qquad (10.258)$$

Thus, the Poisson bracket of two of the angular momentum components is equal to the third. It follows that if two of the components are constants of motion, then so is the third. Also note how this relation is very reminiscent of the Lie bracket between the generators of the rotation group $SO(3)$, see Eq. (4.41). In fact, taking $L_i = J_i$, it *is* the Lie bracket and the angular momentum components may be seen as the generators of rotation. As we shall see in a while, this is not something coincidental. Instead, the angular momentum will be seen to generate rotations in a very explicit manner.

10.4.4   Liouville's theorem

With the introduction of Hamiltonian mechanics, the generalised coordinates and their canonical momenta have very similar roles. After the introduction of the Poisson bracket, their equations of motion are completely symmetric. In what follows, we therefore introduce a new short-hand for denoting both of them at the same time with the notation y^r such that $y^r = q^r$ for $1 \leq r \leq N$ and $y^r = p_{r-N}$ for $N + 1 \leq r \leq 2N$. We will here use indices from the latter part of the alphabet, i.e., r, s, t, to denote indices taking values from one to $2N$ and reserve the earlier part of the alphabet for indices taking values between one and N. With this notation, the Poisson bracket can be written as

$$\{f, g\} = \frac{\partial f}{\partial q^a}\frac{\partial g}{\partial p_a} - \frac{\partial g}{\partial q^a}\frac{\partial f}{\partial p_a} = \frac{\partial f}{\partial y^a}\frac{\partial g}{\partial y^{a+N}} - \frac{\partial g}{\partial y^a}\frac{\partial f}{\partial y^{a+N}} = \frac{\partial f}{\partial y^r}\omega^{rs}\frac{\partial g}{\partial y^s}, \qquad (10.259)$$

where we have introduced $\omega^{rs} = \delta_{r+N,s} - \delta_{r,s+N}$. By construction ω^{rs} is anti-symmetric and has constant components.

Example 10.36 Consider a particle moving in one dimension with coordinate x. Denoting its canonical momentum by p, we introduce the phase space coordinates

$$y^1 = x \quad \text{and} \quad y^2 = p. \qquad (10.260)$$

The Poisson bracket in this system may be described as

$$\{f, g\} = \frac{\partial f}{\partial x}\frac{\partial g}{\partial p} - \frac{\partial f}{\partial p}\frac{\partial g}{\partial x} = \frac{\partial f}{\partial y^1}\frac{\partial g}{\partial y^2} - \frac{\partial f}{\partial y^2}\frac{\partial g}{\partial y^1}. \qquad (10.261)$$

In this case, we find that ω_{rs} has the non-zero entries

$$\omega^{12} = 1 \quad \text{and} \quad \omega^{21} = -1. \qquad (10.262)$$

Let us now consider the time evolution of the phase space coordinates y^r themselves. By construction, we know that

$$\dot{y}^r = \frac{\partial y^r}{\partial y^s}\omega^{st}\frac{\partial \mathcal{H}}{\partial y^t} = \omega^{rt}\frac{\partial \mathcal{H}}{\partial y^t}. \qquad (10.263)$$

This equation describes the phase space flow of the system and we know from Section 3.9.3

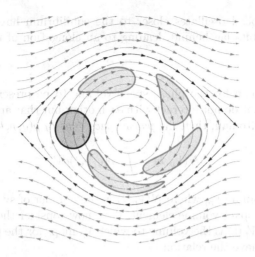

Figure 10.25 The phase space flow of the initially circular region (darker shading) of phase space at different times for a bead moving on a ring in a gravitational field, cf. Fig. 10.24. As time goes by, the shape of the region moves and becomes distorted but retains the same phase space area in accordance with Liouville's theorem.

that a flow preserves volumes if $\nabla \cdot \vec{v} = 0$, where $\vec{v}$ is the flow velocity field. In our case, we indeed find that

$$\partial_r \dot{y}^r = \frac{\partial}{\partial y^r} \omega^{rt} \frac{\partial \mathcal{H}}{\partial y^t} = \omega^{rt} \frac{\partial^2 \mathcal{H}}{\partial y^r \partial y^t} = 0 \tag{10.264}$$

due to the second derivative of the Hamiltonian being symmetric and ω_{rt} being anti-symmetric. The phase space flow is therefore incompressible, i.e., the phase space volume does not change under this flow. This statement is known as *Liouville's theorem*.

Example 10.37 In the case of the harmonic oscillator, the phase space flow is given by

$$x(t) = x_0 \cos(\omega t) + \frac{p_0}{m\omega} \sin(\omega t), \tag{10.265a}$$

$$p(t) = p_0 \cos(\omega t) - m\omega x_0 \sin(\omega t). \tag{10.265b}$$

This is a rotation of the phase space coordinates by an angle ωt and in general areas, in this case the phase space area, are invariant under rotations.

Example 10.38 Liouville's theorem only states that the phase space volume is preserved under the phase space flow. The shape of this volume may still be significantly distorted. Consider the phase space flow in the case of the bead moving on a ring in a gravitational field of Example 10.29. In Fig. 10.25 we show the evolution of a circular region in phase space under the phase space flow. As we can see from this figure, the area of the region remains the same under the flow while the shape does not. In particular, the phase space orbits for which the bead oscillates closer to the equilibrium have a shorter period and therefore the other orbits lag behind, leading to an elongation of the flowing phase space volume.

So what does it mean for the phase space volume to be preserved? If we consider an ensemble of systems that do not interact with each other but that are described by the same phase space and Hamiltonian, then the *phase space density* $\rho(q, p, t)$ is defined such that

$$N(V_p, t) = \int_{V_p} \rho(q, p, t) dV_p \qquad (10.266)$$

for any phase space volume V_p, where $N(V_p, t)$ is the number of systems within V_p at time t and dV_p is the phase space volume element. If we now consider the phase space flow $V_p(t)$ of the volume V_p, i.e., $V_p(t)$ is the volume to which V_p maps for the phase space flow during time t, then we must have the relation

$$N(V_p(t), t) = N(V_p, 0). \qquad (10.267)$$

Since this must hold for any volume and the volume of $V_p(t)$ is the same as the volume of V_p, it follows that

$$\frac{d\rho}{dt} = \frac{\partial \rho}{\partial t} + \frac{\partial \rho}{\partial q^a} \dot{q}^a + \frac{\partial \rho}{\partial p_a} \dot{p}^a = 0. \qquad (10.268)$$

Note that this is just a statement saying that the density along the flow does not change under an incompressible flow. Inserting Hamilton's equations of motion now results in

$$\frac{\partial \rho}{\partial t} = \frac{\partial \mathcal{H}}{\partial q^a} \frac{\partial \rho}{\partial p_a} - \frac{\partial \mathcal{H}}{\partial p_a} \frac{\partial \rho}{\partial q^a} = \{\mathcal{H}, \rho\}. \qquad (10.269)$$

This equation describes how the evolution of the phase space density with time depends on the phase space density itself and the dynamics of the system and is known as *Liouville's equation*. Using phase space distributions and considering their time evolution is a central part in the study of statistical mechanics. In many cases, it will not be possible to consider an ensemble as non-interacting. In such situations, Liouville's equation must be modified and the result will generally contain terms that are non-linear in the phase space density ρ.

Example 10.39 An example of an ensemble with non-interacting systems is given by an ideal gas. In this case, the molecules of the gas are assumed to bounce around in a container with a certain temperature and pressure and without colliding with each other. The motion of each particle is then described independently by the same phase space and Hamiltonian. For most practical purposes, there are so many molecules in any given container that the phase space density ρ can be considered continuous.

Liouville's theorem has another important implication. Let us look at the flow of the phase space region P and define P_k as the volume to which P flows in time $k\tau$, where τ is some fixed time. If all of the phase space regions P_k are disjoint, then the total volume of the first n such regions is given by

$$V = nV_0, \qquad (10.270)$$

where V_0 is the phase space volume of P, since all of the regions have the same volume according to Liouville's theorem. Therefore, if the flow of P is restricted to a region with finite phase space volume, a k such that P_k and P are not disjoint must exist. Since we can select P arbitrarily small as long as it has a finite volume, we can conclude that, with this assumption, a system described by Hamiltonian mechanics will eventually come arbitrarily close to its initial state. This statement is known as the *Poincaré recurrence theorem*.

Example 10.40 Consider the harmonic oscillator and look at a region of phase space P for which the total energy E varies between $E_{\min}$ and $E_{\max}$. Since energy is a constant of motion, it follows that

$$E_{\min} \leq \frac{p^2}{2m} + \frac{kx^2}{2} \leq E_{\max} \qquad (10.271)$$

for any point in parameter space to which P can flow. This describes an elliptic band in phase space and has the finite phase space area $\pi(E_{\max} - E_{\min})/\omega$. It follows that Poincaré's recurrence theorem applies to the harmonic oscillator, which should not be surprising as we know that the harmonic oscillator solutions are periodic.

Note that, while the recurrence theorem states that a system satisfying its assumptions will eventually return to a state arbitrarily close to its original state, the time for this to occur may in practice be prohibitively large.

10.4.5 Canonical transformations

In Lagrangian mechanics, the result of applying Hamilton's principle does not depend on the generalised coordinates chosen to describe the system and so we were free to use any coordinates that could be useful for solving the equations of motion for a particular system. We therefore ask ourselves what the corresponding statement in Hamiltonian mechanics is. Let us briefly summarise the basis of Hamiltonian mechanics and define the condition that we wish to impose on a phase space coordinate transformation in order to keep the same formulation of mechanics.

The basis of Hamiltonian mechanics is the phase space with coordinates q^a and p_a, or equivalently y^r, a function $\mathcal{H}$ that depends on those coordinates, and the Poisson bracket that determines the equations of motion. The Poisson bracket is given by

$$\{f, g\} = \frac{\partial f}{\partial y^r} \omega^{rs} \frac{\partial g}{\partial y^s}. \qquad (10.272)$$

Let us now introduce a different set of coordinates Q^a and P_a, or equivalently Y^r, that describe the same phase space. If we want to keep the same formulation of Hamiltonian mechanics in these coordinates, the corresponding Poisson bracket must be given by

$$\{f, g\}_Y = \frac{\partial f}{\partial Y^r} \omega^{rs} \frac{\partial g}{\partial Y^s}. \qquad (10.273)$$

If we want the theory to remain the same, this new Poisson bracket must coincide with the original one, i.e.,

$$\{f, g\}_Y = \{f, g\}_y \qquad (10.274)$$

for any functions f and g on the phase space. In terms of the coordinates, this condition takes the form

$$\frac{\partial f}{\partial Y^r} \omega^{rs} \frac{\partial g}{\partial Y^s} = \frac{\partial f}{\partial y^r} \omega^{rs} \frac{\partial g}{\partial y^s} = \frac{\partial f}{\partial Y^r} \frac{\partial Y^r}{\partial y^t} \omega^{tu} \frac{\partial Y^s}{\partial y^u} \frac{\partial g}{\partial Y^s}, \qquad (10.275)$$

where we have used the chain rule in the last step. In order for this to be satisfied for all functions f and g we need to impose the condition

$$\omega^{rs} = \frac{\partial Y^r}{\partial y^t} \omega^{tu} \frac{\partial Y^s}{\partial y^u}. \tag{10.276}$$

If this is satisfied, the transformation is called a *canonical transformation*. Alternatively, this can be written on matrix form as

$$\Omega = J^T \Omega J, \tag{10.277}$$

where the matrices Ω and J have ω^{rs} and $\partial Y^s/\partial y^r$, respectively, as their components.

Example 10.41 For a particle of mass m moving in one dimension, we can define the phase space rotation

$$X = x\cos(\alpha) + \frac{p}{m\omega_0}\sin(\alpha), \quad P = p\cos(\alpha) - m\omega_0 x\sin(\alpha), \tag{10.278}$$

where ω_0 is a constant with units $1/\mathsf{T}$ and α is some angle. We find that the matrix J is given by

$$J = \begin{pmatrix} \frac{\partial X}{\partial x} & \frac{\partial X}{\partial p} \\ \frac{\partial P}{\partial x} & \frac{\partial P}{\partial p} \end{pmatrix} = \begin{pmatrix} \cos(\alpha) & \frac{1}{m\omega_0}\sin(\alpha) \\ -m\omega_0\sin(\alpha) & \cos(\alpha) \end{pmatrix} \tag{10.279}$$

and the matrix Ω by construction takes the form

$$\Omega = \begin{pmatrix} 0 & 1 \\ -1 & 0 \end{pmatrix}. \tag{10.280}$$

Inserting this into the right-hand side of Eq. (10.277), we find that the identity is indeed satisfied and that this is a canonical transformation.

It should also be noted that Eq. (10.275) can also be rewritten as

$$\{f, g\}_Y = \frac{\partial f}{\partial Y^r} \frac{\partial g}{\partial Y^s} \{Y^r, Y^s\}_y = \frac{\partial f}{\partial Y^r} \frac{\partial g}{\partial Y^s} \omega^{rs}. \tag{10.281}$$

Thus, another way of writing the requirement for the transformation to be canonical is that the Poisson bracket between the new coordinates with respect to the old satisfies

$$\{Y^r, Y^s\}_y = \omega^{rs}. \tag{10.282a}$$

This implies that the canonical commutation relations

$$\{Q^a, Q^b\}_y = \{P_a, P_b\}_y = 0 \quad \text{and} \quad \{Q^a, P_b\}_y = \delta^a_b \tag{10.282b}$$

must be satisfied.

Example 10.42 Looking at the phase space rotation of Example 10.41, we find that

$$\{X, X\}_{xp} = \{P, P\}_{xp} = 0 \tag{10.283}$$

are trivially fulfilled by virtue of the Poisson bracket being anti-symmetric. For the Poisson bracket between X and P, we find that

$$\{X, P\}_{xp} = \frac{\partial X}{\partial x}\frac{\partial P}{\partial p} - \frac{\partial X}{\partial p}\frac{\partial P}{\partial x} = \cos^2(\alpha) + \frac{m\omega_0}{m\omega_0}\sin^2(\alpha) = 1, \tag{10.284}$$

which also shows that the phase space rotation is a canonical transformation.

When we considered coordinate transformations in the Lagrangian formalism, we found that any transformation of the general coordinates was allowed. In contrast, we have now found that only canonical transformations are allowed in the Hamiltonian formalism and it is therefore instructive to examine how the general coordinate transformation translates to a canonical transformation, since it must be allowed. We therefore consider a general coordinate transformation such that the new coordinates Q^a can be written as functions of the old coordinates q^a and the relation can be inverted, as for the general coordinate transformations in Lagrangian mechanics. We find that

$$\dot{q}^a = \frac{\partial q^a}{\partial Q^b}\dot{Q}^b \implies \frac{\partial \dot{q}^a}{\partial \dot{Q}^b} = \frac{\partial q^a}{\partial Q^b}. \tag{10.285}$$

The canonical momentum related to the new coordinates is then given by

$$P_a = \frac{\partial \mathcal{L}}{\partial \dot{Q}^a} = \frac{\partial \mathcal{L}}{\partial q^b}\frac{\partial q^b}{\partial \dot{Q}^a} + \frac{\partial \mathcal{L}}{\partial \dot{q}^b}\frac{\partial \dot{q}^b}{\partial \dot{Q}^a} = \frac{\partial \mathcal{L}}{\partial \dot{q}^b}\frac{\partial q^b}{\partial Q^a} = p_b\frac{\partial q^b}{\partial Q^a}, \tag{10.286}$$

which results in

$$\frac{\partial P_a}{\partial q^b} = p_c\frac{\partial}{\partial q^b}\frac{\partial q^c}{\partial Q^a} = p_c\frac{\partial Q^d}{\partial q^b}\frac{\partial^2 q^c}{\partial Q^d\partial Q^a} \quad \text{and} \quad \frac{\partial P_a}{\partial p_b} = \frac{\partial q^b}{\partial Q^a}. \tag{10.287}$$

Let us now verify that this transformation is indeed canonical. For the commutation relations of the new phase space coordinates, we find that

$$\{Q^a, Q^b\}_{qp} = \frac{\partial Q^a}{\partial q^c}\frac{\partial Q^b}{\partial p_c} - \frac{\partial Q^b}{\partial q^c}\frac{\partial Q^a}{\partial p_c} = 0, \tag{10.288a}$$

$$\{P_a, P_b\}_{qp} = \frac{\partial P_a}{\partial q^c}\frac{\partial P_b}{\partial p_c} - \frac{\partial P_b}{\partial q^c}\frac{\partial P_a}{\partial p_c} = p_e\delta_b^d\frac{\partial^2 q^e}{\partial Q^d\partial Q^a} - p_e\delta_a^d\frac{\partial^2 q^e}{\partial Q^d\partial Q^b}$$

$$= p_e\left(\frac{\partial^2 q^e}{\partial Q^b\partial Q^a} - \frac{\partial^2 q^e}{\partial Q^a\partial Q^b}\right) = 0, \tag{10.288b}$$

$$\{Q^a, P_b\}_{qp} = \frac{\partial Q^a}{\partial q^c}\frac{\partial P_b}{\partial p_c} - \frac{\partial P_b}{\partial q^c}\frac{\partial Q^a}{\partial p_c} = \frac{\partial Q^a}{\partial q^c}\frac{\partial q^c}{\partial Q_b} = \delta_b^a \tag{10.288c}$$

and the transformations are indeed canonical. It should be noted that, while the generalised coordinates Q^a are functions of q^a only, the corresponding canonical momenta will generally depend on both q^a and p_a as displayed in Eq. (10.286).

Example 10.43 For a particle of mass m moving in two dimensions, we can use either Cartesian or polar coordinates. Starting from the formulation using Cartesian coordinates x^1 and x^2, the polar coordinates are defined through the relations

$$x^1 = \rho\cos(\phi) \quad \text{and} \quad x^2 = \rho\sin(\phi). \tag{10.289}$$

We therefore find that the relation between the canonical momenta in Cartesian and polar coordinates are given by

$$p_\rho = p_1 \frac{\partial x^1}{\partial \rho} + p_2 \frac{\partial x^2}{\partial \rho} = p_1 \cos(\phi) + p_2 \sin(\phi) = \frac{\vec{p} \cdot \vec{x}}{\rho}, \tag{10.290a}$$

$$p_\phi = p_1 \frac{\partial x^1}{\partial \phi} + p_2 \frac{\partial x^2}{\partial \phi} = -p_1 \rho \sin(\phi) + p_2 \rho \cos(\phi) = p_2 x^1 - p_1 x^2, \tag{10.290b}$$

where $\vec{p} = p_1 \vec{e}_1 + p_2 \vec{e}_2$ and we can again identify p_ϕ with the angular momentum relative to the origin.

10.4.6 Phase space flows and symmetries

Just as we considered infinitesimal transformations in the Lagrangian formalism in Section 10.2.4, let us now consider infinitesimal canonical transformations in phase space such that

$$q^a \to Q^a = q^a + \varepsilon \, \delta q^a \quad \text{and} \quad p_a \to P_a = p_a + \varepsilon \, \delta p_a, \tag{10.291}$$

where δq^a and δp_a generally are functions of all of the phase space coordinates. In order for this transformation to be canonical, we obtain the following conditions from the Poisson brackets

$$0 = \{Q^a, Q^b\} \simeq \varepsilon \left(\frac{\partial \delta q^b}{\partial p_a} - \frac{\partial \delta q^a}{\partial p_b} \right), \tag{10.292a}$$

$$0 = \{P_a, P_b\} \simeq \varepsilon \left(\frac{\partial \delta p_a}{\partial q^b} - \frac{\partial \delta p_b}{\partial q^a} \right), \tag{10.292b}$$

$$\delta_b^a = \{Q^a, P_b\} \simeq \delta_b^a + \varepsilon \left(\frac{\partial \delta q^a}{\partial q^b} + \frac{\partial \delta p_b}{\partial p_a} \right), \tag{10.292c}$$

where we have used $\simeq$ to denote equality up to first order in ε. All of these conditions can be satisfied by letting

$$\delta q^a = \frac{\partial G}{\partial p_a} = \{q^a, G\} \quad \text{and} \quad \delta p_a = -\frac{\partial G}{\partial q^a} = \{p_a, G\}, \tag{10.293}$$

where G is an arbitrary function of the phase space coordinates. Now this looks rather familiar, if we let $G = \mathcal{H}$, it is just Hamilton's equations of motion for an infinitesimal time t. In general, it is the infinitesimal form of a continuous phase space transformation with parameter s such that $q_0^a \to q^a(q_0, p_0, s)$ and $p_{0a} \to p_a(q_0, p_0, s)$ satisfying

$$\frac{dq^a}{ds} = \{q^a, G\} \quad \text{and} \quad \frac{dp_a}{ds} = \{p_a, G\} \tag{10.294}$$

as well as the initial conditions $q^a(q_0, p_0, 0) = q_0^a$ and $p_a(q_0, p_0, 0) = p_{0a}$. We say that the function G is the *generator* of this canonical transformation and the functions $q^a(q_0, p_0, s)$ and $p_a(q_0, p_0, s)$ are the *phase space flows* generated by G.

Example 10.44 Comparing Examples 10.37 and 10.41, the phase space flow generated by the harmonic oscillator Hamiltonian is exactly the same as the canonical phase space rotation with $\alpha = \omega t$ and $\omega_0 = \omega$. In fact, we should expect the time evolution of a system to define a canonical transformation as it is equivalent to the phase space flow generated by the Hamiltonian, what we have earlier referred to as the phase space flow. This holds for any Hamiltonian and we establish the Hamiltonian as the generator of time translations as it maps a given set of phase space coordinates to the phase space coordinates describing the system a time t later.

Example 10.45 For a particle moving in one dimension with position x and corresponding canonical momentum p, let us consider the canonical transformation generated by the function $G = p$. The phase space flow generated by G satisfies the differential equations

$$\frac{dx}{ds} = \{x, p\} = 1 \quad \text{and} \quad \frac{dp}{ds} = \{p, p\} = 0. \tag{10.295}$$

The solution to this set of differential equations is

$$x = x_0 + s \quad \text{and} \quad p = p_0. \tag{10.296}$$

For a fixed value of the parameter s, this is a translation in space and we say that the canonical momentum p generates translations in the position x.

If a function F does not change its value under a canonical transformation, then we say that the transformation is a *symmetry* of F. In practice, this means that for infinitesimal canonical transformations under which $F \to F + \varepsilon\, \delta F$ we have

$$\delta F = \frac{\partial F}{\partial y^r}\delta y^r = \frac{\partial F}{\partial y^r}\{y^r, G\} = \frac{\partial F}{\partial y^r}\omega^{rs}\frac{\partial G}{\partial y^s} = \{F, G\} = 0. \tag{10.297}$$

Since the Poisson bracket is anti-symmetric $\{F, G\} = 0$ implies that $\{G, F\} = 0$ and so G is a symmetry of F if F is a symmetry of G. In particular, if we let F be the Hamiltonian $\mathcal{H}$, we have already seen that a constant of motion G satisfies

$$\dot{G} = \{G, \mathcal{H}\} = 0. \tag{10.298}$$

This is the same statement as saying that $\mathcal{H}$ is a symmetry of G, which therefore does not change under the phase space flow generated by $\mathcal{H}$. We have now also seen that this means that G is a symmetry of $\mathcal{H}$ and therefore $\mathcal{H}$ does not change under the phase space flow generated by G. This is the Hamiltonian version of *Noether's theorem*, for every symmetry of the Hamiltonian, the generator of that symmetry is a constant of motion. In addition, the converse is also true and if there exists a constant of motion, then the canonical transformation generated by it is a symmetry of the Hamiltonian.

Example 10.46 For a particle of mass m moving freely in one dimension, the Hamiltonian is given by

$$\mathcal{H} = \frac{p^2}{2m}. \tag{10.299}$$

Since the Hamiltonian does not contain x, the canonical momentum p Poisson commutes with it and is a constant of motion. As we saw in Example 10.45, the canonical transformation generated by p is given by $x \to x + s$ and $p \to p$. Inserting this into the Hamiltonian gives $\mathcal{H} \to \mathcal{H}$ and it the transformation is indeed a symmetry of the Hamiltonian.

Given two functions G_1 and G_2, we can look at and compare their respective phase space flows. In particular, we are interested in what happens if we first follow the flow of G_1 and then the flow of G_2 and compare the result to first following the flow of G_2 and then the flow of G_1. On the infinitesimal level, let us look at following the flows for a parameter distance ε_1 and ε_2, respectively. Looking at the flow from the phase space coordinates y_0^r and first following the flow of G_1, we find that we end up at the point given by

$$y_1^r \simeq y_0^r + \varepsilon_1 \left\{ y^r, G_1 \right\}, \tag{10.300}$$

which is exact to first order in ε_1. Under this flow, we also change the function G_2 according to

$$G_2 \to G_2 + \varepsilon_1 \delta G_2 = G_2 + \varepsilon_1 \left\{ G_2, G_1 \right\}. \tag{10.301}$$

By following the flow of G_2 from y_1^r we will therefore end up at the coordinates

$$\begin{aligned} y_{21}^r &\simeq y_1^r + \varepsilon_2 \left\{ y^r, G_2 + \varepsilon_1 \left\{ G_2, G_1 \right\} \right\} \\ &= y_0^r + \varepsilon_1 \left\{ y^r, G_1 \right\} + \varepsilon_2 \left\{ y^r, G_2 \right\} + \varepsilon_1 \varepsilon_2 \left\{ y^r, \left\{ G_2, G_1 \right\} \right\}. \end{aligned} \tag{10.302}$$

By exchanging the roles of G_1 and G_2 and first following the flow generated by G_2, we find that the difference in the coordinates of the final points is given by

$$y_{21}^r - y_{12}^r = 2\varepsilon_1 \varepsilon_2 \left\{ y^r, \left\{ G_2, G_1 \right\} \right\}. \tag{10.303}$$

In particular, this implies that if the functions G_1 and G_2 Poisson commute, then the flows generated by them commute and it does not matter which flow we follow first.

Example 10.47 A particle moving in a central potential $V(r)$ described in polar coordinates has the Hamiltonian

$$\mathcal{H} = \frac{p_\rho^2}{2m} + \frac{p_\phi^2}{2m\rho^2} + V(\rho). \tag{10.304}$$

Since this Hamiltonian does not depend explicitly on the polar coordinate ϕ, we find that

$$\left\{ p_\phi, \mathcal{H} \right\} = 0 \tag{10.305}$$

and therefore the flow generated by p_ϕ under which $\phi \to \phi + s$, i.e., a rotation by an angle s in the plane of motion, should commute with the time evolution of the system. It therefore does not matter if we first rotate the system and apply the equations of motion and then the rotation or vice versa. This also means that rotating a solution to the equations of motion will result in a solution to the equations of motion that also has had its initial conditions rotated by the same angle, see Fig. 10.26.

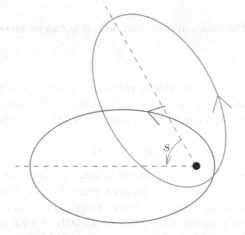

Figure 10.26 Due to the rotational symmetry of the Hamiltonian in the case of a central potential, rotating one solution by an angle s will result in a new solution to the equations of motion. This is here illustrated for the elliptic solutions to the Kepler problem. Both ellipses share the center of the potential as one of their foci and have the same size and eccentricity.

10.5 MANIFOLDS AND CLASSICAL MECHANICS

If you have read Chapter 9, you are likely to have noticed several passages in this chapter where the mathematics would be more conveniently expressed in terms of the calculus on manifolds. From the definition of configuration space, a configuration space was indeed our first example of a manifold, see Example 9.1, to the commutation of phase space flows in Hamiltonian mechanics, remember that two commuting flows imply that their velocity vector fields Lie commute. Indeed you are right and the calculus on manifolds is very well suited to express the ideas of classical mechanics in general. Although a full exposition is beyond the scope of this book, we will briefly discuss its application within both the Lagrangian and the Hamiltonian formalisms.

10.5.1 The Lagrangian formalism revisited

Starting with the most pressing issue, we have already seen that the configuration space of a system can be described by N coordinates q^a and that the coordinates chosen do not really matter as we end up with the same theory regardless. In many situations, such as that of the spherical pendulum, the coordinate system chosen is not necessarily a good global coordinate system in the sense of being a chart on a manifold. Instead, we will often need several charts to cover the full configurations space. In many cases, we can therefore define configuration space as a manifold M describing the configuration of a given system.

Example 10.48 Consider the bead of mass m moving on a ring in a gravitational field. The configuration of the system is given by the position of the bead on the ring, which can be described by an angle θ. We know that this description is 2π periodic and $\theta = \theta_0$ and $\theta = \theta_0 + 2\pi$ correspond to the same configuration. We can avoid covering the same point several times within a chart by considering two different charts as described in Example 9.4. The configuration space of this physical system is therefore just the circle S^1. This is well in line with our intuition, the position of a bead on a ring can be given by providing a position on the ring.

In most of the physical systems we have considered, the Lagrangian $\mathcal{L}$ has been a function of the generalised coordinates q^a as well as their time derivatives $\dot{q}^a$ along some path γ in configuration space that takes time t as its parameter. The tangent vector $\dot{\gamma}$ of this path is expressed as

$$\dot{\gamma}(t) = \dot{q}^a(t)\partial_a \tag{10.306}$$

in any coordinate system. This tangent vector belongs to the point $\gamma(t)$ in the configuration space and is therefore an element of the tangent space $T_{\gamma(t)}M$. Having established that the Lagrangian is a function of the tangent vector components and the point in configuration space at which the tangent vector is located, we identify it with a function on the *tangent bundle* TM, which is the union of all tangent spaces T_pM, i.e., an element of TM is any tangent vector of the manifold M and may be specified by specifying the point p and the tangent vector X at that point in some local coordinate system and we will write this as (p, X). If the point p can be implicitly understood from X, such as when $X = \dot{\gamma}$ is a tangent vector of a curve γ and therefore must be evaluated at the corresponding point of the curve, it is often easier to just denote the element of the tangent bundle by writing the tangent vector itself.

Example 10.49 For a particle of mass m moving freely in three dimensions, the Lagrangian is given by

$$\mathcal{L}(\vec{x}, \dot{\vec{x}}) = \frac{m}{2}\dot{\vec{x}}^2, \tag{10.307}$$

where $\vec{x}$ is the position vector. In an arbitrary coordinate system, this takes the form

$$\mathcal{L}(X) = \frac{m}{2}g(X, X), \tag{10.308}$$

where g is the metric tensor at the point p where X is defined. Evaluated for the tangent vector $\dot{\gamma}$ in a local coordinate system with coordinates q^a, this takes the form

$$\mathcal{L}(\dot{\gamma}) = \frac{m}{2}g_{ab}\dot{q}^a\dot{q}^b, \tag{10.309}$$

which is an expression we recognise from Eq. (10.79).

For a tangent bundle, we also define the projection operator π that is a map from TM to M such that

$$\pi(X) = p, \tag{10.310}$$

where X belongs to the tangent space T_pM, i.e., it maps an element of the tangent bundle to the point in the manifold where the vector is located. A large class of Lagrangians that we have encountered and will encounter will be of the form

$$\mathcal{L}(X) = \frac{1}{2}M(X, X) - V(\pi(X)) = \frac{1}{2}M_{ab}X^aX^b - V(\pi(X)), \tag{10.311}$$

where M is the generalised inertia tensor and V is the potential, which is a function on configuration space. In particular, evaluated for the tangent vector of a curve in local coordinates q^a, this takes the form

$$\mathcal{L}(\dot{\gamma}) = \frac{1}{2} M_{ab} \dot{q}^a \dot{q}^b - V(\gamma(t)),$$
(10.312)

which we again recognise from our earlier discussion.

Example 10.50 Looking at the spherical pendulum in a gravitational field, the Lagrangian is given by

$$\mathcal{L} = \frac{mr_0^2}{2} (\dot{\theta}^2 + \sin^2(\theta)\dot{\varphi}^2) - mgr_0 \cos(\theta)$$
(10.313)

in spherical coordinates. In arbitrary coordinates, this becomes

$$\mathcal{L} = \frac{mr_0^2}{2} g_{ab} \dot{q}^a \dot{q}^b - mgh(\gamma(t)),$$
(10.314)

where $h(p)$ is the height of the point p above the center of the sphere and g_{ab} are the components of the metric tensor on the unit sphere. We note that the generalised inertia tensor in this case is directly proportional to the metric

$$M_{ab} = mr_0^2 g_{ab}.$$
(10.315)

Note: Do not confuse the gravitational field strength g with the metric tensor here!

The above examples are very suggestive in terms of relating the generalised inertia tensor to a metric. Let us examine the properties of the generalised inertia tensor to see if we can extend this argument.

1. The generalised inertia tensor is symmetric. As the anti-symmetric part of M does not contribute to the Lagrangian, we may as well just consider the symmetric part as it contains all of the information about the physics.

2. The generalised inertia tensor is positive definite. If this would not be the case, there would exist directions in which motion resulted from no applied force or where motion was induced in the opposite direction of the applied force. In comparison to the special case of a single particle, this would correspond to zero or negative masses.

These properties are exactly the properties required of a metric tensor and given a Lagrangian of the form in Eq. (10.311) we can identify the generalised inertia tensor as a metric on configuration space. We then say that configuration space is endowed with the *kinematic metric*.

As usual, we define the action $\mathcal{S}$ as the integral of the Lagrangian along the path γ with respect to the curve parameter

$$\mathcal{S}[\gamma] = \int \mathcal{L}(\dot{\gamma})dt.$$
(10.316)

In the case where there is no potential V, this functional exactly corresponds to the functional s_γ in Eq. (9.126) using the kinematic metric. We saw that this functional was precisely the functional whose stationary curves were the geodesics with a constant magnitude tangent vector. As a direct consequence, the solutions to the equations of motion in this case satisfy

$$\nabla_{\dot{\gamma}} \dot{\gamma} = 0,$$
(10.317)

i.e., the geodesic equations, where the connection is the Levi-Civita connection based on the kinematic metric. In particular, since the magnitude of the tangent vector does not change along these geodesics, we find that the kinetic energy

$$T = \frac{1}{2} M(\dot{\gamma}(t), \dot{\gamma}(t)) \tag{10.318}$$

is constant. This is to be expected in the absence of a potential since the Lagrangian does not depend explicitly on time and the kinetic energy is the total energy.

Example 10.51 If we remove the potential from the spherical pendulum in Example 10.50, then the system will evolve along the geodesics of the sphere, since the kinematic metric is proportional to the standard metric on the sphere, with constant speed. As we have seen earlier, these geodesics are the great circles, see Problem 9.33.

When we add a potential V to our Lagrangian the situation changes slightly and we have

$$\mathcal{L}(\dot{\gamma}) = \mathcal{L}_{\text{kin}}(\dot{\gamma}) - V(\gamma), \tag{10.319}$$

where $\mathcal{L}_{\text{kin}}$ is the Lagrangian that we just discussed that contains only the kinetic energy. The Euler–Lagrange equations now take the form

$$\frac{\partial \mathcal{L}}{\partial q^a} - \frac{d}{dt}\frac{\partial \mathcal{L}}{\partial \dot{q}^a} = \frac{\partial \mathcal{L}_{\text{kin}}}{\partial q^a} - \frac{d}{dt}\frac{\partial \mathcal{L}_{\text{kin}}}{\partial \dot{q}^a} - \frac{\partial V}{\partial q^a} = -M_{ab}\nabla_{\dot{\gamma}}\dot{q}^b - \partial_a V = 0. \tag{10.320}$$

This is the component version of the equality

$$M(\partial_a, \nabla_{\dot{\gamma}}\dot{\gamma}) = -dV(\partial_a). \tag{10.321}$$

We now have a relation that on the left-hand side contains the generalised inertia tensor and the change in the velocity, i.e., acceleration, and on the right-hand side has the differential of the potential, i.e., the force. Defining $F = -dV$, we can write this as

$$M_{ab}\nabla_{\dot{\gamma}}\dot{q}^b = F_a, \tag{10.322}$$

which is a generalisation of *Newton's second law*.

Example 10.52 In the case of a particle of mass m moving in three dimensions with a potential $V(\vec{x})$, the generalised inertia tensor is given by $m\delta_{ij}$ in Cartesian coordinates. This implies that the Christoffel symbols of the kinematic metric all vanish and $\nabla_{\dot{\gamma}}\dot{x}^i = \ddot{x}^i$. The generalisation of Newton's second law now becomes

$$m\delta_{ij}\ddot{x}^j = m\ddot{x}^i = F_i, \tag{10.323}$$

which is just its usual form.

10.5.2 The Hamiltonian formalism revisited

The Lagrangian formalism expressed in terms of the calculus on manifolds was concerned with the configuration space as a manifold M and the Lagrangian as a function on the

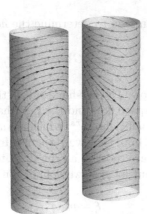

Figure 10.27 The phase space flow of Fig. 10.24 represented on a cylinder, here seen from two different perspectives, to illustrate the fact that the phase space itself is a cylinder, which is a manifold that can be described using the methods developed in Chapter 9.

tangent bundle TM, let us now discuss the properties of the phase space coordinates and the corresponding formulation of Hamiltonian mechanics. The first thing we need to do is to identify the relation of the phase space to the configuration space M. If we assume that configuration space is an N-dimensional manifold with local coordinates q^a, we have seen that the canonical momentum corresponding to q^a is given by

$$p_a = \frac{\partial \mathcal{L}}{\partial \dot{q}^a}. \tag{10.324}$$

Changing coordinates to Q^a, we found that p_a could be expressed in terms of the canonical momenta of the new coordinates as

$$p_a = \frac{\partial \mathcal{L}}{\partial \dot{Q}^b} \frac{\partial \dot{Q}^b}{\partial \dot{q}^a} = P_b \frac{\partial Q^b}{\partial q^a}, \tag{10.325}$$

which is just the transformation rule for the components of a dual vector, cf. Eq. (9.27). The phase space of a system is therefore equivalent to the *cotangent bundle* T^*M of the configuration space, i.e., the union of all cotangent spaces $T_p^* M$. An element of the cotangent bundle is any cotangent vector in configuration space and just as for the tangent bundle we can define a projection π such that $\pi(\xi) = p$ if ξ is a dual vector in $T_p^* M$. As any element in the cotangent bundle can be specified by N coordinates q^a and N components of the canonical momentum p_a, it is in itself also a manifold of dimension $2N$ with the coordinates y^r that we have already introduced.

Example 10.53 The configuration space of a bead of mass m moving on a ring of radius r_0 in a gravitational field with strength g can be described as the circle S^1 that may be parametrised by an angle θ. The phase space is therefore the cotangent bundle of the circle on which we can use the angle θ and the component p of the dual vector $p\,d\theta$ as coordinates. Since $-\infty < p < \infty$, the phase space of the pendulum is a cylinder, see Fig. 10.27.

The Poisson bracket has been seen to determine the flows in phase space according to

$$\dot{y}^r = \{y^r, G\} = \omega^{rs} \frac{\partial G}{\partial y^s}, \tag{10.326}$$

where G is the generator of the flow. The left-hand side of this relation are the components of the tangent vector of the phase space flow and the right-hand side contains the components of the differential of G and the coefficients ω^{rs}, which are anti-symmetric. We first note that the anti-symmetry of ω is something that we recognise from our discussion on differential forms although the indices here are those of a type $(2,0)$ tensor and not a differential form. We will deal with this in time, but for now we define a one-form ξ on the cotangent bundle according to

$$\xi = p_a dq^a. \tag{10.327a}$$

This one-form has the exterior derivative

$$\omega = d\xi = dp_a \wedge dq^a \tag{10.327b}$$

and furthermore

$$\omega \left(\frac{\partial}{\partial q^a}, \frac{\partial}{\partial p_b} \right) = -\delta_a^b. \tag{10.327c}$$

This is exactly the Poisson bracket relation between q^a and p_b up to a sign and in general we define

$$\omega_{rs} = \omega(\partial_r, \partial_s). \tag{10.328}$$

In what follows, we will consider ω as a map from the tangent space of phase space to its cotangent space. Whenever we give only one argument, what is intended will be the one-form

$$\omega(X) = \omega_{rs} X^s dy^r, \tag{10.329}$$

where X is a phase space tangent vector. We now introduce the type $(2,0)$ tensor Ω on phase space that is the inverse of $\omega(X)$, i.e.,

$$\Omega(dy^r, \omega(\partial_s)) = \delta_s^r \tag{10.330}$$

for all ∂_s and dy^r. By construction, we also have the relation

$$\omega \left(\frac{\partial}{\partial q^a} \right) = \left(\frac{\partial q^b}{\partial q^a} \right) dp_b - \left(\frac{\partial p_b}{\partial q^a} \right) dq^b = dp_a \tag{10.331}$$

and therefore

$$\Omega(dq^a, dp_b) = \Omega \left(dq^a, \omega \left(\frac{\partial}{\partial q^b} \right) \right) = \delta_b^a. \tag{10.332}$$

This is exactly the relation imposed by the Poisson bracket and the tensor Ω is a map from one-forms on phase space to tangent vectors on phase space with components given by

$$\Omega(dy^r, dy^s) = \delta_{r+N,s} - \delta_{r,s+N} = \omega^{rs}, \tag{10.333}$$

where the ω^{rs} are the same coefficients as those introduced in Section 10.4. The definition of the flow tangent vector in Eq. (10.326) now takes the form

$$\dot{y}^r = \{y^r, G\} = \Omega(dy^r, dG). \tag{10.334a}$$

Alternatively, this can be written in terms of the two-form ω as

$$\omega(\dot{\gamma}) = dG, \qquad (10.334b)$$

where γ is a flow line of the phase space flow generated by G and $\dot{\gamma}$ its tangent vector. As before, the time evolution of a physical system in phase space is by definition given by the phase space flow generated by the Hamiltonian $\mathcal{H}$, which is given by

$$\mathcal{H}(q, p) = p_a \dot{q}^a - \mathcal{L}, \qquad (10.335)$$

where a $\dot{q}^a$ is everywhere replaced using the inverse of

$$p_a = \frac{\partial \mathcal{L}}{\partial \dot{q}^a}, \qquad (10.336)$$

which is an invertible map from the tangent space to the cotangent space.

Example 10.54 For the bead on the sphere discussed in Example 10.53, the two-form ω is given by

$$\omega = d(p\, d\theta) = dp \wedge d\theta. \qquad (10.337)$$

With the Hamiltonian

$$\mathcal{H} = \frac{p^2}{2mr_0^2} - mr_0 g \cos(\theta), \qquad (10.338)$$

where r_0 is the radius of the sphere, the exterior derivative of the Hamiltonian is given by

$$d\mathcal{H} = \frac{p}{mr_0^2} dp + mr_0 g \sin(\theta) d\theta. \qquad (10.339)$$

For the phase space flow generated by the Hamiltonian, we therefore find that

$$\omega(\dot{\gamma}) = \omega(\dot{\theta}\partial_\theta + \dot{p}\partial_p) = \dot{\theta}\, dp - \dot{p}\, d\theta = d\mathcal{H}. \qquad (10.340)$$

Identification with the exterior derivative of $\mathcal{H}$ gives Hamilton's equations of motion

$$\dot{\theta} = \frac{p}{mr_0^2} \quad \text{and} \quad \dot{p} = -mr_0 g \sin(\theta). \qquad (10.341)$$

The set of canonical transformations is a particular set of coordinate transformations on phase space that leaves the components of the two-form ω invariant. In a way, such coordinate transformations are similar to Cartesian coordinate transformations in a Euclidean space that leave the metric components on the form δ_{ij}. In a more general setting, we can allow any coordinate transformations as long as we transform the components of ω in a proper fashion, i.e., they transform as the components of a type $(0, 2)$ tensor

$$\omega_{r's'} = \frac{\partial y^r}{\partial y'^{r'}} \frac{\partial y^s}{\partial y'^{s'}} \omega_{rs}. \qquad (10.342)$$

The only requirements on ω in a general coordinate system is that it is anti-symmetric, has zero exterior derivative, and is non-degenerate, i.e., that $\omega(X) = 0$ only if $X = 0$, such that the inverse Ω exists. A two-form that satisfies these conditions is called a *symplectic form* and a manifold along with the specification of such a two-form is called a *symplectic manifold*. Many of the properties and theorems of Hamiltonian mechanics can be described more succinctly in terms of the corresponding statements for symplectic manifolds and, just as a system in Lagrangian mechanics is specified by the configuration space and a function on its tangent bundle, a general system in Hamiltonian mechanics is described by a symplectic manifold and a function on it.

Let us consider the phase space flow $y^r(\tau)$ that is generated by the function G such that $y^r(0) = y_0^r$. Since both $y^r(\tau)$ and y_0^r are coordinates describing a point in phase space, this flow can be considered as a one-parameter family of maps $g_\tau(y)$ from phase space to itself, which is invertible since $g_\tau(g_{-\tau}(y)) = y$. The pullback $g_\tau^* \omega$ of the symplectic form ω under this map satisfies

$$g_\tau^* \omega_{rs} = \frac{\partial y^t}{\partial y_0^r} \frac{\partial y^u}{\partial y_0^s} \omega_{tu}|_y \simeq \omega_{rs} + \tau \left(\frac{\partial \dot{y}^t}{\partial y_0^r} \omega_{ts} + \frac{\partial \dot{y}^t}{\partial y_0^s} \omega_{rt} + \dot{y}^t \partial_t \omega_{rs} \right) \tag{10.343}$$

to linear order in τ. Before continuing with this expression, let us derive a few useful relations. Writing Eq. (10.334b) on coordinate form, we find that

$$\omega_{rt} \dot{y}^t = \partial_r G \quad \Longrightarrow \quad \partial_s \omega_{rt} \dot{y}^t = \omega_{rt} \partial_s \dot{y}^t + \dot{y}^t \partial_s \omega_{rt} = \partial_s \partial_r G. \tag{10.344a}$$

Reshuffling now results in

$$\omega_{rt} \partial_s \dot{y}^t = \partial_s \partial_r G - \dot{y}^t \partial_s \omega_{rt}. \tag{10.344b}$$

Furthermore, we have assumed that the exterior derivative of the symplectic form vanishes, which implies that

$$d\omega = \frac{1}{2} (\partial_r \omega_{st}) dy^r \wedge dy^s \wedge dy^t = \frac{1}{2} (\partial_{[r} \omega_{st]}) dy^r \wedge dy^s \wedge dy^t = 0 \tag{10.345}$$

and therefore

$$3\partial_{[r} \omega_{st]} = -\partial_r \omega_{ts} - \partial_s \omega_{rt} + \partial_t \omega_{rs} = 0. \tag{10.346}$$

Returning to the pullback $g_\tau^* \omega$, we find that

$$\frac{d}{d\tau} g_\tau^* \omega_{rs} = \omega_{ts} \partial_r \dot{y}^t + \omega_{rt} \partial_s \dot{y}^t + \dot{y}^t \partial_t \omega_{rs} = \dot{y}^t (-\partial_r \omega_{ts} - \partial_s \omega_{rt} + \partial_t \omega_{rs}) = 0, \tag{10.347}$$

where we have used both of the relations that were just derived. Since the derivative of $g_\tau^* \omega$ is zero and $g_0^* \omega = \omega$, it follows that

$$g_\tau^* \omega = \omega, \tag{10.348}$$

i.e., the pullback of the symplectic form under the phase space flow is equal to the symplectic form itself and we say that the flow preserves the symplectic structure of the manifold.

We are now ready to frame *Liouville's theorem* in a different manner. The first thing we must do is to define what is meant by the phase space volume. After all, our previous discussion on Liouville's theorem essentially regarded all phase space variables y^r as coordinates in a Euclidean space when equating an incompressible flow with the divergence of the flow field. If the phase space is $2N$-dimensional (to show that a symplectic manifold must be of even dimension is left as Problem 10.54), then we can define the *phase space volume form* η as

$$\eta = \omega^N, \tag{10.349}$$

where ω^N denotes the wedge product of N copies of the symplectic form. The volume of the phase space region P is now given by

$$V(P) = \int_P \omega^N. \tag{10.350}$$

We now let $g_\tau(P)$ be the phase space region which P is mapped to under the phase space flow and its volume is given by (see Problem 9.54)

$$V(g_\tau(P)) = \int_{g_\tau(P)} \omega^N = \int_P g_\tau^* \omega^N = \int_P \omega^N = V(P) \tag{10.351}$$

and therefore this flow preserves the phase space volume, which is just the statement of Liouville's theorem. In fact, the above argumentation can be made more general by letting P be a $2k$-dimensional submanifold in phase space and relating the integral of ω^k over P and its phase space flow $g_\tau(P)$ according to (see Problem 9.54)

$$\int_{g_\tau(P)} \omega^k = \int_P g_\tau^* \omega^k = \int_P \omega^k. \tag{10.352}$$

Thus, any integral of ω^k over a $2k$-dimensional submanifold, and not only volumes in the phase space, is preserved under generated phase space flows.

10.6 FIELD THEORY

The discussion in this chapter has been mainly focused on the situations where the configuration space is finite dimensional and can be described by a set of coordinates q^a. As we mentioned when introducing the configuration space, this will not always be the case and we may instead have to consider a more general configuration space with an infinite number of degrees of freedom that may be denoted as $q(a, t)$, where a belongs to some indexing set. When the indexing set can be thought of as a manifold, the degrees of freedom form a field on that manifold and we are then dealing with a *field theory*. In general, we may also have a theory involving several fields on the same manifold and we can then add additional indices to denote which field we are referring to, e.g., $q_i(a, t)$. In what follows, we will assume that we are dealing with a field theory and work exclusively in the Lagrangian formalism.

Example 10.55 We have seen an example of a field theory already in Example 10.11, where we discussed that the configuration space of an oscillating string was given by functions $u(x, t)$ that describe the transversal displacement of the string as a function of the position x on the string. In this case, the index set is the interval $0 < x < \ell$, where ℓ is the length of the string. This is an elementary example of a manifold with x as a global coordinate chart. A fact we have many times ignored is that there are two directions in which the string can oscillate, mainly because the oscillations in different directions decouple to first approximation, but we can include this in the description by using two fields $u_1(x, t)$ and $u_2(x, t)$ that describe the displacement in the different directions, see Fig. 10.28. We can also extend this description to an infinite or half-infinite string by extending the allowed range of values for x.

In a similar fashion, the transversal displacement of a membrane can be described by a field $u(\vec{x}, t)$, where $\vec{x}$ is a two-dimensional vector describing the position on the membrane. The index set in this case is a manifold describing the shape of the membrane itself.

It is not necessary for a field to be a scalar field. In fact, there are many situations where this is not the case. Instead, we will often encounter situations where the fields themselves are assumed to have certain transformation properties when the coordinates on the index set are changed.

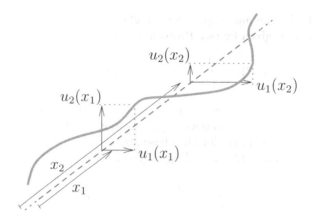

Figure 10.28 The transversal displacements u_1 and u_2 of an oscillating string (solid curve) are functions of the point on the string and are therefore fields on the manifold describing the position on the string (dashed line), i.e., an interval on the real line. For each value of the coordinate x, the fields u_1 and u_2 take the values $u_1(x)$ and $u_2(x)$, respectively.

Example 10.56 The configuration space of electromagnetic theory will turn out not to be the electromagnetic fields themselves, but rather the scalar $\phi(\vec{x}, t)$ and vector $\vec{A}(\vec{x}, t)$ potentials modulo gauge transformations. While the scalar potential is a scalar field, the vector potential transforms as a vector under coordinate changes.

When dealing with Lagrangian field theory, we will generally write down a Lagrangian as a function of the configuration space for a fixed time t as usual. However, as we are dealing with fields on a manifold, the Lagrangian may depend not only on the fields themselves, but also on their derivatives on the manifold. We will consider Lagrangians of the form

$$\mathcal{L} = \int_K L(q, \partial_a q, t) \, dK, \tag{10.353}$$

where K is the index set, dK the volume element on the index set, and L generally is a function of q and its derivatives as well as time t called the *Lagrangian density*. We also assume that the derivatives $\partial_a q$ can be derivatives with respect to any variable that q depends on, including the time t. The action is now given by

$$S = \int_{t_1}^{t_2} \mathcal{L} \, dt = \int_{t_1}^{t_2} \int_K L(q, \partial_a q, t) \, dK \, dt. \tag{10.354}$$

In order to put the time variable t and the index set K on a more equal footing, we can construct a generalised *space-time* manifold $\mathcal{K}$ where we use one set of coordinates from K and time t as coordinates. The volume element in the space-time is given by $d\mathcal{K} = dK \, dt$ and the action can now be written as

$$S = \int_{\mathcal{K}} L \, d\mathcal{K}. \tag{10.355}$$

From this expression, we may use the Euler–Lagrange equations to find the equations of motion for the fields. Since the action is an integral over several variables, these equations of motion will generally be in the form of a set of partial differential equations.

Example 10.57 Consider the example of transversal displacements of a string with tension S and linear density ρ_ℓ. We have already expressed the Lagrangian density for this situation in Example 8.25 as

$$L = \frac{\rho_\ell}{2} u_t(x,t)^2 - \frac{S}{2} u_x(x,t)^2 - f(x,t)u(x,t), \tag{10.356}$$

where $f(x,t)$ is an external force density acting on the string. A point in the generalised space-time manifold in this case is specified by giving a position x on the string and a time t and we saw that the equation of motion resulting from the Lagrangian density is the sourced wave equation

$$u_{tt}(x,t) - c^2 u_{xx}(x,t) = \frac{f(x,t)}{\rho_\ell}, \tag{10.357}$$

where $c^2 = S/\rho_\ell$.

The above example is a special case of a *Klein–Gordon field*, which is one of the most common examples used in both classical field theory as well as in quantum field theory. With a space-time using a position $\vec{x}$ and a time t, the free Klein–Gordon field $\phi(\vec{x},t)$ has the Lagrangian density

$$L_0 = \frac{1}{2} \left[\phi_t^2 - c^2 (\nabla \phi)^2 - m^2 c^4 \phi^2 \right]. \tag{10.358a}$$

Note that this, depending on the dimensions of ϕ, could be multiplied by a dimensionful constant in order for the dimensions to match without altering the equations of motion. In the case of an interacting Klein–Gordon field, we add a source term

$$L = L_0 + \phi J, \tag{10.358b}$$

where J is a function describing the source. The source may be modelled as a known source, depend on other fields, or on ϕ itself. In the case when the source depends on ϕ, the field is said to be *self-interacting*. In the case where the field is not self-interacting, the equation of motion for the Klein–Gordon field is given by

$$\partial_a \frac{\partial L}{\partial (\partial_a \phi)} - \frac{\partial L}{\partial \phi} = \phi_{tt} - c^2 \nabla^2 \phi + m^2 c^4 \phi - J = 0, \tag{10.359}$$

which is an inhomogeneous partial differential equation of the same type that we have encountered in the earlier chapters. In fact, this differential equation, known as the sourced *Klein–Gordon equation*, has exactly the same structure as the differential equation describing the motion of a string with a restoring force proportional to the displacement, see Problem 3.24.

10.6.1 Noether's theorem revisited

In Section 10.2.4, we discussed the relation between quasi-symmetries and constants of motion in the case of a finite-dimensional phase space. When the action is an integral in several variables, the situation changes slightly and instead of constants of motion, we will be dealing with conserved currents, of which we have already seen some examples in Section 8.6.1. In order to generalise to field theory, we consider an infinitesimal transformation

$$Y^a = y^a + \varepsilon \, \delta y^a \quad \text{and} \quad Q_i(Y) = q_i(y) + \varepsilon \, \delta q_i, \tag{10.360}$$

where y^a denotes the set of space-time coordinates, i.e., t and the coordinates on the index set. We have also assumed that there are a number of fields q_i indexed by i. The variation of the action in a space-time region P under this transformation is given by

$$\mathcal{S}' - \mathcal{S} = \int_P \underbrace{[L(Q_i(y), \partial Q_i(y), y) - L(q_i, \partial q_i, y) + \partial_a(L\,\delta y^a)]}_{\equiv \varepsilon \delta L} d\mathcal{K}, \qquad (10.361)$$

where we have also defined the variation δL of the Lagrangian density. By definition, this transformation is a quasi-symmetry of the Lagrangian density if

$$\delta L = \partial_a F^a, \qquad (10.362)$$

where F^a is a set of functions. Expressing δL in terms of the infinitesimal transformation, we find that

$$\delta L = \frac{\partial L}{\partial q_i}\bar{\delta}q_i + \frac{\partial L}{\partial(\partial_a q_i)}\partial_a\bar{\delta}q_i + \partial_a(L\,\delta y^a), \qquad (10.363)$$

where we have introduced the field difference

$$\varepsilon\bar{\delta}q_i = Q_i(y) - q_i(y), \qquad (10.364\text{a})$$

which describes the change in the field at y. By definition, we find that

$$\varepsilon\,\delta q_i = Q_i(y + \varepsilon\,\delta y) - q_i(y) = \varepsilon(\delta y^a \partial_a q_i + \bar{\delta}q_i). \qquad (10.364\text{b})$$

Example 10.58 Looking back at our original version of Noether's theorem, our space-time only contains a time t and no spatial coordinates and the Lagrangian $\mathcal{L}$ coincides with the Lagrangian density L. For any transformation that has a δt that does not depend on t, we find that

$$\begin{aligned}
\delta L &= \frac{\partial L}{\partial q_i}(\delta q_i - \dot{q}_i\delta t) + \frac{\partial L}{\partial\dot{q}_i}(\delta\dot{q}_i - \ddot{q}_i\delta t) + \frac{dL}{dt}\delta t \\
&= \frac{\partial L}{\partial q_i}\delta q_i + \frac{\partial L}{\partial\dot{q}_i}\delta\dot{q}_i + \left(\frac{dL}{dt} - \frac{\partial L}{\partial q_i}\dot{q}_i - \frac{\partial L}{\partial\dot{q}_i}\ddot{q}_i\right)\delta t \\
&= \frac{\partial L}{\partial q_i}\delta q_i + \frac{\partial L}{\partial\dot{q}_i}\delta\dot{q}_i + \frac{\partial L}{\partial t}\delta t.
\end{aligned} \qquad (10.365)$$

This is exactly the form of Eq. (10.97) and the type of transformations we considered in Section 10.2.4.

Using the equations of motion to rewrite Eq. (10.363), we find that

$$\delta L = \partial_a\left(\frac{\partial L}{\partial(\partial_a q_i)}\bar{\delta}q_i + L\delta y^a\right) \qquad (10.366)$$

is always satisfied on-shell. It follows that we have the on-shell relation

$$\delta L = \partial_a\left[\frac{\partial L}{\partial(\partial_a q_i)}\delta q_i - \left(\frac{\partial L}{\partial(\partial_a q_i)}(\partial_b q_i) - \delta_b^a L\right)\delta y^b\right]. \qquad (10.367)$$

If the transformation is a quasi-symmetry of the Lagrangian density, we therefore find that

$$\partial_a J^a = 0, \qquad (10.368\text{a})$$

where we have defined the *conserved current*

$$J^a = \left(\frac{\partial L}{\partial(\partial_a q_i)}(\partial_b q_i) - \delta_b^a L \right) \delta y^b - \frac{\partial L}{\partial(\partial_a q_i)}\delta q_i + F^a. \qquad (10.368b)$$

As discussed in Example 8.19, this is the continuity equation with the time-component of J^a as the density and the spatial components as the currents. Thus, the continuity equation for a density and current given by the transformation is the field theory version of Noether's theorem. In some more generality, if the transformation is not a quasi-symmetry and we instead have

$$\delta L = \partial_a F^a - \kappa, \qquad (10.369)$$

then the function κ is a source term for the corresponding current as we would find $\partial_a J^a = \kappa$.

Example 10.59 Consider the situation where the Lagrangian density does not depend explicitly on time. In this situation, the transformation given by $\delta t = 1$ and all other δy^a and δq_i equal to zero is a symmetry of the Lagrangian density. The conserved current corresponding to this situation is given by

$$J^a = \frac{\partial L}{\partial(\partial_a q_i)}\dot{q}_i - \delta_t^a L, \qquad (10.370)$$

where $\dot{q}_i = \partial_t q_i$. The time-component of this current is given by

$$J^t = \frac{\partial L}{\partial \dot{q}_i}\dot{q}_i - L = H, \qquad (10.371)$$

where H is the *Hamiltonian density*. For many situations, this can be interpreted as the energy density and the spatial components of the current

$$\vec{J} = \frac{\partial L}{\partial(\nabla q_i)}\dot{q}_i \qquad (10.372)$$

can then be interpreted as the energy current.

Example 10.60 For an infinite string with tension S and linear density ρ_ℓ, we can consider transversal oscillations in two orthogonal directions. This is described by the amplitudes $u_1(x,t)$ and $u_2(x,t)$ in the different directions. The resulting Lagrangian density

$$L = \frac{\rho_\ell}{2}[(\partial_t u_1)^2 + (\partial_t u_2)^2] - \frac{S}{2}[(\partial_x u_1)^2 + (\partial_x u_2)^2] \qquad (10.373)$$

is invariant under the transformation

$$\delta u_1 = u_2 \quad \text{and} \quad \delta u_2 = -u_1, \qquad (10.374)$$

with all other variations equal to zero. The corresponding conserved current is given by

$$J^a = -\frac{\partial L}{\partial(\partial_a u_i)}\delta u_i = -\frac{\partial L}{\partial(\partial_a u_1)}u_2 + \frac{\partial L}{\partial(\partial_a u_2)}u_1. \qquad (10.375)$$

The time-component of this current is

$$J^t = \rho_\ell(\dot{u}_2 u_1 - \dot{u}_1 u_2),$$ (10.376)

which is the angular momentum density of the string with respect to its equilibrium position. The spatial component

$$J^x = S[u_2(\partial_x u_1) - u_1(\partial_x u_2)]$$ (10.377)

at position x describes the corresponding torque on the string to the right of x from the string to the left of x. If we introduce an external gravitational field g in the u_2 direction such that the Lagrangian density is instead given by

$$L_{\text{grav}} = L - \rho_\ell g u_2,$$ (10.378)

then the transformation is no longer a symmetry of the Lagrangian density. Instead we find that

$$\delta L_{\text{grav}} = \rho_\ell g u_1$$ (10.379)

and the continuity equation for the angular momentum of the string is instead given by

$$\partial_a J^a = -\rho_\ell g u_1,$$ (10.380)

where the right-hand side represents the external torque density provided by the gravitational field. Note that this is exactly what we expect, since the force density due to the gravitational field is $\rho_\ell g$ and its perpendicular distance to the equilibrium position is u_1.

10.6.2 Symmetries of the wave equation

We will end by looking at the symmetries of the Lagrangian density

$$L = \frac{1}{2}(\rho_\ell q_t^2 - S q_x^2) = \frac{\rho_\ell}{2}(q_t^2 - c^2 q_x^2),$$ (10.381)

which is the Lagrangian density of the Klein–Gordon field with $m = 0$ and $\phi = \sqrt{\rho_\ell} q$. As we have already seen, this Lagrangian density results in the equations of motion

$$q_{tt} - c^2 q_{xx} = 0,$$ (10.382)

i.e., the wave equation for the field $q(x,t)$. The invariance of this Lagrangian density under space-time translations $\delta y^a = k^a$ for constant k^a was discussed already in Example 8.26, where we found the conserved current

$$T_t^t = \frac{\rho_\ell}{2}\left(q_t^2 + c^2 q_x^2\right), \quad T_t^x = -\rho_\ell c^2 q_t q_x$$ (10.383a)

corresponding to time-translational invariance and

$$T_x^t = \rho_\ell q_t q_x, \quad T_x^x = -\frac{\rho_\ell}{2}\left(q_t^2 + c^2 q_x^2\right)$$ (10.383b)

corresponding to invariance under spatial translations. The first of these equations was identified with the continuity equation for energy with $\mathcal{E} = T_t^t$ being the energy density and $j_\mathcal{E} = T_t^x$ being the energy current, while the second was interpreted as the continuity equation for momentum with the momentum density $p = -T^t_x$ and momentum current $j_p = -T^x_x$. Integrating over all of space and assuming that the field q goes to zero sufficiently fast as $x \to \pm\infty$, we find that the total energy and momentum

$$E = \int \mathcal{E}\, dx \quad \text{and} \quad P = \int p\, dx \tag{10.384}$$

are conserved quantities.

Example 10.61 We have already seen that any function on the form $q(x,t) = f(x - ct)$ is a solution to the wave equation in the form of a wave moving to larger values of x as t increases. For such a wave, we find that the energy and momentum densities are given by

$$\mathcal{E} = \rho_\ell c^2 f'^2 \quad \text{and} \quad p = \rho_\ell c f'^2, \tag{10.385}$$

respectively. As a result, it holds that $\mathcal{E} = cp$ in this scenario. The total energy and momenta are given by the integrals of these expressions and hence $E = cP$.

The Lagrangian density in Eq. (10.381) has one more important symmetry that is given by (see Problem 10.57)

$$\delta t = -\frac{x}{c}, \quad \delta x = -ct, \quad \text{and} \quad \delta q = 0. \tag{10.386}$$

The conserved current corresponding to this transformation has the components

$$J^a = \left(\frac{\partial L}{\partial q_a} q_b - \delta^a_b L \right) \delta y^b = T^a_b \delta y^b \tag{10.387a}$$

or, explicitly,

$$J^t = -\frac{x}{c}\mathcal{E} + ctp \quad \text{and} \quad J^x = -\frac{x}{c} j_\mathcal{E} + ct j_p. \tag{10.387b}$$

So what does this conserved current tell us? Integrating cJ^t over all of space and again assuming that the field q vanishes sufficiently fast as $x \to \pm\infty$, we find that

$$\int cJ^t dx = c^2 tP - Ex_E \tag{10.388a}$$

is a conserved quantity, where we have defined the *center of energy*

$$x_E = \frac{1}{E} \int x\mathcal{E}\, dx. \tag{10.388b}$$

Denoting the conserved quantity as $-Ex_0$, we now find that

$$x_E = x_0 + ct\frac{cP}{E}. \tag{10.389}$$

In other words, the center of energy moves at a constant speed determined by the ratio $c^2 P/E$.

Example 10.62 The wave $q(x,t) = f(x - ct)$ discussed in Example 10.61 is moving at speed c towards larger x values. We found that the relation between the total energy and momenta was given by $E = cP$ and can therefore conclude that the center of energy moves according to

$$x_E = x_0 + ct. \tag{10.390}$$

As might be expected, the center of energy therefore also moves towards larger x values with the wave speed c.

Having found the conserved quantity related to the transformation defined by Eq. (10.386), it is natural to ask what kind of continuous transformation it corresponds to. We can do so by calling the continuous transformation parameter θ and solving the coupled differential equations

$$\frac{dt}{d\theta} = \delta t = -\frac{x}{c} \quad \text{and} \quad \frac{dx}{d\theta} = \delta x = -ct \tag{10.391}$$

with the initial condition $t(0) = t_0$ and $x(0) = x_0$. Solving this system results in

$$t(\theta) = t_0 \cosh(\theta) - \frac{x_0}{c} \sinh(\theta) \quad \text{and} \quad x(\theta) = x_0 \cosh(\theta) - ct_0 \sinh(\theta). \tag{10.392}$$

The continuous parameter θ is known as the *rapidity* of the transformation. There is also an alternative form for this transformation using a different parameter v that is defined as

$$v = c \tanh(\theta). \tag{10.393}$$

Using the relation $\cosh^2(\theta) - \sinh^2(\theta) = 1$, we can solve for $\cosh(\theta)$ and find that

$$\cosh(\theta) = \frac{1}{\sqrt{1 - \frac{v^2}{c^2}}} \equiv \gamma \tag{10.394a}$$

and, consequently,

$$\sinh(\theta) = \frac{v}{c} \cosh(\theta) = \gamma \frac{v}{c}. \tag{10.394b}$$

Insertion into Eq. (10.392) now provides the result

$$t = \gamma \left(t_0 - \frac{v}{c^2} x_0 \right) \quad \text{and} \quad x = \gamma \left(x_0 - vt_0 \right). \tag{10.395}$$

Even if we have not covered special relativity in this book, you might recognise this as the *Lorentz transformations*. The invariance under Lorentz transformations will be of particular importance in the study of special relativity, where c will be taken to be the speed of light in vacuum.

10.7 PROBLEMS

Problem 10.1. One end of a rigid homogeneous rod of length ℓ is moving according to

$$\vec{x}_0(t) = \frac{\vec{a}t^2}{2}. \tag{10.396}$$

Determine the motion of the other end if its displacement from the first at time $t = 0$ is given by $\vec{\xi} - \vec{\xi}_0 = \vec{d}_0$ and the rod is rotating with constant angular velocity $\vec{\omega}$.

Problem 10.2. Verify that the center of mass motion $\vec{x}_{cm}(t)$ that was defined in Eq. (10.11) coincides with the motion of the original center of mass $\vec{\xi}_{cm}$ defined by $\vec{x}(\vec{\xi}_{cm}, t)$.

Problem 10.3. A wheel of radius r_0 is rolling without slipping on a road that can be described as the x^1-x^2-plane. Using the wheel's center as reference point and assuming that it moves according to

$$\vec{x}_0(t) = x_0^1(t)\vec{e}_1 + r_0\vec{e}_3, \qquad (10.397)$$

find the angular velocity $\vec{\omega}(t)$ of the wheel and find the velocity $\vec{v}(t)$ of a point on the wheel that is located at a radius r.

Problem 10.4. The *parallel axis theorem* for the moment of inertia tensor states that the moment of inertia tensor I_{ij} of a body relative to the point $\vec{x}_0$ can be written as

$$I_{ij} = I_{ij}^{cm} + I_{ij}^{point}, \qquad (10.398)$$

where I_{ij}^{cm} is the moment of inertia relative to the center of mass and I_{ij}^{point} is the moment of inertia relative to $\vec{x}_0$ for a point particle of mass M placed in the center of mass. Verify that this relation is true starting from the definition of the moment of inertia tensor.

Problem 10.5. Compute the moment of inertia tensor for:

a) A homogeneous sphere of radius r_0 with respect to its center.

b) A homogeneous sphere of radius r_0 with respect to a point on its surface.

c) A homogeneous rod of length ℓ with respect to its center.

d) A homogeneous rod of length ℓ with respect to one of its ends.

e) A homogeneous cube of side length ℓ with respect to its center.

f) A homogeneous cube of side length ℓ with respect to one of its corners.

Assume that each object has a total mass M and express your results in terms of M and the given lengths. *Hint:* You may find the parallel axis theorem (see Problem 10.4) and various symmetry arguments useful.

Problem 10.6. Show that a rigid body does not allow any point in it to have zero velocity unless $\vec{\omega}$ is orthogonal to $\vec{v}_0$.

Problem 10.7. Find a general expression for the acceleration $\vec{a}(t)$ of a point in a rigid body assuming that the reference point has velocity $\vec{v}_0(t)$ and the body rotates with angular velocity $\vec{\omega}(t)$. Verify that you expression results in the relation

$$a = \frac{v^2}{r} \qquad (10.399)$$

when the reference point is taken to be a fixed point and the angular velocity is constant, where a and v are the magnitudes of $\vec{a}$ and $\vec{v}$, respectively, and r is the length of the component of the difference vector $\vec{d}(t) = \vec{x}(t) - \vec{x}_0(t)$ that is orthogonal to $\vec{\omega}$.

Problem 10.8. Use the parallel axis theorem (see Problem 10.4) to verify that the total kinetic energy of a rigid body may equally well be described as the rotational energy around a fixed point or as the translational energy of the center of mass plus the rotational energy with respect to the center of mass, i.e., that

$$\frac{1}{2}I_{ij}^{fix}\omega^i\omega^j = \frac{M}{2}\vec{v}_{cm}^2 + \frac{1}{2}I_{ij}^{cm}\omega^i\omega^j. \qquad (10.400)$$

Problem 10.9. A rigid body in two dimensions rotates around a fixed point. Its center of mass is displaced by a vector $\vec{\delta}_{cm}$ from the fixed point. There is no net external force on the body, but the net torque on the body is τ. Assuming that the moment of inertia of the body is I relative to the fixed point, express the force $\vec{F}_{fix}$ at the fixed point in terms of the given quantities, the total mass M of the body, and its angular velocity ω.

Problem 10.10. A car travelling horizontally at speed v breaks with constant acceleration a. Determine the apparent strength and direction of the gravitational field in the rest frame of the car if the car stops in a distance ℓ. Assume that the gravitational field in the road's rest frame has strength g.

Problem 10.11. Consider an object initially at rest but moving under the influence of a gravitational field $\vec{g} = -g\vec{e}_3$. Write down the corresponding solution $\vec{x}(t)$ to Newton's equations in an inertial frame. Transform your solution to a frame whose origin coincides with that of the original inertial frame, but that rotates with angular velocity $\vec{\omega} = \omega\vec{e}_2$. Verify that the resulting motion in the rotating frame satisfies Newton's equation in that frame when including the Coriolis and centrifugal forces.

Problem 10.12. If the persons A and B on the carousel depicted in Fig. 10.8 wish to throw balls to each other in such a way that they can be caught, then the Coriolis force implies that they should not throw the balls directly at one another, as discussed in the corresponding example. In which direction and with which speed should A throw the ball (in the rotating frame) if it should reach B at time $t = \theta/\omega$? Assume that the persons are located at the radii r_A and r_B, respectively.

Problem 10.13. Determine the external force necessary in order to keep an object of mass m moving radially away from the origin in a frame that rotates with constant angular velocity ω. You may restrict your treatment to the two spatial dimensions in which the frame rotates.

Problem 10.14. By definition, there is no net force on an object which is at rest in an inertial frame. If this object is described from a frame rotating with constant angular velocity $\vec{\omega}$, there will appear to be a centrifugal force acting on the object that would make it accelerate away from the axis of rotation. However, in the rotating frame, the object appears to undergo uniform circular motion, for which the net force should be directed towards the axis of rotation. Verify that the Coriolis force provides exactly the right force in order for this to occur.

Problem 10.15. Identify the configuration space and the corresponding indexing set of the following physical systems:

a) A particle constrained to move along the parabolic surface $\ell_0 x^3 = (x^1)^2 + (x^2)^2$.

b) The transversal movement of a circular drum skin of radius r_0 fixed at the borders.

c) The half-infinite chain of masses connected by springs shown in Fig. 10.29.

Problem 10.16. The configuration space of a system of k particles consists of the particle positions $\vec{x}_r(t)$ for each particle r. The Galilei transformation with velocity $\vec{v} = v^i\vec{e}_i = v\vec{n}$ is given by

$$t \to t \quad \text{and} \quad \vec{x}_r \to \vec{x}_r - \vec{v}t. \tag{10.401}$$

a) Determine δt and $\delta\vec{x}_r$ for infinitesimal Galilei transformation in the direction $\vec{n}$, where the speed v is seen as the transformation parameter.

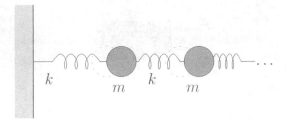

Figure 10.29 A half-infinite chain of masses m connected by springs of spring constant k.

b) Assuming that particle r has the mass m_r and that the potential energy of the system depends only on the particle separations $\vec{x}_{rs} = \vec{x}_r - \vec{x}_s$, we have already seen that the total momentum

$$\vec{P} = \sum_r m_r \dot{\vec{x}}_r \qquad (10.402)$$

is a constant of motion based on translational invariance. Using your result from (a), show that the Galilei transformation is a quasi-symmetry of the Lagrangian and show that the corresponding conserved quantity is

$$M\vec{X} = \vec{P}t - M\vec{x}_{\text{cm}}, \qquad (10.403)$$

where M is the sum of all the particle masses.

Note: The conserved quantity in (b) also follows directly from the fact that the expression for $\vec{P}$ is a total derivative. However, here we wish to show it using Noether's theorem instead.

Problem 10.17. Verify that the equations of motion for the generalised coordinates in the effective Lagrangian in Eq. (10.149) are equivalent to those of the original Lagrangian of Eq. (10.141).

Problem 10.18. Consider the mechanical system discussed in Problem 2.46, where you were asked to determine the generalised inertia tensor for the system. By considering the symmetries of the resulting Lagrangian, find two constants of motion and use them to show that the system can never reach the point $r = 0$ if $\dot{\varphi}(0) \neq 0$.

Problem 10.19. In Example 10.20, we studied a coordinate change for a bead on a rotating ring, which could be regarded as motion on a sphere with a time-dependent holonomic constraint. The original Lagrangian was given by

$$\mathcal{L} = \frac{mr_0^2}{2}[\dot{\theta}^2 + \sin^2(\theta)\dot{\varphi}^2] - mgr_0 \cos(\theta) \qquad (10.404)$$

and the holonomic constraint by

$$\varphi = \omega t. \qquad (10.405)$$

We implemented the parametrisation $\theta = \theta$ and $\varphi = \omega t$ of the constraint surface and showed that it resulted in an effective problem of the form

$$\mathcal{L}_{\text{eff}} = \frac{I}{2}\dot{q}^2 - V_{\text{eff}}(q). \qquad (10.406)$$

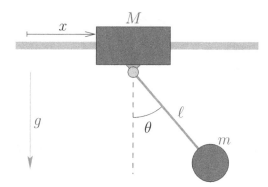

Figure 10.30 A mass m attached to a pendulum of length ℓ with the other end attached to a mass M moving without friction along a horizontal bar. The system is placed in a homogeneous gravitational field $\vec{g}$ and may be described using the generalised coordinates x and θ.

Another parametrisation of the constraint surface is given by

$$\alpha = \theta - \omega_0 t \quad \text{and} \quad \varphi = \omega t. \tag{10.407}$$

Verify that this parametrisation does not result in an effective Lagrangian on the form given in Eq. (10.406) and that we instead get additional terms proportional to $\dot{q}$.

Problem 10.20. A mass M is allowed to slide without friction on a horizontal bar. Attached to the mass is a pendulum with length ℓ and a mass m attached to its other end, see Fig. 10.30, and there is an external gravitational field $\vec{g}$. Using the generalised coordinates x and θ described in the figure,

a) write down an expression for the Lagrangian of the system,

b) deduce the conserved quantities related to invariance under translations in time and in the x-coordinate,

c) find the effective potential and the effective inertia for the θ-coordinate.

Problem 10.21. Consider a non-homogeneous circular wheel of radius r_0, mass m, and moment of inertia I relative to its center of mass. Its center of mass is displaced from its geometrical center by a distance r_1. The wheel rolls without slipping on a horizontal surface in a gravitational field of strength g. Write down an expression for the Lagrangian of the system and derive the corresponding equation of motion.

Problem 10.22. One end of a pendulum of length ℓ is forced to undergo harmonic motion with angular frequency ω and amplitude a while a point mass m is attached to the other end, see Fig. 10.31. Determine the equation of motion for the system when the strength of the external gravitational field is g.

Problem 10.23. For the mass m moving freely on the surface of a sphere of radius r_0 in an external gravitational field of strength g discussed in Example 10.22, find the expressions for the energy E and angular momentum L that result in a circular orbit at $\theta = \theta_0$ and compute the period of the orbit. Discuss the physical interpretation of the limiting behaviour when $\theta_0 \to \pi/2$ and $\theta_0 \to \pi$, respectively.

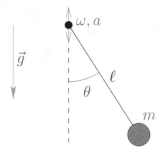

Figure 10.31 A pendulum of length ℓ with a mass m attached to one end and the other end undergoing forced harmonic motion with angular frequency ω and amplitude a. The system is placed in a gravitational field $\vec{g}$ and its configuration can be described by the angle θ.

Problem 10.24. Consider a particle of mass m moving in one dimension with a potential that moves with velocity v_0 such that

$$V(x,t) = V_0(x - v_0 t). \tag{10.408}$$

Show that the infinitesimal transformation given by $\delta t = 1$ and $\delta x = v_0$ is a symmetry of the Lagrangian for this system and determine the corresponding conserved quantity. Verify that, up to an additive constant, the conserved quantity is equal to the total energy in a reference frame that moves with velocity v_0.

Problem 10.25. A particle of mass m is moving in two dimensions subject to a rotating potential energy

$$V(\vec{x}, t) = V_0(R^{-\omega t}\vec{x}), \tag{10.409}$$

where $R^{-\omega t}$ is a rotation by an angle $-\omega t$. Derive the equations of motion and show that the quantity

$$J = E - \omega L \tag{10.410}$$

is a constant of motion, where E is the total energy of the system and $L = m(x^1\dot{x}^2 - x^2\dot{x}^1)$ is its angular momentum. Find the infinitesimal symmetry transformation related to this conserved quantity.

Problem 10.26. A hollow pipe in a gravitational field $\vec{g}$ rotates with angular frequency ω around an axis parallel to the field. The pipe makes a fixed angle θ with the axis and a small ball of mass m is free to move within it, see Fig. 10.32.

a) Find the Lagrangian of the system.

b) Find the effective potential for the coordinate r and the constant of motion that follows from the effective Lagrangian not being explicitly time-dependent. Differentiate the constant of motion in order to find the equation of motion for r.

c) Verify that the special cases $\theta = 0$ and $\theta = \pi/2$ have the expected equations of motion

$$\ddot{r} = -g \quad \text{and} \quad \ddot{r} = r\omega^2, \tag{10.411}$$

respectively.

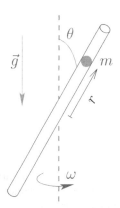

Figure 10.32 A ball of mass m is free to move within a hollow pipe that rotates around an axis parallel to the gravitational field $\vec{g}$ with angular frequency ω. The pipe makes a constant angle θ with this axis and the displacement of the mass from the axis can be described by the coordinate r.

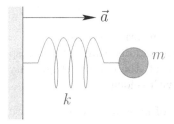

Figure 10.33 One end of a spring with spring constant k is being accelerated with acceleration $\vec{a}$ while the other end is attached to a mass m. We wish to study the resulting motion of the mass.

d) The problem has an unstable equilibrium for any $0 < \theta < \pi/2$. Find an expression for the r-coordinate in this equilibrium and interpret it in terms of a force sum.

Problem 10.27. An object with mass m moving in one dimension is attached to one end of a spring with spring constant k. The other end of the spring starts out at rest and then accelerates with constant acceleration a, see Fig. 10.33.

a) Determine the potential energy of the system as a function of the time t and the displacement x from the minimum of the potential at time $t = 0$.

b) Using the Lagrangian formalism, show that the equations of motion for the mass are equivalent to those that would be obtained if the other end was fixed, but an external gravitational field of strength $g = -a$ was applied.

c) Show that the Lagrangians of the two cases are different, but differ only by a total derivative df/dt, where f is a function of t and x.

d) Show that the infinitesimal transformation given by $\delta t = 1$ and $\delta x = at$ is a quasi-symmetry of the system. Determine the corresponding conserved quantity.

Problem 10.28. A mechanical system consists of a rotationally symmetric wheel of radius

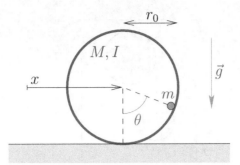

Figure 10.34 A wheel of radius r_0, mass M, and moment of inertia I is rolling without slipping along a horizontal surface. A mass m is allowed to move freely along the perimeter of the wheel and the system is placed in a gravitational field $\vec{g}$. The configuration space of the system is parametrised by the generalised coordinates x and θ.

r_0 with mass M and moment of inertia I relative to its center and a bead of mass m that can move freely along the wheel's perimeter. The wheel is rolling without slipping on a horizontal surface in a gravitational field of strength g, see Fig. 10.34.

a) Write down the Lagrangian of the system using the two configuration space coordinates x and θ given in the figure.

b) Show that the system is invariant under the transformation $x \to x + s$ and find the corresponding conserved quantity J.

c) Use your result from (b) to rewrite the problem as an effective one-dimensional problem for the angle θ. Identify the corresponding effective potential and effective generalised inertia.

Problem 10.29. Consider the motion of a point particle of mass m in two-dimensions constrained to be at a distance $\ell(t)$ from the origin of a Cartesian coordinate system at time t. Using polar coordinates, show that the angular momentum $L = m\ell(t)^2\dot{\phi}$ is conserved and that this implies that the particle must have a larger tangential velocity for smaller $\ell(t)$.

Problem 10.30. A *double Atwood machine* is shown in Fig. 10.35. Assume that the masses are m_1, m_2, and m_3, respectively, and that the pulleys are massless. Describe the configuration space of this system and find the generalised inertia tensor. Also write down the potential energy and find the equations of motion. Determine the condition on the masses in order for the infinitesimal transformation given by $\delta x = 1$ (with all other variations equal to zero) to be a symmetry of the Lagrangian, determine the corresponding conserved quantity, and give its physical interpretation. Repeat the discussion for the transformation $\delta y = 1$.

Problem 10.31. A particle of mass m moves freely on the paraboloid $z = k\rho^2$ given in cylinder coordinates. An external gravitational field of strength g acts in the negative z-direction. Using that the problem has a rotational symmetry, determine the effective potential for the movement in the ρ-direction.

Problem 10.32. The motion of a charged particle with mass m and charge q in an electromagnetic field can be described by the Lagrangian

$$\mathcal{L} = \frac{m}{2}\dot{\vec{x}}^2 - q\phi + q\vec{A}\cdot\dot{\vec{x}},\tag{10.412}$$

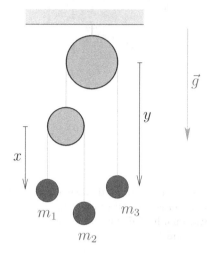

Figure 10.35 A double Atwood machine with two idealised pulleys and three masses m_1, m_2, and m_3. The system is placed in a gravitational field $\vec{g}$ and can be described by the two generalised coordinates x and y.

where ϕ is the scalar potential and $\vec{A}$ the vector potential that appear in electrodynamics. For a static electromagnetic field, the potentials do not depend explicitly on time.

a) Find the canonical momentum $\vec{p} = \partial\mathcal{L}/\partial\dot{\vec{x}}$ and verify that it does not correspond to the physical momentum $m\dot{\vec{x}}$ of the particle.

b) Verify that the equations of motion are given by the *Lorentz force law*

$$m\ddot{\vec{x}} = q(\vec{E} + \dot{\vec{x}} \times \vec{B}), \tag{10.413a}$$

where the electric and magnetic fields are

$$\vec{E} = -\nabla\phi \quad \text{and} \quad \vec{B} = \nabla \times \vec{A}, \tag{10.413b}$$

respectively.

c) Express the Hamiltonian of the system as a function of $\vec{x}$ and the canonical momentum $\vec{p}$.

Problem 10.33. Show that the unbound states of the Kepler problem, i.e., states with $E > 0$, result in the trajectory of the particle being hyperbolic and compute the deflection angle α, the distance of closest approach $\rho_{\min}$ to the center, and the impact parameter d, see Fig. 10.36, in terms of the constants E and L. *Hint:* Like an ellipse, a hyperbola may be described by the relation $\rho = \rho_0/(1 + \varepsilon\cos(\phi))$ in polar coordinates, where ε is the eccentricity. Unlike an ellipse, the eccentricity for a hyperbola is $\varepsilon > 1$.

Problem 10.34. Find the period of oscillations in the radial direction for the harmonic central potential $V(r) = kr^2/2$ without solving the equation of motion or relying on knowledge of its solution. Verify that the result does not depend on the angular momentum L or the energy E.

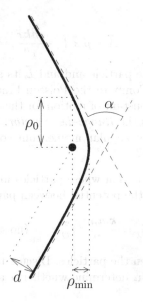

Figure 10.36 A hyperbola can be described by the same formula as an ellipse, but with an eccentricity $\varepsilon > 1$. We define the deflection angle α as the angular difference in the asymptotic directions of the trajectory, the distance of closest approach $\rho_{\min}$ as the minimal distance between the trajectory and the center of the potential, and the impact parameter d as the distance between the asymptotes and the center.

Problem 10.35. By explicitly solving the problem of the harmonic central potential in Cartesian coordinates, verify that the trajectories are ellipses also in this case, but now with the center of the ellipse at the center of the potential instead of one of the foci.

Problem 10.36. For a bead of mass m moving on a rotating ring of radius r_0 in a gravitational field, we have seen that the effective Lagrangian is given by

$$\mathcal{L} = \frac{mr_0^2}{2}\dot{\theta}^2 + \frac{mr_0^2\omega^2}{2}\sin^2(\theta) - mgr_0\cos(\theta). \tag{10.414}$$

Starting from this expression, find the canonical momentum p_θ related to the coordinate θ and express the Hamiltonian $\mathcal{H}$ in terms of θ and p_θ. Verify that Hamilton's equations of motion for this system coincides with the equation of motion derived from the Lagrangian in Lagrangian mechanics. Sketch the phase space flow for the cases $g > r_0\omega^2$, $g < r_0\omega^2$, and $g = r_0\omega^2$.

Problem 10.37. For motion in a central potential, show that the total angular momentum squared $\vec{L}^2$ is a constant of motion by computing its Poisson bracket with the Hamiltonian. We have also seen that given two constants of motion, their Poisson bracket is a new constant of motion. Verify that the Poisson bracket between $\vec{L}^2$ and the individual angular momentum components does not lead to any new conserved quantity.

Problem 10.38. Consider a particle of mass m moving in a central potential

$$V(r) = -\frac{k}{r^n}. \tag{10.415}$$

We define the vector

$$\vec{A} = \vec{p} \times \vec{L} - \frac{mk\vec{x}}{r^n}, \qquad (10.416)$$

where $\vec{p}$ is the momentum of the particle and and $\vec{L}$ its angular momentum relative to the center point of the potential. Compute the Poisson bracket between $\vec{A}$ and the Hamiltonian $\mathcal{H}$ and verify that $\vec{A}$ is a constant of motion for the special case of the Kepler problem ($n = 1$), in which $\vec{A}$ is known as the *Runge–Lenz vector*. Does the Poisson bracket between the Runge–Lenz vector and the angular momentum components provide any additional constants of motion in this case?

Problem 10.39. Imagine a situation where particles interact via harmonic potentials instead of Kepler potentials, i.e., the potential between particles i and j is given by

$$V_{ij} = \frac{km_i m_j}{2} r_{ij}^2, \qquad \text{(no sum)} \qquad (10.417)$$

where r_{ij} is the distance between the particles. Discuss the constrained three-body system for such potentials. In particular, determine whether it allows any Lagrange points and, if so, find their location.

Problem 10.40. Write down the Hamiltonian for a particle of mass m falling in a homogeneous gravitational field g. Sketch the phase space flow lines in the resulting phase space considering the motion to be one-dimensional in the direction of the gravitational field.

Problem 10.41. With the definition of the Poisson bracket in Eq. (10.238), show that it satisfies the relations listed in Eqs. (10.239) to (10.241).

Problem 10.42. For a particle moving on the surface of a sphere (see Example 10.22), we found that Noether's theorem implied that the quantity

$$L = mr_0^2 \sin^2(\theta)\dot{\varphi} \qquad (10.418)$$

is a constant of motion. Verify that this can be derived also in the Hamiltonian formalism by showing that $\{L, \mathcal{H}\} = 0$. Find the corresponding phase space flow generated by L and explicitly verify that the Hamiltonian is invariant under it.

Problem 10.43. A particle moving in a central potential an energy $E < V(\infty)$ and angular momentum $L > 0$ and is therefore in a bound state. Since the angular momentum and energy are conserved, the particle will always be contained in the phase space volume given by $E_{\min} < E < E_{\max} < V(\infty)$ and $0 < L_{\min} < L < L_{\max}$. Show that this phase space volume is finite so that the Poincaré recurrence theorem applies, i.e., the particle will come arbitrarily close to its initial state within a finite time.

Problem 10.44. Explicitly verify that the change from Cartesian to polar coordinates in the two-dimensional plane along with the canonical momentum transformations defined in Eqs. (10.290) define a canonical transformation.

Problem 10.45. Show that the Jacobian

$$J = \begin{vmatrix} \frac{\partial x}{\partial x_0} & \frac{\partial x}{\partial p_0} \\ \frac{\partial p}{\partial x_0} & \frac{\partial p}{\partial p_0} \end{vmatrix} \qquad (10.419)$$

of the phase space flow defined in Example 10.37 is equal to one and the flow therefore satisfies Liouville's theorem.

Problem 10.46. In a *Maxwell–Boltzmann distribution*, the phase space density is taken to be

$$\rho = \rho_0 e^{-\frac{\mathcal{H}}{kT}}, \tag{10.420}$$

where ρ_0 is a normalising constant, T is the temperature, and k is Boltzmann's constant. Show that this is a stationary distribution according to Liouville's equation. *Note:* This is a special case of the Maxwell–Boltzmann distribution being a stationary state also in the case of interacting systems. Indeed, for non-interacting systems, any ensemble with a phase space density that is a function of $\mathcal{H}$ is a stationary distribution.

Problem 10.47. In a general phase space with coordinates q^a and p_a, determine the phase space flow generated by the coordinates q^a. Explicitly write down and physically interpret these flows for

a) a particle of mass m moving in three dimensions described in Cartesian coordinates,

b) a particle of mass m moving in two dimensions described in polar coordinates, and

c) a rigid object with moment of inertia I with respect to the axis it is rotating around described by the rotation angle θ.

Problem 10.48. For a particle moving in a central potential, we have seen that the angular momentum $\vec{L}$ commutes with the Hamiltonian $\mathcal{H}$. As a result, the phase space flow generated by any of the angular momentum components commutes with the time evolution of the system. By considering the change in the angular momentum $\vec{L}$ under the flow generated by $\vec{n} \cdot \vec{L}$, where $\vec{n}$ is an arbitrary unit vector, show that the normal vector of the plane of motion is invariant under this flow only if $\vec{n} \times \vec{L} = 0$ and that it is otherwise results in a rotation of that plane.

Problem 10.49. Consider a two-dimensional motion with generalised coordinates x^1 and x^2 and corresponding canonical momenta p_1 and p_2 such that $\{x^i, p_j\} = \delta^i_j$.

a) Let $L = x^1 p_2 - x^2 p_1$ and show that if $\dot{L} = \dot{p}_1 = 0$, then also $\dot{p}_2 = 0$.

b) Determine the flow generated by L in the phase space. Explicitly verify that this flow defines a canonical transformation for every fixed value of the flow parameter.

c) Show that the quantity $p_1^2 + p_2^2$ is a constant of motion if the Hamiltonian is given by $\mathcal{H} = L^2$.

Problem 10.50. A hollow pipe is allowed to rotate about its center of mass and its moment of inertia about the center of mass is assumed to be I. A bead of mass m is allowed to move without friction within the pipe. The resulting system can be described by the distance r of the bead from the center of mass and the inclination of the pipe θ, see Fig. 10.37.

a) Determine the components of the generalised inertia tensor M_{ab} in this coordinate system.

b) Using the result from (a), determine the Christoffel symbols of the kinematic metric.

c) Placing the system in a gravitational field of strength g, write down the equations of motion resulting from the generalisation of Newton's second law.

Problem 10.51. Starting from the generalisation of Newton's second law in Eq. (10.322) and the Christoffel symbols of Euclidean space in spherical coordinates, write down the equations of motion for a particle of mass m subjected to a force $\vec{F}$ in spherical coordinates.

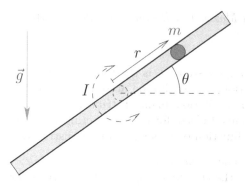

Figure 10.37 A hollow pipe with moment of inertia I rotating around its center of mass and a bead of mass m allowed to move freely within it. The system can be described by the generalised coordinates r and θ.

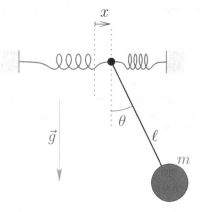

Figure 10.38 A pendulum of length ℓ with a mass m at one end and the other end allowed to move horizontally with a harmonic potential, which may be due to springs that act to restore it to the equilibrium position. The system is placed in a gravitational field $\vec{g}$. The system is described by the generalised coordinates x and θ.

Problem 10.52. Consider the mechanical system shown in Fig. 10.38, where a pendulum of length ℓ has a mass m attached at one of its ends and the other end is allowed to move horizontally subject to a harmonic potential $V(x) = kx^2/2$. The pendulum is placed in a gravitational field of strength g.

a) Show that the configuration space of the system is a cylinder.

b) Write down the Lagrangian for the system and identify the components of the generalised inertia tensor.

c) Due to the potential breaking translational invariance, the canonical momentum p corresponding to the x-coordinate is not conserved. Show that the momentum transfer to the system $\dot{p}$ equals $\delta\mathcal{L}$ up to a total time derivative under the transformation with $\delta x = 1$ and all other variations equal to zero and that $\delta\mathcal{L} = -kx$ in this particular case.

d) What is the physical interpretation of the result in (c)?

Problem 10.53. We introduced the symplectic form ω on phase space as the exterior derivative of the one-form $\xi = p_a dq^a$. Show that this definition is independent of the generalised coordinates q^a, i.e., if we instead use the coordinates Q^a on the configuration space, then $P_a dQ^a = p_a dq^a$.

Problem 10.54. Starting from its definition, show that a symplectic form cannot exist in a manifold with an odd number of dimensions.

Problem 10.55. In general, a manifold may admit several different symplectic forms, resulting in different phase space flows for the same Hamiltonian function. Show that the flow lines in a two-dimensional manifold are the same regardless of the symplectic form imposed, but that the velocity with which they are being followed differs.

Problem 10.56. A common example of an interacting field theory is given by so called ϕ^4 theory where the source J in Eq. (10.358b) is given by

$$J = -\frac{\lambda}{4!}\phi^3 \tag{10.421}$$

(the four in the name ϕ^4 comes from the interaction term being $\phi J \propto \phi^4$). Determine the equations of motion for classical ϕ^4 theory. Assuming that λ is a small parameter and perturbation theory can be used, write down an integral representation for the Green's function of the linearised problem and determine the Feynman rules for the equation of motion.

Problem 10.57. Verify that the infinitesimal Lorentz transformation that was defined in Eq. (10.386) is a symmetry of the Lagrangian density

$$L = \frac{\rho\ell}{2}(q_t^2 - c^2 q_x^2). \tag{10.422}$$

Problem 10.58. In a four-dimensional space-time where we use the coordinates $x^0 = ct$ and x^i for $i = 1, 2, 3$, we can introduce a one-form $\mathcal{A}$ with four components

$$\mathcal{A} = \frac{\phi}{c}dx^0 - A_i dx^i. \tag{10.423}$$

The exterior derivative $\mathcal{F} = d\mathcal{A}$ then has the components

$$\mathcal{F}_{ab} = \partial_a \mathcal{A}_b - \partial_b \mathcal{A}_a. \tag{10.424}$$

Raising and lowering indices with the pseudo-metric η_{ab} that has the components

$$\eta_{00} = 1, \quad \eta_{0i} = \eta_{i0} = 0, \quad \text{and} \quad \eta_{ij} = -\delta_{ij}, \tag{10.425}$$

we can define the Lagrangian density L as

$$L = -\frac{1}{4\mu_0}\mathcal{F}^{ab}\mathcal{F}_{ab} - \mathcal{A}_a J^a, \tag{10.426}$$

where J^a is a source term.

a) Taking the components $\mathcal{A}_a$ as independent fields, show that the equations of motion for these fields can be written in the form

$$\partial_a \mathcal{F}^{ab} = \mu_0 J^b. \tag{10.427}$$

b) Writing out the components of the equations of motion derived in (a), show that the equations of motion are equivalent to half of Maxwell's equations, see Eqs. (2.198), and identify the components of the source term J^a in terms of the charge density ρ and the current density $\vec{J}$.

c) Verify that the other half of Maxwell's equations follow directly from $d\mathcal{F} = d^2\mathcal{A} = 0$.

Note: The pseudo-metric η is central to the formulation of special relativity. In fact, electromagnetism in the form of Maxwell's equations is a relativistic field theory, formulated before Einstein developed special relativity. The inconsistencies between Maxwell's theory of electromagnetism and Galilei invariance played a central role in this development and Einstein's seminal paper was titled "On the electrodynamics of moving bodies" (or, in the original German, "Zur Elektrodynamik bewegter körper").

Problem 10.59. When discussing the symmetries of the wave equation, we found the energy and momentum densities of a wave of the form $q(x,t) = f(x - ct)$.

a) Repeat this computation for a wave of the form $q(x,t) = h(x + ct)$.

b) The general solution to the wave equation is a sum of waves in both directions $q(x,t) = f(x - ct) + h(x + ct)$. Using this, show that $c\,|P| \leq E$.

Problem 10.60. Starting from the Klein–Gordon Lagrangian density in one spatial dimension

a) show that it is symmetric under the general space-time translations and the Lorentz transformation and

b) find the energy and momentum densities $\mathcal{E}$ and p of the Klein–Gordon field.

c) For a plane wave of the form $\phi(x,t) = \sin(kx - \omega t)$, determine the relation between the averages of $\mathcal{E}$ and p over an entire wavelength.

Note: In (c) you will first need to determine the relation between ω and c that must be satisfied for the Klein–Gordon equation to hold.

Reference material

A.1 GROUPS AND CHARACTER TABLES

A small collection of groups with low order and their corresponding character tables.

A.1.1 Cyclic groups

A.1.1.1 C_2

Order: 2
Generators: c
Defining relations: $c^2 = e$
Conjugacy classes: $e = \{e\}$, $C = \{c\}$

Character table of irreps:

C_2	e	C
A	1	1
B	1	−1

A.1.1.2 C_3

Order: 3
Generators: c
Defining relations: $c^3 = e$
Conjugacy classes: $e = \{e\}$, $C^1 = \{c\}$, $C^2 = \{c^2\}$

Character table of irreps:

C_3	e	C^1	C^2
A	1	1	1
E	1	$e^{2\pi i/3}$	$e^{-2\pi i/3}$
E^*	1	$e^{-2\pi i/3}$	$e^{2\pi i/3}$

A.1.1.3 C_{2v}

Order: 4
Generators: c, σ
Defining relations: $c^2 = e$, $\sigma^2 = e$, $\sigma c \sigma c = e$
Conjugacy classes: $e = \{e\}$, $C = \{c\}$, $\sigma = \{\sigma\}$, $\sigma' = \{\sigma c\}$

Character table of irreps:

C_{2v}	e	C	σ	σ'
A_1	1	1	1	1
A_2	1	1	-1	-1
B_1	1	-1	1	-1
B_2	1	-1	-1	1

A.1.2 Dihedral groups

A.1.2.1 D_2

Isomorphic to C_{2v}.

A.1.2.2 D_3

$$\begin{aligned}
\text{Order:} \quad & 6 \\
\text{Generators:} \quad & c, \sigma \\
\text{Defining relations:} \quad & c^3 = e, \sigma^2 = e, c\sigma c\sigma = e \\
\text{Conjugacy classes:} \quad & e = \{e\}, C = \{c, c^2\}, \sigma = \{\sigma, \sigma c, \sigma c^2\}
\end{aligned}$$

Character table of irreps:

D_3	e	$2C$	3σ
A_1	1	1	1
E	2	-1	0
A_2	1	1	-1

A.1.2.3 D_{3h}

$$\begin{aligned}
\text{Order:} \quad & 12 \\
\text{Generators:} \quad & c_3, c_2, \sigma_h \\
\text{Defining relations:} \quad & c_3^3 = e, c_2^2 = e, \sigma_h^2 = e \\
& c_3\sigma_h = \sigma_h c_3, \sigma_h c_2 = c_2\sigma_h, c_2 c_3 c_2 c_3 = e \\
\text{Conjugacy classes:} \quad & e = \{e\}, C_3 = \{c_3, c_3^2\}, C_2 = \{c_2, c_2 c_3, c_2 c_3^2\}, \\
& \sigma_h = \{\sigma_h\}, S_3 = \{\sigma_h c_3, \sigma_h c_3^2\}, \\
& \sigma_v = \{\sigma_h c_2, \sigma_h c_2 c_3, \sigma_h c_2 c_3^2\}
\end{aligned}$$

Character table of irreps:

D_{3h}	e	$2C_3$	$3C_2$	σ_h	$2S_3$	$3\sigma_v$
A_1'	1	1	1	1	1	1
A_2'	1	1	-1	1	1	-1
E'	2	-1	0	2	-1	0
A_1''	1	1	1	-1	-1	-1
A_2''	1	1	-1	-1	-1	1
E''	2	-1	0	-2	1	0

A.1.3 Symmetric groups

A.1.3.1 S_2

Isomorphic to C_2.

A.1.3.2 S_3

Isomorphic to D_3.

A.1.3.3 S_4

$$
\begin{aligned}
\text{Order:} \quad & 24 \\
\text{Generators:} \quad & c_3, c_2, \sigma \\
\text{Defining relations:} \quad & c_3^3 = e, \; c_2^2 = e, \; \sigma^2 = e \\
& c_3 \sigma c_3 \sigma = e, \; \sigma c_2 = c_2 \sigma, \; (c_2 c_3)^3 = e \\
\text{Conjugacy classes:} \quad & e = \{e\}, \; C_2 = \{c_2, c_3 c_2 c_3^2, c_3^2 c_2 c_3\}, \\
& C_3 = \{c_3, c_3^2, c_3 c_2, (c_3 c_2)^2, c_2 c_3 c_2, c_2 c_3^2 c_2, c_2 c_3, (c_2 c_3)^2\}, \\
& S_4 = \{\sigma c_2 c_3, \sigma(c_2 c_3)^2, \sigma c_3^2 c_2 c_3, \sigma c_3 c_2 c_3^2, \sigma c_3 c_2, \sigma(c_3 c_2)^2\}, \\
& \sigma_d = \{\sigma, \sigma c_3, \sigma c_3^2, \sigma c_2, \sigma c_2 c_3 c_2, \sigma c_2 c_3^2 c_2\}
\end{aligned}
$$

Character table of irreps:

S_4	e	$3C_2$	$8C_3$	$6S_4$	$6\sigma_d$
A_1	1	1	1	1	1
A_2	1	1	1	-1	-1
E	2	2	-1	0	0
T_1	3	-1	0	1	-1
T_2	3	-1	0	-1	1

A.2 DIFFERENTIAL OPERATORS IN ORTHOGONAL COORDINATES

A.2.1 General expressions

In a general orthogonal coordinate system in N dimensions, the scale factors are defined according to

$$
h_a = |\vec{E}_a| = \left| \frac{\partial \vec{x}}{\partial y^a} \right|. \tag{A.1}
$$

The Jacobian determinant $\mathcal{J}$ is given by the product of the scale factors

$$
\mathcal{J} = \prod_a h_a. \tag{A.2}
$$

The gradient, divergence, curl, and Laplace operators in an orthogonal coordinate system are given by

$$
\nabla f = \sum_a \frac{1}{h_a} \vec{e}_a \partial_a f, \tag{A.3a}
$$

$$
\nabla \cdot \vec{v} = \frac{1}{\mathcal{J}} \sum_a \partial_a \left(\frac{\tilde{v}_a \mathcal{J}}{h_a} \right), \tag{A.3b}
$$

$$
\nabla \times \vec{v} = \frac{1}{\mathcal{J}} \sum_{abc} \vec{e}_a \varepsilon_{abc} h_a \partial_b (h_c \tilde{v}_c), \tag{A.3c}
$$

$$
\nabla^2 f = \frac{1}{\mathcal{J}} \sum_a \partial_a \left(\frac{\mathcal{J}}{h_a^2} \partial_a f \right). \tag{A.3d}
$$

A.2.2 Cylinder coordinates

The scale factors in cylinder coordinates are given by

$$h_\rho = h_z = 1 \quad \text{and} \quad h_\phi = \rho. \tag{A.4}$$

The gradient, divergence, curl, and Laplace operators in cylinder coordinates are given by

$$\nabla f = \vec{e}_\rho \partial_\rho f + \vec{e}_\phi \frac{1}{\rho} \partial_\phi f + \vec{e}_z \partial_z f, \tag{A.5a}$$

$$\nabla \cdot \vec{v} = \frac{1}{\rho}(\partial_\rho \rho \tilde{v}_\rho + \partial_\phi \tilde{v}_\phi) + \partial_z \tilde{v}_z, \tag{A.5b}$$

$$\nabla \times \vec{v} = \vec{e}_\rho \left(\frac{1}{\rho} \partial_\phi \tilde{v}_z - \partial_z \tilde{v}_\phi \right) + \vec{e}_\phi (\partial_z \tilde{v}_\rho - \partial_\rho \tilde{v}_z) + \frac{1}{\rho} \vec{e}_z (\partial_\rho \rho \tilde{v}_\phi - \partial_\phi \tilde{v}_\rho), \tag{A.5c}$$

$$\nabla^2 f = \frac{1}{\rho} \partial_\rho (\rho \partial_\rho f) + \frac{1}{\rho^2} \partial_\phi^2 f + \partial_z^2 f. \tag{A.5d}$$

A.2.3 Spherical coordinates

The scale factors in spherical coordinates are given by

$$h_r = 1, \quad h_\theta = r, \quad \text{and} \quad h_\varphi = r \sin(\theta). \tag{A.6}$$

The gradient, divergence, curl, and Laplace operators in spherical coordinates are given by

$$\nabla f = \vec{e}_r \partial_r f + \frac{1}{r} \vec{e}_\theta \partial_\theta f + \frac{1}{r \sin(\theta)} \vec{e}_\varphi \partial_\varphi f, \tag{A.7a}$$

$$\nabla \cdot \vec{v} = \frac{1}{r^2} \partial_r (r^2 \tilde{v}_r) + \frac{1}{r \sin(\theta)} \partial_\theta (\sin(\theta) \tilde{v}_\theta) + \frac{1}{r \sin(\theta)} \partial_\varphi \tilde{v}_\varphi, \tag{A.7b}$$

$$\nabla \times \vec{v} = \frac{1}{r \sin(\theta)} \vec{e}_r (\partial_\theta \sin(\theta) \tilde{v}_\varphi - \partial_\varphi \tilde{v}_\theta) + \frac{1}{r} \vec{e}_\theta \left(\frac{1}{\sin(\theta)} \partial_\varphi \tilde{v}_r - \partial_r r \tilde{v}_\varphi \right)$$
$$+ \frac{1}{r} \vec{e}_\varphi (\partial_r r \tilde{v}_\theta - \partial_\theta \tilde{v}_r), \tag{A.7c}$$

$$\nabla^2 f = \frac{1}{r^2} \left[\partial_r (r^2 \partial_r f) + \frac{1}{\sin(\theta)} \partial_\theta (\sin(\theta) \partial_\theta f) + \frac{1}{\sin^2(\theta)} \partial_\varphi^2 f \right]. \tag{A.7d}$$

A.3   SPECIAL FUNCTIONS AND THEIR PROPERTIES

A.3.1   The Gamma function

The gamma function is defined as

$$\Gamma(x) = \int_0^\infty t^{x-1} e^{-t} dt. \tag{A.8}$$

For all non-negative integers n, the gamma function satisfies:

$$\Gamma(n+1) = n! \tag{A.9}$$

For any x, the gamma function satisfies

$$\Gamma(x+1) = x\Gamma(x). \tag{A.10}$$

A.3.2   Bessel functions

A.3.2.1   Bessel functions

The Bessel functions of the first kind $J_\nu(\rho)$ and of the second kind $Y_\nu(\rho)$ are linearly independent solutions to Bessel's differential equation

$$\rho^2 f''(\rho) + \rho f'(\rho) + (\rho^2 - \nu^2) f(\rho) = 0. \tag{A.11}$$

A.3.2.2   Modified Bessel functions

The modified Bessel functions of the first kind $I_\nu(\rho)$ and of the second kind $K_\nu(\rho)$ are linearly independent solutions to Bessel's modified differential equation

$$\rho^2 f''(\rho) + \rho f'(\rho) - (\rho^2 + \nu^2) f(\rho) = 0. \tag{A.12}$$

Relation to the Bessel functions:

$$I_\nu(x) = i^{-\nu} J_\nu(ix) \tag{A.13a}$$

$$K_\nu(x) = \frac{\pi}{2} \frac{I_{-\nu}(x) - I_\nu(x)}{\sin(\pi\nu)} \tag{A.13b}$$

A.3.2.3   Integral representations

For integer values of m:

$$J_m(\rho) = \frac{1}{2\pi} \int_{-\pi}^{\pi} e^{i(\rho\sin(\phi) - m\phi)} d\phi = \frac{1}{\pi} \int_0^\pi \cos(\rho\sin(\phi) - m\phi) d\phi \tag{A.14a}$$

$$Y_m(\rho) = \frac{1}{\pi} \int_0^\pi \sin(\rho\sin(\phi) - m\phi) d\phi = -\frac{1}{\pi} \int_0^\infty [e^{m\tau} + (-1)^m e^{-m\tau}] e^{-\rho\sinh(\tau)} d\tau \tag{A.14b}$$

A.3.2.4   Asymptotic form

For small values of ρ $(0 < \rho \ll \sqrt{\nu+1})$ and non-negative ν:

$$J_\nu(\rho) \simeq \frac{1}{\Gamma(\alpha+1)} \frac{z^\nu}{2^\nu} \tag{A.15}$$

A.3.2.5   Relations among Bessel functions

For integer values of m:

$$J_{-m}(\rho) = (-1)^m J_m(\rho) \tag{A.16}$$

For $Z_\nu(\rho) = J_\nu(\rho)$ or $Z_\nu(\rho) = Y_\nu(\rho)$:

$$Z_\nu(\rho) = \frac{\rho}{2\nu} [Z_{\nu-1}(\rho) + Z_{\nu+1}(\rho)] \tag{A.17a}$$

$$Z_\nu'(\rho) = \frac{1}{2} [Z_{\nu-1}(\rho) - Z_{\nu+1}(\rho)] \tag{A.17b}$$

$$\left(\frac{1}{\rho} \frac{d}{d\rho}\right)^k (\rho^\nu Z_\nu(\rho)) = \rho^{\nu-k} Z_{\nu-k}(\rho) \tag{A.17c}$$

$$\left(\frac{1}{\rho} \frac{d}{d\rho}\right)^k \left(\frac{Z_\nu(\rho)}{\rho^\nu}\right) = (-1)^k \frac{Z_{\nu+k}(\rho)}{\rho^{\nu+k}} \tag{A.17d}$$

For $V_\nu(\rho) = I_\nu(\rho)$ or $V_\nu(\rho) = e^{i\pi\nu} K_\nu(\rho)$:

$$V_\nu(\rho) = \frac{\rho}{2\nu}[V_{\nu-1}(\rho) + V_{\nu+1}(\rho)] \tag{A.18a}$$

$$V'_\nu(\rho) = \frac{1}{2}[V_{\nu-1}(\rho) - V_{\nu+1}(\rho)] \tag{A.18b}$$

A.3.2.6 Expansions

Fourier expansions of $e^{i\rho\sin(\phi)}$ and $e^{i\rho\cos(\phi)}$:

$$e^{i\rho\sin(\phi)} = \sum_{m=-\infty}^{\infty} J_m(\rho)e^{im\phi} \tag{A.19a}$$

$$e^{i\rho\cos(\phi)} = \sum_{m=-\infty}^{\infty} i^m J_m(\rho)e^{im\phi} \tag{A.19b}$$

A.3.2.7 Orthogonality relations

The Bessel functions satisfy the following orthogonality relations:

$$\int_0^1 J_\nu(\alpha_{\nu k}\rho) J_\nu(\alpha_{\nu\ell}\rho)\rho\,d\rho = \frac{\delta_{k\ell}}{2} J_{\nu+1}(\alpha_{\nu k})^2 = \frac{\delta_{k\ell}}{2} J'_\nu(\alpha_{\nu k})^2 \tag{A.20a}$$

$$\int_0^\infty J_\nu(\alpha\rho) J_\nu(\beta\rho)\rho\,d\rho = \frac{1}{\alpha}\delta(\alpha - \beta) \tag{A.20b}$$

The second relation is valid for $\nu > -1/2$ and $\alpha_{\nu k}$ is the kth zero of $J_\nu(\rho)$.

A.3.2.8 Bessel function zeros

The first five zeroes of the first four Bessel functions $J_m(\rho)$ are:

α_{mk}		m		
k	**0**	**1**	**2**	**3**
1	2.40	3.83	5.14	6.38
2	5.52	7.02	8.42	9.76
3	8.65	10.17	11.62	13.02
4	11.79	13.32	14.80	16.22
5	14.93	16.47	17.96	19.41

The first five zeros of the derivatives $J'_m(\rho)$ of the first four Bessel functions are:

α'_{mk}		m		
k	**0**	**1**	**2**	**3**
1	0	1.84	3.05	4.20
2	3.83	5.33	6.71	8.02
3	7.02	8.54	9.97	11.36
4	10.17	11.71	13.17	14.59
5	13.32	14.86	16.35	17.79

A.3.3 Spherical Bessel functions

A.3.3.1 Spherical Bessel functions

The spherical Bessel functions of the first kind $j_n(r)$ and of the second kind $y_n(r)$ are linearly independent solutions to the differential equation

$$r^2 f''(r) + 2r f'(r) + [r^2 - n(n+1)] f(r) = 0. \tag{A.21}$$

A.3.3.2 Relation to Bessel functions

The spherical Bessel functions are related to the Bessel functions as:

$$j_n(r) = \sqrt{\frac{\pi}{2r}} J_{n+1/2}(r) \tag{A.22a}$$

$$y_n(r) = \sqrt{\frac{\pi}{2r}} Y_{n+1/2}(r) = (-1)^{n+1} \sqrt{\frac{\pi}{2r}} J_{-n-1/2}(r) \tag{A.22b}$$

A.3.3.3 Explicit expressions

The first few spherical Bessel functions can be written as:

$$j_0(r) = \frac{\sin(r)}{r} \tag{A.23a}$$

$$j_1(r) = \frac{\sin(r)}{r^2} - \frac{\cos(r)}{r} \tag{A.23b}$$

$$j_2(r) = \left(\frac{3}{r^2} - 1\right) \frac{\sin(r)}{r} - \frac{3\cos(r)}{r} \tag{A.23c}$$

$$j_3(r) = \left(\frac{15}{r^3} - \frac{6}{r}\right) \frac{\sin(r)}{r} - \left(\frac{15}{r^2} - 1\right) \frac{\cos(r)}{r} \tag{A.23d}$$

$$y_0(r) = -\frac{\cos(r)}{r} \tag{A.24a}$$

$$y_1(r) = -\frac{\cos(r)}{r^2} - \frac{\sin(r)}{r} \tag{A.24b}$$

$$y_2(r) = \left(-\frac{3}{r^2} + 1\right) \frac{\cos(r)}{r} - \frac{3\sin(r)}{r} \tag{A.24c}$$

$$y_3(r) = \left(-\frac{15}{r^3} + \frac{6}{r}\right) \frac{\cos(r)}{r} - \left(\frac{15}{r^2} - 1\right) \frac{\sin(r)}{r} \tag{A.24d}$$

A.3.3.4 Rayleigh formulas

The spherical Bessel functions satisfy:

$$j_n(r) = (-r)^n \left(\frac{1}{r} \frac{d}{dr}\right)^n \frac{\sin(r)}{r} \tag{A.25a}$$

$$y_n(r) = -(-r)^n \left(\frac{1}{r} \frac{d}{dr}\right)^n \frac{\cos(r)}{r} \tag{A.25b}$$

A.3.3.5   Relations among spherical Bessel functions

For $f_n(r) = j_n(r)$ and $f_n(r) = y_n(r)$:

$$\left(\frac{1}{r}\frac{d}{dr}\right)^k (r^{n+1}f_n(r)) = r^{n-k+1}f_{n-k}(r) \tag{A.26a}$$

$$\left(\frac{1}{r}\frac{d}{dr}\right)^k (r^{-n}f_n(r)) = (-1)^k r^{-n-k}f_{n+k}(r) \tag{A.26b}$$

A.3.3.6   Orthogonality relations

The spherical Bessel functions satisfy the following orthogonality relations:

$$\int_0^1 j_n(\beta_{nk}r)j_n(\beta_{n\ell}r)r^2 dr = \frac{\delta_{k\ell}}{2}j_{n+1}(\beta_{n\ell})^2 \tag{A.27a}$$

$$\int_0^\infty j_n(\alpha r)j_n(\beta r)r^2 dr = \frac{\pi}{2\alpha^2}\delta(\alpha - \beta) \tag{A.27b}$$

The second relation is valid for $n > -1$ and β_{nk} is the kth zero of $j_n(r)$.

A.3.3.7   Spherical Bessel function zeros

The first five zeroes of the first four spherical Bessel functions $j_\ell(r)$ are:

$\beta_{\ell k}$		ℓ		
k	**0**	**1**	**2**	**3**
1	π	4.49	5.76	6.99
2	2π	7.73	9.10	10.42
3	3π	10.90	12.32	13.70
4	4π	14.07	15.51	16.92
5	5π	17.22	18.69	20.12

The first five zeros of the derivatives $j_\ell'(r)$ of the first four spherical Bessel functions are:

$\beta_{\ell k}'$		ℓ		
k	**0**	**1**	**2**	**3**
1	0	2.08	3.34	4.51
2	4.49	5.94	7.29	8.58
3	7.73	9.21	10.61	11.97
4	10.90	12.40	13.85	15.24
5	14.07	15.58	17.04	18.47

A.3.4   Legendre functions

A.3.4.1   Legendre functions

The Legendre functions are the solutions to Legendre's differential equation

$$\frac{d}{dx}(1 - x^2)\frac{d}{dx}f_\ell(x) + \ell(\ell + 1)f_n(x) = 0. \tag{A.28}$$

The Legendre polynomials $P_\ell(x)$ are the Legendre functions for non-negative integer ℓ that are regular at $x = \pm 1$. The standard normalisation is such that $P_\ell(1) = 1$. The second independent solution to Legendre's differential equation is usually denoted $Q_\ell(x)$ and is not regular at $x = \pm 1$.

A.3.4.2 Rodrigues' formula

The Legendre polynomials can be expressed as:

$$P_\ell(x) = \frac{1}{2^\ell \ell!} \frac{d^\ell}{dx^\ell} (x^2 - 1)^\ell \qquad \text{(A.29)}$$

A.3.4.3 Relation among Legendre polynomials

Bonnet's recursion formula:

$$(\ell + 1)P_{\ell+1}(x) = (2\ell + 1)xP_\ell(x) - P_{\ell-1}(x) \qquad \text{(A.30)}$$

A.3.4.4 Explicit expressions

The first few Legendre polynomials are:

$$P_0(x) = 1 \qquad \text{(A.31a)}$$
$$P_1(x) = x \qquad \text{(A.31b)}$$
$$P_2(x) = \frac{1}{2}(3x^2 - 1) \qquad \text{(A.31c)}$$
$$P_3(x) = \frac{1}{2}(5x^3 - 3x) \qquad \text{(A.31d)}$$
$$P_4(x) = \frac{1}{8}(35x^4 - 30x^2 + 3) \qquad \text{(A.31e)}$$
$$P_5(x) = \frac{1}{8}(63x^5 - 70x^3 + 15x) \qquad \text{(A.31f)}$$

A.3.4.5 Associated Legendre functions

The associated Legendre functions are the solutions to the general Legendre differential equation

$$\frac{d}{dx}(1 - x^2)\frac{d}{dx}f_\ell^m(x) + \left[\ell(\ell + 1) - \frac{m^2}{1 - x^2}\right] f_\ell^m(x) = 0. \qquad \text{(A.32)}$$

The Legendre functions are the special case $m = 0$.
The associated Legendre functions are related to the Legendre polynomials as:

$$P_\ell^m(x) = (-1)^m (1 - x^2)^{m/2} \frac{d^m}{dx^m} P_\ell(x) \qquad \text{(A.33)}$$

By construction, $-\ell \le m \le \ell$.

A.3.4.6 Orthogonality relations

For fixed m:

$$\int_{-1}^{1} P_\ell^m(x)P_k^m\,dx = \frac{2(\ell + m)!\,\delta_{k\ell}}{(2\ell + 1)(\ell - m)!} \qquad \text{(A.34)}$$

For fixed ℓ:

$$\int_{-1}^{1} \frac{P_\ell^m(x)P_\ell^n(x)}{1-x^2}\,dx = \begin{cases} 0 & (m \neq n) \\ \frac{(\ell+m)!}{m(\ell-m)!} & (m = n \neq 0) \\ \infty & (m = n = 0) \end{cases} \qquad (A.35)$$

A.3.5 Spherical harmonics

A.3.5.1 Spherical harmonics

The spherical harmonics are the eigenfunctions of the Sturm–Liouville operator

$$\hat{\Lambda} = \frac{1}{\sin(\theta)}\partial_\theta \sin(\theta)\partial_\theta + \frac{1}{\sin^2(\theta)}\partial_\varphi^2. \qquad (A.36)$$

This operator is equivalent to the generalised Laplace operator on the sphere where θ and φ are the usual spherical coordinates.

A.3.5.2 Expression in terms of associated Legendre functions

The spherical harmonics are of the form:

$$Y_\ell^m(\theta, \varphi) = (-1)^m \sqrt{\frac{(2\ell+1)}{4\pi}\frac{(\ell-m)!}{(\ell+m)!}} P_\ell^m(\cos(\theta))e^{im\varphi} \qquad (A.37)$$

$Y_\ell^m(\theta, \varphi)$ is an eigenfunction of $\hat{\Lambda}$ with eigenvalue $-\ell(\ell+1)$. As for the associated Legendre functions $-\ell \leq m \leq \ell$.
Note: The normalisation is conventional and may differ between different texts!

A.3.5.3 Orthogonality relation

The spherical harmonics satisfy the orthogonality relation:

$$\int_{\theta=0}^{\pi} \int_{\varphi=0}^{2\pi} Y_\ell^m(\theta, \varphi)Y_k^n(\theta, \varphi)^* \sin(\theta)d\theta\,d\varphi = \delta_{\ell k}\delta_{mn} \qquad (A.38)$$

A.3.5.4 Parity

The spherical harmonics satisfy the relation:

$$Y_\ell^m(\pi - \theta, \pi + \varphi) = (-1)^\ell Y_\ell^m(\theta, \varphi) \qquad (A.39)$$

A.3.6 Hermite polynomials

A.3.6.1 Hermite polynomials

The hermite polynomials $H_n(x)$ are the solutions to the Sturm–Liouville problem

$$e^{x^2}\partial_x e^{-x^2}\partial_x H(x) + x^2 H(x) = (2E - 1)H(x) = 2\lambda H(x). \qquad (A.40)$$

This is equivalent to the Hermite equation

$$-H''(x) + 2xH'(x) = (2E - 1)H(x) = 2\lambda H(x). \qquad (A.41)$$

The eigenvalues are given by $\lambda_n = n$ and $E_n = n + 1/2$.

A.3.6.2 Explicit expressions

The first five Hermite polynomials are:

$$H_0(x) = 1 \tag{A.42a}$$
$$H_1(x) = 2x \tag{A.42b}$$
$$H_2(x) = 4x^2 - 2 \tag{A.42c}$$
$$H_3(x) = 8x^3 - 12x \tag{A.42d}$$
$$H_4(x) = 16x^4 - 48x^2 + 12 \tag{A.42e}$$

A.3.6.3 Creation operators

The creation operator $\hat{a}_+$ is defined as

$$\hat{a}_+ = (2x - \partial_x). \tag{A.43}$$

The Hermite polynomials satisfy

$$\hat{a}_+ H_n(x) = H_{n+1}(x). \tag{A.44}$$

The Hermite polynomials can be written

$$H_n(x) = \hat{a}_+^n H_0(x) = \hat{a}_+^n 1. \tag{A.45}$$

A.3.6.4 Hermite functions

The Hermite functions are the solutions to the eigenvalue problem

$$-\psi''(x) + x^2\psi(x) = 2E\psi(x). \tag{A.46}$$

The Hermite functions are given by

$$\psi_n(x) = \frac{e^{-x^2/2}}{\sqrt{2^n n! \sqrt{\pi}}} H_n(x). \tag{A.47}$$

A.3.6.5 Orthogonality relations

The Hermite polynomials satisfy the orthogonality relation:

$$\int_{-\infty}^{\infty} H_n(x) H_m(x) e^{-x^2} dx = 2^n n! \sqrt{\pi} \delta_{nm} \tag{A.48}$$

The Hermite functions satisfy the orthogonality relation:

$$\int_{-\infty}^{\infty} \psi_n(x) \psi_m(x) dx = \delta_{nm} \tag{A.49}$$

A.4 TRANSFORM TABLES

A.4.1 Fourier transform

The Fourier transform of a function $f(x)$ is given by:

$$\tilde{f}(k) = \int_{-\infty}^{\infty} f(x) e^{-ikx} dx \tag{A.50}$$

The inverse Fourier transform is given by:

$$f(x) = \frac{1}{2\pi} \int_{-\infty}^{\infty} \tilde{f}(k) e^{ikx} dk \qquad (A.51)$$

We have adopted a definition where the factor of 2π appears in the inverse Fourier transform. Another common definition is to include $1/\sqrt{2\pi}$ in both the transform and the inverse transform.

A.4.1.1 Transform table

The following is a list of useful Fourier transform relations:

Function $f(x)$	Fourier transform $\tilde{f}(k)$
$f(x - x_0)$	$e^{-ikx_0}\tilde{f}(k)$
$e^{ik_0 x} f(x)$	$\tilde{f}(k - k_0)$
$f(ax)$	$\tilde{f}(k/a)/|a|$
$\dfrac{d^n}{dx^n} f(x)$	$(ik)^n \tilde{f}(k)$
$x^n f(x)$	$i^n \dfrac{d^n}{dk^n} \tilde{f}(k)$
1	$2\pi\delta(k)$
$\delta(x)$	1
$\cos(ax)$	$\pi[\delta(k - a) + \delta(k + a)]$
$\sin(ax)$	$-\pi i[\delta(k - a) - \delta(k + a)]$
$e^{-ax}\theta(x)$	$\dfrac{1}{a + ik}$
e^{-ax^2}	$\sqrt{\dfrac{\pi}{a}} e^{-k^2/4a}$
$e^{-a|x|}$	$\dfrac{2a}{a^2 + k^2}$

Here, $\delta(x)$ is the delta distribution, $\theta(x)$ is the Heaviside function, and $a > 0$.

A.4.2 Laplace transform

The Laplace transform of a function $f(t)$ defined for $t > 0$ is given by

$$F(s) = \int_0^{\infty} e^{-st} f(t)\, dt. \qquad (A.52)$$

A.4.2.1 Transform table

The following is a list of useful Laplace transform relations:

Function $f(t)$	Fourier transform $F(s)$
$\theta(t-a)f(t-a)$	$e^{-as}F(s)$
$e^{bt}f(t)$	$F(s-b)$
$f(at)$	$\dfrac{1}{a}F(s/a)$
$tf(t)$	$-F'(s)$
$f'(t)$	$sF(s)-f(0)$
$f''(t)$	$s^2F(s)-f'(0)-sf(0)$
$\displaystyle\int_0^t f(x)\,dx$	$\dfrac{1}{s}F(s)$
$\displaystyle\int_0^t f_1(x)f_2(t-x)\,dx$	$F_1(s)F_2(s)$
$\delta(t-a)$	e^{-as}
1	$\dfrac{1}{s}$
t^n	$\dfrac{n!}{s^{n+1}}$
e^{-at}	$\dfrac{1}{s+a}$
$\sin(\omega t)$	$\dfrac{\omega}{s^2+\omega^2}$
$\cos(\omega t)$	$\dfrac{s}{s^2+\omega^2}$

Index